# Advanced Calculus

# About the Author

## PATRICK M. FITZPATRICK

(Ph.D., Rutgers University) held post-doctoral positions at the Courant Institute of New York University and the University of Chicago. Since 1975 he has been a member of the Mathematics Department at the University of Maryland at College Park, where he is now Professor of Mathematics and Associate Chairman for Undergraduate Studies. He has also held Visiting Professorships at the University of Paris and the University of Florence. Professor Fitzpatrick's principal research interest, on which he has written more than fifty research articles, is nonlinear functional analysis.

# Advanced Calculus

## A COURSE IN MATHEMATICAL ANALYSIS

**Patrick M. Fitzpatrick**

UNIVERSITY OF MARYLAND

**PWS Publishing Company**

I(T)P  An International Thomson Publishing Company

*Boston • Albany • Bonn • Cincinnati • Detroit • London • Madrid • Melbourne*
*Mexico City • New York • Paris • San Francisco • Singapore • Tokyo • Toronto • Washington*

**PWS PUBLISHING COMPANY**
20 Park Plaza, Boston, MA 02116-4324

Copyright © 1996 by PWS Publishing Company, a division of International Thomson Publishing Inc.

International Thomson Publishing
The trademark ITP is used under license.

*This book is printed on recycled, acid-free paper*

*For more information, contact:*
**PWS Publishing Co.**
**20 Park Plaza**
**Boston, MA 02116**

International Thomson Publishing Europe
Berkshire House I68-I73
High Holborn
London WC1V 7AA
England

Thomas Nelson Australia
102 Dodds Street
South Melbourne, 3205
Victoria, Australia

Nelson Canada
1120 Birchmont Road
Scarborough, Ontario
Canada M1K 5G4

International Thomson Editores
Campos Eliseos 385, Piso 7
Col. Polanco
11560 Mexico D.F., Mexico

International Thomson Publishing GmbH
Königswinterer Strasse 418
53227 Bonn, Germany

International Thomson Publishing Asia
221 Henderson Road
#05-10 Henderson Building
Singapore 0315

International Thomson Publishing Japan
Hirakawacho Kyowa Building, 31
2-2-1 Hirakawacho
Chiyoda Ku, Tokyo 102
Japan

**Library of Congress Cataloging-in-Publication Data**

Fitzpatrick, Patrick
    Advanced calculus : a course in mathematical analysis / Patrick M. Fitzpatrick.
      p.  cm.
    Includes index.
    ISBN 0-534-92612-6
    1.  Calculus.  I.  Title.
QA303.F5812  1995
515.8—dc20                94-41176
                            CIP

Sponsoring Editor: *Steve Quigley*
Production Editor: *Monique A. Calello*
Editorial Assistant: *John Ward*
Marketing Manager: *Marianne Rutter*
Manufacturing Manager: *Ellen Glisker*
Interior Designer: *Catherine Hawkes*
Cover Designer: *Kathleen Wilson*

Interior Illustrator: *Hayden Graphics*
Compositor: *Atlis Graphics & Design, Inc.*
Cover Illustration: *"Cities of the Plain,"* watercolor,
© *Richard C. Karwoski, The Image Bank*
Cover Printer: *Henry N. Sawyer Co., Inc.*
Text Printer: *R.R. Donnelley & Sons–Crawfordsville*

Printed and bound in the United States of America
95 96 97 98 99—10 9 8 7 6 5 4 3 2 1

THIS BOOK IS DEDICATED TO THE MEMORY OF MY FATHER,

MICHAEL JOSEPH FITZPATRICK (1912–1989)

In order to put his system into mathematical form at all, Newton had to devise the concept of differential quotients and propound the laws of motion in the form of differential equations—perhaps the greatest advance in thought that a single individual was ever privileged to make.

**Albert Einstein**
from an essay
*On the one hundredth anniversary of Maxwell's birth*
James Clerk Maxwell: A Commemorative Volume

# Contents

## Continuous Functions and Limits    42

## Differentiation    68

## The Elementary Functions as Solutions of Differential Equations    98

# Integration    113

# The Second Fundamental Theorem and Its Consequences    141

# Approximation by Taylor Polynomials    165

# 9 The Convergence of Sequences and Series of Functions    192

# 10 The Euclidean Space $\mathbb{R}^n$    229

# Continuity, Compactness, and Connectedness     249

# Metric Spaces     270

# Partial Differentiability of Real-Valued Functions of Several Variables     303

## Integration for Functions of Several Variables        428

## Line and Surface Integrals        473

## Consequences of the Field and Positivity Axioms        A1

## Linear Algebra        A7

## Answer Section        A23

# Preface

The goal of this book is to rigorously present the fundamental concepts of mathematical analysis in the clearest, simplest way, within the context of illuminating examples and stimulating exercises. I hope that the student will assimilate a precise understanding of the subject, together with an appreciation of its coherence and significance. The full book is suitable for a year-long course; the first nine chapters are suitable for a one-semester course on functions of a single variable.

Mathematical analysis has been seminal in the development of many branches of science. Indeed, the importance of the applications of the computational algorithms that are a part of the subject often leads to courses in which familiarity with implementing these algorithms is emphasized at the expense of the ideas that underlie the subject. While these techniques are very important, without a genuine understanding of the concepts that are at the heart of these algorithms, it is possible to make only limited use of these computational possibilities.

I have tried to emphasize the unity of the subject. Mathematical analysis is not a collection of isolated facts and techniques, but is, instead, a coherent body of knowledge. Beyond the intrinsic importance of the actual subject, the study of mathematical analysis instills habits of thought that are essential for a proper understanding of many areas of pure and applied mathematics.

In addition to the absolutely essential topics, other important topics have been arranged in such a way that selections can be made without disturbing the coherence of the course. As three examples of such optional topics, I mention the approximation methods for estimating integrals, the Weierstrass Approximation Theorem, and metric spaces. Precise estimates for the errors incurred in the approximation of integrals were always present in the classical courses in mathematical analysis. Nowadays, they do not appear so frequently. In view of the recent growth in computational capability and the attendant need to estimate errors in approximation methods, this topic seems to me worthy of consideration for inclusion in a course. This material is presented in the last section of Chapter 7; subsequent material is independent of this section. An approximation theorem of quite a different flavor is the Weierstrass Approximation Theorem. This stands as one of the singular jewels of classical analysis. It can be presented as a companion to the discussion of approximation of functions by Taylor polynomials; this theorem is proven in the last section of Chapter 8, and again, the subsequent material is independent of this section. The third topic, metric spaces, is more abstract than other topics in the book, but it is an abstraction that is wholly justified by its synthetic power and the applicability of the general theory to important specific problems. On the other hand, however, two of the most important examples of metric spaces, Euclidean spaces and function spaces, are of sufficient independent interest that it can be argued that they deserve a separate, self-contained discussion. The choice I have made is to begin the study of functions of several variables in Chapter 10 with the study of Euclidean space $\mathbb{R}^n$, and then, in Chapter 11, to study functions and mappings between Euclidean spaces. Then I have included a separate chapter, Chapter 12, on metric spaces. The student will have already seen important specific realizations of the general theory, namely the concept of uniform convergence for sequences of functions and the study of subsets of Euclidean space, and with these examples in mind can better appreciate the general theory. The Contraction Mapping Principle is proved and used to establish the fundamental existence result on the solvability of nonlinear differential equations. This serves as a powerful example of the use of brief, general theory to furnish concrete information about specific problems. Once more, none of the subsequent material depends on Chapter 12.

At the beginning of this course it is necessary to establish a base on which the subsequent proofs will be built. It has been my experience that in order to cover, within the allotted time, a substantial amount of analysis, it is not possible to provide a detailed construction of the real numbers starting with a serious treatment of set theory. I have chosen to codify the properties of the real numbers as three groups of axioms. In the Preliminaries, the arithmetic and order properties are codified in the Field and Positivity Axioms; a detailed discussion of the consequences of these axioms, which certainly are familiar to the student, is provided in Appendix A. The least familiar of these axioms, the Completeness Axiom, is presented in the first section of the first chapter, Section 1.1.

In Chapter 2, convergent sequences are studied. Monotonicity and linearity properties of convergent sequences are proven, and the Completeness Axiom is recast as the Bolzano-Weierstrass Theorem and the Nested Interval Theorem for convergent sequences. The material from this chapter is used repeatedly throughout the book. For instance, in Chapter 3, continuity of functions and limits of functions are defined in terms of sequential convergence. The linearity properties of convergent sequences from Chapter 2 immediately imply corresponding linearity properties for continuous functions, limits, derivatives, and later, in Chapter 7, for integrals.

Chapter 4 is devoted to differentiation. In Section 4.5, Darboux's Theorem is proven: It asserts that in order for a real-valued function that is defined on an open interval to be the derivative of another function, it is necessary that the given function possess the intermediate value property. This is the first result regarding the solvability of differential equations.

The students will be familiar with the properties of the logarithmic and trigonometric functions and their inverses, although, most probably, they will not have seen a rigorous analysis of these functions. In Chapter 5, the natural logarithm, the sine, and the cosine functions are introduced as the (unique) solutions of particular differential equations; on the provisional assumption that these equations have solutions, an analytic derivation of the properties of these functions and their inverses is provided. Later, after the differentiability properties of functions defined by integrals and by power series have been established, it is proven that these differential equations do indeed have solutions, and so the provisional assumptions of Chapter 5 are removed.

Integration is studied in Chapters 6 and 7. The integral is defined in terms of Darboux sums, and later its property of being the limit of appropriate sequences of Riemann sums is established. The relationship between integration and differentiation is described in two theorems, which I call the First and the Second Fundamental Theorems of Calculus. This is done to emphasize the distinction between the formula for evaluating the integral of a function that is known to be the derivative of another function and the related, but different, matter of understanding the conditions under which a given function is the derivative of some other function and providing integral representations of solutions of differential equations. The study of the approximation of functions by Taylor polynomials is the subject of Chapter 8. In Chapter 9, we consider a sequence of functions that converges to a limit function and study the way in which the limit function inherits properties possessed by the functions that are the terms of the sequence; the distinction between pointwise and uniform convergence is emphasized. This concludes the study of functions of a single variable.

The study of functions of several variables begins with Chapters 10 and 11, which start with the study of the structure of Euclidean space $\mathbb{R}^n$ and then turn to the manner in which the results about sequences of numbers and functions of a single variable extend to sequences of points in $\mathbb{R}^n$, to functions defined on subsets of Euclidean space and to mappings between such spaces. There is no class of subsets of $\mathbb{R}^n$ that play the same distinguished role with regard to functions of several variables as do intervals with regard to functions of a single variable. For this reason, the general concepts of open, closed, compact, and connected are introduced for subsets of $\mathbb{R}^n$. The notions of compactness and connectedness for a subset of $\mathbb{R}^n$ are motivated by the necessity to extend the Intermediate Value Theorem and the Extreme Value Theorem to functions of several variables. As already mentioned, Chapter 12 is an independent chapter on metric spaces. The material related to differentiation of functions of several variables is covered in Chapters 13 and 14. Emphasis is placed on precise assertions of the way a function may be approximated, in a neighborhood of a point in its domain, by a simpler function.

The study of mappings between Euclidean spaces is the topic of Chapters 15, 16, and 17. Here, and at other points in the book, it is necessary to understand some linear algebra. In Section 15.1, the correspondence between linear mappings from $\mathbb{R}^n$ to $\mathbb{R}^m$ and $m \times n$ matrices is established. As for the other topics that involve linear algebra, in Appendix B the requisite topics in linear algebra are described, and using the cross product of two vectors in $\mathbb{R}^3$, full proofs are provided for the case of vectors and linear mappings in $\mathbb{R}^3$. In Chapter 15, differentiation is studied for mappings between Euclidean spaces: at each point in the domain of a continuously differentiable mapping there is defined the derivative matrix, together with the corresponding linear mapping called the differential. Approximation by linear mappings is studied and the chapter concludes with the Chain Rule for mappings. The Inverse Function Theorem and the

Implicit Function Theorem are the focus of Chapters 16 and 17, respectively. I have made special effort to clearly present these theorems and related material, such as the minimization principle for studying nonlinear systems of equations, not as isolated technical results but as part of the theme of understanding what properties a mapping can be expected to inherit from its linearization. These two theorems are surely the clearest expression of the way that a nonlinear object (a mapping or a system of equations) inherits properties from a linear approximation.

The theory of integration of functions of several variables occupies the last two chapters of the book. In Chapter 18, the integral is first defined for bounded functions defined on generalized rectangles. It is proven that a bounded function defined on a generalized rectangle is integrable if its set of discontinuities has Jordan content 0. Then integration for bounded functions defined on bounded subsets of $\mathbb{R}^n$ is considered, in terms of extensions of such functions to generalized rectangles containing the original domain. Familiar properties of the integral of a function of a single variable (linearity, monotonicity, additivity over domains, and so forth) are established for the integral of functions of several variables. In Section 18.3, Fubini's Theorem on iterated integration is proved and in the last section of Chapter 18, the Change of Variables Theorem for the integral of functions of several variables is proved. In Chapter 19, the book concludes with the study of line and surface integrals and the way in which the First Fundamental Theorem of Calculus for functions of a single variable extends to Green's Formula and Stokes's Formula. I have resisted the temptation to present the general theory of integration on manifolds. In order to make the analytical ideas transparent, rather than present the most general results, emphasis has been placed on a careful treatment of parametrized paths and parametrized surfaces, so that the essentially technical issues associated with patching of surfaces are not present.

I cannot overemphasize the importance of the exercises. To achieve a genuine understanding of the material, it is necessary that the student do many problems. The problems are designed to be challenging and to stimulate the student to carefully reread the relevant sections in order to properly assimilate the material. Many of the problems foreshadow future developments.

## Acknowledgments

Preliminary versions of this book, in note form, have been used in classes by a number of my colleagues. The book has been improved by their comments about the notes and also by suggestions from other colleagues. Accepting sole responsibility for the final manuscript, I warmly thank Professors James Alexander, Stuart Antman, John Benedetto, Ken Berg, Michael Boyle, Joel Cohen, Jeffrey Cooper, Craig Evans, Seymour Goldberg, Paul Green, Denny Gulick, David Hamilton, Chris Jones, Adam Kleppner, John Millson, Umberto Neri, Jacobo Pejsachowicz, Dan Rudolph, Jerome Sather, James Schafer, and Daniel Sweet.

I would like to thank the following reviewers for their comments and criticisms: Bruce Barnes, *University of Oregon;* John Van Eps, *California Polytechnic State University–San Luis Obispo;* Christopher E. Hee, *Eastern Michigan University;* Gordon Melrose, *Old Dominion University;* Claudio Morales, *University of Alabama;* Harold R. Parks, *Oregon State University;* Steven Michael Seubert, *Bowling Green State University;* William Yslas Velez, *University of Arizona;* Clifford E. Weil, *Michigan State University;* and W. Thurmon Whitney, *University of New Haven.*

It is a pleasure to thank Ms. Jaya Nagendra for her excellent typing of various versions of the manuscript, and also to thank the editorial and production personnel at PWS Publishing Company for their considerate and expert assistance in making the manuscript into a book.

I am especially grateful to a teacher of mine and to a student of mine. As an undergraduate at Rutgers University, I was very fortunate to have Professor John Bender as my teacher. He introduced me to mathematical analysis. Moreover, his personal encouragement was what led me to pursue mathematics as a lifetime study. It is not possible to adequately express my debt to him. I also wish to single out for special thanks one of the many students who have contributed to this book. Alan Preis was a great help to me in the final preparation of the manuscript. His assistance and our stimulating discussions in this final phase made what could have been a very tiresome task into a pleasant one.

**Patrick M. Fitzpatrick**

# Preliminaries

## Sets and Functions

For a set $A$, the membership of the element $x$ in $A$ is denoted by $x \in A$ or $x$ in $A$, and the nonmembership of $x$ in $A$ is denoted by $x \notin A$. A member of $A$ is often called a *point* in $A$. Two sets are the same if and only if they have the same members. Frequently sets will be denoted by braces, so that $\{x \mid \text{proposition about } x\}$ is the set of all elements $x$ such that the proposition about $x$ is true.

If $A$ and $B$ are sets, then $A$ is called a *subset* of $B$ if and only if each member of $A$ is a member of $B$, and we denote this by $A \subseteq B$ or by $B \supseteq A$. The *union* of two sets $A$ and $B$, written $A \cup B$, is the set of all elements that belong either to $A$ or to $B$; that is, $A \cup B = \{x \mid x \text{ in } A \text{ or } x \text{ in } B\}$. The word *or* is used here in the nonexclusive sense, so that points that belong to both $A$ and $B$ belong to $A \cup B$. The *intersection* of $A$ and $B$, denoted by $A \cap B$, is the set of all points that belong to both $A$ and $B$; that is, $A \cap B = \{x \mid x \text{ in } A \text{ and } x \text{ in } B\}$. Given sets $A$ and $B$, the *complement* of $A$ in $B$, denoted by $B \backslash A$, is the set of all points in $B$ that are not in $A$. In particular, for a set $B$ and a point $x_0$, $B \backslash \{x_0\}$ denotes the set of points in $B$ that are not equal to $x_0$. The set that has no members is called the *empty set* and is denoted by $\emptyset$.

Given two sets $A$ and $B$, by a *function* from $A$ to $B$ we mean a correspondence that associates with each point in $A$ a point in $B$. Frequently we denote such a function by $f: A \to B$, and for each point $x$ in $A$, we denote by $f(x)$ the point in $B$ that is associated with $x$. We call the set $A$ the *domain* of the function $f: A \to B$ and we define the *image* of $f: A \to B$, denoted by $f(A)$, to be $\{y \mid y = f(x) \text{ for some point } x \text{ in } A\}$. If $f(A) = B$, the function $f: A \to B$ is said to be *onto*. If for each point $y$ in $f(A)$ there is exactly one point $x$ in $A$ such that $y = f(x)$, the function $f: A \to B$ is said to be *one-to-one*. A function $f: A \to B$ that is both one-to-one and onto is said to be *invertible*. For an invertible function $f: A \to B$, for each point $y$ in $f(B)$ there is exactly one point $x$ in $A$ such that $f(x) = y$, and this point is denoted by $f^{-1}(y)$; this correspondence defines the function $f^{-1}: B \to A$, which is called the *inverse function* of the function $f: A \to B$.

---

# The Field Axioms for the Real Numbers

---

In order to rigorously develop analysis, it is necessary to understand the foundation on which it is constructed; this foundation is the set of real numbers, which we will denote by $\mathbb{R}$. Of course, the reader is quite familiar with many properties of the real numbers. However, in order to clarify the basis of our development, it is very useful to codify the properties of $\mathbb{R}$. We will assume that the set of real numbers $\mathbb{R}$ satisfies three groups of axioms: the Field Axioms, the Positivity Axioms, and the Completeness Axiom. A discussion of the Completeness Axiom, which is perhaps the least familiar to the reader, will be deferred until Chapter 1. We will now describe the Field Axioms and the Positivity Axioms and some of their consequences.

For each pair of real numbers $a$ and $b$, a real number is defined that is called the *sum* of $a$ and $b$, written $a + b$, and a real number is defined that is called the *product* of $a$ and $b$, denoted by $ab$. These operations satisfy the following collection of axioms.

## The Field Axioms

*Commutativity of Addition:*   For all real numbers $a$ and $b$,

$$a + b = b + a.$$

*Associativity of Addition:*   For all real numbers $a$, $b$, and $c$,

$$(a + b) + c = a + (b + c).$$

*The Additive Identity:*   There is a real number, which is denoted by 0, such that

$$0 + a = a + 0 = a \quad \text{for all real numbers } a.$$

*The Additive Inverse:*   For each real number $a$, there is a real number $b$ such that

$$a + b = 0.$$

*Commutativity of Multiplication:*   For all real numbers $a$ and $b$,

$$ab = ba.$$

*Associativity of Multiplication:*   For all real numbers $a$, $b$, and $c$,

$$(ab)c = a(bc).$$

*The Multiplicative Identity:*   There is a real number, which is denoted by 1, such that

$$1a = a1 = a \quad \text{for all real numbers } a.$$

*The Multiplicative Inverse:*   For each real number $a \neq 0$, there is a real number $b$ such that

$$ab = 1.$$

*The Distributive Property:*   For all real numbers $a$, $b$, and $c$,

$$a(b + c) = ab + ac.$$

*The Nontriviality Assumption:*

$$1 \neq 0.$$

The Field Axioms are simply a record of the properties that one has always assumed about addition and multiplication of real numbers.

From the Field Axioms it follows* that there is only one number that has the property attributed to 0 in the Additive Identity Axiom. Moreover, it also follows that for each real number $a$,

$$a0 = 0a = 0,$$

and that for any real numbers $a$ and $b$,

$$\text{if} \quad ab = 0, \quad \text{then} \quad a = 0 \quad \text{or} \quad b = 0.$$

The Additive Inverse Axiom asserts that for each real number $a$, there is a solution of the equation

$$a + x = 0.$$

One can show that this equation has only one solution; we denote it by $-a$ and call it the *additive inverse* of $a$. For each pair of numbers $a$ and $b$, we define their *difference*, denoted by $a - b$, by

$$a - b \equiv a + (-b).$$

The Field Axioms also imply that there is only one number having the property attributed to 1 in the Multiplicative Identity Axiom. For a real number $a \neq 0$, the Multiplicative Inverse Axiom asserts that the equation

$$ax = 1$$

has a solution. One can show there is only one solution; we denote it by $a^{-1}$ and call it the *multiplicative inverse* of $a$. We then define for each pair of numbers $a$ and $b \neq 0$ their *quotient,* denoted by $a/b$, as

$$\frac{a}{b} \equiv ab^{-1}.$$

It is an interesting algebraic exercise to verify the implications of the Field Axioms that we have just mentioned and also to verify the following familiar consequences of these same axioms: For any real numbers $a$ and $b \neq 0$,

$$-(-a) = a, \quad (b^{-1})^{-1} = b, \quad \text{and} \quad (-b)^{-1} = -b^{-1}.$$

---

*Verification of these and of subsequent assertions in these Preliminaries is provided in Appendix A.

# The Positivity Axioms for the Real Numbers

In the real numbers there is a natural notion of order: greater than, less than, and so on. A convenient way to codify these properties is by specifying axioms satisfied by the set of positive numbers.

## The Positivity Axioms

There is a set of real numbers, denoted by $\mathcal{P}$, called the set of *positive numbers*. It has the following two properties:

**P1** If $a$ and $b$ are positive, then $ab$ and $a + b$ are also positive.

**P2** For a real number $a$, exactly one of the following three alternatives is true:

$$a \text{ is positive,} \quad -a \text{ is positive,} \quad a = 0.$$

The Positivity Axioms lead in a natural way to an ordering of the real numbers: For real numbers $a$ and $b$, we define $a > b$ to mean that $a - b$ is positive, and $a \geq b$ to mean that $a > b$ or $a = b$. We then define $a < b$ to mean that $b > a$, and $a \leq b$ to mean that $b \geq a$.

Using the Field Axioms and the Positivity Axioms, it is possible to establish the following familiar properties of inequalities (see Appendix A):

(i)  For each real number $a \neq 0$, $a^2 > 0$. In particular, $1 > 0$, since $1 \neq 0$ and $1 = 1^2$.

(ii)  For each positive number $a$, its multiplicative inverse $a^{-1}$ is also positive.

(iii)  If $a > b$, then

$$ac > bc \quad \text{if} \quad c > 0,$$

and

$$ac < bc \quad \text{if} \quad c < 0.$$

## Interval Notation

For a pair of real numbers $a$ and $b$ such that $a < b$, we define

$$(a, b) \equiv \{x \text{ in } \mathbb{R} \mid a < x < b\},$$
$$[a, b] \equiv \{x \text{ in } \mathbb{R} \mid a \leq x \leq b\},$$
$$(a, b] \equiv \{x \text{ in } \mathbb{R} \mid a < x \leq b\},$$

and

$$[a, b) \equiv \{x \text{ in } \mathbb{R} \mid a \leq x < b\}.$$

Moreover, it is convenient to use the symbols $\infty$ and $-\infty$ in the following manner. We define

$$[a, \infty) \equiv \{x \text{ in } \mathbb{R} \mid a \leq x\},$$
$$(-\infty, b] \equiv \{x \text{ in } \mathbb{R} \mid x \leq b\},$$
$$(a, \infty) \equiv \{x \text{ in } \mathbb{R} \mid a < x\},$$
$$(-\infty, b) \equiv \{x \text{ in } \mathbb{R} \mid x < b\},$$

and

$$(-\infty, \infty) \equiv \mathbb{R}.$$

The reader should be very careful to observe that although we have defined, say, $[a, \infty)$, we have *not* defined the symbols $\infty$ and $-\infty$. In particular, we have *not* adjoined additional numbers to $\mathbb{R}$.

It will also be convenient to set $[a, a] \equiv \{a\}$. In general, when we write $[a, b]$ or $(a, b)$, unless another meaning is explicitly mentioned, it is assumed that $a$ and $b$ are real numbers such that $a < b$.

Each of the sets listed above is called an *interval*. In the analysis of functions $f: A \rightarrow \mathbb{R}$, where $A$ is a set of real numbers, a special role is played by those functions that have an interval as their domain $A$. In particular, intervals of the form $(a, b)$, which we call *open intervals,* or of the form $[a, b]$, which we call *closed intervals,* will frequently be the domains of the functions that we will study.

# The Real Numbers

 **The Completeness Axiom: The Natural, Rational, and Irrational Numbers**

A rigorous understanding of mathematical analysis must be based on a proper understanding of the set of real numbers. The purpose of this first chapter is to establish the fundamental properties of the set $\mathbb{R}$ of real numbers.

The properties of addition and multiplication of real numbers have been codified in the Preliminaries as the Field Axioms. The set of real numbers is also equipped with the concept of order, and the properties of order and inequality have been codified in the Preliminaries as the Positivity Axioms. Many interesting properties of the real numbers are consequences of the Field and Positivity Axioms. However, an additional axiom is necessary. To explain why this is so, let us now introduce some special subsets of $\mathbb{R}$.

The subsets of $\mathbb{R}$ that we will now define are the natural numbers, the integers, and the rationals. Of course, these are already familiar to the reader: The natural numbers are the numbers 1, 2, 3, and so on. However, it is necessary to make this statement more precise, and a convenient way of doing so is to first introduce the concept of an *inductive set.*

**DEFINITION**    *A set S of real numbers is said to be inductive provided that*

   (i)  *the number 1 is in S*

*and*

   (ii)  *if the number x is in S, the number x + 1 is also in S.*

The whole set of real numbers $\mathbb{R}$ is inductive. Also, using just the fact that the number 1 is greater than the number 0, it follows that the set $\{x \text{ in } \mathbb{R} \mid x \geq 0\}$ is inductive, as is the set $\{x \text{ in } \mathbb{R} \mid x \geq 1\}$. The set of *natural numbers*, denoted by $\mathbb{N}$, is defined as the intersection of all inductive subsets of $\mathbb{R}$. The set $\mathbb{N}$ itself is inductive. To see this, observe that the number 1 belongs to $\mathbb{N}$, since 1 belongs to every inductive set. Furthermore, if the number $k$ belongs to $\mathbb{N}$, then $k$ belongs to every inductive set; thus, $k + 1$ belongs to every inductive set, and therefore $k + 1$ belongs to $\mathbb{N}$. Thus $\mathbb{N}$ is inductive and, by definition, it is contained in every other inductive set.

We define the set of *integers*, denoted by $\mathbb{Z}$, to be the set of numbers consisting of the natural numbers, their negatives, and the number 0. The set of *rational numbers*, denoted by $\mathbb{Q}$, is defined to be the set of quotients of integers. Now it is a little tedious, but not really difficult, to prove that the set of rational numbers satisfies the Field Axioms and the Positivity Axioms. However, it is not possible to develop calculus using only rational numbers. For instance, it will be necessary to conclude that a polynomial that attains both positive and negative values must also attain the value 0. This is not true if one considers only rational numbers. For example, consider the polynomial defined by $p(x) = x^2 - 2$ for all real numbers $x$. Then $p(0) < 0$ and $p(2) > 0$. However, as has been known since antiquity, there is no rational number $x$ having the property that $x^2 = 2$; that is, there is no rational number $x$ such that $p(x) = 0$. Before giving the classical proof of this assertion, we first note the following two properties of the integers:

- Each rational number $z$ may be expressed as $z = m/n$, where $m$ and $n$ are integers and either $m$ or $n$ is odd.

- If $q$ is an integer and $q^2$ is even, then $q$ is even.

The proofs of these properties are outlined in Exercises 23 and 24 of Section 1.3.

**PROPOSITION 1.1**    There is no rational number whose square equals 2.

**Proof**    We will suppose that the proposition is false and derive a contradiction. Suppose there is a rational number $x$ such that $x^2 = 2$. We may express $x$ as $x = m/n$, where $m$ and $n$ are integers and either $m$ or $n$ is odd. Since $m^2/n^2 = 2$, we have $m^2 = 2n^2$. Thus $m^2$ is even, so $m$ is also even. We now express $m$ as $m = 2k$ where $k$ is an integer. Since $m^2 = 2n^2$, we have $4k^2 = 2n^2$. Thus $n^2$ is even, so $n$ is also even. Hence both $m$ and $n$ are even. But we initially chose these integers so that at least one of them was odd.

The assumption that the proposition is false has led to a contradiction, so the proposition must be true.    ∎

Thus there is no rational number $x$ such that $x^2 = 2$, and hence it is not possible to prove even the simplest geometric result concerning the intersection of the graph of a polynomial and the $x$-axis (that is, points where $x^2 - 2 = 0$) if we restrict ourselves to rational numbers. Worse yet, even the Pythagorean Theorem fails if we restrict ourselves to rational numbers: If $r$ is the length of the hypotenuse of a right-angled triangle whose other two sides have length 1, then $r^2 = 2$, and so the length of the hypotenuse is not a rational number.

We need an additional axiom for the real numbers that at the very least assures us that there are real numbers whose square equals 2. The final axiom will be the Completeness Axiom. In order to state this axiom, we first need the following.

**DEFINITION**    *A nonempty set S of real numbers is said to be bounded above provided that there is a number c having the property that*

$$x \leq c \quad \text{for all } x \text{ in } S.$$

*Such a number c is called an upper bound for S.*

It is clear that if a number $c$ is an upper bound for a set $S$, then every number greater than $c$ is also an upper bound for this set. For a nonempty set $S$ of numbers that is bounded above, among all of the upper bounds for $S$ it is not at all obvious why there should be a smallest, or least, upper bound. In fact, the assertion that there is such a least upper bound will be the final axiom for the real numbers.

## The Completeness Axiom

*Suppose that S is a nonempty set of real numbers that is bounded above. Then among the set of upper bounds for S there is a smallest, or least, upper bound.*

For a nonempty set $S$ of real numbers that is bounded above, the *least upper bound* of $S$, the existence of which is asserted by the Completeness Axiom, will be denoted by l.u.b. $S$. Sometimes the least upper bound of $S$ is also called the *supremum* of $S$ and is denoted by sup $S$.

It is worthwhile to note explicitly that if the number $b$ is an upper bound for the set $S$, then in order to verify that $b =$ l.u.b. $S$ it is necessary to show that $b$ is less than any other upper bound for $S$. This task, however, is equivalent to showing that each number smaller than $b$ is not an upper bound for $S$. This observation will be used frequently, so we will record it here as the following.

**PROPOSITION 1.2**    Suppose that $S$ is a nonempty set of real numbers that is bounded above, and that the number $b$ is an upper bound for $S$. Then the following three assertions are equivalent:

(i)  $b =$ l.u.b. $S$.

(ii)  If $d$ is any upper bound for the set $S$, then $b \leq d$.

(iii)  If $a$ is any number less than $b$, then $a$ is not an upper bound for $S$, so there is a number $x$ in $S$ such that $a < x \leq b$.

At first glance, it is not at all apparent that the Completeness Axiom will help our development of analysis. Perhaps the best way to illustrate the role to be played by this final axiom is to prove that with the Completeness Axiom there is necessarily a real number $x$ such that $x^2 = 2$, and so, in particular, there are real numbers that are not rational. Such numbers are called *irrational.*

In preparation for the proof of the next proposition, it is useful to observe that from the difference of squares formula, $a^2 - b^2 = (a - b)(a + b)$, it follows that for any two nonnegative numbers $a$ and $b$, $a \leq b$ if and only if $a^2 \leq b^2$.

**PROPOSITION 1.3**

Let $c$ be a positive number. Then there is a positive number whose square is $c$.

**Proof**

Define $S = \{x \text{ in } \mathbb{R} \mid x > 0 \text{ and } x^2 < c\}$. First, observe that the set $S$ is nonempty: If $c > 1$, then the number 1 is in $S$, whereas if $c \leq 1$, then the number $c/2$ is in $S$. Second, observe that the set $S$ is bounded above: If the number $x$ is in $S$, then $x$ and $c + 1$ are positive numbers such that $x^2 < c < (c + 1)^2$, so $x < c + 1$. Thus the number $c + 1$ is an upper bound for $S$.

By the Completeness Axiom, the set $S$ has a least upper bound. We define

$$b = \text{l.u.b. } S,$$

and note that $b$ is positive. The Positivity Axiom P2 implies that either $b^2 < c$, $b^2 > c$, or $b^2 = c$. We will argue by contradiction to show that the first two possibilities do not occur and therefore $b^2 = c$.

First suppose that $b^2 < c$. We will choose a suitably small positive number $r$ such that $(b + r)^2$ is also less than $c$. Thus the number $b + r$, which is larger than $b$, belongs to $S$, contradicting the choice of $b$ as an upper bound for $S$. Hence it cannot be the case that $b^2 < c$. To see how to choose such a number $r$, observe that if $r$ is any positive number less than 1,

$$(b + r)^2 = b^2 + 2rb + r^2$$
$$< b^2 + r(2b + 1)$$
$$= c - (c - b^2) + r(2b + 1).$$

Hence $(b + r)^2 < c$ if $r$ is any positive number that is smaller than 1 and also smaller than $(c - b^2)/(2b + 1)$.

Now suppose that $b^2 > c$. We will choose a suitably small positive number $t$ less than $b$ such that $(b - t)^2$ is also greater than $c$. Hence $(b - t)^2 > x^2$ for all $x$ in $S$, and so, because $b - t$ is positive, it follows that $(b - t) > x$ for all $x$ in $S$. This means that the number $b - t$ is an upper bound for $S$, contradicting the choice of $b$ as the least upper bound for $S$. Thus it cannot be the case that $b^2 > c$. To see how to choose such a number $t$, observe that if $t$ is any positive number, then

$$(b - t)^2 = b^2 - 2tb + t^2$$
$$= c + (b^2 - c) - 2tb + t^2$$
$$> c + (b^2 - c) - 2tb.$$

Thus it suffices to choose for $t$ any positive number that is smaller than $b$ and is also smaller than $(b^2 - c)/2b$.

We have shown that it is not possible that $b^2 < c$ or that $b^2 > c$. We therefore conclude that $b^2 = c$. ∎

We will see in Chapter 3 that, in fact, the above proposition is a corollary of a much more general result, called the Intermediate Value Theorem.

For a positive number $c$, there is only one positive number whose square equals $c$. To see why this is so, just observe that if $a$ and $b$ are positive numbers each of whose square is $c$, then $0 = a^2 - b^2 = (a - b)(a + b)$. Since $a + b > 0$, it follows that $a = b$. As usual, we denote the positive number whose square is $c$ by $\sqrt{c}$. We define $\sqrt{0} \equiv 0$.

**DEFINITION**    *A nonempty set $S$ of real numbers is said to be bounded below provided that there is a number b having the property that*

$$b \leq x \quad \text{for all } x \text{ in } S.$$

*Such a number b is called a lower bound for S.*

It is clear that if a number $b$ is a lower bound for a set $S$, then every number less than $b$ is also a lower bound for $S$. We will now use the Completeness Axiom to show that for a nonempty set of numbers $S$ that is bounded below, among the lower bounds for the set there is a *greatest lower bound*, which is denoted by g.l.b. $S$. Sometimes the greatest lower bound of $S$ is called the *infimum* of $S$ and is denoted by inf $S$.

**THEOREM I.4**    Suppose that $S$ is a nonempty set of real numbers that is bounded below. Then among the set of lower bounds for $S$ there is a largest, or greatest, lower bound.

**Proof**    We will consider the set obtained by "reflecting" the set $S$ about the number 0; that is, we will consider the set $T = \{x \text{ in } \mathbb{R} \mid -x \text{ in } S\}$.

For any number $x$, $b \leq x$ if and only if $-x \leq -b$. Thus a number $b$ is a lower bound for $S$ if and only if the number $-b$ is an upper bound for $T$. Since the set $S$ has been assumed to be bounded below, it follows that the set $T$ is bounded above. The Completeness Axiom asserts that there is a least upper bound for $T$, which we will denote by $c$. Since lower bounds of $S$ occur as negatives of upper bounds for $T$, the number $-c$ is the greatest lower bound for $S$.    ∎

The following result will play an important part in our development of integration.

**THEOREM I.5**    **The Dedekind Gap Theorem**    Suppose that $S$ and $T$ are nonempty sets of real numbers having the property that

$$s \leq t \quad \text{for all } s \text{ in } S \text{ and } t \text{ in } T. \tag{1.1}$$

Then
$$\sup S \leq \inf T. \tag{1.2}$$

Moreover, the following three assertions are equivalent:

(i) There is exactly one real number $c$ having the property that

$$s \leq c \leq t \quad \text{for all } s \text{ in } S \text{ and } t \text{ in } T.$$

(ii) $\sup S = \inf T$.

(iii) For each positive number $\epsilon$, there are numbers $s$ in $S$ and $t$ in $T$ such that $t - s < \epsilon$.

**Proof**   From (1.1) it follows that every member of $T$ is an upper bound for the set $S$. Since the number sup $S$ is the smallest upper bound for the set $S$, we have sup $S \leq t$ for all $t$ in $T$. Thus the number sup $S$ is a lower bound for the set $T$. Being a lower bound, it is less than or equal to the greatest lower bound; that is, the inequality (1.2) holds.

We leave the proof of the equivalence of (i) and (ii) as an exercise; we will prove the equivalence of (ii) and (iii).

Suppose that (ii) holds. Let $\epsilon$ be any positive number. Since sup $S$ is the smallest upper bound for $S$, the number sup $S - \epsilon/2$ is not an upper bound for $S$. Thus there is a number $s$ in $S$ such that $s > \sup S - \epsilon/2$. Similarly, since inf $T$ is the largest lower bound for $T$, the number inf $T + \epsilon/2$ is not a lower bound for $T$. Thus there is a number $t$ in $T$ such that $t < \inf T + \epsilon/2$. Hence

$$t - s < (\inf T + \epsilon/2) - (\sup S - \epsilon/2) = \epsilon.$$

This proves assertion (iii).

Finally, we prove that (iii) implies (ii). Suppose that (iii) holds. Let $\epsilon$ be any positive number. Then we can choose numbers $s$ in $S$ and $t$ in $T$ such that $t - s < \epsilon$. But then

$$0 \leq \inf T - \sup S \leq t - s < \epsilon.$$

It follows that sup $S = \inf T$, for otherwise, by choosing $\epsilon = (\inf T - \sup S)/2$, we contradict the preceding inequality. This proves assertion (ii).   ∎

**EXERCISES**

1.   Let $A$ be a set of real numbers that is bounded above, and let $B$ be a nonempty subset of $A$. Prove that sup $A \geq \sup B$. (*Hint:* Define $a = \sup A$ and show that $a$ is an upper bound for $B$.)

2.   Suppose that $A$ is a nonempty set of real numbers that is both bounded above and bounded below, and that inf $A = \sup A$. Prove that the set $A$ consists of exactly one number.

3.   For a positive number $c$, show that if $x$ is any number such that $x^2 = c$, then either $x = \sqrt{c}$ or $x = -\sqrt{c}$.

4.   Prove that the sum of rational numbers is rational. Also prove that the sum of a rational and an irrational number must be irrational.

5.   For a set $A$ of numbers, a member $c$ of $A$ is called the *maximum* of $A$ provided that it is an upper bound for $A$. Prove that a set $A$ of numbers has a maximum if and only if it is bounded above and sup $A$ belongs to $A$. Give an example of a set $A$ of numbers that is nonempty and bounded above but has no maximum.

6.   Prove that $\sqrt{3}$ is not a rational number. (*Hint:* Follow the idea of the proof of Proposition 1.1.)

7.  Let $a$, $b$, and $c$ be real numbers such that $a \neq 0$, and consider the quadratic equation

$$ax^2 + bx + c = 0.$$

Prove that a number $x$ is a solution of this equation if and only if

$$(2ax + b)^2 = b^2 - 4ac.$$

Suppose that $b^2 - 4ac > 0$. Prove that the quadratic equation has exactly two solutions, given by

$$x = \frac{-b + \sqrt{b^2 - 4ac}}{2a} \quad \text{and} \quad x = \frac{-b - \sqrt{b^2 - 4ac}}{2a}.$$

8.  In the preceding exercise, now suppose that $b^2 - 4ac < 0$. Prove that there are no real numbers that are solutions of the quadratic equation.

9.  Define $S = \{x \text{ in } \mathbb{R} \mid x^2 < x\}$. Prove that $\sup S = 1$.

10. Prove that there is a positive number $x$ such that $x^3 = 3$. (*Hint:* Follow the proof of Proposition 1.3.)

## 1.2   The Archimedean Property and the Density of the Rationals and the Irrationals

We will devote this section to a more precise description of the way the natural numbers, the rational numbers, and the irrational numbers are situated in the set of real numbers. By its very definition, the set $\mathbb{N}$ of natural numbers is inductive and is contained in every other inductive set. Thus:

If $A$ is a set of natural numbers that is inductive, then $A = \mathbb{N}$.        (1.3)

Some elementary properties of the natural numbers follow immediately from this observation. For instance, if $n$ is a natural number, then $n \geq 1$; this follows from (1.3) if we observe that since the number 1 is greater than the number 0, the set $A = \{n \text{ in } \mathbb{N} \mid n \geq 1\}$ is an inductive subset of $\mathbb{N}$, and hence is equal to $\mathbb{N}$. Also, the interval $(1, 2)$ contains no natural numbers; this also follows from (1.3) if we observe that the set $\{1\} \cup \{n \text{ in } \mathbb{N} \mid n \geq 2\}$ is an inductive subset of $\mathbb{N}$, and hence is equal to $\mathbb{N}$.

Arguments that are based on the above property (1.3) occur so frequently that it is useful to formalize them as follows.

### Principle of Mathematical Induction

*For each natural number n, let S(n) be some mathematical assertion. Suppose that S(1) is true. Also suppose that whenever k is a natural number such that S(k) is true, then S(k + 1) is also true. Then S(n) is true for every natural number n.*

**Proof**   Define $A = \{k \text{ in } \mathbb{N} \mid S(k) \text{ is true}\}$. The assumptions mean precisely that $A$ is an inductive subset of $\mathbb{N}$. According to (1.3), $A = \mathbb{N}$. Thus $S(n)$ is true for every natural number $n$.                                                                                    ∎

For a set $A$ of numbers, a member of $A$ that is a lower bound for $A$ is called the *minimum* of $A$, and a member of $A$ that is an upper bound for $A$ is called the *maximum* of $A$. It is important to distinguish between *minimum* and *infimum*, and between *maximum* and *supremum*. The Completeness Axiom asserts that a nonempty set $A$ of real numbers that is bounded above always has a supremum. However, in general, the number sup $A$ need not be a member of $A$, so $A$ need not have a maximum. For instance, the interval $(0, 1)$ has no maximum, since its supremum lies outside of the interval. Similarly, although Theorem 1.4 asserts that each nonempty set of real numbers that is bounded below has an infimum, such a set need not have a minimum. However, for sets of natural numbers we can say more. In fact, every nonempty set of natural numbers has a minimum, and every nonempty set of natural numbers that is bounded above has a maximum. In order to verify these two assertions, it is useful first to prove the following.

**LEMMA 1.6**   Suppose that $n$ is a natural number. Then the interval $(n,\ n+1)$ contains no natural numbers.

**Proof**   We will prove this by induction. Let the assertion of the lemma be the inductive statement $S(n)$. We have already observed that $S(1)$ is true. Now suppose that $k$ is a natural number such that $S(k)$ is true. Then we claim that $S(k+1)$ must also be true. Indeed, if $S(k+1)$ is not true, then there is a natural number $m$ such that $k + 1 < m < k + 2$, in which case $m - 1$ is a natural number (see Exercise 9) with $k < m - 1 < k + 1$, which contradicts the inductive hypothesis $S(k)$. Thus $S(k+1)$ is true. By the Principle of Mathematical Induction, $S(n)$ is true for every natural number $n$.                                                ∎

For a natural number $n$, it will be convenient to denote by $\{1, \ldots, n\}$ the set of natural numbers $k$ such that $1 \le k \le n$.

**PROPOSITION 1.7**   Let $n$ be a natural number and let $A \subseteq \{1, \ldots, n\}$ be nonempty. Then the set $A$ has a minimum and a maximum.

**Proof**   We will also prove this by induction. Let the assertion of the proposition be the inductive statement $S(n)$. If $n = 1$, then $A = \{1\}$, so the number 1 is both a minimum and a maximum for the set $A$. Thus $S(1)$ is true.

Now let $k$ be a natural number such that $S(k)$ is true; that is, suppose that every nonempty subset of $\{1, \ldots, k\}$ has a minimum and a maximum. To prove $S(k+1)$, we select a nonempty subset $A$ of, $\{1, \ldots, k + 1\}$, and show that it has a maximum and a minimum. First we will show that $A$ has a maximum. If $k + 1$ belongs to $A$, then $k + 1$ is the maximum of $A$. If $k + 1$ does not belong to $A$, then, by Lemma 1.6, $A \subseteq \{1, \ldots, k\}$, so by the inductive assumption, the set $A$ has a maximum. Now we need to show that $A$ has a minimum. If the set $A \cap \{1, \ldots, k\}$ is empty, then, again using Lemma 1.6, we conclude that $A = \{k + 1\}$, so $k + 1$ is its minimum. Otherwise, the set $A \cap \{1, \ldots, k\}$ is nonempty, so by the inductive assumption, $A \cap \{1, \ldots, k\}$ has a minimum, and this minimum is also a minimum for $A$. Thus $S(k + 1)$ is true. By the Principle of Mathematical Induction, $S(n)$ is true for every natural number $n$. ∎

---

**PROPOSITION 1.8**   Every nonempty set of natural numbers has a minimum.

**Proof**   Let $A$ be a nonempty set of natural numbers. Since the set $A$ is nonempty, we may select a number $n$ in $A$. Then the set $A \cap \{1, \ldots, n\}$ is a nonempty subset of $\{1, \ldots, n\}$, and so, according to Proposition 1.7, it has a minimum. This minimum for $A \cap \{1, \ldots, n\}$ is also a minimum for $A$. ∎

Of course, it is not the case that each nonempty set of natural numbers has a maximum. For example, the set $\mathbb{N}$ itself does not have a maximum, since if $n$ is any natural number, then $n + 1$ is a natural number that is greater than $n$. Moreover, the set $\mathbb{N}$ of natural numbers is not bounded above. It is rather surprising that in order to prove this we will need to use the Completeness Axiom.

---

**THEOREM 1.9**   **The Archimedean Property**   For any number $c$, there is a natural number $n$ that is greater than $c$.

**Proof**   We will suppose that the theorem is false and derive a contradiction. Suppose that there is no natural number that is greater than $c$. Then, using the Positivity Axiom P2, we conclude that

$$n \leq c \quad \text{for every natural number } n.$$

Thus the set $\mathbb{N}$ of natural numbers is bounded above. The Completeness Axiom asserts that $\mathbb{N}$ has a least upper bound. Denote the least upper bound of $\mathbb{N}$ by $b$.

Since $b$ is the smallest upper bound for $\mathbb{N}$, the number $b - 1/2$ is not an upper bound for $\mathbb{N}$. Thus we can choose a positive integer $n$ that is greater than $b - 1/2$. But then $n + 1 > (b - 1/2) + 1 > b$, so $n + 1$ is a natural number that is larger than $b$. This contradicts the choice of $b$ as an upper bound for $\mathbb{N}$. This contradiction proves the result. ∎

The Archimedean Property is often stated as follows: For each pair of positive numbers $a$ and $b$, there is a natural number $n$ such that $na > b$. This is clearly equivalent to the above formulation.

---

**COROLLARY 1.10**   For each positive number $\epsilon$, there is a natural number $n$ such that $1/n < \epsilon$.

**Proof**   Let $\epsilon$ be a positive number. Then the number $1/\epsilon$ is also positive. By the Archimedean Property, there is a natural number $n$ such that $n > 1/\epsilon$, which means that $1/n < \epsilon$. ∎

We conclude this section with a theorem about the distribution in $\mathbb{R}$ of the rational numbers and the irrational numbers.

**DEFINITION**   *A set S of real numbers is said to be dense in* $\mathbb{R}$ *provided that every interval* $I = (a, b)$, *where* $a < b$, *contains a member of S.*

**THEOREM 1.11**   The set of rational numbers and the set of irrational numbers are *both* dense in $\mathbb{R}$.

**Proof**   Let $a$ and $b$ be real numbers such that $a < b$. We need to show that the interval $(a, b)$ contains both a rational and an irrational number. We will consider only the case when $0 \le a < b$, because the other cases follow easily from this one.

First we will show that the interval $(a, b)$ contains a rational number. According to Corollary 1.10, we can choose a natural number $m$ such that $1/m$ is less than the length of the interval; that is, $1/m < b - a$. Again using the Archimedean Property, we can choose another natural number $n$ such that $n/m > b$.

Define $A$ to be the set of natural numbers $k$ such that $k/m < b$. Since $1/m < b - a < b$, the number 1 belongs to the set $A$. Thus $A$ is nonempty. The natural number $n$ was chosen such that $n/m > b$, so every member of $A$ is less than $n$. Proposition 1.7 asserts that the set $A$ has a maximum. Denote the maximum by $j$. We will show that the rational number $j/m$ is in the interval $(a, b)$.

Since $j$ is in $A$, $j/m < b$. Thus it only remains to show that $j/m > a$. However, since $j$ is the maximum of the set $A$, the integer $j + 1$ is not in $A$, and this means that $(j + 1)/m \ge b$. This inequality, together with the inequality $1/m < b - a$, gives us

$$\frac{j}{m} \ge b - \frac{1}{m} > b - (b - a) = a.$$

It remains to find an irrational number in the interval $(a, b)$. Once more using Corollary 1.10, since $b - j/m > 0$, we can choose a natural number $\ell$ such that

$$\frac{\sqrt{2}}{\ell} < b - \frac{j}{m}.$$

Since $\sqrt{2}$ is irrational, the number

$$\frac{j}{m} + \frac{\sqrt{2}}{\ell}$$

is also irrational. (This assertion is a consequence of Exercise 4 of Section 1.1.) The natural number $\ell$ was chosen to be large enough to ensure that this irrational number lies in the interval $(a, b)$.   ∎

**EXERCISES**

1.   Verify that the sets $A = \{n \text{ in } \mathbb{N} \mid n \ge 1\}$ and $B = \{1\} \cup \{n \text{ in } \mathbb{N} \mid n \ge 2\}$ are inductive.
2.   Let $A$ be a nonempty set of real numbers that is bounded below. Prove that the set $A$ has a minimum if and only if the number inf $A$ belongs to $A$.
3.   For each of the following two sets, find the maximum, minimum, infimum, and supremum, if they are defined: (a) $\{1/n \mid n \text{ in } \mathbb{N}\}$, (b) $\{x \text{ in } \mathbb{R} \mid x^2 < 2\}$. Justify your conclusions.
4.   Suppose that the number $a$ has the property that for every natural number $n$, $a \le 1/n$. Prove that $a \le 0$.
5.   Given a real number $a$, define $S = \{x \mid x \text{ in } \mathbb{Q}, x < a\}$. Prove that $a = \sup S$.

6.  Prove that every bounded, nonempty set of natural numbers has a maximum.
7.  In the proof of Theorem 1.11, we considered only the case of an interval of the form $(a, b)$, where $0 \leq a < b$. Use this case to prove the other cases.
8.  Prove that the Archimedean Property is equivalent to Corollary 1.10.
9.  Let $n$ be a natural number greater than 1. Prove that $n - 1$ is also a natural number. (*Hint*: Prove that the set $\{n \mid n = 1 \text{ or } n \text{ in } \mathbb{N} \text{ and } n - 1 \text{ in } \mathbb{N}\}$ is inductive.)

## 1.3  Three Inequalities and Three Algebraic Identities

At the heart of many arguments in analysis lies the problem of estimating the sizes of various quotients, differences, and sums. In order to do so, it is useful to have available a small storehouse of inequalities and algebraic identities. In this section, we will derive three useful inequalities and three useful algebraic identities that will be used frequently throughout this book.

Recall that for a real number $x$, its *absolute value*, denoted by $|x|$, is defined by

$$|x| = \begin{cases} x & \text{if } x \geq 0 \\ -x & \text{if } x < 0. \end{cases}$$

Directly from this definition and from the Positivity Axioms for $\mathbb{R}$, it follows that if $c$ and $d$ are any numbers such that $d$ is nonnegative, then

$$|c| \leq d \quad \text{if and only if} \quad -d \leq c \leq d. \tag{1.4}$$

Given a pair of real numbers $a$ and $b$, we will often need to estimate the size of $|a + b|$. The following important inequality provides estimates, from above and from below, for the size of the absolute value of the sum of two numbers.

**THEOREM 1.12**    For any pair of numbers $a$ and $b$,

$$|a| - |b| \leq |a + b| \leq |a| + |b|. \tag{1.5}$$

**Proof**    We will first prove the right-hand inequality, $|a + b| \leq |a| + |b|$. Using (1.4), we see that this is equivalent to the assertion that

$$-|a| - |b| \leq a + b \leq |a| + |b|.$$

In fact, it follows from (1.4) that

$$-|x| \leq x \leq |x| \quad \text{for all real numbers } x.$$

This is sufficient to prove the preceding inequality.

Now the left-hand inequality in (1.5) is a consequence of the right-hand inequality in (1.5). Indeed,

$$|a| = |(a + b) + (-b)| \leq |a + b| + |b|,$$

since we have just shown that the absolute value of the sum is less than or equal to the sum of the absolute values. Thus, $|a| - |b| \leq |a + b|$.    ■

For later reference, we name each of the inequalities in (1.5).

## The Triangle Inequality

*For any two numbers a and b,*

$$|a + b| \leq |a| + |b|.$$

## The Reverse Triangle Inequality

*For any two numbers a and b,*

$$|a + b| \geq |a| - |b|.$$

For any two numbers $x$ and $y$, we define the *distance* between $x$ and $y$ to be $|x - y|$. We explicitly record the following three properties of distance, the first two of which follow directly from the definition of absolute value, and the last of which follows from the Triangle Inequality if we set $a = x - z$ and $b = z - y$.

**PROPOSITION 1.13**

For any numbers $x$, $y$, and $z$,

$$|x - y| = |y - x|,$$

$$|x - y| \geq 0, \quad \text{and} \quad |x - y| = 0 \quad \text{if and only if} \quad x = y,$$

and

$$|x - y| \leq |x - z| + |z - y|.$$

In the geometric language of distance between points, given a number $a$ and a positive number $r$, the set of numbers $x$ such that $|x - a| < r$ consists of all the numbers whose distance from $a$ is less than $r$. From (1.4) we see that this set consists exactly of the interval $(a - r, a + r)$.

For a natural number $n$ and any number $a$, as usual, we write $a^n$ to denote the product of $a$ multiplied by itself $n$ times. We will need a number of inequalities and algebraic identities involving such powers. We begin with the following important inequality.

## Bernoulli's Inequality

*For each natural number n and each number a such that $a \geq -1$,*

$$(1 + a)^n \geq 1 + na.$$

**Proof** We will give an inductive proof of this inequality. Fix $a \geq -1$. For each natural number $n$, let $S(n)$ be the assertion of the above inequality. Then $S(1)$ is clearly true. Now let $k$ be a natural number such that $S(k)$ is true; that is, $(1 + a)^k \geq 1 + ka$. We need to show that $S(k + 1)$ is true. However,

$$(1 + a)^{k+1} = (1 + a)^k (1 + a) \geq (1 + ka)(1 + a),$$

since $a + 1 \geq 0$ and $S(k)$ is true. Hence, since

$$(1 + ka)(1 + a) = 1 + (k + 1)a + ka^2 \geq 1 + (k + 1)a,$$

we see that $S(k + 1)$ is true. By the Principle of Mathematical Induction, $S(n)$ is true for every natural number $n$. ∎

A simple algebraic identity that we will find very useful is the following.

## Difference of Powers Formula

*For any natural number n and any numbers a and b,*

$$a^n - b^n = (a - b)(a^{n-1} + a^{n-2}b + \cdots + ab^{n-2} + b^{n-1}).$$

It is easy to verify this formula, just by expanding the right-hand side. Indeed,

$$(a - b)(a^{n-1} + a^{n-2}b + \cdots + ab^{n-2} + b^{n-1})$$
$$= a^n + a^{n-1}b + a^{n-2}b^2 + \cdots + a^2b^{n-2} + ab^{n-1}$$
$$- a^{n-1}b - a^{n-2}b^2 - \cdots - a^2b^{n-2} - ab^{n-1} - b^n$$
$$= a^n - b^n.$$

The special case of the Difference of Powers Formula that arises when we set $a = 1$ and $b = r$ becomes, upon rewriting, the following.

## Geometric Sum Formula

*For any natural number n and any number r $\neq$ 1,*

$$1 + r + r^2 + \cdots + r^{n-1} = \frac{1 - r^n}{1 - r}.$$

It will be useful to have a formula that expresses powers of the sum of the numbers $a$ and $b$ in terms of the powers of $a$ and of $b$. In order to state this formula, we need to introduce factorial notation. For each natural number $n$ we define the *factorial* of $n$, denoted by $n!$, as follows: We define $1! \equiv 1$, and if $k$ is any natural number for which $k!$ has been defined, we then define $(k + 1)! \equiv (k + 1)k!$. By the Principle of Mathematical Induction, the symbol $n!$ is defined for all natural numbers $n$. It is convenient to define $0! \equiv 1$. We also need to introduce, for each pair of nonnegative integers $n$ and $k$ such that $n \geq k$, the *binomial coefficient* $\binom{n}{k}$, which is defined by the formula

$$\binom{n}{k} \equiv \frac{n!}{k!(n - k)!}.$$

We have the following formula for $(a + b)^n$, a proof of which is outlined in Exercises 18 and 19.

## The Binomial Formula

*For each natural number n and each pair of numbers a and b,*

$$(a + b)^n = \binom{n}{0}a^n + \binom{n}{1}a^{n-1}b + \binom{n}{2}a^{n-2}b^2 + \cdots + \binom{n}{n-1}ab^{n-1} + \binom{n}{n}b^n.$$

Let us close this chapter by recalling the summation notation. For a natural number $n$ and numbers $a_0, a_1, \ldots, a_n$, we define

$$\sum_{k=0}^{n} a_k \equiv a_0 + a_1 + \cdots + a_n.$$

This notation condenses many formulas. For example, in this summation notation, the three algebraic formulas we have described become:

## The Difference of Powers Formula

$$a^n - b^n = (a - b) \sum_{k=0}^{n-1} a^{n-1-k}b^k.$$

## The Geometric Sum Formula

$$\sum_{k=0}^{n-1} r^k = \frac{1 - r^n}{1 - r} \quad \text{if } r \neq 1.$$

## The Binomial Formula

$$(a + b)^n = \sum_{k=0}^{n} \binom{n}{k}a^{n-k}b^k.$$

---

**EXERCISES**

1.   Let $a$ be a positive number. Prove that if $x$ is a number such that $|x - a| < a/2$, then $x > a/2$.

2.   Let $a$ and $b$ be numbers such that $|a - b| \leq 1$. Prove that $|a| \leq |b| + 1$.

3.   Let $a, b, c,$ and $d$ be numbers such that $|c| \neq |d|$. Prove that

$$\left| \frac{a + b}{c + d} \right| \leq \frac{|a| + |b|}{||c| - |d||}.$$

4.   For a natural number $n$ and any two nonnegative numbers $a$ and $b$, use the Difference of Powers Formula to prove that

$$a \leq b \quad \text{if and only if} \quad a^n \leq b^n.$$

5.   In the case when $a \geq 0$, show that Bernoulli's Inequality is a consequence of the Binomial Formula.

6.   Write out the Difference of Powers Formula explicitly for $n = 2, 3,$ and $4$.

7. Write out the Binomial Formula explicitly for $n = 2, 3$, and 4.

8. For a natural number $n$ and numbers $a$ and $b$ such that $a \geq b \geq 0$, prove that

$$a^n - b^n \geq nb^{n-1}(a - b).$$

9. Using the fact that the square of a real number is nonnegative, prove Cauchy's Inequality: For any numbers $a$ and $b$,

$$ab \leq \frac{1}{2}(a^2 + b^2).$$

Use Cauchy's Inequality to prove that if $a \geq 0$ and $b \geq 0$, then

$$\sqrt{ab} \leq \frac{1}{2}(a + b).$$

10. Let $a, b,$ and $c$ be nonnegative numbers. Prove the following inequalities:

(a)   $ab + bc + ca \leq a^2 + b^2 + c^2.$

(b)   $8abc \leq (a + b)(b + c)(c + a).$

(c)   $abc(a + b + c) \leq a^2b^2 + b^2c^2 + c^2a^2.$

11. (a)   For numbers $a, b,$ and $c$, prove that

$$1 + a + b + c \geq 1 - |a| - |b| - |c|.$$

(b)   For any numbers $c_0, c_1,$ and $c_2$, consider the polynomial defined by

$$p(x) = x^3 + c_2x^2 + c_1x + c_0 \quad \text{for all } x.$$

Prove that there is a positive number $r$ such that

$$p(x) > 0 \quad \text{for all } x \geq r.$$

(*Hint:* Factor out $x^3$ and use part (a).)

12. A function $f: \mathbb{R} \to \mathbb{R}$ is called *strictly increasing* provided that $f(u) > f(v)$ for all numbers $u$ and $v$ such that $u > v$.
   (a)   Define $p(x) = x^3$ for all $x$. Prove that the polynomial $p: \mathbb{R} \to \mathbb{R}$ is strictly increasing.
   (b)   Fix a number $c$ and define $q(x) = x^3 + cx$ for all $x$. Prove that the polynomial $q: \mathbb{R} \to \mathbb{R}$ is strictly increasing if and only if $c \geq 0$. (*Hint:* For $c < 0$, consider the graph to understand why it is not strictly increasing, and then prove it is not increasing.)

13. Let $n$ be a natural number and $a_1, a_2, \cdots, a_n$ be positive numbers. Prove that

$$(1 + a_1)(1 + a_2) \cdots (1 + a_n) \geq 1 + a_1 + a_2 + \cdots + a_n$$

and that

$$(a_1 + a_2 + \cdots + a_n)\left(a_1^{-1} + a_2^{-1} + \cdots + a_n^{-1}\right) \geq n^2.$$

14. Rewrite the Geometric Sum Formula and replace $n$ with $n + 1$ to show that for every natural number $n$,

$$\frac{1}{1 - r} = 1 + r + \cdots + r^n + \frac{r^{n+1}}{1 + r} \quad \text{if} \quad r \neq 1.$$

(a) Use the above formula to find a formula for

$$\frac{1}{1 + x^2} + \frac{1}{(1 + x^2)^2} + \cdots + \frac{1}{(1 + x^2)^n}.$$

(b) Also, show that if $a \neq 0$, then

$$\frac{1}{a} = 1 + (1 - a) + (1 - a)^2 + \frac{(1 - a)^3}{a}.$$

15. Use the Principle of Mathematical Induction to prove the following equalities for each natural number $n$:

(a) $\displaystyle\sum_{k=1}^{n} k = \frac{n(n + 1)}{2}.$

(b) $\displaystyle\sum_{k=1}^{n} k^2 = \frac{n(n + 1)(2n + 1)}{6}.$

16. Let $n$ be a natural number. Find a formula for $\sum_{k=1}^{n} k(k + 1)$.

17. Let $n$ be a natural number. Prove that

$$1^3 + 2^3 + \cdots + n^3 = (1 + 2 + \cdots + n)^2.$$

18. Prove that if $n$ and $k$ are natural numbers such that $k \leq n$, then

$$\binom{n + 1}{k} = \binom{n}{k - 1} + \binom{n}{k}.$$

19. Use the formula in the preceding exercise to provide an inductive proof of the Binomial Formula.

20. Prove that if $n$ is a natural number greater than 1, then $n - 1$ is also a natural number. (*Hint:* Prove that the set $\{n \mid n = 1$ or $n$ in $\mathbb{N}$ and $n - 1$ in $\mathbb{N}\}$ is inductive.)

21. Prove that if $n$ and $m$ are natural numbers such that $n > m$, then $n - m$ is also a natural number. (*Hint:* Prove this by induction on $m$, making use of the preceding exercise.)

22. Let $a$ be a nonzero number and $m$ and $n$ be integers. Prove the following equalities:

(a) $a^{m+n} = a^m a^n$

(b) $(ab)^n = a^n b^n$

23. A natural number $n$ is called *even* if it can be written as $n = 2k$ for some other natural number $k$, and is called *odd* if either $n = 1$ or $n = 2k + 1$ for some other natural number $k$.

(a) Prove that each natural number $n$ is either odd or even.

(b) Prove that if $m$ is a natural number, then $2m > 1$.

(c) Prove that a natural number $n$ cannot be both odd and even. (*Hint:* Recall that if $n$ and $m$ are natural numbers such that $m > n$, then $m - n$ is also a natural number.)

(d)    Suppose that $k_1, k_2, \ell_1,$ and $\ell_2$ are natural numbers such that $\ell_1$ and $\ell_2$ are odd. Prove that if $2^{k_1}\ell_1 = 2^{k_2}\ell_2$, then $k_1 = k_2$ and $\ell_1 = \ell_2$.

24.    (a)    Prove that if $n$ is a natural number, then $2^n > n$.

(b)    Prove that if $n$ is a natural number, then

$$n = 2^{k_0}\ell_0$$

for some odd natural number $\ell_0$ and some nonnegative integer $k_0$. (*Hint:* If $n$ is odd, let $k = 0$ and $\ell = n$; if $n$ is even, let $A = \{k$ in $\mathbb{N} \mid n = 2^k\ell$ for some $\ell$ in $\mathbb{N}.\}$ By (a), $A \subseteq \{1, 2, \ldots, n\}$. Choose $k_0$ to be the maximum of $A$.)

25.    Prove that the preceding two exercises are sufficient to prove the assertions that preceded the proof of the irrationality of $\sqrt{2}$.

26.    A real number of the form $m/2^n$ where $m$ and $n$ are integers, is called a *dyadic rational*. Prove that the set of dyadic rationals is dense in $\mathbb{R}$.

# 2

# Sequences of Real Numbers

## 2.1    The Convergence of Sequences

Two of the central topics in the analysis of real-valued functions of a real variable are the differentiation and integration of functions that have as their domain an interval of real numbers. Sequences of real numbers are also important and, in fact, properties of general functions can be deduced from an understanding of sequences. Accordingly, in this chapter we will study sequences of real numbers, and in Chapter 3 we will turn to the study of general functions.

**DEFINITION**    *A sequence of real numbers is a real-valued function whose domain is the set of natural numbers.*

Since in the first nine chapters we will be considering only sequences of real numbers, we will abbreviate *sequence of real numbers* by writing *sequence*. Also, rather than denoting a sequence with standard functional notation, such as $f : \mathbb{N} \to \mathbb{R}$, it is customary to use subscripts, replacing $f(n)$ with $a_n$, and denoting a sequence by $\{a_n\}$. A natural number $n$ is called an *index* for the sequence, and the number $a_n$ associated with the index $n$ is called the $n$th *term* of the sequence.

Often sequences are defined by presenting an explicit formula. Thus, for example, $\{1/n\}$ denotes the sequence that has, for each index $n$, an $n$th term equal to $1/n$. The sequence $\{1 + (-1)^n\}$ has, for each index $n$, an $n$th term equal to $1 + (-1)^n$, so the $n$th term of this sequence equals 0 if the index $n$ is odd, and equals 2 if the index $n$ is even.

Frequently sequences are defined in a less explicit manner, as in the following example.

**EXAMPLE 2.1** For each natural number $n$, define $a_n$ to be the largest natural number that is less than or equal to $\sqrt{n^3}$. Proposition 1.7 of Chapter 1 implies that every bounded nonempty set of natural numbers has a maximum, so that for each natural number $n$ there is a largest natural number that is less than or equal to $\sqrt{n^3}$. Thus the sequence $\{a_n\}$ is properly defined. We leave it as an exercise for the reader to find the first four terms of this sequence. □

We now give an example of a sequence $\{a_n\}$ that is defined recursively; that is, the sequence is defined by defining the first term $a_1$, then defining $a_{n+1}$ whenever $n$ is a natural number such that the $n$th term $a_n$ is defined. By the Principle of Mathematical Induction, the $n$th term $a_n$ is defined for every natural number $n$, and thus the sequence $\{a_n\}$ is properly defined.

**EXAMPLE 2.2** Define $a_1 = 1$. If $n$ is a natural number such that $a_n$ has been defined, then define

$$a_{n+1} = \begin{cases} a_n + 1/n & \text{if } a_n^2 \le 2 \\ a_n - 1/n & \text{if } a_n^2 > 2. \end{cases}$$

This formula defines the sequence recursively. We leave it as an exercise for the reader to find the first four terms of this sequence. □

**EXAMPLE 2.3** Let $r$ be any number. Define the sequence $\{s_n\}$ by

$$s_n = \sum_{k=1}^{n} r^k \quad \text{for every natural number } n. \qquad \square$$

**EXAMPLE 2.4** Define the sequence $\{s_n\}$ by

$$s_n = \sum_{k=1}^{n} \frac{1}{k} \quad \text{for every natural number } n. \qquad \square$$

The two preceding sequences are formed in the following manner: Given a sequence $\{c_n\}$, define a new sequence $\{s_n\}$ by the formula

$$s_n = \sum_{k=1}^{n} c_k \quad \text{for every natural number } n. \qquad (2.1)$$

Sequences formed in this manner are called *infinite series*.

We will be interested in sequences $\{a_n\}$ that have the following property: "As $n$ gets large, the $a_n$'s approach a fixed number." We make this precise as follows.

**DEFINITION**  *A sequence $\{a_n\}$ is said to converge to the number $a$ provided that for every positive number $\epsilon$ there is a natural number $N$ such that*

$$|a_n - a| < \epsilon \quad \text{for all integers } n \geq N.$$

Now, a given sequence may or may not converge. But if a sequence does converge, it cannot converge to more than one point. Indeed, suppose the sequence $\{a_n\}$ converges to $a$ and to $a'$. Observe that the Triangle Inequality implies that

$$|a - a'| \leq |a_n - a| + |a_n - a'| \quad \text{for every natural number } n. \tag{2.2}$$

Let $\epsilon > 0$. Since the sequence $\{a_n\}$ converges to $a$, we may choose a natural number $N_1$ such that $|a_n - a| < \epsilon$ for all integers $n \geq N_1$. Also, since the sequence $\{a_n\}$ converges to $a'$, we may choose a natural number $N_2$ such that $|a_n - a'| < \epsilon$ for all integers $n \geq N_2$. Choose $n$ to be a natural number greater than both $N_1$ and $N_2$. From inequality (2.2) and the choice of $N_1$ and $N_2$, it follows that

$$|a - a'| \leq |a_n - a| + |a_n - a'| < \epsilon + \epsilon = 2\epsilon.$$

Thus $a = a'$, since otherwise, by letting $\epsilon = |a - a'|/2$ , we contradict the preceding inequality.

If the sequence $\{a_n\}$ converges to the number $a$, we call $a$ the *limit of the sequence* $\{a_n\}$, and write

$$\lim_{n \to \infty} a_n = a.$$

**PROPOSITION 2.1**  The sequence $\{1/n\}$ converges to 0; that is, $\lim_{n \to \infty} 1/n = 0$.

**Proof**  Let $\epsilon > 0$. We need to find a natural number $N$ such that

$$\left| \frac{1}{n} - 0 \right| < \epsilon \quad \text{for all integers } n \geq N;$$

that is, $1/n < \epsilon$ if $n \geq N$. But by the Archimedean Property of $\mathbb{R}$, we may select a natural number $N$ such that $N > 1/\epsilon$. Thus $1/N < \epsilon$, and hence

$$\frac{1}{n} \leq \frac{1}{N} < \epsilon \quad \text{for all integers } n \geq N. \qquad \blacksquare$$

**EXAMPLE 2.5**  The sequence $\{(-1)^n\}$ does not converge. To see this, we argue by contradiction. Suppose that the sequence $\{(-1)^n\}$ converges to a number $a$. Taking $\epsilon = 1$, it follows from the definition of convergence that there is a natural number $N$ such that

$$|(-1)^n - a| < 1 \quad \text{for all integers } n \geq N.$$

In particular, $1 - a \leq |1 - a| = |(-1)^{2N} - a| < 1$, so $a > 0$. On the other hand, we also have $1 + a \leq |1 + a| = |(-1)^{2N+1} - a| < 1$, so $a < 0$. This contradiction shows that the sequence $\{(-1)^n\}$ does not converge. $\qquad \square$

**EXAMPLE 2.6**  The sequence $\{2/n^2 + 4/n + 3\}$ converges to 3; that is,

$$\lim_{n \to \infty} \left[ \frac{2}{n^2} + \frac{4}{n} + 3 \right] = 3.$$

In order to verify this assertion, we choose $\epsilon > 0$. Then we need to find a natural number $N$ such that

$$\left| \frac{2}{n^2} + \frac{4}{n} + 3 - 3 \right| < \epsilon \quad \text{for all integers } n \geq N. \tag{2.3}$$

Observe that

$$\left| \frac{2}{n^2} + \frac{4}{n} + 3 - 3 \right| = \frac{2}{n^2} + \frac{4}{n} \leq \frac{6}{n} \quad \text{for every natural number } n.$$

Now, by the Archimedean Property of $\mathbb{R}$, we may select a natural number $N$ such that $N > 6/\epsilon$. Thus $6/N < \epsilon$, and so

$$\left| \frac{2}{n^2} + \frac{4}{n} + 3 - 3 \right| \leq \frac{6}{n} \leq \frac{6}{N} < \epsilon \quad \text{for all integers } n \geq N,$$

and so (2.3) holds.                                                                                       □

We will soon prove a general result, Theorem 2.5, that will allow us to analyze the previous example, and others like it, in a very simple manner.

**PROPOSITION 2.2**  For any number $c$ such that $|c| < 1$, the sequence $\{c^n\}$ converges to 0; that is,

$$\lim_{n \to \infty} c^n = 0.$$

**Proof**  Let $\epsilon > 0$. We need to find a natural number $N$ such that

$$|c^n - 0| < \epsilon \quad \text{for all integers } n \geq N.$$

Observe that since $0 < |c| < 1$, if we set $d = 1/|c| - 1$, it follows that $|c| = 1/(1 + d)$ and $d$ is positive. Hence, using Bernoulli's Inequality, we obtain the inequality

$$|c|^n = \frac{1}{(1 + d)^n} \leq \frac{1}{1 + nd} \leq \frac{1}{nd} \text{ for every natural number } n. \tag{2.4}$$

Using the Archimedean Property of $\mathbb{R}$, we may choose a natural number $N$ such that $N > 1/\epsilon d$. Consequently, since the numbers $d$ and $\epsilon$ are positive, $1/Nd < \epsilon$, and hence, by inequality (2.4),

$$|c^n - 0| = |c|^n \leq \frac{1}{nd} \leq \frac{1}{Nd} < \epsilon \quad \text{for all integers } n \geq N. \quad ■$$

It is usually the case that after first examining some particular examples of a concept, we then find it useful to prove some general results. We will now prove that the sum of convergent sequences converges to the sum of the limits, the product of convergent sequences converges to the product of the limits, and, when all quotients are defined, the quotient of convergent sequences converges to the quotient of the limits. To do so, it is convenient first to introduce a definition and prove two preliminary lemmas.

**DEFINITION**   *A sequence $\{a_n\}$ is said to be bounded provided that there is a number $M$ such that*

$$|a_n| \leq M \quad \text{for every natural number } n.$$

**LEMMA 2.3**   Every convergent sequence is bounded.

**Proof**   Let $\{a_n\}$ be a sequence that converges to the number $a$. Taking $\epsilon = 1$, it follows from the definition of convergence that we may select a natural number $N$ such that

$$|a_n - a| < 1 \quad \text{for all integers } n \geq N.$$

Thus, using the Reverse Triangle Inequality, we have

$$|a_n| - |a| \leq |a_n - a| < 1 \quad \text{for all integers } n \geq N,$$

from which it follows that

$$|a_n| \leq 1 + |a| \quad \text{for all integers } n \geq N.$$

Define $M = \max\{1 + |a|, |a_1|, \ldots, |a_{N-1}|\}$. Then

$$|a_n| \leq M \quad \text{for every natural number } n.$$

Thus the sequence $\{a_n\}$ is bounded.                                         ■

**LEMMA 2.4**   Suppose that the sequence $\{b_n\}$ converges to the nonzero number $b$. Then there is a natural number $N$ such that

$$|b_n| > \frac{|b|}{2} \quad \text{for all integers } n \geq N.$$

**Proof**   Since $|b|/2$ is positive, we may take $\epsilon = |b|/2$ and use the definition of convergence of a sequence to choose a natural number $N$ such that

$$|b_n - b| < \frac{|b|}{2} \quad \text{for all integers } n \geq N.$$

Thus, using the Reverse Triangle Inequality, we have

$$|b| - |b_n| \leq |b_n - b| < \frac{|b|}{2} \quad \text{for all integers } n \geq N,$$

from which it follows that

$$|b_n| > \frac{|b|}{2} \cdot \quad \text{for all integers } n \geq N.$$                ■

**THEOREM 2.5**    Suppose that the sequence $\{a_n\}$ converges to the number $a$ and that the sequence $\{b_n\}$ converges to the number $b$. Then the sequence $\{a_n + b_n\}$ converges, and

$$(i) \quad \lim_{n \to \infty} [a_n + b_n] = a + b.$$

Also, the sequence $\{a_n b_n\}$ converges, and

$$(ii) \quad \lim_{n \to \infty} [a_n b_n] = ab.$$

Moreover, if $b_n \neq 0$ for all $n$ and $b \neq 0$, then the sequence $\{a_n / b_n\}$ converges, and

$$(iii) \quad \lim_{n \to \infty} \left[ \frac{a_n}{b_n} \right] = \frac{a}{b}.$$

**Proof of (i)**    Let $\epsilon > 0$. We need to find a natural number $N$ such that

$$|(a_n + b_n) - (a + b)| < \epsilon \quad \text{for all integers } n \geq N.$$

In order to do so, we first observe that for every natural number $n$,

$$|(a_n + b_n) - (a + b)| = |(a_n - a) + (b_n - b)|,$$

and hence, by the Triangle Inequality,

$$|(a_n + b_n) - (a + b)| \leq |a_n - a| + |b_n - b|. \tag{2.5}$$

Since the sequence $\{a_n\}$ converges to $a$, we may choose a natural number $N_1$ such that

$$|a_n - a| < \frac{\epsilon}{2} \quad \text{for all integers } n \geq N_1,$$

and since the sequence $\{b_n\}$ converges to $b$, we may choose a natural number $N_2$ such that

$$|b_n - b| < \frac{\epsilon}{2} \quad \text{for all integers } n \geq N_2.$$

Define $N = \max \{N_1, N_2\}$. Then from inequality (2.5) and the choice of $N_1$ and $N_2$, it follows that if $n \geq N$, then

$$|(a_n + b_n) - (a + b)| \leq |a_n - a| + |b_n - b| < \frac{\epsilon}{2} + \frac{\epsilon}{2} = \epsilon.$$

**Proof of (ii)**    Let $\epsilon > 0$. We need to find a natural number $N$ such that

$$|a_n b_n - ab| < \epsilon \quad \text{for all integers } n \geq N.$$

In order to do so, we first observe that for every natural number $n$,

$$a_n b_n - ab = a_n b_n - a_n b + a_n b - ab = a_n (b_n - b) + b(a_n - a),$$

and hence, by the Triangle Inequality,

$$|a_n b_n - ab| \leq |a_n||b_n - b| + |b||a_n - a|. \tag{2.6}$$

Lemma 2.3 asserts that every convergent sequence is bounded. Thus we may choose a number $M$ such that $|a_n| \leq M$ for every natural number $n$. This choice of $M$, together with inequality (2.6), implies that for every natural number $n$,

$$|a_n b_n - ab| \leq M|b_n - b| + |b||a_n - a|. \tag{2.7}$$

Since the sequence $\{b_n\}$ converges to $b$, we may choose a natural number $N_1$ such that

$$|b_n - b| < \frac{\epsilon}{2(M+1)} \quad \text{for all integers } n \geq N_1,$$

and since the sequence $\{a_n\}$ converges to $a$, we may choose a natural number $N_2$ such that

$$|a_n - a| < \frac{\epsilon}{2(|b|+1)} \quad \text{for all integers } n \geq N_2.$$

Let $N = \max\{N_1, N_2\}$. From inequality (2.7) and the choices of $N_1$ and $N_2$, it follows that if $n \geq N$, then

$$|a_n b_n - ab| \leq M|b_n - b| + |b||a_n - a| < \epsilon.$$

**Proof of (iii)**    Using (iii), it is clear that it suffices to prove that the sequence $\{1/b_n\}$ converges to $1/b$. Let $\epsilon > 0$. We need to find a natural number $N$ such that

$$\left| \frac{1}{b_n} - \frac{1}{b} \right| < \epsilon \quad \text{for all integers } n \geq N.$$

In order to do so, first observe that for every natural number $n$,

$$\frac{1}{b_n} - \frac{1}{b} = \frac{b - b_n}{bb_n}.$$

According to Lemma 2.4, we can choose a natural number $N$, such that

$$|b_n| > \frac{|b|}{2} \quad \text{for all integers } n \geq N_1.$$

Thus,

$$\left| \frac{1}{b_n} - \frac{1}{b} \right| = \left| \frac{b - b_n}{bb_n} \right| \leq \frac{2}{|b|^2}|b_n - b| \quad \text{for all integers } n \geq N_1. \tag{2.8}$$

Since the sequence $\{b_n\}$ converges to $b$ and the number $\epsilon|b|^2/2$ is positive, we can choose a natural number $N_2$ such that

$$|b_n - b| < \frac{\epsilon|b|^2}{2} \quad \text{for all integers } n \geq N_2. \tag{2.9}$$

Define $N = \max\{N_1, N_2\}$. From the inequalities (2.8) and (2.9), it follows that if $n \geq N$, then

$$\left| \frac{1}{b_n} - \frac{1}{b} \right| \leq \frac{2}{|b|^2}|b_n - b| < \epsilon. \qquad \blacksquare$$

For convenient reference, it is useful to name the assertions in the preceding theorem. For convergent sequences $\{a_n\}$ and $\{b_n\}$, we have:

**The Sum Property**

$$\lim_{n\to\infty}[a_n+b_n]=\lim_{n\to\infty}a_n+\lim_{n\to\infty}b_n.$$

**The Product Property**

$$\lim_{n\to\infty}[a_nb_n]=\lim_{n\to\infty}a_n\cdot\lim_{n\to\infty}b_n.$$

**The Quotient Property**

*If $b_n\neq 0$ for all natural numbers n, and $b\neq 0$, then*

$$\lim_{n\to\infty}\left[\frac{a_n}{b_n}\right]=\frac{\lim_{n\to\infty}a_n}{\lim_{n\to\infty}b_n}.$$

Furthermore, from the sum and product properties of convergent sequences we also get the following:

**The Difference Property**

$$\lim_{n\to\infty}[a_n-b_n]=\lim_{n\to\infty}a_n-\lim_{n\to\infty}b_n$$

and, more generally,

**The Linearity Property**

*For any two numbers $\alpha$ and $\beta$,*

$$\lim_{n\to\infty}[\alpha a_n+\beta b_n]=\alpha\lim_{n\to\infty}a_n+\beta\lim_{n\to\infty}b_n.$$

We have already shown that

$$\lim_{n\to\infty}\frac{1}{n}=0,\quad\text{and that if}\quad|c|<1,\quad\text{then}\quad\lim_{n\to\infty}c^n=0.$$

It is clear that a constant sequence converges to its constant value. These particular sequence calculations, together with Theorem 2.5, allow us to calculate the limits of many other sequences.

**EXAMPLE 2.7**   The preceding remarks, together with Theorem 2.5, imply that

$$\lim_{n\to\infty}\left[\left(\frac{3}{4}\right)^n+\frac{2}{n}-6\right]=\lim_{n\to\infty}\left[\left(\frac{3}{4}\right)^n\right]+2\lim_{n\to\infty}\left[\frac{1}{n}\right]-\lim_{n\to\infty}[6]=-6.\qquad\square$$

Theorem 2.5 describes the behavior of convergent sequences with respect to addition, multiplication, and division; that is, it relates convergence of sequences to the field properties of $\mathbb{R}$. We now turn to a description of the way in which the limit of a convergent sequence inherits order properties that are possessed by the individual terms of the sequence, in that we will show the relationship between convergence of sequences and the order properties of $\mathbb{R}$.

**LEMMA 2.6**   Suppose that the sequence $\{d_n\}$ converges to the number $d$, and that $d_n \geq 0$ for every natural number $n$. Then $d \geq 0$.

**Proof**   We will suppose that the conclusion is false and derive a contradiction. Indeed, suppose that $d < 0$. Letting $\epsilon = |d|/2$, we see that $\epsilon > 0$ and that $d + \epsilon = d/2 < 0$. Thus, the interval $(d - \epsilon, d + \epsilon)$ consists entirely of negative numbers, so no term of the sequence $\{d_n\}$ belongs to this interval. Therefore, the sequence $\{d_n\}$ cannot converge to $d$. This is a contradiction; thus, $d$ must be nonnegative.   ∎

The preceding lemma asserts that a convergent sequence of nonnegative numbers has a limit that is also nonnegative. It is not always true that a convergent sequence of positive numbers has a limit that is also positive. For instance, $\{1/n\}$ is a sequence of positive numbers that converges to 0.

**THEOREM 2.7**   Let the sequence $\{a_n\}$ converge to $a$, the sequence $\{b_n\}$ converge to $b$, and the sequence $\{c_n\}$ converge to $c$. Suppose that

$$a_n \leq c_n \leq b_n \quad \text{for every natural number } n.$$

Then

$$a \leq c \leq b.$$

**Proof**   The difference property of convergent sequences implies that the sequence $\{b_n - c_n\}$ converges to $b - c$. By assumption, $\{b_n - c_n\}$ is a sequence of nonnegative numbers. The preceding lemma implies that its limit is also nonnegative; that is, $c \leq b$. The very same argument, applied to the sequence $\{c_n - a_n\}$, shows that $a \leq c$.   ∎

**COROLLARY 2.8**   Let $a$ and $b$ be real numbers such that $a < b$. Suppose that the sequence $\{c_n\}$ converges to $c$, and that

$$a \leq c_n \leq b \quad \text{for every natural number } n.$$

Then $a \leq c \leq b$.

**Proof**   Define $a_n = a$ and $b_n = b$ for every natural number $n$. The conclusion follows from the preceding theorem applied to the sequences $\{a_n\}$, $\{b_n\}$, and $\{c_n\}$.   ∎

In the proof of the next theorem, we will use the observation that given a number $\ell$ and a positive number $r$, for any number $x$,

$$|x - \ell| < r \quad \text{if and only if} \quad \ell - r < x < \ell + r.$$

**THEOREM 2.9**   **The Squeezing Principle**   Let $\{a_n\}$, $\{b_n\}$, and $\{c_n\}$ be sequences such that

$$a_n \leq c_n \leq b_n \quad \text{for every natural number } n.$$

Suppose that the sequences $\{a_n\}$ and $\{b_n\}$ converge to the same limit $\ell$. Then the sequence $\{c_n\}$ also converges to $\ell$.

**Proof**    Let $\epsilon > 0$. We need to find a natural number $N$ such that

$$\ell - \epsilon < c_n < \ell + \epsilon \quad \text{for all integers } n \geq N. \tag{2.10}$$

Since the sequence $\{a_n\}$ converges to $\ell$, we can select a natural number $N_1$ such that

$$\ell - \epsilon < a_n < \ell + \epsilon \quad \text{for all integers } n \geq N_1.$$

In particular,

$$\ell - \epsilon < a_n \quad \text{for all integers } n \geq N_1. \tag{2.11}$$

On the other hand, since the sequence $\{b_n\}$ also converges to $\ell$, we can select a natural number $N_2$ such that

$$\ell - \epsilon < b_n < \ell + \epsilon \quad \text{for all integers } n \geq N_2.$$

In particular,

$$b_n < \ell + \epsilon \quad \text{for all integers } n \geq N_2. \tag{2.12}$$

Define $N = \max \{N_1, \ N_2\}$. From inequalities (2.11) and (2.12) and the assumption that $a_n \leq c_n \leq b_n$ for every natural number $n$, it follows that if $n \geq N$, then

$$\ell - \epsilon < a_n \leq c_n \leq b_n < \ell + \epsilon.$$

Thus we have found a natural number $N$ such that (2.10) holds.    ∎

---

**EXERCISES**

1.  Using only the Archimedean Property of $\mathbb{R}$, give a direct "$\epsilon$–$N$" verification of the following limits:

    (a)  $\displaystyle \lim_{n \to \infty} \frac{1}{\sqrt{n}} = 0$
    
    (b)  $\displaystyle \lim_{n \to \infty} \frac{1}{n + 5} = 0$

2.  Using only the Archimedean Property of $\mathbb{R}$, give a direct "$\epsilon$–$N$" verification of the convergence of the following sequences:

    (a)  $\displaystyle \left\{ \frac{2}{\sqrt{n}} + \frac{1}{n} + 3 \right\}$
    
    (b)  $\displaystyle \left\{ \frac{n^2}{n^2 + n} \right\}$

3.  Prove that the sequence $\{a_n\}$ converges to $a$ if and only if the sequence $\{|a_n - a|\}$ converges to 0.

4.  Let the sequence $\{b_n\}$ converge to $b$. Suppose that the sequence $\{a_n\}$ and the number $a$ have the property that there is a number $M$ and a natural number $N$ such that

    $$|a_n - a| \leq M|b_n - b| \quad \text{for all integers } n \geq N.$$

    Prove that the sequence $\{a_n\}$ converges to $a$.

5.  Suppose that the sequence $\{a_n\}$ converges to $a$. Use the Reverse Triangle Inequality to prove that the sequence $\{|a_n|\}$ converges to $|a|$.

6.  Suppose $\{a_n\}$ is a sequence of nonnegative numbers that converges to $a$. Prove that the sequence $\{\sqrt{a_n}\}$ converges to $\sqrt{a}$. (*Hint:* Consider the cases $a = 0$ and $a > 0$ separately.)

7.  If the sequence $\{a_n^2\}$ converges, does the sequence $\{a_n\}$ also necessarily converge?

**8.** (a)   Prove that if $0 < r < 1$, then

$$\lim_{n \to \infty} \sqrt{n}\, r^n = 0.$$

   (*Hint:* Follow the proof that $\lim_{n \to \infty} r^n = 0$.)

   (b)   Prove that if $0 < r < 1$, then

$$\lim_{n \to \infty} n r^n = 0.$$

   (*Hint:* if $a = \sqrt{r}$, then $n r^n = (\sqrt{n}\, a^n)(\sqrt{n}\, a^n)$.)

**9.** Prove that

$$\lim_{n \to \infty} n^{1/n} = 1.$$

   (*Hint:* Define $\alpha_n = n^{1/n} - 1$ and show that $n = (1 + \alpha_n)^n \geq 1 + [n(n-1)/2]\alpha_n^2$.)

**10.** Suppose that the sequence $\{a_n\}$ converges to $a$. Give necessary and sufficient conditions for the sequence $\{(-1)^n a_n\}$ to converge. Prove your assertion.

**11.** Prove that the Archimedean Property of $\mathbb{R}$ is equivalent to the fact that $\lim_{n \to \infty} 1/n = 0$.

**12.** We have proven that $\lim_{n \to \infty} c^n = 0$ if $|c| < 1$. Prove that the sequence $\{c^n\}$ does not converge if $|c| > 1$.

**13.** Let $\{a_n\}$ be a sequence of real numbers. Suppose that for each positive number $c$ there is a natural number $N$ such that

$$a_n > c \quad \text{for all integers } n \geq N.$$

When this is so, the sequence $\{a_n\}$ is said to *converge to infinity*, and we write

$$\lim_{n \to \infty} a_n = \infty.$$

Prove the following:

   (a)   $\displaystyle\lim_{n \to \infty} [n^3 - 4n^2 - 100n] = \infty$   (b)   $\displaystyle\lim_{n \to \infty} \left[ \sqrt{n} - \frac{1}{n^2} + 4 \right] = \infty$

**14.** Discuss the convergence of each of the following sequences:

   (a)   $\{\sqrt{n+1} - \sqrt{n}\}$   (b)   $\{(\sqrt{n+1} - \sqrt{n})\sqrt{n}\}$   (c)   $\{(\sqrt{n+1} - \sqrt{n})n\}$

**15.** Define the sequence $\{s_n\}$ by

$$s_n = \frac{1}{2 \cdot 1} + \frac{1}{3 \cdot 2} + \cdots + \frac{1}{(n+1)(n)} \qquad \text{for every natural number } n.$$

Prove that

$$\lim_{n \to \infty} s_n = 1.$$

**16.** Let the sequences $\{a_n\}$ and $\{b_n\}$ have the property that

$$\lim_{n \to \infty} [a_n^2 + b_n^2] = 0.$$

Prove that

$$\lim_{n \to \infty} a_n = \lim_{n \to \infty} b_n = 0.$$

**17.** Suppose that the sequence $\{a_n\}$ converges to $a$ and that $|a| < 1$. Prove that the sequence $\{(a_n)^n\}$ converges to 0.

**18.** We have proven that the sequence $\{1/n\}$ converges to 0 and that it does not converge to any other number. Use this to prove that none of the following assertions is equivalent to the definition of convergence of a sequence $\{a_n\}$ to the number $a$.

(a)    For some $\epsilon > 0$ there is a natural number $N$ such that

$$|a_n - a| < \epsilon \quad \text{for all integers } n \geq N.$$

(b)    For each $\epsilon > 0$ and each natural number $N$,

$$|a_n - a| < \epsilon \quad \text{for all integers } n \geq N.$$

(c)    There is a natural number $N$ such that for every number $\epsilon > 0$,

$$|a_n - a| < \epsilon \quad \text{for all integers } n \geq N.$$

19.    For the sequence defined in Example 2.2, prove that for every integer $n$, $|a_n - \sqrt{2}| < 1/n$. Use this property to show that the sequence converges to $\sqrt{2}$ .

20.    (The convergence of Cesaro averages.) Suppose that the sequence $\{a_n\}$ converges to $a$. Define the sequence $\{\sigma_n\}$ by

$$\sigma_n = \frac{a_1 + a_2 + \cdots + a_n}{n} \quad \text{for every natural number } n.$$

Prove that the sequence $\{\sigma_n\}$ also converges to $a$.

---

## 2.2    Monotone Sequences, the Bolzano-Weierstrass Theorem, and the Nested Interval Theorem

In Section 2.1 we showed that $\lim_{n \to \infty} 1/n = 0$, and that $\lim_{n \to \infty} c^n = 0$ provided that $|c| < 1$. It is clear that constant sequences converge to their constant value. Thus, using the sum, product, and quotient properties of convergent sequences, we can combine these three examples to obtain further examples of convergent sequences. It will be important to analyze more general sequences. We now turn to the task of providing criteria that are sufficient to determine that a sequence converges, but that do not require any explicit knowledge of the proposed limit.

**DEFINITION**    *A sequence $\{a_n\}$ is said to be monotonically increasing provided that*

$$a_{n+1} \geq a_n \quad \text{for every natural number } n.$$

*A sequence $\{a_n\}$ is said to be monotonically decreasing provided that*

$$a_{n+1} \leq a_n \quad \text{for every natural number } n.$$

*A sequence $\{a_n\}$ is called monotone if it is either monotonically increasing or monotonically decreasing.*

In the preceding section, we proved that if a sequence converges, it must be bounded. Of course, as the sequence $\{(-1)^n\}$ shows, a bounded sequence need not converge. However, in the case of a monotone sequence, there is the following important theorem.

**THEOREM 2.10**   **The Monotone Convergence Theorem**   A monotone sequence converges if and only if it is bounded.

**Proof**   We have already proven that a convergent sequence is bounded, so it remains to be shown that if the sequence $\{a_n\}$ is bounded and monotone, it converges. Let us first suppose that the sequence $\{a_n\}$ is monotonically increasing. Then if we define $S = \{a_n \mid n \in \mathbb{N}\}$, by assumption, the set $S$ is bounded above. According to the Completeness Axiom, $S$ has a least upper bound. Define $a = \text{l.u.b.}\ S$. We claim that the sequence $\{a_n\}$ converges to $a$. Indeed, let $\epsilon > 0$. We need to find a natural number $N$ such that

$$|a_n - a| < \epsilon \quad \text{for all integers } n \geq N;$$

that is,

$$a - \epsilon < a_n < a + \epsilon \quad \text{for all integers } n \geq N. \tag{2.13}$$

Since the number $a$ is an upper bound for the set $S$, we have

$$a_n \leq a < a + \epsilon \quad \text{for every natural number } n. \tag{2.14}$$

On the other hand, since $a$ is the least upper bound for $S$, the number $a - \epsilon$ is not an upper bound for $S$, so there is a natural number $N$ such that $a - \epsilon < a_N$. However, the sequence $\{a_n\}$ is monotonically increasing, so

$$a - \epsilon < a_N \leq a_n \quad \text{for all integers } n \geq N. \tag{2.15}$$

From the inequalities (2.14) and (2.15), we obtain the required inequality (2.13). Thus, the sequence $\{a_n\}$ converges to $a$.

It remains to consider the case when the sequence $\{a_n\}$ is monotonically decreasing. But then the sequence $\{-a_n\}$ is monotonically increasing. By the case just considered, $\{-a_n\}$ converges, and so, using the linearity property of convergent sequences, it follows that $\{a_n\}$ also converges. ∎

**EXAMPLE 2.8**   Define

$$s_n = \sum_{k=1}^{n} \frac{1}{k} \cdot \frac{1}{2^k} \quad \text{for every natural number } n.$$

Then it is clear that the sequence $\{s_n\}$ is monotonically increasing. According to the Monotone Convergence Theorem, $\{s_n\}$ converges if and only if it is bounded. We will show that it is bounded. Indeed, using the Geometric Sum Formula, we see that for every natural number $n$,

$$s_n \leq \frac{1}{2} + \left(\frac{1}{2}\right)^2 + \cdots + \left(\frac{1}{2}\right)^n = \frac{1/2 - (1/2)^{n+1}}{1 - 1/2} \leq 1.$$

Hence the monotone sequence $\{s_n\}$ is bounded, and so it converges. Observe that we have proven that $\{s_n\}$ converges without explicitly identifying the limit. □

**EXAMPLE 2.9** Define

$$s_n = \sum_{k=1}^{n} \frac{1}{k} \quad \text{for every natural number } n.$$

The series $\{s_n\}$ is called the *harmonic series*. Again, we see that the sequence $\{s_n\}$ is monotonically increasing. We claim that it is not bounded and hence not convergent. Indeed, to see this, observe that

$$s_2 = 1 + \frac{1}{2} \geq 1 + \frac{1}{2},$$

and that

$$s_4 = s_2 + \frac{1}{3} + \frac{1}{4} \geq 1 + \frac{1}{2} + \frac{1}{2} = 1 + \frac{2}{2},$$

and that in general we have

$$s_{2^n} \geq 1 + \frac{n}{2} \quad \text{for every natural number } n.$$

From this, using the Archimedean Property of $\mathbb{R}$, it follows that the sequence $\{s_n\}$ is not bounded and hence does not converge. $\square$

**EXAMPLE 2.10** Consider the sequence $\{(1 + 1/n)^n\}$. We claim that this sequence is monotonically increasing. Indeed, using the Binomial Formula from Section 1.3, we see that for all natural numbers $n$,

$$\left(1 + \frac{1}{n}\right)^n = 1 + n \cdot \frac{1}{n} + \frac{n(n-1)}{1 \cdot 2} \frac{1}{n^2} + \cdots + \frac{n(n-1)\cdots(n-n+1)}{1 \cdot 2 \cdots n} \frac{1}{n^n}$$

$$= 1 + 1 + \frac{1}{1 \cdot 2}\left(1 - \frac{1}{n}\right) + \frac{1}{1 \cdot 2 \cdot 3}\left(1 - \frac{1}{n}\right)\left(1 - \frac{2}{n}\right) + \cdots$$

$$+ \frac{1}{1 \cdot 2 \cdots n}\left(1 - \frac{1}{n}\right)\left(1 - \frac{2}{n}\right)\cdots\left(1 - \frac{n-1}{n}\right).$$

From this, using a similar expansion for $(1 + 1/(n+1))^{n+1}$ and comparing the first $n + 1$ terms, we see that for every natural number $n$,

$$\left(1 + \frac{1}{n}\right)^n < \left(1 + \frac{1}{n+1}\right)^{n+1}.$$

Hence the sequence $\{(1 + 1/n)^n\}$ is monotonically increasing. The above expansion and the Geometric Sum Formula imply that for every natural number $n$,

$$2 < \left(1 + \frac{1}{n}\right)^n < 2 + \frac{1}{2} + \cdots + \frac{1}{2^{n-1}} < 3.$$

The Monotone Convergence Theorem implies that the sequence $\{(1 + 1/n)^n\}$ converges. The limit of this sequence is one of the important constants of mathematics and is denoted by $e$. From the above we see that $2 < e \leq 3$. In Chapter 8 we will estimate the number $e$ more precisely. $\square$

**DEFINITION**  *Consider a sequence $\{a_n\}$. Let $\{n_k\}$ be a sequence of natural numbers that is strictly increasing; that is,*

$$n_1 < n_2 < n_3 < \cdots.$$

*Then the sequence $\{b_k\}$ defined by*

$$b_k = a_{n_k} \quad \text{for every natural number } k$$

*is called a subsequence of the sequence $\{a_n\}$.*

Often a subsequence of $\{a_n\}$ is simply denoted by $\{a_{n_k}\}$, it being implicitly understood that $\{n_k\}$ is a strictly increasing sequence of natural numbers and that the $k$th term of the sequence $\{a_{n_k}\}$ is $a_{n_k}$.

A given sequence may or may not be monotone. But, in fact, every sequence has a monotone subsequence. This is not at all obvious. For instance, it is not obvious that the sequence $\{\sin(n)\}$ has a monotone subsequence.

**THEOREM 2.11**  Every sequence has a monotone subsequence.

**Proof**  Consider a sequence $\{a_n\}$. We call a natural number $m$ a *peak index* for the sequence $\{a_n\}$ provided that $a_n \le a_m$ for all integers $n \ge m$.

Either there are only finitely many peak indices for the sequence or there are infinitely many such indices.

*Case 1. There are finitely many peak indices.* Then we may choose a natural number $N$ such that there are no peak indices greater than $N$. We will recursively define a monotonically increasing subsequence of $\{a_n\}$. Indeed, define $n_1 = N + 1$. Now suppose that $k$ is a natural number such that positive integers

$$n_1 < n_2 < \cdots < n_k$$

have been chosen such that

$$a_{n_1} < a_{n_2} < \cdots < a_{n_k}.$$

Since $n_k > N$, the index $n_k$ is not a peak index. Hence there is an index $n_{k+1} > n_k$ such that $a_{n_{k+1}} > a_{n_k}$. Thus we recursively define a strictly increasing sequence of positive integers $\{n_k\}$ having the property that the subsequence $\{a_{n_k}\}$ is monotonically increasing.

*Case 2. There are infinitely many peak indices.* For each natural number $k$, let $n_k$ be the $k$th peak index. From the very definition of peak index it follows that the subsequence $\{a_{n_k}\}$ is monotonically decreasing. ∎

**THEOREM 2.12**  Every bounded sequence has a convergent subsequence.

**Proof** Let $\{a_n\}$ be a bounded sequence. According to the preceding theorem, we may choose a monotone subsequence $\{a_{n_k}\}$. Since $\{a_n\}$ is bounded, so is its subsequence $\{a_{n_k}\}$. Hence $\{a_{n_k}\}$ is a bounded monotone sequence. According to the Monotone Convergence Theorem, $\{a_{n_k}\}$ converges. ∎

There is a slightly more refined version of Theorem 2.12 that we will find very useful.

For a subset $D$ of $\mathbb{R}$, we say that the sequence $\{a_n\}$ is *a sequence in $D$* provided that for every natural number $n$, $a_n$ belongs to $D$.

**THEOREM 2.13** **The Bolzano-Weierstrass Theorem** Let $a$ and $b$ be numbers such that $a < b$. Every sequence in the interval $[a, b]$ has a subsequence that converges to a point in $[a, b]$.

**Proof** Let $\{x_n\}$ be a sequence in $[a, b]$. Then $\{x_n\}$ is bounded. Hence, by the preceding theorem, there is a subsequence $\{x_{n_k}\}$ that converges. But the sequence $\{x_{n_k}\}$ is a sequence in $[a, b]$, and hence, according to Corollary 2.8, its limit is also in $[a, b]$. ∎

The following result should not be surprising.

**PROPOSITION 2.14** Let the sequence $\{a_n\}$ converge to the limit $a$. Then every subsequence of $\{a_n\}$ also converges to the same limit $a$.

**Proof** Let $\{a_{n_k}\}$ be a subsequence of $\{a_n\}$. Let $\epsilon > 0$. We need to find a natural number $N$ such that

$$|a_{n_k} - a| < \epsilon \quad \text{for all integers } k \geq N. \tag{2.16}$$

Since the whole sequence $\{a_n\}$ converges to $a$, we may choose a natural number $N$ such that

$$|a_n - a| < \epsilon \quad \text{for all integers } n \geq N. \tag{2.17}$$

But observe that since $\{n_k\}$ is a strictly increasing sequence of natural numbers,

$$n_j \geq j \quad \text{for every natural number } j.$$

Thus inequality (2.16) follows from inequality (2.17). Hence the subsequence $\{a_{n_k}\}$ also converges to $a$. ∎

The sequence $\{a_{n+1}\}$ is a subsequence of $\{a_n\}$. Hence,

$$\text{if } \lim_{n \to \infty} a_n = a, \quad \text{then} \quad \lim_{n \to \infty} a_{n+1} = a \quad \text{as well.}$$

This simple observation can be quite useful in analyzing sequences that are defined recursively.

**EXAMPLE 2.11**   Define $a_1 = 1$. If $n$ is a natural number for which $a_n$ has been defined, then define

$$a_{n+1} = \frac{1+a_n}{2+a_n}.$$

An induction argument shows that $\{a_n\}$ is a sequence of positive numbers. Moreover, directly from the definition of the sequence, it follows that for every natural number $n$,

$$a_{n+2} - a_{n+1} = \frac{a_{n+1} - a_n}{(2 + a_n)(2 + a_{n+1})}.$$

Since $a_2 < a_1$, the preceding identity and an induction argument show that $\{a_n\}$ is monotonically decreasing. According to the Monotone Convergence Theorem, the sequence $\{a_n\}$ converges. Denote the limit by $a$. From the fact that $\lim_{n\to\infty} a_{n+1} = a$, and also from the sum, product, and quotient properties of convergent sequences, it follows that

$$a = \lim_{n\to\infty} a_{n+1} = \lim_{n\to\infty} \frac{1+a_n}{2+a_n} = \frac{1+a}{2+a}.$$

Thus, $a^2 + a - 1 = 0$ and $a \geq 0$. It follows from the quadratic formula that

$$\lim_{n\to\infty} a_n = \frac{-1 + \sqrt{5}}{2}. \qquad \qquad \square$$

We finish this section with a geometric consequence of the Monotone Convergence Theorem.

**THEOREM 2.15**   **The Nested Interval Theorem**   For each natural number $n$, let $a_n$ and $b_n$ be numbers such that $a_n < b_n$, and consider the interval $I_n = [a_n, b_n]$. Assume that

$$I_{n+1} \subseteq I_n \qquad \text{for every natural number } n. \tag{2.18}$$

Also assume that

$$\lim_{n\to\infty} [b_n - a_n] = 0. \tag{2.19}$$

Then there is exactly one point $x$ that belongs to the interval $I_n$ for all $n$, and both of the sequences $\{a_n\}$ and $\{b_n\}$ converge to this point.

**Proof**   Assumption (2.18) means precisely that for every natural number $n$,

$$a_n \leq a_{n+1} < b_{n+1} \leq b_n.$$

In particular, the sequence $\{a_n\}$ is a monotonically increasing sequence that is bounded above by $b_1$. The Monotone Convergence Theorem implies that the sequence $\{a_n\}$ converges; we will denote its limit by $a$. Observe that $a_n \leq a$ for every natural number $n$. A similar argument shows that the monotonically decreasing sequence $\{b_n\}$ converges to a number that we denote by $b$, and that $b \leq b_n$ for every natural number $n$. Thus

$$a_n \leq a \text{ and } b \leq b_n \qquad \text{for every natural number } n. \tag{2.20}$$

Using assumption (2.19) and the difference property of convergent sequences, we conclude that

$$0 = \lim_{n \to \infty} [b_n - a_n] = b - a.$$

Thus $a = b$. Setting $x = a = b$, it follows from (2.20) that the point $x$ belongs to $I_n$ for every natural number $n$. There can only be one such point, since the existence of two such points would contradict the assumption (2.19) that the lengths of the intervals converge to 0.    ∎

---

**EXERCISES**

1.  Which of the following sequences is monotone? Justify your conclusions.

    (a)  $\left\{ n + \dfrac{(-1)^n}{n} \right\}$    (b)  $\left\{ \dfrac{(-1)^n}{5n} \right\}$    (c)  $\left\{ \dfrac{1}{n^2} + \dfrac{(-1)^n}{3^n} \right\}$

2.  For each of the following sequences, find the peak indices. Justify your conclusions.

    (a)  $\left\{ \dfrac{1}{n} \right\}$    (b)  $\{(-1)^n\}$    (c)  $\{(-1)^n n\}$    (d)  $\left\{ \dfrac{(-1)^n}{n} \right\}$

3.  Prove that the sequence

    $$\left\{ 1 + \frac{1}{2!} + \cdots + \frac{1}{n!} \right\}$$

    converges. (We will see later that it converges to $e - 1$.)

4.  Suppose that the sequence $\{a_n\}$ is monotone. Prove that $\{a_n\}$ converges if and only if $\{a_n^2\}$ converges. Show that this result does not hold without the monotonicity assumption.

5.  Let $\{b_n\}$ be a bounded sequence of nonnegative numbers and $r$ be any number such that $0 \le r < 1$. Define

    $$s_n = b_1 r + b_2 r^2 + \cdots + b_n r^n \qquad \text{for every natural number } n.$$

    Use the Monotone Convergence Theorem to prove that the series $\{s_n\}$ converges.

6.  Let $a > 0$. Prove that the sequence $\{a^{1/n}\}$ converges. Then, by considering the subsequence $\{\sqrt{a^{1/n}}\}$, prove that $\lim_{n \to \infty} a^{1/n} = 1$.

7.  Let $\{a_{n_k}\}$ be a subsequence of the sequence $\{a_n\}$. Show that $n_k \ge k$ for every natural number $k$.

8.  For a positive number $c$, consider the quadratic equation

    $$x^2 - x - c = 0, \qquad x > 0.$$

    Define the sequence $\{x_n\}$ recursively by fixing $x_1 > 0$ and then, if $n$ is a natural number for which $x_n$ has been defined, defining

    $$x_{n+1} = \sqrt{c + x_n}.$$

    Prove that the sequence $\{x_n\}$ converges monotonically to the solution of the above equation.

9.  For a pair of positive numbers $\alpha$ and $\beta$, the number $\sqrt{\alpha\beta}$ is called the *geometric mean* of $\alpha$ and $\beta$, and the number $(\alpha + \beta)/2$ is called the *arithmetic mean* of $\alpha$ and $\beta$. By observing that $(\sqrt{\alpha} - \sqrt{\beta})^2 \ge 0$, show that $(\alpha + \beta)/2 \ge \sqrt{\alpha\beta}$.

10.   For a pair of positive numbers $a$ and $b$, define sequences $\{a_n\}$ and $\{b_n\}$ recursively as follows: Define $a_1 = a$ and $b_1 = b$. If $n$ is a natural number for which $a_n$ and $b_n$ have been defined, define

$$a_{n+1} = \frac{a_n + b_n}{2} \quad \text{and} \quad b_{n+1} = \sqrt{a_n b_n}.$$

11.   (a)   Use the preceding exercise to prove that for every natural number $n$,

$$a_n \geq a_{n+1} \geq b_{n+1} \geq b_n.$$

(b)   From (a), show that the sequences $\{a_n\}$ and $\{b_n\}$ converge. Then show that $\{a_n\}$ and $\{b_n\}$ have the same limit. This common limit is called the *Gauss Arithmetic-Geometric Mean* of $a$ and $b$; it occurs as the value of an important elliptic integral involving $a$ and $b$.

12.   A set $A$ of real numbers is called *compact* if whenever $\{x_n\}$ is a sequence in $A$, there is a subsequence of $\{x_n\}$ that converges to a point in $A$. Using this terminology, the Bolzano-Weierstrass Theorem may be restated as follows: *If $a$ and $b$ are numbers such that $a \leq b$, the interval $[a, b]$ is compact.* Which of the following sets is compact? Justify your conclusions.

(a)   $(0, 1)$          (b)   $[0, 1)$          (c)   $\mathbb{R}$          (d)   $[0, 1] \cup [3, 4]$

# 3

# Continuous Functions and Limits

## 3.1   Continuity

In Chapter 2, we considered real-valued functions that have as their domain the set of natural numbers; that is, we considered sequences of real numbers. We now begin the study of real-valued functions having as their domain a general subset of $\mathbb{R}$. There is a standard notation: For a set of real numbers $D$, by

$$f : D \to \mathbb{R}$$

we denote a function whose domain is $D$, and for each point $x$ in $D$ we denote by $f(x)$ the value that the function assigns to $x$. When we write $f : D \to \mathbb{R}$ we will assume without further mention that $D$ is a set of real numbers.

Two of the concepts that are essential to an analytic description of functions $f : D \to \mathbb{R}$ are *continuity* and *differentiability*. The first five sections of this chapter are devoted to the study of continuity. In the final section we study limits in preparation for the discussion of differentiability, which we will begin in Chapter 4.

**DEFINITION**   *A function $f: D \rightarrow \mathbb{R}$ is said to be continuous at the point $x_0$ in D provided that whenever $\{x_n\}$ is a sequence in D that converges to $x_0$, the image sequence $\{f(x_n)\}$ converges to $f(x_0)$. The function $f: D \rightarrow \mathbb{R}$ is said to be continuous provided that it is continuous at every point in D.*

The definition of continuity of the function $f: D \rightarrow \mathbb{R}$ at the point $x_0$ in $D$ is formulated to make precise the intuitive notion that "if $x$ is a point in $D$ that is close to $x_0$, its image $f(x)$ is close to $f(x_0)$."

**FIGURE 3.1**

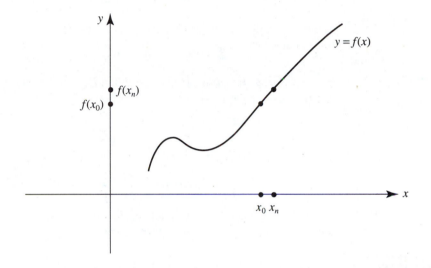

**EXAMPLE 3.1**   For each number $x$, define $f(x) = x^2 - 2x + 4$. Then the function $f: \mathbb{R} \rightarrow \mathbb{R}$ is continuous. To verify this, we select a point $x_0$ in $\mathbb{R}$, and we will show that the function is continuous at $x_0$. Let $\{x_n\}$ be a sequence that converges to $x_0$. By the sum and product properties of convergent sequences,

$$\lim_{n \to \infty} f(x_n) = \lim_{n \to \infty} [x_n^2 - 2x_n + 4] = x_0^2 - 2x_0 + 4 = f(x_0).$$

Thus, $f: \mathbb{R} \rightarrow \mathbb{R}$ is continuous at $x_0$. □

**EXAMPLE 3.2**   Define $f(x) = \sqrt{x}$ for $x \geq 0$. Then the function $f: [0, \infty) \rightarrow \mathbb{R}$ is continuous. To verify this, we select a nonnegative number $x_0$ and let $\{x_n\}$ be a sequence of nonnegative numbers that converges to $x_0$. But then the sequence $\{\sqrt{x_n}\}$ converges to $\sqrt{x_0}$; that is,

$$\lim_{n \to \infty} f(x_n) = \lim_{n \to \infty} \sqrt{x_n} = \sqrt{x_0} = f(x_0).$$

Thus, $f: [0, \infty) \rightarrow \mathbb{R}$ is continuous at $x_0$. □

**EXAMPLE 3.3** Define the function $f: \mathbb{R} \rightarrow \mathbb{R}$ by

$$f(x) = \begin{cases} 1 & \text{if } x \text{ is rational} \\ 0 & \text{if } x \text{ is irrational.} \end{cases}$$

This function is called *Dirichlet's function*. There is no point $x_0$ in $\mathbb{R}$ at which Dirichlet's function is continuous. Indeed, given a point $x_0$ in $\mathbb{R}$, for each natural number $n$, we may, by the density of the rationals and the irrationals (Theorem 1.11), choose a rational number, which we label $u_n$, in the interval $(x_0, x_0 + 1/n)$ and an irrational number, which we label $v_n$, in the interval $(x_0, x_0 + 1/n)$. But for each natural number $n$, $f(u_n) = 1$ and $f(v_n) = 0$, so

$$\lim_{n \to \infty} f(u_n) = 1 \neq 0 = \lim_{n \to \infty} f(v_n).$$

Since both of the sequences $\{u_n\}$ and $\{v_n\}$ converge to $x_0$, it is not possible for $f: \mathbb{R} \rightarrow \mathbb{R}$ to be continuous at $x_0$. □

Given two functions $f: D \rightarrow \mathbb{R}$ and $g: D \rightarrow \mathbb{R}$, we define the *sum* $f + g: D \rightarrow \mathbb{R}$ and the *product* $fg: D \rightarrow \mathbb{R}$ by

$$(f + g)(x) \equiv f(x) + g(x) \quad \text{and} \quad (fg)(x) \equiv f(x)g(x) \quad \text{for all } x \text{ in } D.$$

Moreover, if $g(x) \neq 0$ for all $x$ in $D$, the *quotient* $f/g: D \rightarrow \mathbb{R}$ is defined by

$$(f/g)(x) \equiv \frac{f(x)}{g(x)} \quad \text{for all } x \text{ in } D.$$

The following theorem is an analogue, and also a consequence, of the sum, product, and quotient properties of convergent sequences.

**THEOREM 3.1** Suppose that the functions $f: D \rightarrow \mathbb{R}$ and $g: D \rightarrow \mathbb{R}$ are continuous at the point $x_0$ in $D$. Then the sum

$$f + g: D \rightarrow \mathbb{R} \quad \text{is continuous at } x_0, \tag{3.1}$$

the product

$$fg: D \rightarrow \mathbb{R} \quad \text{is continuous at } x_0, \tag{3.2}$$

and, if $g(x) \neq 0$ for all $x$ in $D$, the quotient

$$f/g: D \rightarrow \mathbb{R} \quad \text{is continuous at } x_0. \tag{3.3}$$

**Proof** Let $\{x_n\}$ be a sequence in $D$ that converges to $x_0$. By the definition of continuity,

$$\lim_{n \to \infty} f(x_n) = f(x_0) \quad \text{and} \quad \lim_{n \to \infty} g(x_n) = g(x_0).$$

The sum property of convergent sequences implies that

$$\lim_{n \to \infty} [f(x_n) + g(x_n)] = f(x_0) + g(x_0), \tag{3.4}$$

and the product property of convergent sequences implies that

$$\lim_{n \to \infty} [f(x_n)g(x_n)] = f(x_0)g(x_0). \tag{3.5}$$

If $g(x) \neq 0$ for all $x$ in $D$, the quotient property of convergent sequences implies that

$$\lim_{n \to \infty} \frac{f(x_n)}{g(x_n)} = \frac{f(x_0)}{g(x_0)}. \tag{3.6}$$

By the definition of continuity, (3.1), (3.2), and (3.3) follow from (3.4), (3.5), and (3.6), respectively.                                                                                     ∎

For a nonnegative integer $k$ and numbers $c_0, c_1, \ldots, c_k$, the function $p \colon \mathbb{R} \to \mathbb{R}$ defined by

$$p(x) = \sum_{i=0}^{k} c_i x^i \quad \text{for all } x \text{ in } \mathbb{R}$$

is called a *polynomial*. If $c_k \neq 0$, $p \colon \mathbb{R} \to \mathbb{R}$ is said to have *degree $k$*.

It is clear that constant functions are continuous. Moreover, the function $f \colon \mathbb{R} \to \mathbb{R}$ defined by $f(x) = x$ for all $x$ in $\mathbb{R}$ is also continuous. From these two simple observations and the sum, product, and quotient properties of continuous functions asserted by Theorem 1, together with an induction argument, we obtain:

**COROLLARY 3.2**   Let $p \colon \mathbb{R} \to \mathbb{R}$ be a polynomial. Then $p \colon \mathbb{R} \to \mathbb{R}$ is continuous. Moreover, if $q \colon \mathbb{R} \to \mathbb{R}$ is also a polynomial and $D = \{x \text{ in } \mathbb{R} \mid q(x) \neq 0\}$, then the quotient $p/q \colon D \to \mathbb{R}$ is continuous.

In addition to forming the sum, product, and quotient of functions, there is another useful way to combine functions: They may be *composed*.

**DEFINITION**   *For functions $f \colon D \to \mathbb{R}$ and $g \colon U \to \mathbb{R}$ such that $f(D)$ is contained in $U$, we define the composition of $f \colon D \to \mathbb{R}$ with $g \colon U \to \mathbb{R}$, denoted by $g \circ f \colon D \to \mathbb{R}$, by the formula*

$$(g \circ f)(x) \equiv g(f(x)) \quad \text{for all } x \text{ in } D.$$

We have the following composition property for continuous functions.

**THEOREM 3.3**   For functions $f \colon D \to \mathbb{R}$ and $g \colon U \to \mathbb{R}$ such that $f(D)$ is contained in $U$, suppose that $f \colon D \to \mathbb{R}$ is continuous at the point $x_0$ in $D$ and $g \colon U \to \mathbb{R}$ is continuous at the point $f(x_0)$. Then the composition

$$g \circ f \colon D \to \mathbb{R}$$

is continuous at $x_0$.

**Proof** Let $\{x_n\}$ be a sequence in $D$ that converges to $x_0$. By the continuity of the function $f: D \to \mathbb{R}$ at the point $x_0$, the sequence $\{f(x_n)\}$ converges to $f(x_0)$. But then $\{f(x_n)\}$ is a sequence in $U$ that converges to $f(x_0)$, so by the continuity of $g: U \to \mathbb{R}$ at the point $f(x_0)$, the sequence $\{g(f(x_n))\}$ converges to $g(f(x_0))$; that is,

$$\lim_{n \to \infty} (g \circ f)(x_n) = (g \circ f)(x_0).$$

Thus, the composition $g \circ f: D \to \mathbb{R}$ is continuous at $x_0$. ∎

**EXAMPLE 3.4** Define $h(x) = \sqrt{1 - x^2}$ for $x$ in $[-1, 1]$. Then $h: [-1, 1] \to \mathbb{R}$ is continuous. This follows immediately from the continuity of polynomials, the continuity of the square-root function, and the composition property of continuous functions. □

**EXERCISES**

1. For a function $f: D \to \mathbb{R}$ and a point $x_0$ in $D$, define $A = \{x \text{ in } D \mid x \geq x_0\}$ and $B = \{x \text{ in } D \mid x \leq x_0\}$. Prove that $f: D \to \mathbb{R}$ is continuous at $x_0$ if and only if $f: A \to \mathbb{R}$ and $f: B \to \mathbb{R}$ are continuous at $x_0$.

2. Define

$$f(x) = \begin{cases} x^2 & \text{if } x \geq 0 \\ x & \text{if } x < 0. \end{cases}$$

Prove that the function $f: \mathbb{R} \to \mathbb{R}$ is continuous.

3. Define

$$f(x) = \begin{cases} 0 & \text{if } 0 \leq x \leq 1 \\ x & \text{if } 1 < x \leq 2. \end{cases}$$

At what points is the function $f: [0, 2] \to \mathbb{R}$ continuous?

4. Suppose that the function $g: \mathbb{R} \to \mathbb{R}$ is continuous and that $g(x) = 0$ if $x$ is rational. Prove that $g(x) = 0$ for all $x$ in $\mathbb{R}$.

5. Define

$$g(x) = \begin{cases} x^2 & \text{if } x \text{ is rational} \\ -x^2 & \text{if } x \text{ is irrational}. \end{cases}$$

At what points is the function $g: \mathbb{R} \to \mathbb{R}$ continuous?

6. Let the function $f: D \to \mathbb{R}$ be continuous. Then define the function $|f|: D \to \mathbb{R}$ by $|f|(x) = |f(x)|$ for $x$ in $D$. Prove that the function $|f|: D \to \mathbb{R}$ also is continuous.

7. Let the continuous function $f: D \to \mathbb{R}$ have nonnegative functional values. Define the function $\sqrt{f}: D \to \mathbb{R}$ by $\sqrt{f}(x) = \sqrt{f(x)}$ for $x$ in $D$. Prove that $\sqrt{f}: D \to \mathbb{R}$ is also continuous.

8. Suppose that the function $f: \mathbb{R} \to \mathbb{R}$ has the property that

$$f(u + v) = f(u) + f(v) \quad \text{for all } u \text{ and } v \text{ in } \mathbb{R}.$$

(a) Prove that if $m = f(1)$, then

$$f(x) = mx \quad \text{for all rational numbers } x.$$

(b) Use (a) to prove that if $f: \mathbb{R} \to \mathbb{R}$ is continuous, then

$$f(x) = mx \quad \text{for all } x \text{ in } \mathbb{R}.$$

## 3.2   The Extreme Value Theorem

For a function $f: D \to \mathbb{R}$, we define

$$f(D) = \{y \text{ in } \mathbb{R} \mid y = f(x) \text{ for some } x \text{ in } D\}$$

and call $f(D)$ the *image* of $f: D \to \mathbb{R}$. We say that a function $f: D \to \mathbb{R}$ attains a *maximum value* provided that its image $f(D)$ has a maximum; that is, there is a point $x_0$ in $D$ such that

$$f(x) \leq f(x_0) \quad \text{for all } x \text{ in } D.$$

We will call such a point in $D$ a *maximizer* of the function $f: D \to \mathbb{R}$. Similarly, the function $f: D \to \mathbb{R}$ is said to attain a *minimum value* provided that its image $f(D)$ has a minimum; a point in $D$ at which this minimum value is attained is called a *minimizer* of $f: D \to \mathbb{R}$.

In general, no assertion can be made concerning the existence of a minimum or maximum value for a function $f: D \to \mathbb{R}$. However, in the case when the domain $D = [a, b]$ and the function $f: [a, b] \to \mathbb{R}$ is continuous, we have the following important theorem.

**THEOREM 3.4**   **The Extreme Value Theorem**   Suppose that the function $f: [a, b] \to \mathbb{R}$ is continuous. Then $f: [a, b] \to \mathbb{R}$ attains both a minimum and a maximum value.

**FIGURE 3.2**

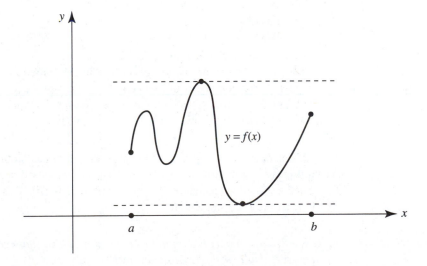

In order to prove this theorem, it is convenient first to prove a weaker result.

**LEMMA 3.5**    Suppose that the function $f:[a, b] \to \mathbb{R}$ is continuous. Then the image of $f:[a, b] \to \mathbb{R}$ is bounded above; that is, there is a number $M$ such that

$$f(x) \leq M \quad \text{for all } x \text{ in } [a, b].$$

**Proof of**    We will argue by contradiction. Assume there that is no such number $M$. Let $n$ be a
**Lemma 3.5**    natural number. Then it is not true that

$$f(x) \leq n \quad \text{for all } x \text{ in } [a, b].$$

Thus, there is a point $x$ in $[a, b]$ at which $f(x) > n$. Choose such a point and label it $x_n$. This defines a sequence $\{x_n\}$ in $[a, b]$ with the property that $f(x_n) > n$ for every natural number $n$. We can employ the Bolzano-Weierstrass Theorem to find a subsequence $\{x_{n_k}\}$ of $\{x_n\}$ that converges to a point $x_0$ in $[a, b]$. Since the function $f:[a, b] \to \mathbb{R}$ is continuous at $x_0$, the image sequence $\{f(x_{n_k})\}$ converges to $f(x_0)$. This contradicts the unboundedness of the sequence $\{f(x_{n_k})\}$. This contradiction proves that the image of $f:[a, b] \to \mathbb{R}$ is bounded above. ∎

**Proof of**    Define $S = f([a, b])$. Then $S$ is a nonempty set of real numbers that, by the preceding
**Theorem 3.4**    lemma, is bounded above. According to the Completeness Axiom, $S$ has a supremum. Define $c = \sup S$. It is necessary to find a point $x$ in $[a, b]$ at which $c = f(x)$.

Let $n$ be a natural number. Then the number $c - 1/n$ is smaller than $c$, and is therefore not an upper bound for the set $S$. Thus there is a point $x$ in $[a, b]$ at which $f(x) > c - 1/n$. Choose such a point and label it $x_n$. From this choice and from the fact that $c$ is an upper bound for $S$, we see that $c - 1/n < f(x_n) \leq c$ for every natural number $n$. Hence the sequence $\{f(x_n)\}$ converges to $c$.

The Bolzano-Weierstrass Theorem asserts that there is a subsequence $\{x_{n_k}\}$ of $\{x_n\}$ that converges to a point $x_0$ in $[a, b]$. Since $f:[a, b] \to \mathbb{R}$ is continuous at $x_0$, $\{f(x_{n_k})\}$ converges to $f(x_0)$. But $\{f(x_{n_k})\}$ is a subsequence of $\{f(x_n)\}$, so $c = f(x_0)$. The point $x_0$ is a maximizer of the function $f:[a, b] \to \mathbb{R}$.

To complete the proof, we observe that the function $-f:[a, b] \to \mathbb{R}$ is also continuous. Consequently, using what we have just proven, we may select a point in $[a, b]$ at which $-f:[a, b] \to \mathbb{R}$ attains a maximum value, and at this point the function $f:[a, b] \to \mathbb{R}$ attains a minimum value. ∎

If one examines the proofs of the preceding lemma and theorem, one sees that the only property of the domain of the function that was used was that each sequence in $[a, b]$ had a subsequence that converged to a point in $[a, b]$. This property is so important that it deserves to be singled out.

**DEFINITION**    *A set $K$ of real numbers is said to be compact provided that every sequence in $K$ has a subsequence that converges to a point in $K$.*

In this new terminology, the Bolzano-Weierstrass Theorem is simply the assertion that if $a$ and $b$ are numbers such that $a < b$, then the set $[a, b]$ is compact.

**THEOREM 3.6**   Let $K$ be a compact nonempty set of real numbers and suppose that the function $f: K \to \mathbb{R}$ is continuous. Then $f: K \to \mathbb{R}$ attains both a minimum and a maximum value.

**Proof**   Exercise.                                                                                                    ∎

---

**EXERCISES**

1. Find a maximizer for each of the following functions:
   (a)  $f: [0, 1] \to \mathbb{R}$, defined by $f(x) = \sqrt{x} + x^{10} + 4$ for $0 \le x \le 1$
   (b)  $g: [-1, 1] \to \mathbb{R}$, defined by $g(x) = -x^{10}(x - 1/4)^{24}$ for $-1 \le x \le 1$
   (c)  $h: [-1, 1] \to \mathbb{R}$, defined by $h(x) = 4 - 2x^3$ for $-1 \le x \le 1$

2. Let $a$ and $b$ be real numbers with $a < b$. Find a continuous function $f: (a, b) \to \mathbb{R}$ having an image that is unbounded above. Also, find a continuous function $f: (a, b) \to \mathbb{R}$ having an image that is bounded above but that does not attain a maximum value.

3. Decide which of the following subsets of $\mathbb{R}$ is compact:
   (a)  $(0, 1)$        (b)  $[0, 1)$        (c)  $[1, 2] \cup [3, 4]$
   (d)  $\mathbb{R}$        (e)  $\mathbb{Q} \cap [0, 1]$        (f)  $\{1, 2, 3, 4\}$

4. Rewrite the proofs of Lemma 3.5 and Theorem 3.4, with $[a, b]$ replaced by a compact set $K$, and thereby prove Theorem 3.6.

5. Suppose that $K$ is a nonempty set of real numbers that is not compact. Prove that there is a monotone sequence in $K$ that does not converge to a point in $K$.

6. Let $K$ be a nonempty set of real numbers that is not compact. Prove that there is a continuous function $g: K \to \mathbb{R}$ that does not attain a maximum value. (*Hint:* By the preceding exercise, we may choose a monotone sequence $\{a_n\}$ in $K$ that does not converge to a point in $K$. If $\{a_n\}$ is bounded, let $a$ be the point to which the sequence converges, and define $g(x) = 1/|x - a|$ for all $x$ in $K$. If $\{a_n\}$ is unbounded, define $g(x) = |x|$ for all $x$ in $K$.)

7. Suppose that the function $f: [0, 1] \to \mathbb{R}$ is continuous, $f(0) > 0$, and $f(1) = 0$. Prove that there is a number $x_0$ in $(0, 1]$ such that $f(x_0) = 0$ and $f(x) > 0$ for $0 \le x < x_0$.

---

## 3.3   The Intermediate Value Theorem

The second important geometric property of the graph of a continuous function that we will establish is that if a continuous function has a domain consisting of an interval, and if its graph contains points that are both above and below a line $y = c$, then, in fact, the function attains the value $c$ on the interval. The heart of the matter lies in the following theorem.

**THEOREM 3.7**   Suppose that the function $f: [a, b] \to \mathbb{R}$ is continuous. Assume also that

$$f(a) < 0 \quad \text{and} \quad f(b) > 0.$$

Then there is a point $x_0$ in the open interval $(a, b)$ at which $f(x_0) = 0$.

**Proof**    We will recursively define a sequence of nested, closed subintervals of $[a, b]$ whose endpoints converge to a point in $[a, b]$ at which $f(x) = 0$.

Let $a_1 = a$ and $b_1 = b$. For a natural number $n$, suppose that the interval $[a_n, b_n]$ contained in $[a, b]$ has been defined such that $f(a_n) \leq 0$ and $f(b_n) > 0$. Consider the midpoint $c_n = (a_n + b_n)/2$.

- If $f(c_n) \leq 0$, define $a_{n+1} = c_n$ and $b_{n+1} = b_n$.
- If $f(c_n) > 0$, define $a_{n+1} = a_n$ and $b_{n+1} = c_n$.

Observe that for each natural number $n$,

$$a \leq a_n \leq a_{n+1} < b_{n+1} \leq b_n \leq b,$$

$$f(a_{n+1}) \leq 0 \quad \text{and} \quad f(b_{n+1}) > 0,$$

and

$$(b_{n+1} - a_{n+1}) = \frac{b_n - a_n}{2}.$$

It follows that $(b_n - a_n) = (b - a)/2^{n-1}$ for all $n$. Thus the sequences $\{a_n\}$ and $\{b_n\}$ satisfy the assumptions of the Nested Interval Theorem (Theorem 2.15 of Chapter 2), so there is a point $x_0$ in $[a, b]$ to which both $\{a_n\}$ and $\{b_n\}$ converge. Since $f: [a, b] \to \mathbb{R}$ is continuous at $x_0$, the image sequences $\{f(a_n)\}$ and $\{f(b_n)\}$ converge to $f(x_0)$. It follows that $f(x_0) \leq 0$, since $\{f(a_n)\}$ is a sequence of nonpositive numbers, and that $f(x_0) \geq 0$, since $\{f(b_n)\}$ is a sequence of nonnegative numbers. Consequently, $f(x_0) = 0$.    ∎

**FIGURE 3.3(a)**

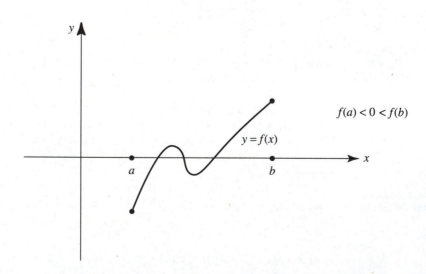

$f(a) < 0 < f(b)$

$y = f(x)$

**FIGURE 3.3(b)**

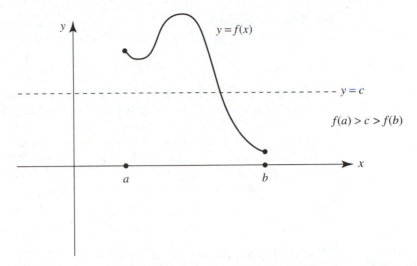

The general Intermediate Value Theorem is obtained from the preceding theorem by a simple algebraic manipulation.

**THEOREM 3.8**  **The Intermediate Value Theorem**  Let the function $f:[a, b] \to \mathbb{R}$ be continuous, and $c$ be a number strictly between $f(a)$ and $f(b)$; that is,

$$f(a) < c < f(b) \quad \text{or} \quad f(b) < c < f(a).$$

Then there is a point $x_0$ in the open interval $(a, b)$ at which $f(x_0) = c$.

**Proof**  First let us suppose that $f(a) < c < f(b)$. Define the function $g:[a, b] \to \mathbb{R}$ by $g(x) = f(x) - c$ for $x$ in $[a, b]$. Then $g:[a, b] \to \mathbb{R}$ is a continuous function such that $g(a) < 0$ and $g(b) > 0$. We may apply the preceding theorem to conclude that there is a point $x_0$ in $(a, b)$ at which $g(x_0) = 0$; that is, $f(x_0) = c$. In the case when $f(b) < c < f(a)$, we define $g:[a, b] \to \mathbb{R}$ by $g(x) = c - f(x)$ for $x$ in $[a, b]$ and follow the same argument.  ■

For a natural number $n$ and real numbers $a_0, a_1, \ldots, a_n$, consider the equation

$$a_0 + a_1 x + \cdots + a_n x^n = 0, \qquad x \text{ in } \mathbb{R}. \tag{3.7}$$

In general, of course, this equation might not have any solution. For instance, the equation

$$1 + x^2 = 0, \qquad x \text{ in } \mathbb{R}$$

has no solution. For $n = 1$, we can easily analyze (3.7). For $n = 2$, the quadratic formula permits us to analyze (3.7). For $n = 3$ and $n = 4$, there are explicit formulas similar to, but slightly more complicated than, the quadratic formula for determining the solutions of (3.7). *However, for $n \geq 5$ there cannot be a formula for determining the solutions of (3.7) for arbitrary choices of the coefficients $a_0, a_1, \ldots, a_n$; this follows from a beautiful theorem of Galois, which, unfortunately, lies outside the scope of this book.* Hence,

---

*See I. N. Herstein, *Topics in Algebra.*

even when the function $f: \mathbb{R} \to \mathbb{R}$ is a polynomial, if its degree is greater than 4, it is usually not possible to explicitly determine the solutions of the equation

$$f(x) = 0, \qquad x \text{ in } \mathbb{R}. \tag{3.8}$$

So one can imagine how difficult it is to determine the solutions of equation (3.8) when $f: \mathbb{R} \to \mathbb{R}$ is defined in terms of, say, trigonometric and exponential functions.

However, the Intermediate Value Theorem is useful in the study of equation (3.8). If $f: \mathbb{R} \to \mathbb{R}$ is continuous and we can find numbers $a$ and $b$ with $a < b$ and $f(a) \cdot f(b) < 0$, then (3.8) has a solution in the open interval $(a, b)$. Moreover, the method of proof that we gave for Theorem 3.7, which is called the *bisection method*, provides a recursive method that, after $n$ steps, determines a subinterval of $[a, b]$ of length $(b - a)/2^{n-1}$ that contains a solution of equation (3.8).

**EXAMPLE 3.5**  Consider the equation

$$x^5 + x + 1 = 0, \qquad x \text{ in } \mathbb{R}.$$

We claim that there is a solution of the above equation. Indeed, define $h: \mathbb{R} \to \mathbb{R}$ by $h(x) = x^5 + x + 1$ for all $x$ in $\mathbb{R}$. Observe that $h(-2) < 0$ and $h(0) > 0$. Thus we may apply the Intermediate Value Theorem to the restriction $h: [-2, 0] \to \mathbb{R}$ to conclude that there is a point $x_0$ in $(-2, 0)$ that is a solution of this equation. $\qquad \square$

There is a slightly more general form of the Intermediate Value Theorem that is of interest because it can be generalized to the situation in which one considers real-valued functions of several real variables. Recall that for real numbers $a$ and $b$ with $a < b$ we have called the sets

$$[a, b], (a, b), (a, b], [a, b), (-\infty, b), (-\infty, b], (a, \infty), \text{ and } [a, \infty)$$

*intervals*. We will also call the empty set, a set consisting of a single member, and the whole set of real numbers *intervals*. It turns out that there is a simple characterization of all intervals: *A subset $I$ of $\mathbb{R}$ is an interval if and only if whenever the points $u$ and $v$ are in $I$ and $u < v$, then the whole interval $[u, v]$ is contained in $I$* (see Exercise 11). This characterization of a general interval will be useful in the proof of the next result.

**THEOREM 3.9**  Let $I$ be an interval and suppose that the function $f: I \to \mathbb{R}$ is continuous. Then its image $f(I)$ is an interval.

**Proof**  Let $y_1$ and $y_2$ be points in the image $f(I)$, with $y_1 < y_2$. We must show that $[y_1, y_2]$ is contained in $f(I)$. Indeed, let $y_1 < c < y_2$. Since $y_1$ and $y_2$ are in $f(I)$, there are points $x_1$ and $x_2$ in $I$ with $f(x_1) = y_1$ and $f(x_2) = y_2$. If we let $J$ be the closed interval having $x_1$ and $x_2$ as endpoints, then $J$ is contained in $I$, since $I$ is an interval. Thus, we can apply the Intermediate Value Theorem to the function $f: J \to \mathbb{R}$ in order to conclude that there is a point $x_0$ in $J$ at which $f(x_0) = c$. Thus $x_0$ belongs to $I$ and $f(x_0) = c$. It follows that $[y_1, y_2]$ is contained in $f(I)$. $\qquad \blacksquare$

In general, it is quite difficult to explicitly determine the image of a function. Consider even the special case when $D = [a, b]$ and the function $f : [a, b] \to \mathbb{R}$ is continuous. According to the Extreme Value Theorem, this function attains maximum and minimum functional values, $m$ and $M$, and the Intermediate Value Theorem implies that it attains every value between $m$ and $M$. Thus to find the image of $f : [a, b] \to \mathbb{R}$, one has to determine the maximum and minimum values of this function. Unless the function is relatively simple, it is not possible to explicitly determine these extreme values.

**EXERCISES**

1.  Prove that there is a solution of the equation

    $$x^9 + x^2 + 4 = 0, \qquad x \text{ in } \mathbb{R}.$$

2.  Prove that there is a solution of the equation

    $$\frac{1}{\sqrt{x + x^2}} + x^2 - 2x = 0, \qquad x > 0.$$

3.  Let $f : [0, \infty) \to \mathbb{R}$ be a continuous function such that $f(0) = 0$ and

    $$f(x) \geq \sqrt{x} \quad \text{for all } x \geq 0.$$

    Show that for each $c > 0$ there is some $x > 0$ such that $f(x) = c$.

4.  For a function $f : D \to \mathbb{R}$, a solution of the equation

    $$f(x) = x, \qquad x \text{ in } D$$

    is called a *fixed-point* of $f : D \to \mathbb{R}$. If $f : [-1, 1] \to \mathbb{R}$ is continuous, $f(-1) > -1$, and $f(1) < 1$, show that $f : [-1, 1] \to \mathbb{R}$ has a fixed-point.

5.  Suppose that the functions $h : [a, b] \to \mathbb{R}$ and $g : [a, b] \to \mathbb{R}$ are continuous. Observe that a solution of the equation

    $$h(x) = g(x), \qquad x \text{ in } [a, b]$$

    corresponds to a point where the graphs intersect. Show that if $h(a) \leq g(a)$ and $h(b) \geq g(b)$, then this equation has a solution.

6.  Suppose that $f : \mathbb{R} \to \mathbb{R}$ is continuous and that its image $f(\mathbb{R})$ is bounded. Prove that there is a solution of the equation

    $$f(x) = x, \qquad x \text{ in } \mathbb{R}.$$

7.  Suppose that the function $f : [a, b] \to \mathbb{R}$ is continuous. For a natural number $n$, let $x_1, \ldots, x_n$ be points in $[a, b]$. Prove that there is a point $z$ in $[a, b]$ at which

    $$f(z) = \frac{f(x_1) + \cdots + f(x_n)}{n}.$$

8.  The proof of Theorem 3.7 has a constructive aspect: At the $n$th stage one has isolated an interval $[a_n, b_n]$ of length $(b - a)/2^{n-1}$ that contains a solution of the equation

    $$f(x) = 0, \qquad x \text{ in } [a, b].$$

    Find an interval of width smaller than $1/8$ that contains a solution of the equation

    $$x^2 = 2, \qquad x > 0.$$

9. Let $p: \mathbb{R} \to \mathbb{R}$ be a polynomial of odd degree. Prove that there is a solution of the equation

$$p(x) = 0, \qquad x \text{ in } \mathbb{R}.$$

10. Suppose that the function $f: [0, 1] \to \mathbb{R}$ is continuous and that its image consists entirely of rational numbers. Prove that $f: [0, 1] \to \mathbb{R}$ is a constant function.

11. Let $I$ be a nonempty set of real numbers that has the property that if $u$ and $v$ are in $I$ with $u < v$, the closed interval $[u, v]$ is contained in $I$. If $I$ is bounded above, define $b = \sup I$; if $I$ is bounded below, define $a = \inf I$. Prove the following:
    (a) If $I$ is unbounded above and below, then $I = \mathbb{R}$.
    (b) If $I$ is bounded below but not above, then $I = (a, \infty)$ or $I = [a, \infty)$.
    (c) If $I$ is bounded above but not below, then $I = (-\infty, b]$ or $I = (-\infty, b)$.
    (d) If $I$ is bounded, then $I$ is one of the sets $[a, b], (a, b), [a, b), (a, b]$.

## 3.4   Images and Inverses

**DEFINITION**   *A function $f: D \to \mathbb{R}$ is said to be one-to-one provided that for each point $y$ in its image $f(D)$, there is exactly one point $x$ in its domain $D$ such that $f(x) = y$.*

It is easy to see that the above definition is equivalent to the assertion that $f: D \to \mathbb{R}$ is one-to-one if when $u$ and $v$ are points in $D$ such that $f(u) = f(v)$, then $u = v$.

For a function $f: D \to \mathbb{R}$ that is one-to-one, by definition, if $y$ is a point in $f(D)$, there is exactly one point $x$ in $D$ such that $f(x) = y$. We will denote this point $x$ by $f^{-1}(y)$, so we have defined the function

$$f^{-1}: f(D) \to \mathbb{R},$$

which we call the *inverse* of the function $f: D \to \mathbb{R}$.

In the analysis of real-valued functions of a single real variable, a particular type of one-to-one function occurs frequently, namely a function whose domain is an interval and that is strictly monotone.

**DEFINITION**   *The function $f: D \to \mathbb{R}$ is called strictly increasing provided that*

$$f(v) > f(u) \quad \text{for all points } u \text{ and } v \text{ in } D \text{ such that } v > u.$$

*The function $f: D \to \mathbb{R}$ is called strictly decreasing provided that*

$$f(v) < f(u) \quad \text{for all points } u \text{ and } v \text{ in } D \text{ such that } v > u.$$

*A function that is either strictly increasing or strictly decreasing is said to be strictly monotone.*

Now it is clear that if $f: D \to \mathbb{R}$ is strictly monotone, it is one-to-one and its inverse $f^{-1}: f(D) \to \mathbb{R}$ is also strictly monotone. We have the following result about the continuity of the inverse.

**THEOREM 3.10**   Let $I$ be an interval and suppose that the function $f: I \to \mathbb{R}$ is strictly monotone. Then the inverse function $f^{-1}: f(I) \to \mathbb{R}$ is continuous.

**Proof**   We will consider the case when $f: I \to \mathbb{R}$ is strictly increasing. Let $y_0 = f(x_0)$ be a point in $f(I)$. To prove that $f^{-1}: f(I) \to \mathbb{R}$ is continuous at $y_0$, we will argue by contradiction. Suppose that $f^{-1}: f(I) \to \mathbb{R}$ is not continuous at $y_0$. Then there is a sequence $\{y_n\}$, with $y_n = f(x_n)$, where $x_n$ is in $I$ for each natural number $n$, such that $\{y_n\}$ converges to $y_0$ but $\{x_n\}$ does not converge to $x_0$. This last assertion means that there is some $\epsilon_0 > 0$ such that $|x_n - x_0| \geq \epsilon_0$ for infinitely many $n$. We may suppose that $x_0 < x_0 + \epsilon_0 \leq x_n$ for infinitely many $n$. Since $I$ is an interval, $x_0 + \epsilon_0$ belongs to $I$, and hence, since $f: I \to \mathbb{R}$ is strictly increasing, $f(x_0) < f(x_0 + \epsilon_0) \leq f(x_n)$ for infinitely many $n$. Thus the sequence $\{f(x_n)\}$ does not converge to $f(x_0)$. This contradiction proves that $f^{-1}: f(I) \to \mathbb{R}$ is continuous at $y_0$.   ∎

The above theorem is remarkable in that we have not assumed that the function $f: I \to \mathbb{R}$ is continuous! In fact, this theorem can be used to establish the following criterion for a strictly monotone function to be continuous.

**THEOREM 3.11**   Let $I$ be an interval and suppose that the function $f: I \to \mathbb{R}$ is strictly monotone. Then the function $f: I \to \mathbb{R}$ is continuous if and only if its image $f(I)$ is an interval.

**Proof**   If the function $f: I \to \mathbb{R}$ is continuous, then, according to the Intermediate Value Theorem, as expressed in Theorem 3.9, its image $f(I)$ is an interval.

To prove the converse, suppose that $J = f(I)$ is an interval. Then $f^{-1}: J \to \mathbb{R}$ is strictly monotone. Thus we may apply the preceding theorem, with $f^{-1}: J \to \mathbb{R}$ playing the role of $f: I \to \mathbb{R}$, to conclude that $(f^{-1})^{-1}: f^{-1}(J) \to \mathbb{R}$ is continuous. However, $(f^{-1})^{-1}: f^{-1}(J) \to \mathbb{R}$ is precisely $f: I \to \mathbb{R}$.   ∎

**PROPOSITION 3.12**   For $n$ a natural number, define

$$f(x) = x^n \quad \text{for all } x \geq 0.$$

Then the function $f: [0, \infty) \to \mathbb{R}$ is strictly increasing and continuous, and has image equal to $[0, \infty)$. Moreover, its inverse $f^{-1}: [0, \infty) \to \mathbb{R}$ is continuous.

**Proof**   Let $0 \leq u < v$. Then, according to the Difference in Powers formula,

$$f(v) - f(u) = (v - u) \sum_{k=0}^{n-1} v^{n-1-k} u^k > 0.$$

Thus $f: [0, \infty) \to \mathbb{R}$ is strictly increasing.

To show that $f([0, \infty)) = [0, \infty)$, let $y > 0$. We may select $x_0$ with $f(x_0) > y$; for instance, if $x_0 = y + 1$, then $f(x_0) = (y + 1)^n > y$. Since $f: [0, \infty) \to \mathbb{R}$ is continuous and $f(0) = 0$, it follows from the Intermediate Value Theorem, applied to the function $f: [0, x_0] \to \mathbb{R}$, that there is a positive number $x$ such that $f(x) = y$.

The continuity of $f^{-1}: [0, \infty) \to \mathbb{R}$ is a consequence of Theorem 3.10.   ∎

For a natural number $n$, if $f^{-1}:[0, \infty) \to \mathbb{R}$ is as in the above proposition, as usual, we define $x^{1/n} \equiv f^{-1}(x)$ for $x \geq 0$. Also, for each positive number $x$ and rational number $r = m/n$, where $m$ and $n$ are integers with $n$ positive, we define $x^r \equiv (x^{1/n})^m$.

1. For $n$ an odd natural number, define $f(x) = x^n$ for all $x$ in $\mathbb{R}$. Prove that the function $f:\mathbb{R} \to \mathbb{R}$ is strictly increasing and $f(\mathbb{R}) = \mathbb{R}$.

2. Find the images of each of the following functions:
   (a)  $f:[0, \infty) \to \mathbb{R}$ defined by $f(x) = 1/(1 + x^2)$ for $x \geq 0$.
   (b)  $h:(0, 1) \to \mathbb{R}$ defined by $h(x) = 1/(x^2 + 8x)$ for $0 < x < 1$.

3. Define
$$f(x) = \begin{cases} x & \text{if } x \leq 0 \\ x + 1 & \text{if } x > 0. \end{cases}$$
   Determine $f^{-1}: f(\mathbb{R}) \to \mathbb{R}$ and prove that $f^{-1}: f(\mathbb{R}) \to \mathbb{R}$ is continuous at 0.

4. (a)  Find a continuous function $f:(0, 1) \to \mathbb{R}$ with image equal to $\mathbb{R}$.
   (b)  Find a continuous function $f:(0, 1) \to \mathbb{R}$ with image equal to $[0, 1]$.
   (c)  Find a continuous function $f:\mathbb{R} \to \mathbb{R}$ that is strictly increasing and has image equal to $(-1, 1)$.

5. Define
$$f(x) = \begin{cases} 1 + x^2 & \text{if } x > 0 \\ 0 & \text{if } x = 0 \\ -(1 + x^2) & \text{if } x < 0. \end{cases}$$
   Show that the function $f:\mathbb{R} \to \mathbb{R}$ is not continuous, but that it has a continuous inverse.

6. Let $D = [0, 1] \cup (2, 3]$ and define $f:D \to \mathbb{R}$ by
$$f(x) = \begin{cases} x & \text{if } 0 \leq x \leq 1 \\ x - 1 & \text{if } 2 < x \leq 3. \end{cases}$$
   Prove that $f:D \to \mathbb{R}$ is continuous. Determine $f^{-1}: f(D) \to \mathbb{R}$ and prove that $f^{-1}: f(D) \to \mathbb{R}$ is not continuous. Does this contradict Theorem 3.10?

7. Let $K$ be a nonempty, compact set of real numbers and suppose that the function $f:K \to \mathbb{R}$ is continuous. Show that its image $f(K)$ is compact.

8. Let the function $f:\mathbb{R} \to \mathbb{R}$ be continuous and suppose that its image $f(\mathbb{R})$ is bounded. Prove that there is a solution of the equation
$$f(x) = x, \qquad x \text{ in } \mathbb{R}.$$
   Now choose a number $a$ with $f(a) > a$ and define the sequence $\{a_n\}$ recursively by defining $a_1 = a$ and $a_{n+1} = f(a_n)$ if $n$ is a natural number for which $a_n$ is defined. If $f:\mathbb{R} \to \mathbb{R}$ is strictly increasing, show that $\{a_n\}$ converges to a solution of the above equation. This method for approximating the solution is called an *iterative method*.

## 3.5 An Equivalent Criterion for Continuity; Uniform Continuity

For a function $f:D \to \mathbb{R}$ and a point $x_0$ in its domain $D$, we have defined $f:D \to \mathbb{R}$ to be continuous at $x_0$ provided that whenever a sequence $\{x_n\}$ in $D$ converges to $x_0$, the image sequence $\{f(x_n)\}$ converges to $f(x_0)$. With this definition of continuity, we have been able to use the results that we proved for sequences to establish properties of continuous functions. In particular, we used the Bolzano-Weierstrass Theorem to prove the Extreme Value Theorem, and the Nested Interval Theorem to prove the Intermediate Value Theorem.

There is an equivalent way of defining continuity that gives us a different perspective from which to view this concept. From this alternative perspective, certain properties of continuous functions can be seen more clearly.

**THEOREM 3.13**   For a function $f: D \to \mathbb{R}$ and a point $x_0$ in its domain $D$, the following two assertions are equivalent:

(i) The function $f: D \to \mathbb{R}$ is continuous at $x_0$.

(ii) For each positive number $\epsilon$ there is a positive number $\delta$ such that

$$|f(x) - f(x_0)| < \epsilon \quad \text{for all points } x \text{ in } D \text{ such that } |x - x_0| < \delta. \qquad (3.9)$$

**FIGURE 3.4**

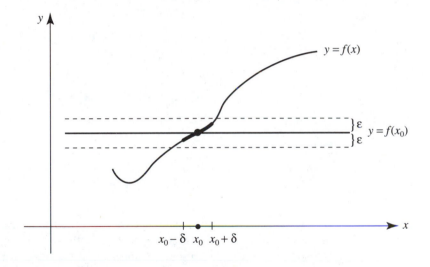

**Proof**   First of all, suppose that $f: D \to \mathbb{R}$ is continuous at $x_0$. We will argue by contradiction to verify criterion (ii). Suppose that (ii) does not hold. Then there is some $\epsilon_0 > 0$ such that for $\epsilon = \epsilon_0$ there is no $\delta > 0$ for which (3.9) holds. Let $n$ be a natural number. Then (3.9) does not hold for $\epsilon = \epsilon_0$ and $\delta = 1/n$. This means precisely that there is a point $x$ in $D$ such that $|x - x_0| < 1/n$ but $|f(x) - f(x_0)| \geq \epsilon_0$. Choose such a point and label it $x_n$. This defines a sequence $\{x_n\}$ in $D$ that converges to $x_0$. But by the continuity of $f: D \to \mathbb{R}$ at $x_0$, $\{f(x_n)\}$ converges to $f(x_0)$. This clearly contradicts the assertion that $|f(x_n) - f(x_0)| \geq \epsilon_0$ for every natural number $n$. Thus (ii) holds.

Now suppose that (ii) holds. We will show that $f: D \to \mathbb{R}$ is continuous at $x_0$. Indeed, let $\{x_n\}$ be a sequence in $D$ that converges to $x_0$. To show that $\{f(x_n)\}$ converges to $f(x_0)$, we let $\epsilon > 0$ and seek a natural number $N$ such that

$$|f(x_n) - f(x_0)| < \epsilon \quad \text{for all integers } n \geq N. \qquad (3.10)$$

But (ii) asserts that we may select $\delta > 0$ such that (3.9) holds. Moreover, since $\{x_n\}$ converges to $x_0$, we can choose a natural number $N$ such that

$$|x_n - x_0| < \delta \quad \text{for all integers } n \geq N. \tag{3.11}$$

Clearly (3.9) and (3.11) imply (3.10). ∎

Observe the geometric meaning of the "$\epsilon$–$\delta$" criterion in the preceding theorem. It asserts that if one chooses a symmetric band of width $2\epsilon$ about the line $y = f(x_0)$, then no matter how small this width is, one can find a corresponding symmetric open interval about $x_0$ such that the graph of the function, restricted to this interval, lies within the chosen band.

**EXAMPLE 3.6**   Define $f(x) = x^3$ for all $x$ in $\mathbb{R}$. We have already proven that the function $f: \mathbb{R} \to \mathbb{R}$ is continuous. However, let us verify the "$\epsilon$–$\delta$" criterion of the preceding theorem at the point $x_0 = 2$. Let $\epsilon > 0$. We must find a $\delta > 0$ such that

$$|x^3 - 8| < \epsilon \quad \text{if} \quad |x - 2| < \delta.$$

But observe that the Difference in Powers formula and the Triangle Inequality imply that

$$|x^3 - 8| = |(x - 2)(x^2 + 2x + 4)|$$
$$\leq |x - 2|[|x|^2 + 2|x| + 4] \quad \text{for all } x \text{ in } \mathbb{R}.$$

However,
$$|x|^2 + 2|x| + 4 \leq 19 \quad \text{for} \quad 1 < x < 3$$

so that
$$|x^3 - 8| \leq 19|x - 2| \quad \text{if} \quad 1 < x < 3. \tag{3.12}$$

Define $\delta = \min\{1, \epsilon/19\}$. If $|x - 2| < \delta$, then $x$ belongs to the interval $(1, 3)$ and $19|x - 2| < \epsilon$, so from (3.12) we see that $|x^3 - 8| < \epsilon$. □

In the preceding example it is clear that for each number $x_0$, the choice of $\delta > 0$ that responds to an $\epsilon > 0$ challenge depends both on $\epsilon$ and on the point $x_0$. Frequently, when $f: D \to \mathbb{R}$ is continuous it happens that the choice of $\delta > 0$ depends only on $\epsilon > 0$ and is independent of the choice of point in $D$. At first glance, this may seem to be a distinction that is too fine. However, we will see when we study integration that this distinction is significant.

**DEFINITION**   *A function $f: D \to \mathbb{R}$ is said to be uniformly continuous provided that for each positive number $\epsilon$ there is a positive number $\delta$ such that*

$$|f(u) - f(v)| < \epsilon \quad \text{for all points } u \text{ and } v \text{ in } D \text{ such that } |u - v| < \delta. \tag{3.13}$$

It is clear that the function $f: D \to \mathbb{R}$ is uniformly continuous if and only if $f: D \to \mathbb{R}$ is continuous and that, moreover, for a point $x_0$ in $D$, if $\epsilon > 0$ is prescribed, the choice of $\delta$ in (3.9) that responds to the $\epsilon$ challenge does not depend on the particular point $x_0$ in $D$.

**EXAMPLE 3.7** Define $f(x) = x^3$ for $x$ in $[0, 20]$. Then the function $f: [0, 20] \to \mathbb{R}$ is uniformly continuous. To see this, observe that for all $u$ and $v$ in $[0, 20]$,

$$|f(u) - f(v)| = |u^2 + uv + v^2||u - v| \leq 1200|u - v|.$$

Hence, for $\epsilon > 0$, if we define $\delta = \epsilon/1200$, then (3.13) holds. $\qquad\square$

**EXAMPLE 3.8** Define $f(x) = 1/x$ for $x$ in $(0, 1)$. The function $f: (0, 1) \to \mathbb{R}$ is continuous. However, it is not uniformly continuous. To verify this, we must find some $\epsilon_0 > 0$ so that for no choice of $\delta > 0$ is it true that

$$|1/u - 1/v| < \epsilon \text{ for all points } u \text{ and } v \text{ in } (0, 1) \text{ such that } |u - v| < \delta.$$

Indeed, set $\epsilon_0 = 1$. Then, given $\delta > 0$, we may find points $u$ and $v$ in the interval $(0, 1)$ with $|u - v| < \delta$ but $|1/u - 1/v| \geq 1$. For instance, choose $u$ to be any point in the interval $(0, 1)$ such that $0 < u < \delta$, then let $v = u/2$. Then $|u - v| = u/2 < \delta$, but $|f(u) - f(v)| = 1/u > 1$. Thus, $f: \mathbb{R} \to \mathbb{R}$ is continuous at $x_0$. $\qquad\square$

**FIGURE 3.5**

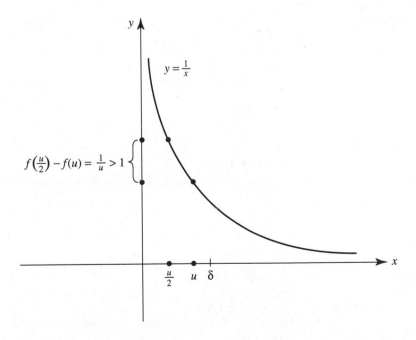

$$f\left(\tfrac{u}{2}\right) - f(u) = \tfrac{1}{u} > 1$$

$y = \frac{1}{x}$

**THEOREM 3.14** Suppose that the function $f: [a, b] \to \mathbb{R}$ is continuous. Then $f: [a, b] \to \mathbb{R}$ is uniformly continuous.

**Proof**   We will argue by contradiction. Suppose that $f:[a, b] \to \mathbb{R}$ is not uniformly continuous. Then there is some $\epsilon_0 > 0$ such that for no $\delta > 0$ is it true that

$$|f(u) - f(v)| < \epsilon_0 \quad \text{for all points } u \text{ and } v \text{ in } [a, b] \text{ such that } |u - v| < \delta.$$

Let $n$ be a natural number. Then there are points $u$ and $v$ in $[a, b]$ such that $|u - v| < 1/n$ but $|f(u) - f(v)| \geq \epsilon_0$. Choose two such points and label them $u_n$ and $v_n$.

This defines two sequences $\{u_n\}$ and $\{v_n\}$ in $[a, b]$. The Bolzano-Weierstrass Theorem can now be invoked to choose a subsequence $\{u_{n_k}\}$ of $\{u_n\}$ that converges to a point $u$ in $[a, b]$. But for each natural number $k$, $|u_{n_k} - v_{n_k}| < 1/n_k \leq 1/k$, so the subsequence $\{v_{n_k}\}$ also converges to $u$. The continuity of $f:[a, b] \to \mathbb{R}$ at $u$ implies that both $\{f(u_{n_k})\}$ and $\{f(v_{n_k})\}$ converge to $f(u)$. Thus, their difference $\{f(u_{n_k}) - f(v_{n_k})\}$ converges to 0. This contradicts the property that for each natural number $k$, $|f(u_{n_k}) - f(v_{n_k})| \geq \epsilon_0$. It follows that $f:[a, b] \to \mathbb{R}$ is uniformly continuous. ∎

An examination of the above proof shows that the only property of the domain of the function $f:[a, b] \to \mathbb{R}$ that was used was the compactness of $[a, b]$. Hence the theorem is true for a continuous function $f: K \to \mathbb{R}$ provided that its domain $K$ is compact. We leave the proof as an exercise.

---

**EXERCISES**

1.  Define $f(x) = x^2$ for all $x$ in $\mathbb{R}$. Verify the "$\epsilon$-$\delta$" criterion for continuity at $x = 2$ and at $x = 50$.
2.  Define $f(x) = \sqrt{x}$ for all $x \geq 0$. Verify the "$\epsilon$-$\delta$" criterion for continuity at $x = 4$ and at $x = 9$.
3.  Define $f(x) = x^3$ for all $x$ in $\mathbb{R}$. Verify the "$\epsilon$-$\delta$" criterion for continuity at each point $x_0$.
4.  Define

$$f(x) = \begin{cases} x & \text{if } x < 0 \\ x + 1 & \text{if } x \geq 0. \end{cases}$$

Use the "$\epsilon$-$\delta$" criterion for continuity to show that $f: \mathbb{R} \to \mathbb{R}$ is not continuous at 0.
5.  Define $g(x) = 6x + 7$ for all $x$ in $\mathbb{R}$. Prove that the function $g: \mathbb{R} \to \mathbb{R}$ is uniformly continuous.
6.  Define $h(x) = 1/(1 + x^2)$ for all $x$ in $\mathbb{R}$. Prove that the function $h: \mathbb{R} \to \mathbb{R}$ is uniformly continuous.
7.  Define $f(x) = x/(x - 1)$ for $x \geq 2$. Prove that the function $f:[2, \infty) \to \mathbb{R}$ is uniformly continuous.
8.  Define $f(x) = 1/x$ for $1/100 \leq x \leq 1$. Prove that the function $f:[1/100, 1] \to \mathbb{R}$ is uniformly continuous.
9.  Define $h(x) = \sqrt{x + 1} - \sqrt{x}$ for $x \geq 0$. Prove that the function $h:[0, \infty) \to \mathbb{R}$ is uniformly continuous.
10. Define $f(x) = x^3$ for all $x$ in $\mathbb{R}$. Prove that the function $f: \mathbb{R} \to \mathbb{R}$ is not uniformly continuous.
11. For an open interval $I = (a, b)$, find a continuous function $f: I \to \mathbb{R}$ that is not uniformly continuous.
12. Prove that the sum of uniformly continuous functions is uniformly continuous. Is the product of uniformly continuous functions uniformly continuous?

13.  A function $f: D \to \mathbb{R}$ is called a *Lipschitz function* if there is some $c \geq 0$ such that

$$|f(u) - f(v)| \leq c|u - v| \quad \text{for all points } u \text{ and } v \text{ in } D.$$

Prove that if $f: D \to \mathbb{R}$ is a Lipschitz function, then it is uniformly continuous.

14.  Define $f(x) = \sqrt{x}$ for $0 \leq x \leq 1$. Prove that the function $f: [0, 1] \to \mathbb{R}$ is uniformly continuous, but that it is not a Lipschitz function.

15.  Define the function $h: [1, 2] \to \mathbb{R}$ as follows: $h(x) = 0$ if the point $x$ in $[1, 2]$ is irrational; $h(x) = 1/n$ if the point $x$ in $[1, 2]$ is rational and $x = m/n$, where $m$ and $n$ are natural numbers having no common positive integer factor other than 1.
     (a)   Prove that $h: [1, 2] \to \mathbb{R}$ fails to be continuous at each rational number in $[1, 2]$.
     (b)   Prove that if $\epsilon > 0$, then the set $\{x \text{ in } [1, 2] \mid h(x) > \epsilon\}$ has only a finite number of points.
     (c)   Use (b) to prove that $h: [1, 2] \to \mathbb{R}$ is continuous at each irrational number in $[1, 2]$.

16.  Suppose that the function $f: (a, b) \to \mathbb{R}$ is uniformly continuous. Prove that $f: (a, b) \to \mathbb{R}$ is bounded.

17.  Suppose that the functions $g: (a, b) \to \mathbb{R}$ and $h: (a, b) \to \mathbb{R}$ are uniformly continuous. Prove that $gh: (a, b) \to \mathbb{R}$ is also uniformly continuous.

18.  A continuous function $f: (a, b) \to \mathbb{R}$ is said to be *continuously extendable* to $[a, b]$ if there is a continuous function whose domain is $[a, b]$ and whose restriction to $(a, b)$ is $f: (a, b) \to \mathbb{R}$. Prove that $f: (a, b) \to \mathbb{R}$ is continuously extendable to $[a, b]$ if and only if $f: (a, b) \to \mathbb{R}$ is uniformly continuous.

19.  Suppose that $f: (a, b) \to \mathbb{R}$ is continuous and monotone. Prove that $f: (a, b) \to \mathbb{R}$ is uniformly continuous if and only if its image $f(a, b)$ is bounded.

20.  Rewrite the proof of Theorem 3.14, with the interval $[a, b]$ replaced by a compact set $K$.

---

## 3.6   Limits

In the preceding sections of this chapter we have studied the properties of continuous functions. We now turn to the study of the behavior of functions near points that are not necessarily in the domain of the given function.

**DEFINITION**   *For a set $D$ of real numbers, the number $x_0$ is called a limit point of $D$ provided that there is a sequence of points in $D \backslash \{x_0\}$ that converges to $x_0$.*

**EXAMPLE 3.9**   For numbers $a$ and $b$ such that $a < b$, both $a$ and $b$ are limit points of the open interval $(a, b)$, although neither point belongs to $(a, b)$. To see that $a$ is a limit point of $(a, b)$, observe that $\{a + (b - a)/2n\}$ is a sequence in $(a, b)$, distinct from $a$, that converges to $a$. Also, every point $x_0$ in $(a, b)$ is also a limit point of $(a, b)$, because for such a point the sequence $\{x_0 + (b - x_0)/2n\}$ is a sequence in $(a, b)$, distinct from $x_0$, that converges to $x_0$.   □

**EXAMPLE 3.10**    Every real number is a limit point of $\mathbb{Q}$, the set of rational numbers. Indeed, let $x_0$ be any real number. Then, by the density of the rational numbers (Theorem 1.11), for each natural number $n$ we may select a rational number $q_n$ in the interval $(x_0, x_0 + 1/n)$. Then $\{q_n\}$ is a sequence of rational numbers, distinct from $x_0$, that converges to $x_0$. A similar argument shows that every real number is also a limit point of the set of irrational numbers.    □

**DEFINITION**    *Given a function $f: D \to \mathbb{R}$ and a limit point $x_0$ of its domain $D$, for a number $\ell$, we write*

$$\lim_{x \to x_0} f(x) = \ell \tag{3.14}$$

*provided that whenever $\{x_n\}$ is a sequence in $D \backslash \{x_0\}$ that converges to $x_0$,*

$$\lim_{n \to \infty} f(x_n) = \ell.$$

*We read (3.14) as "The limit of $f(x)$ as $x$ approaches $x_0$, with $x$ in $D$, equals $\ell$."*

For a function $f: D \to \mathbb{R}$ and point $x_0$ that is a limit point of its domain $D$, if there is a number $\ell$ such that $\lim_{x \to x_0} f(x) = \ell$, we write "$\lim_{x \to x_0} f(x)$ *exists*," and if there is no such number $\ell$ we write "$\lim_{x \to x_0} f(x)$ *does not exist*."

Comparing the definition of *limit* with the definition of *continuity* of a function at a point in its domain, it is not difficult (see Exercise 9) to see that if the number $x_0$ is a limit point of the set of numbers $D$ and also belongs to $D$, then a function $f: D \to \mathbb{R}$ is continuous at $x_0$ if and only if

$$\lim_{x \to x_0} f(x) = f(x_0).$$

Therefore, since we have already provided many examples of continuous functions, we have already computed many limits.

**EXAMPLE 3.11**    We have proven that the quotient of polynomials is continuous at points where the denominator is nonzero, that the square root function is continuous, and that composition of continuous functions is continuous. From this it follows that

$$\lim_{x \to 2} \sqrt{\frac{3x + 3}{x^3 - 4}} = \frac{3}{2}.$$    □

## EXAMPLE 3.12

$$\lim_{x \to 1} \frac{x^2 - 1}{x - 1} = 2. \tag{3.15}$$

To verify this, we let the sequence $\{x_n\}$ converge to 1, with $x_n \neq 1$ for all $n$. Then, by the Difference of Squares Formula, $[x_n^2 - 1]/[x_n - 1] = x_n + 1$ for all $n$. By the sum and product properties of convergent sequences,

$$\lim_{n \to \infty} \frac{x_n^2 - 1}{x_n - 1} = \lim_{n \to \infty} [x_n + 1] = 2,$$

and this proves (3.15).  $\square$

## EXAMPLE 3.13

$$\lim_{x \to 8} \frac{x - 8}{x^{1/3} - 2} = 12. \tag{3.16}$$

To verify this, we let the sequence $\{x_n\}$ converge to 8, with $x_n \neq 8$ for all $n$. Then, by the Difference of Cubes Formula, for each $n$,

$$x - 8 = (x^{1/3})^3 - 2^3 = (x^{1/3} - 2)(x^{2/3} + 2x^{1/3} + 4),$$

so that by the continuity of the $n$th root functions,

$$\lim_{n \to \infty} \frac{x_n - 8}{x_n^{1/3} - 2} = \lim_{n \to \infty} [x_n^{2/3} + 2x_n^{1/3} + 4] = 12. \qquad \square$$

For a function $f: D \to \mathbb{R}$ and a subset $U$ of its domain $D$ having $x_0$ as a limit point, we define

$$\lim_{x \to x_0, x \in U} f(x) = \ell$$

to mean that whenever $\{x_n\}$ is a sequence in $U \setminus \{x_0\}$ that converges to $x_0$, then $\{f(x_n)\}$ converges to $\ell$. If the point $x_0$ is a limit point of the set $\{x \text{ in } D \mid x > x_0\}$ and also of the set $\{x \text{ in } D \mid x < x_0\}$, then

$$\lim_{x \to x_0} f(x) = \ell \quad \text{if and only if} \quad \lim_{x \to x_0, x > x_0} f(x) = \lim_{x \to x_0, x < x_0} f(x) = \ell.$$

We leave the proof of this equivalence as an exercise.

**EXAMPLE 3.14** We will show that

$$\lim_{x \to 0, x > 0} \frac{|x|}{x} = 1 \quad \text{and} \quad \lim_{x \to 0, x < 0} \frac{|x|}{x} = -1 \quad \text{but that} \quad \lim_{x \to 0, x \neq 0} \frac{|x|}{x} \quad \text{does not exist.}$$

Observe that if $x \neq 0$, then

$$\frac{|x|}{x} = \begin{cases} 1 & \text{if } x > 0 \\ -1 & \text{if } x < 0. \end{cases}$$

To verify that $\lim_{x \to 0, x > 0} |x|/x = 1$, just observe that if $\{x_n\}$ is a sequence of positive numbers that converges to 0, then $\{x_n/|x_n|\}$ is a sequence with constant value 1, so $\lim_{n \to \infty} x_n/|x_n| = 1$. Similarly, if $\{x_n\}$ is a sequence of negative numbers that converges to 0, then $\{x_n/|x_n|\}$ is a sequence with constant value $-1$, so $\lim_{n \to \infty} x_n/|x_n| = -1$. Finally, $\lim_{x \to 0} |x|/x$ does not exist, since $\lim_{x \to 0, x < 0} f(x) \neq \lim_{x \to 0, x > 0} f(x)$. $\square$

**EXAMPLE 3.15** Define the function $f: \mathbb{R} \to \mathbb{R}$ by

$$f(x) = \begin{cases} 1 & \text{if } x \text{ is rational} \\ 0 & \text{if } x \text{ is irrational.} \end{cases}$$

This function is called *Dirichlet's function*. There is no point $x_0$ in $\mathbb{R}$ at which $\lim_{x \to x_0} f(x)$ exists. This follows from Example 3.3, in which we showed that there is no point at which the Dirichlet function is continuous. $\square$

The following theorem is an analogue, and also a consequence, of the sum, product, and quotient properties of convergent sequences. A completely similar result was established for continuous functions in Section 3.1.

**THEOREM 3.15** For functions $f: D \to \mathbb{R}$ and $g: D \to \mathbb{R}$, and a limit point $x_0$ of their domains $D$, suppose that

$$\lim_{x \to x_0} f(x) = A \quad \text{and} \quad \lim_{x \to x_0} g(x) = B.$$

Then

$$\lim_{x \to x_0} [f(x) + g(x)] = A + B, \tag{3.17}$$

$$\lim_{x \to x_0} [f(x)g(x)] = AB, \tag{3.18}$$

and, if $B \neq 0$ and $g(x) \neq 0$ for all $x$ in $D$,

$$\lim_{x \to x_0} \frac{f(x)}{g(x)} = \frac{A}{B}. \tag{3.19}$$

**Proof**   Let $\{x_n\}$ be a sequence in $D\backslash\{x_0\}$ that converges to $x_0$. From the definition of limit, it follows that

$$\lim_{n\to\infty} f(x_n) = A \quad \text{and} \quad \lim_{n\to\infty} g(x_n) = B.$$

The sum property of convergent sequences implies that

$$\lim_{n\to\infty} [f(x_n) + g(x_n)] = A + B, \tag{3.20}$$

and the product property of convergent sequences implies that

$$\lim_{n\to\infty} [f(x_n)g(x_n)] = AB. \tag{3.21}$$

If $g(x) \neq 0$ for all $x$ in $D$, the quotient property of convergent sequences implies that

$$\lim_{n\to\infty} \frac{f(x_n)}{g(x_n)} = \frac{A}{B}. \tag{3.22}$$

From the definition of limit, (3.17), (3.18), and (3.19) follow from (3.20), (3.21), and (3.22), respectively.   ∎

We have the following composition property for limits.

**THEOREM 3.16**   For functions $f: D \to \mathbb{R}$ and $g: U \to \mathbb{R}$, suppose that $x_0$ is a limit point of $D$ such that

$$\lim_{x\to x_0} f(x) = y_0, \tag{3.23}$$

and that $y_0$ is a limit point of $U$ such that

$$\lim_{y\to y_0} g(y) = \ell. \tag{3.24}$$

Moreover, suppose that

$$f(D\backslash\{x_0\}) \quad \text{is contained in} \quad U\backslash\{y_0\}. \tag{3.25}$$

Then   $$\lim_{x\to x_0} (g \circ f)(x) = \ell.$$

**Proof**   Let $\{x_n\}$ be a sequence in $D\backslash\{x_0\}$ that converges to $x_0$. From (3.23) it follows that $\{f(x_n)\}$ converges to $y_0$. Set $y_n = f(x_n)$ for each natural number $n$. Then the sequence $\{y_n\}$ converges to $y_0$, and assumption (3.25) implies that $\{y_n\}$ is a sequence in $U\backslash\{y_0\}$. From (3.24) it follows that $\{g(y_n)\}$ converges to $\ell$. Thus,

$$\lim_{n\to\infty} (g \circ f)(x_n) = \ell.$$   ∎

**EXAMPLE 3.16** Suppose that the function $f: \mathbb{R} \to \mathbb{R}$ has the property that

$$\lim_{x\to 0} \frac{f(x) - f(0)}{x} = \ell.$$

From the composition property of limits it follows that if $k$ is any natural number, then

$$\lim_{x\to 0} \frac{f(x^k) - f(0)}{x^k} = \ell.$$

Moreover, for any $c \neq 0$, the composition property also implies that

$$\lim_{x\to 0} \frac{f(cx) - f(0)}{cx} = \ell,$$

and so
$$\lim_{x\to 0} \frac{f(cx) - f(0)}{x} = c \lim_{x\to 0} \frac{f(cx) - f(0)}{cx} = c\ell. \qquad \square$$

---

**EXERCISES**

1. Find the following limits or determine that they do not exist:

   (a) $\lim_{x\to 0} |x|$

   (b) $\lim_{x\to 0, x>0} \frac{x + \sqrt{x}}{2 + \sqrt{x}}$

   (c) $\lim_{x\to 0} \frac{|x|^2}{x}$

   (d) $\lim_{x\to 0} \frac{1}{x}$

2. Prove that

   (a) $\lim_{x\to 1} \frac{x^4 - 1}{x - 1} = 4$

   (b) $\lim_{x\to 1} \frac{\sqrt{x} - 1}{x - 1} = \frac{1}{2}$

3. Define the function $f: \mathbb{R} \to \mathbb{R}$ by $f(x) = x$ if $x \neq 0$, and $f(0) = 4$. Show that $\lim_{x\to 0} f(x) = 0$.

4. Find the following limits or determine that they do not exist:

   (a) $\lim_{x\to 0} \frac{1 + 1/x}{1 + 1/x^2}$

   (b) $\lim_{x\to 0} \frac{1 + 1/x^2}{1 + 1/x}$

   (c) $\lim_{x\to 1} \frac{1 + 1/(x - 1)}{2 + 1/(x - 1)^2}$

5. For functions $f: D \to \mathbb{R}$ and $g: D \to \mathbb{R}$, with $x_0$ a limit point of $D$, suppose that $\lim_{x\to x_0} [f(x) + g(x)]$ exists. Can any conclusion be drawn about $\lim_{x\to x_0} f(x)$?

6. Let $D$ be the set of real numbers consisting of the single number $x_0$. Show that the set $D$ has no limit points. Also show that the set $\mathbb{N}$ of natural numbers has no limit points.

7. Let $D$ be a nonempty subset of $\mathbb{R}$ that is bounded above. Is the supremum of $D$ a limit point of $D$?

8. Explain why, in the definition of $\lim_{x\to x_0} f(x)$, it is necessary to require that $x_0$ be a limit point of $D$.

9. (a) A point $x_0$ in $D$ is said to be an *isolated point* of $D$ provided that there is an $r > 0$ such that the only point of $D$ in the interval $(x_0 - r, x_0 + r)$ is $x_0$ itself. Prove that a point $x_0$ in $D$ is either an isolated point or a limit point of $D$.

   (b) Suppose that $x_0$ is an isolated point of $D$. Prove that every function $f: D \to \mathbb{R}$ is continuous at $x_0$.

(c)   Prove that if the point $x_0$ in $D$ is a limit point of $D$, then a function $f: D \to \mathbb{R}$ is continuous at $x_0$ if and only if $\lim_{x \to x_0} f(x) = f(x_0)$.

**10.**   Suppose the function $f: \mathbb{R} \to \mathbb{R}$ has the property that there is some $M > 0$ such that

$$|f(x)| \leq M|x|^2 \quad \text{for all } x \text{ in } \mathbb{R}.$$

Prove that

$$\lim_{x \to 0} f(x) = 0 \quad \text{and} \quad \lim_{x \to 0} \frac{f(x)}{x} = 0.$$

**11.**   For each number $x$, define $f(x)$ to be the largest integer that is less than or equal to $x$. Graph the function $f: \mathbb{R} \to \mathbb{R}$. Given a number $x_0$, analyze the following limits:

(a)   $\lim_{x \to x_0, x < x_0} f(x)$ 　　　　　　　　　　　　　　(b)   $\lim_{x \to x_0, x > x_0} f(x)$

**12.**   Let $k$ be a natural number. Prove that

$$\lim_{x \to 1} \frac{x^k - 1}{x - 1} = k.$$

**13.**   (A general Squeezing Principle.) Let the number $x_0$ be a limit point of the set $D$ and suppose that the functions $f: D \to \mathbb{R}$, $g: D \to \mathbb{R}$, and $h: D \to \mathbb{R}$ have the property that

$$f(x) \leq g(x) \leq h(x) \quad \text{for all } x \text{ in } D.$$

If $\lim_{x \to x_0} f(x) = \lim_{x \to x_0} h(x) = \ell$, prove that $\lim_{x \to x_0} g(x) = \ell$.

**14.**   (A general Monotone Convergence Principle.) Let $a$ and $b$ be numbers with $a < b$ and set $I = (a, b)$. Suppose that the function $f: I \to \mathbb{R}$ is bounded, and is monotone increasing in the sense that if $u$ and $v$ are in $I$ and $u < v$, then $f(u) \leq f(v)$. Prove that $\lim_{x \to a} f(x)$ exists and that $\lim_{x \to b} f(x)$ exists.

# 4 Differentiation

## 4.1   The Algebra of Derivatives

The simplest type of function $f \colon \mathbb{R} \to \mathbb{R}$ is one whose graph is a line. For such a function, the ratio

$$\frac{f(x_1) - f(x_2)}{x_1 - x_2},$$

where $x_1 \neq x_2$, does not depend on the choice of points $x_1$ and $x_2$. We denote this ratio by $m$ and call $m$ the *slope* of the graph of $f \colon \mathbb{R} \to \mathbb{R}$. So a function $f \colon \mathbb{R} \to \mathbb{R}$ whose graph is a line is completely determined by prescribing its functional value at one point, say at $x_0$, and then prescribing its slope $m$; it is then defined by the formula

$$f(x) = f(x_0) + m(x - x_0) \quad \text{for all } x \text{ in } \mathbb{R}. \tag{4.1}$$

For a function whose graph is not a line, it makes no sense to speak of "the slope of the graph." However, many functions have the property that at certain points on their graph, the graph can be approximated, in a sense that we will soon make precise, by a tangent line. One then defines the slope of the graph at that point to be the slope of the tangent line. The slope will vary from point to point, and when we can determine the

**68**

slope at each point we have very useful information for analyzing the function. This is the basic geometric idea behind differentiation.

To make the above precise, we need to define *tangent line*. For a function $f: I \to \mathbb{R}$, where $I$ is an open interval containing the point $x_0$, observe that for a point $x$ in $I$, with $x \neq x_0$, the slope of the line joining the points $(x_0, f(x_0))$ and $(x, f(x))$ is

$$\frac{f(x) - f(x_0)}{x - x_0}.$$

**FIGURE 4.1**

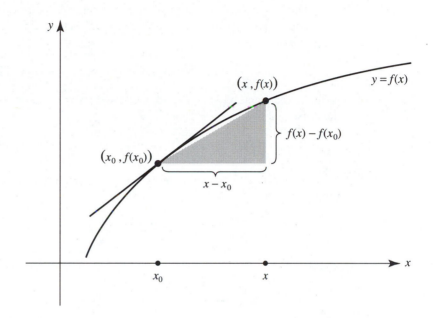

It is reasonable to expect that if there is a tangent line to the graph of $f: I \to \mathbb{R}$ at $(x_0, f(x_0))$, which has a slope $m_0$, then one should have

$$\lim_{x \to x_0} \frac{f(x) - f(x_0)}{x - x_0} = m_0.$$

**DEFINITION**  *Let $I$ be an open interval containing the point $x_0$. Then the function $f: I \to \mathbb{R}$ is said to be differentiable at $x_0$ if*

$$\lim_{x \to x_0} \frac{f(x) - f(x_0)}{x - x_0} \tag{4.2}$$

*exists, in which case we denote this limit by $f'(x_0)$ and call it the derivative of $f: I \to \mathbb{R}$ at $x_0$; that is*

$$f'(x_0) \equiv \lim_{x \to x_0} \frac{f(x) - f(x_0)}{x - x_0}. \tag{4.3}$$

*If the function $f: I \to \mathbb{R}$ is differentiable at every point in $I$, we say that $f: I \to \mathbb{R}$ is differentiable and call the function $f': I \to \mathbb{R}$ the derivative of $f: I \to \mathbb{R}$.*

For a function $f: I \to \mathbb{R}$ that is differentiable at $x_0$, we call the line determined by the equation

$$y = f(x_0) + f'(x_0)(x - x_0) \quad \text{for all } x \text{ in } \mathbb{R}$$

the *tangent line to the graph of $f: I \to \mathbb{R}$ at the point* $(x_0, f(x_0))$.

Observe that if $I$ is an open interval containing the point $x_0$ and the function $f: I \to \mathbb{R}$ is continuous at $x_0$, then no matter what number $m$ is chosen, the line determined by the graph of

$$g(x) = f(x_0) + m(x - x_0) \quad \text{for all } x \text{ in } \mathbb{R}$$

contains the point $(x_0, f(x_0))$ and is a "good approximation to $f: I \to \mathbb{R}$ near $x_0$" in the sense that

$$\lim_{x \to x_0} [f(x) - g(x)] = \lim_{x \to x_0} [f(x) - f(x_0) - m(x - x_0)] = 0.$$

Does there exist a line—that is, a choice of $m$—that has the much better approximation property

$$\lim_{x \to x_0} \frac{f(x) - g(x)}{x - x_0} = 0?$$

Since

$$\frac{f(x) - g(x)}{x - x_0} = \frac{f(x) - f(x_0)}{x - x_0} - m \quad \text{if } x \neq x_0 \text{ is in } I,$$

it is clear that the preceding approximation property holds if and only if the function $f: I \to \mathbb{R}$ is differentiable at $x_0$ and one chooses $m = f'(x_0)$.

We will devote this chapter to the study of differentiation. In Chapters 6 and 7 we will study integration. Once we have understood differentiation and integration, we will prove a version of formula (4.1) for differentiable functions whose graphs are not lines; it is called a Fundamental Theorem of Calculus.*

Observe that since $\lim_{x \to x_0} [x - x_0] = 0$, we cannot use the quotient formula for limits in the determination of differentiability. To overcome this obstacle, in this and the next section we will develop techniques for evaluating limits of the type (4.2), which are referred to as *differentiation rules*. Before turning to these, we will consider some specific examples.

---

*For a differentiable function $f: \mathbb{R} \to \mathbb{R}$ whose derivative is continuous, formula (4.1) becomes

$$f(x) = f(x_0) + \int_{x_0}^{x} f'(t)\, dt \quad \text{for all } x \text{ in } \mathbb{R}.$$

The symbols and the formula will be explained in Chapters 6 and 7.

**EXAMPLE 4.1** Define $f(x) = mx + b$ for all $x$ in $\mathbb{R}$. Then $f: \mathbb{R} \to \mathbb{R}$ is differentiable and

$$f'(x) = m \quad \text{for all } x \text{ in } \mathbb{R}.$$

This is clear since for each $x_0$ in $\mathbb{R}$,

$$\lim_{x \to x_0} \frac{f(x) - f(x_0)}{x - x_0} = \lim_{x \to x_0} \frac{m(x - x_0)}{x - x_0} = m. \qquad \square$$

**EXAMPLE 4.2** Consider the simplest function whose graph is not a line. Define $f(x) = x^2$ for all $x$ in $\mathbb{R}$. Then $f: \mathbb{R} \to \mathbb{R}$ is differentiable and

$$f'(x) = 2x \quad \text{for all } x \text{ in } \mathbb{R}.$$

Indeed, for each $x_0$ in $\mathbb{R}$,

$$\frac{f(x) - f(x_0)}{x - x_0} = \frac{x^2 - x_0^2}{x - x_0} = x + x_0 \quad \text{if } x \neq x_0,$$

and so $\qquad \displaystyle \lim_{x \to x_0} \frac{f(x) - f(x_0)}{x - x_0} = \lim_{x \to x_0} [x + x_0] = 2x_0. \qquad \square$

More generally, we have:

**PROPOSITION 4.1** For a natural number $n$, define $f(x) = x^n$ for all $x$ in $\mathbb{R}$. Then the function $f: \mathbb{R} \to \mathbb{R}$ is differentiable and

$$f'(x) = nx^{n-1} \quad \text{for all } x \text{ in } \mathbb{R}.$$

**Proof** Fix a number $x_0$. Observe that by the Difference of Powers Formula,

$$x^n - x_0^n = (x - x_0)(x^{n-1} + x^{n-2}x_0 + \cdots + xx_0^{n-2} + x_0^{n-1}) \quad \text{for all } x \text{ in } \mathbb{R},$$

and hence

$$\frac{f(x) - f(x_0)}{x - x_0} = x^{n-1} + x^{n-2}x_0 + \cdots + xx_0^{n-2} + x_0^{n-1} \quad \text{if} \quad x \neq x_0.$$

Thus, by the sum and product properties of limits,

$$\lim_{x \to x_0} \frac{f(x) - f(x_0)}{x - x_0} = nx_0^{n-1}. \qquad \blacksquare$$

**EXAMPLE 4.3** Define $f(x) = |x|$ for all $x$ in $\mathbb{R}$. Then $f: \mathbb{R} \to \mathbb{R}$ is not differentiable at $x = 0$. To see this, observe that

$$\lim_{x \to 0, x > 0} \frac{f(x) - f(0)}{x - 0} = \lim_{x \to 0, x > 0} \frac{|x|}{x} = 1$$

while

$$\lim_{x \to 0, x < 0} \frac{f(x) - f(0)}{x - 0} = \lim_{x \to 0, x < 0} \frac{|x|}{x} = -1.$$

Thus $\lim_{x \to 0} [f(x) - f(0)]/[x - 0]$ does not exist. It is easy to see that if $x \neq 0$, then $f: \mathbb{R} \to \mathbb{R}$ is differentiable at $x$, and $f'(x) = 1$ if $x > 0$, while $f'(x) = -1$ if $x < 0$. □

Let us now turn to some general results. We will use the sum, product, and quotient properties of limits to determine formulas for the derivative of the sum, product, and quotient of differentiable functions. Before doing so, we will prove that differentiability implies continuity.

**PROPOSITION 4.2** Let $I$ be an open interval containing the point $x_0$, and suppose that the function $f: I \to \mathbb{R}$ is differentiable at $x_0$. Then $f: I \to \mathbb{R}$ is continuous at $x_0$.

**Proof** Since

$$\lim_{x \to x_0} \frac{f(x) - f(x_0)}{x - x_0} = f'(x_0) \quad \text{and} \quad \lim_{x \to x_0} [x - x_0] = 0,$$

it follows from the product property of limits that

$$\lim_{x \to x_0} [f(x) - f(x_0)] = \lim_{x \to x_0} \left[ \frac{f(x) - f(x_0)}{x - x_0} \cdot (x - x_0) \right] = f'(x_0) \cdot 0 = 0.$$

Thus $\lim_{x \to x_0} f(x) = f(x_0)$, which means that $f: I \to \mathbb{R}$ is continuous at $x_0$. ■

As Example (4.3) shows, it is not true that continuity implies differentiability.

**THEOREM 4.3** Let $I$ be an open interval containing the point $x_0$, and suppose that the functions $f: I \to \mathbb{R}$ and $g: I \to \mathbb{R}$ are differentiable at $x_0$. Then

(i) $f + g: I \to \mathbb{R}$ is differentiable at $x_0$ and

$$(f + g)'(x_0) = f'(x_0) + g'(x_0),$$

(ii) $fg: I \to \mathbb{R}$ is differentiable at $x_0$ and

$$(fg)'(x_0) = f(x_0)g'(x_0) + f'(x_0)g(x_0),$$

and

(iii) if $g(x) \neq 0$ for all $x$ in $I$, then $f/g: I \to \mathbb{R}$ is differentiable at $x_0$ and

$$(f/g)'(x_0) = \frac{g(x_0)f'(x_0) - f(x_0)g'(x_0)}{(g(x_0))^2}.$$

**Proof of (i)**   For $x$ in $I$, with $x \neq x_0$,

$$\frac{(f+g)(x) - (f+g)(x_0)}{x - x_0} = \frac{f(x) - f(x_0)}{x - x_0} + \frac{g(x) - g(x_0)}{x - x_0}.$$

Hence, by the definition of derivative and the sum property of limits,

$$\lim_{x \to x_0} \left[ \frac{(f+g)(x) - (f+g)(x_0)}{x - x_0} \right] = f'(x_0) + g'(x_0).$$

**Proof of (ii)**   For $x$ in $I$, with $x \neq x_0$,

$$\begin{aligned}
\frac{(fg)(x) - (fg)(x_0)}{x - x_0} &= \frac{f(x)g(x) - f(x_0)g(x_0)}{x - x_0} \\
&= \frac{f(x)g(x) - f(x)g(x_0) + f(x)g(x_0) - f(x_0)g(x_0)}{x - x_0} \\
&= f(x)\left[ \frac{g(x) - g(x_0)}{x - x_0} \right] + g(x_0)\left[ \frac{f(x) - f(x_0)}{x - x_0} \right].
\end{aligned}$$

Since differentiability implies continuity, $\lim_{x \to x_0} f(x) = f(x_0)$. Consequently, using the definition of derivative and the addition and product properties for limits,

$$\lim_{x \to x_0} \left[ \frac{(fg)(x) - (fg)(x_0)}{x - x_0} \right] = f(x_0)g'(x_0) + g(x_0)f'(x_0).$$

**Proof of (iii)**   For $x$ in $I$, with $x \neq x_0$,

$$\begin{aligned}
\frac{(f/g)(x) - (f/g)(x_0)}{x - x_0} \\
= \frac{f(x)/g(x) - f(x_0)/g(x_0)}{x - x_0} \\
= \frac{1}{g(x)g(x_0)}\left[ \frac{f(x)g(x_0) - f(x_0)g(x)}{x - x_0} \right] \\
= \frac{1}{g(x)g(x_0)}\left[ \frac{f(x)g(x_0) - f(x_0)g(x_0) + f(x_0)g(x_0) - f(x_0)g(x)}{x - x_0} \right] \\
= \frac{1}{g(x)g(x_0)}\left[ g(x_0)\left\{ \frac{f(x) - f(x_0)}{x - x_0} \right\} - f(x_0)\left\{ \frac{g(x) - g(x_0)}{x - x_0} \right\} \right].
\end{aligned}$$

Since differentiability implies continuity, $\lim_{x \to x_0} g(x) = g(x_0)$. Hence we may use the definition of derivative together with the sum, product, and quotient properties of limits to conclude from the preceding identity that

$$\lim_{x \to x_0} \left[ \frac{(f/g)(x) - (f/g)(x_0)}{x - x_0} \right] = \frac{g(x_0)f'(x_0) - f(x_0)g'(x_0)}{(g(x_0))^2}. \qquad \blacksquare$$

---

**PROPOSITION 4.4**

For an integer $n$, define $\mathcal{O} = \mathbb{R}$ if $n \geq 0$ and $\mathcal{O} = \{x \text{ in } \mathbb{R} \mid x \neq 0\}$ if $n < 0$. Then define

$$f(x) = x^n \quad \text{for all } x \text{ in } \mathcal{O}.$$

The function $f \colon \mathcal{O} \to \mathbb{R}$ is differentiable and

$$f'(x) = nx^{n-1} \quad \text{for all } x \text{ in } \mathcal{O}.$$

**Proof** The case in which $n > 0$ is precisely Proposition 4.1, so we need only consider the case $n < 0$. But if $n < 0$, then

$$f(x) = \frac{1}{x^{-n}} \quad \text{for all } x \text{ in } \mathcal{O},$$

where $-n$ is a natural number. Then from Proposition 4.1 and the quotient formula for derivatives, it follows that $f \colon \mathcal{O} \to \mathbb{R}$ is differentiable and

$$f'(x) = \frac{x^{-n} \cdot 0 - (-n)x^{-n-1} \cdot 1}{(x^{-n})^2}$$

$$= nx^{2n-n-1}$$

$$= nx^{n-1} \quad \text{for } x \text{ in } \mathcal{O}. \qquad \blacksquare$$

---

**COROLLARY 4.5**

For polynomials $p \colon \mathbb{R} \to \mathbb{R}$ and $q \colon \mathbb{R} \to \mathbb{R}$, define $\mathcal{O} = \{x \text{ in } \mathbb{R} \mid q(x) \neq 0\}$. Then the quotient $p/q \colon \mathcal{O} \to \mathbb{R}$ is differentiable.

**Proof** From Proposition 4.1 and parts (i) and (ii) of Theorem 4.3 it follows that both $p \colon \mathcal{O} \to \mathbb{R}$ and $q \colon \mathcal{O} \to \mathbb{R}$ are differentiable. Then part (iii) of Theorem 4.3 implies that $p/q \colon \mathcal{O} \to \mathbb{R}$ is differentiable. $\qquad \blacksquare$

---

**EXERCISES**

1. Use the definition of derivative to compute the derivative of the following functions at $x = 1$:
   (a) $f(x) = \sqrt{x+1}$ for all $x > 0$.
   (b) $f(x) = x^3 + 2x$ for all $x$ in $\mathbb{R}$.
   (c) $f(x) = 1/(1+x^2)$ for all $x$ in $\mathbb{R}$.
2. Define $f(x) = x^3 + 2x + 1$ for all $x$ in $\mathbb{R}$. Find the equation of the tangent line to the graph of $f \colon \mathbb{R} \to \mathbb{R}$ at the point $(2, 13)$.
3. For $m_1$ and $m_2$ distinct real numbers, define

$$f(x) = \begin{cases} m_1 x + 4 & \text{if } x \leq 0 \\ m_2 x + 4 & \text{if } x \geq 0. \end{cases}$$

   Prove that the function $f \colon \mathbb{R} \to \mathbb{R}$ is continuous but not differentiable at $x = 0$.
4. Evaluate the following limits or determine that they do not exist:
   (a) $\lim_{x \to 0} \dfrac{x^2}{x}$
   (b) $\lim_{x \to 1} \dfrac{x^2 - 1}{\sqrt{x} - 1}$
   (c) $\lim_{x \to 0} \dfrac{x - 1}{\sqrt{x} - 1}$
   (d) $\lim_{x \to 2} \dfrac{x^4 - 16}{x - 2}$

**5.** For a natural number $n \geq 2$, define
$$f(x) = \begin{cases} 0 & \text{if } x \leq 0 \\ x^n & \text{if } x > 0. \end{cases}$$
Prove that the function $f: \mathbb{R} \to \mathbb{R}$ is differentiable.

**6.** Suppose that the function $f: \mathbb{R} \to \mathbb{R}$ has the property that
$$-x^2 \leq f(x) \leq x^2 \quad \text{for all } x \text{ in } \mathbb{R}.$$
Prove that $f: \mathbb{R} \to \mathbb{R}$ is differentiable at $x = 0$ and that $f'(0) = 0$.

**7.** Define
$$g(x) = \begin{cases} x^2 & \text{if } x \text{ is rational} \\ -x^2 & \text{if } x \text{ is irrational.} \end{cases}$$
Prove that the function $g: \mathbb{R} \to \mathbb{R}$ is differentiable at $x = 0$.

**8.** Determine the differentiability of each of the following functions at $x = 0$:
(a)   $f(x) = x|x|$   for all $x$ in $\mathbb{R}$.
(b)   $g(x) = \sqrt{|x|^3}$   for all $x$ in $\mathbb{R}$.

**9.** For real numbers $a$ and $b$, define
$$g(x) = \begin{cases} 3x^2 & \text{if } x \leq 1 \\ a + bx & \text{if } x > 1. \end{cases}$$
For what values of $a$ and $b$ is the function $g: \mathbb{R} \to \mathbb{R}$ differentiable at $x = 1$?

**10.** Define
$$f(x) = |x^3(1 - x)| \quad \text{for all } x \text{ in } \mathbb{R}.$$
At what points is the function $f: \mathbb{R} \to \mathbb{R}$ differentiable?

**11.** For a natural number $n$, the Geometric Sum Formula asserts that
$$1 + x + \cdots + x^n = \frac{1 - x^{n+1}}{1 - x} \quad \text{if } x \neq 1.$$
By differentiating, find a formula for
$$1 + x + 2x^2 + \cdots + nx^n,$$
and then for
$$1^2 + 2^2 x + \cdots + n^2 x^{n-1}.$$

**12.** Suppose that the function $g: \mathbb{R} \to \mathbb{R}$ is differentiable at $x = 0$. Also, suppose that for each natural number $n$, $g(1/n) = 0$. Prove that $g(0) = 0$ and $g'(0) = 0$.

**13.** Suppose that the function $f: \mathbb{R} \to \mathbb{R}$ is differentiable at $x_0$. Analyze the limit
$$\lim_{h \to 0} \left[ \frac{f(x_0 + h) - f(x_0 - h)}{h} \right].$$

**14.** Suppose that the function $f: \mathbb{R} \to \mathbb{R}$ is differentiable at $x_0$. Prove that
$$\lim_{x \to x_0} \frac{xf(x_0) - x_0 f(x)}{x - x_0} = f(x_0) - x_0 f'(x_0).$$

**15.** Let the function $g: \mathbb{R} \to \mathbb{R}$ be differentiable. Under what conditions is the function $|g|: \mathbb{R} \to \mathbb{R}$ differentiable at $x = 0$?

**16.** Let the function $f: \mathbb{R} \to \mathbb{R}$ be differentiable at $x = 0$. Prove that
$$\lim_{x \to 0} \frac{f(x^2) - f(0)}{x} = 0.$$

**17.** Define $p(x) = 1 - x + x^3$ for all $x$ in $\mathbb{R}$. Find $p'(1)$. Then write $p(x)$ as the sum of powers of $(x - 1)$ and observe that the coefficient of $(x - 1)$ is $p'(1)$.

**18.** Let the function $h: \mathbb{R} \to \mathbb{R}$ be bounded. Define the function $f: \mathbb{R} \to \mathbb{R}$ by

$$f(x) = 1 + x + x^2 h(x) \quad \text{for all } x \text{ in } \mathbb{R}.$$

Prove that $f(0) = 1$ and $f'(0) = 1$.

## 4.2 Differentiating Inverses and Compositions

Theorem 3.1 asserts that if $I$ is an interval and the function $f: I \to \mathbb{R}$ is strictly monotone, then its inverse is continuous. It is natural to consider the question of the differentiability of the inverse function at the point $y_0 = f(x_0)$ if $f: I \to \mathbb{R}$ is differentiable at $x_0$. The following example is instructive.

**EXAMPLE 4.4**  Define $f(x) = x^3$ for all $x$ in $\mathbb{R}$. We have shown that $f: \mathbb{R} \to \mathbb{R}$ is differentiable, strictly increasing, that $f(\mathbb{R}) = \mathbb{R}$, and that its inverse $f^{-1}: \mathbb{R} \to \mathbb{R}$ is continuous. However, its inverse is not differentiable at $x = 0$. Indeed, if $x \neq 0$, then

$$\frac{f^{-1}(x) - f^{-1}(0)}{x - 0} = \frac{x^{1/3}}{x} = \frac{1}{x^{2/3}},$$

so that

$$\lim_{x \to 0} \frac{f^{-1}(x) - f^{-1}(0)}{x - 0}$$

does not exist. $\qquad \square$

Thus the inverse of a differentiable function need not be differentiable. In the above example, the nondifferentiability of the inverse occurred at a point $f(x_0)$ where $f'(x_0) = 0$. The next theorem proves that it is only at such points that the inverse of a differentiable function can fail to be differentiable.

**THEOREM 4.6**  Let $I$ be an open interval containing the point $x_0$ and let the function $f: I \to \mathbb{R}$ be strictly monotone and continuous. Suppose that $f: I \to \mathbb{R}$ is differentiable at $x_0$ and that $f'(x_0) \neq 0$. Define $J = f(I)$. Then the inverse $f^{-1}: J \to \mathbb{R}$ is differentiable at the point $y_0 = f(x_0)$ and

$$(f^{-1})'(y_0) = \frac{1}{f'(x_0)}.$$

**Proof**  It follows from the Intermediate Value Theorem that $J$ is an open interval containing the point $y_0 = f(x_0)$. For a point $y$ in $J$, with $y \neq y_0$, since

$$y = f(f^{-1}(y)) \quad \text{and} \quad y_0 = f(f^{-1}(y_0)),$$

we have

$$\frac{f^{-1}(y) - f^{-1}(y_0)}{y - y_0} = 1 \Big/ \frac{y - y_0}{f^{-1}(y) - f^{-1}(y_0)} = 1 \Big/ \frac{f(f^{-1}(y)) - f(f^{-1}(y_0))}{f^{-1}(y) - f^{-1}(y_0)}.$$
(4.4)

Since the inverse function is continuous,

$$\lim_{y \to y_0} f^{-1}(y) = f^{-1}(y_0) = x_0,$$

and, by the definition of a derivative,

$$\lim_{x \to x_0} \frac{f(x) - f(x_0)}{x - x_0} = f'(x_0).$$

Thus, using the composition theorem for limits (Theorem 3.16), it follows that

$$\lim_{y \to y_0} \frac{f(f^{-1}(y)) - f(f^{-1}(y_0))}{f^{-1}(y) - f^{-1}(y_0)} = f'(x_0).$$
(4.5)

From (4.4), (4.5), and the quotient formula for limits, it follows that

$$\lim_{y \to y_0} \frac{f^{-1}(y) - f^{-1}(y_0)}{y - y_0} = \frac{1}{f'(x_0)}. \qquad \blacksquare$$

**COROLLARY
4.7**
Let $I$ be an open interval and suppose that the function $f: I \to \mathbb{R}$ is strictly monotone and differentiable with $f'(x) \neq 0$ for all $x$ in $I$. Define $J = f(I)$. Then the inverse function $f^{-1}: J \to \mathbb{R}$ is differentiable and

$$(f^{-1})'(x) = \frac{1}{f'(f^{-1}(x))} \quad \text{for all } x \text{ in } J.$$

**Proof**  Since differentiability implies continuity, the function $f: I \to \mathbb{R}$ is continuous. Hence we may apply the previous theorem at $x$ in $J$, where we have $x = f(f^{-1}(x))$ and $f^{-1}(x)$ plays the role of $x_0$ in the preceding theorem. $\qquad \blacksquare$

**PROPOSITION
4.8**
For a natural number $n$, define $g(x) = x^{1/n}$ for all $x > 0$. Then the function $g: (0, \infty) \to \mathbb{R}$ is differentiable and

$$g'(x) = \frac{1}{n} x^{1/n - 1} \quad \text{for all } x > 0.$$

**Proof** If $f:(0, \infty) \to \mathbb{R}$ is defined by $f(x) = x^n$ for $x > 0$, then, by definition, $g:(0, \infty) \to \mathbb{R}$ is the inverse of $f:(0, \infty) \to \mathbb{R}$. According to Proposition 4.1, $f'(x) = nx^{n-1}$ if $x > 0$. Using Corollary 4.7, we conclude that

$$g'(x) = \frac{1}{f'(g(x))} = \frac{1}{n\left(x^{1/n}\right)^{n-1}} = \frac{1}{n}x^{1/n-1} \quad \text{if } x > 0. \qquad \blacksquare$$

We have shown that the composition of continuous functions is continuous. The composition of differentiable functions is differentiable, and there is a formula for the derivative of the composition. This is the content of the following theorem.

**THEOREM 4.9** **The Chain Rule** Let $I$ be an open interval containing the point $x_0$ and suppose that the function $f: I \to \mathbb{R}$ is differentiable at $x_0$. Let $J$ be an open interval such that $f(I) \subseteq J$, and suppose that $g: J \to \mathbb{R}$ is differentiable at $f(x_0)$. Then the composition $g \circ f: I \to \mathbb{R}$ is differentiable at $x_0$, and

$$(g \circ f)'(x_0) = g'(f(x_0))f'(x_0). \qquad (4.6)$$

**Proof** Define $y_0 = f(x_0)$. For each $x$ in $I$ with $x \neq x_0$, if we let $y = f(x)$, we have

$$\frac{(g \circ f)(x) - (g \circ f)(x_0)}{x - x_0} = \frac{g(y) - g(y_0)}{y - y_0} \cdot \frac{f(x) - f(x_0)}{x - x_0}, \qquad (4.7)$$

provided that $f(x) \neq f(x_0)$. If there is an open interval containing $x_0$ in which $f(x) \neq f(x_0)$ if $x \neq x_0$, then the result follows by taking limits in the above identity and using the composition and product properties of limits. To account for the possibility that there is no such interval, we introduce an auxiliary function $h: J \to \mathbb{R}$ by defining

$$h(y) = \begin{cases} [g(y) - g(y_0)]/[y - y_0] & \text{for } y \text{ in } J \text{ with } y \neq y_0 \\ g'(y_0) & \text{if } y = y_0. \end{cases}$$

Observe that

$$g(y) - g(y_0) = h(y)[y - y_0] \quad \text{for all } y \text{ in } J,$$

so the preceding identity (4.7) can be rewritten as

$$\frac{(g \circ f)(x) - (g \circ f)(x_0)}{x - x_0} = h(f(x))\left[\frac{f(x) - f(x_0)}{x - x_0}\right] \quad \text{for } x \neq x_0 \text{ in } I. \qquad (4.8)$$

From the very definition of $g'(y_0)$ it follows that $h: J \to \mathbb{R}$ is continuous at $y_0$. Furthermore, the differentiability of $f: I \to \mathbb{R}$ at $x_0$ implies the continuity of $f: I \to \mathbb{R}$ at $x_0$, and hence, since the composition of continuous functions is continuous, $h \circ f: I \to \mathbb{R}$ is also continuous at $x_0$. From this, using the product theorem for limits and the identity (4.8), we see that

$$\lim_{x \to x_0} \frac{g(f(x)) - g(f(x_0))}{x - x_0} = h(f(x_0))f'(x_0) = g'(f(x_0))f'(x_0). \qquad \blacksquare$$

**PROPOSITION**
**4.10**

For a rational number $r$, define $h(x) = x^r$ for $x > 0$. Then the function $h: (0, \infty) \to \mathbb{R}$ is differentiable and

$$h'(x) = rx^{r-1} \quad \text{for } x > 0.$$

**Proof**    Since $r$ is a rational number, we can choose integers $m$ and $n$ with $m > 0$ such that $r = m/n$. For $x > 0$, define $g(x) = x^n$ and $f(x) = x^{1/m}$, so that $h(x) = g(f(x))$. According to Proposition 4.4, the function $g: (0, \infty) \to \mathbb{R}$ is differentiable and $g'(x) = nx^{n-1}$ if $x > 0$. On the other hand, according to Proposition 4.8, the function $f: (0, \infty) \to \mathbb{R}$ is differentiable and $f'(x) = (1/m)x^{1/m-1}$ if $x > 0$. From the Chain Rule, it follows that

$$h'(x) = g'(f(x))f'(x)$$
$$= n(x^{1/m})^{n-1}\frac{1}{m}x^{1/m-1}$$
$$= rx^{r-1} \quad \text{if } x > 0. \qquad \blacksquare$$

So far, in the present chapter, all of the results have been concerned with the evaluation of limits of the form

$$\lim_{x \to x_0} \frac{h(x)}{x - x_0}, \tag{4.9}$$

where $I$ is an interval containing the point $x_0$ and the function $h: I \to \mathbb{R}$ has the property that $\lim_{x \to x_0} h(x) = 0$. The reason, of course, that we need these particular results is that since $\lim_{x \to x_0} (x - x_0) = 0$, we cannot use the quotient formula for limits in the analysis of the limit (4.9). We have the same difficulty in evaluating any limit of the form

$$\lim_{x \to x_0} \frac{h(x)}{g(x)} \tag{4.10}$$

if $\lim_{x \to x_0} g(x) = 0$. If $\lim_{x \to x_0} h(x) \neq 0$, the limit (4.10) does not exist. The remaining case, when

$$\lim_{x \to x_0} g(x) = 0 \quad \text{and} \quad \lim_{x \to x_0} h(x) = 0,$$

is often called the *indeterminate case*. We have, however, the following simple but useful result, which is the first of various formulas for computing the limit of quotients known as L'Hôpital's Rules.

**THEOREM 4.11**    Let $I$ be an open interval containing the point $x_0$. Suppose that the functions $g: I \to \mathbb{R}$ and $h: I \to \mathbb{R}$ are differentiable at the point $x_0$ and that

$$g(x_0) = h(x_0) = 0.$$

Assume, moreover, that $g(x) \neq 0$ for $x$ in $I$ with $x \neq x_0$, and that $g'(x_0) \neq 0$. Then

$$\lim_{x \to x_0} \frac{h(x)}{g(x)} = \frac{h'(x_0)}{g'(x_0)}. \tag{4.11}$$

**Proof**   For a point $x$ in $I$, with $x \neq x_0$,

$$\frac{h(x)}{g(x)} = \frac{h(x) - h(x_0)}{g(x) - g(x_0)} = \frac{h(x) - h(x_0)}{x - x_0} \Big/ \frac{g(x) - g(x_0)}{x - x_0}.$$

Thus (4.11) follows from the very definition of a derivative and the quotient formula for limits.  ■

**EXAMPLE 4.5**   Formula (4.11) implies that

$$\lim_{x \to 0} \frac{4x + x^4}{x - x^3} = 4 \quad \text{and} \quad \lim_{x \to 1} \frac{x^2 - 1}{\sqrt{1 + x^2} - \sqrt{2}} = 2\sqrt{2}. \qquad \square$$

We will consider a more refined version of L'Hôpital's Rule in Section 4.4.

---

**EXERCISES**

1.   Suppose that the function $f: (0, \infty) \to \mathbb{R}$ is differentiable and let $c > 0$. Now define $g: (0, \infty) \to \mathbb{R}$ by $g(x) = f(cx)$, for $x > 0$. Prove that the function $g: (0, \infty) \to \mathbb{R}$ is also differentiable and $g'(x) = cf'(cx)$, for $x > 0$.

2.   Define
$$f(x) = \frac{1}{\sqrt{1 + x^2}} \quad \text{for all } x > 0.$$
Prove that $f: (0, \infty) \to \mathbb{R}$ is differentiable, strictly decreasing, and that $f(0, \infty) = (0, 1)$. Find $(f^{-1})'\left(\sqrt{1/5}\right)$.

3.   Let $I$ be an open interval containing the point $x_0$ and let $f: I \to \mathbb{R}$ be continuous, strictly monotone, and differentiable at $x_0$. Assume that $f'(x_0) = 0$. Use the Chain Rule to prove that $f^{-1}: f(I) \to \mathbb{R}$ is not differentiable at $f(x_0)$. Thus Theorem 4.6 cannot be improved.

4.   If the function $f: \mathbb{R} \to \mathbb{R}$ is differentiable at 0, does the limit $\lim_{x \to 0} [f(x) - f(0)]/x^2$ necessarily exist?

5.   Let $f: \mathbb{R} \to \mathbb{R}$ be differentiable and suppose that $h: \mathbb{R} \to \mathbb{R}$ is differentiable, strictly increasing, $h'(x) > 0$ for all $x$ in $\mathbb{R}$, and $h(\mathbb{R}) = \mathbb{R}$. Define $g(x) = f(h^{-1}(x))$ for all $x$ in $\mathbb{R}$. Find $g'(x)$ for $x$ in $\mathbb{R}$.

6.   Suppose that the function $f: \mathbb{R} \to \mathbb{R}$ is differentiable at 0. For real numbers $a$, $b$, and $c$, with $c \neq 0$, prove that
$$\lim_{x \to 0} \frac{f(ax) - f(bx)}{cx} = \frac{a - b}{c} f'(0).$$

7.   Let $I$ be an open interval and suppose that the function $f: I \to \mathbb{R}$ is differentiable and monotone increasing. Prove that $f'(x) \geq 0$ for all $x$ in $I$.

8.   Suppose that the function $f: \mathbb{R} \to \mathbb{R}$ is differentiable and that there is a bounded sequence $\{x_n\}$, with $x_n \neq x_m$ if $n \neq m$, such that $f(x_n) = 0$ for all natural numbers $n$. Prove that there is a number $x_0$ at which $f'(x_0) = 0$. (We will see in the next section that the conclusion holds if $f: \mathbb{R} \to \mathbb{R}$ attains the value 0 at just two distinct points.) (*Hint:* Apply the Bolzano-Weierstrass Theorem.)

9.   Suppose that the function $f: \mathbb{R} \to \mathbb{R}$ is differentiable and $\{x_n\}$ is a strictly increasing, bounded sequence with $f(x_n) \leq f(x_{n+1})$, for all $n$ in $\mathbb{N}$. Prove that there is a number $x_0$ at which $f'(x_0) \geq 0$. (*Hint:* Apply the Monotone Convergence Theorem.)

10.  A function $f: \mathbb{R} \to \mathbb{R}$ is called *even* if

$$f(x) = f(-x) \quad \text{for all } x \text{ in } \mathbb{R},$$

and $f: \mathbb{R} \to \mathbb{R}$ is called *odd* if

$$f(x) = -f(-x) \quad \text{for all } x \text{ in } \mathbb{R}.$$

Prove that if $f: \mathbb{R} \to \mathbb{R}$ is differentiable and odd, $f': \mathbb{R} \to \mathbb{R}$ is even.

## 4.3 The Lagrange Mean Value Theorem and Its Geometric Consequences

We will now prove one of the most useful and geometrically attractive results in calculus, the Lagrange Mean Value Theorem, or simply the Mean Value Theorem. It asserts that if the function $f: [a, b] \to \mathbb{R}$ is continuous and $f: (a, b) \to \mathbb{R}$ is differentiable, then there is a point $x_0$ in the open interval $(a, b)$ with the property that the tangent line to the graph at $(x_0, f(x_0))$ is parallel to the line passing through $(a, f(a))$ and $(b, f(b))$.

To prove the Mean Value Theorem, it is convenient first to prove some preliminary results.

**LEMMA 4.12**   Let $I$ be an open interval containing the point $x_0$ and suppose that the function $f: I \to \mathbb{R}$ is differentiable at $x_0$. If the point $x_0$ is either a maximizer or a minimizer of the function $f: I \to \mathbb{R}$, then $f'(x_0) = 0$.

**FIGURE 4.2**

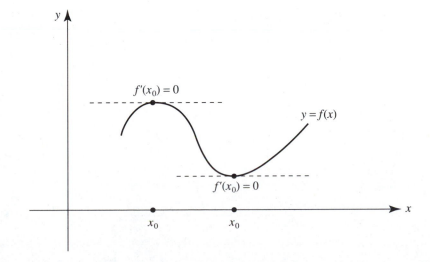

**Proof** Observe that by the very definition of derivative,

$$\lim_{x \to x_0, x < x_0} \frac{f(x) - f(x_0)}{x - x_0} = \lim_{x \to x_0, x > x_0} \frac{f(x) - f(x_0)}{x - x_0} = f'(x_0).$$

First suppose that $x_0$ is a maximizer. Then

$$\frac{f(x) - f(x_0)}{x - x_0} \geq 0 \quad \text{for } x \text{ in } I \text{ with } x < x_0,$$

and hence

$$f'(x_0) = \lim_{x \to x_0, x < x_0} \frac{f(x) - f(x_0)}{x - x_0} \geq 0.$$

On the other hand,

$$\frac{f(x) - f(x_0)}{x - x_0} \leq 0 \quad \text{for } x \text{ in } I \text{ with } x > x_0,$$

and hence

$$f'(x_0) = \lim_{x \to x_0, x > x_0} \frac{f(x) - f(x_0)}{x - x_0} \leq 0.$$

Thus $f'(x_0) = 0$.

In the case when $x_0$ is a minimizer, the same proof applies, with inequalities reversed. ∎

---

**THEOREM 4.13**  **Rolle's Theorem**  Suppose that the function $f: [a, b] \to \mathbb{R}$ is continuous and $f: (a, b) \to \mathbb{R}$ is differentiable. Assume, moreover, that

$$f(a) = f(b) = 0.$$

Then there is a point $x_0$ in the open interval $(a, b)$ at which

$$f'(x_0) = 0.$$

**FIGURE 4.3**

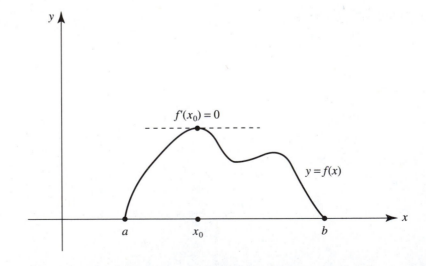

**Proof**   Since $f:[a, b] \to \mathbb{R}$ is continuous, according to the Extreme Value Theorem, it attains both a minimum value and a maximum value on $[a, b]$. Since $f(a) = f(b) = 0$, if both the maximizers and the minimizers occur at the endpoints, then the function $f:[a, b] \to \mathbb{R}$ is identically equal to 0, so $f'(x) = 0$ at every point $x$ in $(a, b)$. Otherwise, the function has either a maximizer or a minimizer at some point $x_0$ in the open interval $I = (a, b)$ and hence, by the preceding lemma, at this point $f'(x_0) = 0$.   ∎

Rolle's Theorem is a special case of the Lagrange Mean Value Theorem, but in fact, the general result follows from Rolle's Theorem by some algebraic manipulations.

**THEOREM 4.14**   **The Lagrange Mean Value Theorem**   Suppose that the function $f:[a, b] \to \mathbb{R}$ is continuous and $f:(a, b) \to \mathbb{R}$ is differentiable. Then there is a point $x_0$ in the open interval $(a, b)$ at which

$$f'(x_0) = \frac{f(b) - f(a)}{b - a}.$$

**FIGURE 4.4**

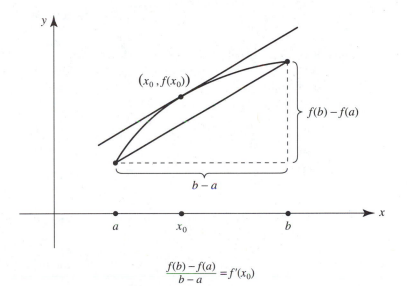

$$\frac{f(b) - f(a)}{b - a} = f'(x_0)$$

**Proof**   The line passing through the points $(a, f(a))$ and $(b, f(b))$ has slope $[f(b) - f(a)]/[b - a]$, so the equation of the line passing through these points is

$$y = g(x) = f(a) + \left[\frac{f(b) - f(a)}{b - a}\right](x - a).$$

Define the auxiliary function $d:[a, b] \to \mathbb{R}$ by $d(x) = f(x) - g(x)$ for all $x$ in $[a, b]$. Then clearly $d(a) = d(b) = 0$. Moreover, from the addition theorems for continuous and differentiable functions, it follows that $d:[a, b] \to \mathbb{R}$ is continuous and $d:(a, b) \to \mathbb{R}$ is differentiable. We apply Rolle's Theorem to select a point $x_0$ in $(a, b)$ at which $d'(x_0) = 0$. But

$$d'(x) = f'(x) - \frac{f(b) - f(a)}{b - a} \quad \text{for} \quad a < x < b,$$

so that
$$f'(x_0) = \frac{f(b) - f(a)}{b - a}.$$
∎

The preceding Mean Value Theorem is one of the most important results in calculus. Observe that its proof depends merely on the definition of the derivative and on the Extreme Value Theorem.

*As a general principle, if we have information about the derivative of a function that we wish to use in order to analyze the function, we should first try to apply the Mean Value Theorem.* The remainder of this section consists of various applications of this strategy.

A function $f: D \to \mathbb{R}$ is said to be *constant* provided that there is some number $c$ such that $f(x) = c$ for all $x$ in $D$.

**LEMMA 4.15**  Let $I$ be an open interval and suppose that the function $f: I \to \mathbb{R}$ is differentiable. Then $f: I \to \mathbb{R}$ is constant if and only if

$$f'(x) = 0 \quad \text{for all } x \text{ in } I.$$

**Proof**  It is clear that if $f: I \to \mathbb{R}$ is constant, then $f'(x) = 0$ for all $x$ in $I$. To prove the converse, let $u$ and $v$ be points in $I$ with $u < v$. Since differentiability implies continuity, the function $f: [u, v] \to \mathbb{R}$ is continuous and, of course, $f: (u, v) \to \mathbb{R}$ is differentiable. According to the Mean Value Theorem, there is a point $x_0$ in $(u, v)$ such that

$$f'(x_0) = \frac{f(v) - f(u)}{v - u}.$$

But $f'(x_0) = 0$, and thus $f(u) = f(v)$. Consequently, the function $f: I \to \mathbb{R}$ is constant. ∎

Two functions $g: I \to \mathbb{R}$ and $h: I \to \mathbb{R}$ are said to *differ by a constant* if there is some number $c$ such that

$$g(x) = h(x) + c \quad \text{for all } x \text{ in } I.$$

Of course, two functions $g: I \to \mathbb{R}$ and $h: I \to \mathbb{R}$ are equal provided that $g(x) = h(x)$ for all $x$ in $I$. Sometimes we say that equal functions are *identically equal*, in order to emphasize that the functions have the same value at all points of their domain. For this reason, we label the following result the *Identity Criterion*.

**PROPOSITION 4.16**  **The Identity Criterion**  Let $I$ be an open interval and let the functions $g: I \to \mathbb{R}$ and $h: I \to \mathbb{R}$ be differentiable. Then these functions differ by a constant if and only if

$$g'(x) = h'(x) \quad \text{for all } x \text{ in } I. \tag{4.12}$$

In particular, these functions are identically equal if and only if (4.12) holds and there is some point $x_0$ in $I$ at which

$$g(x_0) = h(x_0).$$

**Proof**  Define $f = g - h: I \to \mathbb{R}$. According to the differentiation rule for sums, $f: I \to \mathbb{R}$ is differentiable and

$$f'(x) = g'(x) - h'(x) \quad \text{for all } x \text{ in } I.$$

Also, observe that $f: I \to \mathbb{R}$ is constant if and only if the functions $g: I \to \mathbb{R}$ and $h: I \to \mathbb{R}$ differ by a constant. The result now follows from the preceding lemma. ∎

**COROLLARY 4.17**  Let $I$ be an open interval and the function $f: I \to \mathbb{R}$ be differentiable. Suppose that $f'(x) > 0$ for all $x$ in $I$. Then $f: I \to \mathbb{R}$ is strictly increasing.

**Proof**  Let $u$ and $v$ be points in $I$ with $u < v$. Then we may apply the Mean Value Theorem to $f: [u, v] \to \mathbb{R}$ and choose a point $x_0$ in $(u, v)$ at which

$$f'(x_0) = \frac{f(v) - f(u)}{v - u}.$$

Since $f'(x_0) > 0$ and $v - u > 0$, it follows that $f(v) > f(u)$. ∎

By replacing $f: I \to \mathbb{R}$ with $-f: I \to \mathbb{R}$, the above corollary implies that if $f: I \to \mathbb{R}$ has a negative derivative at all $x$ in $I$, then $f: I \to \mathbb{R}$ is strictly decreasing.

The above results give a method for finding intervals on which a differentiable function $f: I \to \mathbb{R}$ is strictly monotonic. The effectiveness of the method depends on being able to find those points at which $f'(x) = 0$. In fact, unless the function $f: I \to \mathbb{R}$ is quite simple, it is usually very difficult to find these points.

**DEFINITION**  *A point $x_0$ in the domain $D$ of a function $f: D \to \mathbb{R}$ is said to be a local maximizer for $f: D \to \mathbb{R}$ provided that there is some $\delta > 0$ such that*

$$f(x) \leq f(x_0) \quad \text{for all } x \text{ in } D \text{ such that } |x - x_0| < \delta.$$

*We call $x_0$ a local minimizer for $f: D \to \mathbb{R}$ provided that there is some $\delta > 0$ such that*

$$f(x) \geq f(x_0) \quad \text{for all } x \text{ in } D \text{ such that } |x - x_0| < \delta.$$

Lemma 4.12 implies that if $I$ is an open interval containing the point $x_0$ and $f: I \to \mathbb{R}$ is differentiable at $x_0$, then for $x_0$ to be either a local minimizer or a local maximizer for $f: I \to \mathbb{R}$, it is necessary that

$$f'(x_0) = 0.$$

However, knowing that $f'(x_0) = 0$ does not guarantee that $x_0$ is either a local maximizer or a local minimizer. For instance, if $f(x) = x^3$ for all $x$ in $\mathbb{R}$, then $f'(0) = 0$, but the point 0 is neither a local maximizer nor a local minimizer. In order to establish criteria that are sufficient for the existence of local maximizers and local minimizers, it is necessary to introduce higher derivatives.

For a differentiable function $f: I \to \mathbb{R}$ that has as its domain an open interval $I$, we say that $f: I \to \mathbb{R}$ has *one derivative* if $f: I \to \mathbb{R}$ is differentiable, and define $f^{(1)}(x) = f'(x)$ for all $x$ in $I$. If the function $f': I \to \mathbb{R}$ itself has a derivative, we say that $f: I \to \mathbb{R}$ has *two derivatives*, or has a *second derivative*, and denote the derivative of $f': I \to \mathbb{R}$ by $f'': I \to \mathbb{R}$ or by $f^{(2)}: I \to \mathbb{R}$. Now let $k$ be a natural number for which we have defined what it means for $f: I \to \mathbb{R}$ to *have $k$ derivatives* and have defined $f^{(k)}: I \to \mathbb{R}$. Then $f: I \to \mathbb{R}$ is said to *have $k+1$ derivatives* if $f^{(k)}: I \to \mathbb{R}$ is differentiable, and we define $f^{(k+1)}: I \to \mathbb{R}$ to be the derivative of $f^{(k)}: I \to \mathbb{R}$. In this context, is it useful to denote $f(x)$ by $f^{(0)}(x)$.

In general, if a function has $k$ derivatives, it does not necessarily have $k+1$ derivatives. For instance, the function $f: \mathbb{R} \to \mathbb{R}$ defined by $f(x) = |x|x$ for all $x$ in $\mathbb{R}$ is differentiable, but does not have a second derivative.

**THEOREM 4.18**  Let $I$ be an open interval containing the point $x_0$ and suppose that the function $f: I \to \mathbb{R}$ has a second derivative. Suppose that

$$f'(x_0) = 0.$$

If $f''(x_0) > 0$, then $x_0$ is a local minimizer of $f: I \to \mathbb{R}$.

If $f''(x_0) < 0$, then $x_0$ is a local maximizer of $f: I \to \mathbb{R}$.

**Proof**  First suppose that $f''(x_0) > 0$. Since

$$f''(x_0) = \lim_{x \to x_0} \frac{f'(x) - f'(x_0)}{x - x_0} > 0,$$

it follows (see Exercise 16) that there is a $\delta > 0$ such that the open interval $(x_0 - \delta, x_0 + \delta)$ is contained in $I$ and

$$\frac{f'(x) - f'(x_0)}{x - x_0} > 0 \quad \text{if} \quad 0 < |x - x_0| < \delta. \tag{4.13}$$

But $f'(x_0) = 0$, so (4.13) amounts to the assertion that if $|x - x_0| < \delta$, then

$$f'(x) > 0 \quad \text{if} \quad x > x_0 \qquad \text{and} \qquad f'(x) < 0 \quad \text{if} \quad x < x_0.$$

Using these two inequalities and the Mean Value Theorem, it follows that

$$f(x) > f(x_0) \quad \text{if} \quad 0 < |x - x_0| < \delta.$$

A similar argument applies when $f''(x_0) < 0$. ∎

The preceding theorem provides no information about $f(x_0)$ as a local extreme point if both $f'(x_0) = 0$ and $f''(x_0) = 0$. As we see from examining functions of the form $f(x) = cx^n$ for all $x$ in $\mathbb{R}$ at $x_0 = 0$, if $f'(x_0) = 0$ and $f''(x_0) = 0$, then $x_0$ may be a local maximizer or a local minimizer or neither.

The geometric consequences of the Mean Value Theorem that we have presented so far certainly conform to one's geometric intuition. However, it is always necessary to be careful. In Section 9.5 we will describe a function $f: \mathbb{R} \to \mathbb{R}$ that is continuous but has the following two properties:

(i) There is no point at which $f: \mathbb{R} \to \mathbb{R}$ is differentiable.

(ii) There is no interval $I$ such that $f: I \to \mathbb{R}$ is monotonic.

It is the existence of such functions that makes it absolutely necessary to root geometrical arguments in firm analytical ground. A less startling, but still somewhat surprising, example is the following function $f: \mathbb{R} \to \mathbb{R}$ that has $f'(0) > 0$ but for which there is no open interval $I$ containing 0 on which $f: I \to \mathbb{R}$ is increasing:

**EXAMPLE 4.6**   Define

$$f(x) = \begin{cases} x - x^2 & \text{if } x \text{ is rational} \\ x + x^2 & \text{if } x \text{ is irrational.} \end{cases}$$

Observe that

$$\left| \frac{f(x) - f(0)}{x - 0} - 1 \right| = |x| \text{ if } x \neq 0,$$

and so

$$f'(0) = \lim_{x \to 0} \frac{f(x) - f(0)}{x - 0} = 1.$$

On the other hand, if $I$ is any open interval containing the point 0, and $u$ is any positive rational number in $I$, then there is a positive irrational number $v$ that is less than $u$ whereas $f(u) > f(v)$. For instance, choose $\alpha$ to be any positive irrational number such that $\alpha < \min\{u, u^2\}$, and observe that $u - \alpha > u - u^2$, so that if $v = u - \alpha$, then $v$ is irrational and $f(v) = u - \alpha + (u - \alpha)^2 > u - u^2 = f(u)$. Thus, $f: I \to \mathbb{R}$ is not increasing.                                                                                      □

In the above example, the function $f: \mathbb{R} \to \mathbb{R}$, although differentiable at $x = 0$, failed even to be continuous at each $x \neq 0$. However, this is not the source of the difficulty. If we work a little harder, we can construct a function $f: \mathbb{R} \to \mathbb{R}$ that is differentiable at every $x$ in $\mathbb{R}$, has $f'(0) > 0$, and yet fails to be increasing on any open interval containing 0. So why does our intuition fail? It fails because it overestimates the consequences of the assumption that $f'(0) > 0$. Even if $f: \mathbb{R} \to \mathbb{R}$ is differentiable at every $x$ in $\mathbb{R}$, the assumption that $f'(0) > 0$ does not imply that there is an open interval containing 0 at each point of which $f'(x) > 0$. The crucial observation is that if $f: \mathbb{R} \to \mathbb{R}$ is differentiable, it need not be the case that $f': \mathbb{R} \to \mathbb{R}$ is continuous, so if $f'(0) > 0$, we cannot conclude that there is a $\delta > 0$ such that $f'(x) > 0$ for all $x$ in $(-\delta, \delta)$.

**EXERCISES**

1.   Sketch the graphs of the following functions. Find the intervals on which they are increasing or decreasing.
   (a)   $f: \mathbb{R} \to \mathbb{R}$ defined by $f(x) = x^3 + ax^2 + bx + c$ for all $x$ in $\mathbb{R}$.
   (b)   $h: (0, \infty) \to \mathbb{R}$ defined by $h(x) = a + b/x$ for $x > 0$, where $a > 0$, $b > 0$.
2.   For real numbers $a, b, c,$ and $d$, define $\mathcal{O} = \{x \mid cx + d \neq 0\}$. Then define

$$f(x) = \frac{ax + b}{cx + d} \quad \text{for all } x \text{ in } \mathcal{O}.$$

Show that if the function $f: \mathcal{O} \to \mathbb{R}$ is not constant, then it fails to have any local maximizers or minimizers. Sketch the graph.
3.   For $c > 0$, prove that the following equation does not have two solutions:

$$x^3 - 3x + c = 0, \qquad 0 < x < 1.$$

**4.** Prove that the following equation has exactly one solution:

$$x^5 + 5x + 1 = 0, \qquad -1 < x < 0.$$

**5.** Prove that the following equation has exactly two solutions:

$$x^4 + 2x^2 - 6x + 2 = 0, \qquad x \text{ in } \mathbb{R}.$$

**6.** For any numbers $a$ and $b$ and an even natural number $n$, show that the following equation has at most two solutions:

$$x^n + ax + b = 0, \qquad x \text{ in } \mathbb{R}.$$

Is this true if $n$ is odd?

**7.** For numbers $a$ and $b$, prove that the following equation has exactly three solutions if and only if $4a^3 + 27b^2 < 0$:

$$x^3 + ax + b = 0, \qquad x \text{ in } \mathbb{R}.$$

**8.** Let $D$ be the set of nonzero real numbers. Suppose that the functions $g: D \to \mathbb{R}$ and $h: D \to \mathbb{R}$ are differentiable and that

$$g'(x) = h'(x) \quad \text{for all } x \text{ in } D.$$

Do the functions $g: D \to \mathbb{R}$ and $h: D \to \mathbb{R}$ differ by a constant? (*Hint:* Is $D$ an interval?)

**9.** Let $n$ be a natural number. Suppose that the function $f: \mathbb{R} \to \mathbb{R}$ is differentiable and that the following equation has at most $n - 1$ solutions:

$$f'(x) = 0, \qquad x \text{ in } \mathbb{R}.$$

Prove that the following equation has at most $n$ solutions:

$$f(x) = 0, \qquad x \text{ in } \mathbb{R}.$$

**10.** Use an induction argument together with the preceding exercise to prove that if $p: \mathbb{R} \to \mathbb{R}$ is a polynomial of degree $n$, then there are at most $n$ solutions of the equation

$$p(x) = 0, \qquad x \text{ in } \mathbb{R}.$$

**11.** Suppose that the function $f: \mathbb{R} \to \mathbb{R}$ is differentiable and

$$\begin{cases} f'(x) = x + x^3 + 2 & \text{for all } x \text{ in } \mathbb{R} \\ f(0) = 5. \end{cases}$$

What is the function $f: \mathbb{R} \to \mathbb{R}$?

**12.** Suppose that the function $g: (-1, 1) \to \mathbb{R}$ is differentiable and

$$\begin{cases} g'(x) = x/\sqrt{1 - x^2} & \text{for} \quad -1 < x < 1 \\ g(0) = 25. \end{cases}$$

What is the function $g: \mathbb{R} \to \mathbb{R}$?

**13.** Let $g: \mathbb{R} \to \mathbb{R}$ and $f: \mathbb{R} \to \mathbb{R}$ be differentiable functions, and suppose that

$$g(x)f'(x) = f(x)g'(x) \quad \text{for all } x \text{ in } \mathbb{R}.$$

If $g(x) \neq 0$ for all $x$ in $\mathbb{R}$, show that there is some $c$ in $\mathbb{R}$ such that $f(x) = cg(x)$ for all $x$ in $\mathbb{R}$.

14. Suppose that $f: \mathbb{R} \to \mathbb{R}$ and $g: \mathbb{R} \to \mathbb{R}$ are each differentiable, and that

$$\begin{cases} f'(x) = g(x) \quad \text{and} \quad g'(x) = -f(x) \quad \text{for all } x \text{ in } \mathbb{R} \\ f(0) = 0 \quad \text{and} \quad g(0) = 1. \end{cases}$$

Prove that

$$[f(x)]^2 + [g(x)]^2 = 1 \quad \text{for all } x \text{ in } \mathbb{R}.$$

(*Hint:* Show that $h(x) = [f(x)]^2 + [g(x)]^2$, for all $x$ in $\mathbb{R}$, defines a constant function.)

15. Let $I$ be an open interval. Suppose that the function $f: I \to \mathbb{R}$ is continuous and that at the point $x_0$ in $I$, $f(x_0) > 0$. Prove that there is a $\delta > 0$ such that $f(x) > 0$ if $|x - x_0| < \delta$.

16. Let $I$ be an open interval containing the point $x_0$ and suppose that the function $g: I \to \mathbb{R}$ is differentiable. Define

$$h(x) = \begin{cases} [g(x) - g(x_0)]/[x - x_0] & \text{if } x \neq x_0 \\ g'(x_0) & \text{if } x = x_0. \end{cases}$$

Show that $h: I \to \mathbb{R}$ is continuous. If $g'(x_0) > 0$, use the preceding exercise to show that there is a $\delta > 0$ such that $[g(x) - g(x_0)]/[x - x_0] > 0$ if $0 < |x - x_0| < \delta$.

17. Suppose that the function $f: \mathbb{R} \to \mathbb{R}$ is differentiable, that $f': \mathbb{R} \to \mathbb{R}$ is continuous at 0, and that $f'(0) > 0$. Prove that there is an open interval $I$ containing 0 such that $f: I \to \mathbb{R}$ is strictly monotonic.

18. Let the function $f: \mathbb{R} \to \mathbb{R}$ have the property that there is a positive number $c$ such that $|f(u) - f(v)| \leq c(u - v)^2$ for all $u, v$ in $\mathbb{R}$. Prove that the function $f: \mathbb{R} \to \mathbb{R}$ is constant.

19. Suppose that $f: \mathbb{R} \to \mathbb{R}$ is differentiable and that there is a positive number $c$ such that

$$f'(x) \geq c \quad \text{for all } x \text{ in } \mathbb{R}.$$

Prove that

$$f(x) \geq f(0) + cx \quad \text{if} \quad x \geq 0 \quad \text{and} \quad f(x) \leq f(0) + cx \quad \text{if} \quad x \leq 0.$$

Use these inequalities to prove that $f(\mathbb{R}) = \mathbb{R}$.

20. Let the function $f: \mathbb{R} \to \mathbb{R}$ have two derivatives and suppose that

$$f(x) \leq 0 \quad \text{and} \quad f''(x) \geq 0 \quad \text{for all } x \text{ in } \mathbb{R}.$$

Prove that $f: \mathbb{R} \to \mathbb{R}$ is constant. (*Hint:* Observe that $f': \mathbb{R} \to \mathbb{R}$ is increasing.)

21. Let the function $f: \mathbb{R} \to \mathbb{R}$ have two derivatives with $f(0) = 0$ and

$$f'(x) \leq f(x) \quad \text{for all } x \text{ in } \mathbb{R}.$$

Is $f(x) = 0$ for all $x$ in $\mathbb{R}$?

## 4.4 The Cauchy Mean Value Theorem and Its Analytic Consequences

The following is a useful extension of the Lagrange Mean Value Theorem.

**THEOREM 4.19**    **The Cauchy Mean Value Theorem**    Suppose that the functions $f: [a, b] \to \mathbb{R}$ and $g: [a, b] \to \mathbb{R}$ are continuous, and that $f: (a, b) \to \mathbb{R}$ and $g: (a, b) \to \mathbb{R}$ are differentiable. Moreover, assume that

$$g'(x) \neq 0 \quad \text{for all } x \text{ in } (a, b).$$

Then there is a point $x_0$ in the open interval $(a, b)$ at which

$$\frac{f(b) - f(a)}{g(b) - g(a)} = \frac{f'(x_0)}{g'(x_0)}. \tag{4.14}$$

**Proof**    Observe that since $g'(x) \neq 0$ for $a < x < b$, it follows from the Lagrange Mean Value Theorem that $g(b) - g(a) \neq 0$. Let us define an auxiliary function $\psi: [a, b] \to \mathbb{R}$ by

$$\psi(x) = f(x) - f(a) - \left[ \frac{f(b) - f(a)}{g(b) - g(a)} \right] [g(x) - g(a)] \quad \text{for} \quad a \leq x \leq b.$$

The addition properties of continuous and differentiable functions imply that $\psi: [a, b] \to \mathbb{R}$ is continuous and $\psi: (a, b) \to \mathbb{R}$ is differentiable. Moreover, we check to see that

$$\psi(a) = \psi(b) = 0.$$

According to Rolle's Theorem, there is a point $x_0$ in the open interval $(a, b)$ at which $\psi'(x_0) = 0$. But

$$\psi'(x_0) = f'(x_0) - \left[ \frac{f(b) - f(a)}{g(b) - g(a)} \right] g'(x_0),$$

and hence we obtain (4.14).    ∎

Observe that if $g(x) = x$ for $a \leq x \leq b$, then the Cauchy Mean Value Theorem reduces to the Lagrange Mean Value Theorem. Also observe that if we were to apply the Lagrange Mean Value Theorem first to $f: [a, b] \to \mathbb{R}$ and then to $g: [a, b] \to \mathbb{R}$, then instead of (4.14) we would have

$$\frac{f(b) - f(a)}{g(b) - g(a)} = \frac{f'(x_1)}{g'(x_2)}$$

for some points $x_1$ and $x_2$ in $(a, b)$. The whole point of Theorem 4.19 is that we can select $x_1 = x_2$. The following is another one of L'Hôpital's Rules.

**THEOREM 4.20** Let $x_0$ be an endpoint of an open interval $I$. Suppose that the functions $f: I \to \mathbb{R}$ and $g: I \to \mathbb{R}$ are differentiable and

$$\lim_{x \to x_0} f(x) = 0 = \lim_{x \to x_0} g(x). \tag{4.15}$$

Moreover, suppose that $g'(x) \neq 0$ for all $x$ in $I$ and

$$\lim_{x \to x_0} \frac{f'(x)}{g'(x)} = \ell. \tag{4.16}$$

Then

$$\lim_{x \to x_0} \frac{f(x)}{g(x)} = \ell. \tag{4.17}$$

**Proof** We will suppose that $I = (x_0, b)$, with $b > x_0$. Extend the functions $f: I \to \mathbb{R}$ and $g: I \to \mathbb{R}$ to $f: [x_0, b) \to \mathbb{R}$ and $g: [x_0, b) \to \mathbb{R}$ by defining $f(x_0) = g(x_0) = 0$. Then (4.15) implies that each of these extensions is continuous at $x_0$. Therefore, since differentiability implies continuity, both $f: [x_0, b) \to \mathbb{R}$ and $g: [x_0, b) \to \mathbb{R}$ are continuous. Let $x$ be in $I$. According to the Cauchy Mean Value Theorem, applied to $f: [x_0, x] \to \mathbb{R}$ and $g: [x_0, x] \to \mathbb{R}$, we can select a point $c = c(x)$ in the interval $(x_0, x)$ at which

$$\frac{f(x)}{g(x)} = \frac{f(x) - f(x_0)}{g(x) - g(x_0)} = \frac{f'(c(x))}{g'(c(x))}. \tag{4.18}$$

Since $\lim_{x \to x_0} c(x) = x_0$, (4.17) follows from (4.18) and (4.16). ∎

**EXAMPLE 4.7** As a simple illustration of Theorem 4.20, we observe that

$$\lim_{x \to 0, x > 0} \frac{x + \sqrt{1 + x} - 1}{x + x^{3/2}} = \frac{3}{2}. \qquad \square$$

We will see many more interesting uses of the above theorem in the context of exponential and trigonometric functions.

**THEOREM 4.21** Let $I$ be an open interval and $n$ be a nonnegative integer, and suppose the function $f: I \to \mathbb{R}$ has $n$ derivatives. Suppose also that at the point $x_0$ in $I$,

$$f^{(k)}(x_0) = 0 \quad \text{for} \quad 0 \leq k \leq n - 1.$$

Then for each point $x \neq x_0$ in $I$, there is a point $z$ strictly between $x$ and $x_0$ at which

$$f(x) = \frac{f^{(n)}(z)}{n!}(x - x_0)^n. \tag{4.19}$$

**Proof**   Define $g(x) = (x - x_0)^n$ for all $x$ in $I$. Then $g^{(k)}(x_0) = 0$ for $0 \le k \le n - 1$, and $g^{(n)}(x_0) = n!$. Let $x$ be a point in $I$, with $x \ne x_0$. We may suppose that $x > x_0$. By applying the Cauchy Mean Value Theorem to the functions $f: [x_0, x] \to \mathbb{R}$ and $g: [x_0, x] \to \mathbb{R}$, we may select a point $x_1$ in $(x_0, x)$ at which

$$\frac{f(x)}{g(x)} = \frac{f(x) - f(x_0)}{g(x) - g(x_0)} = \frac{f'(x_1)}{g'(x_1)}. \tag{4.20}$$

Now apply the Cauchy Mean Value Theorem to $f': [x_0, x_1] \to \mathbb{R}$ and $g': [x_0, x_1] \to \mathbb{R}$ in order to select $x_2$ in $(x_0, x_1)$, at which

$$\frac{f'(x_1)}{g'(x_1)} = \frac{f'(x_1) - f'(x_0)}{g'(x_1) - g'(x_0)} = \frac{f''(x_2)}{g''(x_2)}, \tag{4.21}$$

so that by (4.20),

$$\frac{f(x)}{g(x)} = \frac{f''(x_2)}{g''(x_2)}.$$

Continuing with successively higher derivatives, we obtain a point $x_n$ in $(x_0, x)$ such that

$$\frac{f(x)}{g(x)} = \frac{f^{(n)}(x_n)}{g^{(n)}(x_0)} = \frac{f^{(n)}(x_n)}{n!},$$

and setting $z = x_n$ we obtain (4.19).                                     ∎

---

**EXERCISES**

1.   Evaluate the following limits or determine that they do not exist:

(a)   $\lim_{x \to 0, x > 0} \dfrac{2 - 2(1 + x)^{3/2}}{x}$

(b)   $\lim_{x \to 0, x > 0} \dfrac{1 + x}{x}$

(c)   $\lim_{x \to 1} \dfrac{x^3 - 3x^2 + 3x - 1}{(x - 1)^2}$

(d)   $\lim_{x \to 0, x > 0} \dfrac{2x^2 + x^3}{x|x|}$

2.   Suppose that the function $f: \mathbb{R} \to \mathbb{R}$ has two derivatives, with $f(0) = f'(0) = 0$ and $|f''(x)| \le 1$ if $|x| \le 1$. Prove that $f(x) \le 1/2$ if $x \le 1$.

3.   Let $p: \mathbb{R} \to \mathbb{R}$ be a polynomial of degree no greater than 5. Suppose that at some point $x_0$ in $\mathbb{R}$,

$$p(x_0) = p'(x_0) = \cdots = p^{(5)}(x_0) = 0.$$

Prove that $p(x) = 0$ for all $x$ in $\mathbb{R}$.

4.   Define $f(t) = t^2$ for $0 \le t \le 1$ and $g(t) = t^3$ for $0 \le t \le 1$.

(a)   Find the number $c$ with $0 < c < 1$ at which

$$\frac{f(1) - f(0)}{g(1) - g(0)} = \frac{f'(c)}{g'(c)}.$$

(b)   Show that there does not exist a number $c$ with $0 < c < 1$ at which

$$\begin{cases} f(1) - f(0) = f'(c)(1 - 0) \\ \text{and} \\ g(1) - g(0) = g'(c)(1 - 0). \end{cases}$$

5.   Suppose that the functions $f:[a, b] \to \mathbb{R}$ and $g:[a, b] \to \mathbb{R}$ are continuous, and that $f:(a, b) \to \mathbb{R}$ and $g:(a, b) \to \mathbb{R}$ are differentiable. Also suppose that $|f'(x)| \geq |g'(x)| > 0$ for all $x$ in $(a, b)$. Prove that

$$|f(u) - f(v)| \geq |g(u) - g(v)| \quad \text{for all } u, v \text{ in } [a, b].$$

6.   Suppose that the function $f:(-1, 1) \to \mathbb{R}$ has $n$ derivatives, and $f^{(n)}:(-1, 1) \to \mathbb{R}$ is bounded. Assume also that $f(0) = f'(0) = \cdots = f^{(n-1)}(0) = 0$. Prove that there is a positive number $M$ such that

$$|f(x)| \leq M|x|^n \quad \text{for all } x \text{ in } (-1, 1).$$

7.   Suppose that the function $f:(-1, 1) \to \mathbb{R}$ has $n$ derivatives. Assume that there is a positive number $M$ such that

$$|f(x)| \leq M|x|^n \quad \text{for all } x \text{ in } (-1, 1).$$

Prove that $f(0) = f'(0) = \cdots = f^{(n-1)}(0) = 0$.

8.   Let $I$ be an open interval containing the point $x_0$ and suppose that the function $f:I \to \mathbb{R}$ has two derivatives. Prove that

$$\lim_{h \to 0} \frac{f(x_0 + h) - 2f(x_0) + f(x_0 - h)}{h^2} = f''(x_0).$$

9.   Let $I$ be an open interval and $n$ be a natural number. Suppose that both $f:I \to \mathbb{R}$ and $g:I \to \mathbb{R}$ have $n$ derivatives. Prove that $fg:I \to \mathbb{R}$ has $n$ derivatives, and we have the following formula, called *Leibnitz's formula*:

$$(fg)^{(n)}(x) = \sum_{k=0}^{n} \binom{n}{k} f^{(k)}(x) g^{(n-k)}(x) \quad \text{for all } x \text{ in } I.$$

Write the formula out explicitly for $n = 2$ and $n = 3$.

## 4.5   A Fundamental Differential Equation

For a function $f:I \to \mathbb{R}$ defined on an open interval $I$, consider the following question:

*Does there exist a differentiable function $F:I \to \mathbb{R}$ such that*

$$F'(x) = f(x) \quad \text{for all } x \text{ in } I, \tag{4.22}$$

*and if there is such a function $F:I \to \mathbb{R}$, is it unique?*

A *differential equation* is an equation in which we prescribe some conditions that the derivative or derivatives of a function must satisfy, and from this we wish to find the function. Equation (4.22) is the prototypical differential equation. Differential equations occur throughout mathematics and science. Many of the basic laws of physics are stated as differential equations. As we shall see in Chapter 6, equation (4.22) is the crucial link between differential and integral calculus.

It follows from the first part of the Identity Criterion (Proposition 4.16) that if there is a particular function $F:I \to \mathbb{R}$ that satisfies (4.22), then all of the other functions that satisfy (4.22) are obtained by adding a constant to that particular function. However, it is not at all clear that there are any functions that satisfy (4.22). Consider the following examples.

**EXAMPLE 4.8**

(i) A solution of the differential equation

$$F'(x) = x^7 - 3x + 1 \quad \text{for all } x \text{ in } \mathbb{R}$$

is given by $F(x) = \dfrac{x^8}{8} - \dfrac{3}{2}x^2 + x$ for all $x$ in $\mathbb{R}$.

(ii) A solution of the differential equation

$$F'(x) = \frac{x}{\sqrt{1-x^2}} \quad \text{for } -1 < x < 1$$

is given by $F(x) = -\sqrt{1-x^2}$ for $-1 < x < 1$.

(iii) A solution of the differential equation

$$F'(x) = \frac{1}{1+x^4} \quad \text{for all } x \text{ in } \mathbb{R}$$

cannot be determined on the basis of our present results.    □

In (i) and (ii) above, using the differentiation formulas that we have proven, a solution of the given differential equation was determined by inspection; in (iii), the reader will not be able to find a solution. A more complete analysis of (4.22) will require the development of a new concept, the integral, which we will address in Chapters 6 and 7. For now, we prove a result, due to Darboux, that asserts that if $I$ is an open interval and $f: I \to \mathbb{R}$ fails to have the Intermediate Value Property, then there does not exist a solution of the differential equation (4.22).

**THEOREM 4.22**    **Darboux's Theorem**    Let $I$ be an open interval and suppose that the function $F: I \to \mathbb{R}$ is differentiable. Then the image of the derivative $F': I \to \mathbb{R}$ is an interval; that is, if $u$ and $v$ are points in $I$ and $c$ is strictly between $F'(u)$ and $F'(v)$, then there is a point $x_0$ in $I$ at which $F'(x_0) = c$.

**Proof**    Let $u, v,$ and $c$ be as in the statement of the theorem and we suppose that $u < v$. It is necessary to show that $c$ is in the image of $F': I \to \mathbb{R}$. In order to do this, we define an auxiliary function $h: I \to \mathbb{R}$ by

$$h(x) = F(x) - cx \quad \text{for all } x \text{ in } I.$$

Since $h'(x) = F'(x) - c$, the strategy will be to show that the function $h: (u, v) \to \mathbb{R}$ has a maximizer or minimizer $x_0$. At such a point, by Lemma 4.12, $h'(x_0) = 0$, and so $F'(x_0) = c$.

Since differentiability implies continuity, the function $h: I \to \mathbb{R}$ is continuous. In particular, $h: [u, v] \to \mathbb{R}$ is continuous, and so, according to the Extreme Value Theorem, the function $h: [u, v] \to \mathbb{R}$ attains both maximum and minimum values.

**Case 1:** $F'(u) < F'(v)$. There is a point $x_0$ in $[u, v]$ that is a minimizer for the function $h \colon [u, v] \to \mathbb{R}$. But $h'(u) = F'(u) - c < 0$, so $u$ is not the minimizer, and $h'(v) = F'(v) - c > 0$, so $v$ is not the minimizer. Thus $x_0$ belongs to the open interval $(u, v)$, and so it is a minimizer for $h \colon (u, v) \to \mathbb{R}$.

**Case 2:** $F'(u) > F'(v)$. We argue in the same way to show that if $x_0$ is a maximizer of $h \colon [u, v] \to \mathbb{R}$, then $x_0$ is in the open interval $(u, v)$, so it is a maximizer for $h \colon (u, v) \to \mathbb{R}$. ∎

**EXAMPLE 4.9**   Define the function $f \colon \mathbb{R} \to \mathbb{R}$ by

$$f(x) = \begin{cases} 1 & \text{if } x \geq 0 \\ 0 & \text{if } x < 0. \end{cases}$$

Then the image $f(\mathbb{R})$ consists of two points, so the image is certainly not an interval. It follows from Darboux's theorem that there does not exist a differentiable function $F \colon \mathbb{R} \to \mathbb{R}$ such that

$$F'(x) = f(x) \quad \text{for all } x \text{ in } \mathbb{R}. \qquad \square$$

---

**EXERCISES**

1. Let $F \colon \mathbb{R} \to \mathbb{R}$ be a differentiable function such that $F(1) = 3$ and $F'(x) = 2x + x^3$ for all $x$ in $\mathbb{R}$. Find $F(-1)$.

2. Suppose that the function $h \colon \mathbb{R} \to \mathbb{R}$ has a second derivative and that $h''(x) = 0$ for all $x$ in $\mathbb{R}$. Prove that there are numbers $m$ and $b$ such that

$$h(x) = mx + b \quad \text{for all } x \text{ in } \mathbb{R}.$$

3. Find a function $F \colon \mathbb{R} \to \mathbb{R}$ having a second derivative such that

$$\begin{cases} F''(x) = -x & \text{for all } x \text{ in } \mathbb{R} \\ F(0) = 0 \quad \text{and} \quad F'(0) = 1. \end{cases}$$

   Prove that there is only one such function $F \colon \mathbb{R} \to \mathbb{R}$.

4. Find a differentiable function $F \colon \mathbb{R} \to \mathbb{R}$ such that $F(2) = 4$ and

$$F'(x) = \begin{cases} x & \text{if } x \leq 0 \\ x^2 & \text{if } x > 0. \end{cases}$$

5. Suppose that the function $F \colon \mathbb{R} \to \mathbb{R}$ is differentiable and that for every number $x$, $F'(x)$ is rational. Prove that the graph of $F \colon \mathbb{R} \to \mathbb{R}$ is a line having a rational slope.

6. Let $I$ be an open interval containing the points $a$ and $b$ with $a < b$. Suppose that the function $f \colon I \to \mathbb{R}$ is differentiable and that $f'(a) > 0$. Show that the point $a$ cannot be a maximizer for the function $f \colon [a, b] \to \mathbb{R}$.

7. Let the function $f \colon \mathbb{R} \to \mathbb{R}$ be differentiable with $f'(x) \neq 0$ for each $x$ in $\mathbb{R}$. Prove that $f \colon \mathbb{R} \to \mathbb{R}$ is strictly monotonic.

8. Show that there does not exist a differentiable function $F \colon \mathbb{R} \to \mathbb{R}$ with $F'(x) = 0$ if $x < 0$ and $F'(x) = 1$ if $x \geq 0$, by arguing that such a function would necessarily be constant on $(-\infty, 0)$ and of the form $F(x) = A + Bx$ on $(0, \infty)$, and then deriving a contradiction.

## 4.6 The Notation of Leibnitz

So far, given an open interval $I$ and a differentiable function $f: I \to \mathbb{R}$, we have denoted the function's derivative by

$$f': I \to \mathbb{R},$$

so that $f'(x)$ is the derivative of $f: I \to \mathbb{R}$ at $x$ in $I$. This notation has been completely adequate. However, as we introduce new classes of functions and when we study integration, certain formulas and algorithmic techniques become easier to remember in an alternate notation, due to Leibnitz.

For a differentiable function $f: I \to \mathbb{R}$, we will denote $f'(x)$ by

$$\frac{d}{dx}(f(x)) \quad \text{or} \quad \frac{df}{dx}.$$

If points on the graph of $f: I \to \mathbb{R}$ are denoted by $(x, y)$, we will also denote $f'(x)$ by

$$\frac{dy}{dx} \quad \text{or} \quad y'.$$

The great advantage of the Leibnitzian symbolism is that, when it is properly interpreted, we can treat the symbols $df, dy, dx$, and so on as if they represented members of $\mathbb{R}$ and carry out various algebraic operations, and the resulting formulas will have meaning. We give two examples.

First, for any two real nonzero numbers,

$$\frac{a}{b} = 1 \bigg/ \frac{b}{a}.$$

A corresponding inversion formula for the Leibnitz symbols is

$$\frac{dx}{dy} = 1 \bigg/ \frac{dy}{dx}. \tag{4.23}$$

What significance can be attached to (4.23)? Well, suppose that $f: I \to \mathbb{R}$ is differentiable, with $f'(x) > 0$ for all $x$ in $I$, and we set $y = f(x)$ for $x$ in $I$. According to Theorem 4.6, letting $J = f(I)$, the inverse function $f^{-1}: J \to \mathbb{R}$ is differentiable and

$$(f^{-1})'(f(x)) = \frac{1}{f'(x)} \quad \text{for all } x \text{ in } I. \tag{4.24}$$

If we now let

$$x = f^{-1}(y) \quad \text{for all } y \text{ in } J,$$

then, according to the symbolism introduced above,

$$\frac{dx}{dy} = (f^{-1})'(y) = (f^{-1})'(f(x)),$$

and so, since

$$\frac{dy}{dx} = f'(x),$$

we see that (4.23) may be interpreted as a compact rewriting of (4.24).

As a second example, note that for numbers $a, b \neq 0$, and $c \neq 0$,

$$\frac{a}{b} = \frac{a}{c} \cdot \frac{c}{b}.$$

A corresponding cancellation formula for the Leibnitz symbols is

$$\frac{df}{dx} = \frac{df}{du} \cdot \frac{du}{dx}. \qquad (4.25)$$

Again we seek a suitable interpretation of (4.25). The following is reasonable: Suppose that the functions $f: \mathbb{R} \to \mathbb{R}$ and $u: \mathbb{R} \to \mathbb{R}$ are differentiable. Consider the composition $h = f \circ u: \mathbb{R} \to \mathbb{R}$. According to the Chain Rule, $h: \mathbb{R} \to \mathbb{R}$ is differentiable and

$$h'(x) = f'(u(x))u'(x) \quad \text{for all } x \text{ in } \mathbb{R}. \qquad (4.26)$$

If we substitute

$$\frac{df}{dx} \quad \text{for} \quad \frac{d}{dx}[f(u(x))], \qquad \frac{df}{du} \quad \text{for} \quad f'(u(x)), \quad \text{and} \quad \frac{du}{dx} \quad \text{for} \quad u'(x),$$

$$(4.27)$$

then (4.26) becomes (4.25).

The reader should be careful to observe that in this section *we have not proven anything*. We have simply indicated how (4.23) can reasonably be interpreted as a compact form of (4.24) and (4.25) as a compact symbolic rewriting of (4.26). In fact, not only have we not proven anything, we have used the symbol $d/dt$ ambiguously in (4.25). In spite of this, the symbolism of Leibnitz is extremely useful. We just have to be quite careful to interpret in a precise manner any formulas we are using, then justify this precise interpretation using proven results.

One final notational convention: When $f: I \to \mathbb{R}$ has a second derivative, we denote $f''(x)$ by

$$\frac{d^2}{dx^2}(f(x)), \qquad \frac{d^2 f}{dx^2}, \qquad \frac{d^2 y}{dx^2}, \quad \text{and} \quad y''.$$

---

**EXERCISES**

1.  Give a reasonable interpretation of the formula
$$\frac{d}{dx}(f + g) = \frac{df}{dx} + \frac{dg}{dx}.$$

2.  Is there any way we can justify inequalities and the Leibnitz symbols? For instance, under what conditions does
$$\frac{df}{dx} \geq \frac{dg}{dx} \quad \text{imply} \quad f \geq g?$$

3.  Give a reasonable interpretation of the formula
$$\frac{df}{dr} = \frac{df}{du} \cdot \frac{du}{ds} \cdot \frac{ds}{dr}.$$

4.  Give a reasonable interpretation of the formula
$$\frac{ds}{ds} = 1.$$

# 5

# The Elementary Functions as Solutions of Differential Equations

## 5.1 The Natural Logarithm and Exponential Functions

Given a rational number $r$, we seek a differentiable function $F\colon (0, \infty) \to \mathbb{R}$ such that

$$F'(x) = x^r \quad \text{for all } x > 0.$$

We immediately see that if $r \neq -1$ and $c$ is any real number, then the function $F\colon (0, \infty) \to \mathbb{R}$ defined by

$$F(x) = \frac{x^{r+1}}{r+1} + c \quad \text{for all } x > 0$$

is a solution of the above differential equation, and, by the Identity Criterion (Proposition 4.16), all the solutions are of this form. The exceptional case, $r = -1$, is more interesting. The exceptional problem is to find a differentiable function $F\colon (0, \infty) \to \mathbb{R}$ such that

$$F'(x) = \frac{1}{x} \quad \text{for all } x > 0.$$

Based on the results that we have so far established, we cannot determine whether there is such a function.

In order to extend our stock of differentiable functions, let us provisionally *assume* that there is a differentiable function $F: (0, \infty) \to \mathbb{R}$ such that

$$\begin{cases} F'(x) = 1/x & \text{for all } x > 0 \\ F(1) = 0. \end{cases} \tag{5.1}$$

In Chapter 7 we will prove that there is indeed such a function. It follows from the Identity Criterion that there can be at most one such function.

**THEOREM 5.1**   Let the function $F: (0, \infty) \to \mathbb{R}$ satisfy the differential equation (5.1). Then

   (i)  $F(ab) = F(a) + F(b)$ for all $a, b > 0$.

   (ii)  $F(a^r) = rF(a)$ if $a > 0$ and $r$ is rational.

   (iii)  For each real number $c$ there is a unique positive number $x$ such that $F(x) = c$.

**Proof of (i)**   Fix $a > 0$ and define $g(x) = F(ax) - F(x) - F(a)$ for all $x > 0$. According to the Chain Rule, the function $g: (0, \infty) \to \mathbb{R}$ is differentiable and

$$g'(x) = aF'(ax) - F'(x) = \frac{a}{ax} - \frac{1}{x} = 0 \quad \text{for all } x > 0.$$

The Identity Criterion implies that $g: (0, \infty) \to \mathbb{R}$ is constant. But $g(1) = -F(1) = 0$, so $g(x) = 0$ for all $x > 0$. This proves (i).

**Proof of (ii)**   Define $h(x) = F(x^r) - rF(x)$ for all $x > 0$. Again using the Chain Rule, we see that the function $h: (0, \infty) \to \mathbb{R}$ is differentiable and that

$$h'(x) = rx^{r-1}F'(x^r) - rF'(x) = \frac{rx^{r-1}}{x^r} - \frac{r}{x} = 0 \quad \text{for all } x > 0.$$

Since $h(1) = 0$, it follows from the Identity Criterion that $h(x) = 0$ for all $x > 0$. Thus (ii) is proven.

**Proof of (iii)**   Since $F'(x) = 1/x > 0$ for all $x > 0$, the function $F: (0, \infty) \to \mathbb{R}$ is strictly increasing. Also observe, using (i), that

$$0 = F(1) = F\left(x \cdot \frac{1}{x}\right) = F(x) + F\left(\frac{1}{x}\right) \quad \text{for all } x > 0,$$

so that $F(1/x) = -F(x)$ for all $x > 0$. Thus we may assume $c > 0$, and it suffices to show that there is a solution of the equation

$$F(x) = c, \qquad x > 1. \tag{5.2}$$

Since differentiability implies continuity, the function $F: (0, \infty) \to \mathbb{R}$ is continuous. Consequently, since $F(1) = 0 < c$, according to the Intermediate Value Theorem, to show that equation (5.2) has a solution it will suffice to show that there is a number $x_0 > 1$ such that $F(x_0) > c$. However, $F(2) > F(1) = 0$, and (ii) implies that $F(2^n) = nF(2)$ for every natural number $n$. According to the Archimedean Property, we can choose a natural number $n$ such that $nF(2) > c$, and so, letting $x_0 = 2^n$, we have $F(x_0) > c$. ∎

The function $F: (0, \infty) \to \mathbb{R}$ that satisfies the differential equation (5.1) occurs so frequently in science that it has a special name—it is called the *natural logarithm*—and $F(x)$ is denoted by $\ln x$ for $x > 0$.

From the definition of the natural logarithm and the Chain Rule, it follows that *if I is an open interval and the function $h: I \to \mathbb{R}$ is differentiable and is such that $h(x) > 0$ for all $x$ in $I$, then*

$$\frac{d}{dx}(\ln h(x)) = \frac{h'(x)}{h(x)} \qquad \text{for all } x \text{ in } I. \tag{5.3}$$

Observe that part (iii) of the preceding theorem implies that the function $\ln: (0, \infty) \to \mathbb{R}$ has an inverse function that is defined on all of $\mathbb{R}$. Denote the inverse function by $g: \mathbb{R} \to \mathbb{R}$. Then

$$\begin{cases} g(\ln x) = x & \text{for all } x > 0 \\ \qquad\qquad \text{and} \\ \ln g(x) = x & \text{for all } x \text{ in } \mathbb{R}. \end{cases}$$

Moreover, according to Corollary 4.7, the inverse function $g: \mathbb{R} \to \mathbb{R}$ is differentiable. Since

$$\ln g(x) = x \qquad \text{for all } x \text{ in } \mathbb{R},$$

it follows from the Chain Rule that

$$\frac{g'(x)}{g(x)} = 1 \qquad \text{for all } x \text{ in } \mathbb{R}.$$

Thus $g: \mathbb{R} \to \mathbb{R}$ is a differentiable function that has the property that

$$\begin{cases} g'(x) = g(x) & \text{for all } x \text{ in } \mathbb{R} \\ g(0) = 1. \end{cases} \tag{5.4}$$

Since the function $\ln: (0, \infty) \to \mathbb{R}$ is strictly increasing and its image is all of $\mathbb{R}$, there is a unique solution of the equation

$$\ln x = 1, \qquad x > 0. \tag{5.5}$$

**DEFINITION** *The unique solution of equation (5.5) is denoted by e.*

Recall that if $a$ is a positive number and $b$ is a rational number, we have defined $a^b$. In particular, since $e > 0$, for each rational number $x$ the number $e^x$ is defined.

**PROPOSITION 5.2** Let the function $g: \mathbb{R} \to \mathbb{R}$ and the number $e$ be as defined above. Then

$$g(x) = e^x \qquad \text{for all rational numbers } x. \tag{5.6}$$

**Proof**   Let $x$ be a rational number. Since both $g(x)$ and $e^x$ are positive and the function $\ln\colon (0, \infty) \to \mathbb{R}$ is strictly increasing,

$$g(x) = e^x \quad \text{if and only if} \quad \ln g(x) = \ln e^x.$$

But, by the very definition of the inverse, $\ln g(x) = x$. On the other hand, part (ii) of Theorem 5.1 and the definition of $e$ imply that $\ln e^x = x \ln e = x$.   ∎

Observe that at present, formula (5.6) has no meaning if the number $x$ is irrational, since we have not defined the symbol $a^x$ if $x$ is irrational. For instance, we have not defined the symbol $3^{\sqrt{2}}$. However, since the left-hand side of formula (5.6) is defined for any number $x$, rational or irrational, we now have a natural way to define *irrational powers of positive numbers*.

**DEFINITION**   *For an irrational number $b$, we define $e^b \equiv g(b)$, where $g\colon \mathbb{R} \to \mathbb{R}$ is the inverse of the natural logarithm. More generally, for any positive number $a$, we define $a^b \equiv g(b \ln a)$.*

This definition, together with Proposition 5.2, implies that

$$a^b = g(b \ln a) = e^{b \ln a} \quad \text{for all } a > 0 \text{ and all } b. \tag{5.7}$$

It is not difficult to check that with this extended definition of exponentiation the familiar rules for exponents are still valid (see Exercise 1). Formula (5.4) and the Chain Rule imply that *if $I$ is an open interval and the function $h\colon I \to \mathbb{R}$ is differentiable, then*

$$\frac{d}{dx}\left[e^{h(x)}\right] = e^{h(x)} h'(x) \quad \text{for all } x \text{ in } I. \tag{5.8}$$

We now have two new classes of differentiable functions.

**PROPOSITION 5.3**   Let $a > 0$. Then

$$\frac{d}{dx}[a^x] = a^x \ln a \text{ for all } x \text{ in } \mathbb{R}. \tag{5.9}$$

**Proof**   Using formula (5.8), we have

$$\frac{d}{dx}[a^x] = \frac{d}{dx}[e^{x \ln a}] = a^x \ln a \quad \text{for all } x \text{ in } \mathbb{R}.$$   ∎

**PROPOSITION 5.4**   Let $r$ be any number. Then

$$\frac{d}{dx}[x^r] = rx^{r-1} \quad \text{for all } x > 0. \tag{5.10}$$

**Proof**   Again using formula (5.8), we have

$$\frac{d}{dx}[x^r] = \frac{d}{dx}[e^{r \ln x}]$$

$$= [e^{r \ln x}]\frac{d}{dx}(r \ln x)$$

$$= x^r \left(\frac{r}{x}\right)$$

$$= rx^{r-1} \quad \text{for all } x > 0. \qquad \blacksquare$$

---

**PROPOSITION
5.5**

Let $c$ and $k$ be any real numbers. Then the differential equation

$$\begin{cases} F'(x) = kF(x) & \text{for all } x \text{ in } \mathbb{R} \\ F(0) = c \end{cases} \tag{5.11}$$

has exactly one solution. It is given by the formula

$$F(x) = ce^{kx} \text{ for all } x \text{ in } \mathbb{R}. \tag{5.12}$$

**Proof**   From the differentiation formula (5.8), we see that (5.12) defines a solution of (5.11). It remains to prove uniqueness. Let the function $F\colon \mathbb{R} \to \mathbb{R}$ be a solution of (5.11). Define

$$g(x) = \frac{F(x)}{e^{kx}} \quad \text{for all } x \text{ in } \mathbb{R}.$$

Using the quotient rule for derivatives, we have

$$g'(x) = \frac{ke^{kx}F(x) - ke^{kx}F(x)}{(e^{kx})^2} = 0$$

for all $x$. The Identity Criterion implies that the function $g\colon \mathbb{R} \to \mathbb{R}$ is constant. Since $g(0) = F(0) = c$, $g(x) = c$ for all $x$; that is, $F(x) = ce^{kx}$ for all $x$. Thus, there is exactly one solution of (5.11). $\qquad \blacksquare$

In Section 2.2, we showed that the sequence $\{(1 + 1/n)^n\}$ was monotone increasing and bounded above by 3. In fact,

$$\lim_{n \to \infty} \left[(1 + 1/n)^n\right] = e. \tag{5.13}$$

To see this, observe that from the very definition of the derivative of the natural logarithm at $x = 1$ we have

$$\lim_{n \to \infty} \left[\frac{\ln(1 + 1/n) - \ln 1}{1/n}\right] = \lim_{n \to \infty} \left[n \ln(1 + 1/n)\right] = 1.$$

Therefore, since the exponential function is continuous at $x = 1$,

$$\lim_{n \to \infty} (1 + 1/n)^n = \lim_{n \to \infty} e^{n \ln(1 + 1/n)} = e^1.$$

**EXERCISES**

1.  Let $a > 0$. Prove that for any numbers $x_1$ and $x_2$,
    (a)  $a^{x_1} \cdot a^{x_2} = a^{x_1+x_2}$
    (b)  $(a^{x_1})^{x_2} = a^{x_1 x_2}$

2.  For $a > 0$, show that

    $$\lim_{n \to \infty} n[a^{1/n} - 1] = \ln a.$$

3.  Let $0 < a \le b$. Prove that

    $$\frac{b-a}{b} \le \ln\left[\frac{b}{a}\right] \le \frac{b-a}{a}.$$

4.  Let $a > 0$. Prove that there is a number $k$ such that

    $$a^x = e^{kx} \quad \text{for all } x \text{ in } \mathbb{R}.$$

5.  Use the Mean Value Theorem to show that $e^x > 1 + x$ if $x \ne 0$. Then show that the following equation has exactly one solution:

    $$2e^x = (1 + x)^2, \quad x \text{ in } \mathbb{R}.$$

6.  Show that there is a number $c$ in $(1, e)$ such that

    $$1 = \ln e - \ln 1 = \frac{1}{c}(e - 1).$$

    From this, conclude that $e > 2$.

7.  Use the preceding exercise to prove that the following equation has exactly one solution:

    $$xe^x = 2, \quad 0 < x < 1.$$

8.  For a fixed number $a$, how many solutions does the following equation have?

    $$x \ln x = a, \quad x > 0.$$

9.  Suppose that $h: \mathbb{R} \to \mathbb{R}$ is a differentiable function having the property that

    $$h(a + b) = h(a)h(b) \quad \text{for all } a \text{ and } b \text{ in } \mathbb{R},$$

    and that the function is not identically equal to 0.
    (a)  Using the definition of derivative, prove that

    $$h'(x) = h'(0)h(x) \quad \text{for all } x \text{ in } \mathbb{R}.$$

    (b)  Show that if $k = h'(0)$, then $h(x) = e^{kx}$ for all $x$ in $\mathbb{R}$.

10. Let $a$ be a limit point of the set $U$. Use the continuity of the natural logarithm and the exponential function to show that

    $$\lim_{x \to a} f(x) = \ell \quad \text{if and only if} \quad \lim_{x \to a} e^{f(x)} = e^{\ell}.$$

    Moreover, if $\ell > 0$ and $f(x) > 0$ for all $x$ in $U$, show that

    $$\lim_{x \to a} f(x) = \ell \quad \text{if and only if} \quad \lim_{x \to a} \ln f(x) = \ln \ell.$$

11. By using the preceding exercise, or by other means, analyze the following limits:
    (a)  $\lim_{x \to 0}(1 + x)^{1/x}$

    (b)  $\lim_{x \to 1} \dfrac{\ln x}{x - 1}$

    (c)  $\lim_{n \to \infty} n[e^{b/n} - 1]$

    (d)  $\lim_{x \to 0} \dfrac{a^x - b^x}{x}$ for $a, b > 0$

12.  The *hyperbolic cosine* of $x$, which is denoted by $\cosh x$, and the *hyperbolic sine* of $x$, which is denoted by $\sinh x$, are defined by

$$\cosh x \equiv \frac{e^x + e^{-x}}{2} \quad \text{and} \quad \sinh x \equiv \frac{e^x - e^{-x}}{2} \quad \text{for all } x \text{ in } \mathbb{R}.$$

Given numbers $a$, $\alpha$, and $\beta$, find a solution of the equation

$$\begin{cases} f''(x) - a^2 f(x) = 0 & \text{for all } x \text{ in } \mathbb{R} \\ f(0) = \alpha \quad \text{and} \quad f'(0) = \beta \end{cases}$$

that is of the form

$$f(x) = c_1 \cosh ax + c_2 \sinh ax \quad \text{for all } x \text{ in } \mathbb{R}.$$

## 5.2   The Trigonometric Functions

A function $f : \mathbb{R} \to \mathbb{R}$ is called *periodic*, with *period* $T > 0$, if

$$f(x + T) = f(x) \quad \text{for all } x \text{ in } \mathbb{R}.$$

So far, with the exception of constant functions, we have not encountered any periodic functions. Since periodic phenomena occur in nature (planets, pendulums, and so on) and since the basic functions of trigonometry are periodic, we need to analyze such functions.

In the same way in which the properties of the logarithm and the exponential functions were deduced from a single differential equation, we will now define and analyze the sine and cosine functions with a single differential equation as our starting point. In this section we will study functions $f : \mathbb{R} \to \mathbb{R}$ that have the property that

$$f''(x) + f(x) = 0 \quad \text{for all } x \text{ in } \mathbb{R}. \tag{5.14}$$

**LEMMA 5.6**   Suppose that the function $f : \mathbb{R} \to \mathbb{R}$ is a solution of the differential equation

$$\begin{cases} f''(x) + f(x) = 0 & \text{for all } x \text{ in } \mathbb{R} \\ f(0) = 0 \quad \text{and} \quad f'(0) = 0. \end{cases} \tag{5.15}$$

Then $f(x) = 0$ for all $x$ in $\mathbb{R}$.

**Proof**   Define $g(x) = [f(x)]^2 + [f'(x)]^2$ for all $x$ in $\mathbb{R}$. Observe that

$$g'(x) = 2f(x)f'(x) + 2f'(x)f''(x)$$
$$= 2f'(x)[f(x) + f''(x)]$$
$$= 0 \quad \text{for all } x \text{ in } \mathbb{R}.$$

Thus, by the Identity Criterion, the function $g: \mathbb{R} \to \mathbb{R}$ is constant, and since $g(0) = 0$, $g(x) = 0$ for all $x$ in $\mathbb{R}$. But observe that

$$0 \le [f(x)]^2 \le g(x) \quad \text{for all } x \text{ in } \mathbb{R},$$

so $f(x) = 0$ for all $x$ in $\mathbb{R}$.                                          ∎

For fixed numbers $\alpha$ and $\beta$, consider the differential equation

$$\begin{cases} f''(x) + f(x) = 0 & \text{for all } x \text{ in } \mathbb{R} \\ f(0) = \alpha \quad \text{and} \quad f'(0) = \beta. \end{cases} \tag{5.16}$$

This equation can have at most one solution, since if there were two distinct functions that were solutions of (5.16), we see that their difference would be a solution of (5.15) that is not identically zero, which would contradict Lemma 5.6.

We provisionally *assume* that in the case $\alpha = 1$ and $\beta = 0$, there is a solution of (5.16). In Chapter 9, we will prove that there is such a function. We denote this solution by $C: \mathbb{R} \to \mathbb{R}$. Thus, by definition,

$$\begin{cases} C''(x) + C(x) = 0 & \text{for all } x \text{ in } \mathbb{R} \\ C(0) = 1 \quad \text{and} \quad C'(0) = 0. \end{cases} \tag{5.17}$$

The function $C: \mathbb{R} \to \mathbb{R}$ is *even*; that is,

$$C(-x) = C(x) \quad \text{for all } x \text{ in } \mathbb{R}.$$

This follows from the observation that if we define $f(x) = C(-x)$ for all $x$ in $\mathbb{R}$, then $f: \mathbb{R} \to \mathbb{R}$ is seen to be a solution of the differential equation (5.17), and since there is only one solution, $f(x) = C(x)$ for all $x$ in $\mathbb{R}$.

We define a companion function $S: \mathbb{R} \to \mathbb{R}$ by

$$S(x) = -C'(x) \quad \text{for all } x \text{ in } \mathbb{R}.$$

Thus $S(0) = 0$, since $C'(0) = 0$. Moreover, since $S'(x) = -C''(x) = C(x)$ and $C(0) = 1$, it follows that $S'(0) = 1$. Finally, if we differentiate the first line of (5.17) we see that $C'''(x) + C'(x) = 0$ for all $x$ in $\mathbb{R}$, and hence $S''(x) + S(x) = 0$ for all $x$ in $\mathbb{R}$. Thus

$$\begin{cases} S''(x) + S(x) = 0 & \text{for all } x \text{ in } \mathbb{R} \\ S(0) = 0 \quad \text{and} \quad S'(0) = 1. \end{cases} \tag{5.18}$$

The function $S: \mathbb{R} \to \mathbb{R}$ is *odd*; that is,

$$S(-x) = -S(x) \quad \text{for all } x \text{ in } \mathbb{R}.$$

This follows from the observation that if we define $f(x) = -S(-x)$ for all $x$ in $\mathbb{R}$, then the function $f: \mathbb{R} \to \mathbb{R}$ is seen to be a solution of the differential equation (5.18), and since there is only one solution, $f(x) = S(x)$ for all $x$ in $\mathbb{R}$. For future reference, it is useful to record the formulas

$$S'(x) = C(x) \quad \text{and} \quad C'(x) = -S(x) \quad \text{for all } x \text{ in } \mathbb{R}. \tag{5.19}$$

**THEOREM 5.7**   For all $a$ and $b$,

$$[S(a)]^2 + [C(a)]^2 = 1, \tag{5.20}$$

$$S(a + b) = S(a)C(b) + C(a)S(b), \tag{5.21}$$

$$C(a + b) = C(a)C(b) - S(a)S(b). \tag{5.22}$$

**Proof**   In order to prove (5.20), define $g(x) = [S(x)]^2 + [C(x)]^2 - 1$ for all $x$ in $\mathbb{R}$. Observe that

$$g'(x) = 2S(x)S'(x) + 2C(x)C'(x)$$
$$= 2C(x)[S(x) + S''(x)] = 0 \quad \text{for all } x \text{ in } \mathbb{R}.$$

Thus, by the Identity Criterion, the function $g: \mathbb{R} \to \mathbb{R}$ is constant, and since $g(0) = 0$, it follows that $g(x) = 0$ for all $x$ in $\mathbb{R}$. This proves (5.20).

In order to prove (5.21), fix a real number $b$ and define

$$f(x) = S(x + b) - [S(x)C(b) + C(x)S(b)] \quad \text{for all } x \text{ in } \mathbb{R}.$$

Then $f(0) = 0$ and $f'(0) = 0$. Moreover, for all $x$ in $\mathbb{R}$, $f''(x) + f(x) = 0$, since

$$S''(x + b) + S(x + b) = 0, \quad S''(x) + S(x) = 0, \quad \text{and} \quad C''(x) + C(x) = 0.$$

Thus the function $f: \mathbb{R} \to \mathbb{R}$ is a solution of (5.15), and so, according to Lemma 5.6, $f(x) = 0$ for all $x$ in $\mathbb{R}$. This proves (5.21).

Finally, differentiating the above function $f: \mathbb{R} \to \mathbb{R}$, we have

$$0 = f'(x) = C(x + b) - [C(x)C(b) - S(x)S(b)] \quad \text{for all } x \text{ in } \mathbb{R},$$

so (5.22) is proven.   ■

The above theorem is a statement of the classical trigonometric identities. We call (5.20) the Pythagorean Identity. Observe that as a consequence of the Pythagorean Identity,

$$|S(x)| \leq 1 \quad \text{and} \quad |C(x)| \leq 1 \quad \text{for all } x \text{ in } \mathbb{R}. \tag{5.23}$$

We will now show that the functions $S: \mathbb{R} \to \mathbb{R}$ and $C: \mathbb{R} \to \mathbb{R}$ are periodic. The strategy is to show that there is a smallest positive number $p$ at which $C(p) = 0$, and then to use the addition formulas to prove that these functions have period $T = 4p$.

**LEMMA 5.8**   There is a positive number $x$ at which $C(x) = 0$.

**Proof** From the Mean Value Theorem, it follows that we may select a number $z$ strictly between 0 and 2 such that $S(2) - S(0) = 2C(z)$. Since $S(0) = 0$ and $|S(2)| \leq 1$, we see that $|2C(z)| \leq 1$. Thus, using the identities (5.22) and (5.20), we have

$$C(2z) = [C(z)]^2 - [S(z)]^2 = 2[C(z)]^2 - 1 \leq 0.$$

Hence $C(0) > 0$ and $C(2z) \leq 0$, and so, by the Intermediate Value Theorem, there is a number $x$ between 0 and $2z$ at which $C(x) = 0$. ∎

**THEOREM 5.9** There is a smallest positive number at which $C(x) = 0$.

**Proof** For any continuous function $g: \mathbb{R} \to \mathbb{R}$ for which $g(0) > 0$ and there is a positive number $x$ at which $g(x) = 0$, there is a smallest positive number $x_0$ at which $g(x_0) = 0$ (see Exercise 16). The preceding lemma asserts that at some positive number $x$, $C(x) = 0$. On the other hand, $C(0) = 1$. ∎

**THEOREM 5.10** Let $p$ be the smallest positive number at which $C(x) = 0$. Then the functions $C: \mathbb{R} \to \mathbb{R}$ and $S: \mathbb{R} \to \mathbb{R}$ both have period $4p$.

**Proof** Since $S'(x) = C(x) > 0$ if $0 < x < p$, the function $S: [0, p] \to \mathbb{R}$ is strictly increasing. Hence, since $S(0) = 0$, $S(p) > 0$. Since $C(p) = 0$, the Pythagorean Identity implies that $S(p) = 1$. Thus $C(p) = 0$ and $S(p) = 1$, and so, using the addition formulas (5.21) and (5.22),

$$S(x + p) = C(x) \quad \text{and} \quad C(x + p) = -S(x) \quad \text{for all } x \text{ in } \mathbb{R}. \tag{5.24}$$

Substituting $x + p$ for $x$ in (5.24), we obtain

$$S(x + 2p) = -S(x) \quad \text{and} \quad C(x + 2p) = -C(x) \quad \text{for all } x \text{ in } \mathbb{R}, \tag{5.25}$$

and now, substituting $x + 2p$ for $x$ in (5.25), we see that the functions $S: \mathbb{R} \to \mathbb{R}$ and $C: \mathbb{R} \to \mathbb{R}$ have period $4p$. ∎

As we have already mentioned, none of the functions that we have seen until now have been periodic, except, of course, constant functions.

We define the number $\pi$ to be $2p$, where $p$ is the smallest positive number at which $C(p) = 0$. Hence $S: \mathbb{R} \to \mathbb{R}$ and $C: \mathbb{R} \to \mathbb{R}$ have period $2\pi$. Of course, we need to show that this definition of $\pi$ is in accordance with the usual definition of $\pi$ as the area of a circle of unit radius. Specifically, we must show that the first positive zero of the solution of the differential equation (5.17) occurs at $p$, where $p$ is half the area of a circle of unit radius. To do this, we first need to discuss integration, and so we postpone a justification of our use of the symbol $\pi$ until Chapter 7. However, from now on we will denote $S(x)$ by $\sin x$ and denote $C(x)$ by $\cos x$, for all $x$ in $\mathbb{R}$. The function $S: \mathbb{R} \to \mathbb{R}$ is called the *sine* function; the function $C: \mathbb{R} \to \mathbb{R}$ is called the *cosine* function.

The cosine function was defined to be the unique solution of the differential equation (5.17). In fact, the cosine and sine functions play a central role in the theory of general differential equations. One indication of this is the following.

**THEOREM 5.11**   Let $\alpha$ and $\beta$ be any numbers. Then there is exactly one solution of the differential equation

$$\begin{cases} f''(x) + f(x) = 0 & \text{for all } x \text{ in } \mathbb{R} \\ f(0) = \alpha & \text{and} \quad f'(0) = \beta. \end{cases} \tag{5.26}$$

This solution is defined by

$$f(x) = \alpha \cos x + \beta \sin x \quad \text{for all } x \text{ in } \mathbb{R}. \tag{5.27}$$

**Proof**   The differential equation (5.26) can have at most one solution, since if there were two distinct solutions, their difference would be a solution of the differential equation (5.15) that is not identically zero, which would contradict Lemma 5.6. From (5.17) and (5.18), it follows that formula (5.27) defines a solution of the differential equation (5.26).   ∎

We conclude this section with a discussion of the tangent function. We begin with two observations about the cosine. First, note that the second identity in (5.25) may be rewritten as

$$\cos(x + \pi) = -\cos x \quad \text{for all } x \text{ in } \mathbb{R}.$$

Second, by definition, $\cos \pi/2 = 0$ and $\cos x > 0$ if $0 \leq x < \pi/2$. It follows that $\cos x > 0$ if $-\pi/2 < x < \pi/2$, and that

$$\cos x = 0 \quad \text{if and only if} \quad x = \pi/2 + n\pi \quad \text{for some integer } n.$$

Define $D = \{x \text{ in } \mathbb{R} \mid x \neq \pi/2 + n\pi, n \text{ an integer}\}$. The *tangent* function, with domain $D$, is defined by

$$\tan x = \frac{\sin x}{\cos x} \quad \text{for all } x \text{ in } D.$$

Thus the tangent, being the quotient of differentiable functions, is differentiable, and from the quotient formula for derivatives, together with (5.19), it follows that

$$\frac{d}{dx}[\tan(x)] = \frac{1}{\cos^2 x} \quad \text{if} \quad x \neq \frac{\pi}{2} + n\pi, \quad n \text{ an integer}. \tag{5.28}$$

**THEOREM 5.12**   The function $\tan: (-\pi/2, \pi/2) \to \mathbb{R}$ is strictly increasing, is odd, and has as its image all of $\mathbb{R}$.

**Proof**   Since

$$\frac{d}{dx}[\tan x] = \cos^{-2} x > 0 \quad \text{for} \quad -\frac{\pi}{2} < x < \frac{\pi}{2},$$

the function $\tan: (-\pi/2, \pi/2) \to \mathbb{R}$ is strictly increasing. The oddness of the sine and the evenness of the cosine imply that the tangent is odd.

It remains to be proven that the range is all of $\mathbb{R}$. Since the tangent is odd and $\tan 0 = 0$, it will suffice to show that given $c > 0$ there is a solution of the equation

$$\tan x = c, \quad 0 < x < \frac{\pi}{2}. \tag{5.29}$$

According to the Intermediate Value Theorem, since $\tan 0 = 0$, in order to prove that there is a solution of (5.29) it suffices to find a point $x$ in the interval $(-\pi/2, \pi/2)$ at which $\tan x > c$. However, since the sine is increasing on the interval $[0, \pi/2]$,

$$\tan x = \frac{\sin x}{\cos x} \geq \frac{\sin \pi/4}{\cos x} \quad \text{if} \quad \frac{\pi}{4} \leq x < \frac{\pi}{2}.$$

Moreover, since the cosine is continuous and positive on the interval $(0, \pi/2)$ and $\cos \pi/2 = 0$, we can choose a point $x$ in the interval $(\pi/4, \pi/2)$ at which $\cos x < (\sin \pi/4)/c$. At this point, $\tan x > c$.  ∎

---

**EXERCISES**

1.  Derive formulas for $\cos(a - b)$ and $\sin(a - b)$ in terms of $\sin a$, $\sin b$, $\cos a$, and $\cos b$.

2.  Find a formula for $\sin 3a$ in terms of $\sin a$ and $\cos a$. Use this to calculate $\sin \pi/3$ and $\cos \pi/3$. Also calculate $\sin \pi/6$ and $\cos \pi/4$.

3.  For numbers $a$ and $b$ such that $|a| < 1$, prove that the following equation, which is called *Kepler's equation*, has exactly one solution:

$$x = a \sin x + b, \qquad x \text{ in } \mathbb{R}.$$

4.  Prove that the following equation has exactly one solution:

$$e^{2x} + \cos x + x = 0, \qquad x \text{ in } \mathbb{R}.$$

5.  For numbers $a$ and $b$, define

$$f(x) = \sin x + ax + b \quad \text{for all } x \text{ in } \mathbb{R}.$$

For what values of $a$ is the function $f: \mathbb{R} \to \mathbb{R}$ increasing?

6.  Find the maximum and minimum points of the set $\{\sin x + \cos x \mid x \text{ in } \mathbb{R}\}$.

7.  Using the definition of derivative, prove that

$$\lim_{x \to 0} \frac{\sin x}{x} = 1 \quad \text{and} \quad \lim_{x \to 0} \frac{\cos x - 1}{x} = 0.$$

8.  Let $k$ be a fixed number. Suppose that the function $f: \mathbb{R} \to \mathbb{R}$ is a solution of the differential equation

$$\begin{cases} f''(x) + k^2 f(x) = 0 & \text{for all } x \text{ in } \mathbb{R} \\ f(0) = 0 \quad \text{and} \quad f'(0) = 0. \end{cases}$$

Prove that $f(x) = 0$ for all $x$ in $\mathbb{R}$.

9.  Let $a$, $b$, and $k$ be fixed numbers. Use the preceding exercise to show that the following differential equation has at most one solution:

$$\begin{cases} f''(x) + k^2 f(x) = 0 & \text{for all } x \text{ in } \mathbb{R} \\ f(0) = a \quad \text{and} \quad f'(0) = b. \end{cases}$$

Then verify that if $k \neq 0$, the solution is defined by

$$f(x) = a \cos kx + b \sin kx \quad \text{for all } x \text{ in } \mathbb{R}.$$

10. Let $a$ and $b$ be numbers such that $a^2 + b^2 = 1$. Prove that there exists exactly one number $\theta$ in the interval $[0, 2\pi)$ such that

$$\begin{cases} \cos \theta = a \\ \sin \theta = b. \end{cases}$$

11.  For positive numbers $M$ and $T$, and a number $\theta_0$ in the interval $[0, 2\pi)$, define

$$g(x) = M\sin(Tx + \theta_0) \quad \text{for all } x \text{ in } \mathbb{R}.$$

Graph the function $g: \mathbb{R} \to \mathbb{R}$.

12.  Let $c_1$ and $c_2$ be numbers such that $c_1^2 + c_2^2 = 1$. Define

$$h(x) = c_1 \cos x + c_2 \sin x \quad \text{for all } x \text{ in } \mathbb{R}.$$

Use Exercise 10 and the Addition Formula for the cosine to show that there is a number $\theta_0$ such that

$$h(x) = \cos(x + \theta_0) \quad \text{for all } x \text{ in } \mathbb{R}.$$

13.  Define

$$f(x) = \begin{cases} x^2 \sin(1/x) + x & \text{if } x \neq 0 \\ 0 & \text{if } x = 0. \end{cases}$$

Prove that the function $f: \mathbb{R} \to \mathbb{R}$ is differentiable and that $f'(0) = 1$. Also prove that there is no neighborhood $I$ of $0$ such that the function $f: I \to \mathbb{R}$ is increasing.

14.  Does the function $f: \mathbb{R} \to \mathbb{R}$ defined in the previous exercise have a continuous derivative? Justify your answer.

15.  By using Theorem 4.20 (perhaps repeatedly), verify that:

(a)  $\displaystyle \lim_{x \to 0} \frac{x - \sin x}{x^3} = \frac{1}{6}.$

(b)  $\displaystyle \lim_{x \to 0} \frac{\cos x - 1 + x^2}{x^2} = \frac{1}{2}.$

(c)  $\displaystyle \lim_{x \to 0} \frac{x^3 \sin 1/x}{\sin x} = 0.$

16.  Suppose that the function $g: \mathbb{R} \to \mathbb{R}$ is continuous, $g(0) > 0$, and at some positive number $x_0$, $g(x_0) = 0$. Prove that there is a smallest positive number $p$ at which $g(x) = 0$. (*Hint:* Define $p = \inf\{x \text{ in } \mathbb{R} \mid x > 0, g(x) = 0\}$ and prove that $p > 0$ and $g(p) = 0$.)

## 5.3   The Inverse Trigonometric Functions

The sine, cosine, and tangent functions are all periodic, so none of them has an inverse. However, if we restrict these functions to appropriate intervals, the restrictions will have inverses.

Since $\sin: \left[ -\pi/2, \pi/2 \right] \to \mathbb{R}$ is a strictly increasing continuous function with $\sin -\pi/2 = -1$ and $\sin \pi/2 = 1$, it follows from the Intermediate Value Theorem that for each number $x$ in $[-1, 1]$, there is a unique solution to the equation

$$\sin z = x, \qquad z \text{ in } \left[ -\frac{\pi}{2}, \frac{\pi}{2} \right].$$

We denote this solution by $\arcsin x$, and so we have defined the *arcsine* function, denoted by $\arcsin: [-1, 1] \to \mathbb{R}$, as the inverse of $\sin: \left[ -\frac{\pi}{2}, \frac{\pi}{2} \right] \to \mathbb{R}$.

Since

$$\frac{d}{dx}[\sin x] = \cos x \neq 0 \quad \text{if} \quad -\frac{\pi}{2} < x < \frac{\pi}{2},$$

it follows from Corollary 4.7 that the arcsine function is differentiable and that

$$\frac{d}{dx}[\arcsin x] = \frac{1}{\cos(\arcsin x)} \quad \text{for} \quad -1 < x < 1.$$

However, $\arcsin x$ is in $(-\pi/2, \pi/2)$, so $\cos(\arcsin x) > 0$. Consequently, using the Pythagorean Identity, we have

$$\cos(\arcsin x) = [1 - \sin^2(\arcsin x)]^{1/2} = \sqrt{1 - x^2}.$$

Thus,
$$\frac{d}{dx}[\arcsin x] = \frac{1}{\sqrt{1 - x^2}} \quad \text{if} \quad -1 < x < 1. \tag{5.30}$$

We now turn to the *arccosine* function. Indeed, since $\cos: [0, \pi] \to \mathbb{R}$ is a strictly decreasing continuous function with $\cos 0 = 1$ and $\cos \pi = -1$, it follows from the Intermediate Value Theorem that for each $x$ in $[-1, 1]$ there is a unique solution to the equation

$$\cos z = x, \qquad z \text{ in } [0, \pi]. \tag{5.31}$$

We denote this solution by $\arccos x$, and therefore we have defined the function $\arccos: [-1, 1] \to \mathbb{R}$, which is the inverse of $\cos: [0, \pi] \to \mathbb{R}$. Since

$$\frac{d}{dx}[\cos x] = -\sin x \neq 0 \quad \text{if} \quad 0 < x < \pi,$$

it follows from Corollary 4.7 that

$$\frac{d}{dx}[\arccos x] = -\frac{1}{\sin(\arccos x)} \quad \text{if} \quad -1 < x < 1.$$

However, $\arccos x$ belongs to the interval $(0, \pi)$ when $x$ is in $(-1, 1)$, so $\sin(\arccos x) > 0$. Consequently, using the Pythagorean Identity, we see that

$$\sin(\arccos x) = [1 - \cos^2(\arccos x)]^{1/2} = \sqrt{1 - x^2}.$$

Thus,
$$\frac{d}{dx}[\arccos x] = -\frac{1}{\sqrt{1 - x^2}} \quad \text{if} \quad -1 < x < 1. \tag{5.32}$$

Finally, we consider the *arctangent* function. According to Theorem 5.12, the function $\tan: (-\pi/2, \pi/2) \to \mathbb{R}$ is a strictly increasing function whose image is all of $\mathbb{R}$. It follows that for each number $x$ the equation

$$\tan z = x, \qquad z \text{ in } \left(-\frac{\pi}{2}, \frac{\pi}{2}\right)$$

has a unique solution. We denote this solution by $\arctan x$, and so we have defined $\arctan: \mathbb{R} \to \mathbb{R}$, the inverse of $\tan: (-\pi/2, \pi/2) \to \mathbb{R}$.

From formula (5.28) and Corollary 4.7, we conclude that

$$\frac{d}{dx}[\arctan x] = \cos^2(\arctan x) \quad \text{for all } x \text{ in } \mathbb{R},$$

from which, setting $\alpha = \arctan x$ and using the Pythagorean Identity, it follows that

$$\frac{d}{dx}[\arctan x] = \frac{1}{1+x^2} \quad \text{for all } x \text{ in } \mathbb{R}. \tag{5.33}$$

---

**EXERCISES**

1. Prove that $\arcsin x + \arccos x = \pi/2$ if $-1 \le x \le 1$.
2. Find the unique solution of the differential equation
$$\begin{cases} F'(x) = x/\sqrt{1-x^4}, & -1 < x < 1 \\ F(0) = 1. \end{cases}$$
3. Suppose that $f: \mathbb{R} \to \mathbb{R}$ and $g: \mathbb{R} \to \mathbb{R}$ are periodic functions of period T. Under what conditions is the sum $f + g: \mathbb{R} \to \mathbb{R}$ also periodic? Under what conditions is the composition $f \circ g: \mathbb{R} \to \mathbb{R}$ periodic?
4. Define $h(x) = 4\sin(x/2)$ for all $x$ in $\mathbb{R}$. By restricting the function $h: \mathbb{R} \to \mathbb{R}$ to a suitable interval $[a, b]$ such that $h: [a, b] \to \mathbb{R}$ is strictly increasing and $h([a, b]) = [-4, 4]$, find the inverse of $h: [a, b] \to \mathbb{R}$ and calculate its derivative on the interval $(-4, 4)$.
5. Let $I$ be an open interval and suppose that
$$p(x) = ax^2 + bx + c > 0 \quad \text{for all } x \text{ in } I.$$
Find a solution of the differential equation
$$F'(x) = \frac{1}{p(x)} \quad \text{for all } x \text{ in } I.$$
(*Hint:* Complete the square.)
6. Prove that
$$\arctan v - \arctan u < v - u \quad \text{if} \quad u < v.$$

# 6

# Integration

## 6.1    Motivation for the Definition

For certain functions $f: [a, b] \to \mathbb{R}$, we will define a real number called the *integral* of $f: [a, b] \to \mathbb{R}$ and denoted by $\int_a^b f$. The integral is one of the cornerstones on which mathematical analysis is built. The significance of the integral depends on the context in which it is being considered. For example, the following interpretation of the integral is appropriate in a geometric context:

- For a function $f: [a, b] \to \mathbb{R}$ having the property that $f(x) \geq 0$ for all $x$ in $[a, b]$, the integral $\int_a^b f$ is the area under the graph of $f: [a, b] \to \mathbb{R}$ and above the interval $[a, b]$.

The integral of $f: [a, b] \to \mathbb{R}$ has many other physical interpretations,[*] and the concept of the integral is central in the theory of differential equations, but the geometric interpretation of it as an area is sufficient to motivate our definition.

---

[*] The book *Introduction to Calculus and Analysis*, by R. Courant and F. John (Springer-Verlag, 1989), presents many interesting applications of the integral that arise in problems in physics and engineering.

**FIGURE 6.1**

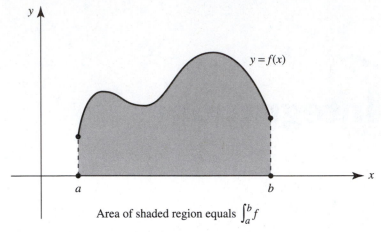

$y = f(x)$

Area of shaded region equals $\int_a^b f$

We will define the notion of the integral in such a manner that the following three properties are satisfied:

(i) If $f(x) = c$ for all $x$ in $(a, b)$, then

$$\int_a^b f = c(b - a).$$

(ii) If $a < c < b$, then

$$\int_a^c f + \int_c^b f = \int_a^b f.$$

(iii) If $f(x) \le g(x)$ for all $x$ in $[a, b]$, then

$$\int_a^b f \le \int_a^b g.$$

If the integral is to have properties (i) and (ii) and the function $f: [a, b] \to \mathbb{R}$ is *piecewise constant*, meaning that there is a natural number $n$ and points $x_0, x_1, \ldots, x_n$ with

$$a = x_0 < x_1 < \cdots < x_{n-1} < x_n = b$$

such that for each index $i$ with $1 \le i \le n$,

$$f(x) = c_i \quad \text{for all } x \text{ in } (x_{i-1}, x_i),$$

then     $$\int_a^b f = c_1(x_1 - x_0) + c_2(x_2 - x_1) + \cdots + c_n(x_n - x_{n-1}).$$

This formula for the integral of a piecewise constant function, together with the monotonicity property (iii), leads to a natural definition of the integral $\int_a^b f$ for a large class of functions $f: [a, b] \to \mathbb{R}$ that includes functions $f: [a, b] \to \mathbb{R}$ that are either continuous or monotone.

Once we have defined the integral and studied its properties, we will see that—as just one example of its importance—it is necessary to understand integration in order to study differential equations. In fact, although the geometric interpretation of the integral as an area is an attractive motivation for the definition, the integral has much greater significance as an analytical tool for describing the solutions of differential equations.

## 6.2   The Definition of the Integral and Criteria for Integrability

Let $a$ and $b$ be real numbers with $a < b$. If $n$ is a natural number and

$$a = x_0 < x_1 < \cdots < x_{n-1} < x_n = b,$$

then $P = \{x_0, \ldots, x_n\}$ is called a *partition* of the interval $[a, b]$. For each index $i$ with $0 \leq i \leq n$, we call $x_i$ a *partition point* of $P$, and if $i \geq 1$ we call the interval $[x_{i-1}, x_i]$ a *subinterval induced by the partition* $P$. The crudest partition of $[a, b]$ occurs when $n = 1$, so that $x_0 = a$ and $x_1 = b$.

Suppose that the function $f : [a, b] \to \mathbb{R}$ is bounded. For each index $i$ with $0 \leq i \leq n$, we define

$$\begin{cases} m_i \equiv \inf\{f(x) \mid x \text{ in } [x_{i-1}, x_i]\} \\ M_i \equiv \sup\{f(x) \mid x \text{ in } [x_{i-1}, x_i]\}. \end{cases} \tag{6.1}$$

**FIGURE 6.2**

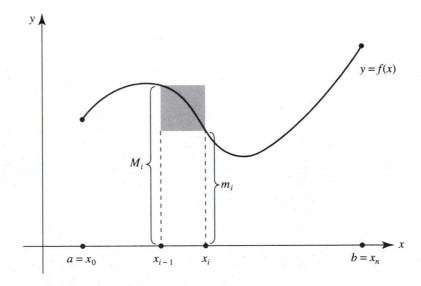

We then define

$$\begin{cases} L(f, P) \equiv \displaystyle\sum_{i=1}^{n} m_i(x_i - x_{i-1}) \\ \\ U(f, P) \equiv \displaystyle\sum_{i=1}^{n} M_i(x_i - x_{i-1}). \end{cases} \tag{6.2}$$

We call $U(f, P)$ the *upper Darboux sum for the function* $f : [a, b] \to \mathbb{R}$ *based on the partition* $P$, and call $L(f, P)$ the *lower Darboux sum for the function* $f : [a, b] \to \mathbb{R}$ *based on the partition P.*

**FIGURE 6.3**

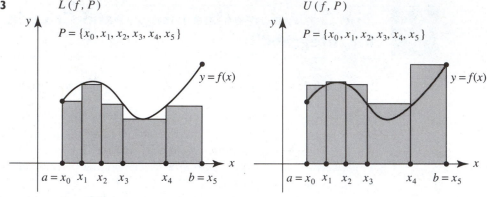

**DEFINITION**     *Suppose that the function* $f : [a, b] \to \mathbb{R}$ *is bounded. Then* $f : [a, b] \to \mathbb{R}$ *is said to be integrable provided that there is exactly one number A that has the property that*

$$L(f, P) \le A \le U(f, P) \quad \text{for every partition P of the interval } [a, b]. \tag{6.3}$$

*We call A the integral of the function* $f : [a, b] \to \mathbb{R}$ *and denote it by* $\int_a^b f$.

It is not immediately clear how one can check that a function is integrable. We will soon see that if the function $f : [a, b] \to \mathbb{R}$ is continuous, then it is integrable. However, even when we have determined that $f : [a, b] \to \mathbb{R}$ is integrable, there still remains the problem of determining the value of the integral. Before beginning the general theory, we will consider two examples.

**EXAMPLE 6.1**    Let the function $f : [a, b] \to \mathbb{R}$ have constant value $c$. Then $f : [a, b] \to \mathbb{R}$ is integrable and

$$\int_a^b f = c(b - a).$$

Indeed, this follows immediately from the definition and the observation that if $P$ is any partition of $[a, b]$, then $L(f, P) = U(f, P) = c(b - a)$.     □

**EXAMPLE 6.2**    Consider the Dirichlet function $f : [0, 1] \to \mathbb{R}$ defined by

$$f(x) = \begin{cases} 0 & \text{if the point } x \text{ in } [0, 1] \text{ is rational} \\ 1 & \text{if the point } x \text{ in } [0, 1] \text{ is irrational.} \end{cases}$$

Let $P = \{x_0, \dots, x_1\}$ be a partition of $[0, 1]$. Since the rationals and the irrationals are dense in $\mathbb{R}$, it follows that for each index $i$ with $1 \leq i \leq n$, if $m_i$ and $M_i$ are as defined by (6.1), then $m_i = 0$ and $M_i = 1$. Hence $L(f, P) = 0$ and $U(f, P) = 1$. It follows that *any* number $A$ in the interval [0, 1] satisfies criterion (6.3). Therefore, the function $f: [0, 1] \to \mathbb{R}$ is not integrable.          $\square$

If one interprets the integral in terms of area, these examples are not surprising. The first is the formula for the area of a rectangle. The second reflects the fact that there is no obvious way to assign an area to a region bounded by a very wild graph.

**LEMMA 6.1**   Suppose that the function $f: [a, b] \to \mathbb{R}$ is bounded and the numbers $m$ and $M$ have the property that

$$m \leq f(x) \leq M \quad \text{for all } x \text{ in } [a, b].$$

Then if $P$ is a partition of the interval $[a, b]$,

$$m(b - a) \leq L(f, P) \leq U(f, P) \leq M(b - a).$$

**Proof**   Let $P = \{x_0, \ldots, x_n\}$. For each index $i$ with $1 \leq i \leq n$, from the very definition of supremum and infimum it follows that

$$m \leq m_i = \inf\{f(x) \mid x \text{ in } [x_{i-1}, x_i]\} \leq \sup\{f(x) \mid x \text{ in } [x_{i-1}, x_1]\} \leq M_i \leq M.$$

Multiply this inequality by $x_i - x_{i-1}$ and sum the resulting $n$ inequalities to obtain

$$m(b - a) = \sum_{i=1}^{n} m(x_i - x_{i-1}) \leq L(f, P) \leq U(f, P)$$

$$\leq \sum_{i=1}^{n} M(x_i - x_{i-1}) = M(b - a). \qquad \blacksquare$$

Given a partition $P$ of the interval $[a, b]$, another partition $P^*$ of $[a, b]$ is called a *refinement* of $P$ if each partition point of $P$ is also a partition point of $P^*$. If $P = \{x_0, \ldots, x_n\}$ and $P^*$ is a refinement of $P$, then for each index $i$ with $1 \leq i \leq n$, $P^*$ induces a partition of the subinterval $[x_{i-1}, x_i]$ that we may denote by $P_i$. Observe that

$$L(f, P^*) = \sum_{i=1}^{n} L(f, P_i) \quad \text{and} \quad U(f, P^*) = \sum_{i=1}^{n} U(f, P_i).$$

**LEMMA 6.2**   **The Refinement Lemma**   Suppose that the function $f: [a, b] \to \mathbb{R}$ is bounded. Let $P$ be a partition of $[a, b]$ and let $P^*$ be a refinement of $P$. Then

$$L(f, P) \leq L(f, P^*) \leq U(f, P^*) \leq U(f, P).$$

**Proof**   Let $P = \{x_0, \ldots, x_n\}$. For each index $i$, let $m_i$ and $M_i$ be as defined by (6.1) and let $P_i$ be the partition of $[x_{i-1}, x_i]$ that is induced by $P^*$. Then Lemma 6.1, applied to the function $f: [x_{i-1}, x_i] \to \mathbb{R}$, yields

$$m_i(x_i - x_{i-1}) \le L(f, P_i) \le U(f, P_i) \le M_i(x_i - x_{i-1}).$$

Sum these $n$ inequalities to obtain

$$L(f, P) \le \sum_{i=1}^{n} L(f, P_i) = L(f, P^*) \le U(f, P^*) = \sum_{i=1}^{n} U(f, P_i) \le U(f, P). \quad \blacksquare$$

Given two partitions $P_1$ and $P_2$ of the interval $[a, b]$, the partition $P^*$ formed by taking the union of the partition points of $P_1$ and of $P_2$ is a *common refinement* of $P_1$ and $P_2$, since $P^*$ is a refinement of both $P_1$ and $P_2$.

**LEMMA 6.3**   Suppose that the function $f : [a, b] \to \mathbb{R}$ is bounded. Then for any two partitions $P_1$ and $P_2$ of $[a, b]$,

$$L(f, P_1) \le U(f, P_2).$$

**Proof**   Let $P^*$ be a common refinement of $P_1$ and $P_2$. From the Refinement Lemma, it follows that

$$L(f, P_1) \le L(f, P^*) \le U(f, P^*) \le U(f, P_2). \quad \blacksquare$$

This leads us to the following very useful theorem.

**THEOREM 6.4**   **The Integrability Criterion**   Suppose that the function $f : [a, b] \to \mathbb{R}$ is bounded. Then $f : [a, b] \to \mathbb{R}$ is integrable if and only if for each positive number $\epsilon$ there is a partition $P$ of the interval $[a, b]$ such that

$$U(f, P) - L(f, P) < \epsilon.$$

**Proof**   Denote by $\mathcal{L}$ the collection of all lower Darboux sums for $f : [a, b] \to \mathbb{R}$ and by $\mathcal{U}$ the collection of all upper Darboux sums for $f : [a, b] \to \mathbb{R}$. Then Lemma 6.3 amounts to the assertion that

$$s \le t \quad \text{whenever } s \text{ is in } \mathcal{L} \text{ and } t \text{ is in } \mathcal{U}. \tag{6.4}$$

Moreover, the function $f : [a, b] \to \mathbb{R}$ is defined to be integrable provided that there is exactly one number $A$ with the property that

$$s \le A \le t \quad \text{whenever } s \text{ is in } \mathcal{L} \text{ and } t \text{ is in } \mathcal{U}. \tag{6.5}$$

According to the Dedekind Gap Theorem, which we proved in Section 1.1, the function $f : [a, b] \to \mathbb{R}$ is integrable if and only if for each positive number $\epsilon$ there are partitions $P_1$ and $P_2$ of the interval $[a, b]$ such that

$$U(f, P_2) - L(f, P_1) < \epsilon. \tag{6.6}$$

This proves the theorem if we can choose $P_1 = P_2$. However, by the Refinement Lemma, if $P$ is a common refinement of $P_1$ and $P_2$, then

$$U(f, P) - L(f, P) \leq U(f, P_2) - L(f, P_1) < \epsilon.$$ ∎

**EXAMPLE 6.3**   Define

$$f(x) = \begin{cases} 7 & \text{if } 1 \leq x < 2 \\ 10 & \text{if } x = 2 \\ -4 & \text{if } 2 < x \leq 3. \end{cases}$$

We will use the Integrability Criterion to prove that the function $f: [1, 3] \to \mathbb{R}$ is integrable. Let $\epsilon > 0$. We must find a partition $P$ of the interval $[1, 3]$ such that $U(f, P) - L(f, P) < \epsilon$.

Observe that the only contribution to the sum $U(f, P) - L(f, P)$ comes from subintervals that contain the point $x = 2$. Define $P = \{1, 2 - \epsilon/30, 2 + \epsilon/30, 3\}$. For $i = 1, 2, 3$, let $m_i$ and $M_i$ be defined by (6.1); we see that $m_1 = M_1, m_3 = M_3, m_2 = -4$, and $M_2 = 10$. Hence

$$U(f, P) - L(f, P) = \frac{14}{15} \epsilon < \epsilon.$$ □

It is not difficult to see that the previous example may be generalized. A function $f: [a, b] \to \mathbb{R}$ is said to be *piecewise constant* if there is a partition $P = \{x_0, \ldots, x_n\}$ of $[a, b]$ such that for each index $i$ with $1 \leq i \leq n$, the function $f: (x_{i-1}, x_i) \to \mathbb{R}$ is constant. A piecewise constant function is integrable (see Exercise 5).

The following example shows that it is possible for $f: [a, b] \to \mathbb{R}$ to have infinitely many discontinuities and still be integrable.

**EXAMPLE 6.4**   Define

$$f(x) = \begin{cases} 1 & \text{if } x = 1/n \text{ for some natural number } n \\ 0 & \text{if the point } x \text{ in } [0, 1] \text{ is not of the above form.} \end{cases}$$

We will again use the Integrability Criterion to prove that the function $f: [0, 1] \to \mathbb{R}$ is integrable. Let $\epsilon > 0$. Observe that the function $f: [\epsilon/2, 1] \to \mathbb{R}$ is piecewise constant, and is therefore integrable. The Integrability Criterion implies that we may choose a partition $P^*$ of the interval $[\epsilon/2, 1]$ such that $U(f, P^*) - L(f, P^*) < \epsilon/2$. Now let $P$ be the partition of the whole interval $[0, 1]$ obtained by adjoining 0 to $P^*$. Then

$$U(f, P) = \frac{\epsilon}{2} + U(f, P^*) \quad \text{and} \quad L(f, P) = L(f, P^*).$$

It follows that $U(f, P) - L(f, P) < \epsilon$. □

**DEFINITION**   *A function $f: D \to \mathbb{R}$ is said to be monotonically increasing provided that*

$$f(x_1) \geq f(x_2) \quad \text{for all points } x_1 \text{ and } x_2 \text{ in } D \text{ such that} \quad x_1 \geq x_2.$$

*This function is said to be monotonically decreasing provided that*

$$f(x_1) \leq f(x_2) \quad \text{for all points } x_1 \text{ and } x_2 \text{ in } D \text{ such that} \quad x_1 \geq x_2.$$

*If a function is either monotonically decreasing or monotonically increasing, it is said to be monotone.*

For a natural number $n$, the partition $P = \{x_0, \ldots, x_n\}$ of the interval $[a, b]$ defined by

$$x_i = a + i\frac{(b-a)}{n} \quad \text{for} \quad 0 \le i \le n$$

is called the *regular partition of* $[a, b]$ *into* $n$ *subintervals*. This partition is characterized by the fact that all of the subintervals induced by the partition have the same length, $(b-a)/n$.

**THEOREM 6.5**    Suppose that the function $f: [a, b] \to \mathbb{R}$ is monotone. Then $f: [a, b] \to \mathbb{R}$ is integrable.

**Proof**    Let us first assume that $f: [a, b] \to \mathbb{R}$ is monotonically increasing. We will apply the Integrability Criterion. Let $\epsilon > 0$. We need to find a partition $P$ of $[a, b]$ such that $U(f, P) - L(f, P) < \epsilon$.

The crucial point is that for each natural number $n$, if $P_n$ is the regular partition of $[a, b]$ into $n$ subintervals, then we have the explicit formula

$$U(f, P_n) - L(f, P_n) = \left(\frac{b-a}{n}\right)(f(b) - f(a)). \tag{6.7}$$

To verify this formula, observe that for each index $i$ with $1 \le i \le n$, since the function $f: [x_{i-1}, x_i] \to \mathbb{R}$ is monotonically increasing,

$$m_i = \inf\{f(x) \mid x \text{ in } [x_{i-1}, x_i]\} = f(x_{i-1}),$$
$$M_i = \sup\{f(x) \mid x \text{ in } [x_{i-1}, x_i]\} = f(x_i),$$

**FIGURE 6.4**

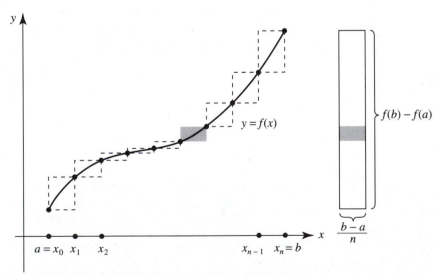

$$U(f, P) - L(f, P) = \frac{[f(b) - f(a)][b - a]}{n}$$

and, by the very definition of a regular partition,

$$x_i - x_{i-1} = \frac{b-a}{n}.$$

Consequently,     $U(f, P_n) - L(f, P_n) = \sum_{i=1}^{n}(M_i - m_i)(x_i - x_{i-1})$

$$= \left(\frac{b-a}{n}\right) \sum_{i=1}^{n}(f(x_i) - f(x_{i-1}))$$

$$= \left(\frac{b-a}{n}\right)(f(b) - f(a)),$$

and thus we have the formula (6.7). By the Archimedean Property of $\mathbb{R}$, we can choose a natural number $n$ such that

$$\left(\frac{b-a}{n}\right)(f(b) - f(a)) < \epsilon.$$

If we set $P = P_n$, we have $U(f, P) - L(f, P) < \epsilon$.

This proves the theorem when $f: [a, b] \to \mathbb{R}$ is monotonically increasing. When $f: [a, b] \to \mathbb{R}$ is monotonically decreasing, the proof proceeds as before, except that (6.7) now becomes

$$U(f, P) - L(f, P) = \left(\frac{b-a}{n}\right)(f(a) - f(b)). \qquad \blacksquare$$

**EXAMPLE 6.5**   According to the previous theorem, if

$$f(x) = e^{x^2} \quad \text{for} \quad 0 \le x \le 1,$$

then since the function $f: [0, 1] \to \mathbb{R}$ is monotonically increasing, it is integrable.   $\square$

**THEOREM 6.6**   Suppose that the function $f: [a, b] \to \mathbb{R}$ is continuous. Then $f: [a, b] \to \mathbb{R}$ is integrable.

**Proof**   Once more, we will apply the Integrability Criterion. Let $\epsilon > 0$. According to Theorem 3.14, the function $f: [a, b] \to \mathbb{R}$ is uniformly continuous. Consequently, since $\epsilon/(b - a) > 0$, we may choose a number $\delta > 0$ such that

$$|f(u) - f(v)| < \frac{\epsilon}{b-a} \quad \text{for all points } u \text{ and } v \text{ in } [a, b] \text{ such that} \quad |u - v| < \delta.$$
$$(6.8)$$

Choose $P = \{x_0, \ldots, x_n\}$ to be a partition of $[a, b]$ with the property that each subinterval $I_i = [x_{i-1}, x_i]$ has length less than $\delta$.

Let $1 \le i \le n$. Since the function $f: I_i \to \mathbb{R}$ is continuous, we may apply the Extreme Value Theorem to choose points $u_i$ and $v_i$ in $I_i$ at which $f: I_i \to \mathbb{R}$ attains maximum and minimum values, respectively. But $I_i$ has length less than $\delta$, so (6.8) implies that

$$M_i - m_i = f(u_i) - f(v_i) < \frac{\epsilon}{b-a}.$$

Consequently,
$$U(f, P) - L(f, P) = \sum_{i=1}^{n} (M_i - m_i)(x_i - x_{i-1})$$

$$< \epsilon/(b - a) \sum_{i=1}^{n} (x_i - x_{i-1})$$

$$= \epsilon. \qquad \blacksquare$$

There is a slight generalization of the preceding theorem that will turn out to be useful.

**COROLLARY 6.7**   Suppose that the function $f: [a, b] \to \mathbb{R}$ is bounded and that $f: (a, b) \to \mathbb{R}$ is continuous. Then $f: [a, b] \to \mathbb{R}$ is integrable.

**Proof**   Again, we will use the Integrability Criterion. Let $\epsilon > 0$. We need to find a partition $P$ of the interval $[a, b]$ such that $U(f, P) - L(f, P) < \epsilon$. Since $f: [a, b] \to \mathbb{R}$ is bounded, we may choose a number $M > 0$ such that

$$-M \leq f(x) \leq M \quad \text{for all } x \text{ in } [a, b].$$

Define $a_0 = a + \epsilon/8M$ and $b_0 = b - \epsilon/8M$. Since the function $f: [a_0, b_0] \to \mathbb{R}$ is continuous, it follows from the preceding theorem that it is integrable, and so, by the Integrability Theorem, there is a partition $P^*$ of the interval $[a_0, b_0]$ such that $U(f, P^*) - L(f, P^*) < \epsilon/2$. Define $P$ to be the partition of the whole interval $[a, b]$ obtained by adjoining the points $a$ and $b$ to $P^*$. The contributions to $U(f, P) - L(f, P)$ from the first and the last subinterval induced by $P$ are each at most equal to the length of the subinterval times $2M$. Thus,

$$U(f, P) - L(f, P) \leq 2M \frac{\epsilon}{8M} + 2M \frac{\epsilon}{8M} + [U(f, P^*) - L(f, P^*)]$$

$$< 2M \frac{\epsilon}{8M} + 2M \frac{\epsilon}{8M} + \frac{\epsilon}{2} = \epsilon. \qquad \blacksquare$$

**EXAMPLE 6.6**   Define

$$f(x) = \begin{cases} \sin(1/x) & \text{if } 0 < x \leq 1 \\ 4 & \text{if } x = 0. \end{cases}$$

Then the function $f: [0, 1] \to \mathbb{R}$ is bounded and $f: (0, 1) \to \mathbb{R}$ is continuous. The preceding corollary implies that the function $f: [0, 1] \to \mathbb{R}$ is integrable. $\square$

For a continuous function $f: [a, b] \to \mathbb{R}$, if we change the functional value at one point, then the new function will not be continuous. In contrast, the property of being integrable is not as sensitive to changes in functional values. It is not difficult to see, for example, that if the function $f: [a, b] \to \mathbb{R}$ is integrable, then by changing its functional values at the endpoints of the interval $[a, b]$ it remains integrable and the value of the integral is unchanged. This observation, together with the preceding theorem, implies that given a bounded, continuous function $f: (a, b) \to \mathbb{R}$, we can define $\int_a^b f$ to be the integral of any extension of the function $f: (a, b) \to \mathbb{R}$ to the interval $[a, b]$.

We will conclude this section with a remark about notation. For an integrable function $f: [a, b] \to \mathbb{R}$, we have denoted the value of the integral by the symbol $\int_a^b f$. The value of the integral is also often denoted by symbols such as $\int_a^b f(x)\, dx$ or $\int_a^b f(t)\, dt$. We will see that this alternate notation, involving Leibnitz symbols, is often quite convenient.

**EXERCISES**

1.  Define $f(x) = x$ for all $x$ in $[0, 1]$. For each natural number $n$, compute $L(f, P_n)$ and $U(f, P_n)$, where $P_n$ is the regular partition of $[0, 1]$ into $n$ subintervals. Then use the Integrability Criterion to show that the function $f: [0, 1] \to \mathbb{R}$ is integrable. (*Hint:* $\sum_{k=1}^n k = n(n+1)/2$.)

2.  Define $f(x) = x^2$ for all $x$ in $[0, 1]$. For each natural number $n$, compute $L(f, P_n)$ and $U(f, P_n)$, where $P_n$ is the regular partition of $[0, 1]$ into $n$ subintervals. Then use the Integrability Criterion to show that the function $f: [0, 1] \to \mathbb{R}$ is integrable. (*Hint:* $\sum_{k=1}^n k^2 = n(n+1)(2n+1)/6$.)

3.  For an interval $[a, b]$ and a positive number $\delta$, show that there is a partition $P = \{x_0, \ldots, x_n\}$ of $[a, b]$ such that each subinterval $[x_{i-1}, x_i]$ has length less than $\delta$.

4.  Suppose that the integrable function $f: [a, b] \to \mathbb{R}$ has the property that $f(x) \geq 0$ for all $x$ in $[a, b]$. Prove that $\int_a^b f \geq 0$.

5.  Prove that a piecewise constant function is integrable.

6.  Define

$$ f(x) = \begin{cases} x & \text{if } 2 \leq x \leq 3 \\ 2 & \text{if } 3 < x \leq 4. \end{cases} $$

Prove that the function $f: [2, 4] \to \mathbb{R}$ is integrable.

7.  Suppose that the integrable function $f: [a, b] \to \mathbb{R}$ has the property that for each rational number $x$ in the interval $[a, b]$, $f(x) = 0$. Prove that $\int_a^b f = 0$.

8.  Suppose that the function $f: [a, b] \to \mathbb{R}$ is bounded and that $f: [a, b] \to \mathbb{R}$ is continuous except at a finite number of points $z_1, \ldots, z_k$ in $[a, b]$. Prove that $f: [a, b] \to \mathbb{R}$ is integrable.

9.  For positive numbers $m$ and $b$, define $f(x) = mx + b$ for all $x$ in $[0, 1]$.
    (a)  From elementary area formulas, show that the area under the graph of $f: [0, 1] \to \mathbb{R}$ and above the $x$ axis is $b + m/2$.
    (b)  For each natural number $n$, compute $L(f, P_n)$ and $U(f, P_n)$, where $P_n$ is the regular partition of $[0, 1]$ into $n$ subintervals. Show that

$$ L(f, P_n) < b + \frac{m}{2} < U(f, P_n). $$

10.  For a point $x$ in the interval $[1, 2]$, define $f(x) = 0$ if $x$ is irrational, and define $f(x) = 1/n$ if $x$ is rational and is expressed as $x = m/n$, for natural numbers $m$ and $n$ having no common positive integer divisor other than 1. Prove that the function $f: [1, 2] \to \mathbb{R}$ is integrable. (*Hint:* First prove that given $\epsilon > 0$ there are only a finite number of points $x$ in the interval $[1, 2]$ at which $f(x) \geq \epsilon$.)

11.  Give a geometric interpretation, in terms of areas, of formula (6.7).

12.  For a point $x$ in the interval $[0, 1]$, define $f(x) = x$ if $x$ is rational and $f(x) = -x$ if $x$ is irrational. Prove that the function $f: [0, 1] \to \mathbb{R}$ is not integrable.

13.  Is the sum of monotone functions also monotone? Is the product of monotone functions also monotone? Justify your answers.

14.  Suppose that $f: [a, b] \to \mathbb{R}$ is a bounded function for which there is a partition $P$ of $[a, b]$ with $L(f, P) = U(f, P)$. Prove that $f: [a, b] \to \mathbb{R}$ is constant.

15. Suppose that the function $f:[a, b] \to \mathbb{R}$ is integrable and there is a positive number $m$ such that $f(x) \geq m$ for all $x$ in $[a, b]$. Show that the reciprocal function $1/f:[a, b] \to \mathbb{R}$ is integrable by proving that for each partition $P$ of the interval $[a, b]$,

$$U(1/f, \ P) - L(1/f, \ P) \leq \frac{1}{m^2}[U(f, \ P) - L(f, \ P)].$$

16. Suppose the continuous function $f:[a, b] \to \mathbb{R}$ has the property that

$$\int_c^d f \leq 0 \quad \text{whenever} \quad a \leq c < d \leq b.$$

Prove that $f(x) \leq 0$ for all $x$ in $[a, b]$. Is this true if we require only integrability of the function?

---

## 6.3 The First Fundamental Theorem of Calculus

In the preceding section, we established a number of criteria that guarantee that a function $f:[a, b] \to \mathbb{R}$ is integrable. However, we have not considered the problem of determining the value of the integral. In this section we will turn to this aspect of integration, and we will prove the truly remarkable First Fundamental Theorem of Calculus.

Let us restate the definition of the value of the integral:

*Suppose that the function $f:[a, b] \to \mathbb{R}$ is integrable. For a real number $A$,*

$$\int_a^b f = A$$

*if and only if*

$$L(f, \ P) \leq A \leq U(f, \ P) \quad \textit{for every partition } P \textit{ of the interval } [a, b]. \quad (6.9)$$

**EXAMPLE 6.7** Suppose that the function $f:[a, b] \to \mathbb{R}$ has constant value $c$ on the open interval $(a, b)$. We have already seen that $f:[a, b] \to \mathbb{R}$ is integrable. To verify that

$$\int_a^b f = c(b - a),$$

observe that if $P = \{x_0, \ldots, x_n\}$ is any partition of $[a, b]$, each subinterval induced by the partition intersects the interval $(a, b)$, and so, for each index $i$ with $1 \leq i \leq n$, $m_i \leq c \leq M_i$. Consequently, $L(f, \ P) \leq c(b - a) \leq U(f, \ P)$, and so, by criterion (6.9), $\int_a^b f = c(b - a)$. $\square$

In order to directly verify criterion (6.9), it is necessary to evaluate or estimate $L(f, \ P)$ and $U(f, \ P)$ for every partition $P$ of $[a, b]$. Even for simple functions, this is usually not so easy. On the other hand, when $P$ is a regular partition of $[a, b]$, the estimation of $L(f, \ P)$ and $U(f, \ P)$ sometimes becomes easier. With this in mind, and because it will be useful later, we prove the following result.

**PROPOSITION**
**6.8**

Suppose that the function $f: [a, b] \to \mathbb{R}$ is bounded. For each natural number $n$, let $P_n$ be a partition of $[a, b]$. Let $A$ be a number having the property that

$$\lim_{n \to \infty} L(f, P_n) = A = \lim_{n \to \infty} U(f, P_n).$$

Then $f: [a, b] \to \mathbb{R}$ is integrable and $\int_a^b = A$.

**Proof**   By the difference property of convergent sequences,

$$\lim_{n \to \infty} [U(f, P_n) - L(f, P_n)] = A - A = 0,$$

and hence the Integrability Criterion implies that $f: [a, b] \to \mathbb{R}$ is integrable. Observe that for each natural number $n$,

$$L(f, P_n) - A \le \int_a^b f - A \le U(f, P_n) - A.$$

It follows from the Squeezing Principle that $\int_a^b f - A = 0$.     ∎

**EXAMPLE 6.8**   We claim that $\int_0^1 x \, dx = 1/2$, and we will verify this claim by applying the preceding proposition. Define $f(x) = x$ for all $x$ in $[0, 1]$. For each natural number $n$, let $P_n$ be the regular partition of $[0, 1]$ into $n$ subintervals. The formula $\sum_{i=1}^m i = m(m + 1)/2$ allows us to explicitly find the Darboux sums. Indeed, since $f: [0, 1] \to \mathbb{R}$ is increasing, it follows that

$$
\begin{aligned}
L(f, P_n) &= \sum_{i=1}^n m_i (x_i - x_{i-1}) \\
&= \sum_{i=1}^n f(x_{i-1}) \frac{1}{n} \\
&= \frac{1}{n} \sum_{i=1}^n \left( \frac{i-1}{n} \right) \\
&= \frac{1}{n^2} \sum_{i=1}^n (i - 1) \\
&= \frac{1}{n^2} \cdot \frac{n(n-1)}{2} \\
&= \frac{1}{2} - \frac{1}{2n}.
\end{aligned}
$$

A similar calculation shows that $U(f, P_n) = 1/2 + 1/2n$. Thus,

$$\lim_{n \to \infty} L(f, P_n) = \lim_{n \to \infty} U(f, P_n) = \frac{1}{2},$$

and so $\int_0^1 x \, dx = 1/2$.     □

**EXAMPLE 6.9**  We will apply Proposition 6.8 to show that $\int_0^1 x^2 \, dx = 1/3$. Define $f(x) = x^2$ for all $x$ in $[0, 1]$. For each natural number $n$, let $P_n$ be the regular partition of $[0, 1]$ into $n$ subintervals. As in the preceding example, there is a formula, namely $\sum_{i=1}^m k^2 = [m(m+1)(2m+1)]/6$, that permits calculation of the Darboux sums. Indeed, since $f : [0, 1] \to \mathbb{R}$ is increasing, it follows that

$$L(f, P_n) = \sum_{i=1}^n m_i(x_i - x_{i-1}) = \sum_{i=1}^n f(x_{i-1})\frac{1}{n}$$

$$= \frac{1}{n} \sum_{i=1}^n \left(\frac{i-1}{n}\right)^2 = \frac{1}{n^3} \sum_{i=1}^n (i-1)^2$$

$$= \frac{1}{n^3}\left[\frac{(n-1)n(2n-1)}{6}\right] = \frac{1}{3} - \frac{1}{2n} + \frac{1}{6n^2}.$$

A similar calculation shows that $U(f, P_n) = 1/3 + 1/2n + 1/6n^2$. Thus,

$$\lim_{n \to \infty} L(f, P_n) = \lim_{n \to \infty} U(f, P_n) = \frac{1}{3},$$

and so $\int_0^1 x^2 \, dx = 1/3$.  □

**EXAMPLE 6.10**  We wish to evaluate the integral $\int_0^1 \sqrt{x} \, dx$. Define $f(x) = \sqrt{x}$ for all $x$ in $[0, 1]$. Then the function $f : [0, 1] \to \mathbb{R}$ is integrable, since it is monotone. But how can one calculate $\int_0^1 \sqrt{x} \, dx$? For the strategy that succeeded in the previous two examples to succeed here as well, it is necessary to find a formula for the sum

$$1 + \sqrt{2} + \sqrt{3} + \cdots + \sqrt{m}.$$

There is no obvious formula.  □

This last example is not atypical. Unless we have very special functions, such as those in Examples 6.8 and 6.9, it is not possible to evaluate $U(f, P)$ or $L(f, P)$, even when $P$ is a regular partition.*

In order to proceed further, a genuinely new idea is needed. The brilliant idea is to replace the problem of calculating the integral with another seemingly different problem, namely the problem of solving the following differential equation:

*Given a function $f : [a, b] \to \mathbb{R}$, find a continuous function $F : [a, b] \to \mathbb{R}$ such that $F : (a, b) \to \mathbb{R}$ is differentiable and*

$$F'(x) = f(x) \quad \text{for all } x \text{ in } (a, b).$$    F.D.E.

(F.D.E. denotes the Fundamental Differential Equation.)

Recall that Section 4.5 was devoted to a discussion of the above problem. We saw there that, for example, when $f : \mathbb{R} \to \mathbb{R}$ is a polynomial, there is a simple explicit solution of the F.D.E. Moreover if $\alpha \neq -1$, $0 < a < b$, and $f(x) = x^\alpha$ for all $x$ in $[a, b]$,

---

*There is a rather ingenious trick for evaluating the integral in Example 6.10, which is due to Fermat (see Exercise 11 of Section 6.4). However, the theory of integration could not proceed by requiring ingenious tricks for each example.

the function defined by $F(x) = x^{\alpha+1}/(\alpha + 1)$ for all $x$ in $[a, b]$ is a solution of the F.D.E.

The first relationship between the F.D.E. and integration is provided by the following theorem.

**THEOREM 6.9**   **The First Fundamental Theorem of Calculus**   Let the function $f: [a, b] \to \mathbb{R}$ be integrable. Suppose that the function $F: [a, b] \to \mathbb{R}$ is continuous, that $F: (a, b) \to \mathbb{R}$ is differentiable, and that

$$F'(x) = f(x) \quad \text{for all } x \text{ in } (a, b).$$

Then
$$\int_a^b f = F(b) - F(a).$$

**Proof**   According to criterion (6.9), we must show that for each partition $P$ of the interval $[a, b]$,

$$L(f, P) \le F(b) - F(a) \le U(f, P). \tag{6.10}$$

Let $P = \{x_0, \ldots, x_n\}$ be a partition of $[a, b]$. Fix an index $i$ with $1 \le i \le n$. We apply the Lagrange Mean Value Theorem to the function $F: [x_{i-1}, x_i] \to \mathbb{R}$ in order to choose a point $c_i$ in the open interval $(x_{i-1}, x_i)$ at which

$$F(x_i) - F(x_{i-1}) = F'(c_i)(x_i - x_{i-1}),$$

so that, since $F'(c_i) = f(c_i)$,

$$F(x_i) - F(x_{i-1}) = f(c_i)(x_i - x_{i-1}). \tag{6.11}$$

Since the point $c_i$ belongs to the interval $[x_{i-1}, x_i]$,

$$m_i = \inf\{f(x) \,|\, x \text{ in } [x_{i-1}, x_i]\} \le f(c_i) \le \sup\{f(x) \,|\, x \text{ in } [x_{i-1}, x_i]\} = M_i. \tag{6.12}$$

Multiplying this last inequality by $x_i - x_{i-1}$ and substituting (6.11), we obtain

$$m_i(x_i - x_{i-1}) \le F(x_i) - F(x_{i-1}) \le M_i(x_i - x_{i-1}).$$

Finally, adding up these $n$ inequalities gives the required inequality (6.10).   ∎

In the following three examples, we will attempt to use the First Fundamental Theorem to evaluate the given integrals.

**EXAMPLE 6.11**   Let $\alpha > 0$. We will evaluate the integral $\int_0^1 x^\alpha \, dx$. Define $f(x) = x^\alpha$ for all $x$ in $[0, 1]$. The function $f \colon [0, 1] \to \mathbb{R}$ is continuous, so it is integrable. In order to apply the First Fundamental Theorem of Calculus, we need to find a continuous function $F \colon [0, 1] \to \mathbb{R}$ such that $F \colon (0, 1) \to \mathbb{R}$ is differentiable and

$$F'(x) = x^\alpha \quad \text{for all } x \text{ in } (0, 1). \tag{6.13}$$

But the function $F \colon [0, 1] \to \mathbb{R}$ defined by

$$F(x) = \frac{x^{\alpha+1}}{\alpha + 1} \quad \text{for all } x \text{ in } [0, 1]$$

has these properties. According to the First Fundamental Theorem,

$$\int_0^1 x^\alpha \, dx = F(1) - F(0) = \frac{1}{\alpha + 1}. \qquad \square$$

Compare the preceding example with Examples 6.8, 6.9, and 6.10.

**EXAMPLE 6.12**   We wish to evaluate $\int_0^1 [1/(1 + x^4)] \, dx$. Define $f(x) = 1/(1 + x^4)$ for all $x$ in $[0, 1]$. Since the function $f \colon [0, 1] \to \mathbb{R}$ is continuous, it is also integrable. In order to apply the First Fundamental Theorem, we need to find a continuous function $F \colon [0, 1] \to \mathbb{R}$ such that $F \colon (0, 1) \to \mathbb{R}$ is differentiable and

$$F'(x) = \frac{1}{1 + x^4} \quad \text{for all } x \text{ in } (0, 1). \tag{6.14}$$

Even if we carefully sift through all of our differentiation results, no solution of (6.14) comes to mind.* Hence we cannot apply the First Fundamental Theorem. $\square$

**EXAMPLE 6.13**   Define

$$f(x) = \begin{cases} 1 & \text{if } 2 \le x < 3 \\ 0 & \text{if } 3 \le x \le 6. \end{cases}$$

There is now no possibility of applying the First Fundamental Theorem in the evaluation of $\int_2^6 f$, because, according to Darboux's Theorem, if $F \colon (2, 6) \to \mathbb{R}$ is differentiable, the function $F' (2, 6) \to \mathbb{R}$ has the Intermediate Value Property. But the function $f \colon (2, 6) \to \mathbb{R}$ does not have the Intermediate Value Property. However, it is not difficult to see that one can apply Proposition 6.8 to show that $\int_2^6 f = 1$. $\square$

The above examples illustrate both the power and the limitations of the First Fundamental Theorem. It replaces the problem of calculating $\int_a^b f$ with the problem of solving the F.D.E. Frequently one can solve the F.D.E., and there are cases when one definitely cannot solve the F.D.E.; sometimes it is not clear. There are a number of techniques of integration that consist of taking an integral $\int_a^b f$ and finding another integral $\int_c^d g$ such that $\int_a^b f = \int_c^d g$ and $\int_c^d g$ can be evaluated by the First Fundamental Theorem. We will consider such techniques in Section 7.3.

---

*There is a precise way to assert and prove that it is not possible to find an "elementary function" that is a solution of (6.14). Unfortunately, to do this requires a background in modern abstract algebra that is outside the scope of this book; see the article "Integration in Finite Terms" by Maxwell Rosenlicht in *American Mathematical Monthly*, Nov. 1972.

As we discussed at the end of Section 6.2, for a continuous bounded function $f: (a, b) \to \mathbb{R}$, the symbol $\int_a^b f$ denotes the integral of any extension of $f: (a, b) \to \mathbb{R}$ to the closed interval $[a, b]$.

**COROLLARY 6.10**   Suppose that the function $F: [a, b] \to \mathbb{R}$ is continuous, that $F: (a, b) \to \mathbb{R}$ is differentiable, and that the derivative $F': (a, b) \to \mathbb{R}$ is both continuous and bounded. Then

$$\int_a^b F'(x)\, dx = F(b) - F(a). \tag{6.15}$$

**Proof**   Define $f(x) = F'(x)$ for all $x$ in $(a, b)$. Since the function $f: (a, b) \to \mathbb{R}$ is both continuous and bounded, any extension of $f: (a, b) \to \mathbb{R}$ to the interval $[a, b]$ defines an integrable function whose integral does not depend on the values $f(a)$ and $f(b)$. The integral of such an extension is what is meant by $\int_a^b F'(x)\, dx$. Formula (6.15) follows directly from the First Fundamental Theorem. ∎

---

**EXERCISES**

1.  Find the values of the following two integrals:

    (a)  $\displaystyle\int_0^1 [x + 1]\, dx$

    (b)  $\displaystyle\int_0^1 [4x + 1]\, dx$

    by applying Proposition 6.8 with the sequence of partitions $\{P_n\}$, where for each $n$, $P_n$ is the regular partition of $[0, 1]$ into $n$ subintervals.

2.  Let $m$ and $b$ be positive numbers. Find the value of $\int_0^1 [mx + b]\, dx$ in the following three ways:

    (a)  Using elementary geometry, interpreting $\int_0^1 [mx + b]\, dx$ as an area.

    (b)  Using Proposition 6.8 where for each natural number $n$, $P_n$ is a regular partition of $[0, 1]$ into $n$ subintervals.

    (c)  Using the First Fundamental Theorem of Calculus.

3.  Use the First Fundamental Theorem of Calculus to evaluate each of the following integrals:

    (a)  $\displaystyle\int_1^2 [1/x^2 + x + \cos x]\, dx$

    (b)  $\displaystyle\int_0^1 x\sqrt{4 - x^2}\, dx$

    (c)  $\displaystyle\int_1^3 x\sqrt{10 - x}\, dx$

    (d)  $\displaystyle\int_0^\pi \cos^2 x\, dx$

---

## 6.4   The Convergence of Darboux Sums and Riemann Sums

For an integrable function $f: [a, b] \to \mathbb{R}$ and a partition $P$ of the interval $[a, b]$, the Refinement Lemma asserts that

$$L(f, P) \leq L(f, P^*) \leq \int_a^b f \leq U(f, P^*) \leq U(f, P)$$

whenever the partition $P^*$ is a refinement of $P$.

Thus, for an integrable function and a given partition, when we refine the partition the associated Darboux sums become better approximations of the value of the integral. It is reasonable to expect that one can make the Darboux sums arbitrarily close to the integral provided that the associated partitions are sufficiently fine. In this section, we will prove that this is so. We will also study a sum called the Riemann sum, which is closely related to the Darboux sum.

**DEFINITION**  *For a partition $P = \{x_0, \ldots, x_n\}$ of the interval $[a, b]$, we define the gap of $P$, which is denoted by $\|P\|$, to be the length of the largest subinterval induced by $P$, that is,*

$$\|P\| \equiv \max_{1 \leq i \leq n} [x_i - x_{i-1}].$$

Observe that for a partition $P$ and a positive number $\alpha$, $\|P\| < \alpha$ if and only if each subinterval induced by the partition $P$ has length less than $\alpha$.

**THEOREM 6.11**  Suppose that the function $f : [a, b] \to \mathbb{R}$ is bounded. Then $f : [a, b] \to \mathbb{R}$ is integrable if and only if for each positive number $\epsilon$ there is a positive number $\delta$ such that

$$U(f, P) - L(f, P) < \epsilon \tag{6.16}$$

whenever $P$ is a partition of $[a, b]$ such that $\|P\| < \delta$.

**Proof**  To show that criterion (6.16) implies the integrability of $f : [a, b] \to \mathbb{R}$, we will use the Integrability Criterion. Let $\epsilon > 0$. According to (6.16), there is a $\delta > 0$ such that $U(f, P) - L(f, P) < \epsilon$ whenever $P$ is a partition of $[a, b]$ such that $\|P\| < \delta$. Thus, if we choose any partition $P$ with $\|P\| < \delta$ the Integrability Criterion is satisfied.

We will now prove the converse. Suppose that $f : [a, b] \to \mathbb{R}$ is integrable. Let $\epsilon > 0$. According to the Integrability Criterion, we may choose a partition $P^* = \{x_0^*, \ldots, x_n^*\}$ of $[a, b]$ such that

$$U(f, P^*) - L(f, P^*) < \frac{\epsilon}{2}. \tag{6.17}$$

Choose $M > 0$ such that

$$-M \leq f(x) \leq M \quad \text{for all } x \text{ in } [a, b]. \tag{6.18}$$

The crucial point in the proof is the following assertion: For any partition $P$ of the interval $[a, b]$, we have the estimate

$$U(f, P) - L(f, P) \leq 2nM\|P\| + [U(f, P^*) - L(f, P^*)]. \tag{6.19}$$

Once this estimate is proven, we define $\delta = \epsilon/4nM$. Then, if $P$ is a partition of $[a, b]$ such that $\|P\| < \delta$,

$$U(f, P) - L(f, P) < 2nM\delta + \frac{\epsilon}{2}$$

$$= \frac{\epsilon}{2} + \frac{\epsilon}{2} = \epsilon.$$

It remains to verify estimate (6.19). Let $P = \{x_0, \ldots, x_m\}$ be a partition of $[a, b]$. As usual, for each index $i$ such that $1 \leq i \leq m$, we set

$$m_i = \inf\{f(x) \mid x \text{ in } [x_{i-1}, x_i]\} \quad \text{and} \quad M_i = \sup\{f(x) \mid x \text{ in } [x_{i-1}, x_i]\}.$$

We now separate the set of indices $\{1, \ldots, m\}$ into two subsets $A$ and $B$. Define $A$ to be the set of indices $i$ in $\{1, \ldots, m\}$ such that the subinterval $(x_{i-1}, x_i)$ contains a partition point of the partition $P^*$. Define $B$ to be the set of indices $i$ in $\{1, \ldots, m\}$ that do not have this property. Now $M_i - m_i \leq 2M$ and $x_i - x_{i-1} \leq \|P\|$ for every index $i$ in $\{1, \ldots, m\}$, so since there are fewer than $n$ indices in the set $A$,

$$\sum_{i \in A} (M_i - m_i)(x_i - x_{i-1}) \leq 2nM \|P\|.$$

On the other hand, if $i$ is an index in the set $B$, then the subinterval $[x_{i-1}, x_i]$ is contained in one of the subintervals $[x_{j-1}^*, x_j^*]$ induced by the partition $P^*$, and so by the Refinement Lemma,

$$\sum_{i \in B} (M_i - m_i)(x_i - x_{i-1}) \leq U(f, P^*) - L(f, P^*).$$

Consequently,

$$U(f, P) - L(f, P) = \sum_{i \in A} (M_i - m_i)(x_i - x_{i-1}) + \sum_{i \in B} (M_i - m_i)(x_i - x_{i-1})$$
$$< 2nM\|P\| + [U(f, P^*) - L(f, P^*)],$$

so (6.19) is proven.                                                                    ∎

**COROLLARY 6.12**   Suppose that the function $f: [a, b] \to \mathbb{R}$ is integrable. If $\{P_n\}$ is any sequence of partitions of $[a, b]$ such that $\lim_{n \to \infty} \|P_n\| = 0$, then

$$\lim_{n \to \infty} U(f, P_n) = \lim_{n \to \infty} L(f, P_n) = \int_a^b f. \tag{6.20}$$

**Proof**   Let $\epsilon > 0$. According to Theorem 6.11, we can choose $\delta > 0$ such that if $P$ is any partition of $[a, b]$, then

$$U(f, P) - L(f, P) < \epsilon \quad \text{if} \quad \|P\| < \delta.$$

On the other hand, since $\lim_{n \to \infty} \|P_n\| = 0$, we can select a natural number $N$ such that $\|P_n\| < \delta$ if $n \geq N$. Thus

$$0 \leq U(f, P_n) - L(f, P_n) < \epsilon \quad \text{for all integers } n \geq N.$$

Moreover, by the definition of the integral,

$$L(f, P_n) \leq \int_a^b f \leq U(f, P_n) \quad \text{for every natural number } n.$$

Consequently, for each integer $n \geq N$,

$$\left| \int_a^b f - L(f, P_n) \right| < \epsilon \quad \text{and} \quad \left| \int_a^b f - U(f, P_n) \right| < \epsilon.$$

This proves (6.20).                                                                    ∎

For convenient future reference, we collect Proposition 6.8 and Corollary 6.12 in the single theorem that follows:

**THEOREM 6.13**   **The Darboux Sum Convergence Criterion**   For a bounded function $f:[a,b] \to \mathbb{R}$ and a real number $A$, the following two assertions are equivalent:

(i)  The function $f:[a,b] \to \mathbb{R}$ is integrable and $\int_a^b f = A$.

(ii)  If $\{P_n\}$ is any sequence of partitions of $[a,b]$ such that $\lim_{n \to \infty} \|P_n\| = 0$, then

$$\lim_{n \to \infty} L(f, P_n) = \lim_{n \to \infty} U(f, P_n) = A.$$

**Proof**   That (i) implies (ii) is the content of Corollary 6.12. To prove the converse, suppose that (ii) holds. For each natural number $n$, let $P_n$ be the regular partition of $[a,b]$ into $n$ subintervals. Then the sequence $\{\|P_n\|\}$ converges to 0, so (ii) implies that

$$\lim_{n \to \infty} L(f, P_n) = \lim_{n \to \infty} U(f, P_n) = A.$$

From this, using Proposition 6.8, we obtain (i).   ∎

**DEFINITION**   *Consider a function $f:[a,b] \to \mathbb{R}$, and let $P = \{x_0, \ldots, x_n\}$ be a partition of the interval $[a,b]$. For each index $i$ such that $1 \le i \le n$, let $c_i$ be a point in the interval $[x_{i-1}, x_i]$. Then the sum*

$$\sum_{i=1}^{n} f(c_i)(x_i - x_{i-1}) \tag{6.21}$$

*is called a Riemann sum for the function $f:[a,b] \to \mathbb{R}$ based on the partition P.*

It is convenient to denote the Riemann sum (6.21) by the symbol $R(f, P)$.* It is clear that if the function $f:[a,b] \to \mathbb{R}$ is bounded, then for each partition $P$ of $[a,b]$,

$$L(f, P) \le R(f, P) \le U(f, P).$$

**COROLLARY 6.14**   Suppose that the function $f:[a,b] \to \mathbb{R}$ is integrable. For each natural number $n$, let $P_n$ be a partition of $[a,b]$ and let $R(f, P_n)$ be a Riemann sum for $f:[a,b] \to \mathbb{R}$ based on $P_n$. Suppose that $\lim_{n \to \infty} \|P_n\| = 0$. Then

$$\lim_{n \to \infty} R(f, P_n) = \int_a^b f.$$

---

*The notation $R(f, P)$ does not explicitly exhibit the dependence of the Riemann sum (6.21) on the choice of the $c_i$'s.

**Proof**   It is clear that for each natural number $n$,

$$L(f, P_n) \le R(f, P_n) \le U(f, P_n).$$

According to Corollary 6.12,

$$\lim_{n\to\infty} L(f, P_n) = \lim_{n\to\infty} U(f, P_n) = \int_a^b f.$$

The conclusion now follows from the Squeezing Principle for convergent sequences.
∎

**EXAMPLE 6.14**   For each natural number $n$, define $P_n$ to be the regular partition of $[0, 1]$ into $n$ subintervals. Consider the Riemann sum for the integral $\int_0^1 \sqrt{x}\, dx$ based on the partition $P_n$ that we obtain by letting $c_i = i/n$ for $1 \le i \le n$. From Corollary 6.14 and the First Fundamental Theorem of Calculus, it follows that

$$\lim_{n\to\infty}\left[\frac{\sqrt{1}+\sqrt{2}+\cdots+\sqrt{n}}{n^{3/2}}\right] = \lim_{n\to\infty}\frac{1}{n}\left[\sqrt{\frac{1}{n}}+\cdots+\sqrt{\frac{n}{n}}\right] = \int_0^1 \sqrt{x}\, dx = \frac{2}{3}. \quad \square$$

We will conclude this section by proving an analogue of the Darboux Sum Convergence Criterion for Riemann sums. In order to do so, we will first prove the following:

**LEMMA 6.15**   Suppose that the function $f: [a, b] \to \mathbb{R}$ is bounded. For each partition $P$ of $[a, b]$ and each positive number $\epsilon$, there are Riemann sums $R(f, P)$ and $R'(f, P)$ for the function $f: [a, b] \to \mathbb{R}$ based on the partition $P$ such that

$$0 \le U(f, P) - R(f, P) < \epsilon \tag{6.22}$$

and

$$0 \le R'(f, P) - L(f, P) < \epsilon. \tag{6.23}$$

**Proof**   Let $P = \{x_0, \ldots, x_n\}$. By definition,

$$U(f, P) = \sum_{k=1}^n M_i(x_i - x_{i-1})$$

where for each index $i$ with $1 \le i \le n$,

$$M_i = \sup\{f(x) \mid x \text{ in } [x_{i-1}, x_i]\}.$$

Fix an index $i$ with $1 \le i \le n$. Since the number $M_i - \epsilon/(b-a)$ is not an upper bound for the set $\{f(x) \mid x \text{ in } [x_{i-1}, x_i]\}$, there is a point $c_i$ in the subinterval $[x_{i-1}, x_i]$ such that $f(c_i) > M_i - \epsilon/(b-a)$. Thus

$$M_i - \epsilon/(b-a) < f(c_i).$$

Multiply this inequality by $(x_i - x_{i-1})$ and sum the resulting $n$ inequalities to obtain

$$U(f, P) - \epsilon = U(f, P) - \epsilon/(b-a)\sum_{k=1}^n(x_i - x_{i-1}) < \sum_{k=1}^n f(c_i)(x_i - x_{i-1}).$$

Define $R(f, P) = \sum_{k=1}^{n} f(c_i)(x_i - x_{i-1})$. This proves (6.22). The proof of (6.23) is entirely similar, except that one chooses a point $c_i'$ in the subinterval $[x_{i-1}, x_i]$ such that $f(c_i') < m_i + \epsilon/(b-a)$. ∎

---

**THEOREM 6.16**   **The Riemann Sum Convergence Criterion**   Suppose that the function $f: [a, b] \to \mathbb{R}$ is bounded, and let $A$ be a number. Then the following two assertions are equivalent:

(i) The function $f: [a, b] \to \mathbb{R}$ is integrable and $\int_a^b f = A$.

(ii) If $\{P_n\}$ is any sequence of partitions of $[a, b]$ such that $\lim_{n \to \infty} \|P_n\| = 0$, and for each natural number $n$, $R(f, P_n)$ is a Riemann sum for $f: [a, b] \to \mathbb{R}$ based on the partition $P_n$, then

$$\lim_{n \to \infty} R(f, P_n) = A.$$

**Proof**   That (i) implies (ii) is the assertion of Corollary 6.14.

We will use the Darboux Sum Convergence Criterion to verify that (ii) implies (i). Suppose that (ii) holds. Let $\{P_n\}$ be a sequence of partitions of $[a, b]$ such that $\lim_{n \to \infty} \|P_n\| = 0$. We must prove that

$$\lim_{n \to \infty} U(f, P_n) = \lim_{n \to \infty} L(f, P_n) = A. \tag{6.24}$$

Fix a natural number $n$. By the preceding lemma, taking $\epsilon = 1/n$, we may choose Riemann sums $R(f, P_n)$ and $R'(f, P_n)$ for the function $f: [a, b] \to \mathbb{R}$ based on the partition $P_n$ such that

$$0 \le U(f, P_n) - R(f, P_n) < \frac{1}{n}$$

and

$$0 \le R'(f, P_n) - L(f, P_n) < \frac{1}{n}.$$

Consequently, for each natural number $n$,

$$R'(f, P_n) - \frac{1}{n} < L(f, P_n) \le U(f, P_n) < R(f, P_n) + \frac{1}{n}. \tag{6.25}$$

Now (ii) implies that

$$\lim_{n \to \infty} R(f, P_n) = \lim_{n \to \infty} R'(f, P_n) = A.$$

Inequality (6.25) and the Squeezing Principle for sequences imply (6.24). According to the Darboux Sum Convergence Criterion, the function $f: [a, b] \to \mathbb{R}$ is integrable and $\int_a^b f = A$. ∎

**EXERCISES**

1. Let $P_1$ and $P_2$ be partitions of $[a, b]$. Show that if $P_1$ is a refinement of $P_2$, then $\|P_1\| \leq \|P_2\|$. Is the converse true?

2. For a fixed positive number $\beta$, find

$$\lim_{n \to \infty} \left[ \frac{1^\beta + 2^\beta + \cdots + n^\beta}{n^{\beta+1}} \right].$$

3. Find

$$\lim_{n \to \infty} \left[ \frac{1}{n+1} + \frac{1}{n+2} + \cdots + \frac{1}{2n} \right].$$

4. Find

$$\lim_{n \to \infty} \left[ \sum_{k=1}^{n} \frac{k}{n^2 + k^2} \right].$$

5. Find

$$\lim_{n \to \infty} \left[ \frac{1}{\sqrt{n \cdot n}} + \frac{1}{\sqrt{n(n+1)}} + \cdots + \frac{1}{\sqrt{n(n+n)}} \right].$$

6. Let $b > 1$. Find the value of the Riemann sum for $\int_1^b [1/\sqrt{x}]\,dx$ that one obtains for the partition $P = \{x_0, \ldots, x_n\}$ of $[1, b]$ by choosing $c_i = [(\sqrt{x_i} + \sqrt{x_{i-1}})/2]^2$ for $1 \leq k \leq n$.

7. Suppose that the function $f: [0, 1] \to \mathbb{R}$ is integrable. Prove that

$$\lim_{n \to \infty} \frac{1}{n} \left[ f\left(\frac{1}{n}\right) + f\left(\frac{2}{n}\right) + \cdots + f\left(\frac{n-1}{n}\right) + f(1) \right] = \int_0^1 f.$$

8. Suppose that the function $f: [a, b] \to \mathbb{R}$ is Lipschitz; that is, there is a number $c$ such that

$$|f(u) - f(v)| \leq c|u - v| \quad \text{for all } u \text{ and } v \text{ in } [a, b].$$

Let $P$ be a partition of $[a, b]$ and $R(f, P)$ be a Riemann sum based on $P$. Prove that

$$\left| R(f, P) - \int_a^b f \right| \leq c\|P\|(b - a).$$

9. Let $p$ and $n$ be natural numbers with $n \geq 2$. Prove that

$$\sum_{k=1}^{n-1} k^p \leq \frac{n^{p+1}}{p+1} \leq \sum_{k=1}^{n} k^p.$$

(*Hint:* Use an induction argument on $n$.)

10. For a natural number $p$, use the preceding exercise and Corollary 6.12 to prove that

$$\int_0^1 x^p\,dx = \frac{1}{p+1}.$$

11. (Fermat's method for computing $\int_1^b x^\beta\,dx$.) Let $b > 1$ and $\beta \neq -1$. Define $f(x) = x^\beta$ for all $x$ in $[1, b]$. For each natural number $n$, let $P_n = \{x_0, \ldots, x_n\}$ be the partition of $[1, b]$ defined by $x_i = b^{i/n}$ for $0 \leq i \leq n$.
    (a) Show that

$$\sum_{i=1}^{n} f(x_i)(x_i - x_{i-1}) = \frac{b^{1/n} - 1}{b^{1/n}} \left[ \frac{1 - (b^{\beta+1})^{\frac{n+1}{n}}}{1 - b^{\frac{\beta+1}{n}}} - 1 \right].$$

(b) Show that

$$\lim_{n \to \infty} \frac{1 - (b^{1/n})^{\beta+1}}{1 - b^{1/n}} = \beta + 1.$$

(c) Use (a) and (b) to show that

$$\int_1^b x^\beta \, dx = \lim_{n \to \infty} \sum_{i=1}^n f(x_i)(x_i - x_{i-1}) = \frac{b^{\beta+1} - 1}{\beta + 1}.$$

## 6.5 Linearity, Monotonicity, and Additivity over Intervals

**THEOREM 6.17**    **Linearity of the Integral**    Suppose that the functions $f : [a, b] \to \mathbb{R}$ and $g : [a, b] \to \mathbb{R}$ are integrable. Then for any two numbers $\alpha$ and $\beta$, the function $\alpha f + \beta g : [a, b] \to \mathbb{R}$ is integrable and

$$\int_a^b [\alpha f + \beta g] = \alpha \int_a^b f + \beta \int_a^b g. \tag{6.26}$$

**Proof**    We will use the Riemann Sum Convergence Criterion to prove the theorem. Let $\{P_n\}$ be a sequence of partitions of the interval $[a, b]$ such that $\lim_{n \to \infty} \|P_n\| = 0$. For each natural number $n$, let $R(\alpha f + \beta g, \ P_n)$ be a Riemann sum for the function $\alpha f + \beta g : [a, b] \to \mathbb{R}$ based on the partition $P_n$. We must show that

$$\lim_{n \to \infty} R(\alpha f + \beta g, \ P_n) = \alpha \int_a^b f + \beta \int_a^b g. \tag{6.27}$$

Fix a natural number $n$. The Riemann sum $R(\alpha f + \beta g, \ P_n)$ is defined by choosing points in the subintervals induced by the partition $P_n$; define $R(f, P_n)$ and $R(g, P_n)$ to be the Riemann sums for the functions $f : [a, b] \to \mathbb{R}$ and $g : [a, b] \to \mathbb{R}$ obtained by making the same choice of points in the subintervals induced by $P_n$. For such a choice of Riemann sums, we have

$$R(\alpha f + \beta g, \ P_n) = \alpha R(f, \ P_n) + \beta R(g, \ P_n). \tag{6.28}$$

The Riemann Sum Convergence Criterion implies that

$$\lim_{n \to \infty} R(f, P_n) = \int_a^b f \quad \text{and} \quad \lim_{n \to \infty} R(g, P_n) = \int_a^b g. \tag{6.29}$$

The linearity property of convergent sequences, together with (6.28) and (6.29), implies that (6.27) holds. ∎

**THEOREM 6.18**   **Monotonicity of the Integral**   Suppose that the functions $f : [a, b] \to \mathbb{R}$ and $g : [a, b] \to \mathbb{R}$ are integrable and that

$$f(x) \leq g(x) \quad \text{for all } x \text{ in } [a, b].$$

Then

$$\int_a^b f \leq \int_a^b g.$$

**Proof**   Let $\{P_n\}$ be any sequence of partitions of $[a, b]$ such that

$$\lim_{n \to \infty} \|P_n\| = 0.$$

For each natural number $n$, let $R(f, P_n)$ and $R(g, P_n)$ be Riemann sums for the functions $f : [a, b] \to \mathbb{R}$ and $g : [a, b] \to \mathbb{R}$, obtained by making the same choice of points in the subintervals induced by $P_n$. Thus,

$$R(f, P_n) \leq R(g, P_n).$$

The order preservation property of convergent sequences and the Riemann Sum Convergence Criterion imply that

$$\int_a^b f = \lim_{n \to \infty} R(f, P_n) \leq \lim_{n \to \infty} R(g, P_n) = \int_a^b g. \qquad \blacksquare$$

**THEOREM 6.19**   **Additivity over Intervals**   Let $a < c < b$. Then the function $f : [a, b] \to \mathbb{R}$ is integrable if and only if both $f : [a, c] \to \mathbb{R}$ and $f : [c, b] \to \mathbb{R}$ are integrable, in which case

$$\int_a^b f = \int_a^c f + \int_c^b f. \qquad (6.30)$$

**Proof**   First we suppose that the functions $f : [a, c] \to \mathbb{R}$ and $f : [c, b] \to \mathbb{R}$ are integrable. For each natural number $n$, let $P_n'$ be the regular partition of the interval $[a, c]$ into $n$ subintervals, let $P_n''$ be the regular partition of the interval $[c, b]$ into $n$ subintervals, and let $P_n$ be the partition of $[a, b]$ obtained from the union of the partition points of $P_n'$ and $P_n''$; observe that

$$L(f, P_n) = L(f, P_n') + L(f, P_n'')$$

and

$$U(f, P_n) = U(f, P_n') + U(f, P_n'').$$

By the Darboux Sum Integrability Criterion,

$$\lim_{n \to \infty} L(f, P_n') = \lim_{n \to \infty} U(f, P_n') = \int_a^c f$$

and

$$\lim_{n \to \infty} L(f, P_n'') = \lim_{n \to \infty} U(f, P_n'') = \int_c^b f.$$

Thus, by the sum property of convergent sequences of numbers,

$$\lim_{n \to \infty} L(f, P_n) = \lim_{n \to \infty} U(f, P_n) = \int_a^c f + \int_c^b f.$$

It follows from Proposition 6.8 that the function $f: [a, b] \to \mathbb{R}$ is integrable and that

$$\int_a^b f = \int_a^c f + \int_c^b f.$$

Now suppose that the function $f: [a, b] \to \mathbb{R}$ is integrable. It is only necessary to show that the functions $f: [a, c] \to \mathbb{R}$ and $f: [c, b] \to \mathbb{R}$ are integrable, since then we have already verified formula (6.30). We will use the Integrability Criterion to show that $f: [a, c] \to \mathbb{R}$ is integrable. Let $\epsilon > 0$. Since $f: [a, b] \to \mathbb{R}$ is integrable, there is a partition $P$ of the interval $[a, b]$ such that $U(f, P) - L(f, P) < \epsilon$. Define $P^*$ to be the refinement of $P$ obtained by adjoining the point $c$ to $P$. The Refinement Lemma implies that

$$U(f, P^*) - L(f, P^*) \leq U(f, P) - L(f, P) < \epsilon.$$

Let $P'$ be the partition that $P^*$ induces on $[a, c]$. Then

$$U(f, P') - L(f, P') \leq U(f, P^*) - L(f, P^*) < \epsilon.$$

Thus $f: [a, c] \to \mathbb{R}$ is integrable. The proof that $f: [c, b] \to \mathbb{R}$ is integrable is the same. ∎

**COROLLARY 6.20** Suppose that the functions $f: [a, b] \to \mathbb{R}$ and $|f|: [a, b] \to \mathbb{R}$ are integrable. Then

$$\left| \int_a^b f(x)\, dx \right| \leq \int_a^b |f(x)|\, dx. \tag{6.31}$$

**Proof** For all $x$ in $[a, b]$,

$$-|f(x)| \leq f(x) \leq |f(x)|.$$

Thus, using the monotonicity and linearity of integration, it follows that

$$-\int_a^b |f(x)|\, dx \leq \int_a^b f(x)\, dx \leq \int_a^b |f(x)|\, dx,$$

which is equivalent to (6.31). ∎

**EXERCISES**

1. Suppose that the function $f: [a, b] \to \mathbb{R}$ has $f(x) = 0$ except at a finite number of points $z_1, \ldots, z_k$ in $[a, b]$. Define $M = \max_{1 \leq i \leq k} |f(z_i)|$. Prove that for any partition $P$ of $[a, b]$, $|R(f, P)| \leq (K + 1)M \|P\|$. Use this to prove that $\int_a^b f = 0$.

2. Suppose that the function $g: [a, b] \to \mathbb{R}$ is integrable and that the function $h: [a, b] \to \mathbb{R}$ has the property that $h(x) = g(x)$ except at a finite number of points in $[a, b]$. Prove that $h: [a, b] \to \mathbb{R}$ is integrable and that

$$\int_a^b g = \int_a^b h.$$

3.  Suppose that the functions $g: [a, b] \to \mathbb{R}$ and $f: [a, b] \to \mathbb{R}$ are continuous. Prove that $\int_a^b [f + g]^2 \geq 0$. Use this to prove that

$$\int_a^b fg \leq \frac{1}{2}\left[\int_a^b f^2 + \int_a^b g^2\right].$$

    (This is the integral version of Cauchy's Inequality.)

4.  Suppose that the functions $g: [a, b] \to \mathbb{R}$ and $f: [a, b] \to \mathbb{R}$ are continuous. Prove that

$$\int_a^b gf \leq \sqrt{\int_a^b g^2}\sqrt{\int_a^b f^2}.$$

    (This is called the Cauchy-Schwarz Inequality for Integrals.) (*Hint:* For each number $\lambda$, define $p(\lambda) = \int_a^b [f - \lambda g]^2$. Show that $p(\lambda)$ is a quadratic polynomial for which $p(\lambda) \geq 0$ for all $\lambda$ and analyze the discriminant of $p(\lambda)$.)

5.  Use the Cauchy-Schwarz Inequality to verify that

$$\int_0^1 \sqrt{1 + x^4}\, dx \leq \sqrt{\frac{6}{5}}$$

    and

$$\int_0^1 \sqrt{1 + x^3}\, dx \leq \sqrt{\frac{5}{2}}.$$

6.  Suppose that the functions $f: [a, b] \to \mathbb{R}$ and $g: [a, b] \to \mathbb{R}$ are continuous. Prove the following integral version of the Triangle Inequality and the Reverse Triangle Inequality:

$$\int_a^b |f| - \int_a^b |g| \leq \int_a^b |f + g| \leq \int_a^b |f| + \int_a^b |g|.$$

7.  Prove that

$$\frac{2}{\pi}x \leq \sin x \leq x \quad \text{if} \quad 0 \leq x \leq \frac{\pi}{2},$$

    and use this to prove that

$$1 \leq \int_0^{\pi/2} \frac{\sin x}{x}\, dx \leq \frac{\pi}{2}.$$

8.  Let $f: [a, b] \to \mathbb{R}$ and $g: [a, b] \to \mathbb{R}$ be continuous functions having the property that $f(x) \leq g(x)$ for all $x$ in $[a, b]$. Prove that

$$\int_a^b f < \int_a^b g$$

    if and only if there is a point $x_0$ in $[a, b]$ at which $f(x_0) < g(x_0)$.

9.  The Monotonicity Property of the integral implies that if the functions $g: [0, \infty) \to \mathbb{R}$ and $h: [0, \infty) \to \mathbb{R}$ are continuous and $g(x) \leq h(x)$ for all $x \geq 0$, then

$$\int_0^x g \leq \int_0^x h \quad \text{for all } x \geq 0.$$

Use this to show that each of the following inequalities implies its successor:

$$\cos x \le 1 \quad \text{if} \quad x \ge 0$$

$$\sin x \le x \quad \text{if} \quad x \ge 0$$

$$1 - \cos x \le \frac{x^2}{2} \quad \text{if} \quad x \ge 0$$

$$x - \sin x \le \frac{x^3}{6} \quad \text{if} \quad x \ge 0.$$

Thus
$$x - \frac{x^3}{6} \le \sin x \le x \quad \text{if} \quad x \ge 0.$$

# 7

# The Second Fundamental Theorem and Its Consequences

## 7.1 The Second Fundamental Theorem of Calculus

In Section 4.5, we first considered the question of whether, given a function $f: (a, b) \to \mathbb{R}$, there is a differentiable function $F: (a, b) \to \mathbb{R}$ such that

$$F'(x) = f(x) \quad \text{for all } x \text{ in } (a, b).$$

We call this equation the Fundamental Differential Equation (F.D.E.). As we have seen in Section 4.5,

(i) For certain functions $f: (a, b) \to \mathbb{R}$, such as polynomials, we can explicitly solve the F.D.E.

(ii) If the function $f: (a, b) \to \mathbb{R}$ does not have the Intermediate Value Property, then Darboux's Theorem asserts that there is no solution of the F.D.E.

(iii) For many functions $f: (a, b) \to \mathbb{R}$, we are unable to decide whether there is a solution of the F.D.E.

This is clearly not a satisfactory situation.

The importance of the Fundamental Differential Equation became evident in the statement of the First Fundamental Theorem of Calculus. However, quite independently of the problem of explicitly evaluating integrals, this equation is of great importance in mathematics. *In particular, it is the first step in the study of general differential equations.* In this section, we will prove that if the function $f:(a, b) \to \mathbb{R}$ is continuous, then the F.D.E. has a solution for which there is an explicit formula. Before proving this result, we derive two preliminary results that are of independent interest.

**THEOREM 7.1**    **The Mean Value Theorem for Integrals**    Suppose that the function $f:[a, b] \to \mathbb{R}$ is continuous. Then there is a point $x_0$ in the interval $[a, b]$ such that

$$\frac{1}{b - a} \int_a^b f = f(x_0).$$

**FIGURE 7.1**

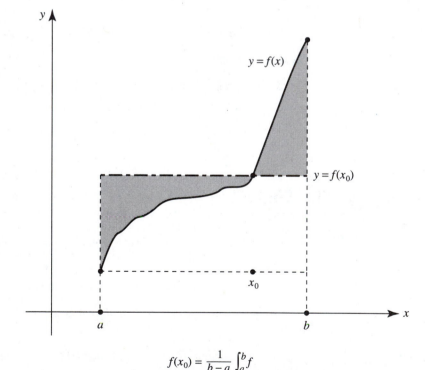

$$f(x_0) = \frac{1}{b - a} \int_a^b f$$

**Proof**    Since the function $f:[a, b] \to \mathbb{R}$ is continuous, we may use the Extreme Value Theorem to choose points $x_m$ and $x_M$ in the interval $[a, b]$ at which $f:[a, b] \to \mathbb{R}$ attains minimum and maximum values, respectively.

The monotonicity property of integration implies that

$$f(x_m)(b-a) \le \int_a^b f(x)\,dx \le f(x_M)(b-a),$$

and so

$$f(x_m) \le \frac{1}{b-a}\int_a^b f(x)\,dx \le f(x_M).$$

Thus, by the Intermediate Value Theorem, there is a point $x_0$ between $x_m$ and $x_M$ such that

$$f(x_0) = \frac{1}{b-a}\int_a^b f. \qquad \blacksquare$$

**PROPOSITION 7.2**

Suppose that the function $f:[a,b] \to \mathbb{R}$ is integrable. Define

$$F(x) = \int_a^x f \quad \text{for all } x \text{ in } [a,b].$$

Then the function $F:[a,b] \to \mathbb{R}$ is continuous.

**Proof** First of all, observe that, according to Theorem 6.19, for each point $x$ in $[a,b]$, the function $f:[a,x] \to \mathbb{R}$ is integrable. Hence, the function $F:[a,b] \to \mathbb{R}$ is properly defined.

Now, since the function $f:[a,b] \to \mathbb{R}$ is integrable, it is bounded. Choose $M > 0$ such that

$$-M \le f(x) \le M \quad \text{for all } x \text{ in } [a,b]. \tag{7.1}$$

We claim that

$$|F(u) - F(v)| \le M|u-v| \quad \text{for all points } u \text{ and } v \text{ in } [a,b]. \tag{7.2}$$

Indeed, let $u$ and $v$ be points in $[a,b]$ with $u < v$. By the additivity over intervals property of the integral,

$$F(v) = \int_a^v f = \int_a^u f + \int_u^v f = F(u) + \int_u^v f,$$

so that

$$F(v) - F(u) = \int_u^v f. \tag{7.3}$$

But (7.1) implies, in particular, that

$$-M \le f(x) \le M \quad \text{if } u \le x \le v,$$

so that, by (7.3) and the monotonicity of the integral,

$$-M(v-u) \le F(v) - F(u) \le M(v-u).$$

Thus, (7.2) holds when $u < v$. Since (7.2) remains unchanged when $u$ and $v$ are interchanged, it also holds when $v < u$.

Finally, (7.2) immediately implies the continuity of the function $F:[a,b] \to \mathbb{R}$.

$\blacksquare$

**EXAMPLE 7.1** Define

$$f(x) = \begin{cases} 2 & \text{if } 0 \leq x \leq 1 \\ x & \text{if } 1 < x \leq 2. \end{cases}$$

Now define

$$F(x) = \int_0^x f \quad \text{if} \quad 0 \leq x \leq 2.$$

By the First Fundamental Theorem of Calculus,

$$F(x) = \begin{cases} 2x & \text{if } 0 \leq x \leq 1 \\ F(1) + \int_1^x f = 3/2 + x^2/2 & \text{if } 1 < x \leq 2. \end{cases}$$

As the preceding theorem predicted, the function $F: [0, 2] \to \mathbb{R}$ is continuous. $\qquad \square$

When we strengthen the assumption in Proposition 7.2 by replacing the integrability of $f: [a, b] \to \mathbb{R}$ with the continuity of $f: [a, b] \to \mathbb{R}$, we obtain one of the most important theorems in mathematics.

**THEOREM 7.3** **The Second Fundamental Theorem of Calculus** Suppose that the function $f: [a, b] \to \mathbb{R}$ is continuous. Define

$$F(x) = \int_a^x f \quad \text{for all } x \text{ in } [a, b].$$

Then

$$F'(x) = f(x) \quad \text{for all } x \text{ in } (a, b).$$

**Proof** We have already verified that the function $F: [a, b] \to \mathbb{R}$ is properly defined and is, in fact, continuous. Let $x_0$ be a point in $(a, b)$. We must show that

$$\lim_{x \to x_0} \frac{F(x) - F(x_0)}{x - x_0} = f(x_0).$$

Let $x$ be a point in $(a, b)$ with $x \neq x_0$. By the additivity over intervals property of the integral,

$$F(x) - F(x_0) = \int_{x_0}^x f \quad \text{if} \quad x > x_0$$

and

$$F(x) - F(x_0) = -\int_x^{x_0} f \quad \text{if} \quad x < x_0.$$

Consequently, applying the Mean Value Theorem for Integrals, we see that we may select a point $c(x)$ between $x_0$ and $x$ such that

$$\frac{F(x) - F(x_0)}{x - x_0} = f(c(x)). \tag{7.4}$$

But the function $f: [a, b] \to \mathbb{R}$ is continuous at $x_0$, so that

$$\lim_{x \to x_0} f(c(x)) = f(x_0).$$

Thus the conclusion follows from (7.4).                                            ∎

   As we have previously mentioned, *the principal importance of the Second Fundamental Theorem of Calculus is that it provides the crucial first step on the road to
studying quite general differential equations.* We shall turn to this aspect in Section 7.2.
Also, as we shall see in Section 7.3, this theorem can be used to verify various classical
techniques for replacing complicated integrals by ones for which we can directly apply
the First Fundamental Theorem of Calculus by inspection. Finally, in Section 7.4, we
will see that the Second Fundamental Theorem is important in analyzing the errors that
arise when we use approximation techniques for estimating integrals.
   In the remainder of this section, we will consider some variations of the Second
Fundamental Theorem.

**COROLLARY**   Suppose that the function $f: [a, b] \to \mathbb{R}$ is continuous. Define
**7.4**

$$H(x) = \int_x^b f \quad \text{for all } x \text{ in } [a, b].$$

Then
$$H'(x) = -f(x) \quad \text{for all } x \text{ in } (a, b).$$

**Proof**   By the additivity over intervals property of the integral,

$$H(x) = \int_a^b f - \int_a^x f = k - F(x) \quad \text{for all } x \text{ in } [a, b],$$

where
$$k = \int_a^b f \quad \text{and} \quad F(x) = \int_a^x f \quad \text{for all } x \text{ in } (a, b).$$

The result now follows from the Second Fundamental Theorem of Calculus.        ∎

   Motivated by the previous result, we extend the meaning of the symbol $\int_a^b f$ as
follows.

**DEFINITION**   *Let c and d be numbers such that c < d. For an integrable function $f: [c, d] \to \mathbb{R}$, we
define*

$$\int_d^c f \equiv -\int_c^d f.$$

With the above definition, the additivity over intervals property of the integral extends
as follows: Let $I$ be an interval and suppose that the function $f: I \to \mathbb{R}$ is continuous.
Then for any three points $x_1$, $x_2$, and $x_3$ in $I$,

$$\int_{x_1}^{x_3} f = \int_{x_1}^{x_2} f + \int_{x_2}^{x_3} f.$$

We leave the proof of this as an exercise.

**COROLLARY
7.5**

Let $I$ be an open interval, and suppose that the function $f: I \to \mathbb{R}$ is continuous. Fix a point $x_0$ in $I$ and define

$$F(x) = \int_{x_0}^{x} f \quad \text{for all } x \text{ in } I.$$

Then the function $F: I \to \mathbb{R}$ is differentiable and

$$F'(x) = f(x) \quad \text{for all } x \text{ in } I.$$

**Proof**    If $x > x_0$, then, according to the Second Fundamental Theorem, $F'(x) = f(x)$. If $x < x_0$, then

$$F(x) = -\int_{x}^{x_0} f,$$

so Corollary 7.4 implies that $F'(x) = f(x)$. Now choose $a$ to be a point in $I$ with $a < x_0$, and define $G(x) = \int_{a}^{x} f$ for $x$ in $I$. Then by the Second Fundamental Theorem of Calculus, $G'(x_0) = f(x_0)$, and since the functions $F: I \to \mathbb{R}$ and $G: I \to \mathbb{R}$ differ by a constant function, $F'(x_0) = f(x_0)$.    ∎

**COROLLARY
7.6**

Let $I$ be an open interval, and suppose that the function $f: I \to \mathbb{R}$ is continuous. Let $J$ be an open interval, and suppose that the function $\varphi: J \to \mathbb{R}$ is differentiable and that $\varphi(J) \subseteq I$. Fix a point $x_0$ in $I$ and define

$$G(x) = \int_{x_0}^{\varphi(x)} f \quad \text{for all } x \text{ in } J.$$

Then the function $G: J \to \mathbb{R}$ is differentiable and

$$G'(x) = f(\varphi(x))\varphi'(x) \quad \text{for all } x \text{ in } J.$$

**Proof**    Define $F(x) = \int_{x_0}^{x} f$ for $x$ in $I$. Then $G = F \circ \varphi: J \to \mathbb{R}$ is the composition of differentiable functions, so the result follows from the Second Fundamental Theorem, as expressed in Corollary 7.5, and the Chain Rule.    ∎

In the Leibnitz notation, the conclusion of Corollary 7.6 may be written as

$$\frac{d}{dx}\left[\int_{x_0}^{\varphi(x)} f(t)\, dt\right] = f(\varphi(x))\varphi'(x) \quad \text{for all } x \text{ in } J.$$

The Second Fundamental Theorem of Calculus was proven under the assumption that the function $f: [a, b] \to \mathbb{R}$ is continuous. In fact, if $f: [a, b] \to \mathbb{R}$ is merely integrable and we define $F(x) = \int_{a}^{x} f$ for all $x$ in $[a, b]$, then $F'(x) = f(x)$ at each point $x$ in $(a, b)$ at which the function $f: [a, b] \to \mathbb{R}$ is continuous (see Exercise 11). Thus, if $f: [a, b] \to \mathbb{R}$ is bounded and $f: (a, b) \to \mathbb{R}$ is continuous, then the function $F: [a, b] \to \mathbb{R}$ defined by $F(x) = \int_{a}^{x} f$ for all $x$ in $[a, b]$ has the property that $F: [a, b] \to \mathbb{R}$ is continuous and $F'(x) = f(x)$ at each point $x$ in $(a, b)$. This slight extension of the Second Fundamental Theorem will be useful in Section 7.3.

**EXERCISES**

1.  Calculate the following derivatives:

    (a)  $\dfrac{d}{dx}\left(\displaystyle\int_0^x x^2 t^2 \, dt\right)$

    (b)  $\dfrac{d}{dx}\left(\displaystyle\int_1^{e^x} \ln t \, dt\right)$

    (c)  $\dfrac{d}{dx}\left(\displaystyle\int_{-x}^x e^{t^2} \, dt\right)$

    (d)  $\dfrac{d}{dx}\left(\displaystyle\int_1^x \cos(x+t) \, dt\right)$

2.  For each of the following integrable functions $f:[a,b] \to \mathbb{R}$ define

    $$F(x) = \int_a^x f(t)\,dt \quad \text{for all } x \text{ in } [a,b]$$

    and find a formula for $F(x)$, $a \le x \le b$, that does not involve integrals.

    (a)  $f:[1,4] \to \mathbb{R}$ defined by

    $$f(x) = \begin{cases} 2 & \text{if } 1 \le x \le 3 \\ 6 & \text{if } 3 < x \le 4. \end{cases}$$

    (b)  $f:[0,2] \to \mathbb{R}$ defined by

    $$f(x) = \begin{cases} x^2 & \text{if } 0 \le x \le 1 \\ x & \text{if } 1 < x \le 2. \end{cases}$$

    (c)  $f:[-1,1] \to \mathbb{R}$ defined by

    $$f(x) = \begin{cases} x & \text{if } -1 \le x < 0 \\ x+1 & \text{if } 0 \le x \le 1. \end{cases}$$

3.  Suppose that the function $f:\mathbb{R} \to \mathbb{R}$ is differentiable. Define the function $H:\mathbb{R} \to \mathbb{R}$ by

    $$H(x) = \int_{-x}^x [f(t) + f(-t)]\,dt \quad \text{for all } x.$$

    Find $H''(x)$.

4.  Suppose that the function $f:\mathbb{R} \to \mathbb{R}$ has a continuous second derivative. Prove that

    $$f(x) = f(0) + f'(0)x + \int_0^x (x-t)f''(t)\,dt \quad \text{for all } x.$$

    (*Hint:* Use the Identity Criterion.)

5.  Suppose that the function $f:\mathbb{R} \to \mathbb{R}$ is continuous. Define

    $$G(x) = \int_0^x (x-t)f(t)\,dt \text{ for all } x.$$

    Prove that $G''(x) = f(x)$ for all $x$.

6.  Define

    $$F(x) = \int_1^x \frac{1}{2\sqrt{t}-1}\,dt \quad \text{for all } x \ge 1.$$

    Prove that if $c > 0$, then there is a unique solution to the equation

    $$F(x) = c, \quad x > 1.$$

7. Suppose that the continuous function $f: [a, b] \to \mathbb{R}$ has the property that $\int_{x_1}^{x_2} f = 0$ whenever $a \leq x_1 < x_2 \leq b$. Prove that $f(x) = 0$ for all $x$ in $[a, b]$.

8. Show that the Mean Value Theorem for Integrals does not hold if we replace the assumption that $f: [a, b] \to \mathbb{R}$ is continuous with the assumption that $f: [a, b] \to \mathbb{R}$ is integrable.

9. For numbers $a_1, \ldots, a_n$ define $p(x) = a_1 x + a_2 x^2 + \cdots + a_n x^n$ for all $x$. Suppose that

$$\frac{a_1}{2} + \frac{a_2}{3} + \cdots + \frac{a_n}{n+1} = 0.$$

Prove that there is some point $x$ in the interval $(0, 1)$ such that $p(x) = 0$.

10. Show that the conclusion of the Mean Value Theorem for Integrals may be strengthened so that we can choose the point $x_0$ to be in $(a, b)$, not just in $[a, b]$.

11. The Second Fundamental Theorem of Calculus has a somewhat more general form than we have stated: *For an integrable function $f: [a, b] \to \mathbb{R}$, we define $F(x) = \int_a^x f$ for all $x$ in $[a, b]$. Then at each point $x_0$ in $(a, b)$ at which the function $f: [a, b] \to \mathbb{R}$ is continuous, the function $F: [a, b] \to \mathbb{R}$ is differentiable and $F'(x_0) = f(x_0)$.* Use the monotonicity property of integration to prove this.

12. Let the function $f: [a, b] \to \mathbb{R}$ be continuous. Suppose that the function $F: [a, b] \to \mathbb{R}$ is continuous, that $F: (a, b) \to \mathbb{R}$ is differentiable, and that $F'(x) = f(x)$ for all $x$ in $(a, b)$. Use the Second Fundamental Theorem of Calculus to prove that

$$\frac{d}{dx} \left[ F(x) - \int_a^x f \right] = 0 \quad \text{for all } x \text{ in } (a, b),$$

and from this derive a new proof of the First Fundamental Theorem of Calculus in the case when $f: [a, b] \to \mathbb{R}$ is assumed to be continuous, not just integrable.

13. Suppose that the functions $f: [a, b] \to \mathbb{R}$ and $g: [a, b] \to \mathbb{R}$ are continuous and $\alpha$ and $\beta$ are any real numbers. Define

$$H(x) = \int_a^x [\alpha f + \beta g] - \alpha \int_a^x [f] - \beta \int_a^x [g] \quad \text{for all } x \text{ in } [a, b].$$

Prove that $H'(x) = 0$ for all $x$ in $(a, b)$ and use this to provide another proof of the linearity of the integral provided that the functions are assumed to be continuous, not just integrable.

## 7.2 The Existence of Solutions of Differential Equations

**PROPOSITION 7.7** Let $I$ be an open interval containing the point $x_0$, and suppose that the function $f: I \to \mathbb{R}$ is continuous. For any number $y_0$, the differential equation

$$\begin{cases} F'(x) = f(x) & \text{for all } x \text{ in } I \\ F(x_0) = y_0 \end{cases}$$

has a unique solution $F: I \to \mathbb{R}$ given by the formula

$$F(x) = y_0 + \int_{x_0}^x f \quad \text{for all } x \text{ in } I.$$

**Proof**    By definition, $F(x_0) = y_0$. The Second Fundamental Theorem of Calculus, as expressed in Corollary 7.5, implies that $F'(x) = f(x)$ for all $x$ in $I$. Thus $F: I \to \mathbb{R}$ is a solution of the differential equation. The Identity Criterion (see Section 4.3) implies that there can be only one solution.    ∎

Recall that in Section 5.1 we *assumed* that there was a differentiable function $F: (0, \infty) \to \mathbb{R}$ that was a solution of the differential equation

$$\begin{cases} F'(x) = 1/x & \text{for all } x > 0 \\ F(1) = 0. \end{cases} \tag{7.5}$$

The Identity Criterion implies that there can be at most one solution. We can now *prove* that there is a solution.

**PROPOSITION 7.8**

Define

$$F(x) = \int_1^x \frac{1}{t} \, dt \quad \text{for all } x > 0.$$

Then the function $F: (0, \infty) \to \mathbb{R}$ is the solution of the differential equation (7.5).

**Proof**    By definition, $F(1) = 0$, and Corollary 7.5 implies that $F'(x) = 1/x$ for all $x > 0$.    ∎

We defined the natural logarithm $\ln: (0, \infty) \to \mathbb{R}$ to be the unique solution of the differential equation (7.5) provided that there is a solution. We now have the explicit integral formula

$$\ln x = \int_1^x \frac{1}{t} \, dt \quad \text{for all } x > 0.$$

We now consider the following prototypical example of a linear differential equation depending only on first derivatives.

*Given a continuous function $h: \mathbb{R} \to \mathbb{R}$ and numbers $a$, $x_0$, and $y_0$, find a differentiable function $F: \mathbb{R} \to \mathbb{R}$ such that*

$$\begin{cases} F'(x) + aF(x) = h(x) & \text{for all } x \\ F(x_0) = y_0. \end{cases} \tag{7.6}$$

**PROPOSITION 7.9**

There is at most one solution of the differential equation (7.6).

**Proof**    Suppose that the functions $F_1: \mathbb{R} \to \mathbb{R}$ and $F_2: \mathbb{R} \to \mathbb{R}$ are both solutions of (7.6). We will show that the difference function $g = F_1 - F_2: \mathbb{R} \to \mathbb{R}$ is identically 0, and this will prove that there is at most one solution of (7.6).

First, observe that

$$g'(x) + ag(x) = 0 \quad \text{for all } x. \tag{7.7}$$

Indeed,

$$
\begin{aligned}
g'(x) + ag(x) &= F_1'(x) - F_2'(x) + a[F_1(x) - F_2(x)] \\
&= [F_1'(x) + aF_1(x)] - [F_2'(x) + aF_2(x)] \\
&= h(x) - h(x) \\
&= 0 \quad \text{for all } x.
\end{aligned}
$$

According to Proposition 5.5,

$$
g(x) = g(0)e^{-ax} \quad \text{for all } x.
$$

But $g(x_0) = 0$, so $g(0) = 0$ and hence $g(x) = 0$ for all $x$. ∎

**THEOREM 7.10** Suppose that the function $h: \mathbb{R} \to \mathbb{R}$ is continuous. Then the differential equation (7.6) has precisely one solution, given by the formula

$$
F(x) = y_0 e^{-a(x-x_0)} + e^{-ax} \int_{x_0}^{x} e^{at} h(t) \, dt \quad \text{for all } x. \tag{7.8}
$$

**Proof** Proposition 7.9 asserts that there is at most one solution of (7.6). It remains to prove that formula (7.8) defines a solution of (7.6), and to do this we need to compute $F': \mathbb{R} \to \mathbb{R}$.

Using the Chain Rule, the product rule for differentiation, and the Second Fundamental Theorem of Calculus, we have

$$
\begin{aligned}
F'(x) &= -ay_0 e^{-a(x-x_0)} - ae^{-ax} \int_{x_0}^{x} e^{at} h(t) \, dt + e^{-ax} e^{ax} h(x) \\
&= -aF(x) + h(x) \quad \text{for all } x,
\end{aligned}
$$

so that $\qquad\qquad F'(x) + aF(x) = h(x) \quad$ for all $x$. ∎

**EXAMPLE 7.2** Find the unique solution of the equation

$$
\begin{cases}
F'(x) - F(x) = x & \text{for all } x \\
F(0) = 2.
\end{cases}
$$

According to Theorem 7.10, the unique solution is given by the formula

$$
F(x) = 2e^x + e^x \int_{0}^{x} e^{-t} t \, dt \quad \text{for all } x.
$$

But the integral in this formula may be evaluated by using the First Fundamental Theorem of Calculus. After doing so, we obtain

$$
F(x) = 3e^x - x - 1 \quad \text{for all } x. \qquad \square
$$

1. Find the unique solution of each of the following differential equations:

   (a) $\begin{cases} F'(x) + F(x) = x & \text{for all } x \\ F(0) = 1. \end{cases}$

   (b) $\begin{cases} F'(x) + 4F(x) = e^x & \text{for all } x \\ F(2) = 31. \end{cases}$

   (c) $\begin{cases} F'(x) + F(x) = x^2 & \text{for all } x \\ F(0) = -1. \end{cases}$

2. For numbers $c$ and $a$, consider the differential equation

   $$\begin{cases} F'(x) = c(a - F(x)) & \text{for all } x. \\ F(0) = 0. \end{cases}$$

   Prove that the unique solution is given by the formula

   $$F(x) = a(1 - e^{-cx}) \quad \text{for all } x.$$

3. For an open interval $I$ containing 0 and a function $f: I \to \mathbb{R}$ that has a continuous derivative, prove that

   $$f(x) = f(0) + \int_0^x f'(t)\, dt \quad \text{for all } x \text{ in } I.$$

   Use this formula to obtain explicit integral representations of the arcsine, the arccosine, and the arctangent.

4. Prove that

   $$\int_1^{12} \frac{1}{x}\, dx = 2 \int_1^2 \frac{1}{x}\, dx + \int_1^3 \frac{1}{x}\, dx.$$

5. Suppose that the function $f: \mathbb{R} \to \mathbb{R}$ is continuous and that

   $$f(x) = \int_0^x f(t)\, dt \quad \text{for all } x.$$

   Prove that $f(x) = 0$ for all $x$.

6. Multiply both sides of the first line of (7.6) by $e^{ax}$, and observe that

   $$\frac{d}{dx}(e^{ax} F(x)) = e^{ax} h(x) \quad \text{for all } x.$$

   Use this to motivate formula (7.8), and also to give another proof of Theorem 7.10.

7. Suppose that the function $g: \mathbb{R} \to \mathbb{R}$ is continuous and that $g(x) > 0$ for all $x$. Define

   $$h(x) = \int_0^x \frac{1}{g(t)}\, dt \quad \text{for all } x,$$

   and let $J = h(\mathbb{R})$. Prove that if $f: J \to \mathbb{R}$ is the inverse of $h: \mathbb{R} \to \mathbb{R}$, then $f: J \to \mathbb{R}$ is a solution of the differential equation

   $$\begin{cases} f'(x) = g(f(x)) & \text{for all } x \text{ in } J \\ f(0) = 0. \end{cases}$$

## 7.3 The Verification of Two Classical Integration Methods

For a continuous function $f:[a, b] \to \mathbb{R}$ such that its restriction $f:(a, b) \to \mathbb{R}$ has a continuous bounded derivative, using the First Fundamental Theorem of Calculus we obtain the formula

$$\int_a^b f'(x)\, dx = f(b) - f(a).$$

This formula may also be derived from the Second Fundamental Theorem of Calculus (see Exercise 12 of Section 7.1).

**THEOREM 7.11** **Integration by Parts** Suppose that the functions $h:[a, b] \to \mathbb{R}$ and $g:[a, b] \to \mathbb{R}$ are continuous and that both $h:(a, b) \to \mathbb{R}$ and $g:(a, b) \to \mathbb{R}$ have continuous bounded derivatives. Then

$$\int_a^b h(x)g'(x)\, dx = h(b)g(b) - h(a)g(a) - \int_a^b g(x)h'(x)\, dx. \qquad (7.9)$$

**Proof** The product function $hg:[a, b] \to \mathbb{R}$ is continuous, $hg:(a, b) \to \mathbb{R}$ is differentiable, and according to the product rule for derivatives,

$$(hg)'(x) = h(x)g'(x) + g(x)h'(x) \quad \text{for all } x \text{ in } (a, b).$$

By the integration formula that preceded the statement of this theorem, we have

$$\int_a^b [hg' + gh'] = \int_a^b (hg)' = h(b)g(b) - h(a)g(a), \qquad (7.10)$$

and, on the other hand, the linearity property of the integral implies that

$$\int_a^b (hg' + gh') = \int_a^b hg' + \int_a^b gh'. \qquad (7.11)$$

Formula (7.9) follows from (7.10) and (7.11). ∎

**EXAMPLE 7.3** By reformulating

$$\int_0^1 xe^x\, dx \quad \text{as} \quad \int_0^1 x\frac{d}{dx}(e^x)\, dx,$$

$$\int_0^\pi x\cos x\, dx \quad \text{as} \quad \int_0^\pi x\frac{d}{dx}(\sin x)\, dx,$$

and

$$\int_1^2 \ln x\, dx \quad \text{as} \quad \int_1^2 \ln x\frac{d}{dx}(x)\, dx,$$

each of the integrals on the left-hand side may be evaluated using integration by parts and the First Fundamental Theorem of Calculus.                                                                ☐

**THEOREM 7.12**   **Integration by Substitution**   Let the function $f: [a, b] \to \mathbb{R}$ be continuous. Suppose that the function $g: [c, d] \to \mathbb{R}$ is also continuous, that $g: (c, d) \to \mathbb{R}$ has a bounded, continuous derivative, and, moreover, that the image of $g: (c, d) \to \mathbb{R}$ is in the interval $(a, b)$. Then

$$\int_c^d f(g(x))g'(x)\, dx = \int_{g(c)}^{g(d)} f(x)\, dx. \tag{7.12}$$

**Proof**   Define the function $H: [c, d] \to \mathbb{R}$ by

$$H(x) = \int_c^x (f \circ g)g' - \int_{g(c)}^{g(x)} f \quad \text{for all } x \text{ in } [c, d].$$

Since the composition of continuous functions is continuous, it follows from Proposition 7.2 that the function $H: [c, d] \to \mathbb{R}$ is continuous. Moreover, from the Second Fundamental Theorem of Calculus and Corollary 7.6, we see that

$$H'(x) = f(g(x))g'(x) - f(g(x))g'(x) = 0 \quad \text{for all } x \text{ in } (c, d).$$

The Identity Criterion implies that the function $H: [c, d] \to \mathbb{R}$ is constant. In particular, since $H(c) = 0$, $H(d) = 0$ also; that is, formula (7.12) holds.                                    ■

For an integrable function $f: [a, b] \to \mathbb{R}$ having the property that $f(x) \geq 0$ for all $x$ in $[a, b]$, we *define* the area bounded by the graph of $f: [a, b] \to \mathbb{R}$ and the $x$-axis to be the integral $\int_a^b f(x)\, dx$. Of course, the integral itself was defined in order to make this definition reasonable.

Recall that in Section 5.2 we provided an analytic definition of the number $\pi$ as follows: The function $\cos x = f(x)$ was defined to be the unique solution of a differential equation

$$\begin{cases} f''(x) + f(x) = 0 & \text{for all } x \\ f(0) = 1 \quad \text{and} \quad f'(0) = 0. \end{cases}$$

Then the number $\pi/2$ was defined to be the smallest positive number at which $\cos x = 0$. Of course, the number $\pi$ has the geometric significance of being the area of a circle of unit radius. The next formula reconciles the analytic definition of $\pi$ with its usual geometric significance.

**PROPOSITION
7.13**

$$\frac{\pi}{4} = \int_0^1 \sqrt{1 - x^2}\, dx.$$

**FIGURE 7.2**

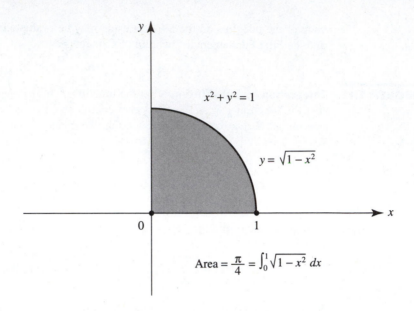

$$\text{Area} = \frac{\pi}{4} = \int_0^1 \sqrt{1-x^2}\, dx$$

**Proof**   First, observe that $\sqrt{1-\sin^2 x} = \cos x$, for $0 \le x \le \pi/2$, since $\cos x \ge 0$ if $0 \le x \le \pi/2$, and $\sin^2 x + \cos^2 x = 1$. Moreover, $\cos^2 x = (1 + \cos 2x)/2$ for all $x$.

Define $g(x) = \sin x$ for $0 \le x \le \pi/2$. Then we may apply Theorem 7.12, together with the two preceding trigonometric identities, to see that since $\sin 0 = 0$ and $\sin \pi/2 = 1$,

$$\int_0^1 \sqrt{1-x^2}\, dx = \int_0^{\pi/2} \sqrt{1-\sin^2 x}\, \cos x\, dx$$

$$= \int_0^{\pi/2} \left[ \frac{1 + \cos 2x}{2} \right] dx.$$

Since

$$\frac{d}{dx}\left( \frac{x}{2} + \frac{\sin 2x}{4} \right) = \frac{1 + \cos 2x}{2} \quad \text{for} \quad 0 \le x \le \frac{\pi}{2},$$

we may apply the First Fundamental Theorem to conclude that

$$\int_0^{\pi/2} \left[ \frac{1 + \cos 2x}{2} \right] dx = \left. \left( \frac{x}{2} + \frac{\sin 2x}{4} \right) \right|_{x=0}^{x=\pi/2} = \frac{\pi}{4}. \qquad \blacksquare$$

**EXAMPLE 7.4**   Consider the integral

$$\int_0^1 e^{x^2}\, dx.$$

Define $f(x) = e^{x^2}$ for all $x$ in $[0, 1]$. If we attempt to write $f : [0, 1] \to \mathbb{R}$ as $f = hg' : [0, 1] \to \mathbb{R}$ so that $\int_0^1 gh'$ can be calculated, we will find that we cannot do so.

Similarly, if we try substitution, we quickly encounter difficulties in choosing a function $g: [c, d] \to \mathbb{R}$ that will make $\int_c^d (f \circ g)g'$ easy to integrate.                    □

The examples we have considered illustrate both the power and the limitations of the techniques of integration by parts and substitution. The object, of course, is to replace one integration problem with another in which we can directly, by inspection, apply the First Fundamental Theorem of Calculus. When we cannot make such a simplification, we need to develop methods for *estimating* integrals. We will turn to this in Section 7.4.*

---

**EXERCISES**

1.   Evaluate the following integrals:

   (a) $\displaystyle\int_1^2 x e^{x^2}\, dx$                    (b) $\displaystyle\int_0^1 (1-x)^2 \sqrt{2+x}\, dx$

   (c) $\displaystyle\int_2^3 x^3 e^{x^2}\, dx$                    (d) $\displaystyle\int_2^\pi x^2 \cos x\, dx$

2.   Evaluate the following integrals:

   (a) $\displaystyle\int_1^e (\ln x)^2\, dx$                    (b) $\displaystyle\int_4^5 \frac{1+x}{1-x}\, dx$

   (c) $\displaystyle\int_4^9 \frac{1}{1-x^2}\, dx$                    (d) $\displaystyle\int_3^4 \left( \frac{1}{x^2 - 2x} + \frac{1}{1+\sqrt{x}} \right) dx$

3.   Prove that for any two natural numbers $n$ and $m$,

$$\int_0^1 x^m (1-x)^n\, dx = \int_0^1 (1-x)^m x^n\, dx.$$

4.   Suppose that the function $f: \mathbb{R} \to \mathbb{R}$ has a continuous second derivative. Fix a number $a$. Prove that

$$\int_a^x f''(t)(x-t)\, dt = -(x-a)f'(a) + f(x) - f(a)    \text{ for all } x.$$

5.   Suppose that the function $f: \mathbb{R} \to \mathbb{R}$ has a continuous second derivative. Prove that for any two numbers $a$ and $b$,

$$\int_a^b x f''(x)\, dx = b f'(b) + f(a) - a f'(a) - f(b).$$

6.   Prove that the area of a circle of radius $r$ is $\pi r^2$.
7.   Calculate the three integrals in Example 7.3.

---

*There is a precise way to assert and prove that for certain functions $f: \mathbb{R} \to \mathbb{R}$, it is not possible to find an "elementary" function $F: \mathbb{R} \to \mathbb{R}$ such that $F'(x) = f(x)$ for all $x$. Unfortunately, to do so requires a background in modern algebra that is outside the scope of this book; see the article "Integration in Finite Terms" by Maxwell Rosenlicht in the *American Mathematical Monthly*, Nov. 1972.

8.  Suppose that the function $f:[0, \infty) \to \mathbb{R}$ is continuous and strictly increasing, and that $f:(0, \infty) \to \mathbb{R}$ is differentiable. Moreover, assume $f(0) = 0$. Consider the formula

$$\int_0^x f + \int_0^{f(x)} f^{-1} = xf(x) \quad \text{for all } x \geq 0.$$

Provide a geometric interpretation of this formula in terms of areas. Then prove this formula. (*Hint:* Differentiate the formula and apply the Identity Criterion.)

9.  Suppose that the function $f:[0, \infty) \to \mathbb{R}$ is continuous and strictly increasing, with $f(0) = 0$ and $f([0, \infty)) = [0, \infty)$. Then define

$$F(x) = \int_0^x f \quad \text{and} \quad G(x) = \int_0^x f^{-1} \quad \text{for all } x \geq 0.$$

(a)  Prove Young's Inequality:

$$ab \leq F(a) + G(b) \quad \text{for all } a \geq 0 \text{ and } b \geq 0.$$

(*Hint:* A sketch will help, as will the formula in Exercise 8.)

(b)  Now use Young's Inequality with $f(x) = x^{p-1}$ for all $x \geq 0$, and $p > 1$ fixed, to prove that if the number $q$ is chosen to have the property that $1/p + 1/q = 1$, then

$$ab \leq \frac{a^p}{p} + \frac{b^q}{q} \quad \text{for all } a \geq 0 \text{ and } b \geq 0.$$

---

## 7.4  The Approximation of Integrals

There are many functions $f:[a, b] \to \mathbb{R}$ that are integrable, but it is not possible by substitution, by integration by parts, or by any other device to reduce the calculation of $\int_a^b f(x)\,dx$ to an application by inspection of the First Fundamental Theorem of Calculus. In such cases, we approximate the value of the integral and we specify a bound for the error that has arisen in the approximation procedure.

The idea is to take a partition $P = \{x_0, \dots, x_n\}$ of $[a, b]$. Then for each index $i$ with $1 \leq i \leq n$, we approximate

$$\int_{x_{i-1}}^{x_i} f(x)\,dx \tag{7.13}$$

by some $A_i$ and define $E_i$ by

$$E_i \equiv \int_{x_{i-1}}^{x_i} f(x)\,dx - A_i. \tag{7.14}$$

Define $A \equiv \sum_{i=1}^n A_i$ and $E \equiv \sum_{i=1}^n E_i$, so that

$$\int_a^b f(x)\,dx - A = E.$$

The $E_i'$s are referred to as the *local errors*; $E$ is called the *global error*. Once we have estimated the local errors, an estimate for the global error usually follows easily. We will look at two approximation methods, the Trapezoid Rule and Simpson's Rule.

For the Trapezoid Rule, consider a partition $P = \{x_0, \ldots, x_n\}$ of the interval $[a, b]$. For each index $i$ with $1 \leq i \leq n$, we approximate the integral $\int_{x_{i-1}}^{x_i} f(x)\, dx$ by

$$A_i \equiv \frac{1}{2}(x_i - x_{i-1})(f(x_{i-1}) + f(x_i)).$$

If $f(x_i) \geq 0$ and $f(x_{i-1}) \geq 0$, the above $A_i$ is simply the area of the trapezoid having vertices $(x_{i-1},\ 0)$, $(x_{i-1},\ f(x_{i-1}))$, $(x_i,\ f(x_i))$, and $(x_i,\ 0)$.

We first estimate the local error.

**THEOREM 7.14**   **The Local Error for the Trapezoid Rule**   Suppose that the function $f : [c, d] \to \mathbb{R}$ is continuous and that its restriction $f : (c, d) \to \mathbb{R}$ has two derivatives. Then there is a point $\zeta$ in the open interval $(c, d)$ at which

$$\int_c^d f(x)\, dx - \frac{(d - c)(f(c) + f(d))}{2} = \frac{-(d - c)^3}{12} f''(\zeta). \tag{7.15}$$

**FIGURE 7.3**

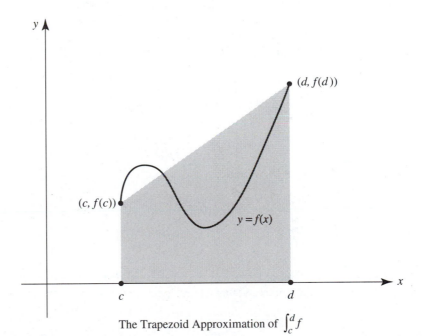

The Trapezoid Approximation of $\int_c^d f$

**Proof**   Let $m = (d + c)/2$ and $t_0 = (d - c)/2$, and define an auxiliary function $E : [0, t_0] \to \mathbb{R}$ by

$$E(t) = \left[ \int_{m-t}^{m+t} f(x)\, dx \right] - t[f(m + t) + f(m - t)] \quad \text{for} \quad 0 \leq t \leq t_0.$$

Since $m + t_0 = d$ and $m - t_0 = c$, the left-hand side of (7.15) is precisely $E(t_0)$. Now define another function $H: [0, t_0] \to \mathbb{R}$ by

$$H(t) = E(t) - \left(\frac{t}{t_0}\right)^3 E(t_0) \quad \text{for} \quad 0 \leq t \leq t_0.$$

It follows from Proposition 7.2 that the function $E: [0, t_0] \to \mathbb{R}$ is continuous, so the function $H: [0, t_0] \to \mathbb{R}$ is also continuous. Moreover, from the Second Fundamental Theorem of Calculus and the Chain Rule, it follows that

$$E'(t) = f(m + t) + f(m - t) - [f(m + t) + f(m - t)] - t[f'(m + t) - f'(m - t)]$$
$$= -t[f'(m + t) - f'(m - t)] \quad \text{for} \quad 0 < t < t_0.$$

Hence

$$H'(t) = -t[f'(m + t) - f'(m - t)] - \frac{3t^2}{t_0^3} E(t_0) \quad \text{for} \quad 0 < t < t_0.$$

It is clear that $H(0) = H(t_0) = 0$. We may apply Rolle's Theorem to the function $H: [0, t_0] \to \mathbb{R}$ to choose a point $t_*$ in $(0, t_0)$ at which $H'(t_*) = 0$; that is,

$$-t_*[f'(m + t_*) - f'(m - t_*)] - \frac{3t_*^2}{t_0^3} E(t_0) = 0.$$

Now apply the Mean Value Theorem to the function $f': [m - t_*, m + t_*] \to \mathbb{R}$ and choose a point $\zeta$ in $(m - t_*, m + t_*)$ at which

$$f'(m + t_*) - f'(m - t_*) = 2t_* f''(\zeta).$$

Substituting the last equality in the preceding equation, we arrive at

$$2t_*^2 \left[ f''(\zeta) + \frac{3}{2t_0^3} E(t_0) \right] = 0.$$

But $t_* \neq 0$, so

$$E(t_0) = -\frac{2t_0^3}{3} f''(\zeta) = \frac{-(d - c)^3}{12} f''(\zeta) \qquad \blacksquare$$

---

**THEOREM 7.15**

**The Global Error for the Trapezoid Rule** Suppose that the function $f: [a, b] \to \mathbb{R}$ is continuous and that its restriction $f: (a, b) \to \mathbb{R}$ has a bounded second derivative. For a partition $P = \{x_0, \ldots, x_n\}$ of the interval $[a, b]$,

$$\int_a^b f = \sum_{i=1}^n \frac{(x_i - x_{i-1})(f(x_{i-1}) + f(x_i))}{2} + E$$

with

$$|E| \leq \frac{M \|P\|^2 (b - a)}{12}, \tag{7.16}$$

where

$$M \equiv \sup\{|f''(x)| \mid x \text{ in } (a, b)\}.$$

**Proof**   For each index $i$ such that $1 \le i \le n$, we may apply the local error estimate for the Trapezoid Rule to choose a point $\zeta_i$ in $(x_{i-1}, x_i)$ at which

$$\int_{x_{i-1}}^{x_i} f = \frac{(x_i - x_{i-1})(f(x_i) + f(x_{i-1}))}{2} - \frac{(x_i - x_{i-1})^3 f''(\zeta_i)}{12}.$$

Sum these $n$ equalities to see that the first formula in (7.16) holds, where

$$E = -\sum_{i=1}^{n} \frac{(x_i - x_{i-1})^3 f''(\zeta_i)}{12}.$$

However, by the Triangle Inequality and the definitions of $\|P\|$ and $M$,

$$|E| \le \frac{\|P\|^2}{12} \sum_{i=1}^{n} (x_i - x_{i-1}) |f''(\zeta_i)|$$

$$\le \frac{\|P\|^2 M}{12} \sum_{i=1}^{n} (x_i - x_{i-1}) = \frac{M\|P\|^2 (b-a)}{12}. \qquad \blacksquare$$

**EXAMPLE 7.5**   We apply the Trapezoid Rule to estimate $\ln 2 = \int_1^2 1/t \, dt$. Define $f(t) = 1/t$ for $1 \le t \le 2$, and note that

$$0 \le f''(t) \le 2 \quad \text{for} \quad 1 \le t \le 2.$$

For a natural number $n$, let $P_n$ be the partition obtained by dividing the interval $[1, 2]$ into $n$ subintervals of equal length. Then (see Exercise 3),

$$\ln 2 = \frac{1}{n}\left[\frac{1}{2} + \frac{n}{n+1} + \frac{n}{n+2} + \cdots + \frac{n}{2n-1} + \frac{1}{4}\right] + E_n,$$

where
$$0 \le E_n \le \frac{1}{6n^2}.$$

Taking $n = 10$, a brief calculation yields .69 as a lower approximation of $\ln 2$, and the error is at most $1/600$; thus, $.69 < \ln 2 < .692$.   $\square$

From the Local Error Estimate for the Trapezoid Rule, we note that the Trapezoid Rule gives the *exact value* of the integral when $f: \mathbb{R} \to \mathbb{R}$ has a line as its graph. Moreover, when

$$f''(x) \ge 0 \quad \text{for all } x \text{ in } (a, b),$$

which means that the function $f: [a, b] \to \mathbb{R}$ is convex, the Trapezoid Rule gives an *upper* approximation to the integral. Each of these observations is clear geometrically.

The second approximation method that we will consider is called Simpson's Rule. Though it is not as easy to motivate geometrically, this rule is, in general, more accurate than the Trapezoid Rule. Given an integrable function $f: [a, b] \to \mathbb{R}$ and a partition $P = \{x_0, \ldots, x_n\}$ of the interval $[a, b]$, for each index $i$ with $1 \le i \le n$, Simpson's Rule approximates the integral $\int_{x_{i-1}}^{x_i} f$ by

$$A_i \equiv \frac{(x_i - x_{i-1})}{6}\left[f(x_{i-1}) + 4f\left(\frac{x_{i-1} + x_i}{2}\right) + f(x_i)\right]. \qquad (7.17)$$

One way to compare approximation methods is to see for what functions the approximation of the integral agrees precisely with the value of the integral. For the Trapezoid Rule, there is exact agreement provided that the function $f\colon [a, b] \to \mathbb{R}$ is a polynomial of degree less than 2; that is, the graph of $f\colon [a, b] \to \mathbb{R}$ is a line. We will show that for Simpson's Rule there is exact agreement provided that the function $f\colon [a, b] \to \mathbb{R}$ is a polynomial of degree less than 4.*

In order to derive the local error estimate for Simpson's Rule it is first convenient to note the following slight extension of Rolle's Theorem.

**THEOREM 7.16** **A Generalized Rolle's Theorem**  Suppose that the function $g\colon [c, d] \to \mathbb{R}$ is continuous. For a natural number $n$, suppose that its restriction $g\colon (c, d) \to \mathbb{R}$ has $n + 1$ derivatives. Let $x_0$ be a point in $(c, d)$ at which

$$g(x_0) = g'(x_0) = \cdots = g^{(n)}(x_0) = 0.$$

Then for each point $x \neq x_0$ in $[c, d]$ at which $g(x) = 0$, there is a point $\zeta$ strictly between $x_0$ and $x$ at which

$$g^{(n+1)}(\zeta) = 0.$$

**Proof**  We will assume that $x > x_0$. If we apply Rolle's Theorem to the function $g\colon [x_0, x] \to \mathbb{R}$, we can choose a point $z_1$ in $(x_0, x)$ at which $g'(z_1) = 0$. Now apply Rolle's Theorem to the function $g'\colon [x_0, z_1] \to \mathbb{R}$ to choose a point $z_2$ in $(x_0, z_1)$ at which $g''(z_2) = 0$. By continuing this procedure $n$ times, we find a point $z_{n+1} = \zeta$ in $(x_0, x)$ at which $g^{n+1}(\zeta) = 0$. ∎

**THEOREM 7.17** **The Local Error for Simpson's Rule**  Suppose that the function $f\colon [c, d] \to \mathbb{R}$ is continuous and that its restriction $f\colon (c, d) \to \mathbb{R}$ has four derivatives. Then there is a point $\zeta$ in the open interval $(c, d)$ at which

$$\int_c^d f(x)\, dx - \frac{(d - c)}{6}\left[ f(c) + 4f\left(\frac{c + d}{2}\right) + f(d) \right] = -\frac{1}{2880}(d - c)^5 f^{(4)}(\zeta).$$

$$(7.18)$$

**Proof**  Let $m = (c + d)/2$ and $t_0 = (d - c)/2$, and define the auxiliary function $E\colon [-t_0, t_0] \to \mathbb{R}$ by

$$E(t) = \int_{m-t}^{m+t} f - \frac{t}{3}[f(m + t) + 4f(m) + f(m - t)] \quad \text{for} \quad -t_0 \leq t \leq t_0.$$

---

*The formula in Simpson's Rule is motivated as follows: For a function $f\colon [c, d] \to \mathbb{R}$, there is a unique quadratic polynomial $p\colon \mathbb{R} \to \mathbb{R}$ that agrees with the function $f\colon [c, d] \to \mathbb{R}$, at $x = c$, $x = (c + d)/2$, and $x = d$, and it can be shown that for this polynomial,

$$\int_c^d p = \frac{(d - c)}{6}\left[ f(c) + 4f\left(\frac{c + d}{2}\right) + f(d) \right].$$

Observe that the left-hand side of (7.18) is $E(t_0)$. Now define the function $H: [-t_0, t_0] \to \mathbb{R}$ by

$$H(t) = E(t) - \left(\frac{t}{t_0}\right)^5 E(t_0) \quad \text{for} \quad -t_0 \le t \le t_0. \tag{7.19}$$

Proposition 7.2 implies that the function $E: [-t_0, t_0] \to \mathbb{R}$ is continuous, so the function $H: [-t_0, t_0] \to \mathbb{R}$ is also continuous. It is clear that $H(0) = H(t_0) = 0$, and we will now show that $H'(0) = H''(0) = 0$ so that we can apply the Generalized Rolle's Theorem. We have the following straightforward calculation of derivatives at each point $t$ in $(-t_0, t_0)$:

$$
\begin{aligned}
E'(t) &= f(m+t) + f(m-t) - \frac{1}{3}[f(m+t) + 4f(m) + f(m-t)] \\
&\quad - \frac{t}{3}[f'(m+t) - f'(m-t)] \\
&= \frac{2}{3}[f(m+t) - 2f(m) + f(m-t)] - \frac{t}{3}[f'(m+t) - f'(m-t)],
\end{aligned}
$$

$$
\begin{aligned}
E''(t) &= \frac{2}{3}[f'(m+t) - f'(m-t)] - \frac{1}{3}[f'(m+t) - f'(m-t)] \\
&\quad - \frac{t}{3}[f''(m+t) + f''(m-t)] \\
&= \frac{1}{3}[f'(m+t) - f'(m-t)] - \frac{t}{3}[f''(m+t) + f''(m-t)],
\end{aligned}
$$

and

$$E'''(t) = -\frac{t}{3}[f'''(m+t) - f'''(m-t)].$$

These calculations, together with (7.19), show that $H(0) = H'(0) = H''(0) = 0$ and that

$$H'''(t) = -\frac{t}{3}[f'''(m+t) - f'''(m-t)] - \frac{60t^2}{t_0^5} E(t_0) \quad \text{for} \quad -t_0 < t < t_0.$$

Consequently, since $H(t_0) = 0$, we can apply the Generalized Rolle's Theorem with $n = 2$ to choose a point $t_*$ in $(0, t_0)$ at which $H'''(t_*) = 0$; that is,

$$-\frac{t_*}{3}[f'''(m+t_*) - f'''(m-t_*)] - \frac{60t_*^2}{t_0^5} E(t_0) = 0. \tag{7.20}$$

Finally, we can apply the Mean Value Theorem to the function $f''': [m - t_*, m + t_*] \to \mathbb{R}$ to choose a point $\zeta$ in $(m - t_*, m + t_*)$ at which

$$f'''(m + t_*) - f'''(m - t_*) = 2t_* f^{(4)}(\zeta).$$

Substituting the last equality in (7.20) leads to

$$t_*^2 \left[ -\frac{2}{3} f^{(4)}(\zeta) - \frac{60}{t_0^5} E(t_0) \right] = 0,$$

so since $t_* \ne 0$, we have

$$E(t_0) = \frac{-t_0^5}{90} f^{(4)}(\zeta). \tag{7.21}$$

However, $t_0 = (d - c)/2$, and since $E(t_0)$ is the left-hand side of (7.18), we see that (7.21) is the same as (7.18). ∎

**COROLLARY 7.18**  For a polynomial $p: \mathbb{R} \to \mathbb{R}$ of degree at most 3 and numbers $c < d$,

$$\int_c^d p(x)\,dx = \frac{1}{6}(d - c)\left[p(c) + 4p\left(\frac{c+d}{2}\right) + p(d)\right].$$

**Proof**  Since $p^{(4)}(\zeta) = 0$ for every number $\zeta$, the result immediately follows from the preceding theorem. ∎

**THEOREM 7.19**  **The Global Error for Simpson's Rule**  Suppose that the function $f: [a, b] \to \mathbb{R}$ is continuous and that its restriction $f: (a, b) \to \mathbb{R}$ has a bounded fourth derivative. Let $P = \{x_0, \ldots, x_n\}$ be a partition of the interval $[a, b]$. Then

$$\int_a^b f = \frac{1}{6}\sum_{i=1}^n (x_i - x_{i-1})\left[f(x_i) + 4f\left(\frac{x_i + x_{i-1}}{2}\right) + f(x_{i-1})\right] + E, \qquad (7.22)$$

with

$$|E| \leq \frac{M\|P\|^4(b - a)}{2880},$$

where

$$M \equiv \sup\{|f^{(4)}(x)| \mid a < x < b\}.$$

**Proof**  Let $i$ be an index such that $1 \leq i \leq n$. We can apply the local error estimate for Simpson's Rule to choose a point $\zeta_i$ in $(x_{i-1}, x_i)$ at which

$$\int_{x_{i-1}}^{x_i} f - \frac{(x_i - x_{i-1})}{6}\left[f(x_i) + 4f\left(\frac{x_i + x_{i-1}}{2}\right) + f(x_{i-1})\right]$$

$$= -\frac{1}{2880}(x_i - x_{i-1})^5 f^{(4)}(\zeta_i).$$

Summing these $n$ equalities, we see that (7.22) holds where

$$E = -\frac{1}{2880}\sum_{i=1}^n (x_i - x_{i-1})^5 f^{(4)}(\zeta_i).$$

However, the Triangle Inequality and the definitions of $\|P\|$ and $M$ yield

$$|E| \leq \frac{\|P\|^4}{2880}\sum_{i=1}^n (x_i - x_{i-1})|f^{(4)}(\zeta_i)|$$

$$\leq \frac{M\|P\|^4}{2880}\sum_{i=1}^n (x_i - x_{i-1})$$

$$= \frac{M\|P\|^4(b - a)}{2880}. \qquad ∎$$

**EXERCISES**

1.  Use the First Fundamental Theorem of Calculus to compute each of the following integrals, then compute the approximations using the Trapezoid Rule and Simpson's Rule with the partition $P = \{c, d\}$. Then compare the actual errors generated against the error estimates provided by Theorems 7.14 and 7.17.

    (a) $\displaystyle\int_1^2 (2x + 3)\, dx$  (b) $\displaystyle\int_0^1 x^2\, dx$

    (c) $\displaystyle\int_0^1 x^4\, dx$

2.  For each of the following integrals, verify Corollary 7.18 by direct computation.

    (a) $\displaystyle\int_0^1 (x^2 + x^3)\, dx$  (b) $\displaystyle\int_2^3 (x + 1)^2\, dx$

3.  Suppose that the function $f:[a, b] \to \mathbb{R}$ is continuous and that its restriction $f: (a, b) \to \mathbb{R}$ has a bounded second derivative. Let $n$ be a natural number and let $P_n$ be the regular partition of $[a, b]$ into $n$ subintervals of equal length. Show that

$$\int_a^b f(x)\, dx = \left(\frac{b-a}{n}\right)\left[\frac{f(a)}{2} + \sum_{k=1}^{n-1} f\left(a + \frac{k}{n}(b-a)\right) + \frac{f(b)}{2}\right] + E,$$

    where $$|E| \le \frac{(b-a)^3}{12n^2}\sup\{|f''(x)| \mid x \text{ in } (a, b)\}.$$

4.  Use Exercise 3 with $n = 3$ to estimate $\ln 4$ and give an upper bound for the error.
5.  Use Exercise 3 with $n = 4$ to estimate $\int_0^1 \sqrt{1 + x^2}\, dx$ and give an upper bound for the error.
6.  Use Simpson's Rule with $P = \{1, 3/2, 2\}$ to estimate $\ln 2$ and give an upper bound for the error.
7.  An approximation rule similar to the Trapezoid Rule is the Midpoint Rule. This rule approximates the integral $\int_c^d f(x)\, dx$ by $(d - c)f([d + c]/2)$. Prove that if the function $f:[c, d] \to \mathbb{R}$ is continuous and its restriction $f:(c, d) \to \mathbb{R}$ has a second derivative, then for some point $\zeta$ in $(c, d)$,

$$\int_c^d f(x)\, dx = (d - c)f\left(\frac{c+d}{2}\right) + \frac{1}{24}(d - c)^3 f''(\zeta).$$

    (*Hint:* Let $m = (c + d)/2$, let $t_0 = (d - c)/2$, and define

$$H(t) = \left[\int_{m-t}^{m+t} f\right] - 2tf(m) - \left(\frac{t}{t_0}\right)^3 \int_{m-t_0}^{m+t_0} f \quad \text{for} \quad -t_0 \le t \le t_0.$$

    Apply the Generalized Rolle's Theorem with $n = 1$ to the function $H:[-t_0, t_0] \to \mathbb{R}$.)
8.  Find a global error estimate for the Midpoint Rule.
9.  Use the local error estimates for the Midpoint Rule (see Exercise 7) and the Trapezoid Rule for the evaluation of

$$\int_{\ln a}^{\ln b} e^x\, dx$$

    to prove that if $0 < a < b$, then

$$\sqrt{ab} < \frac{b-a}{\ln b - \ln a} < \frac{a+b}{2}.$$

(The geometric mean is less than the logarithmic mean, which is less than the arithmetic mean.)

# 8

# Approximation by Taylor Polynomials

## 8.1 Taylor Polynomials and Order of Contact

Polynomials are the simplest kind of functions. Accordingly, for a general function $f: D \to \mathbb{R}$, it is natural to seek a polynomial that is a good approximation of $f: D \to \mathbb{R}$.

**DEFINITION**  *Let $I$ be an open interval containing the point $x_0$. Two functions $f: I \to \mathbb{R}$ and $g: I \to \mathbb{R}$ are said to have contact of order 0 at $x_0$ provided that $f(x_0) = g(x_0)$. For a natural number $n$, the functions $f: I \to \mathbb{R}$ and $g: I \to \mathbb{R}$ are said to have contact of order $n$ at $x_0$ provided that $f: I \to \mathbb{R}$ and $g: I \to \mathbb{R}$ have $n$ derivatives and*

$$f^{(k)}(x_0) = g^{(k)}(x_0) \quad for \quad 0 \le k \le n.$$

**EXAMPLE 8.1** Define $f(x) = \sqrt{2 - x^2}$ and $g(x) = e^{1-x}$ for $0 < x < \sqrt{2}$. Then

$$f(1) = g(1) \quad \text{and} \quad f'(1) = g'(1), \quad \text{but} \quad f''(1) \neq g''(1).$$

Hence the functions $f: (0, \sqrt{2}) \to \mathbb{R}$ and $g: (0, \sqrt{2}) \to \mathbb{R}$ have contact of order 1 at $x_0 = 1$, but do not have contact of order 2 there. At the point $(1, 1)$, which lies on both graphs, the tangent lines to the functions are the same. $\qquad\square$

The following formula is clear: For each pair of nonnegative integers $k$ and $\ell$,

$$\frac{d^k}{dx^k}[(x - x_0)^\ell]\bigg|_{x=x_0} = \begin{cases} k! & \text{if } k = \ell \\ 0 & \text{if } k \neq \ell. \end{cases} \tag{8.1}$$

**PROPOSITION 8.1**

Let $I$ be an open interval containing the point $x_0$ and $n$ be a nonnegative integer. Suppose that the function $f: I \to \mathbb{R}$ has $n$ derivatives. Then there is a unique polynomial of degree at most $n$ that has contact of order $n$ with the function $f: I \to \mathbb{R}$ at $x_0$. This polynomial is defined by the formula

$$p_n(x) = f(x_0) + f'(x_0)(x - x_0) + \cdots + \frac{f^{(k)}(x_0)}{k!}(x - x_0)^k + \cdots + \frac{f^{(n)}(x_0)}{n!}(x - x_0)^n. \tag{8.2}$$

**Proof** If $n = 0$, the result is clear; there is only one constant function whose value at $x_0$ is $f(x_0)$. So suppose that $n \geq 1$. From formula (8.1) it follows that

$$\frac{d^k}{dx^k}[p_n(x)]\bigg|_{x=x_0} = f^{(k)}(x_0) \quad \text{for} \quad 0 \leq k \leq n,$$

so the function $f: I \to \mathbb{R}$ and the polynomial $p_n: I \to \mathbb{R}$ have contact of order $n$ at $x_0$.

It remains to prove uniqueness. However, if we take a general polynomial of degree at most $n$, written in powers of $(x - x_0)$ as

$$p(x) = c_0 + c_1(x - x_0) + \cdots + c_n(x - x_0)^n,$$

then, again from formula (8.1), it is clear that

$$\frac{d^k}{dx^k}[p(x)]\bigg|_{x=x_0} = k!c_k \quad \text{for} \quad 0 \leq k \leq n,$$

so that if the polynomial $p: I \to \mathbb{R}$ has contact of order $n$ with $f: I \to \mathbb{R}$ at $x_0$, we must have $k!c_k = f^{(k)}(x_0)$ for $0 \leq k \leq n$; that is, $p = p_n$. $\qquad\blacksquare$

The polynomial $p_n: \mathbb{R} \to \mathbb{R}$ defined by (8.2) is called the $n$th *Taylor polynomial* for the function $f: I \to \mathbb{R}$ at the point $x_0$.

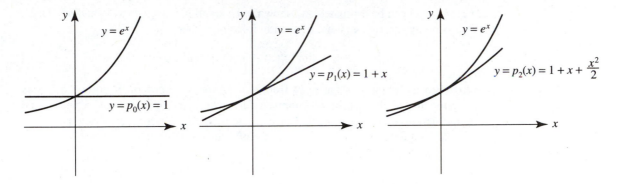

**FIGURE 8.1**

**EXAMPLE 8.2**   Define $f(x) = e^x$ for all $x$. For each natural number $k$, $f^{(k)}(x) = e^x$ for all $x$. Thus the $n$th Taylor polynomial for $f : \mathbb{R} \to \mathbb{R}$ at the point $x = 0$ is defined by

$$p_n(x) = 1 + x + \cdots + \frac{x^n}{n!}.$$   $\square$

**EXAMPLE 8.3**   Define $f(x) = \ln(1 + x)$ for $x > -1$. For each natural number $k$,

$$f^{(k)}(x) = \frac{(-1)^{k+1}(k-1)!}{(1+x)^k} \qquad \text{for all } x > -1.$$

Thus the $n$th Taylor polynomial for $f : (-1, \infty) \to \mathbb{R}$ at $x = 0$ is defined by

$$p_n(x) = x - \frac{x^2}{2} + \cdots + \frac{(-1)^{n+1}}{n} x^n.$$   $\square$

**EXAMPLE 8.4**   Define $f(x) = \cos x$ for all $x$. For each natural number $k$, $f^{(2k)}(x) = (-1)^k \cos x$, and $f^{(2k+1)}(x) = (-1)^{k+1} \sin x$ for all $x$. Thus, for each nonnegative integer $n$, the Taylor polynomials for the cosine function at $x = 0$ are given by

$$p_{2n}(x) = p_{2n+1}(x) = 1 - \frac{x^2}{2!} + \cdots + \frac{(-1)^n}{(2n)!} x^{2n}.$$   $\square$

**EXAMPLE 8.5**   Define $f(x) = \sqrt{x}$ for $x > 0$. For each natural number $k$,

$$f^{(k)}(x) = \frac{1}{2} \left( \frac{1}{2} - 1 \right) \cdots \left( \frac{1}{2} - k + 1 \right) x^{1/2-k} \qquad \text{for all } x > 0.$$

Thus the third Taylor polynomial for the function $f : (0, \infty) \to \mathbb{R}$ at $x = 1$ is

$$p_3(x) = 1 + \frac{1}{2}(x - 1) - \frac{1}{8}(x - 1)^2 + \frac{1}{16}(x - 1)^3.$$   $\square$

For two functions that have a high order of contact at a point, it is reasonable to expect that near this point the difference between them will be small. In particular, if $I$ is an open interval containing the point $x_0$ and $p_n : \mathbb{R} \to \mathbb{R}$ is the $n$th Taylor polynomial for the function $f : I \to \mathbb{R}$ at $x_0$, one expects that for another point $x$ in $I$, the difference

$f(x) - p_n(x)$ can be estimated and shown to be small if $x$ is close to $x_0$ and $n$ is large. What is really surprising is that frequently it happens that

$$\lim_{n \to \infty} [f(x) - p_n(x)] = 0$$

even when the point $x$ is far away from $x_0$. As we will show in Section 8.6, it can also happen that the Taylor polynomials for certain functions do not provide good approximations at any point $x$ other than $x_0$, no matter how large the index $n$ is.

We define $R_n(x) \equiv f(x) - p_n(x)$ for all $x$ in $I$, so that

$$f(x) = p_n(x) + R_n(x) \quad \text{for all } x \text{ in } I,$$

and call $R_n : I \to \mathbb{R}$ the $n$th *remainder*. In Section 8.2, we will begin a rigorous analysis of this remainder.

---

**EXERCISES**

1. For each of the following pairs of functions, determine its highest order of contact at the indicated point:
   (a)  $f(x) = x^2$ and $g(x) = \sin x$ for all $x$; $x_0 = 0$.
   (b)  $f(x) = e^{x^2}$ and $g(x) = 1 + 2x^2$ for all $x$; $x_0 = 0$.
   (c)  $f(x) = \ln x$ and $g(x) = (x - 1)^3 + \ln x$ for all $x > 0$; $x_0 = 1$.

2. Compute the third Taylor polynomial for each of the following functions at the indicated point:
   (a)  $f(x) = \int_0^x (1/1 + t^2)\, dt$ for all $x$; $x_0 = 0$.
   (b)  $f(x) = \sin x$ for all $x$; $x_0 = 0$.
   (c)  $f(x) = \sqrt{2 - x}$ for all $x < 2$; $x_0 = 1$.

3. Define $f(x) = x^6 e^x$ for all $x$. Find the sixth Taylor polynomial for the function $f : \mathbb{R} \to \mathbb{R}$ at $x = 0$.

4. Suppose that the function $f : \mathbb{R} \to \mathbb{R}$ has three derivatives and that the third Taylor polynomial at $x = 0$ is $p_3(x) = 1 + 4x - x^2 + x^3/6$. Show that there is an open interval containing the point 0 such that $f : I \to \mathbb{R}$ is positive, strictly increasing, and strictly concave.

5. Suppose that the function $f : \mathbb{R} \to \mathbb{R}$ has a second derivative and that

$$\begin{cases} f''(x) + f(x) = e^{-x} & \text{for all } x \\ f(0) = 0 \quad \text{and} \quad f'(0) = 2. \end{cases}$$

   Find the fourth Taylor polynomial for $f : \mathbb{R} \to \mathbb{R}$ at $x = 0$.

---

## 8.2   The Lagrange Remainder Theorem

For easy reference, we record again here Theorem 4.21 as the following lemma:

**LEMMA 8.2**   Let $I$ be an open interval and $n$ be a nonnegative integer, and suppose that the function $f: I \to \mathbb{R}$ has $n$ derivatives. Suppose also that at the point $x_0$ in $I$,

$$f^{(k)}(x_0) = 0 \quad \text{for} \quad 0 \le k \le n - 1.$$

Then for each point $x \ne x_0$ in $I$, there is a point $c$ strictly between $x$ and $x_0$ at which

$$f(x) = \frac{f^{(n)}(c)}{n!} (x - x_0)^n.$$

The general remainder theorem is a simple extension of the above lemma.

**THEOREM 8.3**   **The Lagrange Remainder Theorem**   Let $I$ be an open interval containing the point $x_0$ and let $n$ be a nonnegative integer. Suppose that the function $f: I \to \mathbb{R}$ has $n + 1$ derivatives. Then for each point $x$ in $I$, there is a point $c$ strictly between $x$ and $x_0$ such that

$$f(x) = \sum_{k=0}^{n} \frac{f^{(k)}(x_0)}{k!} (x - x_0)^k + \frac{f^{(n+1)}(c)}{(n+1)!} (x - x_0)^{n+1}. \tag{8.3}$$

**Proof**   Consider the $n$th Taylor polynomial for the function $f: I \to \mathbb{R}$ at $x_0$,

$$p_n(x) = \sum_{k=0}^{n} \frac{f^{(k)}(x_0)}{k!} (x - x_0)^k.$$

Since the functions $f: I \to \mathbb{R}$ and $p_n: I \to \mathbb{R}$ have contact of order $n$ at $x_0$, it follows that if we define the function $E: I \to \mathbb{R}$ by

$$E(x) = f(x) - p_n(x) \quad \text{for all } x \text{ in } I,$$

then

$$E(x_0) = E'(x_0) = \cdots = E^{(n)}(x_0) = 0.$$

According to the above lemma, if $x$ is in $I$, then there is a point $c$ strictly between $x$ and $x_0$ such that

$$E(x) = \frac{E^{(n+1)}(c)}{(n+1)!} (x - x_0)^{n+1}.$$

But since $p_n: \mathbb{R} \to \mathbb{R}$ is a polynomial of degree at most $n$,

$$E^{(n+1)}(c) = f^{(n+1)}(c),$$

so from the previous equation it follows that

$$f(x) - p_n(x) = E(x) = \frac{f^{(n+1)}(c)}{(n+1)!} (x - x_0)^{n+1}. \qquad \blacksquare$$

**COROLLARY 8.4**   Suppose that $p: \mathbb{R} \to \mathbb{R}$ is a polynomial of degree at most $n$, and let $x_0$ be any point. Then the $n$th Taylor polynomial for $p: \mathbb{R} \to \mathbb{R}$ at $x_0$ equals $p: \mathbb{R} \to \mathbb{R}$ itself.

**Proof** Fix $x \neq x_0$. According to the Lagrange Remainder Theorem, we may choose a point $c$ strictly between $x$ and $x_0$ such that

$$p(x) - p_n(x) = \frac{p^{(n+1)}(c)}{(n+1)!}(x - x_0)^{n+1}.$$

But $p: \mathbb{R} \to \mathbb{R}$ is a polynomial of degree at most $n$, so $p^{n+1}(c) = 0$. Thus, $p_n(x) = p(x)$. ∎

We will use the Lagrange Remainder Theorem to get a precise estimate of the number $e$. First, we will derive the following crude estimate of $e$.

**LEMMA 8.5**

$$e < 3. \tag{8.4}$$

**Proof** Since the function $\ln: (0, \infty) \to \mathbb{R}$ is strictly increasing,

$$e < 3 \quad \text{if and only if} \quad 1 = \ln e < \ln 3.$$

However, using the integral representation of the natural logarithm, we have

$$\ln 3 = \int_1^3 \frac{1}{t} \, dt$$

$$= \int_1^2 \frac{1}{t} \, dt + \int_2^3 \frac{1}{t} \, dt$$

$$= \int_1^0 \frac{(-1)}{2 - s} \, ds + \int_0^1 \frac{1}{2 + s} \, ds$$

$$= \int_0^1 \frac{4}{4 - s^2} \, ds.$$

But $4/(4 - s^2) > 1$ if $0 < s \leq 1$, so

$$\ln 3 = \int_0^1 \frac{4}{4 - s^2} \, ds > 1. \qquad ∎$$

**THEOREM 8.6** For each natural number $n$ and each nonzero number $x$, there is a point $c$ strictly between $0$ and $x$ such that

$$e^x = 1 + x + \frac{x^2}{2!} + \cdots + \frac{x^n}{n!} + \frac{e^c}{(n+1)!}x^{n+1}. \tag{8.5}$$

In particular,

$$0 < e^x - \left[1 + x + \cdots + \frac{x^n}{n!}\right] < \frac{3}{(n+1)!} \quad \text{if} \quad 0 \leq x \leq 1. \tag{8.6}$$

**Proof**   Formula (8.5) follows directly from the Lagrange Remainder Theorem and Example 8.2. According to the preceding lemma, $e < 3$, and since the function $\exp: \mathbb{R} \to \mathbb{R}$ is strictly increasing, it follows that

$$1 = e^0 \le e^c < e^1 < 3 \quad \text{if} \quad 0 < c < x \le 1.$$

Thus estimate (8.6) follows from (8.5).                                      ∎

---

**PROPOSITION 8.7**   The number $e$ is irrational.

**Proof**   We will argue by contradiction. Suppose that $e$ is rational. Then there are natural numbers $n_0$ and $m_0$ such that $e = n_0/m_0$. Then (8.6), with $x = 1$, becomes

$$0 < \frac{n_0}{m_0} - \left[ 2 + \frac{1}{2!} + \cdots + \frac{1}{n!} \right] \le \frac{3}{(n+1)!} \quad \text{for every natural number } n.$$

Now multiply this inequality by $n!$ to get

$$0 < \frac{n! n_0}{m_0} - n! \left[ 2 + \frac{1}{2!} + \cdots + \frac{1}{n!} \right] \le \frac{3}{n+1} \quad \text{for every natural number } n.$$

However, if $n > 3$ and $n \ge m_0$ the above inequality implies the existence of an integer in the interval $(0, 3/4)$. This contradiction proves that $e$ is irrational.    ∎

---

**THEOREM 8.8**   For each natural number $n$ and each number $x > -1$, there is a number $c$ strictly between $0$ and $x$ such that

$$\ln(1 + x) = x - \frac{x^2}{2} + \cdots + \frac{(-1)^{n+1}}{n} x^n + \frac{(-1)^n}{(n+1)(1+c)^{n+1}} x^{n+1}. \tag{8.7}$$

**Proof**   Formula (8.7) follows from the Lagrange Remainder Theorem and Example 8.3.    ∎

Recall that in Section 2.2, we showed that the sequence

$$\left\{ 1 + \frac{1}{2} + \cdots + \frac{1}{n} \right\}$$

is strictly increasing but unbounded above; that is, the harmonic series diverges. In Section 5.1, we showed that the sequence $\{\ln(n + 1)\}$ is strictly increasing but unbounded above. In fact, these two results are equivalent.

---

**PROPOSITION 8.9**   $\displaystyle \lim_{n \to \infty} \left[ 1 + \frac{1}{2} + \cdots + \frac{1}{n} - \ln(n + 1) \right] = \gamma \quad \text{where} \quad 0 < \gamma \le 1.$

**Proof**   For each natural number $n$, define

$$c_n = 1 + \frac{1}{2} + \cdots + \frac{1}{n} - \ln(n+1).$$

We will show that the sequence $\{c_n\}$ is strictly increasing and bounded above by 1. The conclusion of the proposition will then be a consequence of the Monotone Convergence Theorem.

For each natural number $k$, $\ln(1 + 1/k) = \ln(1 + k) - \ln(k)$. Thus for $n \geq 2$, we may write each $c_n$ as the sum

$$c_n = (1 - \ln 2) + \sum_{k=2}^{n} \left( \frac{1}{k} - \ln\left(1 + \frac{1}{k}\right) \right).$$

Now from formula (8.7), with $n = 1$ and $x = 1/k$, it follows (see Exercise 4) that

$$0 < \frac{1}{k} - \ln\left(1 + \frac{1}{k}\right) < \frac{1}{2k^2} \quad \text{for every natural number } k. \tag{8.8}$$

Thus the sequence $\{c_n\}$ is strictly increasing; it remains to be shown that it is bounded above. In order to do so, we first observe that (8.8) implies that for each integer $n \geq 2$,

$$\sum_{k=2}^{n} \left[ \frac{1}{k} - \ln\left(1 + \frac{1}{k}\right) \right] \leq \sum_{k=2}^{n} \left[ \frac{1}{2k^2} \right]$$

$$\leq \sum_{k=2}^{n} \left[ \frac{1}{2k(k-1)} \right]$$

$$= \frac{1}{2} \sum_{k=2}^{n} \left[ \frac{1}{k-1} - \frac{1}{k} \right]$$

$$= \frac{1}{2} \left[ 1 - \frac{1}{n} \right]$$

$$\leq \frac{1}{2}.$$

On the other hand, (8.8), with $k = 1$, asserts that $1 - \ln 2 < 1/2$, so for every natural number $n \geq 2$,

$$c_n = [1 - \ln 2] + \sum_{k=2}^{n} \left[ \frac{1}{k} - \ln\left(1 + \frac{1}{k}\right) \right] \leq \frac{1}{2} + \frac{1}{2} = 1. \qquad \blacksquare$$

The above number $\gamma$ is called *Euler's constant*. It is unknown whether $\gamma$ is rational or irrational.

---

**EXERCISES**

1.   Prove that

$$1 + \frac{x}{2} - \frac{x^2}{8} < \sqrt{1+x} < 1 + \frac{x}{2} \quad \text{if} \quad x > 0.$$

In particular, show that $1.375 < \sqrt{2} < 1.5$.

**2.** Prove that

$$1 + \frac{x}{3} - \frac{x^2}{9} < (1+x)^{1/3} < 1 + \frac{x}{3} \quad \text{if} \quad x > 0.$$

**3.** Expand the polynomial $p(x) = x^5 - x^3 + x$ in powers of $(x - 1)$.

**4.** Use formula (8.7) to prove that

$$0 < \frac{1}{k} - \ln\left(1 + \frac{1}{k}\right) < \frac{1}{2k^2} \quad \text{for every natural number } k.$$

**5.** Prove that for every pair of numbers $a$ and $b$,

$$|\sin(a + b) - (\sin a + b \cos a)| \leq \frac{b^2}{2}.$$

**6.** Let $I$ be an open interval containing the point $x_0$ and let $n$ be a natural number. Suppose that the function $f: I \to \mathbb{R}$ has $n + 1$ derivatives. Show that the Lagrange Remainder Theorem is equivalent to the following: For each number $h$ such that $x_0 + h$ is in $I$, there is a number $\theta$, strictly between 0 and 1, such that

$$f(x_0 + h) = \sum_{k=0}^{n} \frac{f^{(k)}(x_0)}{k!} h^k + \frac{1}{(n+1)!} f^{(n+1)}(x_0 + \theta h) h^{n+1}.$$

**7.** Using Corollary 8.4, show that if $p: \mathbb{R} \to \mathbb{R}$ is a polynomial of degree $n$ and the number $x_0$ is a root of $p(x)$—that is, $p(x_0) = 0$—then there is a polynomial $q: \mathbb{R} \to \mathbb{R}$ such that $p(x) = (x - x_0)q(x)$.

**8.** A number $x_0$ is said to be a *root of order $k$ of the polynomial* $p: \mathbb{R} \to \mathbb{R}$ provided that $k$ is a natural number such that $p(x) = (x - x_0)^k r(x)$, where $r: \mathbb{R} \to \mathbb{R}$ is a polynomial and $r(x_0) \neq 0$. Prove that $x_0$ is a root of order $k$ of the polynomial $p: \mathbb{R} \to \mathbb{R}$ if and only if

$$p(x_0) = p'(x_0) = \cdots = p^{(k-1)}(x_0) = 0 \quad \text{and} \quad p^{(k)}(x_0) \neq 0.$$

**9.** (a)   For nonnegative integers $n$ and $k$ such that $k \leq n$, show that

$$(1 + x)^n = 1 + \binom{n}{1}x + \binom{n}{2}x^2 + \cdots + \binom{n}{n-1}x^{n-1} + x^n.$$

(b)   Use (a) to prove the Binomial Formula: For each natural number $n$ and any pair of numbers $a$ and $b$,

$$(a + b)^n = \sum_{k=0}^{n} \binom{n}{k} a^k b^{n-k}.$$

**10.** Suppose that each of the functions $f: \mathbb{R} \to \mathbb{R}$ and $g: \mathbb{R} \to \mathbb{R}$ has $n + 1$ continuous derivatives. Prove that $f: \mathbb{R} \to \mathbb{R}$ and $g: \mathbb{R} \to \mathbb{R}$ have contact of order $n$ at 0 if and only if

$$\lim_{x \to 0} \frac{f(x) - g(x)}{x^n} = 0.$$

**11.** Use the Lagrange Remainder Theorem to verify the following criterion for identifying local extreme points: Let $I$ be an open interval containing the point $x_0$ and let $n$ be a natural number. Suppose that the function $f: I \to \mathbb{R}$ has $n + 1$ derivatives and that $f^{(n+1)}: I \to \mathbb{R}$ is continuous. Assume that $f^{(k)}(x_0) = 0$ if $1 \leq k \leq n$, and that $f^{(n+1)}(x_0) \neq 0$.

(a)   If $n + 1$ is even and $f^{(n+1)}(x_0) > 0$, then $x_0$ is a local minimizer.

(b)   If $n + 1$ is even and $f^{(n+1)}(x_0) < 0$, then $x_0$ is a local maximizer.

    (c)   If $n + 1$ is odd, then $x_0$ is neither a local maximizer nor a local minimizer.

**12.** Use Theorem 8.8 to show that if $0 < x \le 1$, then

$$\ln(1 + x) = \sum_{k=1}^{\infty} \frac{(-1)^{k+1}}{k} x^k.$$

**13.** Let $I$ be an open interval containing the point $x_0$, and suppose that the function $f : I \to \mathbb{R}$ has a continuous third derivative with $f'''(x) > 0$ for all $x$ in $I$.

    (a)   Prove that if $x_0 + h$ is in $I$, there is a unique number $\theta = \theta(h)$ in the interval $(0, 1)$ such that

$$f(x_0 + h) = f(x_0) + f'(x_0)h + f''(x_0 + \theta h)\frac{h^2}{2}.$$

    (b)   Prove that

$$\lim_{h \to 0} \theta(h) = \frac{1}{3}.$$

## 8.3  The Convergence of Taylor Polynomials

For a sequence of numbers $\{a_k\}$ that is indexed by the nonnegative integers, we define

$$s_n = \sum_{k=0}^{n} a_k \quad \text{for every nonnegative integer } n,$$

and obtain a new sequence $\{s_n\}$. The sequence $\{s_n\}$ is called the *sequence of partial sums* for the series $\sum_{k=0}^{\infty} a_k$, and $a_k$ is called the $k$th *term* of the series $\sum_{k=0}^{\infty} a_k$. We write

$$\sum_{k=0}^{\infty} a_k = \lim_{n \to \infty} \left[ \sum_{k=1}^{n} a_k \right]$$

if the sequence $\{s_n\}$ converges. If the sequence $\{s_n\}$ does not converge, then we say that the series $\sum_{k=0}^{\infty} a_k$ *diverges*.

    Let $I$ be an open interval containing the point $x_0$ and suppose that the function $f : I \to \mathbb{R}$ has derivatives of all orders. The $n$th Taylor polynomial for $f : I \to \mathbb{R}$ at $x_0$ is defined by

$$p_n(x) = \sum_{k=0}^{n} \frac{f^{(k)}(x_0)}{k!} (x - x_0)^k.$$

In conformity with the above series notation, if $x$ is a point in $I$ at which

$$\lim_{n \to \infty} p_n(x) = f(x), \tag{8.9}$$

we write
$$f(x) = \sum_{k=0}^{\infty} \frac{f^{(k)}(x_0)}{k!} (x - x_0)^k. \tag{8.10}$$

This formula is called a *Taylor series expansion* of the function $f: I \to \mathbb{R}$ about the point $x_0$. By its very definition, (8.10) holds at $x$ if and only if

$$\lim_{n \to \infty} [f(x) - p_n(x)] = 0. \tag{8.11}$$

In this section, we will use the Lagrange Remainder Theorem to determine when $\lim_{n \to \infty}[f(x) - p_n(x)] = 0$. First, we will prove a useful preliminary result.

**LEMMA 8.10**   For any number $c$,

$$\lim_{n \to \infty} \frac{c^n}{n!} = 0.$$

**Proof**   Choose $k$ to be a natural number such that $k \geq 2|c|$. Then if $n \geq k$,

$$0 \leq \left| \frac{c^n}{n!} \right|$$

$$= \left[ \frac{|c|}{1} \cdots \frac{|c|}{k} \right] \left[ \frac{|c|}{k+1} \cdots \frac{|c|}{n} \right]$$

$$\leq |c|^k \left( \frac{1}{2} \right)^{n-k}$$

$$= (2|c|)^k \left( \frac{1}{2} \right)^n.$$

But $\lim_{n \to \infty} (1/2)^n = 0$, and so $\lim_{n \to \infty} c^n/n! = 0$ also. ∎

**THEOREM 8.11**   Let $I$ be an open interval containing the point $x_0$ and suppose that the function $f: I \to \mathbb{R}$ has derivatives of all orders. Suppose also that there are positive numbers $r$ and $M$ such that the interval $[x_0 - r, x_0 + r]$ is contained in $I$ and for every natural number $n$ and every point $x$ in $[x_0 - r, x_0 + r]$,

$$|f^{(n)}(x)| \leq M^n. \tag{8.12}$$

Then
$$f(x) = \sum_{k=0}^{\infty} \frac{f^{(k)}(x_0)}{k!}(x - x_0)^k \quad \text{if} \quad |x - x_0| \leq r. \tag{8.13}$$

**Proof**   The $n$th Taylor polynomial $p_n: \mathbb{R} \to \mathbb{R}$ for $f: I \to \mathbb{R}$ at $x_0$ is defined by

$$p_n(x) = \sum_{k=0}^{n} \frac{f^{(k)}(x_0)}{k!}(x - x_0)^k,$$

and, according to the Lagrange Remainder Theorem, for each point $x$ in $I$, there is a point $c$ strictly between $x$ and $x_0$ such that

$$|f(x) - p_n(x)| = \frac{|f^{(n+1)}(c)|}{(n+1)!}|x - x_0|^{n+1}.$$

In view of inequality (8.12), it follows that for every natural number $n$ and every point $x$ in $[x_0 - r, \ x_0 + r]$,

$$|f(x) - p_n(x)| \leq \frac{M^{n+1}}{(n+1)!}|x - x_0|^{n+1} \leq \frac{c^{n+1}}{(n+1)!}, \qquad (8.14)$$

where $c = Mr$. According to Lemma 8.10, $\lim_{n \to \infty} c^n / n! = 0$. Thus, from (8.14), we see that

$$\lim_{n \to \infty} [f(x) - p_n(x)] = 0 \quad \text{if} \quad |x - x_0| \leq r.$$

This is precisely assertion (8.13). ∎

**COROLLARY**
**8.12**

$$e^x = \sum_{k=0}^{\infty} \frac{x^k}{k!} \quad \text{for all } x. \qquad (8.15)$$

$$\cos x = \sum_{k=0}^{\infty} \frac{(-1)^k}{(2k)!} x^{2k} \quad \text{for all } x. \qquad (8.16)$$

**Proof**  First, we will prove (8.15). Define $f(x) = e^x$ for all $x$, and let $x_0 = 0$. Fix $r > 0$. If we define $M = e^r$, it follows that for every natural number $n$ and every point $x$ in the interval $[-r, r]$,

$$|f^{(n)}(x)| \leq M \leq M^n.$$

According to Theorem 8.11, if $|x| < r$, then

$$f(x) = \sum_{k=0}^{\infty} \frac{x^k}{k!}.$$

But the choice of $r > 0$ was arbitrary, and so the Taylor expansion (8.15) is verified.

The proof of (8.16) is similar. We define $f(x) = \cos x$ for all $x$ and observe that $|f^{(n)}(x)| \leq 1$ for every natural number $n$ and every number $x$. Then the proof proceeds as above. ∎

**EXERCISES**

1. Show that the Taylor expansion of the following functions at the given points converges for all points $x$:
   (a)  $f(x) = \sin x$ at the point $x_0 = 0$.
   (b)  $f(x) = \cos x$ at the point $x_0 = \pi$.
2. Define $f(x) = 1/x$ if $0 < x < 2$.
   (a)  Find $p_n : \mathbb{R} \to \mathbb{R}$, the $n$th Taylor polynomial at $x_0 = 1$.
   (b)  Use the Geometric Sum Formula to show that for every natural number $n$,

   $$f(x) - p_n(x) = \frac{(1-x)^{n+1}}{x} \quad \text{if} \quad 0 < x < 2.$$

   (c)  Use part (b) to prove that

   $$f(x) = \sum_{k=0}^{\infty} \frac{f^{(k)}(1)}{k!}(x-1)^k \quad \text{if} \quad |x - 1| < 1.$$

3. Suppose that the function $F: \mathbb{R} \to \mathbb{R}$ has derivatives of all orders and that

$$\begin{cases} F'(x) - F(x) = 0 & \text{for all } x \\ F(0) = 2. \end{cases}$$

Find a formula for the coefficients of the $n$th Taylor polynomial for $F: \mathbb{R} \to \mathbb{R}$ at $x = 0$. Show that the Taylor expansion converges at every point.

4. Suppose that the function $F: \mathbb{R} \to \mathbb{R}$ has derivatives of all orders and that

$$\begin{cases} F''(x) - F'(x) - F(x) = 0 & \text{for all } x \\ F(0) = 1 \quad \text{and} \quad F'(0) = 1. \end{cases}$$

Find a recursive formula for the coefficients of the $n$th Taylor polynomial for $F: \mathbb{R} \to \mathbb{R}$ at $x = 0$. Show that the Taylor expansion converges at every point.

5. For a pair of numbers $\alpha$ and $\beta$, suppose that the function $f: \mathbb{R} \to \mathbb{R}$ has derivatives of all orders and that

$$f''(x) + \alpha f'(x) + \beta f(x) = 0 \quad \text{for all } x.$$

(a) Show that for every natural number $n$,

$$f^{(n+2)}(x) + \alpha f^{(n+1)}(x) + \beta f^{(n)}(x) = 0 \quad \text{for all } x.$$

(b) Use (a) to show that

$$f(x) = \sum_{k=0}^{\infty} \frac{f^{(k)}(x_0)}{k!}(x - x_0)^k \quad \text{for all } x.$$

## 8.4 A Power Series for the Logarithm

In this section, we will analyze the validity of the Taylor expansion of the natural logarithm. It turns out that in order to do so, it is convenient first to translate the natural logarithm and consider the function $f: (-1, \infty) \to \mathbb{R}$ defined by

$$f(x) = \ln(1 + x) \quad \text{if} \quad x > -1.$$

A direct calculation of derivatives shows that for each natural number $k$,

$$f^{(k)}(x) = \frac{(-1)^{k+1}(k-1)!}{(1+x)^k} \quad \text{for all } x > -1.$$

In particular, the $n$th Taylor polynomial for $f: (-1, \infty) \to \mathbb{R}$ at $x = 0$ is defined by

$$p_n(x) = x - \frac{x^2}{2} + \cdots + \frac{(-1)^{n+1}}{n} x^n. \tag{8.17}$$

Rather than trying to use the Lagrange Remainder Theorem to study the difference $f(x) - p_n(x)$, it is better to jointly exploit the integral formula for the natural logarithm and the Geometric Sum Formula in order to derive a more explicit formula for the difference. Indeed, observe that for each natural number $n$, the Geometric Sum Formula

$$\frac{1}{1-r} = 1 + r + \cdots + r^{n-1} + \frac{r^n}{1-r} \quad \text{if} \quad r \neq 1$$

becomes, if one substitutes $1 - t$ for $r$,

$$\frac{1}{t} = 1 + (1 - t) + \cdots + (1 - t)^{n-1} + \frac{(1 - t)^n}{t} \qquad \text{if} \quad t \neq 0.$$

Thus, using the linearity of integration and the integral representation of the logarithm, we have

$$\ln(1 + x) = \int_1^{1+x} \frac{1}{t} \, dt$$

$$= \int_1^{1+x} [1 + (1 - t) + \cdots + (1 - t)^{n-1}] \, dt + \int_1^{1+x} \frac{(1 - t)^n}{t} \, dt$$

$$= x - \frac{x^2}{2} + \cdots + \frac{(-1)^{n+1}}{n} x^n + \int_1^{1+x} \frac{(1 - t)^n}{t} \, dt$$

$$= p_n(x) + \int_1^{1+x} \frac{(1 - t)^n}{t} \, dt \qquad \text{if} \quad x > -1. \tag{8.18}$$

**THEOREM 8.13**

$$\ln(1 + x) = \sum_{k=1}^{\infty} \frac{(-1)^{k+1} x^k}{k} \qquad \text{if} \quad -1 < x \leq 1. \tag{8.19}$$

**Proof**  Formula (8.18) implies that for each natural number $n$,

$$\ln(1 + x) - \sum_{k=1}^{n} (-1)^{k+1} \frac{x^k}{k} = \int_1^{1+x} \frac{(1 - t)^n}{t} \, dt \qquad \text{if} \quad x > -1.$$

Thus, to verify (8.19), we must show that

$$\lim_{n \to \infty} \left[ \int_1^{1+x} \frac{(1 - t)^n}{t} \, dt \right] = 0 \qquad \text{if} \quad -1 < x \leq 1. \tag{8.20}$$

First, suppose $0 \leq x \leq 1$. Then for each natural number $n$,

$$\left| \int_1^{1+x} \frac{(1 - t)^n}{t} \, dt \right| = \int_1^{1+x} \frac{(t - 1)^n}{t} \, dt$$

$$\leq \int_1^{1+x} (t - 1)^n \, dt$$

$$= \frac{x^{n+1}}{n + 1}$$

$$\leq \frac{1}{n + 1}.$$

Since $\lim_{n\to\infty} 1/n = 0$, we see that (8.20) holds if $0 \leq x \leq 1$.

Now suppose that $-1 < x < 0$. Then for each natural number $n$,

$$\left| \int_1^{1+x} \frac{(1-t)^n}{t}\, dt \right| = \int_{1+x}^1 \frac{(1-t)^n}{t}\, dt$$

$$\leq \frac{1}{1+x} \int_{1+x}^1 (1-t)^n\, dt$$

$$= \left( \frac{1}{1+x} \right) \frac{|x|^{n+1}}{n+1}$$

$$\leq \left( \frac{1}{1+x} \right) \frac{1}{n+1}.$$

Since $\lim_{n\to\infty} 1/n = 0$, we also see that (8.20) holds if $-1 < x < 0$.   ■

In spite of the fact that the function $f(x) = \ln(1+x)$ has derivatives of all orders for all $x > -1$, the Taylor expansion (8.19) is not valid if $x > 1$. Indeed, in view of (8.18) and Bernoulli's Inequality, it follows that if $x > 1$ and $n$ is any natural number, then

$$|\ln(1+x) - p_n(x)| = \int_1^{1+x} \frac{(1-t)^n}{t}\, dt$$

$$\geq \frac{1}{1+x} \int_1^{1+x} (1-t)^n\, dt$$

$$= \frac{1}{1+x} \cdot \frac{x^{n+1}}{n+1}$$

$$= \frac{1}{1+x} \cdot \frac{[1+x-1]^{n+1}}{n+1}$$

$$\geq \frac{1}{1+x} \cdot \frac{1+(n+1)(x-1)}{n+1}$$

$$\geq \frac{x-1}{1+x}.$$

Thus, $\lim_{n\to\infty} [f(x) - p_n(x)] \neq 0$ if $x > 1$.

**EXERCISES**

1. Prove that for each natural number $n$,

$$|\ln(1+x) - p_n(x)| \leq \frac{1}{1+x} \cdot \frac{|x|^{n+1}}{n+1} \quad \text{if} \quad -1 < x \leq 0,$$

and $\qquad |\ln(1+x) - p_n(x)| \leq \frac{x^{n+1}}{n+1} \quad \text{if} \quad 0 \leq x \leq 1.$

Estimate $\ln(1.1)$ with an error of at most $10^{-4}$.

2. Show that with the error estimates in the previous exercise, we need $n = 10{,}000$ to estimate $\ln 2$ with an error bound of $10^{-4}$. How does the identity $\ln 2 = \ln 4/3 - \ln 2/3$ allow us to estimate $\ln 2$ more efficiently?

3. Explain how the identity

$$s = \left(1 + \frac{s-1}{s+1}\right) \Big/ \left(1 - \frac{s-1}{s+1}\right) \quad \text{if} \quad s \neq 0$$

allows us to efficiently compute $\ln(1+x)$ if $-1 < x < 1$ and $x$ is close to $-1$ or $1$.

4. Verify the integral inequalities in the proof of Theorem 8.13.

5. At what points $x$ in the interval $(-1, 1]$ can one use the Lagrange Remainder Theorem to verify the expansion

$$\ln(1+x) = \sum_{k=1}^{\infty} (-1)^{k+1} \frac{x^k}{k!}?$$

## 8.5 The Cauchy Integral Remainder Formula and the Binomial Expansion

If $I$ is an open interval containing the point $x_0$ and the function $f: I \to \mathbb{R}$ is differentiable, then for each point $x$ in $I$, there is a point $c$ strictly between $x$ and $x_0$ such that

$$f(x) = f(x_0) + f'(c)(x - x_0). \tag{8.21}$$

If we further assume that the derivative $f': I \to \mathbb{R}$ is continuous, then, by the First Fundamental Theorem of Calculus,

$$f(x) = f(x_0) + \int_{x_0}^{x} f'(t)\,dt. \tag{8.22}$$

The proof of the Lagrange Remainder Theorem was rooted in the Lagrange Mean Value Theorem as expressed in (8.21). The proof of the following Remainder Theorem will exploit the First Fundamental Theorem of Calculus as expressed in (8.22).

**THEOREM 8.14**  **The Cauchy Integral Remainder Formula**  Let $I$ be an open interval containing the point $x_0$ and $n$ be a natural number. Suppose that the function $f: I \to \mathbb{R}$ has $n+1$ derivatives and that $f^{(n+1)}: I \to \mathbb{R}$ is continuous. Then for each point $x$ in $I$,

$$f(x) = \sum_{k=0}^{n} \frac{f^{(k)}(x_0)}{k!}(x - x_0)^k + \frac{1}{n!}\int_{x_0}^{x} f^{(n+1)}(t)(x-t)^n\,dt. \tag{8.23}$$

**Proof**   By the First Fundamental Theorem of Calculus,

$$f(x) = f(x_0) + \int_{x_0}^{x} f'(t)\,dt. \tag{8.24}$$

Integrating by parts, we see that

$$\int_{x_0}^{x} f'(t)\,dt = -\int_{x_0}^{x} f'(t)\frac{d}{dt}(x-t)\,dt$$

$$= -f'(t)(x-t)\Big|_{t=x_0}^{t=x} + \int_{x_0}^{x} f''(t)(x-t)\,dt \tag{8.25}$$

$$= f'(x_0)(x-x_0) + \int_{x_0}^{x} f''(t)(x-t)\,dt.$$

From (8.24) and (8.25) we obtain (8.23) when $n = 1$. The general formula follows by induction. The inductive step depends on observing that if $1 \le k \le n-1$, then

$$\frac{1}{k!}\int_{x_0}^{x} f^{(k+1)}(t)(x-t)^k\,dt = \frac{-1}{(k+1)!}\int_{x_0}^{x} f^{(k+1)}(t)\frac{d}{dt}[(x-t)^{k+1}]\,dt$$

$$= \frac{1}{(k+1)!} f^{(k+1)}(x_0)(x-x_0)^{k+1}$$

$$+ \frac{1}{(k+1)!}\int_{x_0}^{x} f^{(k+2)}(t)(x-t)^{k+1}\,dt. \qquad \blacksquare$$

Recall from Section 1.3 that for each natural number $n$ and pair of numbers $a$ and $b$, we have the

## Binomial Formula

$$(a+b)^n = \sum_{k=0}^{n} \binom{n}{k} a^{n-k} b^k. \tag{8.26}$$

This formula can be proved directly, using elementary algebra. Another proof can be given by first applying Corollary 8.4 to show that

$$(1+x)^n = \sum_{k=0}^{n} \binom{n}{k} x^k \quad \text{for all } x. \tag{8.27}$$

Then, if $a \ne 0$, substitute $x = b/a$ in (8.27) and multiply by $a^n$ to obtain (8.26).

We will now extend formula (8.27) to the case of exponents that are not necessarily natural numbers. Of course, if $\beta$ is not a nonnegative integer, then the function $(1+x)^\beta$ is not a polynomial, so the right-hand side of (8.27), rather than being a polynomial, will be an infinite series. In order to find an infinite series expansion of $(1+x)^\beta$, it is useful to extend the definition of the binomial coefficients. For each natural number $n$, and each number $\beta$, we define

$$\binom{\beta}{k} \equiv \frac{\beta(\beta - 1) \cdots (\beta - k + 1)}{k!},$$

and define

$$\binom{\beta}{0} \equiv 1.$$

**THEOREM 8.15** **Newton's Binomial Expansion** Let $\beta$ be any real number. Then

$$(1 + x)^\beta = \sum_{k=0}^{\infty} \binom{\beta}{k} x^k \quad \text{if} \quad -1 < x < 1. \tag{8.28}$$

To verify the Binomial Expansion is to prove that

$$\lim_{n \to \infty} \left[ (1 + x)^\beta - \sum_{k=0}^{n} \binom{\beta}{k} x^k \right] = 0 \quad \text{if} \quad -1 < x < 1.$$

The crucial step in the proof is to show that the Binomial Expansion is a Taylor expansion about $x = 0$ so that the Cauchy Integral Remainder Formula provides an integral representation for the remainders. We summarize this step in the following lemma.

**LEMMA 8.16** For any number $\beta$ and any natural number $n$, if $x > -1$, then

$$(1 + x)^\beta - \sum_{k=0}^{n} \binom{\beta}{k} x^k = (n + 1) \binom{\beta}{n + 1} \int_0^x (1 + t)^{\beta - n - 1} (x - t)^n \, dt. \tag{8.29}$$

**Proof** Define the function $f: (-1, \infty) \to \mathbb{R}$ by

$$f(x) = (1 + x)^\beta \quad \text{if} \quad x > -1.$$

Observe that for each natural number $k$,

$$f^{(k)}(x) = \beta(\beta - 1) \cdots (\beta - k + 1)(1 + x)^{\beta - k} \quad \text{if} \quad x > -1,$$

so that

$$\frac{f^{(k)}(x)}{k!} = \binom{\beta}{k}(1 + x)^{\beta - k}$$

and, in particular,

$$\frac{f^{(k)}(0)}{k!} = \binom{\beta}{k}.$$

Thus, the $n$th Taylor polynomial for $f: (-1, \infty) \to \mathbb{R}$ at $x = 0$ is

$$p_n(x) = \sum_{k=0}^{n} \binom{\beta}{k} x^k.$$

According to the Cauchy Integral Remainder Theorem, for each natural number $n$ and each number $x > -1$,

$$f(x) - p_n(x) = \frac{1}{n!} \int_0^x f^{(n+1)}(t)(x - t)^n \, dt$$

$$= \frac{1}{n!} \int_0^x \binom{\beta}{n+1}(n+1)!(1 + t)^{\beta - n - 1}(x - t)^n \, dt$$

$$= (n + 1)\binom{\beta}{n+1} \int_0^x (1 + t)^{\beta - n - 1}(x - t)^n \, dt. \qquad \blacksquare$$

In order to analyze the behavior of the difference (8.29) when $n$ is large, it is convenient first to prove the following lemma, which will also be useful in Chapter 9.

**LEMMA 8.17**   **The Ratio Lemma for Sequences**   Suppose that $\{c_n\}$ is a sequence of nonzero numbers that has the property that

$$\lim_{n \to \infty} \frac{|c_{n+1}|}{|c_n|} = \ell.$$

(i) If $\ell < 1$, then

$$\lim_{n \to \infty} c_n = 0.$$

(ii) If $\ell > 1$, then the sequence $\{c_n\}$ is unbounded.

**Proof**   First, suppose that $0 \le \ell < 1$. Define $\alpha = (\ell + 1)/2$. Since $\ell < \alpha$, we may choose a natural number $N$ such that

$$\frac{|c_{n+1}|}{|c_n|} \le \alpha \quad \text{for all integers } n \ge N.$$

For each natural number $k$, if we successively apply the preceding inequality $k$ times, we see that

$$|c_{N+k}| \le |c_N|\alpha^k.$$

Hence, if we define $M = |c_N|\alpha^{-N}$, it follows that

$$|c_n| \le M\alpha^n \quad \text{for all integers } n \ge N.$$

However, $0 \le \alpha < 1$, so $\lim_{n \to \infty} \alpha^n = 0$. The preceding inequality implies that $\lim_{n \to \infty} c_n = 0$.

Now suppose that $\ell > 1$. Define $\beta = (\ell + 1)/2$. Since $\beta < \ell$, we may choose a natural number $N$ such that

$$\frac{|c_{n+1}|}{|c_n|} \ge \beta \quad \text{for all integers } n \ge N.$$

Hence, for each natural number $k$,

$$|c_{N+k}| \ge |c_N|\beta^k, \qquad (8.30)$$

and since, by Bernoulli's Inequality,

$$\beta^k \ge 1 + k(\beta - 1),$$

(8.30) implies that the sequence $\{c_n\}$ is unbounded. $\qquad \blacksquare$

**LEMMA 8.18** Let $\beta$ be any number. Then

$$\lim_{n \to \infty} n\binom{\beta}{n}x^n = 0 \quad \text{if} \quad |x| < 1.$$

**Proof** Observe that for each natural number $n$,

$$(n+1)\binom{\beta}{n+1}\Big/n\binom{\beta}{n} = \frac{n+1}{n} \cdot \frac{\beta-n}{n+1}.$$

Thus,

$$\lim_{n \to \infty}\left|(n+1)\binom{\beta}{n+1}|x|^{n+1}\Big/n\binom{\beta}{n}|x|^n\right| = |x|.$$

The conclusions follow immediately from the Ratio Lemma. ∎

**Proof of Newton's Binomial Expansion**   First, we consider the case when $-1 < x < 0$. When we write $(x - t) = -(t - x)$ and interchange the limits of integration, formula (8.29) becomes

$$f(x) - p_n(x) = (-1)^{n+1}(n+1)\binom{\beta}{n+1}\int_x^0 \left(\frac{t-x}{1+t}\right)^n (1+t)^{\beta-1}\,dt. \qquad (8.31)$$

But observe that

$$0 \le \left(\frac{t-x}{1+t}\right) \le -x = |x| \quad \text{if} \quad -1 < x \le t \le 0,$$

so for each natural number $n$,

$$0 \le \left(\frac{t-x}{1+t}\right)^n (1+t)^{\beta-1} \le |x|^n \quad \text{if} \quad -1 < x \le t \le 0. \qquad (8.32)$$

From (8.31) and (8.32), it follows that for each natural number $n$,

$$|f(x) - p_n(x)| = (n+1)\binom{\beta}{n+1}\int_x^0 \left(\frac{t-x}{1+t}\right)^n (1+t)^{\beta-1}\,dt$$

$$\le (n+1)\binom{\beta}{n+1}\int_x^0 |x|^n\,dt$$

$$\le (n+1)\binom{\beta}{n+1}|x|^n\int_x^0 dt$$

$$= (n+1)\binom{\beta}{n+1}|x|^{n+1} \quad \text{if} \quad -1 < x < 0. \qquad (8.33)$$

According to Lemma 8.18, if $|x| < 1$,

$$\lim_{n \to \infty}(n+1)\binom{\beta}{n+1}|x|^{n+1} = 0,$$

so from (8.33) we conclude that the Binomial Expansion is valid if $-1 < x < 0$.

It remains to consider the case when $0 < x < 1$. In this case, from (8.29), we obtain

$$|f(x) - p_n(x)| = (n+1)\binom{\beta}{n+1}\int_0^x (1+t)^{\beta-n-1}(x-t)^n\, dt$$

$$= (n+1)\binom{\beta}{n+1}\int_0^x \left[\frac{x-t}{1+t}\right]^n (1+t)^{\beta-1}\, dt$$

$$\leq (n+1)\binom{\beta}{n+1}x^n \int_0^x (1+t)^{\beta-1}\, dt$$

$$= (n+1)\binom{\beta}{n+1}x^{n+1}\left[\frac{(1+x)^\beta - 1}{\beta x}\right]. \qquad (8.34)$$

Again using Lemma 8.18, from (8.34) we conclude that the Binomial Expansion is valid if $0 < x < 1$. ∎

---

**EXERCISES**

1. Verify all of the details in the derivations of the inequalities (8.32), (8.33), and (8.34).
2. Show that for $\beta = -1$, the Binomial Expansion reduces to the Geometric Series.
3. Show that for $\beta$ a natural number, the Binomial Expansion reduces to the Binomial Formula.
4. Prove that if the functions $g\colon [a, b] \to \mathbb{R}$ and $h\colon [a, b] \to \mathbb{R}$ are continuous with $h(x) \geq 0$ for all $x$ in $[a, b]$, then there is a point $c$ in $(a, b)$ such that

$$\int_a^b h(x)g(x)\, dx = g(c)\int_a^b h(x)\, dx.$$

5. Use Exercise 4 to show that the Cauchy Integral Remainder Theorem implies the Lagrange Remainder Theorem if $f^{n+1}\colon I \to \mathbb{R}$ is assumed to be continuous.
6. Apply the Cauchy Integral Remainder Theorem in the analysis of the expansion

$$\ln(1+x) = \sum_{k=1}^\infty (-1)^{k+1}\frac{x^k}{k} \quad \text{if} \quad -1 < x \leq 1.$$

7. Show that for $0 \leq x < 1$, the Lagrange Remainder Theorem can be used to verify the Binomial Expansion.
8. Prove that the Binomial Expansion does not converge if $|x| > 1$.
9. For what values of $r$ does the sequence $\{n^3 r^n\}$ converge?

---

## 8.6   An Infinitely Differentiable Function That Is Not Analytic

We will now present an explicit example of a function $f\colon \mathbb{R} \to \mathbb{R}$ that has derivatives of all orders and yet the only point at which its Taylor expansion about $x = 0$ agrees with its functional value is at $x = 0$.

**THEOREM 8.19**   Define

$$f(x) = \begin{cases} e^{-(1/x^2)} & \text{if } x \neq 0 \\ 0 & \text{if } x = 0. \end{cases}$$

Then the function $f: \mathbb{R} \to \mathbb{R}$ has derivatives of all orders. However, the only point at which

$$f(x) = \sum_{k=0}^{\infty} \frac{f^{(k)}(0)}{k!} x^k \tag{8.35}$$

is at $x = 0$.

**Proof**   To prove the theorem it will suffice to prove that $f: \mathbb{R} \to \mathbb{R}$ has derivatives of all orders and that for each natural number $n$, $f^{(n)}(0) = 0$. Once this is proven, simply observe that the right-hand side of (8.35) is identically zero, and $f(x) = 0$ if and only if $x = 0$.

**Step 1:** We claim that for any polynomial $q: \mathbb{R} \to \mathbb{R}$,

$$\lim_{x \to 0} q\left(\frac{1}{x}\right) e^{-(1/x^2)} = 0. \tag{8.36}$$

In order to verify this, it suffices to show that for each natural number $n$,

$$\lim_{x \to 0} \frac{e^{-(1/x^2)}}{x^n} = 0. \tag{8.37}$$

Indeed, let $n$ be a natural number. According to (8.5),

$$e^b > \frac{b^n}{n!} \quad \text{if} \quad b > 0,$$

so that

$$e^{(1/x^2)} \geq \frac{1}{n! x^{2n}} \quad \text{if} \quad x \neq 0,$$

and hence

$$0 \leq \left| \frac{e^{-(1/x^2)}}{x^n} \right| \leq n! |x|^n \quad \text{if} \quad x \neq 0.$$

This inequality implies (8.37), which in turn implies (8.36).

**Step 2:** We will argue by induction to show that for each natural number $n$, there is a polynomial $q_n: \mathbb{R} \to \mathbb{R}$ such that

$$f^{(n)}(x) = q_n\left(\frac{1}{x}\right) e^{-(1/x^2)} \quad \text{if} \quad x \neq 0. \tag{8.38}$$

Indeed, $f'(x) = (2/x^3)e^{-(1/x^2)}$ if $x \neq 0$, so (8.38) holds when $n = 1$, where $q_1(t) = 2t^3$. Suppose (8.38) holds with $n = k$. Then

$$f^{(k+1)}(x) = \left[ q_k'\left(\frac{1}{x}\right)\left(\frac{-1}{x^2}\right) + q_k\left(\frac{1}{x}\right)\left(\frac{2}{x^3}\right) \right] e^{-(1/x^2)} \quad \text{if} \quad x \neq 0,$$

so (8.38) holds if $n = k + 1$ where $q_{k+1}(t) = q_k'(t)(-t^2) + q_k(t)(2t^3)$. The Principle of Mathematical Induction implies that (8.38) holds for all natural numbers.

From step 2 it follows that the function $f: \mathbb{R} \to \mathbb{R}$ has derivatives of all orders at every point $x \neq 0$. To complete the proof, we will show, again by induction, that for each natural number $n$,

$$f^{(n)}(0) = 0. \tag{8.39}$$

Indeed, if $n = 1$, then, using (8.36) with $q(t) = t$ for all $t$, it follows that

$$\lim_{x \to 0} \frac{f(x) - f(0)}{x} = \lim_{x \to 0} \frac{1}{x} e^{-(1/x^2)} = 0.$$

Now suppose that $k$ is a natural number such that $f^{(k)}(0) = 0$. Then, using (8.38), together with (8.36) with $q(t) = t q_k(t)$ for all $t$, it follows that

$$\lim_{x \to 0} \frac{f^{(k)}(x) - f^{(k)}(0)}{x} = \lim_{x \to 0} \frac{1}{x} q_k\left(\frac{1}{x}\right) e^{-(1/x^2)} = 0,$$

so $f^{(k+1)}(0) = 0$. The Principle of Mathematical Induction implies that for each natural number $n$, $f^{(n)}(0) = 0$.  ∎

A function $g: \mathbb{R} \to \mathbb{R}$ that has derivatives of all orders is said to be *infinitely differentiable*. A function $g: \mathbb{R} \to \mathbb{R}$ that has derivatives of all orders such that

$$g(x) = \sum_{k=0}^{\infty} \frac{g^{(k)}(0) x^k}{k!} \quad \text{for all } x$$

is said to be *analytic*. We have exhibited a function $f: \mathbb{R} \to \mathbb{R}$ that is infinitely differentiable and not analytic.

---

**EXERCISES**

1. Let the function $f: \mathbb{R} \to \mathbb{R}$ be as in Theorem 8.19. Explicitly compute $f''(x)$.
2. Let the function $f: \mathbb{R} \to \mathbb{R}$ be as in Theorem 8.19. Show that there is no positive number $M$ such that for each natural number $n$,

   $$|f^{(n)}(x)| \leq M^n \quad \text{for all } x.$$

3. For $n$ a natural number, a function $f: \mathbb{R} \to \mathbb{R}$ is said to be *n times continuously differentiable* provided that $f: \mathbb{R} \to \mathbb{R}$ has an $n$th derivative and $f^{(n)}: \mathbb{R} \to \mathbb{R}$ is continuous. Define

   $$h(x) = \int_0^x |t|\, dt \quad \text{for all } x.$$

   Show that the function $h: \mathbb{R} \to \mathbb{R}$ is once continuously differentiable but is not twice continuously differentiable. For each natural number $n$, find a function that is $n$ times continuously differentiable but is not $n + 1$ times continuously differentiable.
4. Suppose that the function $g: \mathbb{R} \to \mathbb{R}$ has derivatives of all orders, and that for each natural number $n$, there are positive numbers number $c_n$ and $\delta_n$ such that

   $$|g(x)| \leq c_n |x|^n \quad \text{if} \quad |x| < \delta_n.$$

   Prove that for each natural number $n$, $g^{(n)}(0) = 0$.

## 8.7   The Weierstrass Approximation Theorem

As we have seen in Section 8.6, even if a function has derivatives of all orders, the Taylor polynomial for the function computed at a point $x_0$ in its domain may not provide a good approximation for the function at any point other than $x_0$. Nevertheless, there is the following remarkable theorem.

**THEOREM 8.20**   **The Weierstrass Approximation Theorem**   Let $I$ be a closed and bounded interval and suppose that the function $f: I \to \mathbb{R}$ is continuous. Then for each positive number $\epsilon$, there is a polynomial $p: \mathbb{R} \to \mathbb{R}$ such that

$$|f(x) - p(x)| < \epsilon \quad \text{for all points } x \text{ in } I. \tag{8.40}$$

What is remarkable about this theorem is that there is no assumption about differentiability. For instance, we allow the possibility that there is no point at which the function $f: I \to \mathbb{R}$ is differentiable. Of course, the polynomial that satisfies (8.40) will not, in general, be a Taylor polynomial.

The proof of the Approximation Theorem, which we will present, is due to Bernstein and is quite ingenious. At its roots lie the following three identities:

For each natural number $n$ and any number $x$,

$$\sum_{k=0}^{n} \binom{n}{k} x^k (1 - x)^{n-k} = 1, \tag{8.41}$$

$$\sum_{k=0}^{n} \frac{k}{n} \binom{n}{k} x^k (1 - x)^{n-k} = x, \tag{8.42}$$

and if $n \geq 2$,

$$\sum_{k=0}^{n} \frac{k(k - 1)}{n(n - 1)} \binom{n}{k} x^k (1 - x)^{n-k} = x^2. \tag{8.43}$$

The first identity (8.41) follows from the Binomial Formula, formula (8.26), by setting $a = x$ and $b = 1 - x$. Identities (8.42) and (8.43) are consequences of (8.41). Indeed, if in the identity (8.41) we replace $n$ by $n - 1$ and multiply both sides by $x$, then since

$$\binom{n-1}{k-1} = \frac{k}{n} \binom{n}{k} \quad \text{if} \quad 1 \leq k \leq n,$$

we obtain (8.42). Similarly, if in the identity (8.41) we replace $n$ by $n - 2$ and multiply both sides by $x^2$, then since

$$\binom{n-2}{k-2} = \frac{k(k - 1)}{n(n - 1)} \binom{n}{k} \quad \text{if} \quad 2 \leq k \leq n,$$

we obtain (8.43).

For a natural number $n$, and an integer $k$ such that $0 \leq k \leq n$, it is notationally convenient to define

$$g_k(x) = x^k (1 - x)^{n-k} \quad \text{for all } x \text{ in } \mathbb{R}.$$

**LEMMA 8.21** For each number $x$ and each natural number $n \geq 2$,

$$\sum_{k=0}^{n} \left(x - \frac{k}{n}\right)^2 \binom{n}{k} x^k (1-x)^{n-k} = \frac{x(1-x)}{n}. \qquad (8.44)$$

**Proof** For each integer $n \geq 2$, it follows from (8.41) and (8.42) that

$$\sum_{k=0}^{n} \left(x - \frac{k}{n}\right)^2 \binom{n}{k} g_k(x) = \sum_{k=0}^{n} \left[x^2 - \frac{2xk}{n} + \frac{k^2}{n^2}\right] \binom{n}{k} g_k(x)$$

$$= x^2 \sum_{k=0}^{n} \binom{n}{k} g_k(x) - 2x \sum_{k=0}^{n} \frac{k}{n} \binom{n}{k} g_k(x) + \sum_{k=0}^{n} \frac{k^2}{n^2} \binom{n}{k} g_k(x)$$

$$= x^2 - 2x^2 + \sum_{k=0}^{n} \frac{k^2}{n^2} \binom{n}{k} g_k(x).$$

On the other hand, from (8.42) and (8.43) we have

$$\sum_{k=0}^{n} \frac{k^2}{n^2} \binom{n}{k} g_k(x) = \frac{n-1}{n} \sum_{k=0}^{n} \frac{k^2}{n(n-1)} \binom{n}{k} g_k(x)$$

$$= \frac{n-1}{n} \sum_{k=0}^{n} \left[\frac{k(k-1)}{n(n-1)} + \frac{k}{n(n-1)}\right] \binom{n}{k} g_k(x)$$

$$= \left[\frac{n-1}{n}\right] \left[x^2 + \frac{x}{n-1}\right].$$

Hence,

$$\sum_{k=0}^{n} \left(x - \frac{k}{n}\right)^2 \binom{n}{k} g_k(x) = x^2 - 2x^2 + \left[\frac{n-1}{n}\right]\left[x^2 + \frac{x}{n-1}\right] = \frac{x(1-x)}{n}. \quad \blacksquare$$

**Proof of Weierstrass Approximation Theorem** We will first consider the case when $I = [0, 1]$; the general case follows easily from this case. Let $\epsilon > 0$. According to Theorem 3.14, the function $f: I \to \mathbb{R}$ is uniformly continuous. Hence we can choose $\delta > 0$ such that

$$|f(u) - f(v)| < \frac{\epsilon}{2} \quad \text{for all points } u \text{ and } v \text{ in } I \text{ such that } |u - v| \leq \delta. \qquad (8.45)$$

Also, it follows from the Extreme Value Theorem that the function $f: I \to \mathbb{R}$ is bounded. Thus we can choose a number $M > 0$ such that

$$|f(x)| \leq M \quad \text{for all } x \text{ in } I. \qquad (8.46)$$

Using the Archimedean Property of $\mathbb{R}$, we can select a natural number $n$ such that

$$n > \frac{4M}{\epsilon \delta^2}. \qquad (8.47)$$

Define the polynomial $p: \mathbb{R} \to \mathbb{R}$ by

$$p(x) = \sum_{k=0}^{n} f\left(\frac{k}{n}\right)\binom{n}{k} x^k (1-x)^{n-k} \qquad \text{for all } x.$$

We will show that for this choice of polynomial, the required approximation property (8.40) holds. Indeed, let $x$ be a point in $I$. Then if $0 \le k \le n$, either $|x - k/n| < \delta$ or $|x - k/n| \ge \delta$. If $|x - k/n| < \delta$, it follows from (8.45) that $|f(x) - f(k/n)| < \epsilon/2$. If $|x - k/n| \ge \delta$, then, in view of (8.46),

$$\left| f(x) - f\left(\frac{k}{n}\right) \right| \le 2M \le \frac{2M}{\delta^2}\left(x - \frac{k}{n}\right)^2.$$

Thus, $\qquad \left| f(x) - f\left(\frac{k}{n}\right) \right| \le \frac{\epsilon}{2} + \frac{2M}{\delta^2}\left(x - \frac{k}{n}\right)^2 \qquad \text{for} \quad 0 \le k \le n.$ $\qquad$ (8.48)

From (8.41) it follows that

$$f(x) = \sum_{k=0}^{n} f(x)\binom{n}{k} x^k (1-x)^{n-k},$$

so $\qquad f(x) - p(x) = \sum_{k=0}^{n}\left[ f(x) - f\left(\frac{k}{n}\right) \right]\binom{n}{k} x^k (1-x)^{n-k}.$

Using the Triangle Inequality, (8.48), (8.41), and (8.44), it follows that

$$|f(x) - p(x)| \le \sum_{k=0}^{n} \left| f(x) - f\left(\frac{k}{n}\right) \right| \binom{n}{k} x^k (1-x)^{n-k}$$

$$\le \sum_{k=0}^{n} \left[ \frac{\epsilon}{2} + \frac{2M}{\delta^2}\left(x - \frac{k}{n}\right)^2 \right]\binom{n}{k} x^k (1-x)^{n-k}$$

$$= \frac{\epsilon}{2} + \frac{2M}{\delta^2} \sum_{k=0}^{n}\left(x - \frac{k}{n}\right)^2 \binom{n}{k} x^k (1-x)^{n-k}$$

$$= \frac{\epsilon}{2} + \frac{2M}{n\delta^2} x(1-x)$$

$$< \frac{\epsilon}{2} + \frac{2M}{n\delta^2}.$$

Consequently, since $n$ satisfies (8.47), $|f(x) - p(x)| < \epsilon$.

This proves the theorem if $I = [0, 1]$. Now let $I = [a, b]$. Define $g(t) = a + t(b-a)$ if $0 \le t \le 1$. Then the composite function $f \circ g: [0, 1] \to \mathbb{R}$ is continuous. By what we have just proven, there is a polynomial $q: \mathbb{R} \to \mathbb{R}$ such that $|f(g(t)) - q(t)| < \epsilon$ for all points $t$ in $[0, 1]$. Now define $p(x) = q((x-a)/(b-a))$ for all $x$. Then $p: \mathbb{R} \to \mathbb{R}$ is a polynomial for which the approximation property (8.40) holds. ∎

**EXERCISES**

1.  Show that if $n \geq k \geq 1$, then

$$\frac{k}{n}\binom{n}{k} = \binom{n-1}{k-1}.$$

Use this, together with (8.41) with $n$ replaced by $n-1$, to verify (8.42).

2.  Show that if $n \geq k \geq 2$, then

$$\frac{k(k-1)}{n(n-1)}\binom{n}{k} = \binom{n-2}{k-2}.$$

Use this, together with (8.41) with $n$ replaced by $n-2$, to verify (8.43).

3.  In the proof of the Approximation Theorem, where did we use the fact that $g_k(x) \geq 0$ for all $x$ in $[0, 1]$ and $0 \leq k \leq n$?

4.  Show that the Approximation Theorem does not hold if we replace $I$ by $\mathbb{R}$, by showing that if $f(x) = e^x$ for all $x$, then $f : \mathbb{R} \to \mathbb{R}$ cannot be uniformly approximated by polynomials.

5.  Define $f(x) = |x - 1/2|$ for $0 \leq x \leq 1$. Use the proof of the Approximation Theorem to find an explicit polynomial $p : \mathbb{R} \to \mathbb{R}$ such that $|f(x) - p(x)| < 1/4$ for all $x$ in $[0, 1]$.

6.  Verify the assertion about the composition that was made on the last line of the proof of the Approximation Theorem.

7.  Suppose that the function $h : [-1, 1] \to \mathbb{R}$ is continuous. Prove that there is a sequence $\{p_k : \mathbb{R} \to \mathbb{R}\}$ of polynomials having the property that

$$h(x) = \sum_{k=1}^{\infty} p_k(x) \quad \text{for all points } x \text{ in } [-1, 1].$$

# 9

# The Convergence of Sequences and Series of Functions

## 9.1   Sequences and Series of Numbers

In Chapter 2, we studied the convergence of sequences of numbers. In particular, in Section 2.2 we proved the following important result.

**THEOREM 9.1**   **The Monotone Convergence Theorem**   A monotone sequence of numbers converges if and only if it is bounded.

This is a criterion for convergence that is intrinsic to the sequence itself; it does not require any information about the proposed limit. But the Monotone Convergence Theorem does require that the sequence be monotone. As the sequence $\{(-1)^n\}$ shows, in general it is not true that any bounded sequence converges. We will now establish a criterion for convergence that applies to *all* sequences of numbers.

**DEFINITION**   *A sequence of numbers $\{a_n\}$ is said to be a Cauchy sequence provided that for each positive number $\epsilon$ there is a natural number $N$ such that*

$$|a_n - a_m| < \epsilon \quad if \quad n \geq N \quad and \quad m \geq N.$$

We will prove that a sequence of numbers converges if and only if it is a Cauchy sequence. This too is a criterion for convergence that is intrinsic to the sequence itself; it does not require any information about the proposed limit. Moreover, monotonicity is not required. We will prove this result in stages.

**PROPOSITION 9.2**   Every convergent sequence is Cauchy.

**Proof**   Suppose that $\{a_n\}$ is a sequence that converges to the number $a$. Let $\epsilon > 0$. We need to find a natural number $N$ such that

$$|a_n - a_m| < \epsilon \quad if \quad n \geq N \quad and \quad m \geq N.$$

But since $\{a_n\}$ converges to $a$, we may choose a natural number $N$ such that

$$|a_k - a| < \frac{\epsilon}{2} \quad for\ every\ integer\ k \geq N.$$

Thus, if $n \geq N$ and $m \geq N$, the Triangle Inequality implies that

$$|a_n - a_m| = |(a_n - a) + (a - a_m)|$$
$$\leq |a_n - a| + |a_m - a|$$
$$< \frac{\epsilon}{2} + \frac{\epsilon}{2} = \epsilon. \qquad \blacksquare$$

**LEMMA 9.3**   Every Cauchy sequence is bounded.

**Proof**   Suppose that $\{a_n\}$ is a Cauchy sequence. For $\epsilon = 1$, we can choose a natural number $N$ such that

$$|a_n - a_m| < 1 \quad if \quad n \geq N \quad and \quad m \geq N.$$

In particular, we have

$$|a_n - a_N| < 1 \quad for\ all\ integers\ n \geq N.$$

But, by the Reverse Triangle Inequality,

$$|a_n| - |a_N| \leq |a_n - a_N| \quad for\ every\ natural\ number\ n.$$

Consequently, we see that

$$|a_n| \leq |a_N| + 1 \quad for\ all\ integers\ n \geq N.$$

Define $M = \max \{|a_N| + 1, |a_1|, |a_2|, \ldots, |a_{N-1}|\}$. Then

$$|a_n| \leq M \quad for\ every\ natural\ number\ n. \qquad \blacksquare$$

**THEOREM 9.4** **The Cauchy Convergence Criterion for Sequences** A sequence of numbers converges if and only if it is a Cauchy sequence.

**Proof** According to Proposition 9.2, every convergent sequence is a Cauchy sequence. The converse remains to be proven. Suppose that $\{a_n\}$ is a Cauchy sequence. The preceding lemma asserts that $\{a_n\}$ is bounded. Thus, by the Bolzano-Weierstrass Theorem, $\{a_n\}$ has a subsequence $\{a_{n_k}\}$ that converges to a number $a$.

We claim that the whole sequence $\{a_n\}$ converges to $a$. Indeed, let $\epsilon > 0$. We need to find a natural number $N$ such that

$$|a_n - a| < \epsilon \quad \text{for all integers } n \geq N.$$

Since $\{a_n\}$ is a Cauchy sequence, we can choose a natural number $N$ such that

$$|a_n - a_m| < \frac{\epsilon}{2} \quad \text{if} \quad n \geq N \quad \text{and} \quad m \geq N. \tag{9.1}$$

On the other hand, since the subsequence $\{a_{n_k}\}$ converges to $a$, there is a natural number $K$ such that

$$|a_{n_k} - a| < \frac{\epsilon}{2} \quad \text{if} \quad k \geq K. \tag{9.2}$$

Now choose any natural number $k$ such that $k \geq K$ and $n_k \geq N$. Using the inequalities (9.1) and (9.2), together with the Triangle Inequality, it follows that if $n \geq N$, then

$$|a_n - a| = |(a_n - a_{n_k}) + (a_{n_k} - a)|$$

$$\leq |a_n - a_{n_k}| + |a_{n_k} - a|$$

$$< \frac{\epsilon}{2} + \frac{\epsilon}{2} = \epsilon. \qquad \blacksquare$$

Recall that for a sequence of numbers $\{a_k\}$ that is indexed by the natural numbers, we define

$$s_n = \sum_{k=1}^{n} a_k \quad \text{for every natural number } n$$

and obtain a new sequence $\{s_n\}$. The sequence $\{s_n\}$ is called the *sequence of partial sums* for the series $\sum_{k=1}^{\infty} a_k$, and $a_k$ is called the *k*th *term* of the series $\sum_{k=1}^{\infty} a_k$. We write

$$\sum_{k=1}^{\infty} a_k = \lim_{n \to \infty} \left[ \sum_{k=1}^{n} a_k \right]$$

if the sequence $\{s_n\}$ converges. If the sequence $\{s_n\}$ does not converge, we say that the series $\sum_{k=1}^{\infty} a_k$ *diverges*.*

---

*For a sequence of numbers $\{a_k\}$ indexed by the nonnegative integers, we define $s_n = \sum_{k=0}^{n} a_k$ for every nonnegative integer $n$, and then define

$$\sum_{k=0}^{\infty} a_k = \lim_{n \to \infty} \left[ \sum_{k=0}^{n} a_k \right]$$

if the sequence $\{s_n\}$ converges. This change in the initial index for the terms of a series has no material effect on the theory.

**PROPOSITION**   Suppose that the series $\sum_{k=1}^{\infty} a_n$ converges. Then $\lim_{n\to\infty} a_n = 0$.
**9.5**

**Proof**   Define $\{s_n\}$ to be the sequence of partial sums for the series $\sum_{k=1}^{\infty} a_n$ and define $s$ to be the limit of the sequence $\{s_n\}$. Since $\lim_{n\to\infty} s_n = s$, we also have $\lim_{n\to\infty} s_{n-1} = s$. Thus, by the difference property of convergent sequences,

$$\lim_{n\to\infty} [s_n - s_{n-1}] = 0.$$

However, for each natural number $n \geq 2$, $a_n = s_n - s_{n-1}$ and hence $\lim_{n\to\infty} a_n = 0$. ∎

As we have already seen in Chapter 2, the Harmonic Series

$$\sum_{n=1}^{\infty} \frac{1}{n}$$

does not converge, despite the fact that $\lim_{n\to\infty} 1/n = 0$. Thus the convergence of the sequence of terms $\{a_n\}$ to 0 is a necessary, but not sufficient, condition for the series $\sum_{n=1}^{\infty} a_n$ to converge. The remainder of this section will be devoted to presenting conditions on the terms of a series that are sufficient to ensure that the series converges.
Recall that one of the first results we proved about convergent sequences was that

$$\lim_{n\to\infty} r^n = 0 \quad \text{if} \quad |r| < 1. \tag{9.3}$$

This limit, together with the Geometric Sum Formula, is exactly what is needed to establish the convergence of the Geometric Series $\sum_{k=0}^{\infty} r^k$ provided that $|r| < 1$.

**PROPOSITION**   For a number $r$ such that $|r| < 1$,
**9.6**

$$\sum_{k=0}^{\infty} r^k = \frac{1}{1-r}.$$

**Proof**   The Geometric Sum Formula asserts that for each nonnegative integer $n$,

$$\sum_{k=0}^{n} r^k = \frac{1 - r^{n+1}}{1-r}.$$

But $|r| < 1$, so $\lim_{n\to\infty} r^{n+1} = 0$ and hence, by the linearity property of convergent sequences of numbers,

$$\lim_{n\to\infty} \left[ \sum_{k=0}^{n} r^k \right] = \lim_{n\to\infty} \left[ \frac{1 - r^{n+1}}{1-r} \right] = \frac{1}{1-r}. \qquad \blacksquare$$

Given two sequences $\{a_k\}$ and $\{b_k\}$ and two numbers $\alpha$ and $\beta$, observe that for each natural number $n$,

$$\sum_{k=1}^{n} (\alpha a_k + \beta b_k) = \alpha \sum_{k=1}^{n} a_k + \beta \sum_{k=1}^{n} b_k.$$

Thus, from the linearity property of convergent sequences, it follows that if the two series $\sum_{k=1}^{\infty} a_k$ and $\sum_{k=1}^{\infty} b_k$ are convergent, then so is the series $\sum_{k=1}^{\infty} (\alpha a_k + \beta b_k)$, and moreover,

$$\sum_{k=1}^{\infty} (\alpha a_k + \beta b_k) = \alpha \sum_{k=1}^{\infty} a_k + \beta \sum_{k=1}^{\infty} b_k.$$

We have two principal general criteria for a sequence of numbers to converge, namely the Monotone Convergence Theorem and the Cauchy Convergence Criterion. Applying these criteria to series—that is, to sequences of partial sums—we obtain criteria for the convergence of series. First we examine consequences of the Monotone Convergence Theorem.

**THEOREM 9.7**   Suppose that $\{a_k\}$ is a sequence of nonnegative numbers. Then the series $\sum_{k=1}^{\infty} a_k$ converges if and only if there is a positive number $M$ such that

$$a_1 + \cdots + a_n \leq M \quad \text{for every natural number } n.$$

**Proof**   Since the terms of the series $\sum_{k=1}^{\infty} a_k$ are nonnegative, the sequence of partial sums is monotonically increasing. The Monotone Convergence Theorem asserts that the sequence of partial sums converges if and only if the sequence of partial sums is bounded.    ∎

**COROLLARY 9.8**   **The Comparison Test**   Suppose that $\{a_k\}$ and $\{b_k\}$ are sequences of numbers such that for each natural number $k$,

$$0 \leq a_k \leq b_k.$$

(i) The series $\sum_{k=1}^{\infty} a_k$ converges if the series $\sum_{k=1}^{\infty} b_k$ converges.

(ii) The series $\sum_{k=1}^{\infty} b_k$ diverges if the series $\sum_{k=1}^{\infty} a_k$ diverges.

**Proof**   Observe that for each natural number $n$,

$$\sum_{k=1}^{n} a_k \leq \sum_{k=1}^{n} b_k.$$

The result follows from this inequality and Theorem 9.7.    ∎

**EXAMPLE 9.1**   Consider the series

$$\sum_{k=1}^{\infty} \frac{1}{\sqrt{k}\, 2^k}.$$

Since for every natural number $k$, $1/\sqrt{k}\, 2^k \leq 1/2^k$, and the Geometric Series converges, it follows from the Comparison Test that the series $\sum_{k=1}^{\infty} 1/\sqrt{k}\, 2^k$ also converges.    □

**EXAMPLE 9.2**   Consider the series

$$\sum_{k=1}^{\infty} \frac{1}{2+\sqrt{k}}.$$

Since for each integer $k \geq 2$, $1/(2+\sqrt{k}) \geq 1/2k$, and the harmonic series $\sum_{k=1}^{\infty} 1/k$ diverges, it follows from the Comparison Test that the series $\sum_{k=1}^{\infty} 1/(2+\sqrt{k})$ also diverges.                                                                              □

**COROLLARY 9.9**   **The Integral Test**   Let $\{a_k\}$ be a sequence of nonnegative numbers and suppose that the function $f:[1, \infty) \to \mathbb{R}$ is monotonically decreasing and has the property that

$$f(k) = a_k \quad \text{for every natural number } k.$$

Then the series $\sum_{k=1}^{\infty} a_k$ is convergent if and only if the sequence $\{\int_1^n f(x)dx\}$ is bounded.

**Proof**   Since the function $f:[1, \infty) \to \mathbb{R}$ is monotonically decreasing, its restriction to each bounded interval is integrable. Moreover, for each natural number $k$ and each point $x$ in the interval $[k, k+1]$,

$$a_k = f(k) \geq f(x) \geq f(k+1) = a_{k+1},$$

so the monotonicity property of integration implies that for each natural number $n$,

$$\sum_{k=1}^{n} a_k \geq \int_1^{n+1} f(x)\,dx \geq \sum_{k=2}^{n+1} a_k.$$

**FIGURE 9.1**

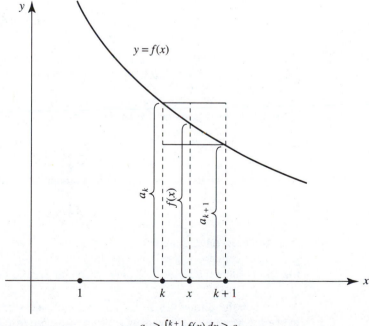

$$a_k \geq \int_k^{k+1} f(x)\,dx \geq a_{k+1}$$

This inequality implies that the sequence of partial sums for the series $\sum_{k=1}^{\infty} a_k$ is bounded if and only if the sequence $\{\int_1^n f(x)\,dx\}$ is bounded. In view of Theorem 9.7, it follows that the series $\sum_{k=1}^{\infty} a_k$ is convergent if and only if the sequence $\{\int_1^n f(x)\,dx\}$ is bounded. ∎

**EXAMPLE 9.3**  Consider the series

$$\sum_{k=1}^{\infty} \frac{1}{(k+1)\ln(k+1)}.$$

Using the First Fundamental Theorem of Calculus, we see that for every natural number $n$,

$$\int_1^n \frac{1}{(x+1)\ln(x+1)}\,dx = \ln(\ln(n+1)) - \ln(\ln 2).$$

Since the sequence $\{\ln(\ln(n+1)) - \ln(\ln 2)\}$ is not bounded, it follows from the Integral Test that the series $\sum_{k=1}^{\infty} 1/[(k+1)\ln(k+1)]$ diverges. □

**COROLLARY 9.10**

**The $p$-Test**  For a positive number $p$, the series

$$\sum_{k=1}^{\infty} \frac{1}{k^p}$$

converges if and only if $p > 1$.

**Proof**  Define $f(x) = x^{-p}$ for $x \geq 1$. The function $f: [1, \infty) \to \mathbb{R}$ is monotone decreasing. For each natural number $n$, the First Fundamental Theorem of Calculus implies that

$$\int_1^n f(x)\,dx = \begin{cases} (n^{1-p} - 1)/(1-p) & \text{if } p \neq 1 \\ \ln n & \text{if } p = 1. \end{cases}$$

Hence the sequence $\{\int_1^n f(x)\,dx\}$ is bounded if and only if $p > 1$. Using the Integral Test, we conclude that the series $\sum_{k=1}^{\infty} 1/k^p$ converges if and only if $p > 1$. ∎

**EXAMPLE 9.4**  Consider the series

$$\sum_{k=1}^{\infty} \frac{k}{e^k}.$$

In Section 8.2, we used the Lagrange Remainder Theorem to prove that

$$e^b > 1 + b + \frac{b^2}{2} + \frac{b^3}{6} > \frac{b^3}{6} \quad \text{if} \quad b > 0.$$

Thus for each natural number $k$, $k/e^k < 6/k^2$. The $p$-Test implies that the series $\sum_{k=1}^{\infty} 1/k^2$ converges, and hence so does the series $\sum_{k=1}^{\infty} 6/k^2$. The Comparison Test now implies that the series $\sum_{k=1}^{\infty} k/e^k$ also converges. □

When the terms of a series fail to be of one sign, it is not possible to directly invoke the Monotone Convergence Theorem in order to check convergence. However, for series whose terms alternate in sign we can indirectly use the Monotone Convergence Criterion to obtain the following convergence test.

**THEOREM 9.11**    **The Alternating Series Test**    Suppose that $\{a_k\}$ is a monotonically decreasing sequence of nonnegative numbers that converges to 0. Then the series

$$\sum_{k=1}^{\infty}(-1)^{k+1}a_k$$

converges.

**Proof**    Define

$$s_n = \sum_{k=1}^{n}(-1)^{k+1}a_k \quad \text{for every natural number } n.$$

In order to prove that the sequence of partial sums $\{s_n\}$ converges, we will first show that the subsequence $\{s_{2n}\}$ converges. Indeed, for each natural number $n$, observe that, since the sequence $\{a_k\}$ is monotonically decreasing,

$$s_{2n+2} - s_{2n} = a_{2n+1} - a_{2n+2} \geq 0,$$

and since the sequence $\{a_k\}$ also consists of nonnegative numbers,

$$s_{2n} = \sum_{k=1}^{n}(a_{2k-1} - a_{2k}) = a_1 - \sum_{k=1}^{n-1}(a_{2k} - a_{2k+1}) - a_{2n} \leq a_1.$$

We conclude that $\{s_{2n}\}$ is monotonically increasing and bounded above by $a_1$. By the Monotone Convergence Theorem, the sequence $\{s_{2n}\}$ converges. Define $s = \lim_{n\to\infty} s_{2n}$. But

$$s_{2n+1} = s_{2n} + a_{2n+1} \quad \text{for every natural number } n.$$

Since $\lim_{n\to\infty} a_n = 0$, it follows that the sequence $\{s_{2n+1}\}$ converges to the same limit $s$.

We claim that the whole sequence $\{s_n\}$ converges to $s$. Indeed, let $\epsilon > 0$. We can choose natural numbers $N_1$ and $N_2$ such that

$$|s_{2n} - s| < \epsilon \quad \text{for all integers } n \geq N_1$$

and

$$|s_{2n+1} - s| < \epsilon \quad \text{for all integers } n \geq N_2.$$

Define $N = \max\{2N_1, \, 2N_2 + 1\}$. Then

$$|s_n - s| < \epsilon \quad \text{for all integers } n \geq N. \qquad \blacksquare$$

**EXAMPLE 9.5** From the Alternating Series Test, it follows that the series

$$\sum_{k=1}^{\infty} \frac{(-1)^{k+1}}{k}$$

converges. In fact, in Section 8.4 we proved that it converged to $\ln 2$. □

For series whose terms are neither of one sign nor alternating in sign, it is natural to apply the Cauchy Convergence Criterion for Sequences to the sequence of partial sums in order to determine if the series converges.

It is sometimes useful, particularly when considering series, to restate the definition of a Cauchy sequence as follows: The sequence $\{s_n\}$ is a Cauchy sequence provided that for each positive number $\epsilon$ there is a natural number $N$ such that for each integer $n \geq N$ and any natural number $k$,

$$|s_{n+k} - s_n| < \epsilon.$$

**THEOREM 9.12** **The Cauchy Convergence Criterion for Series** The series $\sum_{k=1}^{\infty} a_k$ converges if and only if for each positive number $\epsilon$ there is a natural number $N$ such that

$$|a_{n+1} + \cdots + a_{n+k}| < \epsilon \quad \text{for all integers } n \geq N \text{ and all natural numbers } k.$$

**Proof** Apply the Cauchy Convergence Criterion for Sequences to the sequence of partial sums. ∎

**DEFINITION** *The series $\sum_{k=1}^{\infty} a_k$ is said to converge absolutely provided that the series $\sum_{k=1}^{\infty} |a_k|$ converges.*

**COROLLARY 9.13** **The Absolute Convergence Test** The series $\sum_{k=1}^{\infty} a_k$ converges if the series $\sum_{k=1}^{\infty} |a_k|$ converges.

**Proof** From the Triangle Inequality, we conclude that for each pair of natural numbers $n$ and $k$,

$$\left| \sum_{j=n+1}^{n+k} a_j \right| \leq \sum_{j=n+1}^{n+k} |a_j|.$$

Since the series $\sum_{k=1}^{\infty} |a_k|$ converges, it follows from the Cauchy Convergence Criterion for Series that the sequence of partial sums for $\sum_{k=1}^{\infty} |a_k|$ is a Cauchy sequence. The preceding inequality implies that the sequence of partial sums for $\sum_{k=1}^{\infty} a_k$ is also a Cauchy sequence. Once more using the Cauchy Convergence Criterion for Series, it follows that the series $\sum_{k=1}^{\infty} a_k$ converges. ∎

**EXAMPLE 9.6**   The series

$$\sum_{k=1}^{\infty} \frac{\sin k}{k^2}$$

converges. To verify this, first observe that by the $p$-test, with $p = 2$, the series $\sum_{k=1}^{\infty} 1/k^2$ converges. Since for every natural number $k$, $|\sin k| \leq 1$, it follows from the Comparison Test that the series $\sum_{k=1}^{\infty} |\sin k|/k^2$ also converges. The Absolute Convergence Test now implies that the series $\sum_{k=1}^{\infty} \sin k/k^2$ converges.   $\square$

A series that converges but does not converge absolutely is said to *converge conditionally*. The series $\sum_{k=1}^{\infty} (-1)^{k+1}/k$ converges conditionally, since the Harmonic Series diverges, but, by the Alternating Series Test, the series $\sum_{k=1}^{\infty} (-1)^{k+1}/k$ converges.

**THEOREM 9.14**   For the series $\sum_{k=1}^{\infty} a_k$, suppose that there is a number $r$ with $0 \leq r < 1$ and a natural number $N$ such that

$$|a_{n+1}| \leq r|a_n| \quad \text{for all integers } n \geq N. \tag{9.4}$$

Then the series $\sum_{k=1}^{\infty} a_k$ is absolutely convergent.

**Proof**   First, observe that for each natural number $k$, if we apply the inequality (9.4) successively $k$ times, we obtain

$$|a_{N+k}| \leq r^k|a_N|. \tag{9.5}$$

From this inequality and the Geometric Sum Formula, we conclude that for each natural number $k$,

$$|a_1| + \cdots + |a_{N+k}| = |a_1| + \cdots + |a_{N-1}| + |a_N| + \cdots + |a_{N+k}|$$

$$\leq |a_1| + \cdots + |a_{N-1}| + |a_N|[1 + r + \cdots + r^k]$$

$$= |a_1| + \cdots + |a_{N-1}| + |a_N| \left[ \frac{1 - r^{k+1}}{1 - r} \right]$$

$$\leq |a_1| + \cdots + |a_{N-1}| + |a_N| \left[ \frac{1}{1 - r} \right]. \tag{9.6}$$

Define        $M = |a_1| + \cdots + |a_{N-1}| + |a_N| \left[ \dfrac{1}{1 - r} \right].$

Then, from (9.6), it follows that

$$|a_1| + \cdots + |a_n| \leq M \quad \text{for every natural number } n.$$

This means that the sequence of partial sums of the series $\sum_{k=1}^{\infty} |a_k|$ is bounded. According to Theorem 9.7, the series $\sum_{k=1}^{\infty} |a_k|$ converges.   ∎

**COROLLARY 9.15**

**The Ratio Test for Series** For the series $\sum_{k=1}^{\infty} a_k$, suppose that

$$\lim_{n \to \infty} \frac{|a_{n+1}|}{|a_n|} = \ell.$$

(i) If $\ell < 1$, the series $\sum_{n=1}^{\infty} a_n$ converges absolutely.

(ii) If $\ell > 1$, the series $\sum_{n=1}^{\infty} a_n$ diverges.

**Proof** First, suppose that $\ell < 1$. Define $r = (1 + \ell)/2$. Then $\ell < r$, since $\ell < 1$, and so we can choose a natural number $N$ such that

$$\frac{|a_{n+1}|}{|a_n|} < r \quad \text{for all integers } n \geq N.$$

Also, $r < 1$, since $\ell < 1$. The conclusion now follows from Theorem 9.14.

Now suppose that $\ell > 1$. Then it follows from the Ratio Lemma for Sequences (Lemma 8.17) that the sequence $\{a_n\}$ does not converge to 0. Thus the series $\sum_{n=1}^{\infty} a_n$ diverges. ∎

The theory of convergent and divergent series is a broad and deep subject. The present section gives but a brief glimpse of some ways in which the Monotone Convergence Theorem and the Cauchy Convergence Criterion for sequences of numbers can be applied to obtain criteria that are sufficient for a series to converge.

---

**EXERCISES**

1. Examine the following series for convergence:

   (a) $\displaystyle\sum_{k=1}^{\infty} \frac{a^k}{k^p}$ where $a > 0$ and $p > 0$

   (b) $\displaystyle\sum_{k=1}^{\infty} \frac{1}{2k + 3}$

   (c) $\displaystyle\sum_{k=1}^{\infty} \frac{(-1)^k}{k}$

   (d) $\displaystyle\sum_{k=1}^{\infty} \frac{1}{k(k + 1)}$

   (e) $\displaystyle\sum_{k=1}^{\infty} k e^{-k^2}$

(f)   $\displaystyle\sum_{k=1}^{\infty}\left(\frac{k+1}{k^2+1}\right)^3$

(g)   $\displaystyle\sum_{k=1}^{\infty} k\sin\left(\frac{1}{k}\right)$

2.   For any positive number $\alpha$, prove that the series

$$\sum_{k=1}^{\infty}\frac{k^{\alpha}}{e^k}$$

converges.

3.   Fix a positive number $\alpha$ and consider the series

$$\sum_{k=1}^{\infty}\frac{1}{(k+1)[\ln(k+1)]^{\alpha}}.$$

For what values of $\alpha$ does this series converge?

4.   Under the assumptions of the Alternating Series Test, define

$$s = \sum_{k=1}^{\infty}(-1)^{k+1}a_k.$$

Prove that for every natural number $n$,

$$\left| s - \sum_{k=1}^{n}(-1)^{k+1}a_k \right| \le a_{n+1}.$$

5.   Use the Cauchy Convergence Criterion for Series to provide another proof of the Alternating Series Test.

6.   If a sequence converges, then each of its subsequences converges to the same limit, but the convergence of a subsequence does not imply the convergence of the whole sequence. Based on this, prove that

$$\text{if}\qquad \sum_{k=1}^{\infty}a_k = \ell,\qquad\text{then}\qquad \sum_{k=1}^{\infty}(a_{2k}+a_{2k-1}) = \ell,$$

but that the converse does not necessarily hold. (*Hint:* Consider the series $\sum_{k=1}^{\infty}(-1)^k$.)

7.   For the series $\sum_{k=1}^{\infty}a_k$, suppose that there is a number $r$ with $0 \le r < 1$ and a natural number $N$ such that

$$|a_k|^{1/k} < r \qquad \text{for all integers } k \ge N.$$

Prove that $\sum_{k=1}^{\infty}a_k$ converges absolutely. This is known as the Cauchy Root Test.

8.   Suppose that $\sum_{k=1}^{\infty}a_k$ and $\sum_{k=1}^{\infty}b_k$ are series of positive numbers such that

$$\lim_{n\to\infty}\left(\frac{a_k}{b_k}\right) = \ell \qquad\text{and}\qquad \ell > 0.$$

Prove that the series $\sum_{k=1}^{\infty}a_k$ converges if and only if the series $\sum_{k=1}^{\infty}b_k$ converges.

## 9.2 Pointwise Convergence and Uniform Convergence of Sequences of Functions

In Section 9.1, we studied sequences and series of *numbers*. We now turn to the study of sequences of *functions*.

**DEFINITION** *Given a function $f: D \to \mathbb{R}$ and a sequence of functions $\{f_n: D \to \mathbb{R}\}$, the sequence $\{f_n: D \to \mathbb{R}\}$ is said to converge pointwise to $f: D \to \mathbb{R}$ provided that for each point $x$ in $D$,*

$$\lim_{n \to \infty} f_n(x) = f(x).$$

**EXAMPLE 9.7** For each natural number $n$, define

$$f_n(x) = x^n \quad \text{for} \quad 0 \le x \le 1.$$

Since $\{f_n(1)\}$ is a constant sequence, whose constant value is 1, $\lim_{n \to \infty} f_n(1) = 1$. On the other hand,

$$\lim_{n \to \infty} x^n = 0 \quad \text{if} \quad 0 \le x < 1.$$

**FIGURE 9.2**

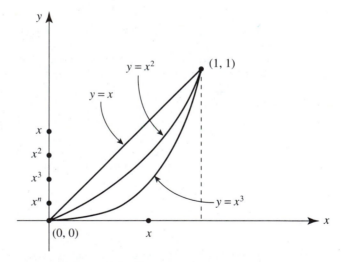

Thus the sequence of functions $\{f_n: [0, 1] \to \mathbb{R}\}$ converges pointwise to the function $f: [0, 1] \to \mathbb{R}$ defined by

$$f(x) = \begin{cases} 1 & \text{if } x = 1 \\ 0 & \text{if } 0 \le x < 1. \end{cases}$$

$\square$

**EXAMPLE 9.8**   For each natural number $n$, define

$$f_n(x) = \sum_{k=0}^{n} \frac{x^k}{k!} \quad \text{for} \quad 0 \le x \le 1.$$

According to formula (8.15), the sequence of functions $\{f_n : [0, 1] \to \mathbb{R}\}$ converges pointwise to the function $f : [0, 1] \to \mathbb{R}$, defined by $f(x) = e^x$ for $0 \le x \le 1$. In infinite series notation, this simply means that

$$e^x = \sum_{k=0}^{\infty} \frac{x^k}{k!} \quad \text{for} \quad 0 \le x \le 1. \qquad \qquad \square$$

**EXAMPLE 9.9**   For each natural number $n$, define

$$f_n(x) = e^{-nx^2} \quad \text{for all } x.$$

FIGURE 9.3

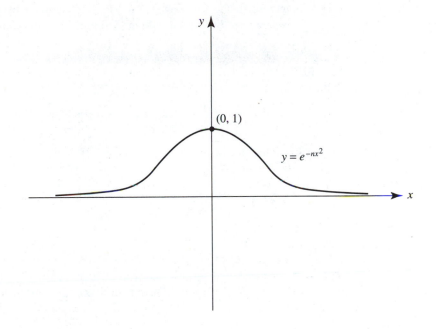

Then $\{f_n(0)\}$ is a constant sequence, whose constant value is 1, so $\lim_{n\to\infty} f_n(0) = 1$. On the other hand, since $e^b > 1 + b$ if $b > 0$, it follows that

$$\frac{1}{e^b} < \frac{1}{1+b} \quad \text{for all } b > 0.$$

Thus, for each natural number $n$ and each $x \ne 0$,

$$0 < f_n(x) < \frac{1}{1+nx^2},$$

so $\lim_{n\to\infty} f_n(x) = 0$ if $x \neq 0$. It follows that the sequence of functions $\{f_n : \mathbb{R} \to \mathbb{R}\}$ converges pointwise to the function $f : \mathbb{R} \to \mathbb{R}$ defined by

$$f(x) = \begin{cases} 0 & \text{if } x \neq 0 \\ 1 & \text{if } x = 0. \end{cases}$$

□

**EXAMPLE 9.10** For each natural number $n$, and each number $x$ with $|x| < 1$, define

$$f_n(x) = \sum_{k=0}^{n} x^k,$$

so that according to the Geometric Sum Formula,

$$f_n(x) = \frac{1 - x^{n+1}}{1 - x}.$$

Since $\lim_{n\to\infty} x^{n+1} = 0$, if $|x| < 1$, the sequence of functions $\{f_n : (-1, 1) \to \mathbb{R}\}$ converges pointwise to the function $f : (-1, 1) \to \mathbb{R}$ defined by $f(x) = 1/(1 - x)$ for $|x| < 1$. We are already familiar with this example; it is the Geometric Series. □

**EXAMPLE 9.11** For each natural number $n$, and each number $x$ in the interval $[0, 1]$, define

$$f_n(x) = \begin{cases} 1 & \text{if } x = k2^{-n} \text{ for some natural number } k \\ 0 & \text{otherwise.} \end{cases}$$

Then we see that the sequence $\{f_n : [0, 1] \to \mathbb{R}\}$ converges pointwise to the function $f : [0, 1] \to \mathbb{R}$ defined by

$$f(x) = \begin{cases} 1 & \text{if } x = k2^{-n}, \text{ for some natural numbers } k \text{ and } n \\ 0 & \text{otherwise.} \end{cases}$$

□

For a sequence of functions $\{f_n : D \to \mathbb{R}\}$ that converges pointwise to the function $f : D \to \mathbb{R}$, we wish to determine the properties of the individual $f_n$'s that are inherited by the limit function $f : D \to \mathbb{R}$. Three natural questions come to mind.

## Question A

*Suppose that each function $f_n : D \to \mathbb{R}$ is continuous. Is the limit function $f : D \to \mathbb{R}$ also continuous?*

## Question B

*If $D = I$ is an open interval and each function $f_n : I \to \mathbb{R}$ is differentiable, is the limit function $f : I \to \mathbb{R}$ also differentiable? If so, does the sequence of derivatives*

$\{f'_n: I \to \mathbb{R}\}$ *converge pointwise to* $f': I \to \mathbb{R}$—*that is, is the derivative of the limit equal to the limit of the derivatives?*

## Question C

*If* $D = [a, b]$ *and each function* $f_n: [a, b] \to \mathbb{R}$ *is integrable, is the limit function* $f: [a, b] \to \mathbb{R}$ *also integrable? If so, is*

$$\lim_{n \to \infty}\left[\int_a^b f_n\right] = \int_a^b f?$$

*That is, is the integral of the limit equal to the limit of the integrals?*

It turns out that the answer to each of the above questions is negative. Example 9.7 shows that the pointwise limit of continuous functions need not be continuous. Example 9.9 shows that the pointwise limit of differentiable functions need not be differentiable. Finally, Example 9.11 exhibits a sequence of integrable functions that converge pointwise to a nonintegrable function.

All is not lost, however. If we strengthen the assumption of pointwise convergence to what we will call *uniform convergence*, then the first and third questions have affirmative answers, and Question *B* has a satisfactory answer. What is of equal importance is that in many interesting situations we can verify uniform convergence.

---

**DEFINITION**  *Given a function* $f: D \to \mathbb{R}$ *and a sequence of functions* $\{f_n: D \to \mathbb{R}\}$, *the sequence* $\{f_n: D \to \mathbb{R}\}$ *is said to converge uniformly to* $f: D \to \mathbb{R}$ *provided that for each positive number* $\epsilon$ *there is a natural number* $N$ *such that*

$$|f(x) - f_n(x)| < \epsilon \quad \text{for all integers } n \geq N \text{ and all points } x \text{ in } D. \tag{9.7}$$

It is clear from the above definition that uniform convergence implies pointwise convergence; however, the converse is not true. To understand the distinction between uniform and pointwise convergence, observe that the sequence $\{f_n: D \to \mathbb{R}\}$ converges pointwise to $f: D \to \mathbb{R}$ provided that for each fixed point $x$ in $D$, the sequence of numbers $\{f_n(x)\}$ converges to the number $f(x)$; thus, for a given point $x$ in $D$ and a positive number $\epsilon$, there is a natural number $N$ such that $|f_n(x) - f(x)| < \epsilon$ for all integers $n \geq N$. *The index* $N$ *that responds to the* $\epsilon$ *challenge may depend on the point* $x$. For uniform convergence, the requirement is that given an $\epsilon > 0$ challenge, we can respond with an index $N$ such that $|f_n(x) - f(x)| < \epsilon$ for all integers $n \geq N$ and for all points $x$ in $D$.

In terms of graphs, the sequence $\{f_n: D \to \mathbb{R}\}$ converges uniformly to $f: D \to \mathbb{R}$ if for each positive number $\epsilon$ there is a natural number $N$ such that if $n \geq N$, the graph of the function $f_n: D \to \mathbb{R}$ lies between the graphs of the functions $f + \epsilon: D \to \mathbb{R}$ and $f - \epsilon: D \to \mathbb{R}$.

**FIGURE 9.4**

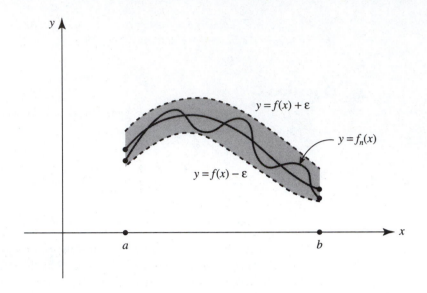

Let us revisit our first two examples of pointwise convergence and analyze them for uniform convergence.

**EXAMPLE 9.12** Let the sequence $\{f_n : [0, 1] \to \mathbb{R}\}$ and the function $f : [0, 1] \to \mathbb{R}$ be as in Example 9.7. The convergence is not uniform. Indeed, for $\epsilon = 1/2$, there is no natural number $N$ having the property that

$$|f_n(x) - f(x)| < \frac{1}{2} \quad \text{for all integers } n \geq N \text{ and all points } x \text{ in } [0, 1],$$

because no matter what natural number $N$ is chosen, by taking $x = (3/4)^{\frac{1}{N+1}}$ we have

$$f_{N+1}(x) - f(x) = \frac{3}{4} > \frac{1}{2}. \qquad \square$$

**EXAMPLE 9.13** Let the sequence $\{f_n : [0, 1] \to \mathbb{R}\}$ and the function $f : [0, 1] \to \mathbb{R}$ be as in Example 9.8. We claim that $\{f_n : [0, 1] \to \mathbb{R}\}$ converges uniformly to $f : [0, 1] \to \mathbb{R}$. To verify this claim, we need the estimate obtained from the Lagrange Remainder Theorem. According to estimate (8.6), for any point $x$ in the interval $[0, 1]$,

$$|f(x) - f_n(x)| \leq \frac{3}{(n + 1)!} \quad \text{for every natural number } n. \tag{9.8}$$

Now let $\epsilon > 0$. By the Archimedean property of $\mathbb{R}$, we can choose a natural number $N$ such that $N > 3/\epsilon$. Thus,

$$|f(x) - f_n(x)| < \epsilon \quad \text{for all integers } n \geq N \text{ and all points } x \text{ in } [0, 1]. \qquad \square$$

In Section 9.1, we proved the Cauchy Convergence Criterion for the convergence of a sequence of numbers. There is a similar criterion for the uniform convergence of a sequence of functions.

**DEFINITION**     *The sequence of functions* $\{f_n: D \to \mathbb{R}\}$ *is said to be uniformly Cauchy provided that for each positive number $\epsilon$ there is a natural number $N$ such that*

$$|f_{n+k}(x) - f_n(x)| < \epsilon \tag{9.9}$$

*for every integer $n \geq N$, every natural number $k$, and every point $x$ in $D$.*

**THEOREM 9.16**   **The Weierstrass Uniform Convergence Criterion**  The sequence of functions $\{f: D \to \mathbb{R}\}$ converges uniformly to a function $f: D \to \mathbb{R}$ if and only if the sequence $\{f_n: D \to \mathbb{R}\}$ is uniformly Cauchy.

**Proof**   Suppose that $\{f_n: D \to \mathbb{R}\}$ converges uniformly to $f: D \to \mathbb{R}$. We will show that $\{f_n: D \to \mathbb{R}\}$ is uniformly Cauchy. Indeed, let $\epsilon > 0$. We can select a natural number $N$ such that

$$|f_n(x) - f(x)| < \frac{\epsilon}{2} \quad \text{for all integers } n \geq N \text{ and every point } x \text{ in } D.$$

Using the Triangle Inequality, it follows that

$$|f_{n+k}(x) - f_n(x)| = |f_{n+k}(x) - f(x) + f(x) - f_n(x)|$$

$$\leq |f_{n+k}(x) - f(x)| + |f_n(x) - f(x)|$$

$$< \frac{\epsilon}{2} + \frac{\epsilon}{2} = \epsilon$$

for every integer $n \geq N$, every natural number $k$, and every point $x$ in $D$. Thus, the sequence $\{f_n: D \to \mathbb{R}\}$ is a uniformly Cauchy sequence.

To prove the converse, we suppose that the sequence of functions $\{f_n: D \to \mathbb{R}\}$ is uniformly Cauchy. Let $x$ be a point in $D$. Then clearly the sequence of real numbers $\{f_n(x)\}$ is a Cauchy sequence and so, by the Cauchy Convergence Criterion for sequences of numbers, $\{f_n(x)\}$ converges. Denote the limit by $f(x)$. This defines a function $f: D \to \mathbb{R}$ that is the only candidate for a function to which $\{f_n: D \to \mathbb{R}\}$ may converge uniformly.

Now let us prove that $\{f_n: D \to \mathbb{R}\}$ does converge uniformly to $f: D \to \mathbb{R}$. Let $\epsilon > 0$. Since $\{f_n: D \to \mathbb{R}\}$ is uniformly Cauchy, we can select a natural number $N$ such that

$$|f_{n+k}(x) - f_n(x)| < \frac{\epsilon}{2} \tag{9.10}$$

for every integer $n \geq N$, every natural number $k$, and every point $x$ in $D$. Let $x$ be a point in $D$. Choose $n \geq N$. Observe that from inequality (9.10) we have

$$f_n(x) - \frac{\epsilon}{2} < f_{n+k}(x) < f_n(x) + \frac{\epsilon}{2} \quad \text{for every natural number } k. \tag{9.11}$$

But
$$\lim_{k \to \infty} f_{n+k}(x) = f(x)$$

so that from (9.11) we obtain

$$f_n(x) - \frac{\epsilon}{2} \le f(x) \le f_n(x) + \frac{\epsilon}{2}.$$

Hence,

$$|f_n(x) - f(x)| < \epsilon \quad \text{for all integers } n \ge N \text{ and all points } x \text{ in } D.$$

It follows that $\{f_n : D \to \mathbb{R}\}$ converges uniformly to $f: D \to \mathbb{R}$. ∎

**EXAMPLE 9.14** For each natural number $n$ and each number $x$ with $|x| \le 1$, define

$$f_n(x) = \sum_{k=1}^{n} \frac{x^k}{k2^k}.$$

Observe, using the Triangle Inequality and the Geometric Sum Formula, that for each pair of natural numbers $n$ and $k$ and each number $x$ with $|x| \le 1$,

$$|f_{n+k}(x) - f_n(x)| \le \frac{|x|^{n+1}}{(n+1)2^{n+1}} + \cdots + \frac{|x|^{n+k}}{(n+k)2^{n+k}}$$

$$\le \frac{1}{2^{n+1}} + \cdots + \frac{1}{2^{n+k}}$$

$$\le \frac{1}{2^n}. \tag{9.12}$$

But $\lim_{n\to\infty}(1/2)^n = 0$, and this, together with the inequality (9.12), implies that the sequence $\{f_n : [-1, 1] \to \mathbb{R}\}$ is uniformly Cauchy. According to the Weierstrass Uniform Convergence Criterion, there is a function $f: [-1, 1] \to \mathbb{R}$ to which the sequence $\{f_n : [-1, 1] \to \mathbb{R}\}$ converges uniformly. ☐

---

**EXERCISES**

1. For each natural number $n$ and each number $x$, define

$$f_n(x) = \frac{1 - |x|^n}{1 + |x|^n}.$$

Find the function $f: \mathbb{R} \to \mathbb{R}$ to which the sequence $\{f_n : \mathbb{R} \to \mathbb{R}\}$ converges pointwise. Prove that the convergence is not uniform.

2. For each natural number $n$ and each number $x \ge 2$, define

$$f_n(x) = \frac{1}{1 + x^n}.$$

Find the function $f: [2, \infty) \to \mathbb{R}$ to which the sequence $\{f_n : [2, \infty) \to \mathbb{R}\}$ converges pointwise. Prove that the convergence is uniform.

3. For each natural number $n$ and each number $x$ in $(0, 1)$, define

$$f_n(x) = \frac{1}{nx + 1}.$$

Find the function $f: (0, 1) \to \mathbb{R}$ to which the sequence $\{f_n : (0, 1) \to \mathbb{R}\}$ converges pointwise. Prove that the convergence is not uniform.

4.  For each natural number $n$ and each number $x$ in $[0, 1]$, define

$$f_n(x) = \frac{x}{nx + 1}.$$

Find the function $f: [0, 1] \to \mathbb{R}$ to which the sequence $\{f_n: [0, 1] \to \mathbb{R}\}$ converges pointwise. Prove that the convergence is uniform.

5.  Determine whether the sequences in Examples 9.9, 9.10, and 9.11 converge uniformly.

6.  Suppose that the sequences $\{f_n: D \to \mathbb{R}\}$ and $\{g_n: D \to \mathbb{R}\}$ converge uniformly to the functions $f: D \to \mathbb{R}$ and $g: D \to \mathbb{R}$, respectively. For any two numbers $\alpha$ and $\beta$, prove that the sequence $\{\alpha f_n + \beta g_n: D \to \mathbb{R}\}$ converges uniformly to the function $\alpha f + \beta g: D \to \mathbb{R}$.

7.  For each natural number $n$, let the function $f_n: \mathbb{R} \to \mathbb{R}$ be bounded. Suppose that the sequence $\{f_n: \mathbb{R} \to \mathbb{R}\}$ converges uniformly to $f: \mathbb{R} \to \mathbb{R}$. Prove that the limit function $f: \mathbb{R} \to \mathbb{R}$ also is bounded.

8.  A number $x$ of the form $\ell/2^k$, where $\ell$ and $k$ are integers, is called a *dyadic rational*. Prove that the dyadic rationals are dense in $\mathbb{R}$. From this, conclude that the limit function in Example 9.11 is not integrable.

9.  For each natural number $n$ and each number $x$ in $(-1, 1)$, define

$$p_n(x) = x + x(1 - x^2) + \cdots + x(1 - x^2)^n.$$

Prove that the sequence $\{p_n: (-1, 1) \to \mathbb{R}\}$ converges pointwise.

10. Let $\{a_n\}$ be a bounded sequence of numbers. For each natural number $n$ and each number $x$, define

$$f_n(x) = a_0 + a_1 x + \frac{a_2 x^2}{2!} + \cdots + \frac{a_n x^n}{n!}.$$

Prove that for each $r > 0$, the sequence of functions $\{f_n: [-r, r] \to \mathbb{R}\}$ is uniformly convergent.

---

**9.3** 
# The Uniform Limit of Continuous Functions, of Integrable Functions, and of Differentiable Functions

We will now provide some affirmative answers to the three questions raised in Section 9.2, by strengthening the assumption of pointwise convergence to that of uniform convergence.

**THEOREM 9.17**    Suppose that $\{f_n: D \to \mathbb{R}\}$ is a sequence of continuous functions that converges uniformly to the function $f: D \to \mathbb{R}$. Then the limit function $f: D \to \mathbb{R}$ is also continuous.

**Proof** Let $x_0$ be a point in $D$. We will use the $\epsilon$–$\delta$ criterion for continuity in order to prove that the function $f: D \to \mathbb{R}$ is continuous at $x_0$. Indeed, let $\epsilon > 0$. Since the sequence $\{f_n: D \to \mathbb{R}\}$ converges uniformly to the function $f: D \to \mathbb{R}$, we can choose a natural number $N$ such that

$$|f_n(x) - f(x)| < \frac{\epsilon}{3} \quad \text{for all integers } n \geq N \text{ and all points } x \text{ in } D. \tag{9.13}$$

Using this inequality with $n = N$ and the Triangle Inequality, we see that

$$|f(x) - f(x_0)| = |f(x) - f_N(x) + f_N(x) - f_N(x_0) + f_N(x_0) - f(x_0)|$$

$$\leq |f(x) - f_N(x)| + |f_N(x) - f_N(x_0)| + |f_N(x_0) - f(x_0)|$$

$$< \frac{\epsilon}{3} + |f_N(x) - f_N(x_0)| + \frac{\epsilon}{3} \quad \text{for all points } x \text{ in } D. \tag{9.14}$$

By assumption, the function $f_N: D \to \mathbb{R}$ is continuous at $x_0$. Hence, we can choose $\delta > 0$ such that

$$|f_N(x) - f_N(x_0)| < \frac{\epsilon}{3} \quad \text{for all points } x \text{ in } D \text{ such that } |x - x_0| < \delta. \tag{9.15}$$

The inequalities (9.14) and (9.15) imply that

$$|f(x) - f(x_0)| < \epsilon \quad \text{for all points } x \text{ in } D \text{ such that } |x - x_0| < \delta.$$

Thus the function $f: D \to \mathbb{R}$ is continuous at the point $x_0$. ∎

**THEOREM 9.18** Suppose that $\{f_n: [a, b] \to \mathbb{R}\}$ is a sequence of integrable functions that converges uniformly to the function $f: [a, b] \to \mathbb{R}$. Then the limit function $f: [a, b] \to \mathbb{R}$ is also integrable. Moreover,

$$\lim_{n \to \infty} \left[ \int_a^b f_n \right] = \int_a^b f.$$

**Proof** We begin with a preliminary observation: It follows directly from the definition of the Darboux sums that if the positive number $\alpha$ and the natural number $n$ have the property that

$$f_n(x) - \alpha \leq f(x) \leq f_n(x) + \alpha \quad \text{for all points } x \text{ in } [a, b], \tag{9.16}$$

then for any partition $P$ of $[a, b]$,

$$L(f_n, P) - \alpha(b - a) \leq L(f, P) \leq U(f, P) \leq U(f_n, P) + \alpha(b - a). \tag{9.17}$$

First, we will use the Integrability Criterion to show that the function $f: [a, b] \to \mathbb{R}$ is integrable. Indeed, let $\epsilon > 0$. Since the sequence $\{f_n: [a, b] \to \mathbb{R}\}$ converges uniformly to the function $f: [a, b] \to \mathbb{R}$ and the number $\epsilon/(b - a)$ is positive, we can choose a natural number $N$ such that (9.16) holds if $n \geq N$ and $\alpha = \epsilon/3(b - a)$, and hence, by the initial observation, for any partition $P$ of the interval $[a, b]$ and any integer $n \geq N$,

$$L(f_n, P) - \frac{\epsilon}{3} \leq L(f, P) \leq U(f, P) \leq U(f_n, P) + \frac{\epsilon}{3}. \tag{9.18}$$

But the function $f_N: [a, b] \to \mathbb{R}$ is integrable, and so, according to the Integrability Criterion, we can choose a partition $P_*$ of $[a, b]$ such that

$$U(f_N, P_*) - L(f_N, P_*) < \frac{\epsilon}{3}. \tag{9.19}$$

Inequalities (9.18) and (9.19) imply that

$$U(f, P_*) - L(f, P_*) < \epsilon.$$

Hence the function $f: [a, b] \to \mathbb{R}$ is integrable.

It remains to be verified that

$$\lim_{n \to \infty} \left[ \int_a^b f_n \right] = \int_a^b f.$$

Let $\epsilon > 0$. We need to find a natural number $N$ such that

$$\left| \int_a^b f_n - \int_a^b f \right| < \epsilon \quad \text{for all integers } n \geq N. \tag{9.20}$$

For $\alpha = \epsilon/2(b - a)$, we may choose a natural number $N$ such that (9.16) holds. But using the linearity and monotonicity properties of the integral, (9.16) implies that for each integer $n \geq N$,

$$\int_a^b f_n - \frac{\epsilon}{2} \leq \int_a^b f \leq \int_a^b f_n + \frac{\epsilon}{2},$$

and so

$$\left| \int_a^b f_n - \int_a^b f \right| < \epsilon. \tag{9.21}$$

∎

The answer to Question $B$ of Section 9.2, regarding the differentiability of the limit of differentiable functions, requires more care than the answers to the other two questions. The uniform limit of differentiable functions need not be differentiable (see Exercise 1). However, there are quite reasonable circumstances under which it is differentiable and the derivative of the limit is the limit of the derivatives.

**THEOREM 9.19**    Let $I$ be an open interval. Suppose that $\{f_n: I \to \mathbb{R}\}$ is a sequence of continuously differentiable functions that has the following two properties:

(i)  The sequence $\{f_n: I \to \mathbb{R}\}$ converges pointwise to the function $f: I \to \mathbb{R}$ and

(ii) The sequence of derivatives $\{f_n': I \to \mathbb{R}\}$ converges uniformly to the function $g: I \to \mathbb{R}$.

Then the function $f: I \to \mathbb{R}$ is continuously differentiable and

$$f'(x) = g(x) \quad \text{for all } x \text{ in } [a, b].$$

**Proof** Fix a point $x_0$ in $I$. According to the First Fundamental Theorem of Calculus, for each natural number $n$ and each point $x$ in $I$,

$$f_n(x) - f_n(x_0) = \int_{x_0}^{x} f_n'. \tag{9.22}$$

Now, Theorem 9.18 implies that for each point $x$ in $I$,

$$\lim_{n \to \infty} \left[ \int_{x_0}^{x} f_n' \right] = \int_{x_0}^{x} g. \tag{9.23}$$

Also, since, by assumption, the sequence $\{f_n: I \to \mathbb{R}\}$ converges pointwise to the function $f: I \to \mathbb{R}$, for each point $x$ in $I$,

$$\lim_{n \to \infty} [f_n(x) - f_n(x_0)] = f(x) - f(x_0). \tag{9.24}$$

From (9.22), (9.23), and (9.24), it follows that

$$f(x) - f(x_0) = \int_{x_0}^{x} g \quad \text{for all } x \text{ in } I. \tag{9.25}$$

By assumption, for each natural number $n$ the function $f_n': I \to \mathbb{R}$ is continuous, so by Theorem 9.17, the uniform limit $g: I \to \mathbb{R}$ also is continuous. From (9.25) and the Second Fundamental Theorem of Calculus, we see that

$$f'(x) = g(x) \quad \text{for all } x \text{ in } I. \quad \blacksquare$$

---

**THEOREM 9.20** Let $I$ be an open interval. Suppose that $\{f_n: I \to \mathbb{R}\}$ is a sequence of continuously differentiable functions that has the following two properties:

(i) The sequence $\{f_n: I \to \mathbb{R}\}$ converges pointwise to the function $f: I \to \mathbb{R}$ and

(ii) The sequence of derivatives $\{f_n': I \to \mathbb{R}\}$ is uniformly Cauchy.

Then the function $f: I \to \mathbb{R}$ is continuously differentiable and for each point $x$ in $I$,

$$\lim_{n \to \infty} f_n'(x) = f'(x).$$

**Proof** The Weierstrass Uniform Convergence Criterion implies that there is a function $g: I \to \mathbb{R}$ to which the sequence $\{f_n': I \to \mathbb{R}\}$ converges uniformly. The conclusion now follows from Theorem 9.19. $\blacksquare$

The property of uniform convergence can frequently be verified. However, there are many interesting cases in which a sequence of functions fails to converge uniformly, but nevertheless the limit function inherits properties possessed by the individual functions in the approximation sequence. We will describe one instance of this: the following verification of a classical formula for $\pi$.

**PROPOSITION
9.21**

**The Newton-Gregory Formula**

$$\frac{\pi}{4} = \int_0^1 \frac{1}{1+x^2}\,dx = \sum_{k=0}^{\infty} \frac{(-1)^k}{2k+1}. \tag{9.26}$$

**Proof**   Since for each number $x$,

$$\frac{d}{dx}(\arctan x) = \frac{1}{1+x^2},$$

it follows from the First Fundamental Theorem of Calculus that

$$\frac{\pi}{4} = \arctan 1 - \arctan 0 = \int_0^1 \frac{1}{1+x^2}\,dx. \tag{9.27}$$

Let $n$ be a natural number. Substituting $-x^2 = r$ in the Geometric Sum Formula, we see that for each number $x$,

$$\frac{1}{1+x^2} = 1 - x^2 + \cdots + (-1)^n x^{2n} + \frac{(-1)^{n+1}x^{2n+2}}{1+x^2},$$

so that

$$\int_0^1 \frac{1}{1+x^2}\,dx = 1 - \frac{1}{3} + \cdots + \frac{(-1)^n}{2n+1} + \int_0^1 \frac{(-1)^{n+1}x^{2n+2}}{1+x^2}\,dx. \tag{9.28}$$

The monotonicity property of the integral gives the estimate

$$\left| \int_0^1 \frac{(-1)^{n+1}x^{2n+2}}{1+x^2}\,dx \right| \leq \int_0^1 x^{2n+2}\,dx = \frac{1}{2n+3},$$

from which it follows that

$$\lim_{n\to\infty} \left[ \int_0^1 \frac{(-1)^{n+1}x^{2n+2}}{1+x^2}\,dx \right] = 0.$$

Thus, (9.26) follows from (9.27) and (9.28).                                    ∎

For each natural number $n$ and each number $x$ in $[0, 1]$, define

$$f_n(x) = \sum_{k=0}^{n}(-1)^k x^{2k},$$

and define $f(x) = 1/(1+x^2)$. The Newton-Gregory Formula may be restated as

$$\lim_{n\to\infty} \left[ \int_0^1 f_n \right] = \int_0^1 f.$$

We proved this without proving that the sequence of functions $\{f_n: [0, 1] \to \mathbb{R}\}$ converges uniformly to the function $f: [0, 1] \to \mathbb{R}$. In fact, we do not even have pointwise convergence on the whole interval $[0, 1]$, since the sequence $\{f_n(1)\}$ does not converge to $f(1)$.

1. For each natural number $n$ and each number $x$ in $(-1, 1)$, define

$$f_n(x) = \sqrt{x^2 + \frac{1}{n}},$$

and define $f(x) = |x|$. Prove that the sequence $\{f_n : (-1, 1) \to \mathbb{R}\}$ converges uniformly to the function $f : (-1, 1) \to \mathbb{R}$. Check that each function $f_n : (-1, 1) \to \mathbb{R}$ is differentiable, whereas the limit function $f : (-1, 1) \to \mathbb{R}$ is not differentiable. Does this contradict Theorem 9.19?

2. For each natural number $n$ and each number $x$ in $[0, 1]$, define

$$f_n(x) = nxe^{-nx^2}.$$

Prove that the sequence $\{f_n : [0, 1] \to \mathbb{R}\}$ converges pointwise to the constant function 0, but that the sequence of integrals $\{\int_0^1 f_n\}$ does not converge to 0. Does this contradict Theorem 9.18?

3. Prove that if $\{f_n : \mathbb{R} \to \mathbb{R}\}$ is a sequence of differentiable functions such that the sequence of derivatives $\{f_n' : \mathbb{R} \to \mathbb{R}\}$ is uniformly convergent and the sequence $\{f_n(0)\}$ is also convergent, then $\{f_n : \mathbb{R} \to \mathbb{R}\}$ is pointwise convergent. Is the assumption that the sequence $\{f_n(0)\}$ converges necessary?

4. Give an example of a sequence of differentiable functions $\{f_n : (-1, 1) \to \mathbb{R}\}$ that converges uniformly but for which $\{f_n'(0)\}$ is unbounded.

5. Under the assumptions of Theorem 9.19, show that for each interval $[\alpha, \beta]$ contained in $I$, the sequence $\{f_n : [\alpha, \beta] \to \mathbb{R}\}$ converges uniformly to $f : [\alpha, \beta] \to \mathbb{R}$.

## 9.4   Power Series

In the study of Taylor series, we began with an infinitely differentiable function; we then constructed a Taylor series, which we analyzed for convergence to the given function. We will now change our point of view. In this section we will *define* a function by a power series expansion and study the properties of such a function.

**DEFINITION**   *Given a sequence of real numbers $\{c_k\}$ indexed by the nonnegative integers, we define the domain of convergence of the series $\sum_{k=0}^{\infty} c_k x^k$ to be the set of all numbers $x$ such that the series $\sum_{k=0}^{\infty} c_k x^k$ converges. Denote the domain of convergence by $D$. We then define a function $f : D \to \mathbb{R}$ by*

$$f(x) = \lim_{n \to \infty} \left[ \sum_{k=0}^{n} c_k x^k \right] = \sum_{k=0}^{\infty} c_k x^k \quad \text{for all } x \text{ in } D. \tag{9.29}$$

*We will refer to (9.29) as a power series expansion, and will call the set $D$ the domain of convergence of the expansion.*

**EXAMPLE 9.15**   Consider the series

$$\sum_{k=0}^{\infty} \frac{(-1)^k x^k}{k+2}. \tag{9.30}$$

Fix a number $x$. Since

$$\lim_{k \to \infty} \left| \frac{(-1)^{k+1} x^{k+1}}{k+3} \middle/ \frac{(-1)^k x^k}{k+2} \right| = |x|,$$

it follows from the Ratio Test for Series that (9.30) converges if $|x| < 1$ and diverges if $|x| > 1$. For $x = 1$, from the Alternating Series Test we conclude that (9.30) converges. For $x = -1$, the Integral Test shows that the series diverges. Thus the domain of convergence of the power series (9.30) is the interval $(-1, 1]$.   □

**EXAMPLE 9.16**   Consider the series

$$\sum_{k=0}^{\infty} k! x^k.$$

For any nonzero number $x$, the terms of the series $\sum_{k=0}^{\infty} k! x^k$ do not converge to 0, and hence the series does not converge. Thus the domain of convergence of the power series $\sum_{k=0}^{\infty} k! x^k$ consists of the single point $x = 0$.   □

**EXAMPLE 9.17**   Consider the series

$$\sum_{k=0}^{\infty} \frac{1}{(1+k!)} x^k.$$

Fix a number $x$. Since

$$\lim_{k \to \infty} \left[ \frac{1}{(1+(k+1)!)} x^{k+1} \middle/ \frac{1}{(1+k!)} x^k \right] = 0,$$

the Ratio Test for Series shows that the domain of convergence of this series is the set of all real numbers.   □

A principal objective of this section is to show that if the function $f \colon (-r, r) \to \mathbb{R}$ is defined by the power series expansion

$$f(x) = \lim_{n \to \infty} \left[ \sum_{k=0}^{n} c_k x^k \right] = \sum_{k=0}^{\infty} c_k x^k \quad \text{for } |x| < r,$$

then $f \colon (-r, r) \to \mathbb{R}$ is differentiable and, moreover,

$$f'(x) = \frac{d}{dx}\left[\lim_{n\to\infty}\sum_{k=0}^{n}c_k x^k\right]$$

$$= \lim_{n\to\infty}\left[\frac{d}{dx}\left[\sum_{k=0}^{n}c_k x^k\right]\right]$$

$$= \lim_{n\to\infty}\left[\sum_{k=1}^{n}kc_k x^{k-1}\right]$$

$$= \sum_{k=1}^{\infty}kc_k x^{k-1} \quad \text{if} \quad |x| < r. \tag{9.31}$$

The above computation is known as *term-by-term differentiation* of a series expansion. It is not at all obvious that the passage from the first line to the second is justified. Once this computation is justified, it follows easily that the function $f:(-r,r)\to\mathbb{R}$ has derivatives of all orders and that term-by-term differentiation of all orders is valid.

For sequences of functions, we have distinguished between pointwise convergence and uniform convergence. It is necessary to make a similar distinction for the convergence of power series.

**DEFINITION** *Let $A$ be a subset of the domain of convergence of the power series $\sum_{k=0}^{\infty}c_k x^k$. Define the function $f: A \to \mathbb{R}$ by*

$$f(x) = \sum_{k=0}^{\infty}c_k x^k \quad \text{for all } x \text{ in } A, \tag{9.32}$$

*and for each natural number $n$, define the function $s_n: A \to \mathbb{R}$ by*

$$s_n(x) = \sum_{k=0}^{n}c_k x^k \quad \text{for all } x \text{ in } A.$$

*The series $\sum_{k=0}^{\infty}c_k x^k$ is said to be uniformly convergent on the set $A$ provided that the sequence of partial sums $\{s_n: A \to \mathbb{R}\}$ converges uniformly to the function $f: A \to \mathbb{R}$.*

**LEMMA 9.22** The power series $\sum_{k=0}^{\infty}c_k x^k$ is uniformly convergent on the set $A$ provided that the following condition holds:

There is a positive number $M$ and a number $\alpha$ with $0 \le \alpha < 1$ such that for each natural number $k$,

$$|c_k x^k| \le M\alpha^k \quad \text{for all } x \text{ in } A. \tag{9.33}$$

**Proof** Define $\{s_n: A \to \mathbb{R}\}$ to be the sequence of partial sums for the series $\sum_{k=0}^{\infty}c_k x^k$ on the set $A$. By the very definition of uniform convergence on a set, we must show that the sequence of functions $\{s_n: A \to \mathbb{R}\}$ is uniformly convergent. However, the Weierstrass Uniform Convergence Criterion asserts that a sequence of functions converges uniformly if and only if the sequence is uniformly Cauchy. Thus it suffices to show that the sequence of partial sums is uniformly Cauchy.

Let $\epsilon > 0$. We need to find a natural number $N$ such that for each integer $n \geq N$ and every natural number $k$,

$$|s_{n+k}(x) - s_n(x)| < \epsilon \quad \text{for all } x \text{ in } A. \tag{9.34}$$

However, from the definition of the partial sums, the Triangle Inequality, and the Geometric Sum Formula, we see that for any pair of natural numbers $k$ and $n$,

$$
\begin{aligned}
|s_{n+k}(x) - s_n(x)| &= |c_{n+k}x^{n+k} + \cdots + c_{n+1}x^{n+1}| \\
&\leq |c_{n+k}x^{n+k}| + \cdots + |c_{n+1}x^{n+1}| \\
&\leq M[\alpha^{n+k} + \cdots + \alpha^{n+1}] \\
&= M\alpha^{n+1}[1 + \cdots + \alpha^{k-1}] \\
&= M\alpha^{n+1}\left[\frac{1 - \alpha^k}{1 - \alpha}\right] \\
&\leq M\alpha^{n+1}\left[\frac{1}{1 - \alpha}\right] \quad \text{for all } x \text{ in } A. \tag{9.35}
\end{aligned}
$$

Since $\lim_{n\to\infty} \alpha^n = 0$, we may choose a natural number $N$ such that

$$\alpha^n < \frac{\epsilon}{M}(1 - \alpha) \quad \text{for all integers } n \geq N.$$

With this choice of $N$ it follows from (9.35) that the required inequality (9.34) holds. ∎

In order to use the above lemma to justify term-by-term differentiation of a power series, it is useful to observe that if $\alpha$ is a number with $0 \leq \alpha < 1$, then there is a number $c$ such that

$$k\alpha^k \leq c(\sqrt{\alpha})^k \quad \text{for every nonnegative integer } k.$$

Indeed, to see this, we write $k\alpha^k = k\sqrt{\alpha}^k \cdot \sqrt{\alpha}^k$ and then choose $c$ to be any upper bound for the sequence $\{k\sqrt{\alpha}^k\}$; there is such an upper bound, because the sequence $\{k\sqrt{\alpha}^k\}$ converges to 0.

**THEOREM 9.23**   Suppose that the nonzero number $x_0$ is in the domain of convergence of the power series $\sum_{k=0}^{\infty} c_k x^k$. Let $r$ be any positive number less than $|x_0|$. Then the interval $[-r, r]$ is in the domain of convergence of the power series $\sum_{k=0}^{\infty} c_k x^k$ and also in the domain of convergence of the derived power series $\sum_{k=1}^{\infty} kc_k x^{k-1}$. Moreover, each of the power series

$$\sum_{k=0}^{\infty} c_k x^k \quad \text{and} \quad \sum_{k=1}^{\infty} kc_k x^{k-1}$$

converges uniformly on the interval $[-r, r]$.

**Proof**   First, we show that the power series $\sum_{k=0}^{\infty} c_k x^k$ converges uniformly on the interval $[-r, r]$. Since the series $\sum_{k=0}^{\infty} c_k x_0^k$ converges, the terms of this series converge to 0, so, in particular, the terms are bounded. Thus we can choose a number $M$ such that

$$|c_k x_0^k| \leq M \quad \text{for every natural number } k.$$

Define $\alpha = r/|x_0|$ and observe that $0 \leq \alpha < 1$. Moreover, writing $x$ as $x = (x/x_0)x_0$, we see that for every natural number $k$,

$$|c_k x|^k \leq M\alpha^k \quad \text{if} \quad |x| \leq r. \tag{9.36}$$

The uniform convergence of the series $\sum_{k=0}^{\infty} c_k x^k$ on the interval $[-r, r]$ now follows from Lemma 9.22.

Now we consider the derived series $\sum_{k=1}^{\infty} k c_k x^{k-1}$. It is convenient to write this series as $\sum_{k=0}^{\infty} a_k x^k$, where $a_k = (k+1)c_{k+1}$ for every nonnegative integer $k$. To show that the series converges uniformly on the interval $[-r, r]$, we will again use Lemma 9.22. In order to do so, observe that for each $x$ in the interval $[-r, r]$ and each natural number $k$,

$$|a_k x^k| = |(k+1)c_{k+1}x^k| = \frac{(k+1)}{k|x_0|} \left| c_{k+1}x_0^{k+1} K \left(\frac{x}{x_0}\right)^k \right|. \tag{9.37}$$

As we observed above, letting $\beta = \sqrt{\alpha}$, there is a number $c$ such that for every natural number $k$,

$$k\alpha^k \leq c\beta^k.$$

Using this and (9.37), we see that for each natural number $k$,

$$|a_k x^k| \leq c'\beta^k \quad \text{if} \quad |x| \leq r,$$

where $c' = 2Mc/|x_0|$. The uniform convergence of the series $\sum_{k=1}^{\infty} k c_k x^{k-1}$ on the interval $[-r, r]$ now follows from Lemma 9.22.   ■

Let $D$ be the domain of convergence of the power series expansion $\sum_{k=0}^{\infty} c_k x^k$. From Theorem 9.23, it follows that $D = \mathbb{R}$ if $D$ is unbounded. If $D$ is bounded, we define

$$r = \sup D,$$

and it follows that

$$(-r, r) \subseteq D \subseteq [r, r].$$

Because of this, we call the number $r$ the *radius of convergence* of the series $\sum_{k=0}^{\infty} c_k x^k$.

We leave it as an exercise for the reader to verify that if the sequence $\{|a_n|^{1/n}\}$ converges to $\alpha$, then $D = \mathbb{R}$ if $\alpha = 0$, and $r = \alpha^{-1}$ if $\alpha > 0$ (see Exercise 14).

**THEOREM 9.24**   Let $r$ be a positive number such that the interval $(-r, r)$ lies in the domain of convergence of the series $\sum_{k=0}^{\infty} c_k x^k$. Define

$$f(x) = \sum_{k=0}^{\infty} c_k x^k \quad \text{if } |x| < r.$$

Then the function $f: (-r, r) \to \mathbb{R}$ has derivatives of all orders. For each natural number $n$,

$$\frac{d^n}{dx^n}[f(x)] = \sum_{k=0}^{\infty} \frac{d^n}{dx^n}[c_k x^k] \quad \text{if} \quad |x| < r,$$

so that, in particular,

$$\frac{f^{(n)}(0)}{n!} = c_n.$$

**Proof**   It is clear that it will be sufficient to prove that

$$f'(x) = \sum_{k=0}^{\infty} \frac{d}{dx}(c_k x^k) \quad \text{if} \quad |x| < r$$

and hence

$$f'(0) = c_1.$$

The general result follows by induction, since according to Theorem 9.23, the derived series also converges on $(-r, r)$.

Choose $R$ to be any positive number less than $r$. Since the series $\sum_{k=0}^{\infty} c_k x^k$ converges at each point between $R$ and $r$, according to Theorem 9.23, each of the series

$$\sum_{k=0}^{\infty} c_k x^k \quad \text{and} \quad \sum_{k=0}^{\infty} k c_k x^{k-1}$$

converges uniformly on the interval $[-R, R]$.

For each natural number $n$, define

$$s_n(x) = \sum_{k=0}^{n} c_k x^k \quad \text{if} \quad |x| < R.$$

Then each of the sequences of functions

$$\{s_n : (-R, R) \to \mathbb{R}\} \quad \text{and} \quad \{s_n' : (-R, R) \to \mathbb{R}\}$$

is uniformly convergent. Theorem 9.20 implies that

$$\lim_{n \to \infty} s_n'(x) = f'(x) \quad \text{if} \quad |x| < R;$$

that is,

$$\sum_{k=0}^{\infty} k c_k x^{k-1} = f'(x) \quad \text{if} \quad |x| < R.$$

Since for each point $x$ in the interval $(-r, r)$ we can choose a positive number $R$ less than $r$ with $|x| < R$, it follows that

$$f'(x) = \sum_{k=0}^{\infty} kc_k x^{k-1} \quad \text{for all points } x \text{ in the interval } (-r, r). \qquad \blacksquare$$

The above theorem implies that a function defined by a power series expansion on the interval $(-r, r)$ coincides with its Taylor series expansion about 0; this is a uniqueness result for the coefficients of a power series expansion.

Recall that in Section 5.2 we *assumed* that the differential equation

$$\begin{cases} F''(x) + F(x) = 0 & \text{for all } x \\ F(0) = 1, \quad F'(0) = 0 \end{cases} \qquad (9.38)$$

had a solution. It is a consequence of Theorem 8.11 that this assumed solution of (9.38) must necessarily have the following Taylor series expansion:

$$F(x) = \sum_{k=0}^{\infty} \frac{(-1)^k}{(2k)!} x^{2k} \quad \text{for all } x. \qquad (9.39)$$

We will now prove that the power series expansion (9.39) does indeed define the unique solution of (9.38).

**THEOREM 9.25** For each number $x$, the series

$$\sum_{k=0}^{\infty} \frac{(-1)^k}{(2k)!} x^{2k}$$

converges. Define

$$F(x) = \sum_{k=0}^{\infty} \frac{(-1)^k}{(2k)!} x^{2k} \quad \text{for all } x. \qquad (9.40)$$

Then the function $F: \mathbb{R} \to \mathbb{R}$ has derivatives of all orders and satisfies the differential equation (9.38).

**Proof** From the Ratio Test for Series, it follows that the domain of convergence of the series $\sum_{k=0}^{\infty} [(-1)^k/(2k)!] x^{2k}$ is the set of all real numbers. Thus, the above function $F: \mathbb{R} \to \mathbb{R}$ is properly defined. Moreover, by Theorem 9.24, it follows that for all $x$,

$$F'(x) = \sum_{k=1}^{\infty} \frac{(-1)^k}{(2k-1)!} x^{2k-1}$$

and

$$F''(x) = \sum_{k=2}^{\infty} \frac{(-1)^k}{(2k-2)!} x^{2k-2} = -F(x).$$

Thus the power series expansion (9.40) defines a function $F: \mathbb{R} \to \mathbb{R}$ that satisfies the differential equation (9.38). $\blacksquare$

For any number $x_0$, the substitution of $x - x_0$ for $x$ reduces the study of power series expansions of the form $\sum_{k=0}^{\infty} c_k (x - x_0)^k$ to the case in which $x_0 = 0$.

1.  Determine the domains of convergence of each of the following power series:

$$\text{(a)} \quad \sum_{k=1}^{\infty} \frac{x^k}{k5^k} \qquad \text{(b)} \quad \sum_{k=1}^{\infty} k!x^k \qquad \text{(c)} \quad \sum_{k=0}^{\infty} \frac{(-1)^k x^{2k-1}}{(2k+1)!}$$

2.  Prove that

$$\frac{1}{(1+x)} = \sum_{k=0}^{\infty} (-1)^k x^k \quad \text{if} \quad |x| < 1.$$

3.  Prove that

$$\frac{1}{(1+x)^2} = \sum_{k=0}^{\infty} (-1)^k k x^{k-1} \quad \text{if} \quad |x| < 1.$$

4.  Prove that

$$\frac{1}{(1+x^2)^2} = \sum_{k=0}^{\infty} (-1)^k k x^{2k-2} \quad \text{if} \quad |x| < 1.$$

5.  Prove that

$$x = \sum_{k=0}^{\infty} \left(1 - \frac{1}{x}\right)^k \quad \text{if} \quad |1 - x| < |x|.$$

6.  Define $f(x) = 1/(1-x)^3$, if $|x| < 1$. Find a power series expansion for the function $f : (-1, 1) \rightarrow \mathbb{R}$.

7.  Suppose that the domain of convergence of the power series $\sum_{k=0}^{\infty} c_k x^k$ contains the interval $(-r, r)$. Define

$$f(x) = \sum_{k=0}^{\infty} c_k x^k \quad \text{if} \quad |x| < r.$$

Let the interval $[a, b]$ be contained in the interval $(-r, r)$. Prove that

$$\int_a^b f(x) \, dx = \sum_{k=0}^{\infty} \frac{c_k}{k+1} [b^{k+1} - a^{k+1}].$$

8.  Obtain a series expansion for the integral

$$\int_0^{1/2} 1/(1 + x^4) \, dx$$

and justify your calculation.

9.  For each number $x$, define

$$h(x) = \sum_{k=0}^{\infty} \frac{x^{2k}}{(2k)!} \quad \text{and} \quad g(x) = \sum_{k=0}^{\infty} \frac{x^{2k+1}}{(2k+1)!}.$$

Prove that for any pair of numbers $\alpha$ and $\beta$, the function

$$f = \alpha h + \beta g : \mathbb{R} \rightarrow \mathbb{R}$$

is a solution of the differential equation

$$\begin{cases} f''(x) - f(x) = 0, & x \text{ in } \mathbb{R} \\ f(0) = \alpha \quad \text{and} \quad f'(0) = \beta. \end{cases}$$

10. Use Bernoulli's Inequality to show that if $\alpha$ is a number with $0 \le \alpha < 1$, then for every natural number $k$,

$$k\alpha^k \le \left[\frac{\sqrt{\alpha}}{1 - \sqrt{\alpha}}\right] (\sqrt{\alpha})^k.$$

11. Prove that if $0 \le \alpha < 1$, then $\lim_{n\to\infty} k\alpha^k = 0$.

12. Rewrite the Geometric Sum Formula as follows: For each natural number $n$,

$$\frac{1}{1-x} - (1 + x + \cdots + x^n) = \frac{x^{n+1}}{1-x} \quad \text{if} \quad x \ne 1.$$

Differentiate this identity to obtain

$$\frac{d}{dx}\left(\frac{1}{1-x}\right) - [1 + 2x + \cdots + nx^{n-1}] = \frac{(n+1)x^n - nx^{n+1}}{(1-x)^2}.$$

Now use this identity and Exercise 11 to directly justify term-by-term differentiation of the Geometric Series.

13. Prove that the series

$$\sum_{k=0}^{\infty} \frac{1}{1 + |x|^k}$$

converges if and only if $|x| > 1$. In particular, show that the series converges at $x = 2$ but not at $x = 1$. Does this contradict Theorem 9.23? (*Hint:* This is not a power series.)

14. Suppose that $\lim_{n\to\infty} |a_n|^{1/n} = \alpha$.
    (a) If $\alpha > 0$, show that $\sum_{k=0}^{\infty} a_n x^n$ converges if $|x| < 1/\alpha$ and diverges if $|x| > 1/\alpha$.
    (b) If $\alpha = 0$, show that $\sum_{k=0}^{\infty} a_n x^n$ converges for all $x \ne 0$.

---

## 9.5   A Continuous Function That Fails at Each Point to Be Differentiable

Weierstrass presented the first example of a continuous function $f: \mathbb{R} \to \mathbb{R}$ that has the remarkable property that there is no point at which it is differentiable. We will analyze such an example, where $f: \mathbb{R} \to \mathbb{R}$ is defined by an expansion

$$f(x) = \sum_{k=1}^{\infty} h_k(x) \quad \text{for all } x$$

and the function $f: \mathbb{R} \to \mathbb{R}$ inherits all of the nondifferentiability possessed by the individual $h_k'$s.

Define $h(x) = |x|$ if $|x| \le 1/2$ and then extend the function $h: [-1/2, 1/2] \to \mathbb{R}$ to $h: \mathbb{R} \to \mathbb{R}$ so that $h: \mathbb{R} \to \mathbb{R}$ has period 1. The function $h: \mathbb{R} \to \mathbb{R}$ is a so-called "saw-toothed" function, the "teeth" of which have base length equal to 1 and depth equal to 1/2.

For each natural number $k$, define the function $h_k: \mathbb{R} \to \mathbb{R}$ by

$$h_k(x) = \left(\frac{1}{4}\right)^{k-1} h(4^{k-1}x) \quad \text{for all } x.$$

Then the function $h_k : \mathbb{R} \to \mathbb{R}$ is also a "saw-toothed" function, the "teeth" of which have base length equal to $1/4^{k-1}$ and depth equal to $1/(2 \cdot 4^{k-1})$. Also, the function $h_k : \mathbb{R} \to \mathbb{R}$ has period $(1/4)^{k-1}$. Furthermore,

$$|h_k(x)| \leq \frac{1}{2} \left( \frac{1}{4} \right)^{k-1} \qquad \text{for all } x. \qquad (9.41)$$

**FIGURE 9.5**

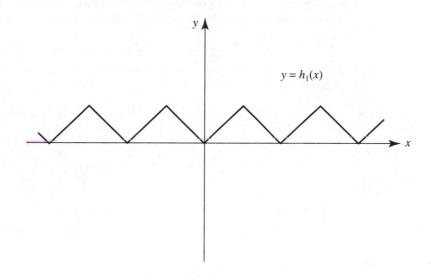

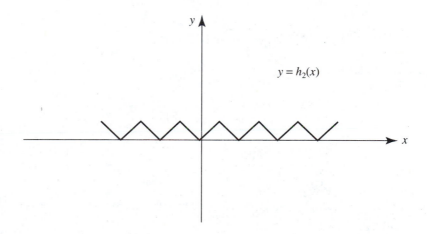

**LEMMA 9.26**   For each number $x$, define

$$f(x) = \lim_{n \to \infty} \left[ \sum_{k=1}^{n} h_k(x) \right] = \sum_{k=1}^{\infty} h_k(x). \qquad (9.42)$$

Then the function $f : \mathbb{R} \to \mathbb{R}$ is continuous.

**Proof** For each number $x$ and each natural number $n$, define

$$f_n(x) = \sum_{k=1}^{n} h_k(x).$$

Since each function $h_k: \mathbb{R} \to \mathbb{R}$ is continuous, each function $f_n: \mathbb{R} \to \mathbb{R}$ is also continuous. We will prove that the sequence of functions $\{f_n: \mathbb{R} \to \mathbb{R}\}$ is uniformly Cauchy. Once this is proven, it follows from the Weierstrass Uniform Convergence Criterion that it converges uniformly. Then, by Theorem 9.17, we may conclude that the limit function is also continuous.

To verify the Cauchy property, observe that from the Triangle Inequality, inequality (9.41), and the Geometric Sum Formula it follows that for any pair of natural numbers $k$ and $n$ and any number $x$,

$$|f_{n+k}(x) - f_n(x)| = |h_{n+k}(x) + \cdots + h_{n+1}(x)|$$

$$\leq |h_{n+k}(x)| + \cdots + |h_{n+1}(x)|$$

$$\leq \left(\frac{1}{2}\right)\left[\left(\frac{1}{4}\right)^{n+k-1} + \cdots + \left(\frac{1}{4}\right)^n\right]$$

$$= \left(\frac{1}{4^n}\right)\left(\frac{1}{2}\right)\left[1 + \cdots + \left(\frac{1}{4}\right)^{k-1}\right]$$

$$\leq \frac{1}{4^n}.$$

Since $\lim_{n \to \infty} 1/4^n = 0$, it follows from this inequality that the sequence of functions $\{f_n: \mathbb{R} \to \mathbb{R}\}$ is uniformly Cauchy. ∎

**THEOREM 9.27** Define the function $f: \mathbb{R} \to \mathbb{R}$ by

$$f(x) = \sum_{k=1}^{\infty} h_k(x) \quad \text{for all } x.$$

Then

(i) The function $f: \mathbb{R} \to \mathbb{R}$ is continuous, but

(ii) There is no point at which the function $f: \mathbb{R} \to \mathbb{R}$ is differentiable.

**Proof** We have already proven that the function $f: \mathbb{R} \to \mathbb{R}$ is continuous.

Let $x_0$ be any number. We will show that $f: \mathbb{R} \to \mathbb{R}$ is not differentiable at $x_0$ by choosing a sequence of numbers $\{x_n\}$, with each $x_n \neq x_0$, that converges to $x_0$ but for which the limit

$$\lim_{n \to \infty} \frac{f(x_n) - f(x_0)}{x_n - x_0}$$

does not exist.

Let $n$ be a natural number. Observe that for an integer $m$, the interval

$$I_m = \left[ m \left(\frac{1}{2}\right)\left(\frac{1}{4}\right)^{n-1}, \ (m+1)\left(\frac{1}{2}\right)\left(\frac{1}{4}\right)^{n-1} \right]$$

has length equal to $1/(2 \cdot 4^{n-1})$ and, moreover, the union of such intervals equals all of $\mathbb{R}$. Thus, we may choose an integer $m$ such that either

(i) the interval $[x_0, \ x_0 + (1/4)^n]$ is contained in $I_m$, or

(ii) the interval $[x_0 - (1/4)^n, \ x_0]$ is contained in $I_m$.

If the first possibility occurs, define $x_n = x_0 + (1/4)^n$; otherwise, define $x_n = x_0 - (1/4)^n$. Now, if $1 \le k \le n$, the function $h_k : \mathbb{R} \to \mathbb{R}$ on the interval $I_m$ is either strictly increasing with slope 1 or strictly decreasing with slope $-1$ (see Exercise 1). Thus,

$$\left| \frac{h_k(x_n) - h_k(x_0)}{x_n - x_0} \right| = 1 \quad \text{if} \quad 1 \le k \le n.$$

On the other hand, since each $h_k : \mathbb{R} \to \mathbb{R}$ has period $(1/4)^{k-1}$, we see that

$$\frac{h_k(x_n) - h_k(x_0)}{x_n - x_0} = 0 \quad \text{if} \quad k > n.$$

Thus,

$$\frac{f(x_n) - f(x_0)}{x_n - x_0} = \sum_{k=1}^{\infty} \frac{h_k(x_n) - h_k(x_0)}{x_n - x_0} = \sum_{k=1}^{n} \frac{h_k(x_n) - h_k(x_0)}{x_n - x_0}$$

is an integer, which is odd if $n$ is odd and is even if $n$ is even. As a consequence, the limit

$$\lim_{n \to \infty} \frac{f(x_n) - f(x_0)}{x_n - x_0}$$

does not exist, so the function $f : \mathbb{R} \to \mathbb{R}$ is not differentiable at the point $x_0$. ∎

In the following exercises, the functions $h_k : \mathbb{R} \to \mathbb{R}$ and $f : \mathbb{R} \to \mathbb{R}$ are those defined above.

---

**EXERCISES**

1. (a)  Show that for each integer $k$,

$$\frac{h_k(u) - h_k(v)}{u - v} = \begin{cases} 1 & \text{if } 0 \le u < v \le (1/2)\,(1/4)^{k-1} \\ -1 & \text{if } -(1/2)\,(1/4)^{k-1} \le u < v \le 0. \end{cases}$$

   (b)  Use (a) to show that for each integer $m$, each natural number $n$, and each natural number $k$ with $1 \le k \le n$,

$$\left| \frac{h_k(u) - h_k(v)}{u - v} \right| = 1$$

   if

$$m\left(\frac{1}{2}\right)\left(\frac{1}{4}\right)^{n-1} \le u < v \le (m+1)\left(\frac{1}{2}\right)\left(\frac{1}{4}\right)^{n-1}.$$

2. Let $r > 0$. Prove that for each number $x_0$ there is an integer $m$ such that either the interval $[x_0, x_0 + r/2]$ is contained in the interval $[mr, (m+1)r]$ or the interval $[x_0 - r/2, x_0]$ is contained in the interval $[mr, (m+1)r]$.

3. Suppose that the function $g: \mathbb{R} \to \mathbb{R}$ has period $T > 0$. Show that for each natural number $n$, the function $g: \mathbb{R} \to \mathbb{R}$ also has period $nT$.

4. For each natural number $n$, define

$$u_n = -\left(\frac{1}{2}\right)\left(\frac{1}{4}\right)^{n-1} \quad \text{and} \quad v_n = \left(\frac{1}{2}\right)\left(\frac{1}{4}\right)^{n-1}.$$

Show that

$$h_k(u_n) = 0 = h_k(v_n) \quad \text{if} \quad k > n,$$

whereas

$$h_k(u_n) > 0 \quad \text{and} \quad h_k(v_n) > 0 \quad \text{if} \quad 1 \le k \le n.$$

Conclude that $f(u_n) > 0$ and $f(v_n) > 0$, and hence that there is no interval $I$ containing the point 0 on which the function $f: I \to \mathbb{R}$ is monotonic.

5. Use Exercise 4 to show that there is no interval $I$ on which the function $f: I \to \mathbb{R}$ is monotonic.

6. Find a function $g: \mathbb{R} \to \mathbb{R}$ that is continuously differentiable, but for which there is no point at which it has a second derivative.

# 10

# The Euclidean Space $\mathbb{R}^n$

## 10.1  The Linear Structure of $\mathbb{R}^n$ and the Inner Product

For each positive integer $n$, we denote by $\mathbb{R}^n$ the set of $n$-tuples of real numbers

$$\mathbf{u} = (u_1, \ldots, u_n),$$

where $u_i$ is a real number for each index $i$ with $1 \leq i \leq n$. We call $u_i$ the $i$th *component of* $\mathbf{u}$ and refer to the members of $\mathbb{R}^n$ as *points in* $\mathbb{R}^n$. Moreover, we shall consider two points $\mathbf{u} = (u_1, \ldots, u_n)$ and $\mathbf{v} = (v_1, \ldots, v_n)$ in $\mathbb{R}^n$ to be *equal* provided that they have the same components; that is,

$$\mathbf{u} = \mathbf{v} \quad \text{if and only if} \quad u_i = v_i \quad \text{for each index } i \text{ with } 1 \leq i \leq n.$$

For any two points $\mathbf{u}$ and $\mathbf{v}$ in $\mathbb{R}^n$, we define the *sum* of $\mathbf{u}$ and $\mathbf{v}$, denoted by $\mathbf{u} + \mathbf{v}$, by the formula

$$\mathbf{u} + \mathbf{v} \equiv (u_1 + v_1, \ldots, u_n + v_n).$$

Also, for each real number $\alpha$, we define the point $\alpha\mathbf{u}$, called the *scalar multiple* of the point $\mathbf{u}$ by $\alpha$, by the formula

$$\alpha\mathbf{u} \equiv (\alpha u_1, \ldots, \alpha u_n).$$

The point in $\mathbb{R}^n$ all of whose components are zero will be denoted by $\mathbf{0}$. In algebraic contexts, it is called *zero*; in geometric contexts, it is called the *origin*. Finally, for each pair of points $\mathbf{u}$ and $\mathbf{v}$ in $\mathbb{R}^n$, we define their *difference*, denoted by $\mathbf{u} - \mathbf{v}$, by the formula

$$\mathbf{u} - \mathbf{v} \equiv \mathbf{u} + (-\mathbf{v}).$$

Of course, $\mathbb{R}^1$ is simply the familiar set of real numbers $\mathbb{R}$, and the definitions of addition of points and of multiplication of points in $\mathbb{R}^1$ by real numbers are exactly the

**FIGURE 10.1**

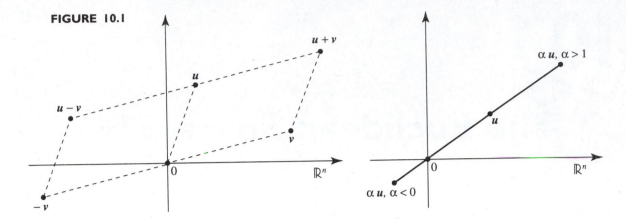

definitions of addition and multiplication of real numbers that we have been using in $\mathbb{R}$. In the Preliminaries, we codified the properties of addition and multiplication in $\mathbb{R}$ as the Field Axioms. Using these Field Axioms and the definition of equality of two points in $\mathbb{R}^n$ as meaning the equality of corresponding components, we obtain the following.

**PROPOSITION 10.1**

Let $\mathbf{u}, \mathbf{v}$, and $\mathbf{w}$ be points in $\mathbb{R}^n$. Then

$$(\mathbf{u} + \mathbf{v}) + \mathbf{w} = \mathbf{u} + (\mathbf{v} + \mathbf{w}),$$
$$\mathbf{u} + \mathbf{0} = \mathbf{u},$$
$$\mathbf{u} - \mathbf{u} = \mathbf{0},$$
$$\mathbf{u} + \mathbf{v} = \mathbf{v} + \mathbf{u},$$

and if $\alpha$ and $\beta$ are real numbers, then

$$\alpha(\mathbf{u} + \mathbf{v}) = \alpha\mathbf{u} + \alpha\mathbf{v},$$
$$(\alpha + \beta)(\mathbf{u}) = \alpha\mathbf{u} + \beta\mathbf{u},$$
$$(\alpha\beta)\mathbf{u} = \alpha(\beta\mathbf{u}).$$

**Proof** Each of these equalities follows from the observation that the Field Axioms for $\mathbb{R}$ imply that we have componentwise equality. ∎

**DEFINITION** *Let $\mathbf{u}$ and $\mathbf{v}$ be points in $\mathbb{R}^n$. The inner product of $\mathbf{u}$ and $\mathbf{v}$, denoted by $\langle \mathbf{u}, \mathbf{v} \rangle$, is defined by the formula*

$$\langle \mathbf{u}, \mathbf{v} \rangle \equiv u_1 v_1 + \cdots + u_n v_n.$$

The inner product is also denoted by $\mathbf{u} \cdot \mathbf{v}$ and is often called the *dot product* or the *scalar product*. The following algebraic properties of the inner product are extensions of the commutative and distributive properties of the addition and multiplication of real numbers.

**PROPOSITION 10.2**

Let $\mathbf{u}$, $\mathbf{v}$, and $\mathbf{w}$ be points in $\mathbb{R}^n$. Then

$$\langle \mathbf{u}, \mathbf{v} \rangle = \langle \mathbf{v}, \mathbf{u} \rangle, \qquad\qquad\qquad \text{(Symmetry)}$$

and if $\alpha$ and $\beta$ are any real numbers, then

$$\langle \alpha\mathbf{u} + \beta\mathbf{w}, \mathbf{v} \rangle = \alpha\langle \mathbf{u}, \mathbf{v} \rangle + \beta\langle \mathbf{w}, \mathbf{v} \rangle. \qquad\qquad \text{(Linearity)}$$

**Proof**    The commutative property of multiplication of real numbers implies that

$$\sum_{i=1}^{n} u_i v_i = \sum_{i=1}^{n} v_i u_i,$$

which is the first identity. The distributive property of addition and multiplication of the real numbers implies that

$$\sum_{i=1}^{n} (\alpha u_i + \beta w_i) v_i = \alpha \sum_{i=1}^{n} u_i v_i + \beta \sum_{i=1}^{n} w_i v_i,$$

which is the second identity.    ■

**DEFINITION**

(i) *Let $\mathbf{w}$ be a point in $\mathbb{R}^n$. Then the norm of $\mathbf{w}$, denoted by $\|\mathbf{w}\|$, is defined by the formula*

$$\|\mathbf{w}\| \equiv \sqrt{\langle \mathbf{w}, \mathbf{w} \rangle} = \sqrt{\sum_{i=1}^{n} w_i^2}.$$

(ii) *Let $\mathbf{u}$ and $\mathbf{v}$ be points in $\mathbb{R}^n$. Then the distance between the points $\mathbf{u}$ and $\mathbf{v}$, denoted by $d(\mathbf{u}, \mathbf{v})$, is defined by the formula*

$$d(\mathbf{u},\mathbf{v}) \equiv \|\mathbf{u} - \mathbf{v}\|.$$

It follows that the norm of the point $\mathbf{w}$ is the distance from the origin to $\mathbf{w}$. Moreover, the distance between the points $\mathbf{u}$ and $\mathbf{v}$ may be expressed in terms of the inner product by the formula

$$d(\mathbf{u},\mathbf{v}) = \|\mathbf{u} - \mathbf{v}\| = (\langle \mathbf{u} - \mathbf{v}, \mathbf{u} - \mathbf{v} \rangle)^{1/2}.$$

When we consider the set $\mathbb{R}^n$ together with the concepts of addition, scalar multiplication, and distance between points that we have introduced so far, it is customary to refer to $\mathbb{R}^n$ as *Euclidean n-space*. Moreover, for a point $\mathbf{u}$ in $\mathbb{R}^n$, it is often convenient to identify with $\mathbf{u}$ all of the *segments* from a point $\mathbf{p}$ in $\mathbb{R}^n$ to the point $\mathbf{p} + \mathbf{u}$ in $\mathbb{R}^n$ and to refer to this collection of segments as the *vector* $\mathbf{u}$. By *the vector $\mathbf{u}$ based at the point* $\mathbf{p}$, we will mean the segment from the point $\mathbf{p}$ to the point $\mathbf{p} + \mathbf{u}$. Addition, scalar multiplication, and the inner product extend to vectors. The norm of the point $\mathbf{u}$ is called the *length* of the vector $\mathbf{u}$; it is equal to the distance between the endpoints of any segment associated with the vector $\mathbf{u}$.

Of course, in the dimensions $n = 1$, $n = 2$, and $n = 3$, the reader is already quite familiar with the geometric meaning of addition and scalar multiplication. The norm and the inner product also have a geometric significance. In the case when $n = 1$, the inner product is just the usual multiplication of real numbers, and the norm of a number is simply its absolute value. In the case when $n = 2$, if $\mathbf{u} = (u_1, u_2)$ is a point in the plane $\mathbb{R}^2$, then the Pythagorean Theorem asserts that the norm of $\mathbf{u}$, which is given by

$$\|\mathbf{u}\| = \sqrt{u_1^2 + u_2^2},$$

is the distance from the origin to the point $\mathbf{u}$. It also is the length of any of the segments associated with the vector $\mathbf{u}$.

**FIGURE 10.2**

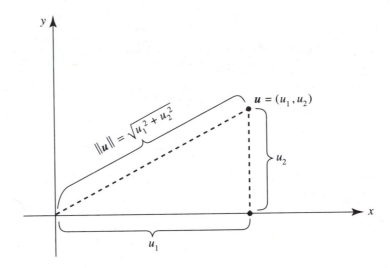

The geometric significance of the inner product of two points (or two vectors) in the plane $\mathbb{R}^2$ is described by the following proposition.

**PROPOSITION 10.3**

Let $\mathbf{u}$ and $\mathbf{v}$ be nonzero vectors in the plane $\mathbb{R}^2$. Then

$$\langle \mathbf{u}, \mathbf{v} \rangle = \|\mathbf{u}\| \|\mathbf{v}\| \cos\theta, \tag{10.1}$$

where $\theta$ is the radian measure of the angle between the vector $\mathbf{u}$ based at the origin and the vector $\mathbf{v}$ based at the origin.

**Proof**    First, recall that if $\mathbf{u}$ is any nonzero point in the plane and $\theta$ is the radian measure of the angle with vertex at $\mathbf{0}$ formed by the points $\mathbf{u}$, $\mathbf{0}$, and $(1, 0)$, then the components of $\mathbf{u}$ may be expressed, in terms of the norm of $\mathbf{u}$ and the angle $\theta$, by the formula

$$\mathbf{u} = (\|\mathbf{u}\| \cos\theta, \|\mathbf{u}\| \sin\theta).$$

Let $\theta_1$ be the radian measure of the angle with vertex at $\mathbf{0}$ formed by the points $\mathbf{u}$, $\mathbf{0}$, and $(1, 0)$; let $\theta_2$ be the radian measure of the angle with vertex at $\mathbf{0}$ formed by the points

$\mathbf{v}$, $\mathbf{0}$, and $(1, 0)$. We may suppose that $\theta_2 > \theta_1$. It follows from the definition of the inner product and the cosine addition formula that

$$\langle \mathbf{u}, \mathbf{v} \rangle = u_1 v_1 + u_2 v_2$$
$$= \|\mathbf{u}\| \|\mathbf{v}\| (\cos \theta_1 \cos \theta_2 + \sin \theta_1 \sin \theta_2)$$
$$= \|\mathbf{u}\| \|\mathbf{v}\| \cos(\theta_2 - \theta_1)$$
$$= \|\mathbf{u}\| \|\mathbf{v}\| \cos \theta,$$

where $\theta$ is the angle with vertex at $\mathbf{0}$ determined by $\mathbf{u}$, $\mathbf{0}$, and $\mathbf{v}$. ∎

From formula (10.1), it follows that for two nonzero vectors $\mathbf{u}$ and $\mathbf{v}$ in the plane $\mathbb{R}^2$, the inner product of $\mathbf{u}$ and $\mathbf{v}$ is zero if and only if vector $\mathbf{u}$ based at the origin is orthogonal (perpendicular) to the vector $\mathbf{v}$ based at the origin. This leads us to make the following definition of orthogonality of two vectors in $\mathbb{R}^n$.

**DEFINITION** *Two vectors $\mathbf{u}$ and $\mathbf{v}$ in $\mathbb{R}^n$ are said to be orthogonal provided that $\langle \mathbf{u}, \mathbf{v} \rangle = 0$.*

**LEMMA 10.4** For two vectors $\mathbf{u}$ and $\mathbf{v}$ in $\mathbb{R}^n$, the following assertions are equivalent:

(i) The vectors $\mathbf{u}$ and $\mathbf{v}$ are orthogonal.

(ii) The Pythagorean Identity holds; that is,

$$\|\mathbf{u} + \mathbf{v}\|^2 = \|\mathbf{u}\|^2 + \|\mathbf{v}\|^2.$$

**Proof** By the definition of the norm,

$$\|\mathbf{u} + \mathbf{v}\|^2 = \langle \mathbf{u} + \mathbf{v}, \mathbf{u} + \mathbf{v} \rangle,$$

and we may use the linearity and the symmetry of the inner product to simplify the right-hand side, to obtain the identity

$$\|\mathbf{u} + \mathbf{v}\|^2 = \|\mathbf{u}\|^2 + \|\mathbf{v}\|^2 + 2\langle \mathbf{u}, \mathbf{v} \rangle.$$

The equivalence of (i) and (ii) follows from this identity. ∎

In the case when the inner product $\langle \mathbf{u}, \mathbf{v} \rangle \neq 0$, it will be very useful to obtain an estimate of the size of $\langle \mathbf{u}, \mathbf{v} \rangle$ in terms of the norms of $\mathbf{u}$ and $\mathbf{v}$. Recall that for any real number $\theta$, $|\cos \theta| \leq 1$. Therefore, it follows from formula (10.1) that if $\mathbf{u}$ and $\mathbf{v}$ are any two vectors in the plane $\mathbb{R}^2$, then

$$|\langle \mathbf{u}, \mathbf{v} \rangle| \leq \|\mathbf{u}\| \cdot \|\mathbf{v}\|. \tag{10.2}$$

Since we have established formula (10.1) only for vectors in the plane $\mathbb{R}^2$, this argument is insufficient to verify inequality (10.2) for vectors $\mathbf{u}$ and $\mathbf{v}$ in $\mathbb{R}^n$ when $n > 2$. Nevertheless, the inequality (10.2) holds for vectors in $\mathbb{R}^n$ even if $n > 2$. Before proving this, it is convenient to establish the following lemma.

**LEMMA 10.5** For vectors $\mathbf{u}$ and $\mathbf{v}$ in $\mathbb{R}^n$, with $\mathbf{v} \neq \mathbf{0}$, define $\lambda = \langle \mathbf{u}, \mathbf{v} \rangle / \langle \mathbf{v}, \mathbf{v} \rangle$. Then the vector $\mathbf{u} - \lambda \mathbf{v}$ is orthogonal to the vector $\mathbf{v}$.

**FIGURE 10.3**

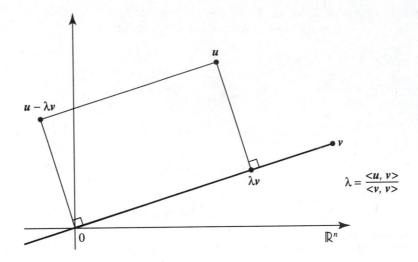

$$\lambda = \frac{\langle u, v \rangle}{\langle v, v \rangle}$$

**Proof**   By the linearity of the inner product and the definition of $\lambda$,

$$\langle \mathbf{u} - \lambda \mathbf{v}, \mathbf{v} \rangle = \langle \mathbf{u}, \mathbf{v} \rangle - \lambda \langle \mathbf{v}, \mathbf{v} \rangle = 0.$$
∎

**THEOREM 10.6**   **The Cauchy-Schwarz Inequality**   For any two vectors $\mathbf{u}$ and $\mathbf{v}$ in $\mathbb{R}^n$,

$$|\langle \mathbf{u}, \mathbf{v} \rangle| \leq \|\mathbf{u}\| \cdot \|\mathbf{v}\|. \tag{10.3}$$

**Proof**   If the vector $\mathbf{v} = \mathbf{0}$, then certainly the Cauchy-Schwarz Inequality holds, since each side of the inequality is 0, so suppose that $\mathbf{v} \neq \mathbf{0}$. Define $\lambda = \langle \mathbf{u}, \mathbf{v} \rangle / \langle \mathbf{v}, \mathbf{v} \rangle$. By Lemma 10.5, the vector $\mathbf{u} - \lambda \mathbf{v}$ is orthogonal to the vector $\mathbf{v}$, and hence is also orthgogonal to $\lambda \mathbf{v}$. Thus, since the inner product of a vector with itself is always nonnegative,

$$0 \leq \langle \mathbf{u} - \lambda \mathbf{v}, \mathbf{u} - \lambda \mathbf{v} \rangle = \langle \mathbf{u} - \lambda \mathbf{v}, \mathbf{u} \rangle = \|\mathbf{u}\|^2 - \frac{\langle \mathbf{u}, \mathbf{v} \rangle^2}{\|\mathbf{v}\|^2}.$$

Multiply this inequality by $\|\mathbf{v}\|^2$ to obtain $\langle \mathbf{u}, \mathbf{v} \rangle^2 \leq \|\mathbf{u}\|^2 \cdot \|\mathbf{v}\|^2$, which is equivalent to the Cauchy-Schwarz Inequality.
∎

In the study of functions of several variables, it is often necessary to estimate the norms of sums and differences of vectors and also to estimate the distance between points. In order to do so, we now extend the most useful of the inequalities that we have already established for real numbers, namely the Triangle and the Reverse Triangle Inequalities. Recall that in Section 1.3, we proved that for any real numbers $a$ and $b$, we have the following upper and lower bounds for $|a + b|$:

$$|a + b| \leq |a| + |b|$$

and

$$|a + b| \geq \big||a| - |b|\big|.$$

The length of a vector $\mathbf{u}$ in $\mathbb{R}^n$ extends the concept of the absolute value of a real number, so the following inequalities are direct extensions of the Triangle Inequality and the Reverse Triangle Inequality.

**THEOREM 10.7**   For two vectors $\mathbf{u}$ and $\mathbf{v}$ in $\mathbb{R}^n$, there are the following estimates for $\|\mathbf{u} + \mathbf{v}\|$:

### The Triangle Inequality

$$\|\mathbf{u} + \mathbf{v}\| \leq \|\mathbf{u}\| + \|\mathbf{v}\|. \tag{10.4}$$

### The Reverse Triangle Inequality:

$$\|\mathbf{u} + \mathbf{v}\| \geq \big|\|\mathbf{u}\| - \|\mathbf{v}\|\big|. \tag{10.5}$$

**Proof**   If we square both sides of inequality (10.4), it is clear that this inequality holds if and only if

$$\|\mathbf{u} + \mathbf{v}\|^2 \leq (\|\mathbf{u}\| + \|\mathbf{v}\|)^2. \tag{10.6}$$

But

$$\|\mathbf{u} + \mathbf{v}\|^2 = \langle \mathbf{u} + \mathbf{v}, \ \mathbf{u} + \mathbf{v} \rangle = \langle \mathbf{u}, \ \mathbf{u} \rangle + 2\langle \mathbf{u}, \ \mathbf{v} \rangle + \langle \mathbf{v}, \ \mathbf{v} \rangle,$$

so (10.6) can be rewritten as

$$\langle \mathbf{u}, \ \mathbf{u} \rangle + 2\langle \mathbf{u}, \ \mathbf{v} \rangle + \langle \mathbf{v}, \ \mathbf{v} \rangle \leq \|\mathbf{u}\|^2 + 2\|\mathbf{u}\|\|\mathbf{v}\| + \|\mathbf{v}\|^2. \tag{10.7}$$

However, the Cauchy-Schwarz Inequality asserts that $|\langle \mathbf{u}, \ \mathbf{v} \rangle| \leq \|\mathbf{u}\|\|\mathbf{v}\|$, so certainly $\langle \mathbf{u}, \ \mathbf{v} \rangle \leq \|\mathbf{u}\|\|\mathbf{v}\|$. This is what is needed to verify inequality (10.7).

The Reverse Triangle Inequality follows from the Triangle Inequality. Indeed, the Triangle Inequality gives

$$\|\mathbf{u}\| = \|(\mathbf{u} + \mathbf{v}) + (-\mathbf{v})\| \leq \|\mathbf{u} + \mathbf{v}\| + \|\mathbf{v}\|,$$

so

$$\|\mathbf{u}\| - \|\mathbf{v}\| \leq \|\mathbf{u} + \mathbf{v}\|.$$

Interchanging $\mathbf{u}$ and $\mathbf{v}$, we also get

$$\|\mathbf{v}\| - \|\mathbf{u}\| \leq \|\mathbf{u} + \mathbf{v}\|.$$

These last two inequalities are equivalent to the Reverse Triangle Inequality.   ∎

1. Consider the two points $\mathbf{u} = (1, 3, -2)$ and $\mathbf{v} = (2, 2, 4)$ in $\mathbb{R}^3$. Find the norm of $\mathbf{u}$ and the norm of $\mathbf{v}$, and show that $\mathbf{u}$ and $\mathbf{v}$ are perpendicular. Show that

$$\|\mathbf{u} + \mathbf{v}\|^2 = \|\mathbf{u}\|^2 + \|\mathbf{v}\|^2.$$

2. Find the maximum value of

$$\frac{x + 2y + 3z}{\sqrt{x^2 + y^2 + z^2}}$$

as $(x, y, z)$ varies among nonzero points in $\mathbb{R}^3$.

3. For a point $\mathbf{u}$ in $\mathbb{R}^n$, show that (a) $\|\mathbf{u}\| = 0$ if and only if $\mathbf{u} = 0$, and that (b) for any number $\alpha$, $\|\alpha\mathbf{u}\| = |\alpha| \|\mathbf{u}\|$.

4. For vectors $\mathbf{u}$ and $\mathbf{v}$ in $\mathbb{R}^n$, verify the identity

$$\|\mathbf{u} - \mathbf{v}\|^2 = \|\mathbf{u}\|^2 + \|\mathbf{v}\|^2 - 2\langle \mathbf{u}, \ \mathbf{v} \rangle.$$

Show that when $n = 2$, this identity is equivalent to the Law of Cosines from trigonometry.

5. Let $\mathbf{u}$ and $\mathbf{v}$ be vectors in $\mathbb{R}^n$. Prove that

$$\langle \mathbf{u}, \ \mathbf{v} \rangle = \frac{\|\mathbf{u} + \mathbf{v}\|^2 - \|\mathbf{u} - \mathbf{v}\|^2}{4}.$$

This identity is called the *Polarization Identity*. Provide a geometric interpretation of this identity when $n = 2$.

6. Let $\mathbf{u}$ and $\mathbf{v}$ be vectors in $\mathbb{R}^n$. Prove that if $\mathbf{u} = 0$ or $\mathbf{v} = \alpha\mathbf{u}$ for some number $\alpha$, then the Cauchy-Schwarz Inequality becomes an equality. Then prove the converse: If $|\langle \mathbf{u}, \ \mathbf{v} \rangle| = \|\mathbf{u}\| \|\mathbf{v}\|$, then either $\mathbf{u} = 0$ or $\mathbf{v} = \alpha\mathbf{u}$ for some number $\alpha$.

7. For $n$ a positive integer and real numbers $a_1, \ldots, a_n$, verify that

$$|a_1 + \cdots + a_n| \leq \sqrt{n}\sqrt{a_1^2 + \cdots + a_n^2}.$$

8. Let $\mathbf{u} = (a, b)$ and $\mathbf{v} = (c, d)$ be nonzero points in the plane $\mathbb{R}^2$, and let $\theta$ be the radian measure of the angle with vertex at $0$ formed by $0$, $\mathbf{u}$, and $\mathbf{v}$.
   (a) Prove that

$$\|\mathbf{u}\|^2 \|\mathbf{v}\|^2 - (\langle \mathbf{u}, \ \mathbf{v} \rangle)^2 = (\|\mathbf{u}\| \|\mathbf{v}\| \sin\theta)^2.$$

   (b) Express the left-hand side of the above equation in components to obtain

$$|ad - bc| = \big|\|\mathbf{u}\| \|\mathbf{v}\| \sin\theta\big|.$$

   (c) Use (b) to verify that $|ad - bc|/2$ is the area of the triangle with vertices $0$, $\mathbf{u}$, and $\mathbf{v}$, and that as a consequence, $|ad - bc|$ is the area of the parallelogram with vertices $0$, $\mathbf{u}$, $\mathbf{u} + \mathbf{v}$, and $\mathbf{v}$.

9. Let $\mathbf{u}$ be a point in $\mathbb{R}^n$, and suppose that $\|\mathbf{u}\| < 1$. Show that if $\mathbf{v}$ is in $\mathbb{R}^n$ and $\|\mathbf{v} - \mathbf{u}\| < 1 - \|\mathbf{u}\|$, then $\|\mathbf{v}\| < 1$.

10. Let $\mathbf{u}$ be a point in $\mathbb{R}^n$ and let $r$ be a positive number. Suppose that the points $\mathbf{v}$ and $\mathbf{w}$ in $\mathbb{R}^n$ are at a distance less than $r$ from the point $\mathbf{u}$. Prove that if $0 \leq t \leq 1$, then the point $t\mathbf{v} + (1 - t)\mathbf{w}$ is also at a distance less than $r$ from $\mathbf{u}$.

11. For points $\mathbf{u}$ and $\mathbf{v}$ in $\mathbb{R}^n$, define the function $p : \mathbb{R} \to \mathbb{R}$ by $p(t) = \|\mathbf{u} + t\mathbf{v}\|^2$ for $t$ in $\mathbb{R}$. Show that $p(t)$ is a quadratic polynomial that attains only nonpositive values. Use this to show that the descriminant is nonpositive, and thus provide another proof of the Cauchy-Schwarz Inequality.

12.  The points $\mathbf{u}_1, \ldots, \mathbf{u}_k$ in $\mathbb{R}^n$ are said to be an *orthonormal set* if $\|\mathbf{u}_i\| = 1$ for $1 \leq i \leq k$, and $\langle \mathbf{u}_i, \mathbf{u}_j \rangle = 0$ if $1 \leq i \leq k$, $1 \leq j \leq k$, and $i \neq j$. Suppose that the points $\mathbf{u}_1, \ldots, \mathbf{u}_k$ in $\mathbb{R}^n$ are an orthonormal set. For $\mathbf{u} = \alpha_1 \mathbf{u}_1 + \cdots + \alpha_k \mathbf{u}_k$, show that

$$\|\mathbf{u}\| = \sqrt{\sum_{i=1}^{k} \alpha_i^2}.$$

13.  Given two continuous functions $f: [0, 1] \to \mathbb{R}$ and $g: [0, 1] \to \mathbf{R}$, we define the *inner product* of $f$ and $g$, denoted by $\langle f, g \rangle$, by the formula

$$\langle f, g \rangle = \int_0^1 f(x)g(x)\, dx.$$

(a)  Verify that this inner product has the properties of the inner product in $\mathbb{R}^n$ listed in Proposition 10.2.

(b)  Follow the proof of the Cauchy-Schwarz Inequality for points in $\mathbb{R}^n$ to prove that

$$\left| \int_0^1 f(x)g(x)\, dx \right| \leq \sqrt{\int_0^1 [f(x)]^2\, dx} \sqrt{\int_0^1 [g(x)]^2\, dx}.$$

## 10.2   Convergence of Sequences in $\mathbb{R}^n$

Recall that in Chapter 2 we studied the concept of convergence of sequences of real numbers. A sequence of real numbers $\{x_k\}$ is defined as *converging* to the real number $x$ provided that *for each positive number $\epsilon$ there is a natural number $K$ such that*

$$|x - x_k| < \epsilon \quad \text{if} \quad k \geq K.$$

We denote the convergence of the sequence $\{x_k\}$ to $x$ by writing

$$\lim_{k \to \infty} x_k = x,$$

and we call $x$ the *limit* of the sequence $\{x_k\}$. It is immediately clear that

$$\lim_{k \to \infty} x_k = x \quad \text{if and only if} \quad \lim_{k \to \infty} |x - x_k| = 0. \tag{10.8}$$

By a *sequence of points in $\mathbb{R}^n$*, we mean a function from the set of natural numbers into $\mathbb{R}^n$. It is customary to denote such a sequence by a symbol such as $\{\mathbf{u}_k\}$, indicating that for each positive integer $k$, the functional value of $k$ is $\mathbf{u}_k$. If $A$ is a subset of $\mathbb{R}^n$, by a *sequence in $A$*, we mean a sequence $\{\mathbf{u}_k\}$ of points in $\mathbb{R}^n$ having the property that $\mathbf{u}_k$ is in $A$ for each index $k$. The aim of this section is to extend the concept of convergence of sequences of real numbers to that of convergence of sequences of points in $\mathbb{R}^n$, and to establish various properties of such sequences.

In Section 10.1, we defined the *distance $d(\mathbf{u}, \mathbf{v})$* between two points $\mathbf{u}$ and $\mathbf{v}$ in $\mathbb{R}^n$ by the formula

$$d(\mathbf{u}, \mathbf{v}) = \|\mathbf{u} - \mathbf{v}\|.$$

In the case when $n = 1$ and $u$ and $v$ are real numbers, the distance formula becomes

$$d(u, v) = |u - v|,$$

so that in view of criterion (10.8) it is reasonable to extend the concept of convergence as follows.

**DEFINITION**

*Let $\{\mathbf{u}_k\}$ be a sequence of points in $\mathbb{R}^n$ and let $\mathbf{u}$ be a point in $\mathbb{R}^n$. Then the sequence $\{\mathbf{u}_k\}$ is said to converge to $\mathbf{u}$ provided that for each positive number $\epsilon$ there is a natural number $K$ such that*

$$d(\mathbf{u}_k, \mathbf{u}) < \epsilon \quad \text{if} \quad k \geq K.$$

In conformity with the notation established for sequences of real numbers, if $\{\mathbf{u}_k\}$ is a sequence of points in $\mathbb{R}^n$, we will denote the convergence of $\{\mathbf{u}_k\}$ to the point $\mathbf{u}$ by writing

$$\lim_{k \to \infty} \mathbf{u}_k = \mathbf{u},$$

and we will call $\mathbf{u}$ the *limit* of the sequence $\{\mathbf{u}_k\}$.

The Triangle Inequality, stated as inequality (10.4), asserts that the length of the sum of two vectors is less than or equal to the sum of the lengths of the individual vectors. It is quite useful to rewrite the Triangle Inequality in an equivalent form related to distance between two points, as follows.

**COROLLARY 10.8**

**Another Version of the Triangle Inequality**   For any points $\mathbf{u}$, $\mathbf{v}$, and $\mathbf{w}$ in $\mathbb{R}^n$,

$$d(\mathbf{u}, \mathbf{v}) \leq d(\mathbf{u}, \mathbf{w}) + d(\mathbf{w}, \mathbf{v}). \tag{10.9}$$

**Proof**   We write $\mathbf{u} - \mathbf{v} = (\mathbf{u} - \mathbf{w}) + (\mathbf{w} - \mathbf{v})$, so that by inequality (10.4),

$$\begin{aligned}
d(\mathbf{u}, \mathbf{v}) &= ||\mathbf{u} - \mathbf{v}|| \\
&= ||(\mathbf{u} - \mathbf{w}) + (\mathbf{w} - \mathbf{v})|| \\
&\leq ||\mathbf{u} - \mathbf{w}|| + ||\mathbf{w} - \mathbf{v}|| \\
&= d(\mathbf{u}, \mathbf{w}) + d(\mathbf{w}, \mathbf{v}).
\end{aligned}$$

■

In order to establish the properties of this extended concept of convergence, rather than directly using the definition, it is convenient to proceed by exploiting what we already know about convergence of sequences of real numbers. We shall proceed based on the following two theorems about real sequences, which were proved in Section 2.1, and which we record again here for future reference.

**THEOREM 10.9**   Let $\{a_k\}$ and $\{b_k\}$ be sequences of real numbers and let $a$ and $b$ be real numbers such that

$$\lim_{k\to\infty} a_k = a \quad \text{and} \quad \lim_{k\to\infty} b_k = b.$$

Then if $\alpha$ and $\beta$ are any real numbers,

$$\lim_{k\to\infty} [\alpha a_k + \beta b_k] = \alpha a + \beta b,$$

$$\lim_{k\to\infty} a_k b_k = ab,$$

and if $b_k \neq 0$ for all $k$ and $b \neq 0$, then

$$\lim_{k\to\infty} \left[\frac{a_k}{b_k}\right] = \frac{a}{b}.$$

**THEOREM 10.10**   **The Squeezing Principle**   Let $\{x_k\}$ be a sequence of real numbers. Suppose that there is a sequence of nonnegative real numbers $\{r_k\}$ that converges to 0 and a real number $x$ such that

$$|x_k - x| \leq r_k \quad \text{for every positive integer } k.$$

Then

$$\lim_{k\to\infty} x_k = x.$$

It follows from the definition of convergence that a sequence $\{\mathbf{u}_k\}$ in $\mathbb{R}^n$ converges to the point $\mathbf{u}$ in $\mathbb{R}^n$ if and only if the sequence of real numbers $\{d(\mathbf{u}_k, \mathbf{u})\}$ converges to 0. Observe that a sequence can have at most one limit, since if $\{\mathbf{u}_k\}$ converges to $\mathbf{u}$ and also to $\mathbf{u}'$, then according to the Triangle Inequality (10.9),

$$0 \leq d(\mathbf{u}, \mathbf{u}') \leq d(\mathbf{u}, \mathbf{u}_k) + d(\mathbf{u}_k, \mathbf{u}') \quad \text{for every positive integer } k.$$

From the Squeezing Principle it follows that $d(\mathbf{u}, \mathbf{u}') = 0$, so $\mathbf{u} = \mathbf{u}'$.

**DEFINITION**   *For each index $i$ with $1 \leq i \leq n$, we define the ith component projection function $p_i \colon \mathbb{R}^n \to \mathbb{R}$ by*

$$p_i(\mathbf{u}) \equiv u_i \quad \text{for} \quad \mathbf{u} = (u_1, \ldots, u_n) \text{ in } \mathbb{R}^n.$$

It follows directly from this definition that

$$\mathbf{u} = (p_1(\mathbf{u}), \ldots, p_n(\mathbf{u})) \quad \text{for } \mathbf{u} \text{ in } \mathbb{R}^n,$$

so a point in $\mathbb{R}^n$ is completely determined by the values of the component projection functions at that point.

We will frequently make use of the following linearity property of the projection function: For each index $i$ with $1 \leq i \leq n$, each pair of points $\mathbf{u}$ and $\mathbf{v}$ in $\mathbb{R}^n$, and each pair of real numbers $\alpha$ and $\beta$,

$$p_i(\alpha\mathbf{u} + \beta\mathbf{v}) = \alpha p_i(\mathbf{u}) + \beta p_i(\mathbf{v}).$$

This property follows from the very definitions of sum and scalar multiple. We will also make use of the inequality

$$|p_i(\mathbf{u})| \leq \|\mathbf{u}\| \quad \text{for each index } i \text{ with } 1 \leq i \leq n,$$

which follows from the definition of $\|\mathbf{u}\|$ in terms of the components of $\mathbf{u}$.

**DEFINITION**   *A sequence of points $\{\mathbf{u}_k\}$ in $\mathbb{R}^n$ is said to converge componentwise to the point $\mathbf{u}$ in $\mathbb{R}^n$ provided that for each index $i$ with $1 \le i \le n$,*

$$\lim_{k \to \infty} p_i(\mathbf{u}_k) = p_i(\mathbf{u}).$$

**THEOREM 10.11**   **The Componentwise Convergence Criterion**   Let $\{\mathbf{u}_k\}$ be a sequence in $\mathbb{R}^n$ and $\mathbf{u}$ be a point in $\mathbb{R}^n$. Then $\{\mathbf{u}_k\}$ converges to $\mathbf{u}$ if and only if $\{\mathbf{u}_k\}$ converges componentwise to $\mathbf{u}$.

**Proof**   First we suppose that the sequence $\{\mathbf{u}_k\}$ converges to $\mathbf{u}$. Fix an index $i$ with $1 \le i \le n$. Then

$$0 \le |p_i(\mathbf{u}_k) - p_i(\mathbf{u})| = |p_i(\mathbf{u}_k - \mathbf{u})| \le \|\mathbf{u}_k - \mathbf{u}\| \quad \text{for each positive integer } k.$$

Since, by definition, the sequence of real numbers $\{\|\mathbf{u}_k - \mathbf{u}\|\}$ converges to 0, it follows from the Squeezing Principle that $\{|p_i(\mathbf{u}_k) - p_i(\mathbf{u})|\}$ also converges to 0; that is, the sequence $\{\mathbf{u}_k\}$ converges componentwise to $\mathbf{u}$.

To prove the converse, suppose that the sequence $\{\mathbf{u}_k\}$ converges componentwise to $\mathbf{u}$. Then, by definition, for each index $i$ with $1 \le i \le n$,

$$\lim_{k \to \infty} p_i(\mathbf{u}_k - \mathbf{u}) = 0.$$

But then by the product and addition properties of convergent real sequences, it follows that

$$\lim_{k \to \infty} [(p_1(\mathbf{u}_k - \mathbf{u}))^2 + \cdots + (p_n(\mathbf{u}_k - \mathbf{u}))^2] = 0.$$

This last assertion means precisely that

$$\lim_{k \to \infty} \|\mathbf{u}_k - \mathbf{u}\|^2 = 0,$$

and hence, by the continuity of the square-root function,

$$\lim_{k \to \infty} \|\mathbf{u}_k - \mathbf{u}\| = 0;$$

that is, the sequence $\{\mathbf{u}_k\}$ converges to $\mathbf{u}$.   ∎

**THEOREM 10.12**   Let $\{\mathbf{u}_k\}$ and $\{\mathbf{v}_k\}$ be sequences in $\mathbb{R}^n$ such that $\{\mathbf{u}_k\}$ converges to the point $\mathbf{u}$ and $\{\mathbf{v}_k\}$ converges to the point $\mathbf{v}$. Then

(i) $\lim_{k \to \infty} \langle \mathbf{u}_k, \mathbf{v}_k \rangle = \langle \mathbf{u}, \mathbf{v} \rangle$,

and for any two real numbers $\alpha$ and $\beta$,

(ii) $\lim_{k \to \infty} [\alpha \mathbf{u}_k + \beta \mathbf{v}_k] = \alpha \mathbf{u} + \beta \mathbf{v}$.

**Proof**  From the Componentwise Convergence Theorem, it follows that for each index $i$ with $1 \leq i \leq n$,

$$\lim_{k \to \infty} p_i(\mathbf{u}_k) = p_i(\mathbf{u}) \quad \text{and} \quad \lim_{k \to \infty} p_i(\mathbf{v}_k) = p_i(\mathbf{v}). \tag{10.10}$$

Using the product property of convergent sequences of real numbers, for each index $i$ with $1 \leq i \leq n$,

$$\lim_{k \to \infty} p_i(\mathbf{u}_k) p_i(\mathbf{v}_k) = p_i(\mathbf{u}) p_i(\mathbf{v}),$$

and hence, by the sum property of convergent real sequences,

$$\lim_{k \to \infty} \sum_{i=1}^{n} p_i(\mathbf{u}_k) p_i(\mathbf{v}_k) = \sum_{i=1}^{n} p_i(\mathbf{u}) p_i(\mathbf{v});$$

that is, (i) holds.

In order to prove (ii), observe that the sum and product properties of convergent real sequences imply that

$$\lim_{k \to \infty} [\alpha p_i(\mathbf{u}_k) + \beta p_i(\mathbf{v}_k)] = \alpha p_i(\mathbf{u}) + \beta p_i(\mathbf{v}),$$

which means that

$$\lim_{k \to \infty} p_i(\alpha \mathbf{u}_k + \beta \mathbf{v}_k) = p_i(\alpha \mathbf{u} + \beta \mathbf{v}).$$

Thus the sequence $\{\alpha \mathbf{u}_k + \beta \mathbf{v}_k\}$ converges componentwise to $\alpha \mathbf{u} + \beta \mathbf{v}$, and so, by the Componentwise Convergence Criterion, the sequence $\{\alpha \mathbf{u}_k + \beta \mathbf{v}_k\}$ converges to the point $\alpha \mathbf{u} + \beta \mathbf{v}$.  ∎

---

**EXERCISES**

1. Let $\{\mathbf{u}_k\}$ be a sequence in $\mathbb{R}^n$ that converges to the point $\mathbf{u}$. Prove that

$$\lim_{k \to \infty} \langle \mathbf{u}_k, \mathbf{v} \rangle = \langle \mathbf{u}, \mathbf{v} \rangle$$

   for every point $\mathbf{v}$ in $\mathbb{R}^n$.

2. Let $\{\mathbf{u}_k\}$ be a sequence in $\mathbb{R}^n$ and $\mathbf{u}$ be a point in $\mathbb{R}^n$. Suppose that for every $\mathbf{v}$ in $\mathbb{R}^n$,

$$\lim_{k \to \infty} \langle \mathbf{u}_k, \mathbf{v} \rangle = \langle \mathbf{u}, \mathbf{v} \rangle.$$

   Prove that $\{\mathbf{u}_k\}$ converges to $\mathbf{u}$. (*Hint:* For each index $i$ with $1 \leq i \leq n$ and each point $\mathbf{u}$ in $\mathbb{R}^n$, $p_i(\mathbf{u}) = \langle \mathbf{u}, \mathbf{e}_i \rangle$, where $\mathbf{e}_i$ is the point in $\mathbb{R}^n$ whose $i$th component is 1 and whose other components are 0.)

3. Suppose that $\{\mathbf{u}_k\}$ is a sequence of points in $\mathbb{R}^n$ that converges to the point $\mathbf{u}$. Prove that the sequence of real numbers $\{\|\mathbf{u}_k\|\}$ converges to $\|\mathbf{u}\|$.

4. Suppose that $\{\mathbf{u}_k\}$ is a sequence of points in $\mathbb{R}^n$ that converges to the point $\mathbf{u}$. Assume also that $\mathbf{u}_k \neq \mathbf{0}$ for all $k$ and that $\mathbf{u} \neq \mathbf{0}$. Define $\mathbf{v}_k = (1/\|\mathbf{u}_k\|)\mathbf{u}_k$ and $\mathbf{v} = (1/\|\mathbf{u}\|)\mathbf{u}$. Prove that the sequence $\{\mathbf{v}_k\}$ converges to $\mathbf{v}$.

5. Suppose that $\{\mathbf{u}_k\}$ is a sequence of points in $\mathbb{R}^n$ that converges to the point $\mathbf{u}$ and that $\|\mathbf{u}\| = r > 0$. Prove that there is a positive integer $K$ such that

$$\|\mathbf{u}_k\| > r/2 \quad \text{if} \quad k \geq K.$$

6. Suppose that $\{\mathbf{u}_k\}$ is a sequence in $\mathbb{R}^n$ that converges to the point $\mathbf{u}$. Let $\mathbf{v}$ be a point in $\mathbb{R}^n$ that is orthogonal to each $\mathbf{u}_k$. Prove that $\mathbf{v}$ is also orthogonal to $\mathbf{u}$.

7. Use the Triangle Inequality for the sum of points in $\mathbb{R}^n$ to give a direct proof of Theorem 10.12.

8. A sequence of points $\{\mathbf{u}_k\}$ in $\mathbb{R}^n$ is called a *Cauchy sequence* provided that for each positive number $\epsilon$ there is a natural number $K$ such that

$$d(\mathbf{u}_k, \mathbf{u}_\ell) < \epsilon \quad \text{if} \quad k \geq K \quad \text{and} \quad \ell \geq K.$$

(a) Prove that $\{\mathbf{u}_k\}$ is a Cauchy sequence if and only if each component sequence is a Cauchy sequence.

(b) Prove that a sequence in $\mathbb{R}^n$ converges if and only if it is a Cauchy sequence. (*Hint:* For sequences of real numbers, this was proved in Section 9.1.)

## 10.3 Interiors, Exteriors, and Boundaries of Subsets of $\mathbb{R}^n$

The first nine chapters of this book were devoted to the study of real-valued functions of a single real variable. In this study, a prominent part was played by those functions that have as their domain an interval of real numbers. In the study of functions that have as their domain a subset of $\mathbb{R}^n$, it turns out that it is necessary to study functions that have as their domain quite general subsets of $\mathbb{R}^n$; there is no special class of subsets of $\mathbb{R}^n$ that play the same distinguished role as do the intervals as subsets of $\mathbb{R}$. As preparation for the study of such functions, in the present section we will consider some special types of subsets of $\mathbb{R}^n$, among which are open subsets and closed subsets.

**DEFINITION**   *Given a point* $\mathbf{u}$ *in* $\mathbb{R}^n$ *and a positive number* $r$, *we call the set*

$$\mathcal{N}_r(\mathbf{u}) \equiv \left\{ \mathbf{v} \text{ in } \mathbb{R}^n \mid d(\mathbf{u}, \mathbf{v}) < r \right\}$$

*the symmetric neighborhood of radius* $r$ *of* $\mathbf{u}$.

**FIGURE 10.4**

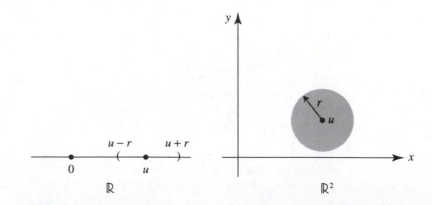

In the case when $n = 1$, the symmetric neighborhood of radius $r$ of a point $u$ in $\mathbb{R}$ is simply the open interval $\{v \text{ in } \mathbb{R} \mid u - r < v < u + r\}$. In the case when $n = 2$, the symmetric neighborhood of radius $r$ of the point $\mathbf{u} = (x_0, y_0)$ in $\mathbb{R}^2$ consists of all points in the plane $\mathbb{R}^2$ that lie inside the circle of radius $r$ centered at the point $(x_0, y_0)$.

**DEFINITION**   *Let $A$ be a subset of $\mathbb{R}^n$. A point $\mathbf{u}$ in $\mathbb{R}^n$ is called an interior point of $A$ provided that there is a symmetric neighborhood of $\mathbf{u}$ that is contained in $A$. The set of all interior points of $A$ is called the interior of $A$, and is denoted by int $A$.*

It is clear that the interior of a set is contained in the set. But there may be points in the set that are not interior points.

**EXAMPLE 10.1**   Let $a$ and $b$ be real numbers with $a < b$. Define $A$ to be the interval $(a, b] = \{u \text{ in } \mathbb{R} \mid a < u \leq b\}$. For a point $u$ in $(a, b)$, define $r$ to be the smaller of the positive numbers $u - a$ and $b - u$. Then $\mathcal{N}_r(u) \subseteq (a, b]$. Thus $u$ is an interior point of $A$. On the other hand, the point $b$ is in the set $A$ but is not an interior point of $A$, since every symmetric neighborhood of the point $b$ contains points greater than $b$. Thus the interior of $A$ is the interval $(a, b)$.   $\square$

**EXAMPLE 10.2**   Let $A$ be the subset of $\mathbb{R}$ consisting of the rational numbers. Then the density of the irrational numbers is equivalent to the assertion that $A$ has no interior points—that is, int $A = \emptyset$.   $\square$

**DEFINITION**   *A subset $A$ of $\mathbb{R}^n$ is said to be open in $\mathbb{R}^n$ provided that every point in $A$ is an interior point of $A$.*

It follows immediately that $\mathbb{R}^n$ is an open subset of $\mathbb{R}^n$ and that the null set $\emptyset$ is also an open subset of $\mathbb{R}^n$. Moreover, if $a$ and $b$ are real numbers with $a < b$, then the interval $(a, b) = \{u \text{ in } \mathbb{R} \mid a < u < b\}$ is easily seen to be open in $\mathbb{R}$. Thus our previous use of the adjective *open* is consistent with the general definition we have just given.

**PROPOSITION 10.13**   Every symmetric neighborhood of a point in $\mathbb{R}^n$ is open in $\mathbb{R}^n$.

**Proof**   Let $\mathbf{u}$ be a point in $\mathbb{R}^n$ and let $r$ be a positive real number. Consider the neighborhood $\mathcal{N}_r(\mathbf{u})$. We must show that every point in $\mathcal{N}_r(\mathbf{u})$ is an interior point of $\mathcal{N}_r(\mathbf{u})$. Let $\mathbf{v}$ be a point in $\mathcal{N}_r(\mathbf{u})$. Define $R = r - d(\mathbf{u}, \mathbf{v})$, and observe that $R$ is positive. We claim that

$$\mathcal{N}_R(\mathbf{v}) \subseteq \mathcal{N}_r(\mathbf{u}). \tag{10.11}$$

Indeed, if $\mathbf{w}$ is in $\mathcal{N}_R(\mathbf{v})$, then

$$d(\mathbf{w}, \mathbf{v}) < R = r - d(\mathbf{u}, \mathbf{v}),$$

**FIGURE 10.5**

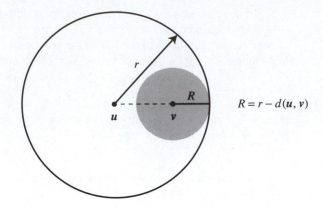

$$R = r - d(\mathbf{u}, \mathbf{v})$$

so using the Triangle Inequality, we have

$$d(\mathbf{w}, \mathbf{u}) \leq d(\mathbf{w}, \mathbf{v}) + d(\mathbf{v}, \mathbf{u})$$
$$< [r - d(\mathbf{u}, \mathbf{v})] + d(\mathbf{v}, \mathbf{u})$$
$$= r.$$

Thus the inclusion (10.11) holds, so $\mathbf{v}$ is an interior point of $\mathcal{N}_r(\mathbf{u})$. ∎

**DEFINITION**    *A subset $A$ of $\mathbb{R}^n$ is said to be closed in $\mathbb{R}^n$ provided that whenever $\{\mathbf{u}_k\}$ is a sequence of points in $A$ that converges to a point $\mathbf{u}$ in $\mathbb{R}^n$, then $\mathbf{u}$ is in $A$.*

**EXAMPLE 10.3**    Let $a$ and $b$ be real numbers with $a < b$. Then the interval $[a, b] = \{u$ in $\mathbb{R} \mid a \leq u \leq b\}$ is closed. This is the content of Corollary 2.8. Thus we have previously used the adjective *closed* in a manner consistent with the general definition that we have just given.    □

**EXAMPLE 10.4**    Define

$$A = \left\{(x, y) \text{ in } \mathbb{R}^2 \mid -1 \leq x \leq 1, -1 \leq y \leq 1\right\}.$$

Then the set $A$ is closed in $\mathbb{R}^2$. This follows from Example 10.3 and the Componentwise Convergence Criterion.    □

**THEOREM 10.14**    **The Complementing Characterization**    A subset of $\mathbb{R}^n$ is open in $\mathbb{R}^n$ if and only if its complement in $\mathbb{R}^n$ is closed in $\mathbb{R}^n$.

**Proof**    First, suppose that $A$ is an open subset of $\mathbb{R}^n$. Thus every point in $A$ is an interior point of $A$, so a sequence in $\mathbb{R}^n \backslash A$ cannot converge to a point in $A$. It follows that a sequence in $\mathbb{R}^n \backslash A$ that converges must converge to a point in $\mathbb{R}^n \backslash A$. Thus $\mathbb{R}^n \backslash A$ is closed in $\mathbb{R}^n$.

To prove the converse, suppose that $A$ is a subset of $\mathbb{R}^n$ such that $\mathbb{R}^n \backslash A$ is closed in $\mathbb{R}^n$. We must show that every point in $A$ is an interior point of $A$. Let $\mathbf{u}$ be a point in $A$. Suppose that $\mathbf{u}$ is not an interior point of $A$. Let $k$ be a natural number. Then the symmetric neighborhood $\mathcal{N}_{1/k}(\mathbf{u})$ is not a subset of $A$, so we can choose a point, which we label $\mathbf{u}_k$, such that

$$\mathbf{u}_k \text{ belongs to } \mathbb{R}^n \backslash A \quad \text{and} \quad d(\mathbf{u}_k, \mathbf{u}) < 1/k.$$

Thus the sequence $\{\mathbf{u}_k\}$ converges to $\mathbf{u}$. But $\mathbb{R}^n \backslash A$ is closed, so $\mathbf{u}$ belongs to $\mathbb{R}^n \backslash A$. This contradiction shows that $\mathbf{u}$ is an interior point of $A$.  ∎

If $A$ and $B$ are subsets of $\mathbb{R}^n$, the *complement in $A$ of $B$*, denoted by $A \backslash B$, is defined by the formula

$$A \backslash B \equiv \{\mathbf{u} \in A \mid \mathbf{u} \notin B\}.$$

Given a subset $A$ of $\mathbb{R}^n$ and $\{A_s\}_{s \in S}$ a collection of subsets of $\mathbb{R}^n$ indexed by a set $S$, from the definition of union, intersection, and complement, it follows that

$$A \backslash \bigcap_{s \in S} A_s = \bigcup_{s \in S} (A \backslash A_s) \quad \text{and} \quad A \backslash \bigcup_{s \in S} A_s = \bigcap_{s \in S} (A \backslash A_s).$$

These formulas are often referred to as *DeMorgan's Laws*.

**THEOREM 10.15**

(i)  The union of a collection of open subsets of $\mathbb{R}^n$ is open in $\mathbb{R}^n$.

(ii)  The intersection of a collection of closed subsets of $\mathbb{R}^n$ is closed in $\mathbb{R}^n$.

**Proof of (i)**   Suppose that $\mathcal{O} = \cup_{s \in S} \mathcal{O}_s$, where each $\mathcal{O}_s$ is an open subset of $\mathbb{R}^n$. We claim that $\mathcal{O}$ is open, that is, that each point of $\mathcal{O}$ is an interior point of $\mathcal{O}$. Indeed, let $\mathbf{u}$ be in $\mathcal{O}$. Then $\mathbf{u}$ is in $\mathcal{O}_s$ for some $s$ in $S$. Since $\mathcal{O}_s$ is open, there is a symmetric neighborhood of $\mathbf{u}$ that is contained in $\mathcal{O}_s$ and is hence contained in $\cup_{s \in S} \mathcal{O}_s = \mathcal{O}$. Thus $\mathbf{u}$ is an interior point of $\mathcal{O}$.

**Proof of (ii)**   Suppose that $C = \bigcap_{s \in S} C_s$, where each $C_s$ is closed in $\mathbb{R}^n$. Then

$$\mathbb{R}^n \backslash C = \mathbb{R}^n \backslash \bigcap_{s \in S} C_s = \bigcup_{s \in S} (\mathbb{R}^n \backslash C_s).$$

From (i) and the Complementing Characterization, it follows that $C$ is closed in $\mathbb{R}^n$.  ∎

It is not always true that the intersection of a collection of open sets is again open. For instance, for each positive integer $k$ the interval of real numbers $(-1/k, 1/k)$ is an open subset of $\mathbb{R}$, yet the intersection of this collection of open sets is the set consisting of the single point 0. But a set consisting of a single point is clearly not open, so the intersection is not open. However, we will now prove that the intersection of a *finite* collection of open sets is open.

**THEOREM 10.16**

(i)  The intersection of a finite collection of open subsets of $\mathbb{R}^n$ is open in $\mathbb{R}^n$.

(ii)  The union of a finite collection of closed subsets of $\mathbb{R}^n$ is closed in $\mathbb{R}^n$.

**Proof of (i)**   Suppose $\mathcal{O} = \bigcap_{i=1}^{k} \mathcal{O}_i$ for some positive integer $k$, where each $\mathcal{O}_i$ is open in $\mathbb{R}^n$. Let $\mathbf{u}$ be a member of $\mathcal{O}$. If $1 \le i \le k$, $\mathbf{u}$ belongs to $\mathcal{O}_i$ and $\mathcal{O}_i$ is open in $\mathbb{R}^n$, so there is a positive number $r_i$ such that $\mathcal{N}_{r_i}(\mathbf{u}) \subseteq \mathcal{O}_i$. Define $r = \min\{r_1, \ldots, r_k\}$. Then $r$ is positive, and $\mathcal{N}_r(\mathbf{u}) \subseteq \bigcap_{i=1}^{k} \mathcal{O}_i = \mathcal{O}$. Thus $\mathbf{u}$ is an interior point of $\mathcal{O}$. Therefore, every point in $\mathcal{O}$ is an interior point of $\mathcal{O}$, so $\mathcal{O}$ is open in $\mathbb{R}^n$.

**Proof of (ii)**   Suppose $\mathcal{C} = \bigcup_{i=1}^{k} \mathcal{C}_i$ for some positive integer $k$, where each $\mathcal{C}_i$ is closed in $\mathbb{R}^n$. Observe, by DeMorgan's Laws, that $\mathbb{R}^n \backslash \mathcal{C} = \bigcap_{i=1}^{k} (\mathbb{R}^n \backslash \mathcal{C}_i)$. From (i) and the Complementing Characterization, it follows that $\bigcap_{i=1}^{k} \mathcal{C}_i$ is closed in $\mathbb{R}^n$.   ∎

---

**DEFINITION**   *Let $A$ be a subset of $\mathbb{R}^n$.*

(i) *A point $\mathbf{u}$ in $\mathbb{R}^n$ is called an exterior point of $A$ provided that there is a symmetric neighborhood of $\mathbf{u}$ that is contained in $\mathbb{R}^n \setminus A$. The set of all exterior points of $A$ is called the exterior of $A$, and is denoted by ext $A$.*

(ii) *A point $\mathbf{u}$ in $\mathbb{R}^n$ is called a boundary point of $A$ provided that each symmetric neighborhood of $\mathbf{u}$ contains a point in $A$ and also contains a point not in $A$. The set of all boundary points of $A$ is called the boundary of $A$, and is denoted by bd $A$.*

It is clear that given a subset $A$ of $\mathbb{R}^n$ and any point $\mathbf{u}$ in $\mathbb{R}^n$, either there is a symmetric neighborhood of $\mathbf{u}$ that is contained in $A$, or there is a symmetric neighborhood of $\mathbf{u}$ that is contained in $\mathbb{R}^n \setminus A$, or every symmetric neighborhood of $\mathbf{u}$ contains a point in $A$ and also contains a point in $\mathbb{R}^n \setminus A$; furthermore, these possibilities are mutually exclusive. This means precisely that $\mathbb{R}^n$ is decomposed into the following disjoint union:

$$\mathbb{R}^n = \text{int } A \cup \text{ext } A \cup \text{bd } A. \tag{10.12}$$

Directly from the definitions of interior, exterior, and boundary, we see that if $A$ is a subset of $\mathbb{R}^n$, then

$$\text{int } A = \text{ext}\,(\mathbb{R}^n \setminus A) \quad \text{and} \quad \text{bd } A = \text{bd}\,(\mathbb{R}^n \setminus A). \tag{10.13}$$

---

**THEOREM 10.17**   Let $A$ be a subset of $\mathbb{R}^n$. Then

(i) *$A$ is open in $\mathbb{R}^n$ if and only if $A \cap \text{bd } A = \emptyset$;*

(ii) *$A$ is closed in $\mathbb{R}^n$ if and only if bd $A \subseteq A$.*

**Proof of (i)**   First, let us suppose that $A$ is open in $\mathbb{R}^n$. Then $A = \text{int } A$. Thus, since int $A \cap \text{bd } A = \emptyset$, it follows that $A \cap \text{bd } A = \emptyset$. To prove the converse, suppose that $A \cap \text{bd } A = \emptyset$. Then since $A \cap \text{ext } A = \emptyset$ and int $A \subseteq A$, it follows from the decomposition (10.12) that $A = \text{int } A$; that is, $A$ is an open subset of $\mathbb{R}^n$.

**Proof of (ii)**   The Complementing Characterization asserts that $A$ is closed in $\mathbb{R}^n$ if and only if $\mathbb{R}^n \backslash A$ is open in $\mathbb{R}^n$. However, from (i) it follows that $\mathbb{R}^n \backslash A$ is open in $\mathbb{R}^n$ if and only if $(\mathbb{R}^n \backslash A) \cap \text{bd}\,(\mathbb{R}^n \backslash A) = \emptyset$. Since $\text{bd}\,A = \text{bd}\,(\mathbb{R}^n \backslash A)$, we conclude that $A$ is closed in $\mathbb{R}^n$ if and only if $(\mathbb{R}^n \backslash A) \cap \text{bd}\,A = \emptyset$. This proves (ii), since clearly $\text{bd}\,A \cap (\mathbb{R}^n \backslash A) = \emptyset$ if and only if $\text{bd}\,A \subseteq A$. ∎

We conclude this chapter by introducing a construction called the *Cartesian product*, which builds subsets of $\mathbb{R}^n$ from subsets of $\mathbb{R}$.

---

**DEFINITION**   *For each index $i$ with $1 \leq i \leq n$, let $A_i$ be a subset of $\mathbb{R}$. The Cartesian product of $A_1, A_2, \ldots, A_n$, denoted by $A_1 \times A_2 \times \cdots \times A_n$, is the subset of $\mathbb{R}^n$ defined by the formula*

$$A_1 \times A_2 \times \cdots \times A_n \equiv \left\{ \mathbf{u} = (u_1, \ldots, u_i, \ldots, u_n) \text{ in } \mathbb{R}^n \mid u_i \text{ in } A_i \text{ for } 1 \leq i \leq n \right\}.$$

If each $A_i$ is a bounded open interval $I_i = (a_i, b_i)$, the Cartesian product is called an *open generalized rectangle*; if each $A_i$ is a bounded closed interval $I_i = [a_i, b_i]$, the Cartesian product is called a *closed generalized rectangle*. We leave it as an exercise to verify that an open generalized rectangle is an open subset of $\mathbb{R}^n$ and that a closed generalized rectangle is a closed subset of $\mathbb{R}^n$. In fact, the Cartesian product of any $n$ open subsets of $\mathbb{R}$ is open in $\mathbb{R}^n$ and the Cartesian product of any $n$ closed subsets of $\mathbb{R}$ is closed in $\mathbb{R}^n$. (See Exercises 10 and 11.)

---

**EXERCISES**

1. Determine which of the following subsets of $\mathbb{R}$ are open in $\mathbb{R}$, closed in $\mathbb{R}$, or neither open nor closed in $\mathbb{R}$. Justify your conclusions.
   (a) $A = (0, \infty)$
   (b) $A = \mathbb{Q}$, the set of rational numbers
   (c) $A = \{u \text{ in } \mathbb{R} \mid u^2 > 4\}$
   (d) $A = \{u \text{ in } \mathbb{R} \mid u^2 \geq 4\}$
   (e) $A = [0, \infty)$

2. Determine which of the following subsets $A$ of $\mathbb{R}^2$ are open in $\mathbb{R}^2$, closed in $\mathbb{R}^2$, or neither open nor closed in $\mathbb{R}^2$. Justify your conclusions.
   (a) $A = \{\mathbf{u} = (x, y) \mid x^2 > y\}$
   (b) $A = \{\mathbf{u} = (x, y) \mid x^2 + y^2 = 1\}$
   (c) $A = \{\mathbf{u} = (x, y) \mid x \text{ is rational}\}$
   (d) $A = \{\mathbf{u} = (x, y) \mid x \geq 0, \ y \geq 0\}$

3. Let $r$ be a positive number, and define $\mathcal{O} = \{\mathbf{u} \text{ in } \mathbb{R}^n \mid \|\mathbf{u}\| > r\}$. Prove that $\mathcal{O}$ is open in $\mathbb{R}^n$ by showing that every point in $\mathcal{O}$ is an interior point of $\mathcal{O}$. (*Hint:* For $\mathbf{u}$ in $\mathcal{O}$, define $R = \|\mathbf{u}\| - r$ and show that $\mathcal{N}_R(\mathbf{u}) \subseteq \mathcal{O}$.)

4. Let $r$ be a positive number, and define $F = \{\mathbf{u} \text{ in } \mathbb{R}^n \mid \|\mathbf{u}\| \leq r\}$. Use the Componentwise Convergence Criterion to prove that $F$ is closed.

5. Let $r$ be a positive number, and define $\mathcal{O} = \{\mathbf{u} \text{ in } \mathbb{R}^n \mid \|\mathbf{u}\| > r\}$. Prove that $\mathcal{O}$ is open in $\mathbb{R}^n$ by showing that its complement is closed in $\mathbb{R}^n$.

6. Let $r$ be a positive number, and define $F = \{\mathbf{u} \text{ in } \mathbb{R}^n \mid \|\mathbf{u}\| = r\}$. Prove that $F$ is closed in $\mathbb{R}^n$ by using the Componentwise Convergence Criterion together with the sum and product properties of convergent real sequences.

7.  Let $A$ be a subset of $\mathbb{R}^n$ and $\mathbf{w}$ be a point in $\mathbb{R}^n$. The *translate* of $A$ by $\mathbf{w}$ is denoted by $\mathbf{w} + A$ and is defined by

$$\mathbf{w} + A \equiv \{\mathbf{w} + \mathbf{u} \mid \mathbf{u} \text{ in } A\}.$$

   (a)  Show that $A$ is open if and only if $\mathbf{w} + A$ is open.
   (b)  Show that $A$ is closed if and only if $\mathbf{w} + A$ is closed.

8.  For $r$ a positive number and $\mathbf{u}$ a point in $\mathbb{R}^n$, define $A = \mathcal{N}_r(\mathbf{u})$. Show that int $A = A$, that bd $A = \{\mathbf{v} \text{ in } \mathbb{R}^n \mid d(\mathbf{u}, \mathbf{v}) = r\}$ and that ext $A = \{\mathbf{v} \text{ in } \mathbb{R}^n \mid d(\mathbf{u}, \mathbf{v}) > r\}$.

9.  Let $A$ and $B$ be subsets of $\mathbb{R}^n$ with $A \subseteq B$.
   (a)  Prove that int $A \subseteq$ int $B$.
   (b)  Is it necessarily true that bd $A \subseteq$ bd $B$?

10. For each index $i$ with $1 \leq i \leq n$, let $F_i$ be a closed subset of $\mathbb{R}$. Prove that the Cartesian product

$$F_1 \times F_2 \times \cdots \times F_n$$

   is a closed subset of $\mathbb{R}^n$.

11. For each index $i$ with $1 \leq i \leq n$, let $\mathcal{O}_i$ be an open subset of $\mathbb{R}$. Prove that the Cartesian product

$$\mathcal{O}_1 \times \mathcal{O}_2 \times \cdots \times \mathcal{O}_n$$

   is an open subset of $\mathbb{R}^n$.

12. For a subset $A$ of $\mathbb{R}^n$, the *closure* of $A$, denoted by cl $A$, is defined by

$$\text{cl } A = \text{int } A \cup \text{bd } A.$$

   Prove that $A \subseteq$ cl $A$, and that $A =$ cl $A$ if and only if $A$ is closed in $\mathbb{R}^n$.

13. Let $A$ be a subset of $\mathbb{R}^n$.
   (a)  Prove that int $A$ is an open subset of $\mathbb{R}^n$.
   (b)  Use (a) to show that ext $A$ is also an open subset of $\mathbb{R}^n$.
   (c)  Use (a) and (b), together with the decomposition (10.12), to show that bd $A$ is a closed subset of $\mathbb{R}^n$.

# 11

# Continuity, Compactness, and Connectedness

## 11.1   Continuity of Functions and Mappings

Recall that a function $f: A \to \mathbb{R}$ whose domain $A$ is a subset of $\mathbb{R}$ has been defined to be *continuous at the point $x$* in $A$ provided that whenever a sequence $\{x_k\}$ in $A$ converges to $x$, the image sequence $\{f(x_k)\}$ converges to $f(x)$. Moreover, the function $f: A \to \mathbb{R}$ has been defined to be *continuous* provided that it is continuous at each point in its domain. We studied such functions in Chapter 3. In the present chapter, we will study more general functions of the form $F: A \to \mathbb{R}^m$ where $A$ is a subset of $\mathbb{R}^n$, and where $m$ may be greater than 1. It turns out that it is sometimes useful to distinguish between the case when $m = 1$ and the case when $m > 1$; there are a number of particular results that hold only in the case when $m = 1$. To emphasize this distinction, we will call general functions $F: A \to \mathbb{R}^m$ *mappings* and reserve the use of the word *function* for the case when $m = 1$ — that is, when the range is $\mathbb{R}$.

**DEFINITION**   *Let $A$ be a subset of $\mathbb{R}^n$.*

(i)  *A mapping $F: A \to \mathbb{R}^m$ is said to be continuous at the point $\mathbf{u}$ in $A$ provided that whenever a sequence $\{\mathbf{u}_k\}$ in $A$ converges to $\mathbf{u}$, the image sequence $\{F(\mathbf{u}_k)\}$ converges to $F(\mathbf{u})$.*

(ii)  *A mapping $F: A \to \mathbb{R}^m$ is said to be continuous provided that it is continuous at every point in its domain.*

**249**

**PROPOSITION
11.1**

For each index $i$ with $1 \leq i \leq n$, the $i$th component projection function $p_i \colon \mathbb{R}^n \to \mathbb{R}$ is continuous.

**Proof**    Let $\mathbf{u}$ be a point in $\mathbb{R}^n$. Suppose that $\{\mathbf{u}_k\}$ is a sequence in $\mathbb{R}^n$ that converges to $\mathbf{u}$. Then, by the Componentwise Convergence Criterion,

$$\lim_{k \to \infty} \{p_i(\mathbf{u}_k)\} = p_i(\mathbf{u}).$$

Thus the function $p_i \colon \mathbb{R}^n \to \mathbb{R}$ is continuous at the point $\mathbf{u}$. Since $\mathbf{u}$ was an arbitrarily chosen point in $\mathbb{R}^n$, the function $p_i \colon \mathbb{R}^n \to \mathbb{R}$ is continuous.    ∎

**EXAMPLE 11.1**    Define the functions $f \colon \mathbb{R}^3 \to \mathbb{R}$, $g \colon \mathbb{R}^3 \to \mathbb{R}$ and $h \colon \mathbb{R}^3 \to \mathbb{R}$ by

$$f(x, y, z) = x, \quad g(x, y, z) = y, \quad \text{and} \quad h(x, y, z) = z \quad \text{for } (x, y, z) \text{ in } \mathbb{R}^3.$$

Proposition 11.1 implies that these three functions are continuous.    □

We will first establish a number of results for real-valued functions of several real variables, and then we will turn to the study of general mappings. We begin with the following extension of Theorem 3.1.

**THEOREM 11.2**    Let $A$ be a subset of $\mathbb{R}^n$ that contains the point $\mathbf{u}$, and suppose that the functions $h \colon A \to \mathbb{R}$ and $g \colon A \to \mathbb{R}$ are both continuous at $\mathbf{u}$. For any real numbers $\alpha$ and $\beta$, the function

$$\alpha h + \beta g \colon A \to \mathbb{R}$$

is continuous at $\mathbf{u}$. Also, the product

$$h \cdot g \colon A \to \mathbb{R}$$

is continuous at $\mathbf{u}$. Moreover, if $g(\mathbf{v}) \neq 0$ for all $\mathbf{v}$ in $A$, then the quotient

$$\frac{h}{g} \colon A \to \mathbb{R}$$

is continuous at $\mathbf{u}$.

**Proof**    Let $\{\mathbf{u}_k\}$ be a sequence in the set $A$ that converges to the point $\mathbf{u}$. Since the function $h \colon A \to \mathbb{R}$ is continuous at the point $\mathbf{u}$, the image sequence $\{h(\mathbf{u}_k)\}$ converges to $h(\mathbf{u})$. Similarly, the continuity of the function $g \colon A \to \mathbb{R}$ at the point $\mathbf{u}$ implies that the sequence $\{g(\mathbf{u}_k)\}$ converges to $g(\mathbf{u})$. From the sum, product, and quotient properties of convergent sequences of real numbers, it follows that

$$\lim_{k \to \infty} (\alpha h + \beta g)(\mathbf{u}_k) = (\alpha h + \beta g)(\mathbf{u}),$$

$$\lim_{k \to \infty} (hg)(\mathbf{u}_k) = (hg)(\mathbf{u}),$$

and

$$\lim_{k \to \infty} \left(\frac{h}{g}\right)(\mathbf{u}_k) = \left(\frac{h}{g}\right)(\mathbf{u}).$$

These three sequential limits are precisely what is required to prove the theorem.    ∎

**EXAMPLE 11.2**   Define the function $f: \mathbb{R}^3 \to \mathbb{R}$ by

$$f(x, y, z) = xz + y^3 \quad \text{for } (x, y, z) \text{ in } \mathbb{R}^3.$$

Since this function is obtained from products and sums of component projection functions, it follows from Proposition 11.1 and Theorem 11.2 that it is continuous.   □

Given a mapping $F: A \to \mathbb{R}^m$, if $B$ is a subset of the domain $A$, the *image* of the set $B$ under the mapping $F: A \to \mathbb{R}^m$, denoted by $F(B)$, is defined by the formula

$$F(B) \equiv \{\mathbf{v} \text{ in } \mathbb{R}^m \mid \mathbf{v} = F(\mathbf{u}) \text{ for some } \mathbf{u} \text{ in } B\}.$$

**THEOREM 11.3**   Let $A$ be a subset of $\mathbb{R}^n$ that contains the point $\mathbf{u}$. Suppose that the mapping $G: A \to \mathbb{R}^m$ is continuous at the point $\mathbf{u}$. Let $B$ be a subset of $\mathbb{R}^m$ with $G(A) \subseteq B$, and suppose that the mapping $H: B \to \mathbb{R}^k$ is continuous at the point $G(\mathbf{u})$. Then the composition

$$H \circ G: A \to \mathbb{R}^k$$

is continuous at $\mathbf{u}$.

**Proof**   Let $\{\mathbf{u}_k\}$ be a sequence in $A$ that converges to the point $\mathbf{u}$. Since the mapping $G: A \to \mathbb{R}^m$ is continuous at $\mathbf{u}$, it follows that the image sequence $\{G(\mathbf{u}_k)\}$ converges to $G(\mathbf{u})$. But then $\{G(\mathbf{u}_k)\}$ is a sequence in $B$ that converges to the point $G(\mathbf{u})$. The continuity of the mapping $H: B \to \mathbb{R}^k$ at the point $G(\mathbf{u})$ implies that the sequence $\{H(G(\mathbf{u}_k))\}$ converges to $H(G(\mathbf{u}))$; that is, the sequence $\{(H \circ G)(\mathbf{u}_k)\}$ converges to $(H \circ G)(\mathbf{u})$.   ■

**EXAMPLE 11.3**   Define the function $f: \mathbb{R}^2 \to \mathbb{R}$ by

$$f(x, y) = x^2 y + e^{xy+1} \text{ for } (x, y) \text{ in } \mathbb{R}^2.$$

Then the function $f: \mathbb{R}^2 \to \mathbb{R}$ is continuous. To see this, we observe that

$$f = p_1 \cdot p_1 \cdot p_2 + h \circ (p_1 \cdot p_2): \mathbb{R}^2 \to \mathbb{R},$$

where $h: \mathbb{R} \to \mathbb{R}$ is defined by $h(x) = e^{x+1}$ for $x$ in $\mathbb{R}$. Since products, sums, and compositions of continuous maps are again continuous, it follows that the function $f: \mathbb{R}^2 \to \mathbb{R}$ is continuous.   □

**EXAMPLE 11.4**   Define the function $f: \mathbb{R}^n \to \mathbb{R}$ by

$$f(\mathbf{u}) = \|\mathbf{u}\| \quad \text{for } \mathbf{u} \text{ in } \mathbb{R}^n.$$

Then the function $f: \mathbb{R}^n \to \mathbb{R}$ is continuous. To see this, we observe that

$$f = h \circ (p_1 p_1 + \cdots + p_n p_n): \mathbb{R}^n \to \mathbb{R},$$

where $h(x) = \sqrt{x}$ for each nonnegative number $x$. Since products, sums, and compositions of continuous maps are again continuous, it follows that the function $f: \mathbb{R}^n \to \mathbb{R}$ is continuous.   □

**DEFINITION**

*Given a mapping $F: A \to \mathbb{R}^m$ where $A$ is a subset of $\mathbb{R}^n$, and an index $i$ with $1 \le i \le n$, we define the function $F_i: A \to \mathbb{R}$ to be the composition of $F: A \to \mathbb{R}^m$ with the ith component projection. We call the function $F_i: A \to \mathbb{R}$ the ith component function of the mapping $F: A \to \mathbb{R}^m$. Thus*

$$F(\mathbf{u}) = (F_1(\mathbf{u}), \ldots, F_m(\mathbf{u})) \quad \text{for } \mathbf{u} \text{ in } A,$$

*and the mapping $F: A \to \mathbb{R}^m$ is said to be represented by its component functions as*

$$F = (F_1, \ldots, F_m): A \to \mathbb{R}^m. \tag{11.1}$$

**EXAMPLE 11.5**   Let $\mathcal{O}$ be the set of all nonzero points in $\mathbb{R}^n$. Define the mapping $F: \mathcal{O} \to \mathbb{R}^n$ by

$$F(\mathbf{u}) = \mathbf{u}/\|\mathbf{u}\|^2 \quad \text{for } \mathbf{u} \text{ in } \mathcal{O}.$$

Then the representation of the mapping in component functions is

$$F(\mathbf{u}) = (u_1/\|\mathbf{u}\|^2, \ldots, u_n/\|\mathbf{u}\|^2) \quad \text{for } \mathbf{u} \text{ in } \mathbb{R}^n.$$

In the case when $n = 3$, this component representation may be written as

$$F(x, y, z) = \left( \frac{x}{x^2 + y^2 + z^2}, \frac{y}{x^2 + y^2 + z^2}, \frac{z}{x^2 + y^2 + z^2} \right) \quad \text{for } (x, y, z) \text{ in } \mathcal{O}.$$

$\square$

Just as we have the Componentwise Convergence Criterion for the convergence of sequences in Euclidean space, we also have the following simple and useful criterion for the continuity of a mapping.

**THEOREM 11.4**   **The Componentwise Continuity Criterion**   Let $A$ be a subset of $\mathbb{R}^n$ that contains the point $\mathbf{u}$, and consider the mapping

$$F = (F_1, \ldots, F_m): A \to \mathbb{R}^m.$$

Then the mapping $F: A \to \mathbb{R}^m$ is continuous at $\mathbf{u}$ if and only if each of its component functions $F_i: A \to \mathbb{R}$ is continuous at $\mathbf{u}$.

**Proof**   This result follows immediately from the Componentwise Convergence Theorem, since if $\{\mathbf{u}_k\}$ is a sequence in $A$ that converges to the point $\mathbf{u}$, then the image sequence $\{F(\mathbf{u}_k)\}$ converges to $F(\mathbf{u})$ if and only if for each index $i$ with $1 \le i \le n$, the sequence $\{F_i(\mathbf{u}_k)\}$ converges to $F_i(\mathbf{u})$. $\blacksquare$

The Componentwise Continuity Criterion provides the following extension of the first assertion of Theorem 11.2.

**COROLLARY 11.5**   Let $A$ be a subset of $\mathbb{R}^n$ that contains the point $\mathbf{u}$, and suppose that the mappings $H: A \to \mathbb{R}^m$ and $G: A \to \mathbb{R}^m$ are both continuous at the point $\mathbf{u}$. Then for any real numbers $\alpha$ and $\beta$, the mapping

$$\alpha H + \beta G: A \to \mathbb{R}^m$$

is also continuous at $\mathbf{u}$.

**Proof**   This result follows from the Componentwise Continuity Criterion and Theorem 11.2 if we observe that for each index $i$ with $1 \le i \le n$, the $i$th component function of the mapping $\alpha H + \beta G: A \to \mathbb{R}^m$ is the function $\alpha H_i + \beta G_i: A \to \mathbb{R}$.  ∎

Recall that in Section 3.5, we proved that for real-valued functions of a single real variable there was an "$\epsilon$–$\delta$" criterion for continuity of a function at a point that was equivalent to the sequential definition. We will now establish a similar criterion for general mappings.

**THEOREM 11.6**   Let $A$ be a subset of $\mathbb{R}^n$ that contains the point $\mathbf{u}$. Then the following two assertions about a mapping $F: A \to \mathbb{R}^m$ are equivalent:

(i) The mapping $F: A \to \mathbb{R}^m$ is continuous at the point $\mathbf{u}$.

(ii) For each positive number $\epsilon$ there is a positive number $\delta$ such that

$$d(F(\mathbf{v}),\ F(\mathbf{u})) < \epsilon \quad \text{if } \mathbf{v} \text{ is in } A \text{ and } d(\mathbf{v}, \mathbf{u}) < \delta. \tag{11.2}$$

**FIGURE 11.1**

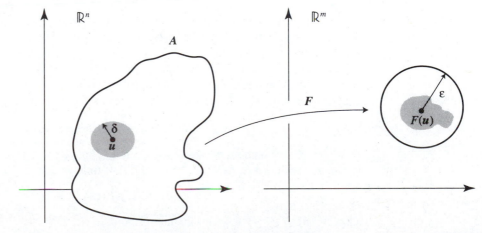

**Proof**  First, suppose that (i) holds. To verify (ii), we will suppose the contrary and derive a contradiction. Indeed, if (ii) does not hold, then there is some $\epsilon_0 > 0$ such that for no positive number $\delta$ is it true that

$$d(F(\mathbf{v}), F(\mathbf{u})) < \epsilon_0 \quad \text{for each } \mathbf{v} \text{ in } A \text{ such that } d(\mathbf{v}, \mathbf{u}) < \delta. \tag{11.3}$$

In particular, if $k$ is a positive integer, then (11.3) fails to hold when $\delta = 1/k$. This means that there is a point in $A$, which we label $\mathbf{u}_k$, with $d(\mathbf{u}_k, \mathbf{u}) < 1/k$ while $d(F(\mathbf{u}_k), F(\mathbf{u})) \geq \epsilon_0$. This defines a sequence $\{\mathbf{u}_k\}$ in $A$ that converges to the point $\mathbf{u}$, whose image sequence $\{F(\mathbf{u}_k)\}$ does not converge to $F(\mathbf{u})$. This contradicts (i).

To prove the converse, suppose that (ii) holds. To verify (i), we choose $\{\mathbf{u}_k\}$ to be a sequence in $A$ that converges to the point $\mathbf{u}$. We must show that the image sequence $\{F(\mathbf{u}_k)\}$ converges to $F(\mathbf{u})$. Let $\epsilon > 0$. According to (ii), we may choose a positive number $\delta$ such that

$$d(F(\mathbf{v}), F(\mathbf{u})) < \epsilon \quad \text{for each } \mathbf{v} \text{ in } A \text{ such that } d(\mathbf{v}, \mathbf{u}) < \delta. \tag{11.4}$$

Moreover, since the sequence $\{\mathbf{u}_k\}$ is a sequence in $A$ that converges to $\mathbf{u}$, we can select a positive integer $K$ such that $d(\mathbf{u}_k, \mathbf{u}) < \delta$ if $k \geq K$. Thus, using (11.4), we conclude that

$$d(F(\mathbf{u}_k), F(\mathbf{u})) < \epsilon \quad \text{if} \quad k \geq K.$$

Hence the image sequence $\{F(\mathbf{u}_k)\}$ converges to the point $F(\mathbf{u})$.  ∎

The preceding characterization of continuity of a mapping at a point leads to the following useful characterization of continuity on the whole domain for mappings that have as their domain an open subset of $\mathbb{R}^n$.

**THEOREM 11.7**  Let $\mathcal{O}$ be an open subset of $\mathbb{R}^n$, and consider the mapping $F: \mathcal{O} \to \mathbb{R}^m$. Then the following assertions are equivalent:

(i)  The mapping $F: \mathcal{O} \to \mathbb{R}^m$ is continuous.

(ii)  $F^{-1}(V)$ is an open subset of $\mathbb{R}^n$ whenever $V$ is an open subset of $\mathbb{R}^m$.

**Proof**  First, suppose that (i) holds. Let $V$ be an open subset of $\mathbb{R}^m$. We wish to show that $F^{-1}(V)$ is open in $\mathbb{R}^n$, which by definition means that every point in $F^{-1}(V)$ is an interior point. Let $\mathbf{u}$ be a point in $F^{-1}(V)$. Then $F(\mathbf{u})$ belongs to $V$, and $V$ is open in $\mathbb{R}^m$, so there is some positive number $\epsilon$ such that $\mathcal{N}_\epsilon(F(\mathbf{u})) \subseteq V$. Since the mapping $F: \mathcal{O} \to \mathbb{R}^m$ is continuous at the point $\mathbf{u}$, it follows from the "$\epsilon$–$\delta$" characterization of continuity provided in Theorem 11.6 that we can select a positive number $\delta$ such that

$$d(F(\mathbf{v}), F(\mathbf{u})) < \epsilon \quad \text{if } \mathbf{v} \text{ is in } \mathcal{O} \text{ and } d(\mathbf{v}, \mathbf{u}) < \delta. \tag{11.5}$$

However, by assumption, the set $\mathcal{O}$ is open in $\mathbb{R}^n$, so we can select a positive number $r$ less than $\delta$ such that $\mathcal{N}_r(\mathbf{u}) \subseteq \mathcal{O}$. From (11.5), it follows that

$$F(\mathcal{N}_r(\mathbf{u})) \subseteq \mathcal{N}_\epsilon(F(\mathbf{u})) \subseteq V.$$

Thus $\mathcal{N}_r(\mathbf{u}) \subseteq F^{-1}(V)$, so $\mathbf{u}$ is an interior point of $F^{-1}(V)$.

To prove the converse, suppose that (ii) holds. Let $\mathbf{u}$ be a point in $\mathcal{O}$. To show that the mapping $F: \mathcal{O} \to \mathbb{R}^m$ is continuous at the point $\mathbf{u}$, we will verify the "$\epsilon$–$\delta$" characterization of continuity asserted in Theorem 11.6. Let $\epsilon > 0$. Symmetric neighborhoods are open, so $\mathcal{N}_\epsilon(F(\mathbf{u}))$ is open in $\mathbb{R}^m$. From (ii) it follows that $F^{-1}(\mathcal{N}_\epsilon(F(\mathbf{u})))$ is open in $\mathbb{R}^n$. Thus $\mathbf{u}$, which belongs to $F^{-1}(\mathcal{N}_\epsilon(F(\mathbf{u})))$, is an interior point of $F^{-1}(\mathcal{N}_\epsilon(F(\mathbf{u})))$, so we can choose a positive number $\delta$ with $\mathcal{N}_\delta(\mathbf{u}) \subseteq F^{-1}(\mathcal{N}_\epsilon(F(\mathbf{u})))$. This means that $F(\mathcal{N}_\delta(\mathbf{u})) \subseteq \mathcal{N}_\epsilon(F(\mathbf{u}))$. Thus the mapping $F: \mathcal{O} \to \mathbb{R}^m$ satisfies the "$\epsilon$–$\delta$" characterization of continuity at the point $\mathbf{u}$. ∎

**COROLLARY 11.8**    Let the function $f: \mathbb{R}^n \to \mathbb{R}$ be continuous, and let $c$ be a real number. Then each of the sets

$$\{\mathbf{u} \text{ in } \mathbb{R}^n \mid f(\mathbf{u}) < c\} \quad \text{and} \quad \{\mathbf{u} \text{ in } \mathbb{R}^n \mid f(\mathbf{u}) > c\}$$

is open in $\mathbb{R}^n$, while each of the sets

$$\{\mathbf{u} \text{ in } \mathbb{R}^n \mid f(\mathbf{u}) \leq c\} \quad \text{and} \quad \{\mathbf{u} \text{ in } \mathbb{R}^n \mid f(\mathbf{u}) \geq c\}$$

is closed in $\mathbb{R}^n$.

**Proof**    By observing that the sets $\{v \text{ in } \mathbb{R} \mid v < c\}$ and $\{v \text{ in } \mathbb{R} \mid v > c\}$ are both open in $\mathbb{R}$, from the characterization of continuity asserted in Theorem 11.7 it follows that both $\{\mathbf{u} \text{ in } \mathbb{R}^n \mid f(\mathbf{u}) < c\}$ and $\{\mathbf{u} \text{ in } \mathbb{R}^n \mid f(\mathbf{u}) > c\}$ are open in $\mathbb{R}^n$. But observe that

$$\{\mathbf{u} \text{ in } \mathbb{R}^n \mid f(\mathbf{u}) \geq c\} = \mathbb{R}^n \setminus \{\mathbf{u} \text{ in } \mathbb{R}^n \mid f(\mathbf{u}) < c\}$$

and

$$\{\mathbf{u} \text{ in } \mathbb{R}^n \mid f(\mathbf{u}) \leq c\} = \mathbb{R}^n \setminus \{\mathbf{u} \text{ in } \mathbb{R}^n \mid f(\mathbf{u}) > c\},$$

so from the Complementing Characterization of openness and closedness, it follows that both $\{\mathbf{u} \text{ in } \mathbb{R}^n \mid f(\mathbf{u}) \leq c\}$ and $\{\mathbf{u} \text{ in } \mathbb{R}^n \mid f(\mathbf{u}) \geq c\}$ are closed in $\mathbb{R}^n$. ∎

**EXAMPLE 11.6**    Define

$$\mathcal{O} = \{(x,\, y,\, z) \text{ in } \mathbb{R}^3 \mid 2x + 3y + 4z > 0\}.$$

Then $\mathcal{O}$ is an open subset of $\mathbb{R}^3$. This follows from Corollary 11.8 if we observe that the function $f: \mathbb{R}^3 \to \mathbb{R}$, defined by

$$f(x,\, y,\, z) = 2x + 3y + 4z \quad \text{for} \quad (x,\, y,\, z) \text{ in } \mathbb{R}^3,$$

is continuous. □

**EXAMPLE 11.7**    Fix positive numbers $a$ and $b$ with $a < b$ and define

$$\mathcal{O} = \{\mathbf{u} \text{ in } \mathbb{R}^n \mid a < \|\mathbf{u}\| < b\}.$$

Then $\mathcal{O}$ is an open subset of $\mathbb{R}^n$. This follows from Corollary 11.8 if we first observe that since the function $f: \mathbb{R}^n \to \mathbb{R}$ defined by

$$f(\mathbf{u}) = \|\mathbf{u}\| \quad \text{for } \mathbf{u} \text{ in } \mathbb{R}^n$$

is continuous, both $\{\mathbf{u} \text{ in } \mathbb{R}^n \mid \|\mathbf{u}\| > a\}$ and $\{\mathbf{u} \text{ in } \mathbb{R}^n \mid \|\mathbf{u}\| < b\}$ are open subsets of $\mathbb{R}$. Hence $\mathcal{O}$, being the intersection of these two sets, is also open in $\mathbb{R}^n$. □

1.  Define the function $f: \mathbb{R}^2 \to \mathbb{R}$ by

$$f(x, y) = \cos(x + y) + x^2 y^2 \quad \text{for } (x, y) \text{ in } \mathbb{R}^2.$$

Prove that $f: \mathbb{R}^2 \to \mathbb{R}$ is continuous.

2.  Define $\mathcal{O} = \{(x, y, z) \text{ in } \mathbb{R}^3 \mid (x, y, z) \neq (0, 0, 0)\}$, and define the function $f: \mathcal{O} \to \mathbb{R}$ by

$$f(x, y, z) = \frac{x}{x^2 + y^2 + z^2} \quad \text{for } (x, y, z) \text{ in } \mathcal{O}.$$

Prove that the function $f: \mathcal{O} \to \mathbb{R}$ is continuous.

3.  Fix a point $\mathbf{v}$ in $\mathbb{R}^n$ and define the function $f: \mathbb{R}^n \to \mathbb{R}$ by

$$f(\mathbf{u}) = \langle \mathbf{u}, \mathbf{v} \rangle \quad \text{for } \mathbf{u} \text{ in } \mathbb{R}^n.$$

Prove that the function $f: \mathbb{R}^n \to \mathbb{R}$ is continuous.

4.  Suppose that the function $f: \mathbb{R}^n \to \mathbb{R}$ is continuous and that $f(\mathbf{u}) > 0$ if the point $\mathbf{u}$ in $\mathbb{R}^n$ has at least one rational component. Prove that $f(\mathbf{u}) \geq 0$ for all points $\mathbf{u}$ in $\mathbb{R}^n$.

5.  Use Corollary 11.8 to show that each of the following sets is open in $\mathbb{R}^2$:
    (a)  $\{(x, y) \text{ in } \mathbb{R}^2 \mid y > 0\}$
    (b)  $\{(x, y) \text{ in } \mathbb{R}^2 \mid x^2/5 + y^2/4 < 1\}$
    (c)  $\{(x, y) \text{ in } \mathbb{R}^2 \mid y > x^2\}$
    (d)  $\{(x, y) \text{ in } \mathbb{R}^2 \mid 1 < x^2 + y^2 < 2\}$

6.  Suppose that the functions $f: \mathbb{R}^n \to \mathbb{R}$ and $g: \mathbb{R}^n \to \mathbb{R}$ are both continuous. Prove that the set $\{\mathbf{u} \text{ in } \mathbb{R}^n \mid f(\mathbf{u}) = g(\mathbf{u}) = 0\}$ is closed in $\mathbb{R}^n$.

7.  Let $\mathcal{O}$ be an open subset of $\mathbb{R}^n$ and suppose that the function $f: \mathcal{O} \to \mathbb{R}$ is continuous. If $a$ and $b$ are numbers with $a < b$, prove that the set

$$\{\mathbf{u} \text{ in } \mathcal{O} \mid a < f(\mathbf{u}) < b\}$$

is open in $\mathbb{R}^n$.

8.  Show that the set $\{\mathbf{u} \text{ in } \mathbb{R}^n \mid u_n > 0\}$ is open in $\mathbb{R}^n$.

9.  Use Corollary 11.8 to show that if $\mathbf{u}$ is a point in $\mathbb{R}^n$ and $r$ is a positive number, then the set $\{\mathbf{v} \text{ in } \mathbb{R}^n \mid d(\mathbf{u}, \mathbf{v}) \leq r\}$ is closed in $\mathbb{R}^n$.

10. Suppose that the function $f: \mathbb{R}^n \to \mathbb{R}$ is continuous and that $c$ is a real number. Prove that the set $\{\mathbf{v} \text{ in } \mathbb{R}^n \mid f(\mathbf{v}) = c\}$ is closed in $\mathbb{R}^n$.

11. Let $\mathcal{O}$ be an open subset of $\mathbb{R}^n$ and suppose that the function $f: \mathcal{O} \to \mathbb{R}$ is continuous. Suppose that $\mathbf{u}$ is a point in $\mathcal{O}$ at which $f(\mathbf{u}) > 0$. Prove that there is a symmetric neighborhood $\mathcal{N}$ of $\mathbf{u}$ such that $f(\mathbf{v}) > f(\mathbf{u})/2$ for all $\mathbf{v}$ in $\mathcal{N}$.

12. Let $A$ be a subset of $\mathbb{R}^n$. The *characteristic function* of the set $A$ is the function $f: \mathbb{R}^n \to \mathbb{R}$ defined by

$$f(\mathbf{u}) = \begin{cases} 1 & \text{if } \mathbf{u} \in A \\ 0 & \text{if } \mathbf{u} \notin A. \end{cases}$$

Prove that this characteristic function is continuous at each interior point of $A$ and at each exterior point of $A$, but that it fails to be continuous at each boundary point of $A$.

13. Give a direct proof of Corollary 11.5 without quoting the Componentwise Continuity Criterion.

# 11.2   Compactness and the Extreme Value Theorem

In Chapter 3, we proved the Extreme Value Theorem, which asserts that if $I$ is a closed, bounded interval of real numbers and the function $f: I \to \mathbb{R}$ is continuous, then $f: I \to \mathbb{R}$ attains both a maximum and a minimum value. The proof of this theorem depended precisely on the fact that every sequence in $I$ has a subsequence that converges to a point in $I$. It is useful to isolate this property.

Given a sequence $\{\mathbf{x}_k\}$ in $\mathbb{R}^n$, and $\{k_j\}$ a strictly increasing sequence of positive integers, the sequence $\{\mathbf{x}_{k_j}\}$ is called a *subsequence* of $\{\mathbf{x}_k\}$. Observe that if the sequence $\{\mathbf{x}_k\}$ converges to the point $\mathbf{x}$ in $\mathbb{R}^n$, then each subsequence $\{\mathbf{x}_{k_j}\}$ also converges to $\mathbf{x}$, since this property has already been established for real sequences, and so if $\lim_{k \to \infty} d(\mathbf{x}_k, \mathbf{x}) = 0$, then $\lim_{j \to \infty} d(\mathbf{x}_{k_j}, \mathbf{x}) = 0$ also.

**DEFINITION**   *Let $A$ be a subset of $\mathbb{R}^n$. Then $A$ is said to be compact provided that every sequence in $A$ has a subsequence that converges to a point in $A$.*

The Bolzano-Weierstrass Theorem (Theorem 2.13) is the assertion that if $a$ and $b$ are real numbers such that $a \leq b$, then the closed, bounded interval $[a, b]$ is compact. The subset $A = (0, 1]$ of $\mathbb{R}$ is not compact, since the sequence $\{1/k\}$ is in $A$, but no subsequence of $\{1/k\}$ converges to a point in $A$, since every subsequence of $\{1/k\}$ converges to the point 0, and 0 does not belong to $(0, 1]$. Also, the subset $A = [0, \infty)$ of $\mathbb{R}$ is not compact since the sequence $\{k\}$ is in $A$, and yet no subsequence of $\{k\}$ converges, since every subsequence is unbounded. So certainly no subsequence converges to a point in $A$.

It is useful to characterize the compact subsets of $\mathbb{R}^n$. As a first step, we will establish two conditions that are necessary in order for a subset of $\mathbb{R}^n$ to be compact.

**DEFINITION**   *A subset $A$ of $\mathbb{R}^n$ is said to be bounded provided that there is a number $M$ such that*

$$\|\mathbf{u}\| \leq M \quad \text{for all points } \mathbf{u} \text{ in } A.$$

**THEOREM 11.9**   A compact subset of $\mathbb{R}^n$ is bounded and closed in $\mathbb{R}^n$.

**Proof**   Let $A$ be a compact subset of $\mathbb{R}^n$. First we show that $A$ is closed in $\mathbb{R}^n$. Let $\{\mathbf{u}_k\}$ be a sequence in $A$ that converges to the point $\mathbf{u}$ in $\mathbb{R}^n$. Then every subsequence of $\{\mathbf{u}_k\}$ also converges to $\mathbf{u}$. By the definition of compactness, some subsequence converges to a point in $A$. Thus $\mathbf{u}$ belongs to $A$; hence $A$ is closed.

To prove that $A$ is bounded, we will assume the contrary and derive a contradiction. Suppose that $A$ is not bounded. Then for each natural number $k$, it is not true that

$$\|\mathbf{u}\| \leq k \quad \text{for all points } \mathbf{u} \text{ in } A.$$

So we can select a point in $A$, which we label $\mathbf{u}_k$, such that $\|\mathbf{u}_k\| > k$. Since $A$ is compact, a subsequence $\{\mathbf{u}_{k_j}\}$ of the sequence $\{\mathbf{u}_k\}$ converges to a point $\mathbf{u}$ that belongs to $A$. But then

$$\|\mathbf{u}_{k_j}\| > k_j \geq j \quad \text{for all positive integers } j.$$

Thus the real sequence $\{\|\mathbf{u}_{k_j}\|\}$ is unbounded but converges to the norm of $\mathbf{u}$. This is a contradiction; hence $A$ is bounded. ∎

It turns out that for a subset $A$ of $\mathbb{R}^n$ to be compact, it is both necessary and sufficient that $A$ be bounded and closed in $\mathbb{R}^n$. In order to prove sufficiency, the crucial result is the following theorem:

**THEOREM 11.10**

Every bounded sequence in $\mathbb{R}^n$ has a convergent subsequence.

**Proof**   We will prove the theorem by induction. The case when $n = 1$ is precisely the assertion of Theorem 2.12.

Now suppose that $n$ is a positive integer and every bounded sequence in $\mathbb{R}^n$ has a convergent subsequence. We must show that every bounded sequence in $\mathbb{R}^{n+1}$ has a convergent subsequence.

Let $\{\mathbf{u}_k\}$ be a bounded sequence in $\mathbb{R}^{n+1}$. Fix a positive integer $k$. Define $x_k$ to be the $(n+1)$st component of $\mathbf{u}_k$, and write

$$\mathbf{u}_k = (\mathbf{v}_k, x_k),$$

where $\mathbf{v}_k$ is the point in $\mathbb{R}^n$ whose $i$th component is equal to the $i$th component of $\mathbf{u}_k$ for each index $i$ with $1 \le i \le n$. This defines two sequences: the sequence of real numbers $\{x_k\}$ and the sequence $\{\mathbf{v}_k\}$ in $\mathbb{R}^n$.

Now $\{\mathbf{v}_k\}$ is a bounded sequence in $\mathbb{R}^n$, and $\{x_k\}$ is a bounded sequence of real numbers. By the induction assumption, a subsequence of $\{\mathbf{v}_k\}$ converges to a point $\mathbf{v}$ in $\mathbb{R}^n$. The corresponding subsequence of $\{x_k\}$ itself has a further subsequence that converges to a real number $x$. From the Componentwise Convergence Criterion, it follows that the subsequence of $\{\mathbf{u}_k\}$ corresponding to this last subsequence converges to the point $\mathbf{u} = (\mathbf{v}, x)$ in $\mathbb{R}^{n+1}$.

The principle of mathematical induction implies that this theorem is true for every positive integer $n$. ∎

**THEOREM 11.11**

A subset of $\mathbb{R}^n$ is compact if and only if it is bounded and closed in $\mathbb{R}^n$.

**Proof**   First, according to Theorem 11.9, every compact subset of $\mathbb{R}^n$ is bounded and closed in $\mathbb{R}^n$.

To prove the converse, let $A$ be a subset of $\mathbb{R}^n$ that is bounded and closed in $\mathbb{R}^n$. Suppose that $\{\mathbf{u}_k\}$ is a sequence in $A$. Since the sequence $\{\mathbf{u}_k\}$ is bounded, it follows from Theorem 11.10 that there is a subsequence $\{\mathbf{u}_{k_j}\}$ of $\{\mathbf{u}_k\}$ that converges to a point $\mathbf{u}$ in $\mathbb{R}^n$. But $\{\mathbf{u}_{k_j}\}$ is itself a sequence in $A$, and since $A$ is closed, $\mathbf{u}$ belongs to the set $A$. Thus $A$ is compact. ∎

We now turn to the study of continuous mappings that have as their domain a compact subset of $\mathbb{R}^n$.

**THEOREM 11.12**

Let $A$ be a subset of $\mathbb{R}^n$, and suppose that the mapping $F: A \to \mathbb{R}^m$ is continuous. If the domain $A$ is compact, then the image $F(A)$ is also compact.

**Proof**   Let $\{\mathbf{u}_k\}$ be a sequence in $F(A)$. For each positive integer $k$, choose a point $\mathbf{v}_k$ in $A$ with $\mathbf{u}_k = F(\mathbf{v}_k)$. Since $A$ is compact, there is a subsequence $\{\mathbf{v}_{k_j}\}$ that converges to a point $\mathbf{v}$ in $A$. But the mapping $F: A \to \mathbb{R}^m$ is continuous at the point $\mathbf{v}$. Thus $\{\mathbf{u}_{k_j}\} = \{F(\mathbf{v}_{k_j})\}$ converges to the point $F(\mathbf{v})$ in $F(A)$. Hence every sequence in $F(A)$ has a subsequence that converges to a point in $F(A)$. By definition, this means that $F(A)$ is compact.   ∎

**LEMMA 11.13**   Every nonempty compact subset of $\mathbb{R}$ has a smallest and a largest member.

**Proof**   Let $A$ be a subset of $\mathbb{R}$ that is compact. According to Theorem 11.9, the set $A$ is bounded and closed in $\mathbb{R}$. Since $A$ is bounded, by the Completeness Axiom for $\mathbb{R}$, $A$ has a least upper bound. Denote the least upper bound of $A$ by $b$. Then $x \leq b$ for all $x$ in $A$. On the other hand, if $k$ is any positive integer, then $b - 1/k$ is not an upper bound of $A$, so we can choose a point in $A$, which we label $x_k$, with $b - 1/k < x_k \leq b$. From the Squeezing Principle, it follows that the sequence $\{x_k\}$ converges to $b$. But $A$ is a closed subset of $\mathbb{R}$, so $b$ belongs to $A$. The number $b$ is the largest member of $A$.

A similar proof, with *greatest lower bound* replacing *least upper bound*, shows the existence of a smallest member of $A$.   ∎

**THEOREM 11.14**   **The Extreme Value Theorem**   Let $A$ be a nonempty compact subset of $\mathbb{R}^n$, and suppose that the function $f: A \to \mathbb{R}$ is continuous. Then the function $f: A \to \mathbb{R}$ attains a smallest and a largest value.

**Proof**   From Theorem 11.12, it follows that the image $f(A)$ is compact. According to Lemma 11.13, $f(A)$ has both a smallest and a largest member.   ∎

The above corollary is sharp in the following precise sense: If $A$ is a subset of $\mathbb{R}^n$ that is not compact, then there is a continuous function $f: A \to \mathbb{R}$ that fails to attain a smallest value. Indeed, if $A$ is not compact, then either $A$ is unbounded or $A$ is not closed in $\mathbb{R}^n$. If $A$ is unbounded, then the function $f: A \to \mathbb{R}$ defined by $f(\mathbf{u}) = -\|\mathbf{u}\|$ is continuous, but certainly fails to attain a smallest value. On the other hand, if $A$ is not closed, then there is a sequence in $A$ that converges to a point $\mathbf{v}$ in $\mathbb{R}^n$, but $\mathbf{v}$ does not belong to the set $A$. In this case, the continuous function $f: A \to \mathbb{R}$ defined by $f(\mathbf{u}) = \|\mathbf{u} - \mathbf{v}\|$ fails to attain a smallest value.

Recall that in order to prove that a continuous function $f: [a, b] \subseteq \mathbb{R} \to \mathbb{R}$ was integrable, we needed to prove that each such function is uniformly continuous. In the study of integration for functions of several real variables, a similar result will be needed. To that end, we introduce the concept of uniform continuity for general mappings.

**DEFINITION**   *Let $A$ be a subset of $\mathbb{R}^n$. A mapping $F: A \to \mathbb{R}^m$ is called uniformly continuous provided that for each positive number $\epsilon$ there is a positive number $\delta$ such that*

$$d(F(\mathbf{u}), \ F(\mathbf{v})) < \epsilon \quad \textit{if } \mathbf{u} \textit{ and } \mathbf{v} \textit{ belong to } A \textit{ and } d(\mathbf{u}, \mathbf{v}) < \delta.$$

Using the "$\epsilon$–$\delta$" characterization of continuity asserted in Theorem 11.6, it is clear that a uniformly continuous mapping is continuous. As we have already seen in the case of functions of a single real variable, the converse is false. However, for mappings that have a domain that is compact, we have the following theorem.

**THEOREM**   Let $A$ be a subset of $\mathbb{R}^n$ and suppose that the mapping $F: A \to \mathbb{R}^m$ is continuous. If $A$
**11.15**   is compact, then the mapping $F: A \to \mathbb{R}^m$ is uniformly continuous.

**Proof**   By assuming that the conclusion is false, we will derive a contradiction. Suppose that
the mapping $F: A \to \mathbb{R}^m$ is not uniformly continuous. Then there is a positive number
$\epsilon_0$ such that for no positive number $\delta$ is it true that

$$d(F(\mathbf{u}), F(\mathbf{v})) < \epsilon_0 \quad \text{for all points } \mathbf{u} \text{ and } \mathbf{v} \text{ in } A \text{ such that } d(\mathbf{u}, \mathbf{v}) < \delta.$$

Let $k$ be a positive integer. Then the previous assertion fails to hold when $\delta = 1/k$. This
means that there are two points in $A$, which we label $\mathbf{u}_k$ and $\mathbf{v}_k$, such that $d(\mathbf{u}_k, \mathbf{v}_k) < 1/k$,
while $d(F(\mathbf{u}_k), F(\mathbf{v}_k)) \geq \epsilon_0$. This defines the two sequences $\{\mathbf{u}_k\}$ and $\{\mathbf{v}_k\}$.

Since $A$ is compact, a subsequence $\{\mathbf{u}_{k_j}\}$ of the sequence $\{\mathbf{u}_k\}$ converges to a point
$\mathbf{u}$ in $A$. By the Triangle Inequality,

$$d(\mathbf{u}, \mathbf{v}_{k_j}) \leq d(\mathbf{u}, \mathbf{u}_{k_j}) + d(\mathbf{u}_{k_j}, \mathbf{v}_{k_j}) \leq d(\mathbf{u}, \mathbf{u}_{k_j}) + \frac{1}{j}$$

for every positive integer $j$. Thus the sequence $\{\mathbf{v}_{k_j}\}$ also converges to the same point
$\mathbf{u}$. However, the mapping $F: A \to \mathbb{R}^m$ is continuous at the point $\mathbf{u}$, so both of the image
sequences $\{F(\mathbf{u}_{k_j})\}$ and $\{F(\mathbf{v}_{k_j})\}$ converge to the same point $F(\mathbf{u})$. This contradicts the
fact that

$$d(F(\mathbf{u}_{k_j}), F(\mathbf{u})) + d(F(\mathbf{u}), F(\mathbf{v}_{k_j})) \geq d(F(\mathbf{u}_{k_j}), F(\mathbf{v}_{k_j})) \geq \epsilon_0$$

for every positive integer $j$. This contradiction proves that the mapping $F: A \to \mathbb{R}^m$ is
uniformly continuous.   ■

**EXERCISES**

1. Determine which of the following subsets of $\mathbb{R}$ is compact. Justify your conclusions.
   (a) $\{x \text{ in } [0, 1] \mid x \text{ is rational}\}$
   (b) $\{x \text{ in } \mathbb{R} \mid x^2 > x\}$
   (c) $\{x \text{ in } \mathbb{R} \mid e^x - x^2 \leq 2\}$

2. Let $\mathbf{u}$ be a point in $\mathbb{R}^n$, and let $r$ be a positive number. Prove that the set

$$\{\mathbf{v} \text{ in } \mathbb{R}^n \mid d(\mathbf{u}, \mathbf{v}) \leq r\}$$

is compact.

3. Can a symmetric neighborhood in $\mathbb{R}^n$ be compact?

4. Let $A$ be a subset of $\mathbb{R}^n$ and the function $f: A \to \mathbb{R}^m$ be continuous. If $A$ is bounded,
   is $f(A)$ bounded?

5. Suppose that the function $f: \mathbb{R}^n \to \mathbb{R}$ is continuous, and $f(\mathbf{u}) \geq \|\mathbf{u}\|$ for every point
   $\mathbf{u}$ in $\mathbb{R}^n$. Prove that $f^{-1}([0, 1])$ is compact.

6. Suppose that $A$ is a subset of $\mathbb{R}^n$ that is not compact. Show that there is a continuous
   function $f: A \to \mathbb{R}$ that does not attain a largest value.

7. Let $A$ and $B$ be compact subsets of $\mathbb{R}$. Define $K = \{(x, y) \text{ in } \mathbb{R}^2 \mid x \text{ in } A, y \text{ in } B\}$.
   Prove that $K$ is compact.

8. Let $A$ be a subset of $\mathbb{R}^n$ that is compact, and let $\mathbf{v}$ be a point in $\mathbb{R}^n \backslash A$. Prove that there
   is a point $\mathbf{u}_0$ in the set $A$ such that

$$d(\mathbf{u}_0, \mathbf{v}) \leq d(\mathbf{u}, \mathbf{v}) \quad \text{for all points } \mathbf{u} \text{ in } A.$$

Is this point unique?

9. A mapping $F: \mathbb{R}^n \to \mathbb{R}^m$ is said to be *Lipschitz* if there is a number $C$ such that

$$d(F(\mathbf{u}), F(\mathbf{v})) \leq C d(\mathbf{u}, \mathbf{v}) \quad \text{for all points } \mathbf{u} \text{ and } \mathbf{v} \text{ in } \mathbb{R}^n.$$

The number $C$ is called a *Lipschitz constant* for the mapping. Show that a Lipschitz mapping is uniformly continuous.

10. Is the product of two real-valued uniformly continuous functions again uniformly continuous?

## 11.3 Connectedness and the Intermediate Value Theorem

Recall that a subset $I$ of $\mathbb{R}$ is defined to be an *interval* if it has the property that for any two points in $I$, every point between these two points also lies in $I$. The Intermediate Value Theorem, as expressed in Theorem 3.9, asserts that if $I$ is an interval and the function $f: I \to \mathbb{R}$ is continuous, then its image $f(I)$ is also an interval. We devote this section to studying those subsets $A$ of $\mathbb{R}^n$ that have the property that if the function $f: A \to \mathbb{R}$ is continuous, then its image $f(A)$ is an interval. The relevant concepts are connectedness and pathwise connectedness. We will begin with the latter.

**DEFINITION** *Let $a$ and $b$ be real numbers with $a < b$. Then a continuous mapping $\gamma: [a, b] \to \mathbb{R}^n$ will be called a parametrized path. The domain of $\gamma: [a, b] \to \mathbb{R}^n$ is called the parameter space and the image of $\gamma: [a, b] \to \mathbb{R}^n$ is called a path.*

It is important to distinguish between a path and a parametrized path: A path is a *subset* of $\mathbb{R}^n$, whereas a parametrized path is a *mapping*. Furthermore, any path will be the image of several different parametrized paths.

**FIGURE 11.2**

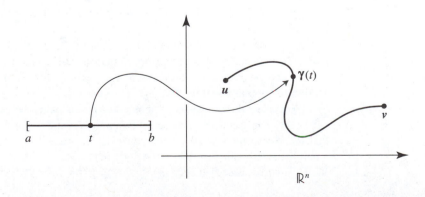

**EXAMPLE 11.8** The upper half-circle

$$\{(x,\ y)\ \text{in}\ \mathbb{R}^2\ |\ x^2 + y^2 = 1,\ y \geq 0\}$$

is a path that is the image of the parametrized path $\gamma\colon [-1, 1] \to \mathbb{R}^2$ defined by $\gamma(t) = (t, \sqrt{1 - t^2})$ for $-1 \leq t \leq 1$. This upper half-circle is also the image of the parametrized path $\gamma\colon [0, \pi] \to \mathbb{R}^2$ defined by $\gamma(t) = (-\cos t, \sin t)$ for $0 \leq t \leq \pi$. □

**FIGURE 11.3**

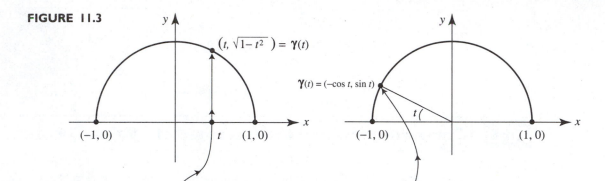

Also note that every path is the image of a parametrized path having as its parameter space the interval $[0, 1]$. Indeed, if a path is the image of a parametrized path $\gamma\colon [a,\ b] \to \mathbb{R}^n$, then it is also the image of the parametrized path $\bar{\gamma}\colon [0, 1] \to \mathbb{R}^n$ defined by

$$\bar{\gamma}(t) = \gamma((1 - t)a + tb) \quad \text{for} \quad 0 \leq t \leq 1.$$

**DEFINITION**

(i) *Let $A$ be a subset of $\mathbb{R}^n$ that contains the points $\mathbf{u}$ and $\mathbf{v}$. By a path in $A$ joining $\mathbf{u}$ and $\mathbf{v}$, we mean the image of a parametrized path $\gamma\colon [a,\ b] \to \mathbb{R}^n$ with $\gamma(a) = \mathbf{u}$, $\gamma(b) = \mathbf{v}$, such that the image of $\gamma\colon [a,\ b] \to \mathbb{R}^n$ is contained in $A$.*

(ii) *A subset $A$ of $\mathbb{R}^n$ is said to be pathwise connected provided that every pair of points in $A$ can be joined by a path in $A$.*

**THEOREM 11.16**

A subset of $\mathbb{R}$ is pathwise connected if and only if it is an interval.

**Proof** First, suppose that $I$ is an interval. Let $u$ and $v$ be points in $I$. We may suppose that $u < v$. Since $I$ is an interval, the closed interval $[u, v]$ is a subset of $I$, so we can define $[u, v]$ to be the parameter space and define $\gamma(t) = t$ for $u \leq t \leq v$ to obtain a path in $I$ joining the points $u$ and $v$.

To prove the converse, suppose that $I$ is a pathwise connected subset of $\mathbb{R}$. To verify that $I$ is an interval, we select two points $u$ and $v$ in $I$ with $u < v$. We must show that the closed interval $[u, v]$ is a subset of $I$. Since $I$ is pathwise connected, there is a parametrized path $\gamma[a,\ b] \to I$ with $\gamma(a) = u$ and $\gamma(b) = v$. Now the Intermediate Value Theorem asserts that $[\gamma(a),\ \gamma(b)] \subseteq \gamma([a,\ b])$ so $[u, v] \subseteq I$. ∎

**EXAMPLE 11.9**   The unit sphere $S$ in $\mathbb{R}^3$ is defined to be the set

$$S = \{\mathbf{u} = (x,\ y,\ z) \text{ in } \mathbb{R}^3 \mid x^2 + y^2 + z^2 = 1\}.$$

**FIGURE 11.4**

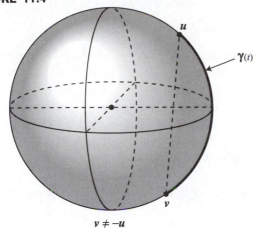

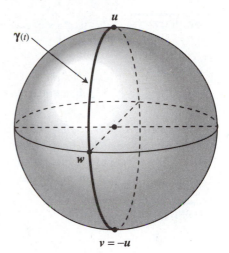

$v \neq -u$          $v = -u$

The sphere is pathwise connected.

The unit sphere is pathwise connected. To verify this, let $\mathbf{u}$ and $\mathbf{v}$ be points in $S$. If $\mathbf{u}$ and $\mathbf{v}$ are not antipodal points—that is, if $\mathbf{v} \neq -\mathbf{u}$—then we can check that $(1 - t)\mathbf{u} + t\mathbf{v} \neq 0$ for $0 \leq t \leq 1$, so

$$\gamma(t) = \frac{(1 - t)\mathbf{u} + t\mathbf{v}}{\|(1 - t)\mathbf{u} + t\mathbf{v}\|} \quad \text{for} \quad 0 \leq t \leq 1$$

defines a parametrized path joining $\mathbf{u}$ and $\mathbf{v}$. It remains to consider the case when $\mathbf{u}$ and $\mathbf{v}$ are antipodal; that is, $\mathbf{v} = -\mathbf{u}$. In this case, choose a point $\mathbf{w}$ in $S$ that is perpendicular to $\mathbf{u}$. We can check that $\cos t\, \mathbf{u} + \sin t\, \mathbf{w}$ lies in $S$ for each real $t$, and so

$$\gamma(t) = \cos t\, \mathbf{u} + \sin t\, \mathbf{w} \quad \text{for} \quad 0 \leq t \leq \pi$$

defines a parametrized path $\gamma \colon [0, \pi] \to \mathbb{R}$ joining the points $\mathbf{u}$ and $-\mathbf{u}$. We leave as an exercise the verification that the image of each of these parametrized paths lies in $S$.   $\square$

**THEOREM 11.17**   Let $A$ be a subset of $\mathbb{R}^n$, and suppose that the mapping $F \colon A \to \mathbb{R}^m$ is continuous. If $A$ is pathwise connected, then its image $F(A)$ is also pathwise connected.

**Proof**   Let $\mathbf{u}$ and $\mathbf{v}$ be points in $F(A)$. We must find a path in $F(A)$ joining $\mathbf{u}$ and $\mathbf{v}$. Choose points $\mathbf{x}$ and $\mathbf{y}$ in $A$ such that $F(\mathbf{x}) = \mathbf{u}$ and $F(\mathbf{y}) = \mathbf{v}$. Since by assumption the domain $A$ is pathwise connected, it follows that there is a parametrized path $\gamma \colon [a, b] \to \mathbb{R}^n$ with $\gamma(a) = \mathbf{x}$, $\gamma(b) = \mathbf{y}$, and $\gamma([a, b]) \subseteq A$. Since the composition of continuous mappings is continuous, it follows that the composition $F \circ \gamma \colon [a, b] \to \mathbb{R}^m$ is a parametrized path in $F(A)$ joining $\mathbf{u}$ and $\mathbf{v}$.   ∎

**THEOREM 11.18**   Let $A$ be a pathwise connected subset of $\mathbb{R}^n$, and suppose that the function $f \colon A \to \mathbb{R}$ is continuous. Then its image $f(A)$ is an interval.

**Proof**    According to Theorem 11.17, $f(A)$ is a pathwise connected subset of $\mathbb{R}$. It then follows from Theorem 11.16 that $f(A)$ is an interval.    ■

**DEFINITION**    *Let **u** and **v** be two points in $\mathbb{R}^n$. Then the parametrized path $\gamma: [0, 1] \to \mathbb{R}^m$ defined by*

$$\gamma(t) = (1 - t)\mathbf{u} + t\mathbf{v} \quad for \quad 0 \le t \le 1$$

*is called the parametrized segment from **u** to **v**. The image of this parametrized segment is called the segment between **u** and **v**.*

Here, again, it is important to emphasize that a parametrized segment is a mapping, whereas a segment is a set of points. The parametrized segment from **u** to **v** is not equal to the parametrized segment from **v** to **u**, although the images of these parametrized segments are the same, namely the segment between the points **u** and **v**.

**DEFINITION**    *A subset $A$ of $\mathbb{R}^n$ is called convex provided that the segment between any two points in $A$ is a subset of $A$.*

**FIGURE 11.5**

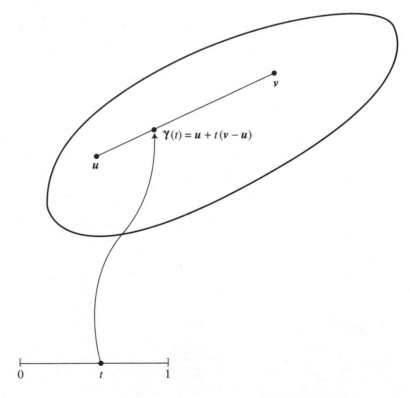

Observe that, by definition, a subset $A$ of $\mathbb{R}$ is convex if and only if it is an interval.

**THEOREM**    Every convex subset of $\mathbb{R}^n$ is pathwise connected.
**11.19**

**Proof** Let $A$ be a convex subset of $\mathbb{R}^n$. Let $\mathbf{u}$ and $\mathbf{v}$ be points in $A$. By the very definition of convexity, the parametrized path $\gamma\colon [0, 1] \to \mathbb{R}^n$, defined by $\gamma(t) = (1 - t)\mathbf{u} + t\mathbf{v}$, has the property that its image is a subset of $A$. Thus $A$ is pathwise connected. ∎

An immediate consequence of the above theorem is that $\mathbb{R}^n$ itself is pathwise connected.

**EXAMPLE 11.10** Let $C$ be a convex subset of $\mathbb{R}^2$ and suppose that the function $f\colon C \to \mathbb{R}$ is continuous. Then the graph of the function

$$G = \{\mathbf{u} = (x,\, y,\, z) \mid (x,\, y) \text{ in } C \text{ and } z = f(x,\, y)\}$$

is pathwise connected. One way to verify this is to observe that the mapping $F\colon C \to \mathbb{R}^3$, defined by $F(x,\, y) = (x,\, y,\, f(x,\, y))$ for $(x,\, y)$ in $C$, is continuous and its image is $G$. Thus, since $C$ is pathwise connected, it follows from Theorem 11.17 that $G$ is also pathwise connected. □

**COROLLARY 11.20** Every symmetric neighborhood in $\mathbb{R}^n$ is pathwise connected.

**Proof** Let $\mathbf{u}$ be a point in $\mathbb{R}^n$ and $r$ be a positive real number. According to Theorem 11.19, in order to show that the symmetric neighborhood $\mathcal{N}_r(\mathbf{u})$ is pathwise connected, it will suffice to show that $\mathcal{N}_r(\mathbf{u})$ is convex. So let $\mathbf{v}$ and $\mathbf{w}$ be points in $\mathcal{N}_r(\mathbf{u})$, and let $t$ be a real number such that $0 \le t \le 1$. We will show that $(1 - t)\mathbf{v} + t\mathbf{w}$ belongs to $\mathcal{N}_r(\mathbf{u})$; that is,

$$\|(1 - t)\mathbf{v} + t\mathbf{w} - \mathbf{u}\| < r.$$

Indeed, since

$$(1 - t)\mathbf{v} + t\mathbf{w} - \mathbf{u} = (1 - t)(\mathbf{v} - \mathbf{u}) + t(\mathbf{w} - \mathbf{u}),$$

the Triangle Inequality implies that

$$
\begin{aligned}
\|(1 - t)\mathbf{v} + t\mathbf{w} - \mathbf{u}\| &= \|(1 - t)(\mathbf{v} - \mathbf{u}) + t(\mathbf{w} - \mathbf{u})\| \\
&\le \|(1 - t)(\mathbf{v} - \mathbf{u})\| + \|t(\mathbf{w} - \mathbf{u})\| \\
&= (1 - t)\|(\mathbf{v} - \mathbf{u})\| + t\|(\mathbf{w} - \mathbf{u})\| \\
&< (1 - t)r + tr \\
&= r.
\end{aligned}
$$
∎

**DEFINITION** *A subset $A$ of $\mathbb{R}^n$ is said to have the Intermediate Value Property provided that every continuous function $f\colon A \to \mathbb{R}$ has an interval as its image.*

Theorem 11.17 asserts that every pathwise connected subset of $\mathbb{R}^n$ has the Intermediate Value Property. It turns out that there are other subsets of $\mathbb{R}^n$ that also have the Intermediate Value Property. We will now characterize subsets of $\mathbb{R}^n$ that have this property.

**DEFINITION**    *Let A be a subset of $\mathbb{R}^n$. Two open subsets $\mathcal{U}$ and $\mathcal{V}$ of $\mathbb{R}^n$ are said to separate the set A provided that*

$$A \cap \mathcal{U} \text{ and } A \cap \mathcal{V} \text{ are nonempty,}$$

$$(A \cap \mathcal{U}) \cup (A \cap \mathcal{V}) = A,$$

*and*    $$(A \cap \mathcal{U}) \cap (A \cap \mathcal{V}) = \varnothing.$$

**DEFINITION**    *A subset A of $\mathbb{R}^n$ is said to be connected provided that there do not exist two open subsets of $\mathbb{R}^n$ that separate A.*

The following theorem justifies the introduction of the concept of connectedness.

**THEOREM 11.21**    A subset of $\mathbb{R}^n$ is connected if and only if it has the Intermediate Value Property.

**Proof**    First, suppose that $A$ is a subset of $\mathbb{R}^n$ that is connected. We will show that $A$ has the Intermediate Value Property. Indeed, let the function $f: A \to \mathbb{R}$ be continuous. To show that its image $f(A)$ is an interval, we will suppose otherwise and derive a contradiction. If $f(A)$ is not an interval, then there are points $\mathbf{u}$ and $\mathbf{v}$ in $A$ and a real number $c$ such that

$$f(\mathbf{u}) < c < f(\mathbf{v}),$$

but $c$ does not belong to the image $f(A)$. Define

$$A_1 = f^{-1}(-\infty, c) \quad \text{and} \quad A_2 = f^{-1}(c, \infty).$$

Observe that neither $A_1$ nor $A_2$ is empty, since $\mathbf{u}$ belongs to $A_1$ and $\mathbf{v}$ belongs to $A_2$. The sets $A_1$ and $A_2$ are disjoint, and furthermore $A_1 \cup A_2 = A$, since the number $c$ is not in $f(A)$.

We will find two subsets $\mathcal{U}$ and $\mathcal{V}$ of $\mathbb{R}^n$, each of which is open in $\mathbb{R}^n$, that have the property that

$$A \cap \mathcal{U} = A_1 \quad \text{and} \quad A \cap \mathcal{V} = A_2.$$

Once this is done, from the above-mentioned properties of the sets $A_1$ and $A_2$ it will follow that $\mathcal{U}$ and $\mathcal{V}$ separate $A$. This contradicts the assumption that $A$ is connected.

Let $\mathbf{u}$ be a point in $A_1$. Since $f(\mathbf{u}) < c$, it follows from the continuity of $f: A \to \mathbb{R}$ and the "$\epsilon$-$\delta$" characterization of continuity provided in Theorem 11.6 that we can choose a positive number $r = r(\mathbf{u})$ such that $f(\mathbf{v}) < c$ if $\mathbf{v}$ is in $\mathcal{N}_r(\mathbf{u}) \cap A$. Define $\mathcal{U}$ to be the union of these symmetric neighborhoods $\mathcal{N}_r(\mathbf{u})$, as $\mathbf{u}$ varies in $A_1$. Then $\mathcal{U}$ is open in $\mathbb{R}^n$, since it is the union of open subsets in $\mathbb{R}^n$, and it is clear that $A \cap \mathcal{U} = A_1$. Similarly, for each point $\mathbf{v}$ in $A_2$ we may choose a symmetric neighborhood whose intersection with $A$ is contained in $A_2$. The union of such neighborhoods defines an open subset $\mathcal{V}$ of $\mathbb{R}^n$ whose intersection with $A$ equals $A_2$.

To prove the converse, suppose that every continuous function $f: A \to \mathbb{R}$ has the Intermediate Value Property. We will show that $A$ is connected by assuming otherwise and deriving a contradiction. Suppose that $A$ is not connected. Then there are two subsets $\mathcal{U}$ and $\mathcal{V}$ of $\mathbb{R}^n$, each of which is open in $\mathbb{R}^n$, that separate $A$. Define the function $f: A \to \mathbb{R}$ by

$$f(\mathbf{u}) = \begin{cases} 0 & \text{if } \mathbf{u} \text{ is in } \mathcal{U} \cap A \\ 1 & \text{if } \mathbf{u} \text{ is in } \mathcal{V} \cap A. \end{cases}$$

Then the function $f: A \to \mathbb{R}$ certainly fails to have the Intermediate Value Property, since it attains exactly two functional values, 0 and 1. On the other hand, the function $f: A \to \mathbb{R}$ is continuous. Indeed, to verify continuity, just observe that since both $\mathcal{U}$ and $\mathcal{V}$ are open subsets of $\mathbb{R}^n$, it follows from the definition of $f: A \to \mathbb{R}$ that for each point $\mathbf{u}$ in $A$ there is a symmetric neighborhood $\mathcal{N}_r(\mathbf{u})$ such that $f: A \to \mathbb{R}$ is constant on $A \cap \mathcal{N}_r(\mathbf{u})$. Thus this function is certainly continuous at the point $\mathbf{u}$. The existence of this continuous function whose image is not an interval contradicts the assumption that $A$ has the Intermediate Value Property. Thus $A$ must be connected.            ∎

**COROLLARY
11.22**

Every pathwise connected subset of $\mathbb{R}^n$ is connected.

**Proof**    Let $A$ be a pathwise connected subset of $\mathbb{R}^n$. Theorem 11.18 asserts that $A$ has the Intermediate Value Property, and so, by Theorem 11.21, $A$ must be connected.            ∎

Theorem 11.16 asserts that a subset of $\mathbb{R}$ is an interval if and only if it is pathwise connected. In particular, by Corollary 11.22, each interval is connected. In fact, it is not difficult to show that each connected subset of $\mathbb{R}$ is an interval (see Exercise 7). Thus for a *subset* of $\mathbb{R}$, there is no distinction between being an interval, being pathwise connected, and being connected. For this reason, there is no need to introduce the concepts of connectedness and pathwise connectedness in the study of real-valued functions of a single real variable.

It is reasonable to ask about the distinction between connectedness and pathwise connectedness for subsets of $\mathbb{R}^n$ if $n > 1$. Corollary 11.22 asserts that every pathwise-connected subset of $\mathbb{R}^n$ is connected. The converse is not true; there are connected subsets of $\mathbb{R}^n$ that are not pathwise connected.

**EXAMPLE 11.11**    We will describe a subset of the plane $\mathbb{R}^2$ that is connected but not pathwise connected. It is the union of two pathwise connected sets. First, define $K = \{(x, y) \text{ in } \mathbb{R}^2 \mid x = 0, \ -1 \leq y \leq 1\}$. Observe that $K$ is convex, so it is pathwise connected. Now define $G = \{(x, y) \text{ in } \mathbb{R}^2 \mid 0 < x \leq 1, \ y = \sin 1/x\}$. Then $G$ is also pathwise connected, since it is the graph of a continuous function whose domain is an interval. We define $A = K \cup G$.

**FIGURE 11.6**

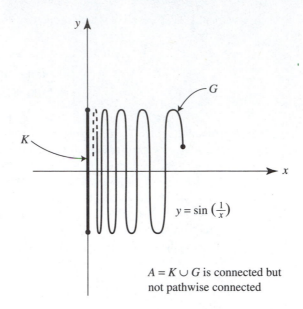

$y = \sin\left(\frac{1}{x}\right)$

$A = K \cup G$ is connected but
not pathwise connected

It turns out that the set $A$ is connected. But $A$ is not pathwise connected, since it is not possible to find a path in the set $A$ that joins a point in $K$ to a point in $G$. The details of verifying these assertions are left as exercises (see Exercises 16 and 17).    □

We will study connectedness more generally in the context of our study of metric spaces, which we will undertake in Chapter 12. In particular, in Section 12.5, we will prove that an open subset of $\mathbb{R}^n$ is connected if and only if it is pathwise connected. So it is not so surprising that the preceding example of a subset of the plane that is connected but not pathwise connected had to be a little wild.

We defined pathwise connectedness in terms of paths joining points. Intuitively, a path seems to be a familiar geometric object. As always, however, care is needed in framing our geometric intuition as a precise mathematical assertion. With respect to pathwise connectedness, the warning note is sounded by a famous example of G. Peano, who proved that the square $S = \{(x,\ y)$ in $\mathbb{R}^2 \mid |x| \leq 1,\ |y| \leq 1\}$ is a path; that is, he discovered a continuous mapping $\gamma \colon [0, 1] \to \mathbb{R}^2$ whose image is the square $S$.[*]

---

**EXERCISES**

1.  Let $I$ be an interval of real numbers, and suppose that the function $f \colon I \to \mathbb{R}$ is continuous. The *graph* of this function is the subset of $\mathbb{R}^2$ defined by

$$G = \{(x,\ y) \text{ in } \mathbb{R}^2 \mid x \text{ in } I,\ y = f(x)\}.$$

Show that $G$ is pathwise connected. Is $G$ convex?

2.  Let $A$ and $B$ be pathwise connected subsets of $\mathbb{R}^n$ whose intersection $A \cap B$ is nonempty. Prove that the union $A \cup B$ is also pathwise connected.

---

[*]See details in the excellent book *Introduction to Topology and Modern Analysis* by George Simmons (McGraw-Hill, 1963).

3.   Let $a$ and $b$ be positive real numbers. Use Exercises 1 and 2 to show that the ellipse

$$\{(x,\ y)\ \text{in}\ \mathbb{R}^2 \mid x^2/a + y^2/b = 1\}$$

is pathwise connected.

4.   Show that the set $\{(x,\ y)\ \text{in}\ \mathbb{R}^2 \mid x^2 + y^4 = 1\}$ is pathwise connected.

5.   Let $Q$ be the set of rational numbers. Show that $\mathbb{Q}$ is not connected.

6.   Show that the set $S = \{(x,\ y)\ \text{in}\ \mathbb{R}^2 \mid \text{either}\ x\ \text{or}\ y\ \text{is rational}\}$ is pathwise connected.

7.   Show that a connected subset of $\mathbb{R}$ must be an interval.

8.   Let $\mathbf{u}$ be a point in $\mathbb{R}^n$ and let $r$ be a positive number. Show that the set $\{\mathbf{v}\ \text{in}\ \mathbb{R}^n \mid d(\mathbf{v},\ \mathbf{u}) \le r\}$ is convex.

9.   Let $A$ be a connected subset of $\mathbb{R}^3$. Suppose that the points $(0,\ 0,\ 1)$ and $(4,\ 3,\ 0)$ are in $A$.
     (a)  Prove that there is a point in $A$ whose second component is 2.
     (b)  Prove that there is a point in $A$ whose norm is 4.

10.  Let $\mathbf{u}$ be a fixed point in $\mathbb{R}^n$, and let $c$ be a fixed real number. Prove that each of the following three sets is convex:

$$\{\mathbf{v}\ \text{in}\ \mathbb{R}^n \mid \langle \mathbf{v},\ \mathbf{u} \rangle > c\} \qquad \{\mathbf{v}\ \text{in}\ \mathbb{R}^n \mid \langle \mathbf{v},\ \mathbf{u} \rangle = c\} \qquad \{\mathbf{v}\ \text{in}\ \mathbb{R}^n \mid \langle \mathbf{v},\ \mathbf{u} \rangle < c\}$$

11.  Let $A$ and $B$ be convex subsets of $\mathbb{R}^n$. Prove that the intersection $A \cap B$ is also convex. Is it true that the intersection of two pathwise connected subsets of $\mathbb{R}^n$ is also pathwise connected?

12.  Given a point $\mathbf{u}$ in $\mathbb{R}^n$ and a point $\mathbf{v}$ in $\mathbb{R}^m$, we define the point $(\mathbf{u},\ \mathbf{v})$ to be the point in $\mathbb{R}^{n+m}$ whose first $n$ components coincide with the components of $\mathbf{u}$ and whose last $m$ components coincide with those of $\mathbf{v}$. Suppose that $A$ is a subset of $\mathbb{R}^n$ and that $F: A \to \mathbb{R}^m$ is continuous. The *graph* $G$ of this mapping is defined by

$$G = \{(\mathbf{u},\ \mathbf{v})\ \text{in}\ \mathbb{R}^{n+m} \mid \mathbf{u}\ \text{in}\ A,\ \mathbf{v} = F(\mathbf{u})\}.$$

Show that if $A$ is pathwise connected, then $G$ is also pathwise connected. (*Hint:* Compose.)

13.  Use Exercise 2 and Exercise 12 to show that the set $\{(x,\ y,\ z)\ \text{in}\ \mathbb{R}^3 \mid 2x^2 + y^2 + z^2 = 1\}$ is pathwise connected.

14.  Suppose that $A$ is a subset of $\mathbb{R}^n$ that fails to be connected, and let $\mathcal{U}$ and $\mathcal{V}$ be open subsets of $\mathbb{R}^n$ that separate $A$. Suppose that $B$ is a subset of $A$ that is connected. Prove that either $B \subseteq \mathcal{U}$ or $B \subseteq \mathcal{V}$.

15.  Let $K$ be a compact subset of $\mathbb{R}^n$, and suppose that $\mathcal{O}$ is an open subset of $\mathbb{R}^n$ that contains $K$. Prove that there is some positive number $r$ such that for any point $\mathbf{u}$ in $K$, $\mathcal{N}_r(\mathbf{u}) \subseteq \mathcal{O}$.

16.  Use Exercises 14 and 15 to show that the set $A$ defined in Example 11.11 is connected.

17.  Show that the set $A$ defined in Example 11.11 is not pathwise connected. (*Hint:* Let $\mathbf{u} = (0,\ 1)$ and $\mathbf{v} = (1,\ \sin 1)$. Suppose that there is a parametrized path $\gamma: [0,\ 1] \to \mathbb{R}^2$ in $A$ joining $\mathbf{u}$ to $\mathbf{v}$. Define $t_*$ to be that supremum of the points $t$ in $[0,\ 1]$ such that $\gamma$ maps the interval $[0,\ t]$ into $K$. Show that the second component of $\gamma$ is not continuous at the point $t_*$.)

18.  Which subsets of $\mathbb{R}$ are both compact and connected?

19.  Show that the parametrized paths described in Example 11.9 have images in $S$.

20.  Let $K$ be a compact subset of $\mathbb{R}^n$. Prove that $K$ is not connected if and only if there are nonempty, disjoint subsets $A$ and $B$ of $K$, with $A \cup B = K$ and a positive number $\epsilon$ such that $d(\mathbf{u},\ \mathbf{v}) > \epsilon$ for all $\mathbf{u}$ in $A$ and all $\mathbf{v}$ in $B$. Is the assumption of compactness necessary for the existence of such an $\epsilon$?

# Metric Spaces

## 12.1  Open Sets, Closed Sets, and Sequential Convergence

The first nine chapters of this book were devoted to the study of real-valued functions of a single real variable. Chapters 10 and 11 have been devoted to the consideration of Euclidean spaces and mappings between these spaces. It was observed in the late nineteenth century that a number of concepts that are useful in this study could be isolated and then studied in the more abstract context of metric spaces. This direction of thought clarifies the basis of many of the arguments that we have used; more important, it permits us to understand those concepts that may be generalized for use in the study of diverse problems in mathematics. In this first section, we will define a *metric space* and give a number of examples of metric spaces. We will also extend to metric spaces the concepts of openness and closedness that were introduced in Chapter 11 for Euclidean spaces. In Section 12.2, we will introduce the notion of a Cauchy sequence in a metric space and will call a metric space *complete* provided that every Cauchy sequence in the space converges to a point in that space. Examples of complete metric spaces will be considered. An important theorem called the Contraction Mapping Principle will be proven. In Section 12.3, we will review some of the results established earlier regarding the solutions of differential equations. Then we will use the Contraction Mapping Principle to prove a fundamental theorem about the solvability of a nonlinear differential equation. The final two sections will be devoted to extending results on continuity, compactness, and connectedness that have been established in Chapter 11 in the context of Euclidean spaces.

**DEFINITION**   *A set X is called a metric space if for any two points p and q in X there is defined a real number d(p, q), called the distance between p and q, such that the following three properties are satisfied:*

*Nonnegativity:*

$$d(p, q) > 0 \quad if \quad p \neq q; \quad d(p, p) = 0$$

*Symmetry:*

$$d(p, q) = d(q, p)$$

*The Triangle Inequality:*

$$d(p, q) \leq d(p, w) + d(w, q) \quad for\ all\ w\ in\ X$$

The above function $d: X \times X \to [0, \infty)$ is called a *metric on X*. It is possible to have more than one metric on a set. When we call $X$ a metric space, we will suppose that a fixed metric has been prescribed. Some of the most important metric spaces in the study of classical analysis are described in the following three theorems.

**THEOREM 12.1**   For any two real numbers $p$ and $q$, define

$$d(p, q) = |p - q|.$$

Then $d$ is a metric on $\mathbb{R}$.

**Proof**   Proposition 1.13 is precisely the assertion that $d(p, q) = |p - q|$ defines a metric.   ∎

**THEOREM 12.2**   For any two points $p$ and $q$ in Euclidean $n$-space $\mathbb{R}^n$, define

$$d(p, q) = \sqrt{\sum_{i=1}^{n} (p_i - q_i)^2}.$$

Then $d$ is a metric on $\mathbb{R}^n$.

**Proof**   The nonnegativity property follows from the observation that the sum of squares of real numbers is always nonnegative and is 0 if and only if all of the numbers are 0. The symmetry property is clear. The Triangle Inequality has already been established as inequality (10.9).   ∎

**THEOREM 12.3**   Define $C([a, b], \mathbb{R})$ to be the set of all continuous functions $f : [a, b] \to \mathbb{R}$, and for any two functions $f$ and $g$ in $C([a, b], \mathbb{R})$, define

$$d(f, g) = \max \{|f(x) - g(x)| \mid x \text{ in } [a, b]\}.$$

Then $d$ is a metric on $C([a, b], \mathbb{R})$.

**Proof**   First, observe that if $f$ and $g$ are in $C([a, b], \mathbb{R})$, then the function $|f - g| : [a, b] \to \mathbb{R}$ is continuous. According to the Extreme Value Theorem, the function $|f - g| : [a, b] \to \mathbb{R}$ has a maximum value. Thus $d(f, g)$ is properly defined.

Choose a point $x_0$ in the interval $[a, b]$ with $d(f, g) = |f(x_0) - g(x_0)|$. Then

$$0 \le |f(x) - g(x)| = |g(x) - f(x)| \le |g(x_0) - f(x_0)| \quad \text{for all } x \text{ in } [a, b],$$

so $d(f, g) = d(g, f) \ge 0$, and $d(f, g) = 0$ if and only if $f(x) = g(x)$ for all $x$ in $[a, b]$. It remains to verify the Triangle Inequality. Let $h$ be in $C([a, b], \mathbb{R})$. By the Triangle Inequality in $\mathbb{R}$, and the definition of the metric,

$$|f(x) - g(x)| \le |f(x) - h(x)| + |h(x) - g(x)|$$
$$\le d(f, h) + d(h, g) \quad \text{for all } x \text{ in } [a, b].$$

Thus we obtain the Triangle Inequality, $d(f, g) \le d(f, h) + d(h, g)$. ∎

It is useful to observe that for any two functions $f$ and $g$ in $C([a, b], \mathbb{R})$ and any positive number $r$ that

$$d(f, g) < r$$

if and only if        $f(x) - r < g(x) < f(x) + r \quad \text{for all } x \text{ in } [a, b].$

**FIGURE 12.1**

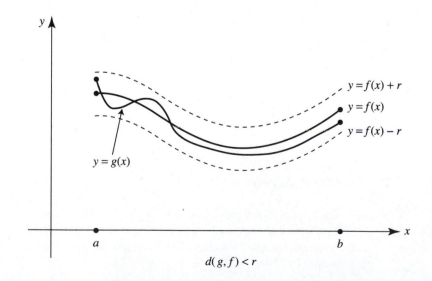

**THEOREM 12.4**   Let $X$ be any set. For any two points $p$ and $q$ in $X$, define

$$d(p, q) = \begin{cases} 0 & \text{if } p = q \\ 1 & \text{if } p \neq q. \end{cases}$$

This defines a metric on $X$ that is called the *discrete metric*.

**Proof**   The nonnegativity and symmetry properties and the Triangle Inequality follow immediately from the definition.   ∎

It is quite clear that the discrete metric on $\mathbb{R}^n$ and the Euclidean metric on $\mathbb{R}^n$ are distinct metrics.

Every subset $Y$ of a metric space $X$ is a metric space in its own right, where given two points $p$ and $q$ in $Y$, the distance between $p$ and $q$ is the same as the distance between these points when they are considered as points in the set $X$. Thus every subset of $\mathbb{R}$, every subset of Euclidean $n$-space $\mathbb{R}^n$, and every subset of $C([a, b], \mathbb{R})$ is a metric space. When we consider a subset $Y$ of a metric space $X$ as a metric space with the metric inherited from the metric on $X$, we will refer to the *subspace $Y$* of the metric space $X$.

**DEFINITION**   *Let $X$ be a metric space.*

(i) *For a point $p$ in $X$ and a positive number $r$, the set*

$$\mathcal{N}_r(p) \equiv \{q \text{ in } X \mid d(q, p) < r\}$$

*is called the symmetric neighborhood of $p$ in $X$ of radius $r$.*

(ii) *Given a subset $A$ of $X$, a point $p$ in $A$ is called an interior point of $A$ if some symmetric neighborhood of $p$ in $X$ is contained in $A$. The set of all interior points of $A$ is called the interior of $A$, and is denoted by int $A$.*

(iii) *A subset $\mathcal{O}$ of $X$ is called open in $X$ if every point in $\mathcal{O}$ is an interior point of $\mathcal{O}$.*

Observe that the definitions we have given of symmetric neighborhood, interior point, and openness for a general metric space generalize the corresponding concepts that we have already considered for Euclidean $n$-space.

**EXAMPLE 12.1**   Consider the metric space $C([0, 1], \mathbb{R})$. Given a function $f$ in $C([0, 1], \mathbb{R})$ and a positive number $r$, we see that the symmetric neighborhood of $f$ in $C([0, 1], \mathbb{R})$ of radius $r$ consists of those continuous functions $g: [0, 1] \to \mathbb{R}$ such that

$$f(x) - r < g(x) < f(x) + r \quad \text{for all } x \text{ in } [0, 1]. \qquad \square$$

**EXAMPLE 12.2**   Let $X$ be any set considered as a metric space with the discrete metric. Given a point $p$ in $X$ and a positive number $r$, it follows directly from the definition of the discrete metric that the neighborhood of $p$ in $X$ of radius $r$ consists of the whole set $X$ if $r \geq 1$, and consists of the single point $p$ if $r < 1$. $\qquad \square$

**EXAMPLE 12.3** Let $h$ and $g$ be in $C([0, 1], \mathbb{R})$, with $h(x) < g(x)$ for all $x$ in $[0, 1]$. Define $\mathcal{O} = \{f$ in $C([0, 1], \mathbb{R}) \mid h(x) < f(x) < g(x)$ for all $x$ in $[0, 1]\}$. Then $\mathcal{O}$ is open. Indeed, let $f$ be in $\mathcal{O}$. Let $r_1$ and $r_2$ be the minimum values of the functions $f - h \colon [0, 1] \to \mathbb{R}$ and $g - f \colon [0, 1] \to \mathbb{R}$, respectively, and then let $r$ be the smaller of the positive numbers $r_1$ and $r_2$. Then we see that $\mathcal{N}_r(f) \subseteq \mathcal{O}$. Thus $\mathcal{O}$ is open.  $\square$

**PROPOSITION 12.5**

Let $X$ be a metric space. Then every symmetric neighborhood in $X$ is open in $X$.

**Proof**  Let $p$ be a point in $X$ and $r$ be a positive number. Consider the symmetric neighborhood $\mathcal{N}_r(p)$. We will show that every point in $\mathcal{N}_r(p)$ is an interior point of $\mathcal{N}_r(p)$. Let $q$ be a point in $\mathcal{N}_r(p)$. Define $R = r - d(p, q)$, and observe that $R$ is positive. We claim that $\mathcal{N}_R(q) \subseteq \mathcal{N}_r(p)$. Indeed, by the Triangle Inequality, if $x$ is an element of $X$, then

$$d(x, p) \le d(x, q) + d(q, p),$$

so that if $d(x, q) < R = r - d(q, p)$, then $d(x, p) < r$. Thus $\mathcal{N}_R(q) \subseteq \mathcal{N}_r(p)$, so $q$ is an interior point of $\mathcal{N}_r(p)$.  ∎

For a set $X$, by a *sequence in $X$*, we mean a function $f \colon \mathbb{N} \to X$. It is customary to denote sequences by symbols such as $\{p_k\}$, $\{q_k\}$, and so on. If $X$ is a metric space, we define the concept of convergence of a sequence as follows.

**DEFINITION**  *Let $X$ be a metric space. A sequence $\{p_k\}$ in $X$ is said to converge to a point $p$ in $X$ provided that for each positive number $\epsilon$ there is a natural number $N$ such that*

$$d(p_k, p) < \epsilon \quad \text{if} \quad k \ge N.$$

We see that a sequence $\{p_k\}$ in $X$ converges to the point $p$ in $X$ if and only if the real sequence $\{d(p_k, p)\}$ converges to 0. We call $p$ the *limit* of the sequence $\{p_k\}$. Observe that a sequence can have at most one limit, since if $\{p_k\}$ converges to $p$ and also to $p'$, then according to the Triangle Inequality,

$$0 \le d(p, p') \le d(p, p_k) + d(p_k, p') \quad \text{for all positive integers } k.$$

From the Squeezing Principle, it follows that $d(p, p') = 0$, so $p' = p$.

Given a sequence $\{p_k\}$ in $X$ and $\{k_j\}$ a strictly increasing sequence of positive integers, the sequence $\{p_{k_j}\}$ is called a *subsequence* of $\{p_k\}$. Observe that if $\{p_k\}$ converges to $p$, then each subsequence of $\{p_{k_j}\}$ also converges to $p$, since, by our results for real sequences, if $\lim_{k \to \infty} d(p_k, p) = 0$, then $\lim_{j \to \infty} d(p_{k_j}, p) = 0$.

Of course, in the Euclidean space $\mathbb{R}^n$, the above definition of convergence coincides with the concept of convergence that we have described in Section 10.2. The Componentwise Convergence Criterion directly reduces the property of convergence of a sequence in $\mathbb{R}^n$ to the convergence of each of the component sequences.

**EXAMPLE 12.4**  If $X$ is any set that we consider as a metric space with the discrete metric, then a sequence $\{p_k\}$ in $X$ converges to the point $p$ in $X$ if and only if there is some index $N$ such that $p_k = p$ for all $k \ge N$. This follows from the observation that $d(p_k, p) < 1$ if and only if $p_k = p$.  $\square$

**EXAMPLE 12.5**   Consider the metric space $C([a, b], \mathbb{R})$. Let $\{f_n\}$ be a sequence of functions in $C([a, b], \mathbb{R})$ and let $f$ be a fixed function in $C([a, b], \mathbb{R})$. In Chapter 9 we described two types of convergence for sequences of functions. The sequence of functions $\{f_k: [a, b] \to \mathbb{R}\}$ was defined to *converge pointwise* to the function $f: [a, b] \to \mathbb{R}$ provided that for each point $x$ in $[a, b]$ the sequence of real numbers $\{f_k(x)\}$ converges to the number $f(x)$. Also, the sequence of functions $\{f_k: [a, b] \to \mathbb{R}\}$ was defined to *converge uniformly* to the function $f: [a, b] \to \mathbb{R}$ provided that for each $\epsilon > 0$ there is an index $N$ such that

$$f(x) - \epsilon < f_k(x) < f(x) + \epsilon \quad \text{for all } x \text{ in } [a, b] \text{ and all } k \geq N.$$

It is easy to see that this inequality holds if and only if

$$d(f_k, f) < \epsilon \quad \text{for all } k \geq N.$$

Thus the sequence of functions converges uniformly if and only if it converges as a sequence in the metric space $C([a, b], \mathbb{R})$. For this reason, this metric is often referred to as the *uniform metric* on $C([a, b], \mathbb{R})$.                                          □

**DEFINITION**   *Let $X$ be a metric space. Then a subset $C$ of $X$ is said to be closed in $X$ if whenever $\{p_k\}$ is a sequence in $C$ that converges to a point $p$ in $X$, then $p$ belongs to $C$.*

Again, the concept of closedness defined above is a generalization of the concept we have defined in Euclidean $n$-space. We considered examples of closed subsets of $\mathbb{R}^n$ in Section 10.3.

**EXAMPLE 12.6**   In the metric space $C([a, b], \mathbb{R})$, consider the set $A$ consisting of all functions whose functional values are nonnegative. Then $A$ is a closed subset of $C([a, b], \mathbb{R})$. This follows from the observation that uniform convergence implies pointwise convergence and the fact that the set of nonnegative real numbers is a closed subset of $\mathbb{R}$.                                          □

For any two sets $A$ and $B$, the *complement* of $A$ in $B$, denoted by $B \backslash A$, is defined by

$$B \backslash A \equiv \{p \in B \mid p \notin A\}.$$

If $B$ is any set and $\{A_s \mid s \in S\}$ is a collection of sets, then from the definitions of union, intersection, and complement, it follows that

$$B \backslash \bigcap_{s \in S} A_s = \bigcup_{s \in S} (B \backslash A_s) \quad \text{and} \quad B \backslash \bigcup_{s \in S} A_s = \bigcap_{s \in S} (B \backslash A_s).$$

These two formulas are frequently referred to as *deMorgan's laws*.

**THEOREM 12.6**   **The Complementing Characterization**   Let $X$ be a metric space and $A$ be a subset of $X$. Then $A$ is open in $X$ if and only if its complement in $X$ is closed in $X$.

**Proof**   First, suppose that $A$ is open in $X$. Then every point in $A$ is an interior point of $A$, so a sequence in $X \backslash A$ cannot converge to a point in $A$. It follows that a sequence in $X \backslash A$ that converges to a point in $X$ must converge to a point in $X \backslash A$. Thus $X \backslash A$ is closed.

We will now prove the converse. Suppose that $X \backslash A$ is closed. We must show that every point in $A$ is an interior point of $A$. Let $p$ be a point in $A$. Suppose that $p$ is not an interior point of $A$. Let $k$ be a positive integer. Then $\mathcal{N}_{1/k}(p)$ is not a subset of $A$, so we can choose a point in $X \backslash A$, which we label $p_k$, such that $d(p_k, p) < 1/k$. Thus $\{p_k\}$ converges to $p$. But $X \backslash A$ is closed, so $p$ is an element of $X \backslash A$. This contradiction shows that $p$ is an interior point of $A$. ∎

**THEOREM 12.7**   Let $X$ be a metric space.

(i) The union of a collection of open subsets of $X$ is open in $X$.

(ii) The intersection of a collection of closed subsets of $X$ is closed in $X$.

**Proof of (i)**   Suppose that $\mathcal{O} = \bigcup_{s \in S} \mathcal{O}_s$, where each $\mathcal{O}_s$ is open in $X$. Let $p$ be a point in $\mathcal{O}$. We must show that $p$ is an interior point of $\mathcal{O}$. But $p$ belongs to some $\mathcal{O}_s$, so since $\mathcal{O}_s$ is open in $X$, there is a neighborhood $\mathcal{N}_r(p)$ in $X$ of $p$ that is contained in $\mathcal{N}_r(q)$. Thus $\mathcal{N}_r(p)$ is also contained in $\mathcal{O}$, so $p$ is an interior point of $\mathcal{O}$.

**Proof of (ii)**   Suppose that $C = \bigcap_{s \in S} C_s$, where each $C_s$ is closed in $X$. Then $X \backslash C = X \backslash \bigcap_{s \in S} C_s = \bigcup_{s \in S} (X \backslash C_s)$. From (i) and the Complementing Characterization, it follows that $C$ is closed in $X$. ∎

As we have already noted in the context of Euclidean spaces, the intersection of a collection of open sets need not be open, nor need the union of a collection of closed sets be closed. However, for finite collections of sets, we have the following theorem.

**THEOREM 12.8**   Let $X$ be a metric space.

(i) The intersection of a finite collection of open subsets of $X$ is open in $X$.

(ii) The union of a finite collection of closed subsets of $X$ is closed in $X$.

**Proof of (i)**   Suppose $\mathcal{O} = \bigcap_{i=1}^k \mathcal{O}_i$ for some positive integer $k$, where each $\mathcal{O}_i$ is open in $X$. Let $p$ be an element of $\mathcal{O}$. If $1 \le i \le k$, $p$ is an element of $\mathcal{O}_i$ and $\mathcal{O}_i$ is open in $X$, so there exists a positive number $r_i$ with $\mathcal{N}_{r_i}(p) \subseteq \mathcal{O}_i$. Define $r = \min \{r_1, \ldots, r_k\}$. Then $r$ is positive, and $\mathcal{N}_r(p) \subseteq \bigcap_{i=1}^k \mathcal{O}_i = \mathcal{O}$. Thus $p$ is an interior point of $\mathcal{O}$, and $\mathcal{O}$ is open in $X$.

**Proof of (ii)**   Suppose $C = \bigcup_{i=1}^k C_i$ for some positive integer $k$, where each $C_i$ is closed in $X$. Observe that $X \backslash C = \bigcap_{i=1}^k (X \backslash C_i)$. From (i) and the Complementing Characterization, it follows that $C$ is closed in $X$. ∎

**EXERCISES**

1. Let $A = \{f \text{ in } C([0, 1], \mathbb{R}) \mid f(x) \geq 0 \text{ for all } x \text{ in } [0, 1]\}$. Prove that $A$ is a closed subset of $C([0, 1], \mathbb{R})$, but that $A$ is not open in $C([0, 1], \mathbb{R})$.

2. Let $X = C([0, 1], \mathbb{R})$. Find $d(f, g)$ for each of the following pairs of functions:
   (a) $f(x) = x$ and $g(x) = \cos x$ for $x$ in $[0, 1]$
   (b) $f(x) = 4x^3$ and $g(x) = 6x^2 - 3x$ for $x$ in $[0, 1]$

3. Let $\{f_k\}$ be the sequence in $C([0, 1], \mathbb{R})$ defined by

   $$f_k(x) = (1 - x)x^k \quad \text{for } x \text{ in } [0, 1] \text{ and each positive integer } k.$$

   Prove that the sequence converges pointwise to the function whose constant value is 0. Is the sequence $\{f_k\}$ a convergent sequence in the metric space $C([0, 1], \mathbb{R})$?

4. For each positive integer $k$, define the function $f_k : [0, 1] \to \mathbb{R}$ by $f_k(x) = \cos(x/k)$ for $x$ in $[0, 1]$. Show that the sequence $\{f_k\}$ converges in the metric space $C([0, 1], \mathbb{R})$.

5. Suppose that $X$ is a metric space that contains the point $p$ and $r$ is a positive number. Prove that the set $\{q \text{ in } X \mid d(p, q) \leq r\}$ is closed in $X$.

6. Verify the assertions made in Example 12.5.

7. Verify the assertions made in Example 12.6.

8. For any two points in the plane $\mathbb{R}^2$, define

   $$d^*(p, q) = |p_1 - q_1| + |p_2 - q_2|.$$

   (a) Show that $d^*$ defines a metric on $\mathbb{R}^2$.
   (b) Compare a symmetric neighborhood of $(0, 0)$ in this metric with a symmetric neighborhood of $(0, 0)$ in the Euclidean metric.
   (c) Show that a sequence in $\mathbb{R}^2$ converges with respect to the above metric if and only if it converges with respect to the Euclidean metric.

9. Let $X$ be any set considered as a metric space with the discrete metric. With this metric, show that every subset of $X$ is both open and closed in $X$.

10. For a metric space $X$ and a positive number $r$, can one have $\mathcal{N}_r(p) = \mathcal{N}_r(q)$ and yet $p \neq q$? Can this happen in $\mathbb{R}^n$ with the Euclidean metric?

11. For any two functions $f$ and $g$ in $C([a, b], \mathbb{R})$, define

    $$d^*(f, g) = \int_a^b |f(x) - g(x)| \, dx.$$

    (a) Prove that this defines a metric on $C([a, b], \mathbb{R})$.
    (b) Prove the following inequality relating this metric and the uniform metric:

    $$d^*(f, g) \leq (b - a) d(f, g).$$

    (c) Compare the concepts of convergence of a sequence of functions in this metric and in the uniform metric.

## 12.2  Completeness and the Contraction Mapping Principle

**DEFINITION**   *Let $X$ be a metric space. A sequence $\{p_k\}$ in $X$ is said to be a Cauchy sequence provided that for each positive number $\epsilon$ there is a natural number $N$ such that*

$$d(p_k, p_\ell) < \epsilon \quad \text{if} \quad k \geq N \quad \text{and} \quad \ell \geq N.$$

Observe that the preceding concept is a direct generalization of the concept of a Cauchy sequence of real numbers, which we considered in Section 9.1.

**PROPOSITION 12.9**

Every convergent sequence is a Cauchy sequence.

**Proof**   Let $X$ be a metric space. Suppose that $\{p_k\}$ is a sequence in $X$ that converges to the point $p$ in $X$. Let $\epsilon > 0$. We can choose a positive integer $N$ such that $d(p_k, p) < \epsilon/2$ if $k \geq N$. From the Triangle Inequality it follows that

$$d(p_k, p_\ell) \leq d(p_k, p) + d(p_\ell, p) < \frac{\epsilon}{2} + \frac{\epsilon}{2} = \epsilon \quad \text{if} \quad k \geq N \quad \text{and} \quad \ell \geq N.$$

Thus the sequence $\{p_k\}$ is Cauchy. ∎

In Section 9.1, we proved the Cauchy Convergence Criterion, which asserts that a sequence of real numbers converges if and only if it is a Cauchy sequence. The point of this equivalence is that it often happens that we have a sequence of real numbers that we wish to prove converges, but there is not sufficient information to identify the proposed limit. In such a case the Cauchy Convergence Criterion is useful, since it is a criterion that is intrinsic to the sequence itself and requires no knowledge of the proposed limit. It will be useful to discover other metric spaces in which every Cauchy sequence converges to a point in the space.

**EXAMPLE 12.7**   Consider a sequence $\{\mathbf{u}_k\}$ in Euclidean space $\mathbb{R}^n$. As in the case of sequences of real numbers, we claim that this sequence is a Cauchy sequence if and only if it converges to a point $\mathbf{u}$ in $\mathbb{R}^n$. Since in any metric space every convergent sequence is a Cauchy sequence, to verify this assertion we need only show that if $\{\mathbf{u}_k\}$ is a Cauchy sequence, then it converges. But observe that if $\mathbf{u} = (u_1, \ldots, u_n)$ and $\mathbf{v} = (v_1, \ldots, v_n)$ are points in $\mathbb{R}^n$, then

$$|u_i - v_i| \leq \|\mathbf{u} - \mathbf{v}\| \quad \text{for each index } i \text{ with } 1 \leq i \leq n.$$

Thus, if $\{\mathbf{u}_k\}$ is a Cauchy sequence in $\mathbb{R}^n$ and $1 \leq i \leq n$, then the sequence of $i$th components is Cauchy, and hence, by the Cauchy Convergence Criterion, this sequence of $i$th components converges to some number $u_i$. Define $\mathbf{u}$ to be the point in $\mathbb{R}^n$ whose $i$th component is $u_i$. Then from the Componentwise Convergence Criterion it follows that the sequence $\{\mathbf{u}_k\}$ converges to the point $\mathbf{u}$. □

**EXAMPLE 12.8**   Consider a sequence $\{f_k\}$ in $C([a, b], \mathbb{R})$. From the definition of the metric on $C([a, b], \mathbb{R})$, it follows that the sequence $\{f_k\}$ is a Cauchy sequence in $C([a, b], \mathbb{R})$ if and only if for each positive number $\epsilon$ there is a natural number $N$ such that

$$|f_k(x) - f_\ell(x)| < \epsilon \quad \text{for all } x \text{ in } [a, b], \text{ if } k \geq N \text{ and } \ell \geq N.$$

This is a concept that we have considered in Section 9.2: We refer to a sequence which has this property as *uniformly Cauchy*. In fact, in that section we established the Weierstrass Uniform Convergence Criterion (Theorem 9.16), which asserts that a sequence in $C([a, b], \mathbb{R})$ is uniformly Cauchy if and only if it converges uniformly to a continuous function $f : [a, b] \to \mathbb{R}$. □

**DEFINITION**   *A metric space X is said to be complete provided that every Cauchy sequence in X converges to a point in X.*

The discussion that preceded this definition can be conveniently summarized in the statement of the following theorem.

**THEOREM 12.10**   The metric spaces $\mathbb{R}$, $\mathbb{R}^n$, and $C([a, b], \mathbb{R})$ are complete.

A subspace of a complete metric space is not necessarily complete. For instance, if $X$ is the subspace of $\mathbb{R}$ consisting of the interval $(0, 2)$, then $X$ is not complete. To see this, observe that the sequence $\{1/k\}$ is a Cauchy sequence in $X$ that does not converge to a point in $X$, since it converges to the point 0, which is not in $X$. However, there is the following criterion for deciding which subspaces of a complete metric space are also complete.

**THEOREM 12.11**   Let $X$ be a complete metric space and $Y$ be a subspace of $X$. Then $Y$ is a complete metric space if and only if $Y$ is a closed subset of $X$.

**Proof**   First, suppose that $Y$ is a closed subset of $X$. Let $\{p_k\}$ be a Cauchy sequence in $Y$. Then $\{p_k\}$ is a Cauchy sequence in $X$, and since $X$ is complete, $\{p_k\}$ converges to a point $p$ in $X$. But $Y$ is a closed subset of $X$, so $p$ belongs to $Y$. Thus $Y$ is complete.

To prove the converse, suppose that $Y$ is a complete metric space. Let $\{p_k\}$ be a sequence in $Y$ that converges to the point $p$ in $X$. From Proposition 12.9 it follows that $\{p_k\}$ is a Cauchy sequence. But $Y$ is complete, so $\{p_k\}$ converges to a point in $Y$. Since a sequence can converge to at most one point, $p$ belongs to $Y$. Thus $Y$ is a closed subset of $X$.   ∎

**COROLLARY 12.12**   Every closed subset of $\mathbb{R}$, of $\mathbb{R}^n$, and of $C([a, b], \mathbb{R})$ is a complete metric space.

**Proof**   The result follows from Theorems 12.10 and 12.11.   ∎

**DEFINITION**   *Let X and Y be metric spaces.*

*(i) A mapping $T: X \to Y$ is said to be a Lipschitz mapping provided that there is some nonnegative number c, called a Lipschitz constant for the mapping, such that*

$$d(T(p), T(q)) \leq cd(p, q) \quad \text{for all points p and q in X.}$$

*(ii) A Lipschitz mapping $T: X \to Y$ that has Lipschitz constant less than 1 is called a contraction.*

**EXAMPLE 12.9**    Let $I$ be an open interval in $\mathbb{R}$ and suppose that the function $f: I \to \mathbb{R}$ is differentiable. We claim that $f: \mathbb{R} \to \mathbb{R}$ is Lipschitz with Lipschitz constant $c$ if and only if

$$|f'(x)| \leq c \quad \text{for all } x \text{ in } I.$$

To verify this, first suppose that this inequality holds. Then if $u$ and $v$ are points in $I$ with $u \neq v$, it follows from the Lagrange Mean Value Theorem that there is some point $z$ between $u$ and $v$ such that

$$f(u) - f(v) = f'(z)[u - v],$$

so $|f(u) - f(v)| \leq c|u - v|$. The converse follows from the very definition of a derivative as the limit of difference quotients.    □

There is a useful generalization of the above example to mappings between Euclidean spaces. Since the extension uses the concept of a partial derivative, we will postpone the discussion of this extension until we have established the appropriate extension of the Mean Value Theorem for such mappings.

**DEFINITION**    *Let X be a metric space. A point p in X is called a fixed-point for the mapping $T: X \to X$ provided that*

$$T(p) = p.$$

We will be interested in finding assumptions on a mapping that ensure the existence of fixed-points of the mapping. Of course, a mapping may or may not have any fixed-points. For instance, the mapping $T: \mathbb{R} \to \mathbb{R}$ defined by $T(x) = x + 1$ certainly has no fixed-points.

For real-valued functions of a single real variable, a fixed-point of the function corresponds to a point where the graph of the function intersects the line $y = x$. This observation provides the geometric insight for the following example.

**EXAMPLE 12.10**    Suppose that the function $f: [a, b] \to \mathbb{R}$ is continuous and that the image $f([a, b])$ is contained in $[a, b]$. Then $f: [a, b] \to \mathbb{R}$ has a fixed-point. This follows from the Intermediate Value Theorem when we observe that if we define $g(x) = f(x) - x$ for $x$ in $[a, b]$, then $g(a) \geq 0$ and $g(b) \leq 0$, so $g(x_0) = 0$ for some $x_0$ in $[a, b]$, which means that $f(x_0) = x_0$.    □

The above result generalizes to mappings between Euclidean spaces as follows: If $K$ is a subset of $\mathbb{R}^n$ that is compact and convex, and the mapping $T: K \to K$ is continuous, then $T: K \to K$ has a fixed-point. This is called Brouwer's Fixed-Point Theorem. Unfortunately, its proof lies outside the scope of this book.*

We will now prove a theorem called the Contraction Mapping Principle, which has many important applications. In addition to being useful, this theorem is particularly interesting because its proof requires merely the definition of a complete metric space

---

*A proof of this theorem can be found in Milnor's book *Topology from the Differentiable Viewpoint* (University Press of Virginia, 1965).

FIGURE 12.2

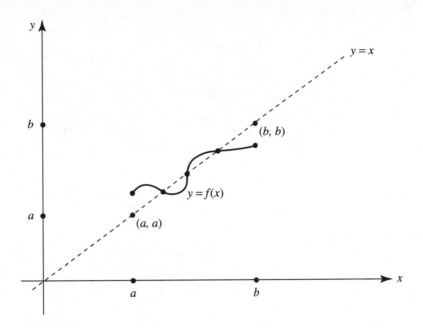

and two results that we have already established in Chapters 1 and 2. The first is that if $c$ is a real number, then

$$\lim_{k \to \infty} c^k = 0 \quad \text{if} \quad |c| < 1;$$

the second is the

## Geometric Sum Formula:

$$\sum_{k=0}^{n} c^k = \frac{1 - c^{n+1}}{1 - c} \quad \text{if} \quad c \neq 1.$$

**THEOREM 12.13**

**The Contraction Mapping Principle** Let $X$ be a complete metric space, and suppose that the mapping $T: X \to X$ is a contraction. Then the mapping $T: X \to X$ has exactly one fixed-point.

**Proof** Let $c$ be a number with $0 \leq c < 1$ that is a Lipschitz constant for the mapping $T: X \to X$; that is,

$$d(T(p), T(q)) \leq cd(p, q) \quad \text{for all points } p \text{ and } q \text{ in } X.$$

Select some point in $X$ and label it $p_0$. Now define the sequence $\{p_k\}$ inductively by setting $p_1 = T(p_0)$; then if $k$ is a positive integer such that $p_k$ is defined, set $p_{k+1} = T(p_k)$. This sequence is properly defined, since $T(X)$ is a subset of $X$. We will show that the sequence $\{p_k\}$ converges to a fixed-point of the mapping $T: X \to X$.

First, observe that from the definition of the sequence and the definition of the Lipschitz constant $c$, it follows that

$$d(p_2, p_1) = d(T(p_1), T(p_0)) \leq cd(T(p_0), p_0),$$

**FIGURE 12.3**

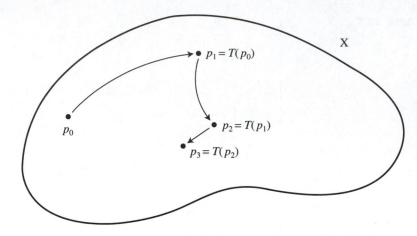

$$d(p_{k+1}, p_k) \le cd(p_k, p_{k-1}), \ 0 < c < 1$$

and that     $d(p_{k+1}, p_k) = d(T(p_k), T(p_{k-1})) \le cd(p_k, p_{k-1})$    if    $k \ge 2$.

Using these two inequalities, an induction argument implies that

$$d(p_{k+1}, p_k) \le c^k d(T(p_0), p_0) \quad \text{for every positive integer } k.$$

Hence, if $m$ and $k$ are positive integers with $m > k$, from the Triangle Inequality and the Geometric Sum Formula it follows that

$$\begin{aligned}
d(p_m, p_k) &\le d(p_m, p_{m-1}) + d(p_{m-1}, p_{m-2}) + \cdots + d(p_{k+1}, p_k) \\
&\le [c^{m-1} + c^{m-2} + \cdots + c^k]d(T(p_0), p_0) \\
&= c^k[1 + c + \cdots + c^{m-1-k}]d(T(p_0), p_0) \\
&= c^k \frac{[1 - c^{m-k}]}{1 - c} d(T(p_0), p_0).
\end{aligned}$$

Since $c$ is between 0 and 1,

$$d(p_m, p_k) \le \frac{c^k}{1 - c} d(T(p_0), p_0) \quad \text{if} \quad m > k. \tag{12.1}$$

But $\lim_{k \to \infty} c^k = 0$, and hence from (12.1) we conclude that $\{p_k\}$ is a Cauchy sequence.

By assumption, the metric space $X$ is complete. Thus there is a point $p$ in $X$ to which the sequence $\{p_k\}$ converges. But

$$d(T(p_k), T(p)) \le cd(p_k, p) \quad \text{for every positive integer } k, \tag{12.2}$$

so from the Squeezing Principle it follows that the image sequence $\{T(p_k)\}$ converges to the point $T(p)$. However, since $T(p_k) = p_{k+1}$ for each $k$, the sequence $\{T(p_k)\}$ is a subsequence of the sequence $\{p_k\}$, so $T(p) = p$. Thus the mapping $T: X \to X$ has at least one fixed-point.

It remains to check that there is only one fixed-point. But if $p$ and $q$ are points in $X$ such that $T(p) = p$ and $T(q) = q$, then

$$0 \leq d(p, q) = d(T(p), T(q)) \leq cd(p, q),$$

so since $0 \leq c < 1$, we must have $d(p, q) = 0$; that is, $p = q$. Thus there is exactly one fixed-point. ∎

The above proof of the Contraction Mapping Principle actually proves much more than the mere *existence* of a unique fixed-point. *It provides an iterative method for approximating the fixed-point.* Indeed, under the assumptions of the above theorem, what has been proven is not only that the mapping $T: X \to X$ has exactly one fixed-point $p_*$, but also that if $p_0$ is any point in $X$, then the sequence $\{p_k\}$ is defined recursively by setting $p_1 = T(p_0)$ and, if $k$ is a positive integer such that $p_k$ is defined, defining $p_{k+1} = T(p_k)$ converges to $p_*$. Moreover, if $c$ is a number with $0 \leq c < 1$ that is a Lipschitz constant for the mapping $T: X \to X$, the following error bounds hold:

$$d(p_*, p_k) \leq \frac{c^k}{1-c} d(T(p_0), p_0) \quad \text{for every positive integer } k. \qquad (12.3)$$

**EXERCISES**

1. Show that none of the following mappings $f: X \to X$ have a fixed-point, and explain why the Contraction Mapping Principle is not contradicted:
   (a) $X = (0, 1) \subseteq \mathbb{R}$, and $f(x) = x/2$ for $x$ in $X$.
   (b) $X = \mathbb{R}$, and $f(x) = x + 1$ for $x$ in $X$.
   (c) $X = \{(x, y) \text{ in } \mathbb{R}^2 \mid x^2 + y^2 = 1\}$, and $f(x, y) = (-y, x)$ for $(x, y)$ in $X$.
2. Fix $\alpha$ a positive real number, let $X = [0, 1] \subseteq \mathbb{R}$, and define $f: X \to \mathbb{R}$ by

$$f(x) = \alpha x(1 - x) \quad \text{for } x \text{ in } X.$$

   (a) For what values of $\alpha$ does the mapping $f: X \to \mathbb{R}$ have the property that $f(X) \subseteq X$?
   (b) For what values of $\alpha$ does the mapping $f: X \to \mathbb{R}$ have the property that $f(X) \subseteq X$ and $f: X \to X$ is a contraction?
3. Define the function $f: [1, \infty) \to \mathbb{R}$ by

$$f(x) = 1 + \sqrt{x} \quad \text{for } x \geq 1.$$

   Show that this function has exactly one fixed-point.
4. Fill in the details of Example 12.9.
5. Suppose that $p: \mathbb{R} \to \mathbb{R}$ is a polynomial. Show that $p: \mathbb{R} \to \mathbb{R}$ is Lipschitz if and only if the degree of the polynomial is less than 2.
6. Suppose that both of the functions $g: \mathbb{R} \to \mathbb{R}$ and $h: \mathbb{R} \to \mathbb{R}$ are Lipschitz. Is the product of these functions Lipschitz?
7. Suppose that the function $f: [a, b] \to \mathbb{R}$ is continuous. Prove that $f: [a, b] \to \mathbb{R}$ is Lipschitz if and only if $f: (a, b) \to \mathbb{R}$ is Lipschitz.
8. (a) Define $f(x) = \sqrt{x}$ for $x \geq 0$. Show that the function $f: [0, \infty) \to \mathbb{R}$ is continuous but is not Lipschitz.
   (b) Define $f(x) = |x|$ for all real numbers $x$. Show that the function $f: \mathbb{R} \to \mathbb{R}$ is Lipschitz but not differentiable.
9. For each positive integer $k$, define $f_k(x) = x^k$ for $0 \leq x \leq 1$. Is the sequence $\{f_k: [0, 1] \to \mathbb{R}\}$ a Cauchy sequence in the metric space $C([0, 1], \mathbb{R})$?

10. For each positive integer $k$, define $f_k(x) = e^{x/k}$ for $0 \le x \le 1$. Is the sequence $\{f_k: [0, 1] \to \mathbb{R}\}$ a Cauchy sequence in the metric space $C([0, 1], \mathbb{R})$?

11. For each positive integer $k$, define $f_k(x) = \cos(x/k)$ for $0 \le x \le 1$. Is the sequence $\{f_k: [0, 1] \to \mathbb{R}\}$ a Cauchy sequence in the metric space $C([0, 1], \mathbb{R})$?

12. Verify that the inequality (12.1) implies that the sequence $\{p_k\}$ is a Cauchy sequence.

13. Verify that the inequality (12.1) implies inequality (12.3).

14. Let $X$ be a metric space. Suppose that $\{p_k\}$ is a sequence in $X$ with the property that

$$d(p_{k+1}, p_k) \le 1/k \quad \text{for all natural numbers } k.$$

Is the sequence a Cauchy sequence? (*Hint:* Let $X = \mathbb{R}$ and let $p_k = \sum_{i=1}^{k} 1/i$ for each positive integer $k$.)

15. Let $U$ be a symmetric neighborhood in $\mathbb{R}^n$. By explicitly finding a Cauchy sequence in $U$ that does not converge to a point in $U$, show that $U$ is not complete.

16. Let $X$ be a subset of $\mathbb{R}^n$ and suppose that the mapping $T: X \to \mathbb{R}^m$ is Lipschitz. Prove that $T(X)$ is bounded if $X$ is bounded. Is this result true if the mapping is only assumed to be continuous?

17. Let $X$ be a complete metric space containing the point $p_0$ and let $r$ be a positive real number. Define $K = \{p \text{ in } X \mid d(p, p_0) \le r\}$. Suppose that $T: K \to X$ is Lipschitz with Lipschitz constant $c$. Suppose also that $cr + d(T(p_0), p_0) \le r$. Prove that $T(K) \subseteq K$ and that $T: K \to K$ has a fixed-point.

---

## 12.3 The Existence Theorem for Nonlinear Differential Equations

The aim of this section is to use the Contraction Mapping Principle to prove an important theorem regarding the existence of solutions of certain nonlinear differential equations. We begin by recalling some results about differential equations that we established earlier. In Section 4.5 we first encountered the following question regarding the solvability of a differential equation.

*Suppose that $I$ is an open interval of real numbers that contains the point $x_0$. Then given a number $y_0$ and a function $h: I \to \mathbb{R}$, does there exist a differentiable function $f: I \to \mathbb{R}$ that is a solution of the following fundamental differential equation?*

$$\begin{cases} f'(x) = h(x) & \text{for all } x \text{ in } I \\ f(x_0) = y_0. \end{cases} \tag{12.4}$$

In general, there may not be any solutions of this equation. For instance, according to Darboux's theorem, Theorem 4.22, if the function $h: I \to \mathbb{R}$ fails to have the Intermediate Value Property, then there is no solution of equation (12.4). However, if the function $h: I \to \mathbb{R}$ is continuous, then we have the following precise result about the solvability of equation (12.4), which was proven as Proposition 7.7, and is a variant of the Second Fundamental Theorem of Calculus.

**THEOREM**
**12.14**
Let $I$ be an open interval of real numbers that contains the point $x_0$, let $y_0$ be a real number, and suppose that the function $h: I \rightarrow \mathbb{R}$ is continuous. Then there is exactly one differentiable function $f: I \rightarrow \mathbb{R}$ that is a solution of the differential equation (12.4), and it is given by the formula

$$f(x) = y_0 + \int_{x_0}^{x} h(t) \, dt \quad \text{for all } x \text{ in } I. \tag{12.5}$$

Formula (12.5) represents the solution of equation (12.4) as an integral. Thus for each point $x$ in $I$, the actual value of $f(x)$ is the limit of Riemann sums. It is, of course, a properly defined function; that is, it associates a definite value with each point $x$ in $I$. Sometimes, of course, it is possible to simplify this representation, using a technique such as integration by parts or integration by substitution, and thus to represent the solution in terms of more familiar functions. However, it is often not possible to represent the solution explicitly in terms of the "elementary functions" $x^k$, $\sin x$, $\ln x$, and so on. In such a case, it is necessary to use an approximation technique, such as those described in Section 7.4, in order to obtain more precise information about the actual functional values of the solution of the differential equation (12.4).

**EXAMPLE 12.11**   Consider the differential equation

$$\begin{cases} f'(x) = 1/(1 + x^4) & \text{for all } x \text{ in } \mathbb{R} \\ f(0) = 1. \end{cases}$$

Define $h(x) = 1/(1 + x^4)$ for each real number $x$ to obtain a continuous function $h: \mathbb{R} \rightarrow \mathbb{R}$. Theorem 12.14 implies that this differential equation has a unique solution $f: \mathbb{R} \rightarrow \mathbb{R}$, defined by the formula

$$f(x) = 1 + \int_0^x \frac{1}{1 + t^4} \, dt \quad \text{for all } x \text{ in } \mathbb{R}.$$

We point out that one cannot explicitly evaluate the functional values of this solution by inspection. For instance,

$$f(2) = 1 + \int_0^2 \frac{1}{1 + t^4} \, dt,$$

and it is necessary to use some approximation technique, such as those described in Section 7.4, in order to approximate $f(2)$.                                                    □

For certain choices of the function $h: I \rightarrow \mathbb{R}$, we might be able to recall a differentiable function $g: I \rightarrow \mathbb{R}$ having the property that

$$g'(x) = h(x) \quad \text{for all } x \text{ in } I.$$

When we can recall such a function, the solution of the differential equation (12.4) is given by

$$f(x) = y_0 - g(x_0) + g(x) \quad \text{for all } x \text{ in } I.$$

**EXAMPLE 12.12**   Consider the differential equation

$$\begin{cases} f'(x) = 1/(1+x^2) & \text{for all } x \text{ in } \mathbb{R} \\ f(1) = 2. \end{cases}$$

Recall that we proved that the arctangent function has the property that

$$\frac{d}{dx}\arctan x = \frac{1}{1+x^2} \quad \text{for all } x \text{ in } \mathbb{R}.$$

Therefore, the unique solution of the above differential equation is given by the formula

$$f(x) = 2 - \arctan 1 + \arctan x = 2 - \frac{\pi}{4} + \arctan x \quad \text{for all } x \text{ in } \mathbb{R}. \qquad \square$$

In Section 7.2, we considered a differential equation that was more general than the differential equation (12.4). Specifically, we introduced an additional parameter $b$ and sought a differentiable function $f: I \to \mathbb{R}$ that is a solution of the equation

$$\begin{cases} f'(x) = bf(x) + h(x) & \text{for all } x \text{ in } I \\ f(x_0) = y_0. \end{cases} \tag{12.6}$$

In fact, there is a trick that reduces this equation to the type we have just considered. The trick* is to multiply the first equation in (12.6) by $e^{-bx}$ and see that the equation then becomes

$$\begin{cases} \dfrac{d}{dx}[e^{-bx}f(x)] = e^{-bx}h(x) & \text{for all } x \text{ in } I \\ f(x_0) = y_0. \end{cases} \tag{12.7}$$

If the function $h: I \to \mathbb{R}$ is continuous, it follows from Theorem 12.14 that there is a unique solution of (12.6) that is given by the formula

$$f(x) = e^{b(x-x_0)}y_0 + \int_{x_0}^{x} e^{b(x-t)}h(t)\,dt \quad \text{for all } x \text{ in } I. \tag{12.8}$$

**EXAMPLE 12.13**   Consider the differential equation

$$\begin{cases} f'(x) = 2f(x) + x & \text{for all } x \text{ in } \mathbb{R} \\ f(0) = 1. \end{cases}$$

Since the function $h: \mathbb{R} \to \mathbb{R}$ defined by $h(x) = x$ for all $x$ is continuous, it follows from the above discussion that this differential equation has a unique solution $f: \mathbb{R} \to \mathbb{R}$ given by the formula

$$f(x) = e^{2x} + \int_{0}^{x} e^{2(x-t)}t\,dt \quad \text{for all } x \text{ in } \mathbb{R}.$$

Integrating by parts and using the First Fundamental Theorem of Calculus, this last formula reduces to

$$f(x) = \frac{5}{4}e^{2x} - \frac{x}{2} - \frac{1}{4} \quad \text{for all } x \text{ in } \mathbb{R}. \qquad \square$$

---

*The trick is called *multiplication by an integrating factor*.

We will now consider much more general differential equations. Suppose that $\mathcal{O}$ is an open subset of the plane $\mathbb{R}^2$ that contains the point $(x_0, y_0)$ and that the function $g: \mathcal{O} \to \mathbb{R}$ is continuous. The problem is to find an open interval $I$ that contains the point $x_0$ and a differentiable function $f: I \to \mathbb{R}$ such that

$$\begin{cases} f'(x) = g(x, f(x)) & \text{for all } x \text{ in } I \\ f(x_0) = y_0. \end{cases} \tag{12.9}$$

This differential equation contains both the differential equation (12.4) and equation (12.6) as particular cases. Indeed, defining $\mathcal{O} = \{(x, y) \mid x \text{ in } I, \ y \text{ in } \mathbb{R}\}$, in the case when $g(x, y) = h(x)$ for $(x, y)$ in $\mathcal{O}$, the differential equation (12.9) reduces to the differential equation (12.4), whereas in the case when $g(x, y) = by + h(x)$ for $(x, y)$ in $\mathcal{O}$, the differential equation (12.9) reduces to the differential equation (12.6).

For a general continuous function $g: \mathcal{O} \to \mathbb{R}$ the study of equation (12.9) can be quite delicate. First of all, as the next example illustrates, there may be more than one solution.

**EXAMPLE 12.14**   Let $\mathcal{O} = \mathbb{R}^2$, let $(x_0, y_0) = (0, 0)$, and define $g(x, y) = 3y^{2/3}$. Then equation (12.9) becomes the equation

$$\begin{cases} f'(x) = 3[f(x)]^{2/3} & \text{for all } x \text{ in } I \\ f(0) = 0. \end{cases} \tag{12.10}$$

It is clear that the function $f: \mathbb{R} \to \mathbb{R}$ that is identically 0 is a solution of equation (12.10). But there is another solution. It is not difficult to check that the function $f: \mathbb{R} \to \mathbb{R}$ defined by

$$f(x) = \begin{cases} 0 & \text{if } x < 0 \\ x^3 & \text{if } x \geq 0 \end{cases}$$

is also a solution of the differential equation (12.10).                    □


Another difficulty that can arise in the study of equation (12.9) is that the interval $I$ on which the solution is defined may be small. This is illustrated in the following example.


**EXAMPLE 12.15**   Let $\mathcal{O} = \mathbb{R}^2$, let $(x_0, y_0) = (0, 0)$, and define $g(x, y) = 1 + y^2$. Then equation (12.9) becomes the equation

$$\begin{cases} f'(x) = 1 + (f(x))^2 & \text{for all } x \text{ in } I \\ f(0) = 0. \end{cases} \tag{12.11}$$

We claim that there is a unique solution of this differential equation on the interval $I = (-\pi/2, \pi/2)$ and that there is no solution on any larger interval. To analyze the differential equation (12.11), first suppose that $I$ is an interval containing the point 0 and that the differentiable function $f: I \to \mathbb{R}$ is a solution of (12.11). Then

$$\frac{f'(x)}{1 + [f(x)]^2} = 1 \quad \text{for all } x \text{ in } I,$$

which, in view of the Chain Rule and the formula for the derivative of the arctangent function, means that

$$\frac{d}{dx}[\arctan{(f(x))} - x] = 0 \quad \text{for all } x \text{ in } I.$$

Since $\arctan{f(0)} = \arctan{0} = 0$, it follows from the Identity Criterion that

$$\arctan{f(x)} = x \quad \text{for all } x \text{ in } I.$$

But the image of the arctangent function is the interval $(-\pi/2, \pi/2)$, so it follows that $I \subseteq (-\pi/2, \pi/2)$ and that $f(x) = \tan{x}$ for all $x$ in $I$.

This argument shows that if there is an interval containing the point 0 and a differentiable function $f: I \to \mathbb{R}$ that is a solution of the differential equation (12.11), then $I \subseteq (-\pi/2, \pi/2)$ and $f(x) = \tan{x}$ for all $x$ in $I$. A straightforward computation of derivatives shows that if we define $I = (-\pi/2, \pi/2)$ and $f: I \to \mathbb{R}$ by $f(x) = \tan{x}$, then this function is a solution of the differential equation (12.11). $\qquad\square$

The purpose of the above example is to show that even when $g: \mathbb{R}^2 \to \mathbb{R}$ is as simple a function as a second degree polynomial in $y$, there are restrictions on the size of the neighborhood $I$ of $x_0$ on which there is a solution of (12.11) (see Exercise 6).

We now turn to the analysis of the differential equation (12.9) for fairly general functions $g: \mathcal{O} \to \mathbb{R}$. Once more, the Second Fundamental Theorem of Calculus will be essential to our analysis. The following lemma establishes the equivalence between the differential equation and an associated integral equation.

**LEMMA 12.15**    **The Equivalence Lemma**    Let $\mathcal{O}$ be an open subset of the plane $\mathbb{R}^2$ that contains the point $(x_0, y_0)$ and suppose that the function $g: \mathcal{O} \to \mathbb{R}$ is continuous. Let $I$ be a neighborhood of the point $x_0$ and suppose that the function $f: I \to \mathbb{R}$ has the property that

$$(x, f(x)) \text{ is in } \mathcal{O} \quad \text{for all } x \text{ in } I.$$

Then the following two assertions are equivalent:

(i) The function $f: I \to \mathbb{R}$ is differentiable and is a solution of the differential equation

$$\begin{cases} f'(x) = g(x, f(x)) & \text{for all } x \text{ in } I \\ f(x_0) = y_0. \end{cases} \tag{12.12}$$

(ii) The function $f: I \to \mathbb{R}$ is continuous and is a solution of the integral equation

$$f(x) = y_0 + \int_{x_0}^{x} g(t, f(t))\, dt \quad \text{for all } x \text{ in } I. \tag{12.13}$$

**Proof**    Define the function $h: I \to \mathbb{R}$ by

$$h(x) = g(x, f(x)) \quad \text{for all } x \text{ in } I$$

and observe that the function $h: I \to \mathbb{R}$ is continuous, since it is the composition of continuous functions.

First, suppose that (i) holds. By the very definition of the function $h: I \to \mathbb{R}$, the differential equation (12.12) can be written as

$$\begin{cases} f'(x) = h(x) & \text{for all } x \text{ in } I \\ f(x_0) = y_0. \end{cases}$$

Theorem 12.14 implies that

$$f(x) = y_0 + \int_{x_0}^{x} h(t)\, dt = y_0 + \int_{x_0}^{x} g(t,\, f(t))\, dt \quad \text{for all } x \text{ in } I,$$

so (ii) holds.

Conversely, if (ii) holds, then clearly $f(x_0) = y_0$. Moreover, the Second Fundamental Theorem of Calculus implies that

$$\frac{d}{dx}\left[ \int_{x_0}^{x} g(t,\, f(t))\, dt \right] = g(x,\, f(x)) \quad \text{for all } x \text{ in } I.$$

From the integral equation (12.13), we conclude that the function $f: I \to \mathbb{R}$ is a differentiable function that is a solution of the differential equation (12.12).   ∎

**THEOREM 12.16**   **The Existence Theorem**   Let $\mathcal{O}$ be an open subset of the plane $\mathbb{R}^2$ that contains the point $(x_0,\, y_0)$. Suppose that the function $g: \mathcal{O} \to \mathbb{R}^2$ is continuous and that there is a positive number $M$ such that

$$|g(x,\, y_1) - g(x,\, y_2)| \leq M|y_1 - y_2| \quad \text{for all points } (x,\, y_1) \text{ and } (x,\, y_2) \text{ in } \mathcal{O}.$$
$$(12.14)$$

Then there is an open interval $I$ containing the point $x_0$ such that the differential equation

$$\begin{cases} f'(x) = g(x,\, f(x)) & \text{for all } x \text{ in } I \\ f(x_0) = y_0 \end{cases}$$
$$(12.15)$$

has exactly one solution.

**Proof**   For $\ell$ a positive number, define $I_\ell$ to be the closed interval $[x_0 - \ell,\, x_0 + \ell]$. We will show that $\ell$ can be chosen so that there is exactly one continuous function $f: I_\ell \to \mathbb{R}$ with

$$f(x) = y_0 + \int_{x_0}^{x} g(s,\, f(s))\, ds \quad \text{for all } x \text{ in } I_\ell.$$
$$(12.16)$$

Once such an $\ell$ is chosen, it follows from the Equivalence Lemma that there is exactly one solution of the differential equation (12.15) on the interval $I = (x_0 - \ell,\, x_0 + \ell)$.

Since $\mathcal{O}$ is open, we can choose positive numbers $a$ and $b$ such that the rectangle $R = [x_0 - a,\, x_0 + a] \times [y_0 - b,\, y_0 + b]$ is contained in $\mathcal{O}$. Now for each positive number $\ell$ with $\ell \leq a$, define $X_\ell$ to be the subspace of the metric space $C(I_\ell,\, \mathbb{R})$ consisting of those continuous functions $f: I_\ell \to \mathbb{R}$ that have the property that

$$|f(x) - y_0| \leq b \quad \text{for all } x \text{ in } I_\ell.$$

Observe that $X_\ell$ consists of the continuous functions $f: I_\ell \to \mathbb{R}$ whose graphs lie in the rectangle $I_\ell \times [y_0 - b, \ y_0 + b]$.

**FIGURE 12.4**

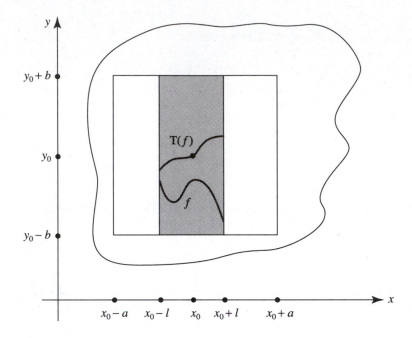

$$T(f)(x) = y_0 + \int_{x_0}^{x} g(t, f(t)) dt$$

For a function $f$ in $X_\ell$, define the function $T(f)$ in $C(I_\ell, \mathbb{R})$ by

$$T(f)(x) = y_0 + \int_{x_0}^{x} g(t, \ f(t)) \, dt \quad \text{for all } x \text{ in } I_\ell.$$

Observe that a solution of the integral equation (12.16) is simply a fixed-point of the mapping $T: X_\ell \to C(I_\ell, \mathbb{R})$.

The strategy of the proof is as follows: Since $C(I_\ell, \mathbb{R})$ is a complete metric space and $X_\ell$ is a closed subset of $C(I_\ell, \mathbb{R})$, it follows from Corollary 12.12 that $X_\ell$ is also a complete metric space. We will show that if $\ell$ is chosen sufficiently small, then

$$T(X_\ell) \subseteq X_\ell \quad \text{and} \quad T: X_\ell \to X_\ell \quad \text{is a contraction.} \tag{12.17}$$

Hence it will follow from the Contraction Mapping Principle that the mapping $T: X_\ell \to X_\ell$ has a unique fixed-point.

In order to choose $\ell$ so that $T(X_\ell) \subseteq X_\ell$, we first choose any positive number $K$ such that

$$|g(x, \ y)| \le K \quad \text{for all points } (x, \ y) \text{ in the rectangle } R.$$

The Extreme Value Theorem permits us to make such a choice. Then, if $f$ is any function in $X_\ell$ and the point $x$ belongs to $I_\ell$,

$$|T(f)(x) - y_0| = \left| \int_{x_0}^x g(t,\ f(t))\,dt \right| \le \ell K,$$

so
$$T(X_\ell) \subseteq X_\ell \quad \text{provided that} \quad \ell K \le b. \tag{12.18}$$

Now observe that if $f_1$ and $f_2$ are any two functions in $X_\ell$ and the point $x$ is in $I_\ell$, then, by assumption (12.14),

$$|g(x,\ f_1(x)) - g(x,\ f_2(x))| \le M|f_1(x) - f_2(x)| \le Md(f_1,\ f_2);$$

hence, using the linearity and monotonicity properties of the integral, it follows that

$$|T(f_1)(x) - T(f_2)(x)| = \left| \int_{x_0}^x [g(t,\ f_1(t)) - g(t,\ f_2(t))]\,dt \right|$$

$$\le \left| \int_{x_0}^x Md(f_1,\ f_2)\,dt \right|$$

$$\le |x - x_0|Md(f_1,\ f_2)$$

$$\le \ell Md(f_1,\ f_2).$$

This last inequality, together with inequality (12.18), implies that for $O \le l \le a$,

$$T: X_\ell \to X_\ell \quad \text{is a contraction provided that} \quad \ell K \le b \quad \text{and} \quad \ell M < 1. \tag{12.19}$$

Hence we may apply the Contraction Mapping Principle to find the unique fixed-point of $T$; that is, the unique solution of the integral equation (12.16) and also of the differential equation (12.15).   ∎

## EXERCISES

1. For each of the following differential equations, use Theorem 12.14 to write an integral formula for the solution and, if possible, write the solution in terms of elementary functions:

   (a) $\begin{cases} f'(x) = x \cos x & \text{for all } x \text{ in } \mathbb{R} \\ f(0) = 1 \end{cases}$

   (b) $\begin{cases} f'(x) = 1 + x^3 & \text{for all } x \text{ in } \mathbb{R} \\ f(1) = 4 \end{cases}$

   (c) $\begin{cases} f'(x) = e^{x^2} & \text{for all } x \text{ in } \mathbb{R} \\ f(0) = 0 \end{cases}$

2. For each of the following differential equations, use formula (12.8) to write an integral formula for the solution and, if possible, write the solution in terms of elementary functions:

   (a) $\begin{cases} f'(x) = f(x) + 1 & \text{for all } x \text{ in } \mathbb{R} \\ f(0) = 1 \end{cases}$

   (b) $\begin{cases} f'(x) = -f(x) + 2 + x & \text{for all } x \text{ in } \mathbb{R} \\ f(0) = 1 \end{cases}$

   (c) $\begin{cases} f'(x) = 2f(x) + e^x & \text{for all } x \text{ in } \mathbb{R} \\ f(1) = 0 \end{cases}$

3. Follow the analysis in Example 12.15 to explicitly find the maximal interval $I$ about 0 on which we can solve the following differential equations:

(a) $\begin{cases} f'(x) = 1 - [f(x)]^2 & \text{for all } x \text{ in } I \\ f(0) = 0 \end{cases}$

(b) $\begin{cases} f'(x) = xf(x) & \text{for all } x \text{ in } I \\ f(0) = 1 \end{cases}$

(c) $\begin{cases} f'(x) = [f(x)]^2 & \text{for all } x \text{ in } I \\ f(0) = -1 \end{cases}$

4. Verify the details of Example 12.14.
5. Verify the details of Example 12.15.
6. Let $\epsilon$ be a positive number. Consider the differential equation

$$\begin{cases} f'(x) = (1/\epsilon)(1 + (f(x))^2) & \text{for all } x \text{ in } I \\ f(0) = 0. \end{cases}$$

Find the length of the maximal interval on which this differential equation has a solution.

7. Verify the integral inequalities that were asserted in the proof of the Existence Theorem.
8. Follow the proof of the Existence Theorem to prove the following: *Let $a$, $b$, $M$, and $K$ be positive real numbers, let $(x_0, y_0)$ be a point in the plane $\mathbb{R}^2$, and let $R$ be the closed rectangle $[x_0 - a, x_0 + a] \times [y_0 - b, y_0 + b]$. Suppose that the function $g: R \to \mathbb{R}$ is continuous and that $aM < 1$ and $aK \le b$, where*

$$|g(x, y)| \le K \quad \text{for all } (x, y) \text{ in } R$$

*and* $\quad |g(x, y_1) - g(x, y_2)| \le M|y_1 - y_2| \quad$ *for all $(x, y_1)$, $(x, y_2)$ in $R$.*

*Then there is a unique solution of the differential equation*

$$\begin{cases} f'(x) = g(x, f(x)) & \text{for all } x \text{ in } (x_0 - a, x_0 + a) \\ f(x_0) = y_0. \end{cases}$$

9. Use Exercise 8 to determine a value of $r$ such that the differential equation

$$\begin{cases} f'(x) = \sin(xf(x)) & \text{for all } x \text{ in } (-r, r) \\ f(0) = 1 \end{cases}$$

has a unique solution.

10. Under the assumptions of Exercise 8, show that if $f_0(x) = y_0$ for all $x$ in $[x_0 - a, x_0 + a]$ and the sequence of functions $\{f_k : (x_0 - a, x_0 + a) \to \mathbb{R}\}$ is defined recursively by the formula

$$f_{k+1}(x) = y_0 + \int_{x_0}^x g(s, f_k(s)) \, ds \quad \text{for all } x \text{ in } (x_0 - a, x_0 + a),$$

then the sequence $\{f_k : (x_0 - a, x_0 + a) \to \mathbb{R}\}$ converges uniformly to the solution of the differential equation.

11. Under the assumptions of Exercise 8, let $\{f_k\}$ be as defined in Exercise 10. Show that

$$|f_k(x) - f(x)| \le \frac{(aM)^k}{1 - aM} aL$$

for all $x$ in $(x_0 - a, x_0 + a)$ and all natural numbers $k$, where $f: I \to \mathbb{R}$ is a solution of equation (12.9). (*Hint:* Use the estimate (12.3).)

# 12.4   Continuous Mappings Between Metric Spaces

We now turn to the study of functions $f: X \to Y$, where $X$ and $Y$ are general metric spaces. In this generality, *functions* are often referred to as *mappings*. As we will see, many of the results for continuous mappings between Euclidean spaces carry over, almost word for word, to mappings between general metric spaces.

Given a mapping $f: X \to Y$, if $A$ is a subset of $X$, we define

$$f(A) = \{q \text{ in } Y \mid q = f(p) \text{ for some point } p \text{ in } A\},$$

and call $f(A)$ the *image* of the set $A$ under the mapping $f: X \to Y$. Also, if $B$ is a subset of $Y$, we define

$$f^{-1}(B) = \{p \text{ in } X \mid f(p) \text{ in } B\}$$

and call $f^{-1}(B)$ the *preimage* of the set $B$ under the mapping $f: X \to Y$.

Since both $X$ and $Y$ are metric spaces, the concept of convergence of a sequence is defined in both $X$ and $Y$. This leads us to the following natural concept of continuity.

**DEFINITION**   *Let $X$ and $Y$ be metric spaces.*

  (i) *A mapping $f: X \to Y$ is said to be continuous at the point $p$ in $X$ provided that whenever a sequence $\{p_k\}$ in $X$ converges to $p$, the image sequence $\{f(p_k)\}$ converges to $f(p)$.*

  (ii) *A mapping $f: X \to Y$ is said to be continuous provided that it is continuous at every point in $X$.*

The above definition extends the definition of continuity that we gave in Chapter 3 for real-valued functions of a real variable and in Chapter 11 for mappings between Euclidean spaces.

For mappings $f: X \to Y$ between general metric spaces, it does not make sense to consider sums and products of mappings, since there may be no addition or multiplication defined on the set $Y$. However, for real-valued mappings—that is, for mappings from a metric space into $\mathbb{R}$—there is a direct extension of Theorem 11.2 concerning the continuity of sums, products, and quotients of continuous mappings.

Given a metric space $X$ and real-valued mappings $f$, $g: X \to \mathbb{R}$, we define the *product* $fg: X \to \mathbb{R}$ and the *sum* $f + g: X \to \mathbb{R}$ by

$$(fg)(p) \equiv f(p)g(p) \quad \text{and} \quad (f + g)(p) \equiv f(p) + g(p) \quad \text{for all } p \text{ in } X.$$

If $g(p) \notin 0$ for all $p$ in $X$, we define the *quotient* $f/g: X \to \mathbb{R}$ by

$$\left(\frac{f}{g}\right)(p) \equiv \frac{f(p)}{g(p)} \quad \text{for all } p \text{ in } X.$$

**THEOREM 12.17**

Let $X$ be a metric space and $p$ be a point in $X$. Suppose that the mappings $f: X \to \mathbb{R}$ and $g: X \to \mathbb{R}$ are both continuous at the point $p$. Then for any real numbers $\alpha$ and $\beta$, the function

$$\alpha f + \beta g: X \to \mathbb{R}$$

is continuous at $p$. Also, the product

$$f \cdot g: X \to \mathbb{R}$$

is continuous at $p$. Moreover, if $g(q) \neq 0$ for all $q$ in $X$, then the quotient

$$\frac{f}{g}: X \to \mathbb{R}$$

is continuous at $p$.

**Proof**

Let $\{p_k\}$ be a sequence in $X$ that converges to $p$. It follows from the definition of continuity that the real sequences $\{f(p_k)\}$ and $\{g(p_k)\}$ converge to $f(p)$ and $g(p)$, respectively. By the sum, product, and quotient properties of convergent sequences of real numbers,

$$\lim_{k \to \infty} (f + g)(p_k) = (f + g)(p),$$

$$\lim_{k \to \infty} (fg)(p_k) = (fg)(p),$$

and

$$\lim_{k \to \infty} \left(\frac{f}{g}\right)(p_k) = \left(\frac{f}{g}\right)(p).$$

These three equalities prove the theorem. ■

We also have the following theorem concerning the composition of continuous mappings, which generalizes Theorem 3.3 and Theorem 11.3.

**THEOREM 12.18**

Let $X$, $Y$, and $Z$ be metric spaces and let $p$ be a point in $X$. Suppose that the mapping $f: X \to Y$ is continuous at $p$ and the mapping $g: Y \to Z$ is continuous at $f(p)$. Then the composition $g \circ f: X \to Z$ is continuous at $p$.

**Proof**

Let $\{p_k\}$ be a sequence in $X$ that converges to $p$. Then $\{f(p_k)\}$ converges to $f(p)$, since the mapping $f: X \to Y$ is continuous at $p$. Thus the sequence $\{g(f(p_k))\}$ converges to $\{g(f(p))\}$, since $\{f(p_k)\}$ is a sequence in $Y$ that converges to $f(p)$, and the mapping $g: Y \to Z$ is continuous at $f(p)$. Hence $\{(g \circ f)(p_k)\}$ converges to $(g \circ f)(p)$. ■

For real-valued functions of a single real variable, we proved in Theorem 3.13 that there is an "$\epsilon$–$\delta$" criterion for continuity of a function at a point that is equivalent to the sequential convergence definition. For mappings between Euclidean spaces, this criterion was established in Theorem 11.6. In fact, as we will now show, this "$\epsilon$–$\delta$" criterion for continuity of a mapping at a point holds in general metric spaces.

**THEOREM 12.19**   Let $X$ and $Y$ be metric spaces, let $p$ be a point in $X$, and consider the mapping $f \colon X \to Y$. Then the following two assertions are equivalent:

(i)  The mapping $f \colon X \to Y$ is continuous at the point $p$.

(ii)  For each positive number $\epsilon$ there is a positive number $\delta$ such that

$$d(f(p), f(q)) < \epsilon \quad \text{for each point } q \text{ in } X \text{ with } d(p, q) < \delta. \qquad (12.20)$$

**Proof**   First, suppose that $f \colon X \to Y$ is continuous at $p$. To verify (12.20), we will suppose the contrary and derive a contradiction. If (12.20) does not hold, then there is some $\epsilon_0 > 0$ such that for no positive number $\delta$ is it true that $f(\mathcal{N}_\delta(p)) \subseteq \mathcal{N}_{\epsilon_0}(f(p))$. In particular, if $k$ is a positive integer, then it is not true that $f(\mathcal{N}_{1/k}(p)) \subseteq \mathcal{N}_{\epsilon_0}(f(p))$. This means that there is a point in $X$, which we label $p_k$, such that $d(p, p_k) < 1/k$ while $d(f(p), f(p_k)) \geq \epsilon_0$. This defines a sequence $\{p_k\}$ in $X$ that converges to $p$, whose image sequence $\{f(p_k)\}$ does not converge to $f(p)$. This contradicts the continuity of the mapping $f \colon X \to Y$ at the point $p$. Thus (12.20) holds.

To prove the converse, suppose that (12.20) holds. Let $\{p_k\}$ be a sequence in $X$ that converges to $p$. We must show that $\{f(p_k)\}$ converges to $f(p)$. Let $\epsilon > 0$. According to (12.20), we can choose a positive number $\delta$ such that $f(\mathcal{N}_\delta(p)) \subseteq \mathcal{N}_\epsilon(f(p))$. Moreover, since the sequence $\{p_k\}$ converges to $p$, we can select a positive integer $N$ such that $p_k$ is in $\mathcal{N}_\delta(p)$ if $k \geq N$; hence $f(p_k)$ is in $\mathcal{N}_\epsilon(f(p))$ if $k \geq N$. Thus the sequence $\{f(p_k)\}$ converges to $f(p)$. By definition, this means that $f \colon X \to Y$ is continuous at the point $p$.   ∎

Finally, we will use the preceding characterization of continuity of a mapping at a point to provide the following criterion for a mapping to be continuous on all of its domain.

**THEOREM 12.20**   Let $X$ and $Y$ be metric spaces and consider the mapping $f \colon X \to Y$. Then the following assertions are equivalent:

(i)  *The mapping $f \colon X \to Y$ is continuous.*

(ii)  *$f^{-1}(V)$ is open in $X$ whenever $V$ is open in $Y$.*

**Proof**   First, suppose that (i) holds; that is, the mapping $f: X \to Y$ is continuous. Let $V$ be an open subset of $Y$. We wish to show that $f^{-1}(V)$ is open in $X$. Let $p$ be a point in $f^{-1}(V)$; we must show that $p$ is an interior point of $f^{-1}(V)$. But $f(p)$ is a point in $V$, which is open in $Y$, so there is some positive number $r$ with $\mathcal{N}_r(f(p)) \subseteq V$. Since $f: X \to Y$ is continuous at the point $p$, it follows from Theorem 12.19 that we can select a positive number $\delta$ with $f(\mathcal{N}_\delta(p)) \subseteq \mathcal{N}_r(f(p)) \subseteq V$. Thus $\mathcal{N}_\delta(p) \subseteq f^{-1}(V)$. So $p$ is an interior point of $f^{-1}(V)$. Since $p$ was arbitrarily chosen, every point in $f^{-1}(V)$ is an interior point. By definition, this means that $f^{-1}(V)$ is open.

To prove the converse, suppose that (ii) holds. Let $p$ be a point in $X$. To show that $f: X \to Y$ is continuous at $p$, we use the "$\epsilon$–$\delta$" characterization of continuity asserted in Theorem 12.19. Let $\epsilon > 0$. According to Proposition 12.5, $\mathcal{N}_\epsilon(f(p))$ is open in $Y$. From (ii) it follows that $f^{-1}(\mathcal{N}_\epsilon(f(p)))$ is open in $X$. Thus the point $p$ in $f^{-1}(\mathcal{N}_\epsilon(f(p)))$ is an interior point of $f^{-1}(\mathcal{N}_\epsilon(f(p)))$, so we can choose a positive number $\delta$ with $\mathcal{N}_\delta(p) \subseteq f^{-1}(\mathcal{N}_\epsilon(f(p)))$. This means that $f(\mathcal{N}_\delta(p)) \subseteq \mathcal{N}_\epsilon(f(p))$. Thus the function $f: X \to Y$ satisfies the "$\epsilon$–$\delta$" characterization of continuity at the point $p$.   ∎

---

**EXERCISES**

1. Let $X$ be a metric space. Prove the following version of the Reverse Triangle Inequality:

$$|d(p, x) - d(x, q)| \le d(p, q) \quad \text{for all points } p, q, \text{ and } x \text{ in } X.$$

2. Let $X$ be a metric space and $p_*$ be a point in $X$. Define the mapping $f: X \to \mathbb{R}$ by

$$f(p) = d(p, p_*) \quad \text{for all } p \text{ in } X.$$

   Use Exercise 1 to prove that $f: X \to \mathbb{R}$ is continuous.

3. Let $X$ be a metric space and let the mapping $f: X \to \mathbb{R}$ be continuous. Let $p$ be a point in $X$ at which $f(p) > 0$. Use Theorem 12.19 to show that there is some positive number $r$ such that $f(q) > 0$ for all $q$ in $\mathcal{N}_r(p)$.

4. Given a mapping $f: X \to Y$ and a subset $Z$ of $X$, the *restriction* of $f: X \to Y$ to $Z$ is the function $g: Z \to Y$ defined by $g(x) = f(x)$ for $x$ in $Z$. Though $f: X \to Y$ and $g: Z \to Y$ are distinct functions when $X \ne Z$, it is customary to denote $g: Z \to Y$ by $f: Z \to Y$. Give an example of a function $f: X \to Y$, a subspace $Z$ of $X$, and a point $p$ in $Z$ such that $f: X \to Y$ is not continuous at $p$ while $f: Z \to Y$ is continuous at $p$.

5. Let $X$ and $Y$ be metric spaces and let $p$ be a point in $X$. Show that a mapping $f: X \to Y$ is continuous at $p$ if and only if there is a neighborhood $\mathcal{N}_r(p)$ of $p$ in $X$ such that $f: \mathcal{N}_r(p) \to Y$ is continuous at $p$. Compare this result with that of Exercise 4.

6. Suppose that $X$ is a metric space and that the mapping $f: \mathbb{R} \to X$ is continuous. Let $C$ be a closed subset of $X$ and let $f(x)$ belong to $C$ if $x$ is rational. Prove that $f(\mathbb{R}) \subseteq C$.

7. Let $X$ and $Y$ be metric spaces and suppose that the mapping $f: X \to Y$ is continuous. Is it true that if $\mathcal{O}$ is open in $X$, then $f(\mathcal{O})$ is open in $Y$?

8. Let $X$ and $Y$ be metric spaces. Prove that $f: X \to Y$ is continuous if and only if $f^{-1}(C)$ is closed in $X$ whenever $C$ is closed in $Y$.

9. Let $X$ be a metric space. A subset $D$ of $X$ is said to be *dense* in $X$ if every point in $X$ is the limit of a sequence in $D$. Formulate the Weierstrass Approximation Theorem as an assertion of denseness in the metric space $C([a, b], \mathbb{R})$.

**10.** Let $X = C([a, b], \mathbb{R})$, and define the function $\psi: X \to \mathbb{R}$ by

$$\psi(f) = \int_a^b f(x)\, dx \quad \text{for each } f \text{ in } X.$$

(a)   Prove that

$$|\psi(f) - \psi(g)| \le (b-a)d(f, g) \quad \text{for all } f \text{ and } g \text{ in } X.$$

(b)   Use the above inequality to verify that $\psi: X \to \mathbb{R}$ is continuous.

---

## 12.5   Compactness and Connectedness

We have already defined what it means for a subset of Euclidean space $\mathbb{R}^n$ to be compact and we have established the Extreme Value Theorem for continuous real-valued functions on compact subsets of $\mathbb{R}^n$. This was the content of Section 11.2. In Section 11.3, we defined what it means for a subset of Euclidean space to be connected, we described certain criteria for verifying connectedness, and we extended the Intermediate Value Theorem to continuous real-valued functions on connected subsets of Euclidean $n$-space. It turns out that there are straightforward extensions of the concepts of compactness and connectedness for general metric spaces, and we will consider these in the present section.

**DEFINITION**   *A metric space $X$ is said to be compact provided that every sequence in $X$ has a subsequence that converges to a point in $X$.*

This extends the definition of compactness that we gave in Section 11.2 when $X$ is a subspace of $\mathbb{R}^n$. The Extreme Value Theorem generalizes to continuous real-valued functions on compact metric spaces and, as we will now see, the proof of the general result is almost the same as the proof for compact subspaces of Euclidean space.

**PROPOSITION 12.21**   Let $X$ and $Y$ be metric spaces. Suppose that the mapping $f: X \to Y$ is continuous. If $X$ is compact, then $f(X)$ is also compact.

**Proof**   Let $\{p_k\}$ be a sequence in $f(X)$. For each natural number $k$, let $q_k$ be a point in $X$ such that $p_k = f(q_k)$. Since $X$ is compact, there is a subsequence $\{q_{k_j}\}$ in $X$ that converges to some point $q$. The mapping $f: X \to Y$ is continuous at the point $q$; thus the image sequence $\{p_{k_j}\} = \{f(q_{k_j})\}$ converges to the point $f(q)$ in $f(X)$. Hence every sequence in $f(X)$ has a subsequence that converges to a point in $f(X)$. By definition, this means that $f(X)$ is compact.   ∎

**THEOREM 12.22**   **The Extreme Value Theorem**   Let $X$ be a nonempty compact metric space and let the function $f: X \to \mathbb{R}$ be continuous. Then $f: X \to \mathbb{R}$ attains both a minimum and a maximum value on $X$.

**Proof**  From the preceding proposition it follows that $f(X)$ is compact. However, Lemma 11.13 asserts that a nonempty compact subset of $\mathbb{R}$ has both a largest and a smallest member. ∎

For $X$ a subspace of $\mathbb{R}^n$, we proved in Section 11.2 that $X$ is compact if and only if $X$ is a closed, bounded subset of $\mathbb{R}^n$. For general metric spaces there is no such simple characterization of compactness. In particular, as the following example shows, it is not true that every closed, bounded subset of $C([a, b], \mathbb{R})$ is compact.

**EXAMPLE 12.16**  Let $K$ be the subset of $C([0, 1], \mathbb{R})$ consisting of all continuous functions $f: [0, 1] \to \mathbb{R}$ such that

$$|f(x)| \leq 1 \quad \text{for all } x \text{ in } [0, 1].$$

We leave it as an exercise for the reader to verify that $K$ is a closed and bounded subset of $C([0, 1], \mathbb{R})$. However, $K$ is not a compact metric space. To establish this assertion, it is necessary to find a sequence in $K$ that has the property that no subsequence converges uniformly to a function in $K$. For each positive integer $k$ define the function $f_k: [0, 1] \to \mathbb{R}$ by

$$f_k(x) = x^k \quad \text{for } x \text{ in } [0, 1].$$

Then define the function $f: [0, 1] \to \mathbb{R}$ by

$$f(x) = \begin{cases} 0 & \text{if } 0 \leq x < 1 \\ 1 & \text{if } x = 1. \end{cases}$$

Since $\lim_{k \to \infty} x^k = 0$ if $0 \leq x < 1$, it follows that the sequence $\{f_k: [0, 1] \to \mathbb{R}\}$ converges *pointwise* to the function $f: [0, 1] \to \mathbb{R}$. Hence every subsequence of $\{f_k: [0, 1] \to \mathbb{R}\}$ also converges *pointwise* to the function $f: [0, 1] \to \mathbb{R}$. But the function $f: [0, 1] \to \mathbb{R}$ is not continuous, and hence is not in $K$. Thus there is no subsequence of $\{f_k: [0, 1] \to \mathbb{R}\}$ that converges in the metric space $C([0, 1], \mathbb{R})$ (that is, converges uniformly) to a function in $K$. □

There is a theorem, called the Arzela-Ascoli Theorem, that characterizes the compact subspaces of $C([0, 1], \mathbb{R})$. Unfortunately, this theorem lies outside the scope of this book.*

**DEFINITION**  *A metric space $X$ is said to have the Intermediate Value Property provided that every continuous function $f: X \to \mathbb{R}$ has an interval as its image.*

In Section 11.3, we proved that a subset of $\mathbb{R}^n$ has the Intermediate Value Property if and only if it is connected. We wish to characterize those general metric spaces that have the Intermediate Value Property. This amounts to discovering the appropriate concept of connectedness for a general metric space.

---

*The excellent book *Introduction to Topology and Modern Analysis* by Simmons (McGraw-Hill, 1963) has a very clear discussion of this and of related matters.

**DEFINITION**   *Let $X$ be a metric space and $\mathcal{U}$ and $\mathcal{V}$ be open subsets of $X$. Then $\mathcal{U}$ and $\mathcal{V}$ are said to separate $X$ provided that*

$$\mathcal{U} \text{ and } \mathcal{V} \text{ are nonempty,}$$

$$\mathcal{U} \cup \mathcal{V} = X,$$

*and*

$$\mathcal{U} \cap \mathcal{V} = \emptyset.$$

**DEFINITION**   *A metric space $X$ is said to be connected provided that there do not exist two open subsets of $X$ that separate $X$.*

The following theorem justifies the introduction of the concept of connectedness.

**THEOREM**   A metric space $X$ is connected if and only if it has the Intermediate Value Property.
**12.23**

**Proof**   First, suppose that the metric space $X$ is connected. We will show that $X$ has the Intermediate Value Property. Indeed, let $f: X \to \mathbb{R}$ be continuous. To show that the image $f(X)$ is an interval, we will suppose otherwise and derive a contradiction. Indeed, if $f(X)$ is not an interval, then there is a real number $c$ and points $u$ and $v$ in $X$ such that

$$f(u) < c < f(v)$$

but $c$ is not in $f(X)$. Define

$$X_1 = f^{-1}(-\infty, c) \quad \text{and} \quad X_2 = f^{-1}(c, \infty).$$

Since both $(-\infty, c)$ and $(c, \infty)$ are open subsets of $\mathbb{R}$ and the function $f: X \to \mathbb{R}$ is continuous, it follows from Theorem 12.20 that both $X_1$ and $X_2$ are open subsets of $X$. Also, observe that neither $X_1$ nor $X_2$ is empty, since $u$ belongs to $X_1$ and $v$ belongs to $X_2$. Moreover, $X = X_1 \cup X_2$, since $c$ is not in $f(X)$. Finally, it is clear that $X_1$ and $X_2$ are disjoint. Thus $X_1$ and $X_2$ separate the metric space $X$, which contradicts the assumption that $X$ is connected. This contradiction shows that $X$ has the Intermediate Value Property.

To prove the converse, suppose that every continuous function $f: X \to \mathbb{R}$ has the Intermediate Value Property. We will show that $X$ is connected by assuming otherwise and deriving a contradiction. Suppose that $X$ is not connected. Then there is a pair of open subsets $\mathcal{U}$ and $\mathcal{V}$ of $X$ that separate $X$. Define the function $f: X \to \mathbb{R}$ by

$$f(p) = \begin{cases} 0 & \text{if } p \in \mathcal{U} \\ 1 & \text{if } p \in \mathcal{V} \end{cases}$$

Then the function $f: X \to \mathbb{R}$ certainly fails to have the Intermediate Value Property, since it attains exactly two functional values, namely 0 and 1. We also claim that the function $f: X \to \mathbb{R}$ is continuous. To see this it will, according to Theorem 12.20, be sufficient to check that the inverse images under $f: X \to \mathbb{R}$ of open subsets of $\mathbb{R}$ are open in $X$. But from the definition of the function $f: X \to \mathbb{R}$, it is clear that the inverse image of any subset of $\mathbb{R}$ is either $\emptyset$, $X$, $\mathcal{U}$, or $\mathcal{V}$, and hence it certainly is open. The existence of this continuous function whose image is not an interval contradicts the assumption that $X$ has the Intermediate Value Property. Thus $X$ must be connected. ∎

The definition that we have given of a connected metric space is formally different from the definition of a connected subset of $\mathbb{R}^n$ that was given in Section 11.3. However, the definitions are equivalent. This follows from the fact that each of them is equivalent to the Intermediate Value Property. However, it also follows by a direct argument. To understand this direct argument, we will finish this chapter with a brief discussion of the relative concepts of *open* and *closed*.

Suppose that $Y$ is a metric space and that $X$ is a subspace of $Y$. Consider a subset $A$ of $X$. Then, of course, $A$ is also a subset of $Y$. Now, considering $A$ as a subset of $X$, we can ask whether $A$ is open in the metric space $X$. On the other hand, considering $A$ as a subset of $Y$, we can ask whether $A$ is open in the metric space $Y$. In general, the answers to these two questions are different. As the following example shows, the concept of openness is *relative*; it depends on the designation of an ambient metric space.

**EXAMPLE 12.17**    Let $Y = \mathbb{R}$ and $X = [0, 1]$. Then $A = (1/2, 1]$ is not an open subset of the metric space $Y$, since every neighborhood of 1 in $Y$ contains points not in $(1/2, 1]$. On the other hand, $(1/2, 1]$ is open in the metric space $X$, since $\{x \text{ in } X \mid d(x, 1) < r\} = (r, 1] \subseteq (1/2, 1]$ if $0 < r < 1/2$.    □

The following theorem provides a description, in the case when the metric space $X$ is a subspace of the metric space $Y$, of the open subsets of $X$ in terms of the open subsets of $Y$.

**THEOREM 12.24**    Let $X$ be a subspace of the metric space $Y$ and let $A$ be a subset of $X$. Then $A$ is open in $X$ if and only if $A = X \cap \mathcal{O}$, where $\mathcal{O}$ is open in $Y$.

**Proof**    First, suppose that $A$ is open in $X$. Let $p$ be a point in $A$. Then there is a neighborhood of $p$ in $X$ that is contained in $X$; that is there is a positive number $r = r(p)$ such that

$$\{q \text{ in } X \mid d(p, q) < r\} \subseteq A. \tag{12.21}$$

Now the neighborhood of the point $p$, in $Y$, of radius $r$ is an open subset of $Y$, and hence the union of such neighborhoods, as $p$ varies in $A$, is also an open subset of $Y$, that is,

$$\mathcal{O} = \bigcup_{p \in A} \{q \text{ in } Y \mid d(q, p) < r\}$$

is an open subset of $Y$. From the inclusion (12.21) it is clear that $A = X \cap \mathcal{O}$.

To prove the converse, suppose that $A = X \cap \mathcal{O}$, where $\mathcal{O}$ is an open subset of $Y$. Let $p$ be a point in $A$. Then $p$ also belongs to $\mathcal{O}$, and is therefore an interior point, relative to $Y$, of $\mathcal{O}$. Thus we may select a positive number $r$ with $\{q \text{ in } Y \mid d(q, p) < r\} \subseteq \mathcal{O}$. Thus $\{q \text{ in } X \mid d(q, p) < r\} \subseteq X \cap \mathcal{O} = A$. Hence $p$ is in the interior, relative to $X$, of $A$. Thus every point in $A$ is an interior point, relative to $X$, of $A$. By definition, this means that $A$ is open in $X$.    ∎

It follows immediately from the above theorem that the concept of connectedness introduced for subsets of Euclidean $n$-space in Section 11.3 is consistent with the general definition of connectedness given in the present section.

As the following example illustrates, the concept of closedness is also a *relative* concept.

**EXAMPLE 12.18** Let $Y = \mathbb{R}$ and $X = (0, 2]$. Consider the set $A = (0, 1]$. Then $A$ is not a closed subset of the metric space $Y$, since the sequence $\{1/2k\}$ is a sequence in $A$ that converges to a point in $Y$ that does not lie in $A$. However, $A$ is a closed subset of the metric space $X$, since it is clear that if a sequence in $A$ converges to a point in $X$, then that point must be in $A$. □

Using the Complementing Criterion and Theorem 12.24, we can show that if the metric space $X$ is a subspace of the metric space $Y$, then a subset $A$ of $X$ is closed in $X$ if and only if there is a closed subset $C$ of $Y$ such that $A = X \cap C$ (see Exercise 8).

---

**EXERCISES**

1. Let $X$ be a compact metric space. Prove that the subspace $K$ of $X$ is compact if and only if $K$ is a closed subset of $X$.

2. Let $X$ be a metric space and let $K \subseteq X$ be a compact subspace. Let $p$ be a point in $X \backslash K$. Prove that there is a point $q_0$ in $K$ that is closest to $p$ of all points in $K$, in the sense that

$$d(p, q_0) \le d(p, q) \quad \text{for all points } q \text{ in } K.$$

   (*Hint:* Define the function $f: K \to \mathbb{R}$ by $f(q) = d(q, p)$ for all $q$ in $K$. Show that $f: K \to \mathbb{R}$ is continuous.) Is this point unique?

3. Show that if $X$ is a metric space and $\mathcal{U}$ and $\mathcal{V}$ are open subsets of $X$ that separate $X$, then both $\mathcal{U}$ and $\mathcal{V}$ are also closed in $X$.

4. Prove that a metric space $X$ is disconnected if and only if there is a subset $D$ of $X$ that is both open and closed in $X$, with $D \ne \phi$ and $D \ne X$.

5. Let $X$ be a compact metric space. Prove that $X$ is disconnected if and only if there are nonempty subsets $A$ and $B$ of $X$ and a positive number $\epsilon$, with $A \cap B = \phi$, $A \cup B = X$, and

$$d(p, q) > \epsilon \quad \text{for all } p \text{ in } A \text{ and } q \text{ in } B.$$

   Is compactness necessary?

6. Let $X$ be a metric space. For each positive integer $k$, let $F_k$ be a nonempty compact subspace of $X$, and suppose that $F_{k+1} \subseteq F_k$. Prove that the intersection $\bigcap_{k=1}^{\infty} F_k$ is nonempty. (*Hint:* For each positive integer $k$, choose a point $p_k$ in $F_k$. A subsequence of the sequence $\{p_k\}$ converges to a point $p$ in $X$. Where does $p$ lie?)

7. Use Theorem 12.24 to prove directly that for a subspace $X$ of $\mathbb{R}^n$ the general definition of connectedness given in the present section coincides with that already given in Section 11.3.

8. Using the Complementing Criterion and Theorem 12.24, show that if the metric space $X$ is a subspace of the metric space $Y$, then a subset $A$ of $X$ is closed in $X$ if and only if there is a closed subset $C$ of $Y$ such that $A = X \cap C$.

9. A metric space $X$ is defined to be *pathwise connected* provided that for any two points $p$ and $q$ in $X$, there is a continuous function $\gamma: [0, 1] \to X$ such that $\gamma(0) = p$ and $\gamma(1) = q$.
   (a) Prove that a pathwise connected metric space has the Intermediate Value Property. (*Hint:* Follow the corresponding proof in Section 11.3.)
   (b) Prove that a pathwise connected metric space is connected.

**10.** We have defined what it means for a subset $X$ of $\mathbb{R}^n$ to be convex. By checking the corresponding discussion in Section 11.3, do the following exercises:
   (a)  Define what it means for a subset $X$ of $C([a, b], \mathbb{R})$ to be convex.
   (b)  Prove that convex subsets of $C([a, b], \mathbb{R})$ are pathwise connected.
   (c)  Prove that symmetric neighborhoods in $C([a, b], \mathbb{R})$ are convex.
   (d)  Prove that symmetric neighborhoods in $C([a, b], \mathbb{R})$ are connected.

**11.** Suppose that $X$ is a set consisting of more than one point, considered as a metric space with the discrete metric. Show that $X$ is not connected.

**12.** Let $X$ be a compact metric space. Define $C(X, \mathbb{R})$ to be the set of all continuous functions $f: X \to \mathbb{R}$, and for two functions $f$ and $g$ in $C(X, \mathbb{R})$, define

$$d(f, g) = \max \{|f(p) - g(p)| \,|\, p \text{ in } X\}.$$

Prove that $d$ defines a metric on $C(X, \mathbb{R})$.

# 13

# Partial Differentiability of Real-Valued Functions of Several Variables

## 13.1    Limits

For $I$ an open interval of real numbers, a function $f: I \to \mathbb{R}$ has been defined to be *differentiable at the point $x_0$ in $I$* provided that the limit

$$\lim_{x \to x_0} \frac{f(x) - f(x_0)}{x - x_0} \tag{13.1}$$

exists. This limit, which is denoted by $f'(x_0)$, is called the *derivative* of the function $f: I \to \mathbb{R}$ at the point $x_0$. Moreover, if the function $f: I \to \mathbb{R}$ is differentiable at every point in $I$, then the function $f: I \to \mathbb{R}$ is said to be *differentiable* and the function $f': I \to \mathbb{R}$ is called the *derivative* of the function $f: I \to \mathbb{R}$. The study of the relationship between a function and its derivative is, of course, one of the important topics in the analysis of real-valued functions of a single real variable.

In the present chapter we will turn to the study of differentiation for real-valued functions of several real variables, and we will extend to the case of functions of several variables many of the results that we obtained earlier for functions of a single variable. We begin this study, in this section, by extending the definitions of limit point of a set and of limit of a function to the case of several variables.

Given a subset $A$ of $\mathbb{R}^n$ and a point $\mathbf{x}_0$ in $\mathbb{R}^n$, recall that $A\backslash\{\mathbf{x}_0\}$ denotes the set $\{\mathbf{x} \text{ in } A \mid \mathbf{x} \neq \mathbf{x}_0\}$. In particular, $A\backslash\{\mathbf{x}_0\} = A$ if the point $\mathbf{x}_0$ does not belong to $A$.

**DEFINITION**    *Let $A$ be a subset of $\mathbb{R}^n$ and let $\mathbf{x}_0$ be a point in $\mathbb{R}^n$. Then $\mathbf{x}_0$ is said to be a limit point of the set $A$ provided that there is a sequence in $A\backslash\{\mathbf{x}_0\}$ that converges to $\mathbf{x}_0$.*

**DEFINITION**    *Let $A$ be a subset of $\mathbb{R}^n$ and let $\mathbf{x}_0$ be a limit point of $A$. Given a function $f: A \to \mathbb{R}$ and a real number $\ell$, we write*

$$\lim_{\mathbf{x}\to\mathbf{x}_0} f(x) = \ell \tag{13.2}$$

*provided that whenever $\{\mathbf{x}_k\}$ is a sequence in $A\backslash\{\mathbf{x}_0\}$ that converges to $\mathbf{x}_0$, then the image sequence $\{f(\mathbf{x}_k)\}$ converges to $\ell$.*

We read (13.2) as: "The limit of $f(\mathbf{x})$ as $\mathbf{x}$ approaches $\mathbf{x}_0$ equals $\ell$."

As in the case of real-valued functions of a single real variable, it is easy to see that if $A$ is a subset of $\mathbb{R}^n$ and $\mathbf{x}_0$ is a limit point of $A$ that also belongs to $A$, then the function $f: A \to \mathbb{R}$ is continuous at $\mathbf{x}_0$ if and only if

$$\lim_{\mathbf{x}\to\mathbf{x}_0} f(\mathbf{x}) = f(\mathbf{x}_0). \tag{13.3}$$

However, in Section 11.1 we studied continuous functions of several variables, and so, in view of the equivalence of continuity and (13.3), we already have at our disposal a number of ways of analyzing limits.

**EXAMPLE 13.1**    Consider the point $(1, 2)$ in the plane $\mathbb{R}^2$. Then

$$\lim_{(x,y)\to(1,2)} [x^2 y + e^{xy+1}] = 2 + e^3.$$

This follows from the fact that the function $f: \mathbb{R}^2 \to \mathbb{R}$ defined by

$$f(x, y) = x^2 y + e^{xy+1} \quad \text{for } (x, y) \text{ in } \mathbb{R}^2$$

has already been shown to be continuous.      □

**EXAMPLE 13.2** Let $\mathbf{x}_0$ be a point in $\mathbb{R}^n$. Then

$$\lim_{\mathbf{x} \to \mathbf{x}_0} \|\mathbf{x}\| = \|\mathbf{x}_0\|.$$

This follows from the fact that the function $f: \mathbb{R}^n \to \mathbb{R}$ defined by

$$f(\mathbf{x}) = \|\mathbf{x}\| \quad \text{for } \mathbf{x} \text{ in } \mathbb{R}^n$$

has already been shown to be continuous. $\qquad\qquad\qquad\qquad\qquad\qquad\qquad$ □

The following theorem follows directly from the definition of limit by using the sum, product, and quotient properties of convergent sequences of real numbers.

**THEOREM 13.1** Let $A$ be a subset of $\mathbb{R}^n$ and let $\mathbf{x}_0$ be a limit point of $A$. Suppose that the functions $f: A \to \mathbb{R}$ and $g: A \to \mathbb{R}$ and the real numbers $\ell_1$ and $\ell_2$ have the property that

$$\lim_{\mathbf{x} \to \mathbf{x}_0} f(\mathbf{x}) = \ell_1 \quad \text{and} \quad \lim_{\mathbf{x} \to \mathbf{x}_0} g(\mathbf{x}) = \ell_2.$$

Then

$$\lim_{\mathbf{x} \to \mathbf{x}_0} [f(\mathbf{x}) + g(\mathbf{x})] = \ell_1 + \ell_2,$$

$$\lim_{\mathbf{x} \to \mathbf{x}_0} [f(\mathbf{x})g(\mathbf{x})] = \ell_1\ell_2,$$

and, if $g(\mathbf{x}) \neq 0$ for all $\mathbf{x}$ in $A$ and $\ell_2 \neq 0$, then

$$\lim_{\mathbf{x} \to \mathbf{x}_0} [f(\mathbf{x})/g(\mathbf{x})] = \ell_1/\ell_2.$$

**Proof** Let $\{\mathbf{x}_k\}$ be a sequence in $A \setminus \{\mathbf{x}_0\}$ that converges to $\mathbf{x}_0$. Since

$$\lim_{\mathbf{x} \to \mathbf{x}_0} f(\mathbf{x}) = \ell_1 \quad \text{and} \quad \lim_{\mathbf{x} \to \mathbf{x}_0} g(\mathbf{x}) = \ell_2,$$

it follows that $\qquad \lim_{k \to \infty} f(\mathbf{x}_k) = \ell_1 \quad \text{and} \quad \lim_{k \to \infty} g(\mathbf{x}_k) = \ell_2.$

Using the sum, product, and quotient properties of convergent sequences of real numbers, we conclude that

$$\lim_{k \to \infty} [f(\mathbf{x}_k) + g(\mathbf{x}_k)] = \ell_1 + \ell_2,$$

$$\lim_{k \to \infty} [f(\mathbf{x}_k)g(\mathbf{x}_k)] = \ell_1\ell_2,$$

and, if $g(\mathbf{x}_k) \neq 0$ for each positive integer $k$ and $\ell_2 \neq 0$, then

$$\lim_{k \to \infty} [f(\mathbf{x}_k)/g(\mathbf{x}_k)] = \ell_1/\ell_2.$$

The calculation of these three sequential limits is precisely what is required in order to prove the theorem. $\qquad\qquad\qquad\qquad\qquad\qquad\qquad\qquad\qquad\qquad\qquad$ ■

One of the interesting but subtle features of many problems in mathematical analysis is the necessity of studying limits of quotients that are of the form

$$\lim_{\mathbf{x} \to \mathbf{x}_0} [f(\mathbf{x})/g(\mathbf{x})]$$

in the case when both

$$\lim_{\mathbf{x}\to\mathbf{x}_0} f(\mathbf{x}) = 0 \quad \text{and} \quad \lim_{\mathbf{x}\to\mathbf{x}_0} g(\mathbf{x}) = 0.$$

Such limits occur frequently. For instance, the limit (13.1), which is the very definition of the derivative for a function of a single real variable, is of this form provided that $f: I \to \mathbb{R}$ is continuous at the point $x_0$. Limits of this form also occur prominently in the study of functions of several real variables. To properly understand how to treat such limits, it is necessary to study differentiation of functions of several real variables. We will study this in the succeeding sections of this chapter. For now, let us consider some examples of this type of limit.

**EXAMPLE 13.3**    Let $(x_0, y_0)$ be a point in the plane $\mathbb{R}^2$. Consider the limit

$$\lim_{(x,y)\to(x_0,y_0)} \frac{xy}{x^2 + y^2}. \tag{13.4}$$

Define $f(x, y) = xy/(x^2 + y^2)$ if $(x, y) \neq (0, 0)$. Since polynomials are continuous,

$$\lim_{(x,y)\to(x_0,y_0)} xy = x_0 y_0 \quad \text{and} \quad \lim_{(x,y)\to(x_0,y_0)} x^2 + y^2 = x_0^2 + y_0^2,$$

so if $(x_0, y_0) \neq (0, 0)$, it follows that

$$\lim_{(x,y)\to(x_0,y_0)} \frac{xy}{x^2 + y^2} = \frac{x_0 y_0}{x_0^2 + y_0^2}.$$

But the limit (13.4) does not exist if $(x_0, y_0) = (0, 0)$. To see this, first observe that the sequence $\{(1/k, 1/k)\}$ converges to the point $(0, 0)$, and since $f(1/k, 1/k) = 1/2$ for each natural number $k$, it follows that the image sequence $\{f(1/k, 1/k)\}$ converges to $1/2$. On the other hand, the sequence $\{(1/k, 0)\}$ also converges to the point $(0, 0)$, and since $f(1/k, 0) = 0$ for each natural number $k$, it follows that the image sequence $\{f(1/k, 0)\}$ converges to 0. Thus the limit (13.4) does not exist if $(x_0, y_0) = (0, 0)$. $\square$

**EXAMPLE 13.4**    Consider the following limit:

$$\lim_{(x,y)\to(0,0)} \frac{x^3}{x^2 + y^2}. \tag{13.5}$$

Define $f(x, y) = x^3/(x^2 + y^2)$ if $(x, y) \neq (0, 0)$. By observing that the sequence $\{(0, 1/k)\}$ converges to $(0, 0)$ and that $f(0, 1/k) = 0$ for each natural number $k$, we see that the only possible value of the limit is 0. To verify that the limit is indeed 0, it is necessary to make some estimates of the size of $f(x, y)$. Indeed, if $x \neq 0$, then

$$\left| \frac{x^3}{x^2 + y^2} \right| \leq \left| \frac{x^3}{x^2} \right| = |x|,$$

and therefore

$$\left| \frac{x^3}{x^2 + y^2} \right| \leq |x| \quad \text{if} \quad (x, y) \neq (0, 0),$$

since this estimate also clearly holds if $x = 0$ and $y \neq 0$. Now suppose that the sequence $\{(x_k, y_k)\}$ converges to $(0, 0)$, with each $(x_k, y_k) \neq (0, 0)$. Then the sequence $\{x_k\}$

converges to 0, so from the preceding estimate and the Squeezing Principle it follows that the image sequence $\{f(x_k, \ y_k)\}$ converges to 0. Thus

$$\lim_{(x,y)\to(0,0)} \frac{x^3}{x^2 + y^2} = 0.$$    □

**EXAMPLE 13.5**   We claim that

$$\lim_{(x,y)\to(0,0)} \frac{\sin(x^2 + y^2)}{x^2 + y^2} = 1. \tag{13.6}$$

To verify this, let the sequence $\{(x_k, \ y_k)\}$ converge to $(0, \ 0)$ with each $(x_k, \ y_k) \neq (0, \ 0)$. Define $t_k = x_k^2 + y_k^2$ for each $k$ and observe that $\{t_k\}$ is a sequence of nonzero real numbers that converges to 0. Since $\sin 0 = 0$, the derivative of the sine is the cosine, and $\cos 0 = 1$, it follows that

$$\lim_{k\to\infty} \frac{\sin(x_k^2 + y_k^2)}{x_k^2 + y_k^2} = \lim_{k\to\infty} \frac{\sin(t_k) - \sin 0}{t_k - 0} = 1.$$

This proves (13.6).    □

For functions of a single real variable, the sequential definition of limit had an equivalent "$\epsilon$–$\delta$" formulation. The following is a similar formulation for limits of functions of several real variables.

**THEOREM 13.2**   Let $A$ be a subset of $\mathbb{R}^n$ and $\mathbf{x}_0$ be a limit point of $A$. For a function $f \colon A \to \mathbb{R}$ and a real number $\ell$, the following two assertions are equivalent:

(i)
$$\lim_{\mathbf{x}\to\mathbf{x}_0} f(\mathbf{x}) = \ell.$$

(ii)   For each positive number $\epsilon$, there is a positive number $\delta$ such that

$$|f(\mathbf{x}) - \ell| < \epsilon \text{ if } \mathbf{x} \text{ is in } A \text{ and } 0 < d(\mathbf{x}, \mathbf{x}_0) < \delta.$$

**Proof**   First, suppose that (i) holds. To verify (ii), we will suppose the contrary and derive a contradiction. Indeed, if (ii) does not hold, then there is some $\epsilon_0 > 0$ such that for no positive number $\delta$ is it true that

$$|f(\mathbf{x}) - \ell| < \epsilon_0 \text{ if } \mathbf{x} \text{ is in } A \text{ and } 0 < d(\mathbf{x}, \mathbf{x}_0) < \delta. \tag{13.7}$$

In particular, if $k$ is a positive integer, then (13.7) fails to hold when $\delta = 1/k$. This means that there is a point in $A$, which we label $\mathbf{x}_k$, with $0 < d(\mathbf{x}_k, \mathbf{x}_0) < 1/k$ while $|f(\mathbf{x}_k) - \ell| \geq \epsilon_0$. This defines a sequence $\{\mathbf{x}_k\}$ in $A \backslash \{\mathbf{x}_0\}$ that converges to the point $\mathbf{x}_0$, whose image sequence $\{f(\mathbf{x}_k)\}$ does not converge to $\ell$. This contradicts (i).

To prove the converse, suppose that (ii) holds. To verify (i), we choose $\{\mathbf{x}_k\}$ to be a sequence in $A \backslash \{\mathbf{x}_0\}$ that converges to the point $\mathbf{x}_0$. We must show that the image sequence $\{f(\mathbf{x}_k)\}$ converges to $\ell$. Let $\epsilon > 0$. According to (ii), we can choose a positive number $\delta$ such that

$$|f(\mathbf{x}) - \ell| < \epsilon \quad \text{if } \mathbf{x} \text{ is in } A \text{ and } 0 < d(\mathbf{x}, \mathbf{x}_0) < \delta. \tag{13.8}$$

Moreover, since the sequence $\{\mathbf{x}_k\}$ is a sequence in $A \setminus \{\mathbf{x}_0\}$ that converges to $\mathbf{x}_0$, we can select a positive integer $K$ such that $0 < d(\mathbf{x}_k, \mathbf{x}_0) < \delta$ if $k \geq K$. Thus, using (13.8), we conclude that

$$|f(\mathbf{x}_k) - \ell| < \epsilon \quad \text{if} \quad k \geq K.$$

Hence the image sequence $\{f(\mathbf{x}_k)\}$ converges to $\ell$. ∎

---

**EXERCISES**

1. Prove that
$$\lim_{(x,y)\to(0,0)} \frac{x^3 y}{x^2 + y^4} = 0.$$

2. Analyze the following limits:

   (a) $\displaystyle\lim_{h\to 0} \frac{e^h - 1}{h}$
   
   (b) $\displaystyle\lim_{t\to 0} \frac{\sin t^2}{t}$

3. Analyze the following limits:

   (a) $\displaystyle\lim_{(x,y)\to(0,0)} \frac{x + y}{x^2 + y^2}$

   (b) $\displaystyle\lim_{(x,y,z)\to(0,0,0)} \frac{x + y + z}{x^2 + y^2 + z^2}$

   (c) $\displaystyle\lim_{(x,y)\to(0,0)} \frac{e^{x^2+y^2} - 1}{x^2 + y^2}$

4. Let $m$ and $n$ be positive integers. Show that the limit
$$\lim_{(x,y)\to(0,0)} \frac{x^n y^m}{x^2 + y^2}$$
exists if and only if $m + n > 2$.

5. Give an example of a subset $A$ of $\mathbb{R}$ and a point $x$ in $A$ that is not a limit point of the set $A$.

6. Let $A$ be a subset of $\mathbb{R}^n$ and let $\mathbf{x}$ be a point in $\mathbb{R}^n$. Show that $\mathbf{x}$ is a limit point of $A$ if and only if every symmetric neighborhood of $\mathbf{x}$ contains a point of $A$ that is not equal to $\mathbf{x}$.

7. Let $A$ be a subset of $\mathbb{R}^n$ and let the point $\mathbf{x}_0$ in $\mathbb{R}^n$ be a limit point of $A$. Suppose that the function $g : A \to \mathbb{R}$ is bounded; that is, there is a number $c$ such that
$$|g(\mathbf{x})| \leq c \quad \text{for all } \mathbf{x} \text{ in } A.$$
Prove that if $\lim_{\mathbf{x}\to\mathbf{x}_0} f(\mathbf{x}) = 0$, then $\lim_{\mathbf{x}\to\mathbf{x}_0} [g(\mathbf{x}) f(\mathbf{x})] = 0$.

8. Let $A$ be a subset of $\mathbb{R}^n$. The function $f : \mathbb{R}^n \to \mathbb{R}$ defined by
$$f(\mathbf{x}) = \begin{cases} 1 & \text{if } \mathbf{x} \in A \\ 0 & \text{if } \mathbf{x} \notin A \end{cases}$$
is called the *characteristic function* of $A$. Let $\mathbf{x}_0$ be a point in $\mathbb{R}^n$. Show that $\lim_{\mathbf{x}\to\mathbf{x}_0} f(\mathbf{x})$ exists if and only if $\mathbf{x}_0$ is either an interior point or an exterior point of the set $A$.

9. Give an example of a function $f : \mathbb{R}^n \to \mathbb{R}$ such that $\lim_{\mathbf{x}\to 0} f(\mathbf{x}) = 0$ but that $\lim_{\mathbf{x}\to 0} f(\mathbf{x})/\|\mathbf{x}\| \neq 0$.

10. For a function $f : \mathbb{R}^n \to \mathbb{R}$ and a positive integer $m$, show that
$$\lim_{\mathbf{x}\to 0} f(\mathbf{x})/\|\mathbf{x}\|^{m+1} = 0 \quad \text{implies that} \quad \lim_{\mathbf{x}\to 0} f(\mathbf{x})/\|\mathbf{x}\|^m = 0,$$
but that the converse does not hold.

11. Let $A$ be a subset of $\mathbb{R}^n$ and suppose that $\mathbf{0}$ is a limit point of $A$. Suppose that the function $f: A \to \mathbb{R}$ has the property that there is a positive number $c$ such that

$$f(\mathbf{x}) \geq c\|\mathbf{x}\|^2 \quad \text{for all } \mathbf{x} \text{ in } A,$$

and that the function $g: A \to \mathbb{R}$ has the property that

$$\lim_{\mathbf{x} \to 0} g(\mathbf{x})/\|\mathbf{x}\|^2 = 0.$$

Prove that there is a positive number $r$ such that

$$f(\mathbf{x}) - g(\mathbf{x}) \geq (c/2)\|\mathbf{x}\|^2 \quad \text{for all } \mathbf{x} \text{ in } A \text{ with } 0 < \|\mathbf{x}\| < r.$$

12. Let $A$ be a subset of $\mathbb{R}^n$. Show that $A$ is closed if and only if it contains all of its limit points.

13. Let $A$ and $B$ be subsets of $\mathbb{R}^n$ with $A \subseteq B$. Suppose that the point $\mathbf{x}_0$ in $\mathbb{R}^n$ is a limit point of $A$. Given a function $f: B \to \mathbb{R}$, we define its *restriction* to the set $A$ to be the function $\bar{f}: A \to \mathbb{R}$ defined by $\bar{f}(\mathbf{x}) = f(\mathbf{x})$ for all $\mathbf{x}$ in $A$. Find an example of a function for which $\lim_{\mathbf{x} \to \mathbf{x}_0} \bar{f}(\mathbf{x})$ exists but $\lim_{\mathbf{x} \to \mathbf{x}_0} f(\mathbf{x})$ does not exist. (*Note:* Frequently, for notational simplicity, the function $\bar{f}: A \to \mathbb{R}$ is simply denoted by $f: A \to \mathbb{R}$, but then a new notational device needs to be invented to distinguish the two limits.)

## 13.2  Partial Derivatives

Consider a real-valued function of several real variables $f: \mathbb{R}^n \to \mathbb{R}$, together with two points $\mathbf{u}$ and $\mathbf{v}$ in $\mathbb{R}^n$. Suppose that we want to compare $f(\mathbf{u})$ with $f(\mathbf{v})$. When $n = 1$ and the function $f: \mathbb{R} \to \mathbb{R}$ is differentiable, we can use the Lagrange Mean Value Theorem to compare these two values. When $n > 1$, the following restriction procedure is natural. Look at the parametrized segment from $\mathbf{u}$ to $\mathbf{v}$—that is, the parametrized path $\gamma: [0, 1] \to \mathbb{R}$ defined by

$$\gamma(t) = \mathbf{u} + t(\mathbf{v} - \mathbf{u}) = t\mathbf{v} + (1 - t)\mathbf{u} \quad \text{for} \quad 0 \leq t \leq 1.$$

Then consider the composition of the function $f: \mathbb{R}^n \to \mathbb{R}$ with this parametrized path, which is the function $\psi: [0, 1] \to \mathbb{R}$ defined by

$$\psi(t) = f(\mathbf{u} + t(\mathbf{v} - \mathbf{u})) \quad \text{for} \quad 0 \leq t \leq 1. \tag{13.9}$$

Then $\psi(0) = f(\mathbf{u})$ and $\psi(1) = f(\mathbf{v})$. Thus to compare $f(\mathbf{u})$ with $f(\mathbf{v})$ is to compare $\psi(0)$ with $\psi(1)$. If we can determine that $\psi: [0, 1] \to \mathbb{R}$ is continuous and that $\psi: (0, 1) \to \mathbb{R}$ is differentiable, then we can apply the Lagrange Mean Value for functions of a single variable to compare $f(\mathbf{u})$ with $f(\mathbf{v})$. Thus it is necessary to investigate the properties of the function $f: \mathbb{R}^n \to \mathbb{R}$ that will allow us to conclude that the above auxiliary function $\psi: [0, 1] \to \mathbb{R}$ is continuous and that $\psi: (0, 1) \to \mathbb{R}$ is differentiable, and that will allow us to compute $\psi': (0, 1) \to \mathbb{R}$.

We can regard the above function $\psi: [0, 1] \to \mathbb{R}$ as being the restriction of the function $f: \mathbb{R}^n \to \mathbb{R}$ to the line segment between the points $\mathbf{u}$ and $\mathbf{v}$, together with the placing of a coordinate system on this line segment. In the case when $n = 2$, the graph

**FIGURE 13.1**

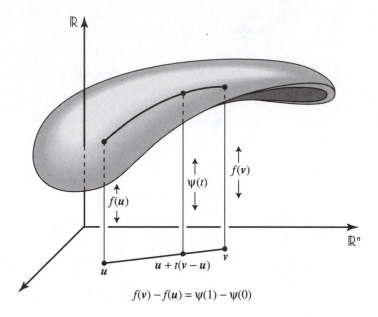

$$f(v) - f(u) = \psi(1) - \psi(0)$$

of $\psi\colon [0, 1] \to \mathbb{R}$ is obtained by intersecting the graph of $f\colon \mathbb{R}^2 \to \mathbb{R}$ with the plane that is parallel to the $z$-axis and contains the segment joining $\mathbf{u}$ and $\mathbf{v}$. For this reason, we will refer to the function $\psi\colon [0, 1] \to \mathbb{R}$ as a *section* of the function $f\colon \mathbb{R}^n \to \mathbb{R}$.

In order to analyze the differentiability of the function $\psi\colon (0, 1) \to \mathbb{R}$ at the point $t_0$, we change variables by setting $\mathbf{x} = \mathbf{u} + t_0(\mathbf{v} - \mathbf{u})$, $\mathbf{p} = \mathbf{v} - \mathbf{u}$, and $s = t - t_0$; then

$$\frac{\psi(t) - \psi(t_0)}{t - t_0} = \frac{f(\mathbf{x} + s\mathbf{p}) - f(\mathbf{x})}{s},$$

and therefore

$$\psi'(t_0) = \lim_{s \to 0} \frac{f(\mathbf{x} + s\mathbf{p}) - f(\mathbf{x})}{s}, \tag{13.10}$$

provided that the limit exists. The strategy of looking at sections of a function, together with formula (13.10), motivates the introduction of the following concept of partial derivative.

**DEFINITION**

*Let $\mathcal{O}$ be an open subset of $\mathbb{R}^n$ that contains the point $\mathbf{x}$ and let $i$ be an index with $1 \leq i \leq n$. A function $f\colon \mathcal{O} \to \mathbb{R}$ is said to have a partial derivative with respect to its $i$th component at the point $\mathbf{x}$ provided that*

$$\lim_{t \to 0} \frac{f(\mathbf{x} + t\mathbf{e}_i) - f(\mathbf{x})}{t}$$

*exists. If this limit exists, we denote its value by $\partial f/\partial x_i (\mathbf{x})$ and call it the partial derivative of $f\colon \mathcal{O} \to \mathbb{R}$ with respect to the $i$th component at the point $\mathbf{x}$.*

The geometric meaning of $\partial f/\partial x_i\,(\mathbf{x})$ is as follows: Choose a positive number $r$ such that the symmetric neighborhood $\mathcal{N}_r(\mathbf{x})$ is contained in $\mathcal{O}$, and consider the section defined by

$$\psi(t) = f(\mathbf{x} + t\mathbf{e}_i) \quad \text{for} \quad |t| < r.$$

Then $f:\mathcal{O} \to \mathbb{R}$ has a partial derivative with respect to its $i$th component at the point $\mathbf{x}$ precisely when there is a tangent line to the graph of this section at the point on the graph corresponding to $t = 0$, at which point the slope of this tangent is the number

$$\psi'(0) = \partial f/\partial x_i\,(\mathbf{x}).$$

**FIGURE 13.2**

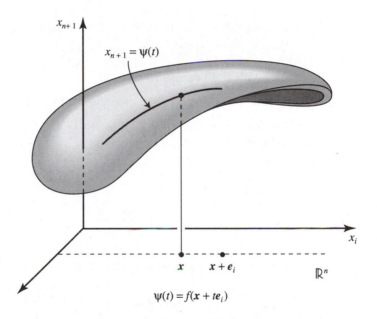

$$\psi(t) = f(x + te_i)$$

Thus the existence of $\partial f/\partial x_i(\mathbf{x})$ is equivalent to the differentiability of a function of a single real variable, so we can immediately use the single-variable differentiation results to obtain addition, product, and quotient rules for partial derivatives.

**DEFINITION**   *Let $\mathcal{O}$ be an open subset of $\mathbb{R}^n$. Then the function $f:\mathcal{O} \to \mathbb{R}$ is said to have first-order partial derivatives provided that for each index $i$ with $1 \le i \le n$, the function $f:\mathcal{O} \to \mathbb{R}$ has a partial derivative with respect to its $i$th component at every point in $\mathcal{O}$.*

It is clear that for each index $i$ with $1 \le i \le n$, the $i$th component projection function

$$p_i:\mathbb{R}^n \to \mathbb{R}$$

has first-order partial derivatives, and that if $j$ is an index with $1 \leq j \leq n$ and $\mathbf{x}$ is any point in $\mathbb{R}^n$, then

$$\partial p_i / \partial x_j (\mathbf{x}) = \begin{cases} 1 & \text{if } i = j \\ 0 & \text{if } i \neq j. \end{cases}$$

Hence sums, products, and suitable quotients of projection functions have first-order partial derivatives. This means that quotients of polynomials in the component variables, if their denominators are nonzero, define functions having first-order partial derivatives.

Naturally, if points in $\mathbb{R}^n$ have their components described in a notation without indices, we make a corresponding change in the notation for partial derivatives. So, for instance, if a function $f : \mathbb{R}^3 \to \mathbb{R}$ is given and points in $\mathbb{R}^3$ are written as $(x, y, z)$, then $\partial f / \partial y (x_0, y_0, z_0)$ will denote the partial derivative of $f : \mathbb{R}^3 \to \mathbb{R}$ with respect to the second component at the point $(x_0, y_0, z_0)$. Likewise, if a function $f : \mathbb{R}^4 \to \mathbb{R}$ is given and points in $\mathbb{R}^4$ are written as $(u, v, w, t)$, then $\partial f / \partial t (u_0, v_0, w_0, t_0)$ will denote the partial derivative of $f : \mathbb{R}^4 \to \mathbb{R}$ with respect to the fourth component at the point $(u_0, v_0, w_0, t_0)$.

**EXAMPLE 13.6**   Define

$$f(x, y, z) = xyz + e^{xy^2} \quad \text{for } (x, y, z) \text{ in } \mathbb{R}^3.$$

Let us formally check that the function $f : \mathbb{R}^3 \to \mathbb{R}$ has a partial derivative with respect to its second component at each point in $\mathbb{R}^3$. Choose a point $(x_0, y_0, z_0)$ in $\mathbb{R}^3$. Since

$$(x_0, y_0, z_0) + t\mathbf{e}_2 = (x_0, y_0 + t, z_0) \text{ for each real number } t$$

we must show that

$$\lim_{t \to 0} \frac{f(x_0, y_0 + t, z_0) - f(x_0, y_0, z_0)}{t}$$

exists. This is precisely the same as fixing the components $x_0$ and $z_0$, defining $g(y) = x_0 y z_0 + e^{x_0 y^2}$ for $y$ in $\mathbb{R}$, and calculating $g'(y_0)$. Hence

$$\frac{\partial f}{\partial y}(x_0, y_0, z_0) = g'(y_0) = x_0 z_0 + 2y_0 x_0 e^{x_0 y_0^2}. \qquad \square$$

The above example illustrates a general method for actually calculating specific first-order partial derivatives.

Proposition 4.2 asserts that if $I$ is an open interval of real numbers and the function $f : I \to \mathbb{R}$ is differentiable at the point $x$ in $I$, then the function $f : I \to \mathbb{R}$ is also continuous at the point $x$. This assertion is usually expressed as "differentiability implies continuity." The proof of this assertion is quite straightforward: If $h \neq 0$ and the point $x + h$ belongs to $I$, write the difference $f(x + h) - f(x)$ as

$$f(x + h) - f(x) = \frac{f(x + h) - f(x)}{h} \cdot h,$$

so that from the product rule for limits,

$$\lim_{h \to 0} [f(x + h) - f(x)] = f'(x) \cdot 0 = 0.$$

Thus the function $f : I \to \mathbb{R}$ is continuous at the point $x$.

There is an important difference between functions of a single real variable and functions of several real variables. For $n > 1$, a function $f : \mathbb{R}^n \to \mathbb{R}$ that has first-order partial derivatives need not be continuous. The following example shows what can occur.

**EXAMPLE 13.7** Define

$$f(x, y) = \begin{cases} xy/(x^2 + y^2) & \text{if } (x, y) \neq (0, 0) \\ 0 & \text{if } (x, y) = (0, 0). \end{cases}$$

For $(x, y) \neq (0, 0)$, there is a neighborhood of $(x, y)$ on which the restriction of $f : \mathbb{R}^2 \to \mathbb{R}$ is a quotient of polynomials whose denominator does not vanish. Thus, $\partial f / \partial x (x, y)$ and $\partial f / \partial y (x, y)$ exist; moreover, a short computation gives

$$\frac{\partial f}{\partial x}(x, y) = \frac{y^3 - x^2 y}{(x^2 + y^2)^2} \quad \text{and} \quad \frac{\partial f}{\partial y}(x, y) = \frac{x^3 - y^2 x}{(x^2 + y^2)^2}.$$

On the other hand, at $(x, y) = (0, 0) = \mathbf{0}$, we observe that for each number $t$,

$$f(\mathbf{0} + t\mathbf{e}_1) = f(t, 0) = 0,$$

so that $\quad \dfrac{\partial f}{\partial x}(0, 0) = \lim_{t \to 0} \dfrac{f(\mathbf{0} + t\mathbf{e}_1) - f(\mathbf{0})}{t} = \lim_{t \to 0} \dfrac{f(t, 0) - f(0, 0)}{t} = 0.$

A similar calculation shows that $\partial f / \partial y (0, 0) = 0$. Thus the function $f : \mathbb{R}^2 \to \mathbb{R}$ has first-order partial derivatives. Yet this function is not continuous at the point $\mathbf{0}$. To see this, observe that the sequence $\{(1/k, 1/k)\}$ converges to $(0, 0)$ and that $f(1/k, 1/k) = 1/2$ for each $k$, so the image sequence $\{f(1/k, 1/k)\}$ converges to $1/2$. But $1/2 \neq f(0, 0)$ so the function $f : \mathbb{R}^2 \to \mathbb{R}$ is not continuous at the point $(0, 0)$. $\qquad \square$

But things are not as bad as the above example makes them seem. We will show that if $\mathcal{O}$ is an open subset of $\mathbb{R}^n$ and the function $f : \mathcal{O} \to \mathbb{R}$ has first-order partial derivatives, then, if we assume in addition that $\partial f / \partial x_i : \mathcal{O} \to \mathbb{R}$ is continuous for each index $i$ with $1 \leq i \leq n$, the basic results of the single-variable theory carry over. Since this additional assumption will play an important part later, it is useful to isolate it with the following definition.

**DEFINITION** *Let $\mathcal{O}$ be an open subset of $\mathbb{R}^n$. Then a function $f : \mathcal{O} \to \mathbb{R}$ is said to be continuously differentiable provided that it has first-order partial derivatives and that for each index $i$ with $1 \leq i \leq n$, the function $\partial f / \partial x_i : \mathcal{O} \to \mathbb{R}$ is continuous.*

By calculating each of the first-order partial derivatives of the function $f : \mathbb{R}^3 \to \mathbb{R}$ defined in Example 13.6, we see that this function is continuously differentiable. The function $f : \mathbb{R}^2 \to \mathbb{R}$ defined in Example 13.7 is not continuously differentiable, since neither of its two first-order partial derivatives is continuous at the point $(0, 0)$ (see Exercise 3).

We now turn to second-order partial derivatives. Given an open subset $\mathcal{O}$ of $\mathbb{R}^n$ and an index $i$ with $1 \leq i \leq n$, if the function $f : \mathcal{O} \to \mathbb{R}$ has a partial derivative with respect to its $i$th component at each point in $\mathcal{O}$, then the function $\partial f / \partial x_i : \mathcal{O} \to \mathbb{R}$ is defined and we can ask whether this new function itself has first-order partial derivatives.

Fix an index $j$ with $1 \leq j \leq n$. If the function $\partial f/\partial x_i : \mathcal{O} \to \mathbb{R}$ has a partial derivative with respect to its $j$th component at the point $\mathbf{x}$ in $\mathcal{O}$, we use

$$\frac{\partial^2 f}{\partial x_j \partial x_i}(\mathbf{x}) \quad \text{to denote} \quad \frac{\partial}{\partial x_j}\left[\frac{\partial f}{\partial x_i}\right](\mathbf{x}).$$

Naturally, when $n = 2$ or $3$ and points are labeled without subscripts, we use a more suggestive notation for second partial derivatives, for example

$$\frac{\partial^2 f}{\partial x \partial y}, \frac{\partial^2 f}{\partial z^2}, \frac{\partial^2 f}{\partial y \partial z}, \text{ etc.}$$

**DEFINITION**  *Let $\mathcal{O}$ be an open subset of $\mathbb{R}^n$ and consider a function $f : \mathcal{O} \to \mathbb{R}$:*

(i) *The function $f : \mathcal{O} \to \mathbb{R}$ is said to have second-order partial derivatives provided that it has first-order partial derivatives and that for each index $i$ with $1 \leq i \leq n$, the function $\partial f/\partial x_i : \mathcal{O} \to \mathbb{R}$ also has first-order partial derivatives.*

(ii) *The function $f : \mathcal{O} \to \mathbb{R}$ is said to have continuous second-order partial derivatives provided that it has second-order partial derivatives and that for each pair of indices $i$ and $j$ with $1 \leq i \leq n$ and $1 \leq j \leq n$, the function $\partial^2 f/\partial x_i \partial x_j : \mathcal{O} \to \mathbb{R}$ is continuous.*

It turns out that every continuously differentiable function is continuous, and that every function with continuous second-order partial derivatives is continuously differentiable. However, in order to prove these assertions, it will be necessary first to prove a Mean Value Theorem for functions of several variables. We will prove such a Mean Value Theorem in Section 13.3.

We close this section with a very useful result about the equality of certain second-order partial derivatives.

**THEOREM 13.3**  Let $\mathcal{O}$ be an open subset of $\mathbb{R}^n$ and suppose that the function $f : \mathcal{O} \to \mathbb{R}$ has continuous second-order partial derivatives. For any two indices $i$ and $j$ with $1 \leq i \leq n$ and $1 \leq j \leq n$ and any point $\mathbf{x}$ in $\mathcal{O}$,

$$\frac{\partial^2 f}{\partial x_i \partial x_j}(\mathbf{x}) = \frac{\partial^2 f}{\partial x_j \partial x_i}(\mathbf{x}). \tag{13.11}$$

In order to prove this theorem, it will clarify matters if we first isolate the following lemma.

**LEMMA 13.4**  Let $\mathcal{U}$ be an open subset of the plane $\mathbb{R}^2$ that contains that point $(x_0, y_0)$ and suppose that the function $f : \mathcal{U} \to \mathbb{R}$ has second-order partial derivatives. Then there are points $(x_1, y_1)$ and $(x_2, y_2)$ in $\mathcal{U}$ at which

$$\frac{\partial^2 f}{\partial x \partial y}(x_1, y_1) = \frac{\partial^2 f}{\partial y \partial x}(x_2, y_2).$$

**Proof of the Lemma**  Since $\mathcal{U}$ is open, we can choose a positive number $r$ such that if we define the intervals of real numbers $I$ and $J$ by $I = (x_0 - 2r, x_0 + 2r)$ and $J = (y_0 - 2r, y_0 + 2r)$, then the rectangle $I \times J$ is contained in $\mathcal{U}$.

The idea of the proof is to express

$$f(x_0 + r, y_0 + r) - f(x_0 + r, y_0) - f(x_0, y_0 + r) + f(x_0, y_0)$$

as a difference, in two different ways: first as the difference

$$[f(x_0 + r, y_0 + r) - f(x_0 + r, y_0)] - [f(x_0, y_0 + r) - f(x_0, y_0)] \qquad (13.12)$$

and then as the difference

$$[f(x_0 + r, y_0 + r) - f(x_0, y_0 + r)] - [f(x_0 + r, y_0) - f(x_0, y_0)]. \qquad (13.13)$$

Then we will use the Mean Value Theorem for functions of a single real variable to express (13.12) and (13.13) as second-order partial derivatives of the function $f : \mathcal{U} \to \mathbb{R}$.

First we will analyze the difference (13.12). Define the auxiliary function $\varphi : I \to \mathbb{R}$ by

$$\varphi(x) = f(x, y_0 + r) - f(x, y_0) \qquad \text{for } x \text{ in } I.$$

Since $f : \mathcal{U} \to \mathbb{R}$ has a partial derivative with respect to its first component, the function $\varphi : I \to \mathbb{R}$ is differentiable. Thus we may apply the Lagrange Mean Value Theorem to the restriction of the function $\varphi : I \to \mathbb{R}$ to the closed interval $[x_0, x_0 + r]$ to select a point $x_1$ in the open interval $(x_0, x_0 + r)$ such that

$$\frac{\varphi(x_0 + r) - \varphi(x_0)}{r} = \varphi'(x_1);$$

that is,

$$\frac{\varphi(x_0 + r) - \varphi(x_0)}{r} = \frac{\partial f}{\partial x}(x_1, y_0 + r) - \frac{\partial f}{\partial x}(x_1, y_0). \qquad (13.14)$$

With this point $x_1$ fixed, define another auxiliary function $\alpha : J \to \mathbb{R}$ by

$$\alpha(y) = \frac{\partial f}{\partial x}(x_1, y) \qquad \text{for } y \text{ in } J.$$

We can apply the Mean Value Theorem to the restriction of the function $\alpha : J \to \mathbb{R}$ to the closed interval $[y_0, y_0 + r]$ to select a point $y_1$ in the open interval $(y_0, y_0 + r)$ such that

$$\frac{\alpha(y_0 + r) - \alpha(y_0)}{r} = \frac{\partial^2 f}{\partial y \partial x}(x_1, y_1). \qquad (13.15)$$

From (13.14) and (13.15), we obtain

$$\varphi(x_0 + r) - \varphi(x_0) = r^2 \frac{\partial^2 f}{\partial y \partial x}(x_1, y_1). \qquad (13.16)$$

However, $\varphi(x_0 + r) - \varphi(x_0)$ equals the difference (13.12), and hence we have

$$[f(x_0 + r, y_0 + r) - f(x_0 + r, y_0)] - [f(x_0, y_0 + r) - f(x_0, y_0)] = r^2 \frac{\partial^2 f}{\partial y \partial x}(x_1, y_1).$$
$$(13.17)$$

In order to analyze the difference (13.13), we now repeat the same argument applied to the auxiliary function $\psi: J \to \mathbb{R}$ defined by

$$\psi(y) = f(x_0 + r, \, y) - f(x_0, \, y) \quad \text{for } y \text{ in } J.$$

From this it will follow that we can select a point $(x_2, \, y_2)$ in the rectangle $I \times J$ such that

$$[f(x_0 + r, \, y_0 + r) - f(x_0, \, y_0 + r)] - [f(x_0 + r, \, y_0) - f(x_0, \, y_0)] = r^2 \frac{\partial^2 f}{\partial x \partial y}(x_2, \, y_2).$$

(13.18)

From the equality of the left-hand sides of (13.17) and (13.18) follows the equality of the right-hand sides, so the lemma is proven. ∎

**Proof of Theorem 13.3**
We will prove the theorem when $n = 2$, and leave the general case to the reader (Exercise 16). Let $(x_0, \, y_0)$ be a point in $\mathcal{O}$. Choose a positive number $r$ such that the symmetric neighborhood $\mathcal{N}_r(x_0, \, y_0)$ is contained in $\mathcal{O}$. Let $k$ be a positive integer. Then we can apply the lemma with $\mathcal{U} = \mathcal{N}_{r/k}(x_0, \, y_0)$ and select points $(x_k, \, y_k)$ and $(u_k, \, v_k)$ in $\mathcal{N}_{r/k}(x_0, \, y_0)$ at which

$$\frac{\partial^2 f}{\partial x \partial y}(x_k, \, y_k) = \frac{\partial^2 f}{\partial y \partial x}(u_k, \, v_k).$$

(13.19)

But, by assumption, the function $\partial^2 f/\partial x \partial y : \mathcal{O} \to \mathbb{R}$ is continuous at $(x_0, \, y_0)$, as is the function $\partial^2 f/\partial y \partial x : \mathcal{O} \to \mathbb{R}$. Since both of the sequences $\{(x_k, \, y_k)\}$ and $\{(u_k, \, v_k)\}$ converge to the point $(x_0, \, y_0)$, it follows that

$$\lim_{k \to \infty} \left[ \frac{\partial^2 f}{\partial x \partial y}(x_k, \, y_k) \right] = \frac{\partial^2 f}{\partial x \partial y}(x_0, \, y_0)$$

and

$$\lim_{k \to \infty} \left[ \frac{\partial^2 f}{\partial y \partial x}(u_k, \, v_k) \right] = \frac{\partial^2 f}{\partial y \partial x}(x_0, \, y_0).$$

In view of (13.19), we conclude that

$$\frac{\partial^2 f}{\partial x \partial y}(x_0, \, y_0) = \frac{\partial^2 f}{\partial y \partial x}(x_0, \, y_0). \qquad ∎$$

Observe that in Lemma 13.4 we required only that the function $f: \mathcal{O} \to \mathbb{R}$ have second-order partial derivatives. On the other hand, in Theorem 13.3, we required that the second-order partial derivatives be continuous. This extra assumption is necessary. The following is an example of a function $f: \mathcal{O} \to \mathbb{R}$ that has second-order partial derivatives, and yet we do not have equality of $\partial^2 f/\partial x \partial y$ and $\partial^2 f/\partial y \partial x$ at all points.

**EXAMPLE 13.8**   Define the function $f: \mathbb{R}^2 \to \mathbb{R}$ by

$$f(x, y) = \begin{cases} xy(x^2 - y^2)/(x^2 + y^2) & \text{if } (x, y) \neq (0, 0) \\ 0 & \text{if } (x, y) = (0, 0). \end{cases}$$

Calculations, which we leave to the reader (see Exercise 13), show that the function $f: \mathbb{R}^2 \to \mathbb{R}$ has second-order partial derivatives, but that

$$\frac{\partial^2 f}{\partial y \partial x}(0, 0) = -1 \quad \text{while} \quad \frac{\partial^2 f}{\partial x \partial y}(0, 0) = 1. \qquad \square$$

**Remark on Notation**   There are other notations in common use for partial derivatives. For instance, when the function $f: \mathbb{R}^3 \to \mathbb{R}$ has first-order partial derivatives, $\partial f/\partial x(x, y, z)$ is often denoted by $f_x(x, y, z)$. Also, when the function $f: \mathbb{R}^3 \to \mathbb{R}$ has second-order partial derivatives,

$$f_{xx}(x, y, z), \quad \frac{\partial^2 f}{\partial x \partial x}(x, y, z), \quad \text{and} \quad \frac{\partial^2 f}{\partial x^2}(x, y, z)$$

are used to denote the same quantity. Moreover, for $\mathcal{O}$ an open subset of $\mathbb{R}^n$, a continuously differentiable function $f: \mathcal{O} \to \mathbb{R}$ is often said to be $C^1$, and if $f: \mathcal{O} \to \mathbb{R}$ has continuous second-order partial derivatives, it is said to be $C^2$.

---

**EXERCISES**

1. Calculate the first-order partial derivatives of the following functions:
   (a)   $f(x, y, z) = x + yz + xy + x \sin(yz)$ for $(x, y, z)$ in $\mathbb{R}^3$.
   (b)   $f(x, y, z) = (\sin(x^2 y^2))/(1 + x^2 + y^3)$ for $(x, y, z)$ in $\mathbb{R}^3$.
   (c)   $f(x, y, z) = \sqrt{1 + \cos^2(xy)}$ for $(x, y, z)$ in $\mathbb{R}^3$.
2. Prove that the function defined in Example 13.6 is continuously differentiable.
3. For the function $f: \mathbb{R}^2 \to \mathbb{R}$ defined in Example 13.7, show that neither the function $\partial f/\partial x: \mathbb{R}^2 \to \mathbb{R}$ nor the function $\partial f/\partial y: \mathbb{R}^2 \to \mathbb{R}$ is continuous at the point $(0, 0)$.
4. Suppose that the function $g: \mathbb{R}^2 \to \mathbb{R}$ has the property that

$$|g(x, y)| \leq x^2 + y^2 \quad \text{for all } (x, y) \text{ in } \mathbb{R}^2.$$

   Prove that $g: \mathbb{R}^2 \to \mathbb{R}$ has partial derivatives with respect to both $x$ and $y$ at the point $(0, 0)$.
5. Suppose that the function $f: \mathbb{R}^2 \to \mathbb{R}$ has first-order partial derivatives and that

$$\frac{\partial f}{\partial x}(x, y) = \frac{\partial f}{\partial y}(x, y) = 0 \quad \text{for all } (x, y) \text{ in } \mathbb{R}^2.$$

   Prove that the function $f: \mathbb{R}^2 \to \mathbb{R}$ is constant; that is, there is some number $c$ such that

$$f(x, y) = c \quad \text{for all } (x, y) \text{ in } \mathbb{R}^2.$$

   (*Hint:* First show that the restriction of $f: \mathbb{R}^2 \to \mathbb{R}$ to a line parallel to one of the coordinate axes is constant.)

6. Define

$$g(x, y) = \begin{cases} x^2 y^4/(x^2 + y^2) & \text{if } (x, y) \neq (0, 0) \\ 0 & \text{if } (x, y) = (0, 0). \end{cases}$$

Prove that the function $g: \mathbb{R}^2 \to \mathbb{R}$ has first-order partial derivatives. Is the function $g: \mathbb{R}^2 \to \mathbb{R}$ continuously differentiable?

7. (An extension of the Chain Rule) Let $\mathcal{O}$ be an open subset of $\mathbb{R}^n$ that contains the point $\mathbf{x}$ and suppose that the function $f: \mathcal{O} \to \mathbb{R}$ has a partial derivative with respect to the $i$th component at the point $\mathbf{x}$. Let $I$ be an open interval in $\mathbb{R}$ with $f(\mathcal{O}) \subseteq I$ and let the function $g: I \to \mathbb{R}$ have a derivative at the point $f(\mathbf{x})$. Prove that the composition $g \circ f: \mathcal{O} \to \mathbb{R}$ has a partial derivative with respect to the $i$th component at the point $\mathbf{x}$, and that

$$\frac{\partial}{\partial x_i}(g \circ f)(\mathbf{x}) = g'(f(\mathbf{x}))\frac{\partial f}{\partial x_i}(\mathbf{x}).$$

8. Suppose that the function $g: \mathbb{R} \to \mathbb{R}$ is differentiable. Use Exercise 7 to calculate the first-order partial derivatives of the following functions:

   (a) The function $h: \mathbb{R}^2 \to \mathbb{R}$ defined by

   $$h(x, y) = g(xy^2 + 1) \quad \text{for } (x, y) \text{ in } \mathbb{R}^2.$$

   (b) The function $h: \mathbb{R}^2 \to \mathbb{R}$ defined by

   $$h(u, v) = g(4u + 7v) \quad \text{for } (u, v) \text{ in } \mathbb{R}^2.$$

   (c) The function $h: \mathbb{R}^2 \to \mathbb{R}$ defined by

   $$h(t, s) = g(t - s) \quad \text{for } (t, s) \text{ in } \mathbb{R}^2.$$

9. Suppose that the function $g: \mathbb{R} \to \mathbb{R}$ has a second derivative. Calculate the second-order partial derivatives of the functions defined in Exercise 8.

10. Suppose that the functions $\varphi: \mathbb{R} \to \mathbb{R}$ and $\psi: \mathbb{R} \to \mathbb{R}$ have continuous second-order partial derivatives. Define

$$f(x, y) = \varphi(x - y) + \psi(x + y) \quad \text{for } (x, y) \text{ in } \mathbb{R}^2.$$

Prove that

$$\frac{\partial^2 f}{\partial x^2}(x, y) - \frac{\partial^2 f}{\partial y^2}(x, y) = 0 \quad \text{for all } (x, y) \text{ in } \mathbb{R}^2.$$

11. A function $f: \mathbb{R}^2 \to \mathbb{R}$ is called *harmonic* provided that it has second-order partial derivatives and

$$\frac{\partial^2 f}{\partial x^2}(x, y) + \frac{\partial^2 f}{\partial y^2}(x, y) = 0 \quad \text{for all } (x, y) \text{ in } \mathbb{R}^2.$$

Which of the following functions is harmonic?
   (a) $f(x, y) = e^x \cos y$ for $(x, y)$ in $\mathbb{R}^2$
   (b) $g(x, y) = x^2 - y^2$ for $(x, y)$ in $\mathbb{R}^2$
   (c) $h(x, y) = x^2 - y^3$ for $(x, y)$ in $\mathbb{R}^2$

12. Given a pair of functions $\phi: \mathbb{R}^2 \to \mathbb{R}$ and $\psi: \mathbb{R}^2 \to \mathbb{R}$, it is often useful to know whether there exists some continuously differentiable function $f: \mathbb{R}^2 \to \mathbb{R}$ such that

$$\partial f/\partial x(x, y) = \phi(x, y) \quad \text{and} \quad \partial f/\partial y(x, y) = \psi(x, y) \quad \text{for all } (x, y) \text{ in } \mathbb{R}^2.$$

Such a function $f: \mathbb{R}^2 \to \mathbb{R}$ is called a *potential function* for the pair of functions $(\phi, \psi)$.

(a)   Show that if a potential function exists for the pair $(\phi, \psi)$, then this potential is uniquely determined up to an additive constant—that is, the difference of any two potentials is constant.

(b)   Show that if there is a potential function for the pair of continuously differentiable functions $\phi: \mathbb{R}^2 \to \mathbb{R}$ and $\psi: \mathbb{R}^2 \to \mathbb{R}$, then

$$\frac{\partial \psi}{\partial x}(x, y) = \frac{\partial \phi}{\partial y}(x, y) \quad \text{for all } (x, y) \text{ in } \mathbb{R}^2.$$

13.   Consider the function defined in Example 13.8.

    (a)   Show that

$$\frac{\partial f}{\partial y}(x, 0) = x \quad \text{for all } x \text{ in } \mathbb{R}$$

and that

$$\frac{\partial f}{\partial x}(0, y) = -y \quad \text{for all } y \text{ in } \mathbb{R}.$$

    (b)   From part (a), conclude that

$$\frac{\partial^2 f}{\partial x \partial y}(0, 0) = 1 \quad \text{and} \quad \frac{\partial^2 f}{\partial y \partial x}(0, 0) = -1.$$

14.   Let $\mathcal{U}$ be an open subset of the plane $\mathbb{R}^2$ that contains the point $(x_0, y_0)$. Prove that there is a positive number $r$ such that $(x, y)$ is in $\mathcal{U}$ if $|x - x_0| < 2r$ and $|y - y_0| < 2r$.

15.   Verify equation (13.18) by following the argument used in the proof of Lemma 13.4 to establish (13.17).

16.   Let the function $f: \mathcal{O} \to \mathbb{R}$ satisfy the assumptions of Theorem 13.3 and let $i$ and $j$ be indices with $1 \le i \le n$ and $1 \le j \le n$. Choose a positive number $r$ such that the symmetric neighborhood $\mathcal{N}_r(\mathbf{x})$ is contained in $\mathcal{O}$. Define

$$g(s, t) = f(\mathbf{x} + t\mathbf{e}_i + s\mathbf{e}_j) \quad \text{for} \quad s^2 + t^2 < r^2.$$

    (a)   Verify that $g: \mathcal{N}_r(0, 0) \to \mathbb{R}$ has continuous second-order partial derivatives and that

$$\frac{\partial^2 g}{\partial s \partial t}(0, 0) = \frac{\partial^2 f}{\partial x_j \partial x_i}(\mathbf{x}) \quad \text{and} \quad \frac{\partial^2 g}{\partial t \partial s}(0, 0) = \frac{\partial^2 f}{\partial x_i \partial x_j}(\mathbf{x}).$$

    (b)   Use (a) to show that the general case of Theorem 13.3 follows from the case when $n = 2$.

17.   Suppose that the function $f: \mathbb{R}^2 \to \mathbb{R}$ has continuous second-order partial derivatives and let $(x_0, y_0)$ be a point in $\mathbb{R}^2$. Prove that for each $\epsilon > 0$ there exists a $\delta > 0$ such that if $0 < |h| < \delta$ and $0 < |k| < \delta$, then

$$\left| \frac{f(x_0 + h, y_0 + k) - f(x_0 + h, y_0) - f(x_0, y_0 + k) + f(x_0, y_0)}{hk} - \frac{\partial^2 f}{\partial x \partial y}(x_0, y_0) \right| < \epsilon.$$

(*Hint:* Follow the idea of the proof of Lemma 13.4.)

## 13.3 The Mean Value Theorem and Directional Derivatives

In the study of real-valued functions of a single variable, a prominent role is played by the Lagrange Mean Value Theorem. For convenient reference, we restate this theorem, which was proved in Section 4.3.

**THEOREM 13.5**    **The Lagrange Mean Value Theorem**    Suppose that the function $f:[a, b] \rightarrow \mathbb{R}$ is continuous and that $f:(a, b) \rightarrow \mathbb{R}$ is differentiable. Then there is a point $c$ in the open interval $(a, b)$ at which

$$f(b) - f(a) = f'(c)(b - a).$$

One of the goals of this section is to extend this theorem to the case of functions of several real variables. As we have already noted at the beginning of Section 13.2, the strategy for making this extension is to try to reduce the general case to the single-variable case by analyzing the restriction of the function to segments in the domain.

**LEMMA 13.6**    **The Mean Value Lemma**    Let $\mathcal{O}$ be an open subset of $\mathbb{R}^n$ and let $i$ be an index with $1 \leq i \leq n$. Suppose that the function $f: \mathcal{O} \rightarrow \mathbb{R}$ has a partial derivative with respect to its $i$th component at each point in $\mathcal{O}$. Let $\mathbf{x}$ be a point in $\mathcal{O}$ and let $a$ be a real number such that the segment between the points $\mathbf{x}$ and $\mathbf{x} + a\mathbf{e}_i$ lies in $\mathcal{O}$. Then there is a number $\theta$ with $0 < \theta < 1$ such that

$$f(\mathbf{x} + a\mathbf{e}_i) - f(\mathbf{x}) = \frac{\partial f}{\partial x_i}(\mathbf{x} + \theta a\mathbf{e}_i)a. \tag{13.20}$$

**Proof**    Since $\mathcal{O}$ is open in $\mathbb{R}^n$, we can select an open interval of real numbers $I$ that contains the numbers $0$ and $a$, such that for each $t$ in $I$ the point $\mathbf{x} + t\mathbf{e}_i$ belongs to $\mathcal{O}$. Define the function $\phi: I \rightarrow \mathbb{R}$ by $\phi(t) = f(\mathbf{x} + t\mathbf{e}_i)$ for each $t$ in $I$. Then the partial differentiability of the function $f: \mathcal{O} \rightarrow \mathbb{R}$ with respect to its $i$th component implies that at each point $t$ in $I$,

$$\phi'(t) = \frac{\partial f}{\partial x_i}(\mathbf{x} + t\mathbf{e}_i).$$

It follows that the function $\phi: I \rightarrow \mathbb{R}$ is differentiable. Thus we can apply the Lagrange Mean Value Theorem for functions of a single variable to the restriction of the function $\phi: I \rightarrow \mathbb{R}$ to the closed interval $[0, a]$ to obtain a point $\theta$ with $0 < \theta < 1$ such that

$$\phi(a) - \phi(0) = \phi'(\theta a)a,$$

which, in view of the definition of the function $\phi: I \rightarrow \mathbb{R}$ and the calculation of $\phi'(t)$, can be rewritten as (13.20). ∎

**PROPOSITION**
**13.7**

**The Mean Value Proposition**   Let $\mathbf{x}$ be a point in $\mathbb{R}^n$ and let $r$ be a positive number. Suppose that the function $f: \mathcal{N}_r(\mathbf{x}) \to \mathbb{R}$ has first-order partial derivatives. Then if the point $\mathbf{x} + \mathbf{h}$ belongs to $\mathcal{N}_r(\mathbf{x})$, there are points $\mathbf{z}_1, \mathbf{z}_2, \ldots, \mathbf{z}_n$ in $\mathcal{N}_r(\mathbf{x})$ such that

$$f(\mathbf{x} + \mathbf{h}) - f(\mathbf{x}) = \sum_{i=1}^{n} h_i \frac{\partial f}{\partial x_i}(\mathbf{z}_i), \tag{13.21}$$

and

$$\|\mathbf{x} - \mathbf{z}_i\| < \|\mathbf{h}\| \quad \text{for each index } i \text{ with } 1 \le i \le n.$$

**Proof**   We will prove the result with $n = 3$. From this, it will be clear that the general result is also true. The trick is to expand the difference $f(\mathbf{x} + \mathbf{h}) - f(\mathbf{x})$. We have

$$\begin{aligned}
f(\mathbf{x} + \mathbf{h}) - f(\mathbf{x}) &= f(x_1 + h_1, \, x_2 + h_2, \, x_3 + h_3) - f(x_1, \, x_2, \, x_3) \\
&= f(x_1 + h_1, \, x_2 + h_2, \, x_3 + h_3) - f(x_1 + h_1, \, x_2 + h_2, \, x_3) \\
&\quad + f(x_1 + h_1, \, x_2 + h_2, \, x_3) - f(x_1 + h_1, \, x_2, \, x_3) \\
&\quad + f(x_1 + h_1, \, x_2, \, x_3) - f(x_1, \, x_2, \, x_3).
\end{aligned}$$

We apply the Mean Value Lemma to each of these differences to find numbers $\theta_1, \theta_2$, and $\theta_3$ in the open interval $(0, \, 1)$ with

$$\begin{aligned}
f(\mathbf{x} + \mathbf{h}) - f(\mathbf{x}) &= \frac{\partial f}{\partial x_3}(x_1 + h_1, \, x_2 + h_2, \, x_3 + \theta_3 h_3) h_3 \\
&\quad + \frac{\partial f}{\partial x_2}(x_1 + h_1, \, x_2 + \theta_2 h_2, \, x_3) h_2 \\
&\quad + \frac{\partial f}{\partial x_1}(x_1 + \theta_1 h_1, \, x_2, \, x_3) h_1.
\end{aligned}$$

Setting $\mathbf{z}_3 = (x_1 + h_1, \, x_2 + h_2, \, x_3 + \theta_3 h_3)$, $\mathbf{z}_2 = (x_1 + h_1, \, x_2 + \theta_2 h_2, \, x_3)$, and $\mathbf{z}_1 = (x_1 + \theta_1 h_1, \, x_2, \, x_3)$, the result follows.   ∎

For $\mathcal{O}$ an open subset of $\mathbb{R}^n$ containing the point $\mathbf{x}$, a function $f: \mathcal{O} \to \mathbb{R}$ has been defined to have a partial derivative with respect to the $i$th component at the point $\mathbf{x}$ provided that the limit

$$\lim_{t \to 0} \frac{f(\mathbf{x} + t\mathbf{e}_i) - f(\mathbf{x})}{t}$$

exists. We now turn to an analysis of this limit when the point $\mathbf{e}_i$ is replaced by a general nonzero point $\mathbf{p}$ in $\mathbb{R}^n$.

**DEFINITION** *Let $\mathcal{O}$ be an open subset of $\mathbb{R}^n$ that contains the point $\mathbf{x}$. Consider a function $f : \mathcal{O} \to \mathbb{R}$ and a nonzero point $\mathbf{p}$ in $\mathbb{R}^n$. If the limit*

$$\lim_{t \to 0} \frac{f(\mathbf{x} + t\mathbf{p}) - f(\mathbf{x})}{t}$$

*exists, we call this limit the directional derivative of the function $f : \mathcal{O} \to \mathbb{R}$ in the direction $\mathbf{p}$ at the point $\mathbf{x}$, and denote it by*

$$\frac{\partial f}{\partial \mathbf{p}}(\mathbf{x}).$$

Observe that if $\mathbf{p} = \mathbf{e}_i$, then

$$\frac{\partial f}{\partial \mathbf{p}}(\mathbf{x}) = \frac{\partial f}{\partial x_i}(\mathbf{x}).$$

**THEOREM 13.8** **The Directional Derivative Theorem** Let $\mathcal{O}$ be an open subset of $\mathbb{R}^n$ and suppose that the function $f : \mathcal{O} \to \mathbb{R}$ is continuously differentiable. Then for each point $\mathbf{x}$ in $\mathcal{O}$ and each nonzero point $\mathbf{p}$ in $\mathbb{R}^n$, the function $f : \mathcal{O} \to \mathbb{R}$ has a directional derivative in the direction $\mathbf{p}$ at the point $\mathbf{x}$ that is given by the formula

$$\frac{\partial f}{\partial \mathbf{p}}(\mathbf{x}) = \sum_{i=1}^{n} p_i \frac{\partial f}{\partial x_i}(\mathbf{x}). \tag{13.22}$$

**Proof** Since $\mathcal{O}$ is an open subset of $\mathbb{R}^n$, we can choose a positive number $r$ such that the symmetric neighborhood $\mathcal{N}_r(\mathbf{x})$ is contained in $\mathcal{O}$. Then from the Mean Value Proposition, we see that if $t$ is any number with $|t| \|\mathbf{p}\| < r$, there are $n$ points $\mathbf{z}_1, \ldots, \mathbf{z}_n$ such that

$$f(\mathbf{x} + t\mathbf{p}) - f(\mathbf{x}) = \sum_{i=1}^{n} t p_i \frac{\partial f}{\partial x_i}(\mathbf{z}_i) \tag{13.23}$$

and $\qquad \|\mathbf{z}_i - \mathbf{x}\| \le |t| \|\mathbf{p}\| \quad$ for each index $i$ with $1 \le i \le n$. $\qquad$ (13.24)

We may rewrite (13.23) as

$$\frac{f(\mathbf{x} + t\mathbf{p}) - f(\mathbf{x})}{t} = \sum_{i=1}^{n} p_i \frac{\partial f}{\partial x_i}(\mathbf{z}_i) \quad \text{for } t \neq 0. \tag{13.25}$$

Since $\partial f / \partial x_i : \mathcal{O} \to \mathbb{R}$ is continuous at the point $\mathbf{x}$ for each index $i$ with $1 \le i \le n$, it follows from (13.24) and (13.25) that

$$\lim_{t \to 0} \frac{f(\mathbf{x} + t\mathbf{p}) - f(\mathbf{x})}{t} = \lim_{t \to 0} \sum_{i=1}^{n} p_i \frac{\partial f}{\partial x_i}(\mathbf{z}_i) = \sum_{i=1}^{n} p_i \frac{\partial f}{\partial x_i}(\mathbf{x}).$$

This proves formula (13.22). ∎

In view of formula (13.22) we introduce the following definition.

**DEFINITION**   *Let $\mathcal{O}$ be an open subset of $\mathbb{R}^n$ that contains the point $\mathbf{x}$ and suppose that the function $f: \mathcal{O} \to \mathbb{R}$ has first-order partial derivatives at $\mathbf{x}$. We define the derivative vector of the function $f: \mathcal{O} \to \mathbb{R}$ at the point $\mathbf{x}$, denoted by $\mathbf{D}f(\mathbf{x})$, to be the point in $\mathbb{R}^n$ given by*

$$\mathbf{D}f(\mathbf{x}) = \left( \frac{\partial f}{\partial x_1}(\mathbf{x}), \frac{\partial f}{\partial x_2}(\mathbf{x}), \ldots, \frac{\partial f}{\partial x_n}(\mathbf{x}) \right).$$

In vector notation, formula (13.22) can be written as

$$\frac{d}{dt}[f(\mathbf{x}+t\mathbf{p})]\bigg|_{t=0} = \frac{\partial f}{\partial \mathbf{p}}(\mathbf{x}) = \langle \mathbf{D}f(\mathbf{x}), \mathbf{p} \rangle. \tag{13.26}$$

It is also useful to observe a slight extension of (13.26): Replacing the point $\mathbf{x}$ with the point $\mathbf{x}+t\mathbf{p}$, it follows that

$$\frac{d}{dt}[f(\mathbf{x}+t\mathbf{p})] = \langle \mathbf{D}f(\mathbf{x}+t\mathbf{p}), \mathbf{p} \rangle, \tag{13.27}$$

provided that the segment between $\mathbf{x}$ and $\mathbf{x}+t\mathbf{p}$ lies in $\mathcal{O}$.

**THEOREM 13.9**   **The Mean Value Theorem**   Let $\mathcal{O}$ be an open subset of $\mathbb{R}^n$ and suppose that the function $f: \mathcal{O} \to \mathbb{R}$ is continuously differentiable. If the segment joining the points $\mathbf{x}$ and $\mathbf{x}+\mathbf{h}$ lies in $\mathcal{O}$, then there is a number $\theta$ with $0 < \theta < 1$ such that

$$f(\mathbf{x}+\mathbf{h}) - f(\mathbf{x}) = \langle \mathbf{D}f(\mathbf{x}+\theta\mathbf{h}), \mathbf{h} \rangle. \tag{13.28}$$

**Proof**   Since $\mathcal{O}$ is an open subset of $\mathbb{R}^n$, we can select an open interval of real numbers $I$, which contains the numbers 0 and 1, such that $\mathbf{x}+t\mathbf{h}$ belongs to $\mathcal{O}$ for each $t$ in $I$. Define

$$\phi(t) = f(\mathbf{x}+t\mathbf{h}) \quad \text{for each } t \text{ in } I.$$

Using the slight generalization of the Directional Derivative Theorem stated as formula (13.27) we see that

$$\phi'(t) = \langle \mathbf{D}f(\mathbf{x}+t\mathbf{h}), \mathbf{h} \rangle \quad \text{for each } t \text{ in } I. \tag{13.29}$$

Thus we may apply the Lagrange Mean Value Theorem for functions of a single real variable to the restriction of the function $\phi: I \to \mathbb{R}$ to the closed interval $[0, 1]$ in order to select a number $\theta$ with $0 < \theta < 1$ such that

$$\phi(1) - \phi(0) = \phi'(\theta).$$

Using (13.29) and the definition of $\phi: [0, 1] \to \mathbb{R}$, it is clear that this formula is a restatement of (13.28). ∎

In the case when $\mathbf{p}$ is a point in $\mathbb{R}^n$ of norm 1, a directional derivative in the direction $\mathbf{p}$ can be interpreted as a *rate of change*. To see this, let $\mathcal{O}$ be an open subset of $\mathbb{R}^n$ that contains the point $\mathbf{x}$ and suppose that the function $f: \mathcal{O} \to \mathbb{R}$ is continuously differentiable. Then if the point $\mathbf{p}$ is of norm 1 and $t$ is a positive real number,

$$t = \|t\mathbf{p}\|,$$

so if $t$ is positive and sufficiently small,

$$\frac{f(\mathbf{x}+t\mathbf{p})-f(\mathbf{x})}{t} = \frac{f(\mathbf{x}+t\mathbf{p})-f(\mathbf{x})}{\|t\mathbf{p}\|}.$$

In view of this, if the norm of $\mathbf{p}$ is 1, it is reasonable to call $\partial f/\partial\mathbf{p}(\mathbf{x})$ the *rate of change* of the function $f\colon \mathcal{O} \to \mathbb{R}$ in the direction $\mathbf{p}$ at the point $\mathbf{x}$.

**COROLLARY 13.10**  Let $\mathcal{O}$ be an open subset of $\mathbb{R}^n$ that contains the point $\mathbf{x}$ and suppose that the function $f\colon \mathcal{O} \to \mathbb{R}$ is continuously differentiable. If $\mathbf{D}f(\mathbf{x}) \neq \mathbf{0}$, then the direction of norm 1 at the point $\mathbf{x}$ in which the function $f\colon \mathcal{O} \to \mathbb{R}$ is increasing the fastest is the direction $\mathbf{p}_0$ defined by

$$\mathbf{p}_0 = \frac{\mathbf{D}f(\mathbf{x})}{\|\mathbf{D}f(\mathbf{x})\|}. \tag{13.30}$$

**Proof**  Using formula (13.26) and the Cauchy-Schwarz Inequality, it follows that if $\mathbf{p}$ is any point in $\mathbb{R}^n$ of norm 1, then

$$\left|\frac{\partial f}{\partial\mathbf{p}}(\mathbf{x})\right| = |\langle \mathbf{D}f(\mathbf{x}), \mathbf{p}\rangle| \leq \|\mathbf{D}f(\mathbf{x})\| \cdot \|\mathbf{p}\| = \|\mathbf{D}f(\mathbf{x})\|. \tag{13.31}$$

On the other hand, if $\mathbf{p}_0$ is defined by (13.30), then $\mathbf{p}_0$ has norm 1, and using (13.26), it follows that

$$\frac{\partial f}{\partial\mathbf{p}_0}(\mathbf{x}) = \langle \mathbf{D}f(\mathbf{x}), \mathbf{p}_0\rangle = \left\langle \mathbf{D}f(\mathbf{x}), \frac{\mathbf{D}f(\mathbf{x})}{\|\mathbf{D}f(\mathbf{x})\|}\right\rangle = \|\mathbf{D}f(\mathbf{x})\|.$$

This calculation, together with inequality (13.31), implies that if $\mathbf{p}$ has norm 1, then

$$\frac{\partial f}{\partial\mathbf{p}}(\mathbf{x}) \leq \frac{\partial f}{\partial\mathbf{p}_0}(\mathbf{x}). \qquad \blacksquare$$

**EXAMPLE 13.9**  Define

$$f(x, y) = e^{x^2 - y^2} \quad \text{for } (x, y) \text{ in } \mathbb{R}^2.$$

The function $f\colon \mathbb{R}^2 \to \mathbb{R}$ is continuously differentiable. A short calculation shows that

$$\frac{\partial f}{\partial x}(1, 1) = 2 \quad \text{and} \quad \frac{\partial f}{\partial y}(1, 1) = -2.$$

Thus $\mathbf{D}f(1, 1) = (2, -2)$, so the direction in which the function $f\colon \mathbb{R}^2 \to \mathbb{R}$ is increasing the fastest at the point $(1, 1)$ is given by the vector $(1/\sqrt{2}, -1/\sqrt{2})$.  $\square$

In Section 13.2, we mentioned that a continuously differentiable function is continuous. We can now prove this assertion.

**THEOREM 13.11**  Let $\mathcal{O}$ be an open subset of $\mathbb{R}^n$ and suppose that the function $f\colon \mathcal{O} \to \mathbb{R}$ is continuously differentiable. Then the function $f\colon \mathcal{O} \to \mathbb{R}$ is continuous.

**Proof**    Let $\mathbf{x}$ be a point in $\mathcal{O}$. We need to show that the function $f: \mathcal{O} \to \mathbb{R}$ is continuous at $\mathbf{x}$. We will directly apply the sequential definition of continuity. First, since $\mathbf{x}$ is an interior point of $\mathcal{O}$, we can select a positive number $r$ such that the symmetric neighborhood $\mathcal{N}_r(\mathbf{x})$ is contained in $\mathcal{O}$.

Let $\{\mathbf{x}_k\}$ be a sequence in $\mathcal{N}_r(\mathbf{x})$ that converges to $\mathbf{x}$. For each natural number $k$, set $\mathbf{h}_k = \mathbf{x}_k - \mathbf{x}$ and apply the Mean Value Theorem to select a number $\theta_k$ with $0 < \theta_k < 1$ such that

$$f(\mathbf{x}_k) - f(\mathbf{x}) = f(\mathbf{x} + \mathbf{h}_k) - f(\mathbf{x}) = \langle \mathbf{D}f(\mathbf{x} + \theta_k \mathbf{h}_k), \mathbf{h}_k \rangle. \tag{13.32}$$

Now observe that

$$\lim_{k \to \infty} \mathbf{h}_k = \mathbf{0} \quad \text{and} \quad \lim_{k \to \infty} [\mathbf{x} + \theta_k \mathbf{h}_k] = \mathbf{x}.$$

Since $f: \mathcal{O} \to \mathbb{R}$ is continuously differentiable, it follows that

$$\lim_{k \to \infty} \mathbf{D}f(\mathbf{x} + \theta_k \mathbf{h}_k) = \mathbf{D}f(\mathbf{x}).$$

Thus, since (13.32) holds for all positive integers $k$, we conclude that

$$\lim_{k \to \infty} [f(\mathbf{x}_k) - f(\mathbf{x})] = \langle \mathbf{D}f(\mathbf{x}), \mathbf{0} \rangle = 0,$$

which means that the image sequence $\{f(\mathbf{x}_k)\}$ converges to $f(\mathbf{x})$.    ∎

**COROLLARY 13.12**    Let $\mathcal{O}$ be an open subset of $\mathbb{R}^n$ and suppose that the function $f: \mathcal{O} \to \mathbb{R}$ has continuous second-order partial derivatives. Then the function $f: \mathcal{O} \to \mathbb{R}$ is continuously differentiable.

**Proof**    For each index $i$ with $1 \leq i \leq n$, the function $\partial f / \partial x_i : \mathcal{O} \to \mathbb{R}$ is continuously differentiable and hence, by Theorem 13.11, it is continuous. This is precisely what it means for the function $f: \mathbb{R}^n \to \mathbb{R}$ to be continuously differentiable.    ∎

**Remark on Notation**    Because of the formula that expresses the rate of change as the inner-product of the derivative vector with the direction, frequently the derivative vector is called the *gradient*. Moreover, it is often denoted by $\nabla f(\mathbf{x})$ rather than $\mathbf{D}f(\mathbf{x})$. With this notational device, formula (13.26), for instance, becomes

$$\frac{d}{dt}[f(\mathbf{x} + t\mathbf{p})]\bigg|_{t=0} = \frac{\partial f}{\partial \mathbf{p}}(\mathbf{x}) = \langle \nabla f(\mathbf{x}), \mathbf{p} \rangle.$$

**EXERCISES**

1. For each of the following functions, find the derivative vector $\mathbf{D}f(\mathbf{x})$ for those points $\mathbf{x}$ in $\mathbb{R}^n$ where it is defined:
   - (a)  $f(\mathbf{x}) = e^{\|\mathbf{x}\|^2}$ for $\mathbf{x}$ in $\mathbb{R}^n$
   - (b)  $f(x, y) = \sin(xy)/\sqrt{x^2 + y^2 + 1}$ for $(x, y)$ in $\mathbb{R}^2$
   - (c)  $f(\mathbf{x}) = 1/\|\mathbf{x}\|^2$ for $\mathbf{x}$ in $\mathcal{O} = \{\mathbf{x} \text{ in } \mathbb{R}^3 \mid \mathbf{x} \neq \mathbf{0}\}$

2. Assume that the functions $f: \mathbb{R}^n \to \mathbb{R}$ and $g: \mathbb{R}^n \to \mathbb{R}$ are continuously differentiable. Find a formula for $\mathbf{D}(fg)(\mathbf{x})$ in terms of $\mathbf{D}f(\mathbf{x})$ and $\mathbf{D}g(\mathbf{x})$.

3. Suppose that the functions $f: \mathbb{R}^n \to \mathbb{R}$ and $g: \mathbb{R} \to \mathbb{R}$ are continuously differentiable. Find a formula for $\mathbf{D}(g \circ f)(\mathbf{x})$ in terms of $\mathbf{D}f(\mathbf{x})$ and $g'(f(\mathbf{x}))$.

4.  Suppose that the function $f: \mathbb{R}^n \to \mathbb{R}$ has first-order partial derivatives and that the point $\mathbf{x}$ in $\mathbb{R}^n$ is a local minimizer for $f: \mathbb{R}^n \to \mathbb{R}$, meaning that there is a positive number $r$ such that

$$f(\mathbf{x} + \mathbf{h}) \geq f(\mathbf{x}) \quad \text{if} \quad d(\mathbf{x}, \mathbf{x} + \mathbf{h}) < r.$$

Prove that $\mathbf{D}f(\mathbf{x}) = \mathbf{0}$.

5.  Show that in the case when $\mathbf{h} = a\mathbf{e}_i$, the mean value formula (13.28) is the same as the mean value formula (13.20).

6.  Define the function $f: \mathbb{R}^3 \to \mathbb{R}$ by

$$f(x, y, z) = xyz + x^2 + y^2 \quad \text{for } (x, y, z) \text{ in } \mathbb{R}^3.$$

The Mean Value Theorem implies that there is a number $\theta$ with $0 < \theta < 1$ for which

$$f(1, 1, 1) - f(0, 0, 0) = \frac{\partial f}{\partial x}(\theta, \theta, \theta) + \frac{\partial f}{\partial y}(\theta, \theta, \theta) + \frac{\partial f}{\partial z}(\theta, \theta, \theta).$$

Find the value of $\theta$.

7.  Suppose that the function $f: \mathbb{R}^2 \to \mathbb{R}$ has first-order partial derivatives and that $f(0, 0) = 1$, while

$$\frac{\partial f}{\partial x}(x, y) = 2 \quad \text{and} \quad \frac{\partial f}{\partial y}(x, y) = 3 \quad \text{for all } (x, y) \text{ in } \mathbb{R}^2.$$

Prove that

$$f(x, y) = 1 + 2x + 3y \quad \text{for all } (x, y) \text{ in } \mathbb{R}^2.$$

8.  Suppose that the function $f: \mathbb{R}^n \to \mathbb{R}$ is continuously differentiable. Let $\mathbf{x}$ be a point in $\mathbb{R}^n$. For $\mathbf{p}$ a nonzero point in $\mathbb{R}^n$ and $\alpha$ a nonzero real number, show that

$$\frac{\partial f}{\partial (\alpha \mathbf{p})}(\mathbf{x}) = \alpha \frac{\partial f}{\partial \mathbf{p}}(\mathbf{x}).$$

9.  Define the function $f: \mathbb{R}^2 \to \mathbb{R}$ by

$$f(x, y) = \begin{cases} (x/|y|)\sqrt{x^2 + y^2} & \text{if } y \neq 0 \\ 0 & \text{if } y = 0. \end{cases}$$

(a)  Prove that the function $f: \mathbb{R}^2 \to \mathbb{R}$ is not continuous at the point $(0, 0)$.

(b)  Prove that the function $f: \mathbb{R}^2 \to \mathbb{R}$ has directional derivatives in all directions at the point $(0, 0)$.

(c)  Prove that if $c$ is any number, then there is a vector $\mathbf{p}$ of norm 1 such that

$$\frac{\partial f}{\partial \mathbf{p}}(0, 0) = c.$$

(d)  Does (c) contradict Corollary 13.10?

10.  Consider the following assertions for a function $f: \mathbb{R}^2 \to \mathbb{R}$:

(a)  The function $f: \mathbb{R}^2 \to \mathbb{R}$ is continuously differentiable.

(b)  The function $f: \mathbb{R}^2 \to \mathbb{R}$ has directional derivatives in all directions at each point in $\mathbb{R}^2$.

(c)  The function $f: \mathbb{R}^2 \to \mathbb{R}$ has first-order partial derivatives at each point in $\mathbb{R}^2$. Explain the implications between these assertions.

11. Suppose that the function $f: \mathbb{R}^n \to \mathbb{R}$ is continuously differentiable. Define $K = \{\mathbf{x} \text{ in } \mathbb{R}^n \mid \|\mathbf{x}\| \leq 1\}$.

    (a) Prove that there is a point $\mathbf{x}$ in $K$ at which the function $f: K \to \mathbb{R}$ attains a smallest value.

    (b) Now suppose also that if $\mathbf{p}$ is any point in $\mathbb{R}^n$ of norm 1, then $\langle \mathbf{D}f(\mathbf{p}), \mathbf{p} \rangle > 0$. Show that the minimizer $\mathbf{x}$ in (a) has norm less than 1.

# Local Approximation of Real-Valued Functions

## First-Order Approximation, Tangent Planes, and Affine Functions

Let $\mathcal{O}$ be an open subset of Euclidean space $\mathbb{R}^n$ and suppose that we wish to analyze the behavior of the real-valued function $f: \mathcal{O} \to \mathbb{R}$ in a neighborhood of the point $\mathbf{x}$ in $\mathcal{O}$. (For instance, we might want to establish that the point $\mathbf{x}$ is a local maximizer or a local minimizer for the function.) One strategy for doing this is to choose another function $g: \mathcal{O} \to \mathbb{R}$ that is simpler than the given function $f: \mathcal{O} \to \mathbb{R}$ and is a good approximation of $f: \mathcal{O} \to \mathbb{R}$ near the point $\mathbf{x}$. Then we can see what properties the function $f: \mathcal{O} \to \mathbb{R}$ inherits from the simpler function $g: \mathcal{O} \to \mathbb{R}$.

For this strategy to be effective, it is first necessary to clearly identify the type of function we consider "simple." In this chapter, we will consider two types of simple functions—the affine functions and the quadratic functions—and sums of such functions. It is then necessary to clearly define what is meant by "good approximation." For a natural number $k$, two functions $f: \mathcal{O} \to \mathbb{R}$ and $g: \mathcal{O} \to \mathbb{R}$ are said to be $k$th-*order* *approximations* of one another at the point $\mathbf{x}$ in $\mathcal{O}$ provided that

$$\lim_{h\to 0} \frac{f(\mathbf{x}+\mathbf{h}) - g(\mathbf{x}+\mathbf{h})}{\|\mathbf{h}\|^k} = 0.$$

In this first section, we will define what is meant by an affine function and show that for a continuously differentiable function $f:\mathcal{O} \to \mathbb{R}$, at each point $\mathbf{x}$ in $\mathcal{O}$ there is an affine function that is a first-order approximation. For functions of two variables, the graph of this affine approximation is the tangent plane. The remaining two sections of this chapter are concerned with functions $f:\mathcal{O} \to \mathbb{R}$ that have continuous second-order partial derivatives. In Section 14.2, the second-order partial derivatives of $f:\mathcal{O} \to \mathbb{R}$ at the point $\mathbf{x}$ are organized into an $n \times n$ matrix, called the *Hessian matrix*, and a formula for second-order directional derivatives is established. We also associate a quadratic function with each $n \times n$ matrix and obtain estimates for the values attained by quadratic functions. In Section 14.3, we show how to find a function $g:\mathcal{O} \to \mathbb{R}$ that is the sum of an affine function and a quadratic function and is a second-order approximation to $f:\mathcal{O} \to \mathbb{R}$ at the point $\mathbf{x}$. This permits us to provide a criterion for deciding when a point $\mathbf{x}$ is a local maximizer or minimizer for the function $f:\mathcal{O} \to \mathbb{R}$.

Suppose that $I$ is an open interval of real numbers and that the function $f:I \to \mathbb{R}$ is differentiable. By definition, this means that if $x$ is a point in $I$, then

$$\lim_{h\to 0} \frac{f(x+h) - f(x)}{h} = f'(x).$$

If we rewrite the difference

$$\frac{f(x+h) - f(x)}{h} - f'(x) = \frac{f(x+h) - [f(x) + f'(x)h]}{h},$$

the above definition of derivative may be rewritten as

$$\lim_{h\to 0} \frac{f(x+h) - [f(x) + f'(x)h]}{h} = 0. \tag{14.1}$$

Moreover, from the Lagrange Remainder Theorem, which we proved in Chapter 8, it follows that if $f:I \to \mathbb{R}$ has a continuous second derivative, then

$$\lim_{h\to 0} \frac{f(x+h) - [f(x) + f'(x)h + (1/2)f''(x)h^2]}{h^2} = 0. \tag{14.2}$$

In this chapter, we wish to establish results analogous to the approximation formulas (14.1) and (14.2) for functions of *several* real variables. It is useful to introduce the following definition.

**DEFINITION**    *Let $\mathcal{O}$ be an open subset of $\mathbb{R}^n$ that contains the point $\mathbf{x}$. For a positive integer $k$, two functions $f:\mathcal{O} \to \mathbb{R}$ and $g:\mathcal{O} \to \mathbb{R}$ are said to be kth-order approximations of one another at the point $\mathbf{x}$ provided that*

$$\lim_{h\to 0} \frac{f(\mathbf{x}+\mathbf{h}) - g(\mathbf{x}+\mathbf{h})}{\|\mathbf{h}\|^k} = 0. \tag{14.3}$$

**EXAMPLE 14.1**   Define $f(h) = e^h$ for each number $h$. Then $f(0) = f'(0) = f''(0) = 1$. From formula (14.1), at the point $x = 0$, we see that the first-degree polynomial $g: \mathbb{R} \to \mathbb{R}$ defined by $g(h) = 1 + h$ is a first-order approximation of the function $f: \mathbb{R} \to \mathbb{R}$ at the point $x = 0$. On the other hand, from formula (14.2) at the point $x = 0$, we see that the second-degree polynomial $q: \mathbb{R} \to \mathbb{R}$, defined by $q(h) = 1 + h + (1/2)h^2$, is a second-order approximation of the function $f: \mathbb{R} \to \mathbb{R}$ at the point $x = 0$.   □

The following theorem provides an extension to functions of several variables of the approximation formula (14.1).

**THEOREM 14.1**   **The First-Order Approximation Theorem**   Let $\mathcal{O}$ be an open subset of $\mathbb{R}^n$ and suppose that the function $f: \mathcal{O} \to \mathbb{R}$ is continuously differentiable. Let $\mathbf{x}$ be a point in $\mathcal{O}$. Then

$$\lim_{\mathbf{h} \to 0} \frac{f(\mathbf{x} + \mathbf{h}) - [f(\mathbf{x}) + \langle \mathbf{D}f(\mathbf{x}), \mathbf{h} \rangle]}{\|\mathbf{h}\|} = 0. \tag{14.4}$$

**Proof**   Since $\mathbf{x}$ is an interior point of $\mathcal{O}$, we can choose a positive number $r$ such that the symmetric neighborhood $\mathcal{N}_r(\mathbf{x})$ is contained in $\mathcal{O}$. Fix a nonzero point $\mathbf{h}$ in $\mathbb{R}^n$ with $\|\mathbf{h}\| < r$. Then the point $\mathbf{x} + \mathbf{h}$ belongs to $\mathcal{N}_r(\mathbf{x})$ and so, by the Mean Value Theorem, we can select a number $\theta$ with $0 < \theta < 1$ such that

$$f(\mathbf{x} + \mathbf{h}) - f(\mathbf{x}) = \langle \mathbf{D}f(\mathbf{x} + \theta\mathbf{h}), \mathbf{h} \rangle.$$

Thus    $f(\mathbf{x} + \mathbf{h}) - f(\mathbf{x}) - \langle \mathbf{D}f(\mathbf{x}), \mathbf{h} \rangle = \langle \mathbf{D}f(\mathbf{x} + \theta\mathbf{h}) - \mathbf{D}f(\mathbf{x}), \mathbf{h} \rangle$,

so that, using the Cauchy-Schwarz Inequality, we obtain the estimate

$$|f(\mathbf{x} + \mathbf{h}) - f(\mathbf{x}) - \langle \mathbf{D}f(\mathbf{x}), \mathbf{h} \rangle| \le \|\mathbf{D}f(\mathbf{x} + \theta\mathbf{h}) - \mathbf{D}f(\mathbf{x})\| \cdot \|\mathbf{h}\|.$$

Dividing this estimate by $\|\mathbf{h}\|$, we obtain

$$\frac{|f(\mathbf{x} + \mathbf{h}) - [f(\mathbf{x}) + \langle \mathbf{D}f(\mathbf{x}), \mathbf{h} \rangle]|}{\|\mathbf{h}\|} \le \|\mathbf{D}f(\mathbf{x} + \theta\mathbf{h}) - \mathbf{D}f(\mathbf{x})\|. \tag{14.5}$$

But the function $f: \mathcal{O} \to \mathbb{R}$ has been assumed to be continuously differentiable, so

$$\lim_{\mathbf{h} \to 0} \|\mathbf{D}f(\mathbf{x} + \theta\mathbf{h}) - \mathbf{D}f(\mathbf{x})\| = 0,$$

and thus (14.4) follows from the estimate (14.5).   ∎

For a continuously differentiable function $f: \mathcal{O} \to \mathbb{R}$ whose domain $\mathcal{O}$ is an open subset of the plane $\mathbb{R}^2$ and a point $(x_0, y_0)$ in $\mathcal{O}$, if we denote a general point in $\mathcal{O}$ by $(x, y)$ and set $\mathbf{h} = (x - x_0, y - y_0)$, it is clear that $\mathbf{h}$ approaches 0 if and only if $(x, y)$ approaches $(x_0, y_0)$ and that $\|\mathbf{h}\| = \sqrt{(x - x_0)^2 + (y - y_0)^2}$. Hence the approximation property (14.4) may be rewritten as

$$\lim_{(x,y) \to (x_0, y_0)} \frac{f(x, y) - [f(x_0, y_0) + \partial f/\partial x(x_0, y_0)(x - x_0) + \partial f/\partial y(x_0, y_0)(y - y_0)]}{\sqrt{(x - x_0)^2 + (y - y_0)^2}} = 0.$$

$$\tag{14.6}$$

This last formula has a geometric interpretation involving the existence of a tangent plane. To describe this, we make the following definition.

**DEFINITION**   *Let $\mathcal{O}$ be an open subset of the plane $\mathbb{R}^2$ and suppose that the function $f: \mathcal{O} \to \mathbb{R}$ is continuous at the point $(x_0, y_0)$ in $\mathcal{O}$. By the tangent plane to the graph of $f: \mathcal{O} \to \mathbb{R}$ at the point $(x_0, y_0, f(x_0, y_0))$, we mean the graph of a function $\psi: \mathbb{R}^2 \to \mathbb{R}$ of the form*

$$\psi(x, y) = a + b(x - x_0) + c(y - y_0) \quad \text{for } (x, y) \text{ in } \mathbb{R}^2,$$

*where $a$, $b$, and $c$ are real numbers, which has the property that*

$$\lim_{(x,y) \to (x_0,y_0)} \frac{f(x, y) - \psi(x, y)}{\sqrt{(x - x_0)^2 + (y - y_0)^2}} = 0. \tag{14.7}$$

A continuous function of two variables $f: \mathcal{O} \to \mathbb{R}$ can have directional derivatives in all directions at the point $(x_0, y_0)$ in $\mathcal{O}$ without having a tangent plane at the point $(x_0, y_0, f(x_0, y_0))$ (see Exercises 17, 18, and 19); such examples occur because the definition of tangent plane requires that the limit (14.7) exist independently of the way in which the point $(x, y)$ approaches $(x_0, y_0)$. However, for continuously differentiable functions, the approximation property (14.6) is exactly what is required in order to prove the following corollary.

**COROLLARY 14.2**   Suppose that $\mathcal{O}$ is an open subset of the plane $\mathbb{R}^2$ that contains the point $(x_0, y_0)$, and that the function $f: \mathcal{O} \to \mathbb{R}$ is continuously differentiable. Then there is a tangent plane to the graph of the function $f: \mathcal{O} \to \mathbb{R}$ at the point $(x_0, y_0, f(x_0, y_0))$. This tangent plane is the graph of the function $\psi: \mathbb{R}^2 \to \mathbb{R}$ defined for $(x, y)$ in $\mathbb{R}^2$ by

$$\psi(x, y) = f(x_0, y_0) + \frac{\partial f}{\partial x}(x_0, y_0)(x - x_0) + \frac{\partial f}{\partial y}(x_0, y_0)(y - y_0). \tag{14.8}$$

**Proof**   For a general point $(x, y)$ in $\mathcal{O}$, set

$$h = (x, y) - (x_0, y_0),$$

and observe that

$$\langle \mathbf{D}f(x_0, y_0), \mathbf{h} \rangle = \frac{\partial f}{\partial x}(x_0, y_0)(x - x_0) + \frac{\partial f}{\partial y}(x_0, y_0)(y - y_0).$$

Since $f: \mathcal{O} \to \mathbb{R}$ is continuously differentiable, the First-Order Approximation Theorem implies that

$$\lim_{(x,y) \to (x_0,y_0)} \frac{f(x, y) - [f(x_0, y_0) + \partial f/\partial x(x_0, y_0)(x - x_0) + \partial f/\partial y(x_0, y_0)(y - y_0)]}{\sqrt{(x - x_0)^2 + (y - y_0)^2}} = 0.$$

But then

$$\lim_{(x,y) \to (x_0,y_0)} \frac{f(x, y) - \psi(x, y)}{\sqrt{(x - x_0)^2 + (y - y_0)^2}} = 0;$$

that is, the graph of the function $\psi : \mathbb{R}^2 \to \mathbb{R}$ is the tangent plane to the graph of the function $f : \mathcal{O} \to \mathbb{R}$ at the point $(x_0, \, y_0, \, f(x_0, \, y_0))$.    ∎

We can reason geometrically to see why the the tangent plane described in the preceding corollary is necessarily described by the equation (14.8).* Indeed, suppose that $\mathcal{O}$ is an open subset of the plane $\mathbb{R}^2$ and consider the function $f : \mathcal{O} \to \mathbb{R}$. At the point $(x_0, \, y_0)$ in $\mathcal{O}$, we look for a plane that is tangent to the graph of $f : \mathcal{O} \to \mathbb{R}$ at the point $(x_0, \, y_0, \, f(x_0, \, y_0))$. If the function $f : \mathcal{O} \to \mathbb{R}$ has first-order partial derivatives at $(x_0, \, y_0)$, then from the definition of partial derivative and the meaning, in the case of functions of a single real variable, of the derivative as the slope of the tangent line, it follows that the vectors

$$\mathbf{T}_1 = \left(1, \, 0, \, \frac{\partial f}{\partial x}(x_0, \, y_0)\right) \quad \text{and} \quad \mathbf{T}_2 = \left(0, 1, \frac{\partial f}{\partial y}(x_0, \, y_0)\right)$$

should be parallel to the proposed tangent plane. Thus the proposed tangent plane should have

$$\boldsymbol{\eta} = \mathbf{T}_1 \times \mathbf{T}_2 = (-\partial f / \partial x(x_0, \, y_0), \, -\partial f / \partial y(x_0, \, y_0), \, 1) \tag{14.9}$$

as a normal vector.

**FIGURE 14.1**

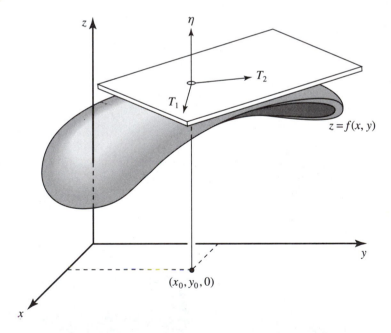

$$z = f(x, y)$$

$$(x_0, y_0, 0)$$

---

*The appendix on linear algebra includes a discussion of the properties of planes in $\mathbb{R}^3$ and the cross product of two vectors in $\mathbb{R}^3$.

The plane that passes through the point $(x_0, y_0, f(x_0, y_0))$ and is normal to $\eta$ consists of all points $(x, y, z)$ in $\mathbb{R}^3$ that satisfy the equation

$$\langle \eta, (x - x_0, y - y_0, z - f(x_0, y_0)) \rangle = 0,$$

and it is clear that this means that the point $(x, y, z)$ in $\mathbb{R}^3$ lies on the graph of the function defined by equation (14.8).

**EXAMPLE 14.2** Let $a$, $b$, and $c$ be positive numbers and let $\mathcal{O}$ be the open subset of the plane $\mathbb{R}^2$ defined by $\mathcal{O} = \{(x, y) \text{ in } \mathbb{R}^2 \mid (x/a)^2 + (y/b)^2 < 1\}$. Define the function $f : \mathcal{O} \to \mathbb{R}$ by

$$f(x, y) = c\sqrt{1 - (x/a)^2 - (y/b)^2} \quad \text{for } (x, y) \text{ in } \mathcal{O}.$$

The graph of this function is the upper half of an ellipsoid. Fix a point $(x_0, y_0)$ in $\mathcal{O}$ and let $z_0 = f(x_0, y_0)$. Since the function $f : \mathcal{O} \to \mathbb{R}$ is continuously differentiable, according to Corollary 14.2 there is a tangent plane to the graph of this function at the point $(x_0, y_0, z_0)$. A brief calculation of partial derivatives shows that

$$\frac{\partial f}{\partial x}(x_0, y_0) = -\frac{c^2 x_0}{z_0 a^2} \quad \text{and} \quad \frac{\partial f}{\partial y}(x_0, y_0) = -\frac{c^2 y_0}{z_0 b^2}.$$

Thus the tangent plane to the ellipsoid at the point $(x_0, y_0, z_0)$ consists of all the points $(x, y, z)$ such that

$$z = z_0 - \frac{c^2 x_0}{z_0 a^2}(x - x_0) - \frac{c^2 y_0}{z_0 b^2}(y - y_0). \qquad \square$$

The First-Order Approximation Theorem is also useful from another, less geometric, perspective. It enables us to approximate rather complicated functions by simpler ones, and to assert precisely the manner in which the functions are close to one another. Of course, the simplest type of function is a constant function. The next two simplest types of functions are the linear functions and the affine functions, which are defined as follows.

**DEFINITION**  A function $g : \mathbb{R}^n \to \mathbb{R}$ is said to be *affine* if it is defined by

$$g(\mathbf{u}) = c + \sum_{i=1}^{n} a_i u_i \quad \text{for } \mathbf{u} \text{ in } \mathbb{R}^n$$

where $c$ and the $a_i'$'s are prescribed numbers. If $c = 0$, the function is called *linear*.

**COROLLARY 14.3**  Let $\mathcal{O}$ be an open subset of $\mathbb{R}^n$ that contains the point $\mathbf{x}$ and suppose that the function $f : \mathcal{O} \to \mathbb{R}$ is continuously differentiable. Then there is an affine function that is a first-order approximation of $f : \mathcal{O} \to \mathbb{R}$ at the point $\mathbf{x}$, namely the function $g : \mathbb{R}^n \to \mathbb{R}$ defined by

$$g(\mathbf{u}) = f(\mathbf{x}) + \langle \mathbf{D}f(\mathbf{x}), \mathbf{u} - \mathbf{x} \rangle \quad \text{for } \mathbf{u} \text{ in } \mathbb{R}^n.$$

**Proof** Observe that the function $g: \mathbb{R}^n \to \mathbb{R}$ is affine and that

$$g(\mathbf{x} + \mathbf{h}) = f(\mathbf{x}) + \langle \mathbf{D}f(\mathbf{x}), \mathbf{h} \rangle \quad \text{for } \mathbf{x} + \mathbf{h} \text{ in } \mathbb{R}^n.$$

The First-Order Approximation Theorem asserts that the functions $f: \mathcal{O} \to \mathbb{R}$ and $g: \mathcal{O} \to \mathbb{R}$ are first-order approximations of one another at the point $\mathbf{x}$. ∎

**EXAMPLE 14.3** Define

$$f(x, y) = \sin(x - y - y^2) \quad \text{for } (x, y) \text{ in } \mathbb{R}^2.$$

The function $f: \mathbb{R}^2 \to \mathbb{R}$ is continuously differentiable. Computing partial derivatives at the point $(0, 0)$, we find that the affine function that is a first-order approximation of $f: \mathbb{R}^2 \to \mathbb{R}$ at the point $(0, 0)$ is defined by

$$\psi(x, y) = x - y \quad \text{for } (x, y) \text{ in } \mathbb{R}^2.$$

Computing partial derivatives at the point $(\pi, 0)$, we find that the affine function that is a first-order approximation of $f: \mathbb{R}^2 \to \mathbb{R}$ at the point $(\pi, 0)$ is given by

$$\psi(x, y) = \pi - x + y \quad \text{for } (x, y) \text{ in } \mathbb{R}^2. \qquad \square$$

**EXAMPLE 14.4** It follows from the First-Order Approximation Theorem that

$$\lim_{(x,y) \to (0,0)} \frac{\sqrt{1 + 4x + x^2 - y^2} - [1 + 2x]}{\sqrt{x^2 + y^2}} = 0.$$

Indeed, to see this, define

$$f(x, y) = \sqrt{1 + 4x + x^2 - y^2} \quad \text{for } x^2 + y^2 < r,$$

where $r$ is a positive number so small that $1 + 4x + x^2 - y^2$ is positive if $x^2 + y^2 < r$. Then the function $f: \mathcal{N}_r(0, 0) \to \mathbb{R}$ is continuously differentiable. The evaluation of this limit follows by applying the First-Order Approximation Theorem at the point $(0, 0)$. $\square$

---

**EXERCISES**

1. Define

$$f(x, y) = e^{2x+4y+1} \quad \text{for } (x, y) \text{ in } \mathbb{R}^2.$$

Find the equation of the tangent plane to the graph of the function $f: \mathbb{R}^2 \to \mathbb{R}$ at the point $(0, 0, e)$.

2. Define

$$f(x, y) = x^2 - xy + 2y^2 + x \quad \text{for } (x, y) \text{ in } \mathbb{R}^2.$$

At what points on the graph of the function $f: \mathbb{R}^2 \to \mathbb{R}$ is the tangent plane parallel to the $x$-$y$ plane?

3. Let $a$, $b$, and $c$ be positive numbers. The set of points $(x, y, z)$ in $\mathbb{R}^3$ such that

$$(x/a)^2 + (y/b)^2 - (z/c)^2 = 1$$

is called a *hyperboloid*. Find the equation of the tangent plane to this hyperboloid at a point $(x_0, y_0, z_0)$ on the hyperboloid with $z_0$ positive.

4.  Define
$$f(x, y) = 2y + x^2 + xy \quad \text{for } (x, y) \text{ in } \mathbb{R}^2.$$
Find the affine function that is a first-order approximation to the function $f \colon \mathbb{R}^2 \to \mathbb{R}$ at the point $(0, 0)$.

5.  Define
$$f(x, y) = e^{\sin(x-y)} \quad \text{for } (x, y) \text{ in } \mathbb{R}^2.$$
Find the affine function that is a first-order approximation to the function $f \colon \mathbb{R}^2 \to \mathbb{R}$ at the point $(0, 0)$.

6.  Define
$$f(x, y, z) = x^2 + y^2 + z \quad \text{for } (x, y, z) \text{ in } \mathbb{R}^3.$$
Find the affine function that is a first-order approximation to the function $f \colon \mathbb{R}^3 \to \mathbb{R}$ at the point $(0, 0, 0)$.

7.  Fix a point $\mathbf{x}$ in $\mathbb{R}^n$. Let $\mathbf{c}$ be a point in $\mathbb{R}^n$ and define the function $\psi \colon \mathbb{R}^n \to \mathbb{R}$ by
$$\psi(\mathbf{u}) = \langle \mathbf{c}, \mathbf{u} - \mathbf{x} \rangle \quad \text{for } \mathbf{u} \text{ in } \mathbb{R}^n.$$
    (a)  Show that the function $\psi \colon \mathbb{R}^n \to \mathbb{R}$ is affine.
    (b)  Now show that given any affine function $\psi \colon \mathbb{R}^n \to \mathbb{R}$, it is possible to choose points $\mathbf{x}$ and $\mathbf{c}$ in $\mathbb{R}^n$ so that $\psi \colon \mathbb{R}^n \to \mathbb{R}$ has the above form.

8.  Suppose that the functions $g \colon \mathbb{R}^2 \to \mathbb{R}$ and $h \colon \mathbb{R}^2 \to \mathbb{R}$ are both continuously differentiable. Define the function $f \colon \mathbb{R}^2 \to \mathbb{R}$ by
$$f(x, y) = xg(x, y) + yh(x, y) \quad \text{for } (x, y) \text{ in } \mathbb{R}^2.$$
Find the affine function that is a first-order approximation to $f \colon \mathbb{R}^2 \to \mathbb{R}$ at $(0, 0)$.

9.  Suppose that the functions $f \colon \mathbb{R}^2 \to \mathbb{R}$ and $h \colon \mathbb{R}^2 \to \mathbb{R}$ are continuously differentiable. Find necessary and sufficient conditions for these functions to be first-order approximations of each other at the point $(0, 0)$.

10.  Let $a$, $b$ and $c$ be real numbers. Prove that
$$\lim_{(x,y) \to (0,0)} \frac{ax^2 + bxy + y^2}{\sqrt{x^2 + y^2}} = 0.$$

11.  Prove that
$$\lim_{(x,y) \to (0,0)} \frac{\sin(2x + 2y) - 2x - 2y}{\sqrt{x^2 + y^2}} = 0.$$

12.  Prove that
$$\lim_{(x,y) \to (0,0)} \frac{(1 + 2x + y^2)^{3/2} - 1 - 3x}{\sqrt{x^2 + y^2}} = 0.$$

13.  Let $a$ be a real number. Prove that if
$$\lim_{(x,y) \to (0,0)} \frac{ax}{\sqrt{x^2 + y^2}} = 0,$$
then $a = 0$.

14.  Let $a$, $b$, and $c$ be real numbers. Prove that if
$$\lim_{(x,y) \to (0,0)} \frac{c + ax + by}{\sqrt{x^2 + y^2}} = 0,$$
then $c = a = b = 0$.

15.    Define

$$f(x, y) = x^2 + x + y \quad \text{and} \quad g(x, y) = x + y \quad \text{for } (x, y) \text{ in } \mathbb{R}^2.$$

Prove that
$$\lim_{(x,y) \to (0,0)} \frac{f(x, y) - g(x, y)}{\sqrt{x^2 + y^2}} = 0.$$

but that
$$\lim_{(x,y) \to (0,0)} \frac{f(x, y) - g(x, y)}{x^2 + y^2} \neq 0.$$

16.    Suppose that the function $f: \mathbb{R}^2 \to \mathbb{R}$ is continuous. Let $a$ and $b$ be any real numbers. Prove that

$$\lim_{(x,y) \to (0,0)} [f(x, y) - (f(0, 0) + ax + by)] = 0.$$

Is it true that
$$\lim_{(x,y) \to (0,0)} \frac{f(x, y) - [f(0, 0) + ax + by]}{\sqrt{x^2 + y^2}} = 0?$$

17.    Let $\mathcal{O}$ be an open subset of the plane $\mathbb{R}^2$ and suppose that the function $f: \mathcal{O} \to \mathbb{R}$ is continuous at the point $(x_0, y_0)$ in $\mathcal{O}$. For numbers $a, b,$ and $c$, define

$$\psi(x, y) = a + b(x - x_0) + c(y - y_0) \quad \text{for } (x, y) \text{ in } \mathbb{R}^2.$$

Assume that the graph of $\psi: \mathbb{R}^2 \to \mathbb{R}$ is tangent to the graph of $f: \mathcal{O} \to \mathbb{R}$ at the point $(x_0, y_0, f(x_0, y_0))$.
(a)    Show that $a = f(x_0, y_0)$.
(b)    Show that $f: \mathcal{O} \to \mathbb{R}$ has first-order partial derivatives at $(x_0, y_0)$ and that $b = \partial f/\partial x(x_0, y_0), c = \partial f/\partial y(x_0, y_0)$.
(c)    Use (a) and (b) to show that there can be only one tangent plane.

18.    Define

$$f(x, y) = \begin{cases} \sin(y^2/x) \cdot \sqrt{x^2 + y^2} & \text{if } x \neq 0 \\ 0 & \text{if } x = 0. \end{cases}$$

(a)    Show that the function $f: \mathbb{R}^2 \to \mathbb{R}$ is continuous at the point $(0, 0)$ and has directional derivatives in every direction at $(0, 0)$.
(b)    Show that there is no plane that is tangent to the graph of $f: \mathbb{R}^2 \to \mathbb{R}$ at the point $(0, 0, f(0, 0))$.

19.    Suppose that the continuous function $f: \mathbb{R}^2 \to \mathbb{R}$ has a tangent plane at the point $(x_0, y_0, f(x_0, y_0))$. Prove that the function $f: \mathbb{R}^2 \to \mathbb{R}$ has directional derivatives in all directions at the point $(x_0, y_0)$.

---

## 14.2    Quadratic Functions, Hessian Matrices, and Second Derivatives

For a function of several variables, it often is necessary to determine those points at which the function attains maximum and minimum values. For functions of a single variable, we developed the second-derivative test in Chapter 4. We wish to discover the appropriate correspondent of this test for functions of several variables. We will do this in Section 14.3. In preparation for this, in the present section we will find a formula for the second derivative of sections of functions of several variables and provide estimates of the values attained by quadratic functions.

Recall that by an $n \times n$ *matrix* we mean a rectangular array of real numbers consisting of $n$ rows and $n$ columns. If such an $n \times n$ matrix is denoted by $\mathbf{A}$, we will write

$$\mathbf{A} = [a_{ij}],$$

where for each pair of indices $i$ and $j$ with $1 \le i \le n$ and $1 \le j \le n$, $a_{ij}$ denotes the number in the $i$th row and $j$th column of the matrix $\mathbf{A}$; we call $a_{ij}$ the $ij$th entry of the matrix.

**DEFINITION**   *Given any $n \times n$ matrix $\mathbf{A} = [a_{ij}]$ and a point $\mathbf{x}$ in $\mathbb{R}^n$, by the symbol $\mathbf{Ax}$ we denote the point in $\mathbb{R}^n$ that for each index $i$ with $1 \le i \le n$ has $i$th component equal to the inner product of the $i$th row of $\mathbf{A}$ with $\mathbf{x}$. Thus*

$$\mathbf{Ax} \equiv \mathbf{y},$$

*where* $$y_i \equiv \sum_{j=1}^{n} a_{ij} x_j \quad \text{for each index } i \text{ with } 1 \le i \le n.$$

Since the point $\mathbf{Ax}$ has the $i$th component given by the inner product of $\mathbf{x}$ with the $i$th row of the matrix $\mathbf{A}$, if we denote by $\mathbf{A}_i$ the $i$th row of the matrix $\mathbf{A}$, then

$$\mathbf{Ax} = (\langle \mathbf{A}_1, \mathbf{x} \rangle, \dots, \langle \mathbf{A}_i, \mathbf{x} \rangle, \dots, \langle \mathbf{A}_n, \mathbf{x} \rangle).$$

**DEFINITION**   *Let $\mathbf{A} = [a_{ij}]$ be an $n \times n$ matrix. The function $Q \colon \mathbb{R}^n \to \mathbb{R}$ defined by*

$$Q(\mathbf{x}) \equiv \langle \mathbf{Ax}, \mathbf{x} \rangle \quad \text{for } \mathbf{x} \text{ in } \mathbb{R}^n$$

*is called the quadratic function associated with the matrix $\mathbf{A}$.*

Observe that

$$Q(\mathbf{x}) = \sum_{i=1}^{n} \sum_{j=1}^{n} a_{ij} x_j x_i \quad \text{for } \mathbf{x} \text{ in } \mathbb{R}^n,$$

so $Q(\mathbf{x})$ is a linear combination of $x_j x_i$'s; hence the name *quadratic function*.

**EXAMPLE 14.5**   The $2 \times 2$ matrix $\mathbf{A} = \begin{bmatrix} a & b \\ b & c \end{bmatrix}$ has associated with it the quadratic function $Q \colon \mathbb{R}^2 \to \mathbb{R}$ defined by

$$Q(x, y) = ax^2 + 2bxy + cy^2 \quad \text{for } (x, y) \text{ in } \mathbb{R}^2. \qquad \square$$

**EXAMPLE 14.6**    Consider the $3 \times 3$ matrix $\mathbf{A}$ defined by

$$\mathbf{A} = \begin{bmatrix} \lambda_1 & 0 & 0 \\ 0 & \lambda_2 & 0 \\ 0 & 0 & \lambda_3 \end{bmatrix}.$$

Associated with the matrix $\mathbf{A}$ is the quadratic function $Q \colon \mathbb{R}^3 \to \mathbb{R}$ defined by

$$Q(x, y, z) = \lambda_1 x^2 + \lambda_2 y^2 + \lambda_3 z^2 \quad \text{for } (x, y, z) \text{ in } \mathbb{R}^3. \qquad \square$$

**DEFINITION**    *Let $\mathcal{O}$ be an open subset of $\mathbb{R}^n$ and suppose that the function $f \colon \mathcal{O} \to \mathbb{R}$ has second-order partial derivatives. The Hessian matrix of the function $f \colon \mathcal{O} \to \mathbb{R}$ at the point $\mathbf{x}$ in $\mathcal{O}$, denoted by*

$$\mathbf{D}^2 f(\mathbf{x}),$$

*is defined to be the $n \times n$ matrix that for each pair of indices $i$ and $j$ with $1 \le i \le n$ and $1 \le j \le n$ has the $ij$th entry defined by*

$$(\mathbf{D}^2 f(\mathbf{x}))_{ij} \equiv \frac{\partial^2 f}{\partial x_j \partial x_i}(\mathbf{x}).$$

Observe that for each index $i$ with $1 \le i \le n$, the $i$th row of the Hessian matrix $\mathbf{D}^2 f(\mathbf{x})$ is the derivative vector of the function $\partial f / \partial x_i \colon \mathcal{O} \to \mathbb{R}$ at the point $\mathbf{x}$. Also observe that in view of the equality of cross–partial derivatives asserted in Theorem 13.3, it follows that the Hessian matrix $\mathbf{D}^2 f(\mathbf{x})$ is *symmetric*, that is, the $ij$th entry equals the $ji$th entry, provided that the function $f \colon \mathcal{O} \to \mathbb{R}$ has continuous second-order partial derivatives.

**EXAMPLE 14.7**    Let $\mathcal{O}$ be an open subset of $\mathbb{R}^2$ containing the point $(x_0, y_0)$ and suppose that the function $f \colon \mathcal{O} \to \mathbb{R}$ has second-order partial derivatives. Then the Hessian matrix of $f \colon \mathcal{O} \to \mathbb{R}$ at $(x_0, y_0)$ is given by

$$\mathbf{D}^2 f(x_0, y_0) = \begin{bmatrix} \partial^2 f / \partial x \partial x(x_0, y_0) & \partial^2 f / \partial y \partial x(x_0, y_0) \\ \partial^2 f / \partial x \partial y(x_0, y_0) & \partial^2 f / \partial y \partial y(x_0, y_0) \end{bmatrix}. \qquad \square$$

**EXAMPLE 14.8**    Define

$$f(x, y) = x^2 y + y^2 + 1 \quad \text{for } (x, y) \text{ in } \mathbb{R}^2.$$

A short computation of second-order derivatives shows that at the point $(x_0, y_0)$ in $\mathbb{R}^2$,

$$\mathbf{D}^2 f(x_0, y_0) = \begin{bmatrix} 2y_0 & 2x_0 \\ 2x_0 & 2 \end{bmatrix}.$$

In particular, at the point $(1, 1)$,

$$\mathbf{D}^2 f(1, 1) = \begin{bmatrix} 2 & 2 \\ 2 & 2 \end{bmatrix}. \qquad \square$$

**EXAMPLE 14.9**   Let $\mathcal{O}$ be an open subset of $\mathbb{R}^3$ and suppose that the function $f: \mathcal{O} \to \mathbb{R}$ has second-order partial derivatives. Then the Hessian matrix of $f: \mathcal{O} \to \mathbb{R}$ at the point $(x_0, y_0, z_0)$ in $\mathcal{O}$ is given by

$$\mathbf{D}^2 f(x_0, y_0, z_0)$$

$$= \begin{bmatrix} \partial^2 f/\partial x \partial x(x_0, y_0, z_0) & \partial^2 f/\partial y \partial x(x_0, y_0, z_0) & \partial^2 f/\partial z \partial x(x_0, y_0, z_0) \\ \partial^2 f/\partial x \partial y(x_0, y_0, z_0) & \partial^2 f/\partial y \partial y(x_0, y_0, z_0) & \partial^2 f/\partial z \partial y(x_0, y_0, z_0) \\ \partial^2 f/\partial x \partial z(x_0, y_0, z_0) & \partial^2 f/\partial y \partial z(x_0, y_0, z_0) & \partial^2 f/\partial z \partial z(x_0, y_0, z_0) \end{bmatrix}.$$

$\square$

**EXAMPLE 14.10**   Define

$$f(x, y, z) = \sin(xyz) + e^{x+y} + z^2 \quad \text{for } (x, y, z) \text{ in } \mathbb{R}^3.$$

A short computation of second-order derivatives shows that at the point $(0, 0, 0)$,

$$\mathbf{D}^2 f(0, 0, 0) = \begin{bmatrix} 1 & 1 & 0 \\ 1 & 1 & 0 \\ 0 & 0 & 2 \end{bmatrix}.$$

$\square$

The following theorem explains how the Hessian matrix enters into second derivative calculations of the sections of a function of several variables.

**THEOREM 14.4**   Let $\mathcal{O}$ be an open subset of $\mathbb{R}^n$ that contains the point $\mathbf{x}$ and suppose that the function $f: \mathcal{O} \to \mathbb{R}$ has continuous second-order partial derivatives. Choose a positive number $r$ such that the symmetric neighborhood $\mathcal{N}_r(\mathbf{x})$ is contained in $\mathcal{O}$. Then if $\|\mathbf{h}\| < r$ and $|t| \leq 1$,

$$\frac{d}{dt}[f(\mathbf{x} + t\mathbf{h})] = \langle \mathbf{D}f(\mathbf{x} + t\mathbf{h}), \mathbf{h} \rangle \tag{14.10}$$

and

$$\frac{d^2}{dt^2}[f(\mathbf{x} + t\mathbf{h})] = \langle \mathbf{D}^2 f(\mathbf{x} + t\mathbf{h})\mathbf{h}, \mathbf{h} \rangle. \tag{14.11}$$

**Proof**   Let $I$ be an open interval of real numbers that contains the points 0 and 1 and is such that the point $\mathbf{x} + t\mathbf{h}$ belongs to $\mathcal{O}$ if $t$ belongs to $I$. Define

$$\phi(t) = f(\mathbf{x} + t\mathbf{h}) \quad \text{for } t \text{ in } I.$$

The Directional Derivative Theorem implies that if $t$ is in $I$, then

$$\phi'(t) = \frac{d}{dt}[f(\mathbf{x} + t\mathbf{h})] = \langle \mathbf{D}f(\mathbf{x} + t\mathbf{h}), \mathbf{h} \rangle = \sum_{i=1}^{n} h_i \frac{\partial f}{\partial x_i}(\mathbf{x} + t\mathbf{h}).$$

However, for each index $i$ with $1 \leq i \leq n$, we can again apply the Directional Derivative

Theorem to the partial derivative $\partial f/\partial x_i \colon \mathcal{O} \to \mathbb{R}$, and hence by differentiating each side of the preceding equality, we see that

$$\phi''(t) = \frac{d}{dt}[\phi'(t)] = \sum_{i=1}^{n} h_i \frac{d}{dt}\left[\frac{\partial f}{\partial x_i}(\mathbf{x}+t\mathbf{h})\right]$$

$$= \sum_{i=1}^{n} h_i \left\langle \mathbf{D}\left[\frac{\partial f}{\partial x_i}\right](\mathbf{x}+t\mathbf{h}), \mathbf{h}\right\rangle$$

$$= \sum_{i=1}^{n} \left\langle \mathbf{D}\left[\frac{\partial f}{\partial x_i}\right](\mathbf{x}+t\mathbf{h}), \mathbf{h}\right\rangle h_i$$

$$= \langle \mathbf{D}^2 f(\mathbf{x}+t\mathbf{h})\mathbf{h}, \mathbf{h}\rangle. \qquad \blacksquare$$

The preceding formula (14.11) will be very useful in the next section for establishing a second-derivative criterion for extreme points of a function of several variables. But we will also need some estimates of the sizes of the values attained by quadratic functions. The remainder of this section will be devoted to obtaining these estimates.

**DEFINITION**   *The norm of an $n \times n$ matrix* $\mathbf{A} = [a_{ij}]$, *denoted by* $\|\mathbf{A}\|$, *is defined by*

$$\|\mathbf{A}\| \equiv \sqrt{\sum_{j=1}^{n}\sum_{i=1}^{n} a_{ij}^2}.$$

Observe that if we define the point $\mathbf{A}_i$ in $\mathbb{R}^n$ to be the $i$th row of the $n \times n$ matrix $\mathbf{A}$, then the square of the norm of $\mathbf{A}$ may be written as

$$\|\mathbf{A}\|^2 = \|\mathbf{A}_1\|^2 + \|\mathbf{A}_2\|^2 + \cdots + \|\mathbf{A}_n\|^2.$$

The above definition of the norm of a matrix is introduced because with this definition of the norm, we have the following useful variant of the Cauchy-Schwarz Inequality.

**THEOREM 14.5**   **A Generalized Cauchy-Schwarz Inequality**   Let $\mathbf{A}$ be an $n \times n$ matrix and let $\mathbf{u}$ be a point in $\mathbb{R}^n$. Then

$$\|\mathbf{Au}\| \le \|\mathbf{A}\| \cdot \|\mathbf{u}\|. \qquad (14.12)$$

**Proof**   Squaring both sides of (14.12), it is clear that this inequality holds if and only if

$$\|\mathbf{Au}\|^2 \le \|\mathbf{A}\|^2 \|\mathbf{u}\|^2. \qquad (14.13)$$

If for each index $i$ with $1 \le i \le n$ we let the point $\mathbf{A}_i$ in $\mathbb{R}^n$ be the $i$th row of $\mathbf{A}$, then

$$\mathbf{Au} = (\langle \mathbf{A}_1, \mathbf{u}\rangle, \dots, \langle \mathbf{A}_n, \mathbf{u}\rangle),$$

so that

$$\|\mathbf{Au}\|^2 = (\langle \mathbf{A}_1, \mathbf{u}\rangle)^2 + \cdots + (\langle \mathbf{A}_n, \mathbf{u}\rangle)^2.$$

Thus, by the standard Cauchy-Schwarz Inequality,

$$\|\mathbf{Au}\|^2 \leq \|\mathbf{A}_1\|^2 \|\mathbf{u}\|^2 + \cdots + \|\mathbf{A}_n\|^2 \|\mathbf{u}\|^2$$
$$= (\|\mathbf{A}_1\|^2 + \cdots + \|\mathbf{A}_n\|^2) \|\mathbf{u}\|^2$$
$$= \|\mathbf{A}\|^2 \|\mathbf{u}\|^2.$$

We have verified inequality (14.13), and hence also inequality (14.12).   ∎

**COROLLARY 14.6**   Let $\mathbf{A}$ be an $n \times n$ matrix, let $Q: \mathbb{R}^n \to \mathbb{R}$ be the quadratic function associated with $\mathbf{A}$, and let $\mathbf{u}$ be a point in $\mathbb{R}^n$. Then

$$|Q(\mathbf{u})| \leq \|\mathbf{A}\| \|\mathbf{u}\|^2. \qquad (14.14)$$

**Proof**   By definition,

$$|Q(\mathbf{u})| = |\langle \mathbf{Au}, \mathbf{u} \rangle|.$$

Thus, if we first use the standard Cauchy-Schwarz Inequality and then the Generalized Cauchy-Schwarz Inequality, it follows that

$$|Q(\mathbf{u})| = |\langle \mathbf{Au}, \mathbf{u} \rangle|$$
$$\leq \|\mathbf{Au}\| \cdot \|\mathbf{u}\|$$
$$\leq \|\mathbf{A}\| \|\mathbf{u}\|^2.$$   ∎

**DEFINITION**   *An $n \times n$ matrix $\mathbf{A}$ is said to be positive definite provided that*

$$\langle \mathbf{Au}, \mathbf{u} \rangle > 0 \quad \text{for all nonzero points } \mathbf{u} \text{ in } \mathbb{R}^n$$

*and is said to be negative definite provided that*

$$\langle \mathbf{Au}, \mathbf{u} \rangle < 0 \quad \text{for all nonzero points } \mathbf{u} \text{ in } \mathbb{R}^n.$$

It is possible to give precise criteria in terms of the entries of a matrix that determine when it is positive definite or negative definite. The simplest case is the $2 \times 2$ case.

**PROPOSITION 14.7**   The $2 \times 2$ symmetric matrix

$$\mathbf{A} = \begin{bmatrix} a & b \\ b & c \end{bmatrix}$$

is positive definite if and only if

$$a > 0 \quad \text{and} \quad ac - b^2 > 0.$$

**Proof**   Observe that the quadratic function $Q: \mathbb{R}^2 \to \mathbb{R}$ associated with the matrix **A** has the form

$$Q(x, y) = ax^2 + 2bxy + cy^2 \quad \text{for } (x, y) \text{ in } \mathbb{R}^2.$$

For points $(x, y)$ with $y \neq 0$, set $t = x/y$ and $p(t) = at^2 + 2bt + c$. Observe that

$$Q(x, y) = y^2[a(x/y)^2 + 2b(x/y) + c] = y^2 p(t).$$

The second-degree polynomial $p(t)$ is positive for all $t$ if and only if $a > 0$ and $ac - b^2 > 0$. In the case when $y = 0$, observe that $a > 0$ if and only if $Q(x, 0) = ax^2 > 0$ for all $x \neq 0$.  ∎

A similar argument shows that the matrix **A** in Proposition 14.7 is negative definite if and only if

$$a < 0 \quad \text{and} \quad ac - b^2 > 0.$$

---

**PROPOSITION 14.8**   Let **A** be an $n \times n$ positive definite matrix. Then there is a positive number $c$ such that

$$Q(\mathbf{u}) = \langle \mathbf{Au}, \mathbf{u} \rangle \geq c\|\mathbf{u}\|^2 \quad \text{for all } \mathbf{u} \text{ in } \mathbb{R}^n.$$

**Proof**   Since the quadratic function $Q: \mathbb{R}^n \to \mathbb{R}$ is a function that is the sum and product of component projection functions, it is continuous. On the other hand, the unit sphere $S = \{\mathbf{u} \text{ in } \mathbb{R}^n \mid \|\mathbf{u}\| = 1\}$ is a closed and bounded subset of $\mathbb{R}^n$. According to the corollary of the Extreme Value Theorem, there is a point in $S$ that is a minimizer for the restriction of the quadratic function to $S$. Define $c$ to be the value of the quadratic function at this minimizer. Observe that $c$ is positive, since we have assumed that the matrix **A** is positive definite, and that

$$Q(\mathbf{u}) \geq c \quad \text{for all points } \mathbf{u} \text{ in } S. \tag{14.15}$$

Now, for all points **u** in $\mathbb{R}^n$ and all real numbers $\lambda$, $\mathbf{A}(\lambda\mathbf{u}) = \lambda\mathbf{Au}$, so

$$Q(\lambda\mathbf{u}) = \lambda^2 Q(\mathbf{u}). \tag{14.16}$$

Moreover, note that if **u** is any nonzero point in $\mathbb{R}^n$, then

$$Q(\mathbf{u}) = Q\left(\|\mathbf{u}\| \frac{\mathbf{u}}{\|\mathbf{u}\|}\right).$$

From equality (14.16), it follows that

$$Q(\mathbf{u}) = \|\mathbf{u}\|^2 Q\left(\frac{\mathbf{u}}{\|\mathbf{u}\|}\right).$$

But $\mathbf{u}/\|\mathbf{u}\|$ is a point in $S$, so by inequality (14.15),

$$Q(\mathbf{u}) \geq c\|\mathbf{u}\|^2.$$

It is clear that this inequality also holds if $\mathbf{u} = \mathbf{0}$.  ∎

**EXERCISES**

1.  Define

$$f(x, y) = e^{xy} + x^2 + 2xy \quad \text{for } (x, y) \text{ in } \mathbb{R}^2.$$

(a)   Define $\phi: \mathbb{R} \to \mathbb{R}$ by $\phi(t) = f(2t, 3t)$ for $t$ in $\mathbb{R}$. Calculate $\phi''(0)$ directly.
(b)   Find the Hessian matrix of the function $f: \mathbb{R}^2 \to \mathbb{R}$ at the point $(0, 0)$ and use formula (14.11) to calculate

$$\phi''(0) = \frac{d^2}{dt^2}[f(2t, 3t)]\Big|_{t=0}.$$

2.  Define

$$f(x, y) = \cos(xy) + x^2 y \quad \text{for } (x, y) \text{ in } \mathbb{R}^2.$$

(a)   Define $\phi: \mathbb{R} \to \mathbb{R}$ by $\phi(t) = f(1 - t, t/2)$ for $t$ in $\mathbb{R}$. Calculate $\phi''(0)$ directly.
(b)   Find the Hessian matrix of the function $f: \mathbb{R}^2 \to \mathbb{R}$ at the point $(1, 0)$ and use formula (14.11) to calculate

$$\phi''(0) = \frac{d^2}{dt^2}[f(1 - t, t/2)]\Big|_{t=0}.$$

3.  Let $\mathcal{O} = \{(x, y, z) \text{ in } \mathbb{R}^3 \mid xyz > -1\}$, and define $g: \mathcal{O} \to \mathbb{R}$ by

$$g(x, y, z) = \sqrt{1 + xyz} \quad \text{for } (x, y, z) \text{ in } \mathcal{O}.$$

(a)   Define $\phi: \mathbb{R} \to \mathbb{R}$ by $\phi(t) = g(3t, 1 - t, t)$ for $t$ in $\mathbb{R}$. Calculate $\phi''(0)$ directly.
(b)   Find the Hessian matrix of the function $g: \mathcal{O} \to \mathbb{R}$ at the point $(0, 1, 0)$ and use formula (14.11) to calculate

$$\phi''(0) = \frac{d^2}{dt^2}[g(3t, 1 - t, t)]\Big|_{t=0}.$$

4.  Suppose that the function $f: \mathbb{R}^2 \to \mathbb{R}$ has continuous second-order partial derivatives, and at the origin $(0, 0)$, suppose that

$$\frac{\partial f}{\partial x}(0, 0) = 0 \quad \text{and} \quad \frac{\partial f}{\partial y}(0, 0) = 0.$$

Let $\mathbf{h}$ be a nonzero point in the plane $\mathbb{R}^2$ and suppose that

$$\langle \mathbf{D}^2 f(0, 0)\mathbf{h}, \mathbf{h} \rangle > 0.$$

Use the single-variable theory to prove that there is some positive number $r$ such that

$$f(t\mathbf{h}) > f(0, 0) \quad \text{if} \quad 0 < |t| < r.$$

5.  In Exercise 4, suppose in fact that

$$\langle \mathbf{D}^2 f(0, 0)\mathbf{h}, \mathbf{h} \rangle > 0 \quad \text{for every nonzero point } \mathbf{h} \text{ in } \mathbb{R}^2.$$

Explain why this is not sufficient to directly conclude that the origin is a local minimizer of the function $f: \mathbb{R}^2 \to \mathbb{R}$.

6.  Let $a$, $b$, and $c$ be real numbers and for each number $t$, define $p(t) = at^2 + 2bt + c$.
(a)   Show that $p(t) > 0$ for every number $t$ if and only if $a > 0$ and $ac - b^2 > 0$.
(b)   Show that $p(t) < 0$ for every number $t$ if and only if $a < 0$ and $ac - b^2 > 0$.

7.  Suppose that $\mathbf{A}$ is a $3 \times 3$ symmetric matrix that is positive definite. Show that each of the following four properties hold, and from each of them obtain information about the entries of the matrix $\mathbf{A}$.
(a)   $\langle \mathbf{Ae}_1, \mathbf{e}_1 \rangle > 0$, $\langle \mathbf{Ae}_2, \mathbf{e}_2 \rangle > 0$, and $\langle \mathbf{Ae}_3, \mathbf{e}_3 \rangle > 0$.
(b)   $\langle \mathbf{Au}, \mathbf{u} \rangle > 0$ for all nonzero $\mathbf{u} = (h, k, 0)$; $h$, $k$ in $\mathbb{R}$.

(c)   $\langle \mathbf{Au}, \mathbf{u} \rangle > 0$ for all nonzero $\mathbf{u} = (0, h, k)$; $h, k$ in $\mathbb{R}$.

(d)   $\langle \mathbf{Au}, \mathbf{u} \rangle > 0$ for all nonzero $\mathbf{u} = (h, 0, k)$; $h, k$ in $\mathbb{R}$.

8.   For each of the following quadratic functions, find a $2 \times 2$ matrix with which it is associated.

(a)   $h(x, y) = x^2 - y^2$   for $(x, y)$ in $\mathbb{R}^2$.

(b)   $g(x, y) = x^2 + 8xy + y^2$   for $(x, y)$ in $\mathbb{R}^2$.

9.   By making a suitable choice of the matrix $\mathbf{A}$, show that the Generalized Cauchy-Schwarz Inequality contains the standard Cauchy-Schwarz Inequality as a special case.

10.   Find a $3 \times 3$ matrix that is associated with the quadratic function $Q: \mathbb{R}^3 \to \mathbb{R}$ defined by

$$Q(x, y, z) = x^2 - y^2 + 3xy + yz - z^2 \quad \text{for } (x, y, z) \text{ in } \mathbb{R}^3.$$

11.   Define the function $Q: \mathbb{R} \to \mathbb{R}$ by $Q(x) = x^4$. Observe that

$$Q(x) > 0 \quad \text{for all } x \neq 0.$$

Show that there is no positive number $c$ such that

$$Q(x) \geq cx^2 \quad \text{for all } x \neq 0.$$

Explain why this does not contradict Proposition 14.8.

---

## 14.3  Second-Order Approximation and the Second-Derivative Test

---

**DEFINITION**   *Let $A$ be a subset of $\mathbb{R}^n$ that contains the point $\mathbf{x}$ and consider the function $f: A \to \mathbb{R}$:*

(i) *The point $\mathbf{x}$ is called a local maximizer for the function $f: A \to \mathbb{R}$ provided that there is some positive number $r$ such that*

$$f(\mathbf{x} + \mathbf{h}) \leq f(\mathbf{x}) \quad \text{if } \mathbf{x} + \mathbf{h} \text{ is in } A \text{ and } \|\mathbf{h}\| < r.$$

(ii) *The point $\mathbf{x}$ is called a local minimizer for the function $f: A \to \mathbb{R}$ provided that there is some positive number $r$ such that*

$$f(\mathbf{x} + \mathbf{h}) \geq f(\mathbf{x}) \quad \text{if } \mathbf{x} + \mathbf{h} \text{ is in } A \text{ and } \|\mathbf{h}\| < r.$$

(iii) *The point $\mathbf{x}$ is called a local extreme point for the function $f: A \to \mathbb{R}$ provided that it is either a local minimizer or a local maximizer for $f: A \to \mathbb{R}$.*

We immediately find the following necessary condition for a point to be a local extreme point for a function.

**PROPOSITION 14.9**   Let $\mathcal{O}$ be an open subset of $\mathbb{R}^n$ that contains the point $\mathbf{x}$ and suppose that the function $f\colon \mathcal{O} \to \mathbb{R}$ has first-order partial derivatives. If the point $\mathbf{x}$ is a local extreme point for the function $f\colon \mathcal{O} \to \mathbb{R}$, then

$$\mathbf{D}f(\mathbf{x}) = \mathbf{0}. \tag{14.17}$$

**Proof**   Since $\mathbf{x}$ is an interior point of $\mathcal{O}$, we can choose a positive number $r$ such that the symmetric neighborhood $\mathcal{N}_r(\mathbf{x})$ is contained in $\mathcal{O}$. Fix an index $i$ with $1 \le i \le n$ and define the function $\phi\colon (-r, r) \to \mathbb{R}$ by

$$\phi(t) = f(\mathbf{x} + t\mathbf{e}_i) \quad \text{for} \quad |t| < r.$$

Then the point 0 is an extreme point of the function $\phi\colon (-r, r) \to \mathbb{R}$, so

$$\phi'(0) = \frac{\partial f}{\partial x_i}(\mathbf{x}) = 0.$$

But this holds for each index $i$ with $1 \le i \le n$, which means that (14.17) holds.   ■

Observe that in order to search for local extreme points, we must first find those points $\mathbf{x}$ in $\mathcal{O}$ at which

$$\mathbf{D}f(\mathbf{x}) = \mathbf{0}. \tag{14.18}$$

However, equation (14.18) is a system of $n$ scalar equations in $n$ real unknowns. Unless the function $f\colon \mathcal{O} \to \mathbb{R}$ has a very simple form, it is not possible to find explicit solutions of (14.18). This should not be so surprising since, in fact, even for a differentiable function of a single variable $f\colon \mathbb{R} \to \mathbb{R}$, unless $f\colon \mathbb{R} \to \mathbb{R}$ is very simple, it is not possible to explicitly find all of the numbers $x$ that are solutions of the equation

$$f'(x) = 0.$$

Moreover, there may exist points $\mathbf{x}$ in $\mathcal{O}$ at which $\mathbf{D}f(\mathbf{x}) = \mathbf{0}$, but which are not extreme points for the function $f\colon \mathcal{O} \to \mathbb{R}$.

**EXAMPLE 14.11**   Consider the functions $f\colon \mathbb{R}^2 \to \mathbb{R}$, $g\colon \mathbb{R}^2 \to \mathbb{R}$, and $h\colon \mathbb{R}^2 \to \mathbb{R}$ defined by

$$f(x, y) = \quad x^2 + y^2 \quad \text{for } (x, y) \text{ in } \mathbb{R}^2$$
$$g(x, y) = -x^2 - y^2 \quad \text{for } (x, y) \text{ in } \mathbb{R}^2$$
$$h(x, y) = \quad x^2 - y^2 \quad \text{for } (x, y) \text{ in } \mathbb{R}^2.$$

Each of these functions is continuously differentiable, and we observe that for each of them the origin $(0, 0)$ is the only point in the plane where the derivative vector is zero. We see that the point $(0, 0)$ is a local minimizer for $f\colon \mathbb{R}^2 \to \mathbb{R}$ and a local maximizer for $g\colon \mathbb{R}^2 \to \mathbb{R}$, but fails to be a local extreme point for $h\colon \mathbb{R}^2 \to \mathbb{R}$.   □

In the above example, it is only because the functions were so simple that we were able to determine their behavior near the point $(0, 0)$. In order to analyze more complicated functions, we need to organize the set of second-order partial derivatives in a manner that will allow us to formulate a "second-derivative test" for functions of several variables. We will see that the right way to do this is to arrange the second-order derivatives in the Hessian matrix and then to examine the quadratic function associated with this matrix.

In Chapter 8, we considered the approximation of a function of a single variable by a Taylor polynomial and obtained estimates for the difference between the function and its polynomial approximation. The Lagrange Remainder Theorem provides such estimates, and we will need the following special case of this theorem.

**THEOREM 14.10**

Let $I$ be an open interval of real numbers and suppose that the function $f: I \to \mathbb{R}$ has a second derivative. Then for each pair of points $x$ and $x + h$ in the interval $I$, there is a number $\theta$ with $0 < \theta < 1$ such that

$$f(x + h) = f(x) + f'(x)h + \frac{1}{2}f''(x + \theta h)h^2. \tag{14.19}$$

From this result about functions of a single variable and the derivative calculations for functions of several variables that we obtained in Section 14.1, we obtain the following theorem.

**THEOREM 14.11**

Let $\mathcal{O}$ be an open subset of $\mathbb{R}^n$ and suppose that the function $f: \mathcal{O} \to \mathbb{R}$ has continuous second-order partial derivatives. Then for each pair of points $\mathbf{x}$ and $\mathbf{x} + \mathbf{h}$ in $\mathcal{O}$ that have the property that the segment between these points also lies in $\mathcal{O}$, there is a number $\theta$ with $0 < \theta < 1$ such that

$$f(\mathbf{x} + \mathbf{h}) = f(\mathbf{x}) + \langle \mathbf{D}f(\mathbf{x}), \mathbf{h} \rangle + \frac{1}{2}\langle \mathbf{D}^2 f(\mathbf{x} + \theta \mathbf{h})\mathbf{h}, \mathbf{h} \rangle. \tag{14.20}$$

**Proof** Choose $I$ to be an open interval of real numbers containing both $0$ and $1$ such that $\mathbf{x} + t\mathbf{h}$ belongs to $\mathcal{O}$ if $t$ is in $I$. Then define the function $\psi: I \to \mathbb{R}$ by

$$\psi(t) = f(\mathbf{x} + t\mathbf{h}) \quad \text{for } t \text{ in } I.$$

Observe that Theorem 14.4 implies that the function $\psi: I \to \mathbb{R}$ has a second derivative and that we have the following formulas for the first and second derivatives:

$$\psi'(t) = \langle \mathbf{D}f(\mathbf{x} + t\mathbf{h}), \mathbf{h} \rangle \quad \text{and} \quad \psi''(t) = \langle \mathbf{D}^2 f(\mathbf{x} + t\mathbf{h})\mathbf{h}, \mathbf{h} \rangle \quad \text{for } t \text{ in } I. \tag{14.21}$$

We now apply Theorem 14.10 to the function $\psi: I \to \mathbb{R}$ with $x = 0$ and $h = 1$ to choose a number $\theta$ with $0 < \theta < 1$ such that

$$\psi(1) = \psi(0) + \psi'(0) + \frac{1}{2}\psi''(\theta), \tag{14.22}$$

an equality that, by substituting the values of $\psi(1)$ and $\psi(0)$ and using the above formulas for $\psi'(0)$ and $\psi''(\theta)$, is seen to be precisely formula (14.20). ∎

**THEOREM 14.12**

**The Second-Order Approximation Theorem** Let $\mathcal{O}$ be an open subset of $\mathbb{R}^n$ that contains the point $\mathbf{x}$ and suppose that the function $f: \mathcal{O} \to \mathbb{R}$ has continuous second-order partial derivatives. Then

$$\lim_{\mathbf{h} \to 0} \frac{f(\mathbf{x} + \mathbf{h}) - [f(\mathbf{x}) + \langle \mathbf{D}f(\mathbf{x}), \mathbf{h} \rangle + \frac{1}{2}\langle \mathbf{D}^2 f(\mathbf{x})\mathbf{h}, \mathbf{h} \rangle]}{\|\mathbf{h}\|^2} = 0. \tag{14.23}$$

**Proof**    Since the point $\mathbf{x}$ is an interior point of $\mathcal{O}$, we can choose a positive number $r$ such that the symmetric neighborhood $\mathcal{N}_r(\mathbf{x})$ is contained in $\mathcal{O}$. It is convenient to define

$$R(\mathbf{h}) = f(\mathbf{x}+\mathbf{h}) - \left[ f(\mathbf{x}) + \langle \mathbf{D}f(\mathbf{x}), \mathbf{h} \rangle + \frac{1}{2}\langle \mathbf{D}^2 f(\mathbf{x})\mathbf{h}, \mathbf{h} \rangle \right] \quad \text{for} \quad \|\mathbf{h}\| < r.$$

We must show that

$$\lim_{\mathbf{h}\to 0} \frac{R(\mathbf{h})}{\|\mathbf{h}\|^2} = 0. \tag{14.24}$$

Fix the point $\mathbf{h}$ in $\mathbb{R}^n$ with $0 < \|\mathbf{h}\| < r$. Using Theorem 14.11, we can choose a number $\theta$ with $1 < \theta < 1$ such that

$$f(\mathbf{x}+\mathbf{h}) = f(\mathbf{x}) + \langle \mathbf{D}f(\mathbf{x}), \mathbf{h} \rangle + \frac{1}{2}\langle \mathbf{D}^2 f(\mathbf{x}+\theta\mathbf{h})\mathbf{h}, \mathbf{h} \rangle,$$

so that
$$R(\mathbf{h}) = \frac{1}{2}\langle \mathbf{D}^2 f(\mathbf{x}+\theta\mathbf{h})\mathbf{h}, \mathbf{h} \rangle - \frac{1}{2}\langle \mathbf{D}^2 f(\mathbf{x}+\mathbf{h})\mathbf{h}, \mathbf{h} \rangle$$
$$= \frac{1}{2}\langle [\mathbf{D}^2 f(\mathbf{x}+\theta\mathbf{h}) - \mathbf{D}^2 f(\mathbf{x}+\mathbf{h})]\mathbf{h}, \mathbf{h} \rangle. \tag{14.25}$$

We can use this formula and the Generalized Cauchy-Schwarz Inequality to obtain

$$|R(\mathbf{h})| \le \frac{1}{2}\|\mathbf{D}^2 f(\mathbf{x}+\theta\mathbf{h}) - \mathbf{D}^2 f(\mathbf{x})\| \|\mathbf{h}\|^2.$$

Dividing this estimate by $\|\mathbf{h}\|^2$, we obtain

$$\frac{|R(\mathbf{h})|}{\|\mathbf{h}\|^2} \le \frac{1}{2}\|\mathbf{D}^2 f(\mathbf{x}+\theta\mathbf{h}) - \mathbf{D}^2 f(\mathbf{x})\|. \tag{14.26}$$

But the function $f:\mathcal{O} \to \mathbb{R}$ has been assumed to have continuous second-order partial derivatives, so

$$\lim_{\mathbf{h}\to 0} \|\mathbf{D}^2 f(\mathbf{x}+\theta\mathbf{h}) - \mathbf{D}^2 f(\mathbf{x})\| = 0,$$

and hence (14.24) follows from the estimate (14.26).    ∎

---

**THEOREM 14.13**    **The Second-Derivative Test**    Let $\mathcal{O}$ be an open subset of $\mathbb{R}^n$ that contains the point $\mathbf{x}$ and suppose that the function $f:\mathcal{O} \to \mathbb{R}$ has continuous second-order partial derivatives. Assume that

$$\mathbf{D}f(\mathbf{x}) = 0.$$

(i)  If the Hessian matrix $\mathbf{D}^2 f(\mathbf{x})$ is positive definite, then the point $\mathbf{x}$ is a strict local minimizer of the function $f:\mathcal{O} \to \mathbb{R}$.

(ii)  If the Hessian matrix $\mathbf{D}^2 f(\mathbf{x})$ is negative definite, then the point $\mathbf{x}$ is a strict local maximizer of the function $f:\mathcal{O} \to \mathbb{R}$.

**Proof**  We need only consider case (i), since case (ii) follows from (i) if we replace $f$ with $-f$. So suppose that the Hessian matrix $\mathbf{D}^2 f(\mathbf{x})$ is positive definite. Since the point $\mathbf{x}$ is an interior point of $\mathcal{O}$, we can choose a positive number $r$ such that the symmetric neighborhood $\mathcal{N}_r(\mathbf{x})$ is contained in $\mathcal{O}$.

The strategy of the proof is to write the difference $f(\mathbf{x} + \mathbf{h}) - f(\mathbf{x})$ as

$$f(\mathbf{x} + \mathbf{h}) - f(\mathbf{x}) = Q(\mathbf{h}) + R(\mathbf{h}), \tag{14.27}$$

where $Q: \mathbb{R}^n \to \mathbb{R}$ is a positive definite quadratic function and

$$\lim_{\mathbf{h} \to 0} \frac{R(\mathbf{h})}{\|\mathbf{h}\|^2} = 0. \tag{14.28}$$

Indeed, if we define

$$R(\mathbf{h} = f(\mathbf{x} + \mathbf{h}) - \left[ f\mathbf{x} + \langle \mathbf{D}f(\mathbf{x}), \mathbf{h} \rangle + \frac{1}{2} \langle \mathbf{D}^2 f(\mathbf{x})\mathbf{h}, \mathbf{h} \rangle \right] \quad \text{for} \quad \|\mathbf{h}\| < r, \tag{14.29}$$

then the Second-Order Approximation Theorem asserts that (14.28) holds. Moreover, if we define $Q: \mathbb{R}^n \to \mathbb{R}$ to be the quadratic function associated with one-half the Hessian matrix $\mathbf{D}^2 f(\mathbf{x})$, then this quadratic function is positive definite. Finally, since $\mathbf{D}f(\mathbf{x}) = \mathbf{0}$, we may rewrite (14.29) to obtain (14.27).

Since the quadratic function $Q: \mathbb{R}^n \to \mathbb{R}$ is positive definite, we can use Proposition 14.8 to choose a positive number $c$ such that

$$Q(\mathbf{h}) \geq c\|\mathbf{h}\|^2 \quad \text{for all } \mathbf{h} \text{ in } \mathbb{R}^n.$$

On the other hand, using (14.28) it follows that we can choose a positive number $\delta$ less than $r$ such that

$$\frac{|R(\mathbf{h})|}{\|\mathbf{h}\|^2} < \frac{c}{2} \quad \text{if} \quad 0 < \|\mathbf{h}\| < \partial.$$

Combining these two estimates, it follows from (14.27) that

$$f(\mathbf{x} + \mathbf{h}) - f(\mathbf{x}) = Q(\mathbf{h}) + R(\mathbf{h}) \geq c\|\mathbf{h}\|^2 + R(\mathbf{h}) > \frac{c}{2}\|\mathbf{h}\|^2 \quad \text{if} \quad 0 < \|\mathbf{h}\| < \partial,$$

so the point $\mathbf{x}$ is a strict local minimizer of the function $f: \mathcal{O} \to \mathbb{R}$.  ∎

In Proposition 14.7, we characterized the positive definite, symmetric $2 \times 2$ matrices; hence it is interesting to record the following special case of the Second-Derivative Test.

**COROLLARY 14.14**  Let $\mathcal{O}$ be an open subset of the plane $\mathbb{R}^2$ that contains the point $(x_0, y_0)$ and suppose that the function $f: \mathcal{O} \to \mathbb{R}$ has continuous second-order partial derivatives. Assume that

$$\frac{\partial f}{\partial x}(x_0, y_0) = 0 \quad \text{and} \quad \frac{\partial f}{\partial y}(x_0, y_0) = 0.$$

Suppose also that

$$\frac{\partial^2 f}{\partial x^2}(x_0, y_0) > 0$$

and that
$$\frac{\partial^2 f}{\partial x^2}(x_0, y_0) \cdot \frac{\partial^2 f}{\partial y^2}(x_0, y_0) - \left[\frac{\partial^2 f}{\partial x \partial y}(x_0, y_0)\right]^2 > 0.$$

Then the point $(x_0, y_0)$ is a strict local minimizer for that function $f: \mathcal{O} \to \mathbb{R}$.

We will conclude this section with some comments about first-order and second-order approximation. For simplicity of notation, we will look at the point $\mathbf{0}$. Let $\mathcal{O}$ be an open subset of $\mathbb{R}^n$ containing $\mathbf{0}$ and consider the function $f: \mathcal{O} \to \mathbb{R}$. If the function is continuously differentiable, then the First-Order Approximation Theorem asserts that the affine function $\psi: \mathbb{R}^n \to \mathbb{R}$ defined by

$$\psi(\mathbf{x}) = f(\mathbf{0}) + \langle \mathbf{D}f(\mathbf{0}), \mathbf{x} \rangle \quad \text{for } \mathbf{x} \text{ in } \mathbb{R}^n$$

is a good approximation to the function $f: \mathbb{R}^n \to \mathbb{R}$ near $\mathbf{0}$, in the precise sense that

$$\lim_{\mathbf{x} \to \mathbf{0}} \frac{f(\mathbf{x}) - \psi(\mathbf{x})}{\|\mathbf{x}\|} = 0.$$

The Second-Order Approximation Theorem asserts that if the function $f: \mathcal{O} \to \mathbb{R}$ has continuous second-order partial derivatives, then by adding to the affine function $\psi: \mathbb{R}^n \to \mathbb{R}$ one-half the quadratic function associated with the Hessian matrix $\mathbf{D}^2 f(\mathbf{0})$, we obtain the function $\phi: \mathbb{R}^n \to \mathbb{R}$ defined by

$$\phi(\mathbf{x}) = f(\mathbf{0}) + \langle \mathbf{D}f(\mathbf{0}), \mathbf{x} \rangle + \frac{1}{2}\langle \mathbf{D}^2 f(\mathbf{0})\mathbf{x}, \mathbf{x} \rangle \quad \text{for } \mathbf{x} \text{ in } \mathbb{R}^n,$$

which is a much better approximation to the function $f: \mathcal{O} \to \mathbb{R}$, in the precise sense that

$$\lim_{\mathbf{x} \to \mathbf{0}} \frac{f(\mathbf{x}) - \phi(\mathbf{x})}{\|\mathbf{x}\|^2} = 0.$$

**EXAMPLE 14.12**  Define
$$f(x, y) = \sin(x - y + x^2 + xy + x^2 y) \quad \text{for } (x, y) \text{ in } \mathbb{R}^2.$$

A short computation of partial derivatives shows that
$$\mathbf{D}f(\mathbf{0}) = (1, -1)$$

and that
$$\mathbf{D}^2 f(\mathbf{0}) = \begin{bmatrix} 2 & 1 \\ 1 & 0 \end{bmatrix}.$$

Thus the First-Order Approximation Theorem implies that
$$\lim_{(x,y) \to (0,0)} \frac{\sin(x - y + x^2 + xy + x^2 y) - [x - y]}{\sqrt{x^2 + y^2}} = 0,$$

and the Second-Order Approximation Theorem implies that
$$\lim_{(x,y) \to (0,0)} \frac{\sin(x - y + x^2 + xy + x^2 y) - [x - y + x^2 + xy]}{x^2 + y^2} = 0. \qquad \square$$

1. Analyze the local extrema of the following functions:
   (a)  $f(x, y) = e^{x^2 - 4y + y^2}$   for $(x, y)$ in $\mathbb{R}^2$
   (b)  $g(x, y, z) = e^{x^2 - 4y + y^2} + z^2$   for $(x, y, z)$ in $\mathbb{R}^3$
   (c)  $f(x, y) = (x^2 + y^2)e^{x^2 + y^2}$   for $(x, y)$ in $\mathbb{R}^2$
   (d)  $f(x, y) = x^3 y^2 (6 - x - y)$   for $(x, y)$ in $\mathbb{R}^2$ with $x > 0$ and $y > 0$

2. Find necessary and sufficient conditions for a $2 \times 2$ symmetric matrix to be negative definite. Use this information to state and prove a sufficient condition for a point to be a local maximizer for a function of two variables.

3. Define $K$ to be the closed rectangle $\{(x, y)$ in $\mathbb{R}^2 \mid -1 \leq x \leq 1, -\pi \leq y \leq \pi\}$ and define the function $f: K \to \mathbb{R}$ by $f(x, y) = xe^{-x} \cos y$ for $(x, y)$ in $K$. Find the largest and smallest functional values of the function $f: K \to \mathbb{R}$. (*Hint:* Analyze the behavior on the boundary of $K$ separately.)

4. Show that the point $(-1, 1)$ is the minimizer of the function $f: \mathbb{R}^2 \to \mathbb{R}$ defined by

$$f(x, y) = (2x + 3y)^2 + (x + y - 1)^2 + (x + 2y - 2)^2 \quad \text{for } (x, y) \text{ in } \mathbb{R}^2.$$

5. Define

$$f(x, y) = x^2 + y^2 - 3xy \quad \text{for } (x, y) \text{ in } \mathbb{R}^2.$$

Explain why the function $f: \mathbb{R}^2 \to \mathbb{R}$ has no local extreme points.

6. Analyze the local extreme points of the function $f: \mathbb{R}^2 \to \mathbb{R}$ defined by

$$f(x, y) = \cos(x + y) + \sin(x + y) \quad \text{for } (x, y) \text{ in } \mathbb{R}^2.$$

7. Suppose that the function $f: \mathbb{R}^n \to \mathbb{R}$ has continuous second-order partial derivatives. Let $\mathbf{x}$ be a point in $\mathbb{R}^n$ at which $\mathbf{D}f(\mathbf{x}) = 0$. Assume also that there are points $\mathbf{u}$ and $\mathbf{v}$ in $\mathbb{R}^n$ at which

$$\langle \mathbf{D}^2 f(\mathbf{x})\mathbf{u}, \mathbf{u} \rangle > 0 \quad \text{and} \quad \langle \mathbf{D}^2 f(\mathbf{x})\mathbf{v}, \mathbf{v} \rangle < 0.$$

Show that the point $\mathbf{x}$ is neither a local maximum nor a local minimum of the function $f: \mathbb{R}^n \to \mathbb{R}$.

8. Suppose that the function $f: \mathbb{R}^n \to \mathbb{R}$ has continuous second-order partial derivatives. Let $\mathbf{x}$ be a point in $\mathbb{R}^n$ at which $\mathbf{D}f(\mathbf{x}) = 0$ and such that all of the entries of the Hessian matrix $\mathbf{D}^2 f(\mathbf{x})$ are also 0. By giving specific examples, show that it is possible for the point $\mathbf{x}$ to be a local maximum, a local minimum, or neither.

9. Show that if the function $f: \mathbb{R}^n \to \mathbb{R}$ has continuous second-order partial derivatives and if at the point $\mathbf{x}$ in $\mathbb{R}^n$, $\mathbf{D}f(\mathbf{x}) = 0$ with $\mathbf{D}^2 f(\mathbf{x})$ positive definite, then there are positive numbers $c$ and $\delta$ such that

$$f(\mathbf{x} + \mathbf{h}) - f(\mathbf{x}) \geq c\|\mathbf{h}\|^2 \quad \text{if} \quad \|\mathbf{h}\| < \delta.$$

10. Complete the proof of the Second-Derivative Test by proving that if the Hessian is negative definite at a point, then the point is a strict local maximum.

11. Calculate the following limits by applying the First-Order or Second-Order Approximation Theorem:

    (a)  $\displaystyle \lim_{(x,y) \to (0,0)} \frac{\sin(x + xy - y) - [x + y]}{\sqrt{x^2 + y^2}}$

    (b)  $\displaystyle \lim_{(s,t) \to (0,0)} \frac{s^2 t}{\sqrt{s^2 + t^2}}$

(c) $\displaystyle\lim_{(x,y)\to(0,0)} \frac{e^{x-y} - 1 - x + y}{x^2 + y^2}$

(d) $\displaystyle\lim_{(x,y)\to(0,0)} \frac{\cos(x - y + xy) - [1 - 1/2x^2 + xy - 1/2y^2]}{x^2 + y^2}$

# 15

# Approximating Nonlinear Mappings by Linear Mappings

## 15.1 Linear Mappings and Matrices

We now turn to the study of nonlinear mappings $\mathbf{F}: \mathcal{O} \to \mathbb{R}^m$, where $\mathcal{O}$ is an open subset of $\mathbb{R}^n$ and $n$ and $m$ are positive integers. In order to study such general mappings, we return to a strategy that we have often used before: We approximate such mappings by mappings of a much simpler form. Here, we consider linear mappings to be the simpler ones.

In this first section, we will consider linear mappings and the correspondence between linear mappings and matrices. In particular, we will establish the equivalence between the invertibility of a linear mapping $\mathbf{T}: \mathbb{R}^n \to \mathbb{R}^n$ and the invertibility of the $n \times n$ matrix with which it is associated. Section 15.2 is devoted to describing the way in which nonlinear mappings may be approximated by linear mappings; the crucial concepts are the derivative matrix and the differential. In Section 15.3, we will study a generalization of the Chain Rule, established in Chapter 4 for differentiating the composition of functions, to a formula for the differential of the composition of general nonlinear mappings.

**DEFINITION**   *A mapping* $\mathbf{T}: \mathbb{R}^n \to \mathbb{R}^m$ *is said to be linear provided that for each pair of points* $\mathbf{u}$ *and* $\mathbf{v}$ *in* $\mathbb{R}^n$ *and each pair of numbers* $\alpha$ *and* $\beta$,

$$\mathbf{T}(\alpha\mathbf{u} + \beta\mathbf{v}) = \alpha\mathbf{T}(\mathbf{u}) + \beta\mathbf{T}(\mathbf{v}). \tag{15.1}$$

**EXAMPLE 15.1**   For each index $i$ such that $1 \le i \le n$, the $i$th component projection function

$$p_i: \mathbb{R}^n \to \mathbb{R}$$

is linear. This follows directly from the definitions of addition and scalar multiplication, since for each pair of points $\mathbf{u}$ and $\mathbf{v}$ in $\mathbb{R}^n$ and each pair of numbers $\alpha$ and $\beta$,

$$p_i(\alpha\mathbf{u} + \beta\mathbf{v}) = \alpha u_i + \beta v_i = \alpha p_i(\mathbf{u}) + \beta p_i(\mathbf{v}). \qquad \square$$

**PROPOSITION 15.1**   For a point $\mathbf{a}$ in $\mathbb{R}^n$, define the mapping $\mathbf{T}: \mathbb{R}^n \to \mathbb{R}$ by

$$\mathbf{T}(\mathbf{x}) = \langle \mathbf{a}, \mathbf{x} \rangle \quad \text{for all } \mathbf{x} \text{ in } \mathbb{R}^n. \tag{15.2}$$

Then the mapping $\mathbf{T}: \mathbb{R}^n \to \mathbb{R}$ is linear. Moreover, for each linear mapping $\mathbf{T}: \mathbb{R}^n \to \mathbb{R}$ there is a unique point $\mathbf{a}$ in $\mathbb{R}^n$ such that $\mathbf{T}: \mathbb{R}^n \to \mathbb{R}$ is defined by formula (15.2).

**Proof**   Given a point $\mathbf{a}$ in $\mathbb{R}^n$, as we have previously noted, the inner product has the property that for each pair of points $\mathbf{u}$ and $\mathbf{v}$ in $\mathbb{R}^n$ and each pair of numbers $\alpha$ and $\beta$,

$$\langle \mathbf{a}, \alpha\mathbf{u} + \beta\mathbf{v} \rangle = \alpha \langle \mathbf{a}, \mathbf{u} \rangle + \beta \langle \mathbf{a}, \mathbf{v} \rangle.$$

This is exactly what it means for the mapping $\mathbf{T}: \mathbb{R}^n \to \mathbb{R}$ to be linear.

Now suppose that $\mathbf{T}: \mathbb{R}^n \to \mathbb{R}$ is any linear mapping. For each index $i$ such that $1 \le i \le n$, define $a_i = \mathbf{T}(\mathbf{e}_i)$ and then define the point $\mathbf{a}$ in $\mathbb{R}^n$ by $\mathbf{a} = (a_1, \ldots, a_n)$. Then, by the linearity of $\mathbf{T}: \mathbb{R}^n \to \mathbb{R}$,

$$\begin{aligned}
\mathbf{T}(\mathbf{x}) &= \mathbf{T}(x_1\mathbf{e}_1 + \cdots + x_n\mathbf{e}_n) \\
&= x_1\mathbf{T}(\mathbf{e}_1) + \cdots + x_n\mathbf{T}(\mathbf{e}_n) \\
&= x_1 a_1 + \cdots + x_n a_n \\
&= \langle \mathbf{a}, \mathbf{x} \rangle \quad \text{for all } \mathbf{x} \text{ in } \mathbb{R}^n. \qquad \blacksquare
\end{aligned}$$

Observe that for an index $i$ such that $1 \le i \le n$, the $i$th component projection function $p_i: \mathbb{R}^n \to \mathbb{R}$ is defined by formula (15.2), where $\mathbf{a} = \mathbf{e}_i$.

**EXAMPLE 15.2**   A linear transformation $\mathbf{T}: \mathbb{R}^3 \to \mathbb{R}$ is completely determined by specifying three numbers $a$, $b$, and $c$ and defining

$$\mathbf{T}(x, y, z) = ax + by + cz \quad \text{for all } (x, y, z) \text{ in } \mathbb{R}^3. \qquad \square$$

For linear mappings $\mathbf{T}: \mathbb{R}^n \to \mathbb{R}^m$, where $m$ is now an integer greater than 1, there is a natural generalization of Proposition 15.1. To understand what this generalization is, for a linear mapping $\mathbf{T}: \mathbb{R}^n \to \mathbb{R}^m$ and an index $i$ such that $1 \le i \le m$, recall that the function

$$\mathbf{T}_i = p_i \circ \mathbf{T}: \mathbb{R}^n \to \mathbb{R}$$

is called the $i$th *component function* of the mapping $\mathbf{T}: \mathbb{R}^n \to \mathbb{R}^m$ and that

$$\mathbf{T}(\mathbf{x}) = (\mathbf{T}_1(\mathbf{x}), \dots, \mathbf{T}_m(\mathbf{x})) \quad \text{for all } \mathbf{x} \text{ in } \mathbb{R}^n.$$

**PROPOSITION 15.2**

A mapping $\mathbf{T}: \mathbb{R}^n \to \mathbb{R}^m$ is linear if and only if each of its component functions is linear.

**Proof**   For each pair of points $\mathbf{u}$ and $\mathbf{v}$ in $\mathbb{R}^n$ and each pair of numbers $\alpha$ and $\beta$, by the very definition of equality in $\mathbb{R}^m$,

$$\mathbf{T}(\alpha\mathbf{u} + \beta\mathbf{v}) = \alpha\mathbf{T}(\mathbf{u}) + \beta\mathbf{T}(\mathbf{v}) \qquad (15.3)$$

if and only if for each index $i$ such that $1 \le i \le m$,

$$p_i(\mathbf{T}(\alpha\mathbf{u} + \beta\mathbf{v})) = p_i(\alpha\mathbf{T}(\mathbf{u}) + \beta\mathbf{T}(\mathbf{v})). \qquad (15.4)$$

But we have already observed that the component projection functions are linear, so for each index $i$ such that $1 \le i \le m$,

$$p_i(\alpha\mathbf{T}(\mathbf{u}) + \beta\mathbf{T}(\mathbf{v})) = \alpha p_i(\mathbf{T}(\mathbf{u})) + \beta p_i(\mathbf{T}(\mathbf{v})).$$

Thus the identity (15.3) holds if and only if for each index $i$ such that $1 \le i \le m$,

$$\mathbf{T}_i(\alpha\mathbf{u} + \beta\mathbf{v}) = \alpha\mathbf{T}_i(\mathbf{u}) + \beta\mathbf{T}_i(\mathbf{v}). \qquad (15.5)$$

The equivalence of (15.3) and (15.5) is exactly what is required in order to prove the proposition. ■

**EXAMPLE 15.3**   A mapping $\mathbf{T}: \mathbb{R}^2 \to \mathbb{R}^2$ may be written as

$$\mathbf{T}(x, y) = (g(x, y), h(x, y)) \quad \text{for } (x, y) \text{ in } \mathbb{R}^2,$$

where $g: \mathbb{R}^2 \to \mathbb{R}$ and $h: \mathbb{R}^2 \to \mathbb{R}$ are the component functions. Proposition 15.2 implies that the mapping $\mathbf{T}: \mathbb{R}^2 \to \mathbb{R}^2$ is linear if and only the functions $g: \mathbb{R}^2 \to \mathbb{R}$ and $h: \mathbb{R}^2 \to \mathbb{R}$ are linear. From the characterization of linear functions just established in Proposition 15.1, it follows that $\mathbf{T}: \mathbb{R}^2 \to \mathbb{R}^2$ is linear if and only if there are numbers $a$, $b$, $c$, and $d$ for which

$$\mathbf{T}(x, y) = (ax + by, cx + dy) \quad \text{for all } (x, y) \text{ in } \mathbb{R}^2. \qquad \square$$

Let $m$ and $n$ be positive integers. Recall that by an $m \times n$ *matrix* we mean a rectangular array of real numbers consisting of $m$ rows and $n$ columns. If such an $m \times n$ matrix is denoted by $\mathbf{A}$, we will write

$$\mathbf{A} = [a_{ij}],$$

where for each pair of indices $i$ and $j$ such that $1 \leq i \leq m$ and $1 \leq j \leq n$, $a_{ij}$ denotes the number in the $i$th row and $j$th column of the matrix $\mathbf{A}$; we call $a_{ij}$ the $ij$th *entry* of the matrix and sometimes, for convenience, denote it by $(\mathbf{A})_{ij}$.

**DEFINITION**   *For an $m \times n$ matrix $\mathbf{A} = [a_{ij}]$ and a point $\mathbf{x}$ in $\mathbb{R}^n$, by the symbol $\mathbf{A}\mathbf{x}$ we denote the point in $\mathbb{R}^m$ that, for each index $i$ such that $1 \leq i \leq m$, has $i$th component equal to the inner product of the $i$th row of $\mathbf{A}$ with $\mathbf{x}$. Thus*

$$\mathbf{A}\mathbf{x} \equiv \mathbf{y},$$

*where* $\qquad y_i \equiv \displaystyle\sum_{j=1}^{n} a_{ij} x_j \quad$ *for each index $i$ such that $1 \leq i \leq m$.*

The fundamental correspondence between matrices and linear mappings is described in the next theorem.

**THEOREM 15.3**   For an $m \times n$ matrix $\mathbf{A}$, define the mapping $\mathbf{T} \colon \mathbb{R}^n \to \mathbb{R}^m$ by

$$\mathbf{T}(\mathbf{x}) = \mathbf{A}\mathbf{x} \quad \text{for all } \mathbf{x} \text{ in } \mathbb{R}^n. \tag{15.6}$$

Then the mapping $\mathbf{T} \colon \mathbb{R}^n \to \mathbb{R}^m$ is linear. Moreover, for each linear mapping $\mathbf{T} \colon \mathbb{R}^n \to \mathbb{R}^m$, there is a unique $m \times n$ matrix $\mathbf{A} = [a_{ij}]$ such that $\mathbf{T} \colon \mathbb{R}^n \to \mathbb{R}^m$ is defined by (15.6); for each pair of indices $i$ and $j$ such that $1 \leq i \leq m$, $1 \leq j \leq n$, the $ij$th entry of the matrix $\mathbf{A}$ is determined by the formula

$$a_{ij} = \langle \mathbf{T}(\mathbf{e}_j), \mathbf{e}_i \rangle. \tag{15.7}$$

**Proof**   Since a mapping is linear if and only if each of its component functions is linear, to verify that formula (15.6) defines a linear mapping, we must show that for each index $i$ such that $1 \leq i \leq m$, the component function $p_i \circ \mathbf{T} \colon \mathbb{R}^n \to \mathbb{R}$ is linear. But, by the very definition of $\mathbf{A}\mathbf{x}$,

$$(p_i \circ \mathbf{T})(\mathbf{x}) = \langle \mathbf{A}_i, \mathbf{x} \rangle \quad \text{for all } \mathbf{x} \text{ in } \mathbb{R}^n,$$

where $\mathbf{A}_i$ is the $i$th row of $\mathbf{A}$, and so, by Proposition 15.1, each component function is linear.

Now suppose that $\mathbf{T} \colon \mathbb{R}^n \to \mathbb{R}^m$ is any linear mapping. Then each component function is linear and hence, by Proposition 15.1, for each index $i$ such that $1 \leq i \leq m$ we can select a point $\mathbf{A}_i$ in $\mathbb{R}^n$ such that the $i$th component function is defined by

$$(p_i \circ \mathbf{T})(\mathbf{x}) = \langle \mathbf{A}_i, \mathbf{x} \rangle \quad \text{for all } \mathbf{x} \text{ in } \mathbb{R}^n. \tag{15.8}$$

Define $\mathbf{A}$ to be the $m \times n$ matrix whose $i$th row is $\mathbf{A}_i$. Then (15.8) is equivalent to (15.6).

Observe that if $\mathbf{A} = [a_{ij}]$, then for each pair of indices $i$ and $j$ such that $1 \leq i \leq m$ and $1 \leq j \leq n$, the $ij$th entry of the matrix $\mathbf{A}$ is the $j$th component of the $i$th row; that is, $a_{ij} = \langle \mathbf{A}_i, \mathbf{e}_j \rangle$. Hence, letting $\mathbf{x} = \mathbf{e}_j$ in (15.8), we have

$$a_{ij} = \langle \mathbf{A}_i, \mathbf{e}_j \rangle = (p_i \circ \mathbf{T})(\mathbf{e}_j) = \langle \mathbf{T}(\mathbf{e}_j), \mathbf{e}_i \rangle. \qquad \blacksquare$$

**DEFINITION**  *For a linear mapping $\mathbf{T} \colon \mathbb{R}^n \to \mathbb{R}^m$, the $m \times n$ matrix $\mathbf{A}$ such that*

$$\mathbf{T}(\mathbf{x}) = \mathbf{A}\mathbf{x} \quad \text{for all } \mathbf{x} \text{ in } \mathbb{R}^n$$

*is said to be the matrix associated with the mapping $\mathbf{T} \colon \mathbb{R}^n \to \mathbb{R}^m$.*

**EXAMPLE 15.4**  Suppose that the mapping $\mathbf{T} \colon \mathbb{R}^2 \to \mathbb{R}^3$ is linear and that

$$\mathbf{T}(1, 0) = (-4, 1/2, 1) \quad \text{and} \quad \mathbf{T}(0, 1) = (0, 6, 10).$$

From formula (15.7), it follows that the $3 \times 2$ matrix associated with this mapping is the matrix $\mathbf{A}$ given by

$$\mathbf{A} = \begin{pmatrix} -4 & 0 \\ 1/2 & 6 \\ 1 & 10 \end{pmatrix}.$$

Thus, by formula (15.6) we see that

$$\mathbf{T}(x, y) = (-4x, x/2 + 6y, x + 10y) \quad \text{for all } (x, y) \text{ in } \mathbb{R}^2. \qquad \square$$

**EXAMPLE 15.5**  Define the mapping $\mathbf{T} \colon \mathbb{R}^3 \to \mathbb{R}^2$ by

$$\mathbf{T}(x, y, z) = (x, z) \quad \text{for all } (x, y, z) \text{ in } \mathbb{R}^3.$$

Then this mapping is linear and is associated with the $2 \times 3$ matrix $\mathbf{A}$ defined by

$$\mathbf{A} = \begin{pmatrix} 1 & 0 & 0 \\ 0 & 0 & 1 \end{pmatrix}. \qquad \square$$

**EXAMPLE 15.6**  For a number $\theta$, define the $2 \times 2$ matrix $\mathbf{A}$ by

$$\mathbf{A} = \begin{pmatrix} \cos\theta & -\sin\theta \\ \sin\theta & \cos\theta \end{pmatrix}.$$

Let $\mathbf{T} \colon \mathbb{R}^2 \to \mathbb{R}^2$ be the linear mapping associated with the matrix $\mathbf{A}$. This linear mapping rotates points in the plane by $\theta$ radians about the origin. To see why this is so, we note that if $(x, y)$ is a point in $\mathbb{R}^2$ that is written in polar coordinates as $(r\cos\phi, r\sin\phi)$, then a direct calculation, using just the addition properties of the sine and cosine, shows that its image $\mathbf{T}(x, y)$ is the point $(r\cos(\theta + \phi), r\sin(\theta + \phi))$.

**FIGURE 15.1**

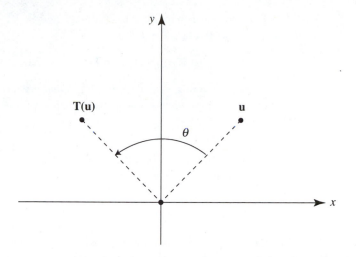

**EXAMPLE 15.7**   In the plane $\mathbb{R}^2$, consider the line $\ell = \{(x,\ y)$ in $\mathbb{R}^2 \mid y = x\}$. For a point $(x,\ y)$ in $\mathbb{R}^2$, define $\mathbf{T}(x,\ y)$ to be the point on the line $\ell$ that is closest to $(x,\ y)$. This defines a mapping $\mathbf{T}\colon \mathbb{R}^2 \to \mathbb{R}^2$ that is linear. To see why it is linear, note that from the geometric significance of the inner product we have the following formula: Let $(x_0,\ y_0) = (1/\sqrt{2},\ 1/\sqrt{2})$, so that $(x_0,\ y_0)$ is a point of length 1 on $\ell$. From a simple trigonometric argument using Proposition 10.3, it follows that

$$\mathbf{T}(x,\ y) = \langle (x,\ y),\ (x_0,\ y_0) \rangle (x_0,\ y_0)$$

$$= \left( \frac{x}{2} + \frac{y}{2},\ \frac{x}{2} + \frac{y}{2} \right) \quad \text{for all } (x,\ y) \text{ in } \mathbb{R}^2.$$

**FIGURE 15.2**

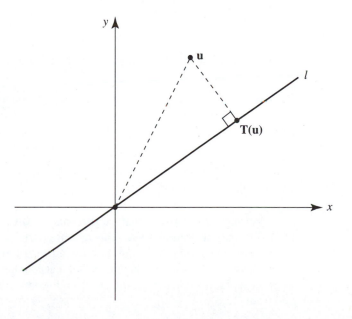

For two linear mappings $\mathbf{T}: \mathbb{R}^n \to \mathbb{R}^m$ and $\mathbf{S}: \mathbb{R}^n \to \mathbb{R}^m$ and two numbers $\alpha$ and $\beta$, define the mapping $\alpha \mathbf{T} + \beta \mathbf{S}: \mathbb{R}^n \to \mathbb{R}^m$ by

$$(\alpha \mathbf{T} + \beta \mathbf{S})(\mathbf{x}) \equiv \alpha \mathbf{T}(\mathbf{x}) + \beta \mathbf{S}(\mathbf{x}) \quad \text{for all } \mathbf{x} \text{ in } \mathbb{R}^n.$$

Such a mapping is called a *linear combination* of the mappings $\mathbf{T}$ and $\mathbf{S}$. It immediately follows that a linear combination of linear mappings is again a linear mapping.

**DEFINITION**

*For two $m \times n$ matrices $\mathbf{A}$ and $\mathbf{B}$ and two numbers $\alpha$ and $\beta$, the matrix $\alpha \mathbf{A} + \beta \mathbf{B}$ is the $m \times n$ matrix whose $ij$th entry $(\alpha \mathbf{A} + \beta \mathbf{B})_{ij}$ is defined by*

$$(\alpha \mathbf{A} + \beta \mathbf{B})_{ij} \equiv \alpha a_{ij} + \beta b_{ij}.$$

This definition of the linear combination of matrices was made so that the matrix associated with the linear combinations of two linear mappings $\mathbf{T}$ and $\mathbf{S}$ will be the same linear combination of the matrices associated with $\mathbf{T}$ and $\mathbf{S}$, respectively. We will formulate this as the next proposition; the proof follows directly from the linearity of the projection functions and formula (15.8).

**PROPOSITION 15.4**

Suppose that the linear mapping $\mathbf{T}: \mathbb{R}^n \to \mathbb{R}^m$ is associated with the $m \times n$ matrix $\mathbf{A}$ and the linear mapping $\mathbf{S}: \mathbb{R}^n \to \mathbb{R}^m$ is associated with the $m \times n$ matrix $\mathbf{B}$. Then for each pair of numbers $\alpha$ and $\beta$, the linear mapping $\alpha \mathbf{T} + \beta \mathbf{S}: \mathbb{R}^n \to \mathbb{R}^m$ is associated with the $m \times n$ matrix $\alpha \mathbf{A} + \beta \mathbf{B}$.

Certain linear mappings can be composed. Specifically, given a linear mapping $\mathbf{T}: \mathbb{R}^n \to \mathbb{R}^m$ and a linear mapping $\mathbf{S}: \mathbb{R}^m \to \mathbb{R}^k$, the composition $\mathbf{S} \circ \mathbf{T}: \mathbb{R}^n \to \mathbb{R}^k$ is defined by

$$(\mathbf{S} \circ \mathbf{T})(\mathbf{x}) \equiv \mathbf{S}(\mathbf{T}(\mathbf{x})) \quad \text{for all } \mathbf{x} \text{ in } \mathbb{R}^n.$$

It is not difficult to check that the composition is again a linear mapping.

**DEFINITION**

*For positive integers $n$, $m$, and $k$, let $\mathbf{A} = [a_{ij}]$ be an $m \times n$ matrix and $\mathbf{B} = [b_{ij}]$ be a $k \times m$ matrix. Then the product matrix $\mathbf{BA}$ is defined to be the $k \times n$ matrix whose $ij$th entry $(\mathbf{BA})_{ij}$ is defined by*

$$(\mathbf{BA})_{ij} \equiv \sum_{\ell=1}^{m} b_{i\ell} a_{\ell j}. \tag{15.9}$$

Formula (15.9) asserts that the $ij$th entry of $\mathbf{BA}$ is the inner product of the $i$th row of $\mathbf{B}$ with the $j$th column of $\mathbf{A}$. This definition of matrix product is made so that the matrix associated with the composition of linear mappings will be the product of the matrices associated with the mappings that make up the composition. We formulate this as the next proposition; here again, the proof follows from the linearity of the projections and from (15.8).

**PROPOSITION
15.5**
For natural numbers $n$, $m$, and $k$, suppose that the linear mapping $\mathbf{T}\colon \mathbb{R}^n \to \mathbb{R}^m$ is associated with the $m \times n$ matrix $\mathbf{A}$ and the linear mapping $\mathbf{S}\colon \mathbb{R}^m \to \mathbb{R}^k$ is associated with the $k \times m$ matrix $\mathbf{B}$. Then the composite mapping $\mathbf{S} \circ \mathbf{T}\colon \mathbb{R}^n \to \mathbb{R}^k$ is associated with the $k \times n$ product matrix $\mathbf{BA}$.

**COROLLARY
15.6**
For natural numbers $n$, $m$, $k$, and $\ell$, let $\mathbf{A}$ be an $m \times n$ matrix, $\mathbf{B}$ be a $k \times m$ matrix, and $\mathbf{C}$ be an $\ell \times k$ matrix. Then

$$\mathbf{C}(\mathbf{BA}) = (\mathbf{CB})\mathbf{A}. \qquad (15.10)$$

**Proof**
Let the linear mapping $\mathbf{T}\colon \mathbb{R}^n \to \mathbb{R}^m$ be associated with the $m \times n$ matrix $\mathbf{A}$, the linear mapping $\mathbf{S}\colon \mathbb{R}^m \to \mathbb{R}^k$ be associated with the $k \times m$ matrix $\mathbf{B}$, and the linear mapping $\mathbf{L}\colon \mathbb{R}^k \to \mathbb{R}^\ell$ be associated with the $\ell \times k$ matrix $\mathbf{C}$. Proposition 15.5 implies that the mapping $\mathbf{S} \circ \mathbf{T}\colon \mathbb{R}^n \to \mathbb{R}^k$ is associated with the matrix $\mathbf{BA}$ and that the mapping $\mathbf{L} \circ (\mathbf{S} \circ \mathbf{T})\colon \mathbb{R}^n \to \mathbb{R}^\ell$ is associated with the matrix $\mathbf{C}(\mathbf{BA})$. Similarly, the mapping $(\mathbf{L} \circ \mathbf{S}) \circ \mathbf{T}\colon \mathbb{R}^n \to \mathbb{R}^\ell$ is associated with the matrix $(\mathbf{CB})\mathbf{A}$. But for each point $\mathbf{x}$ in $\mathbb{R}^n$,

$$((\mathbf{L} \circ \mathbf{S}) \circ \mathbf{T})(\mathbf{x}) = \mathbf{L}(\mathbf{S}(\mathbf{T}(\mathbf{x}))) = (\mathbf{L} \circ (\mathbf{S} \circ \mathbf{T}))(\mathbf{x}),$$

which simply means that the mappings $(\mathbf{L} \circ \mathbf{S}) \circ \mathbf{T}\colon \mathbb{R}^n \to \mathbb{R}^\ell$ and $\mathbf{L} \circ (\mathbf{S} \circ \mathbf{T})\colon \mathbb{R}^n \to \mathbb{R}^\ell$ are equal. Thus their associated matrices are equal; that is, (15.10) holds. ∎

The property of matrix multiplication stated in formula (15.10) is called the *associative property* of matrix multiplication. For two matrices $\mathbf{A}$ and $\mathbf{B}$, the product matrix $\mathbf{BA}$ is defined only when the number of columns in $\mathbf{B}$ equals the number of rows in $\mathbf{A}$. Thus $\mathbf{AB}$ need not be defined when $\mathbf{BA}$ is defined. Moreover, when $n = m = k$, so that both $\mathbf{BA}$ and $\mathbf{AB}$ are properly defined $n \times n$ matrices, in general it is *not* true that $\mathbf{AB} = \mathbf{BA}$. In the language of algebra, this means that matrix multiplication is not commutative.

**EXAMPLE 15.8**   Define the $2 \times 2$ matrices $\mathbf{A}$ and $\mathbf{B}$ by

$$\mathbf{A} = \begin{pmatrix} 1 & 0 \\ 2 & 3 \end{pmatrix} \quad \text{and} \quad \mathbf{B} = \begin{pmatrix} 0 & 1 \\ 0 & 1 \end{pmatrix}.$$

Then
$$\mathbf{AB} = \begin{pmatrix} 0 & 1 \\ 0 & 5 \end{pmatrix}, \quad \text{whereas} \quad \mathbf{BA} = \begin{pmatrix} 2 & 3 \\ 2 & 3 \end{pmatrix}. \qquad \square$$

Recall that given any two sets $A$ and $B$ and a mapping $F\colon A \to B$, the *image* of this mapping, which is denoted by $F(A)$, is defined to be the subset of $B$ consisting of all points $b$ in $B$ for which there is a point $a$ in $A$ such that $b = F(a)$. This mapping is said to be *one-to-one* provided that for each point $b$ in $F(A)$ there is exactly one point $a$ in $A$ such that $F(a) = b$. Moreover, the mapping $F\colon A \to B$ is said to be *onto* provided that its image $F(A)$ equals $B$. Finally, the mapping $F\colon A \to B$ is said to be *invertible* provided that it is both one-to-one and onto. Given an invertible mapping $F\colon A \to B$, for each point $b$ in $B$ define $F^{-1}(b)$ to be the unique point $a$ in $A$ such that $F(a) = b$. This defines the mapping $F^{-1}\colon B \to A$, which is called the *inverse mapping* of the mapping $F\colon A \to B$.

An invertible linear mapping $\mathbf{T} \colon \mathbb{R}^n \to \mathbb{R}^n$ has an inverse that also is linear. To verify this, we must show that for each pair of points $\mathbf{u}$ and $\mathbf{v}$ in $\mathbb{R}^n$ and each pair of numbers $\alpha$ and $\beta$,

$$\mathbf{T}^{-1}(\alpha \mathbf{u} + \beta \mathbf{v}) = \alpha \mathbf{T}^{-1}(\mathbf{u}) + \beta \mathbf{T}^{-1}(\mathbf{v}),$$

which, by the very definition of inverse, means that

$$\mathbf{T}(\alpha \mathbf{T}^{-1}(\mathbf{u}) + \beta \mathbf{T}^{-1}(\mathbf{v})) = \alpha \mathbf{u} + \beta \mathbf{v}.$$

This last equality, however, follows from the linearity of the mapping $\mathbf{T} \colon \mathbb{R}^n \to \mathbb{R}^n$ and the definition of inverse.

The *identity mapping* on $\mathbb{R}^n$, which is denoted by $\mathbf{Id} \colon \mathbb{R}^n \to \mathbb{R}^n$, is the mapping that maps each point to itself; that is, $\mathbf{Id}(\mathbf{x}) \equiv \mathbf{x}$ for all points $\mathbf{x}$ in $\mathbb{R}^n$. The $n \times n$ matrix that is associated with the identity mapping is the matrix that has 1's on the diagonal and 0's elsewhere; it is denoted by $\mathbf{I}_n$ and called the *identity matrix*.

Just as there is a concept of invertibility for mappings, there is also a concept of invertibility for $n \times n$ matrices.

**DEFINITION**  *An $n \times n$ matrix $\mathbf{A}$ is said to be invertible provided that there is an $n \times n$ matrix $\mathbf{B}$ having the property that*

$$\mathbf{AB} = \mathbf{I}_n \quad and \quad \mathbf{BA} = \mathbf{I}_n.$$

There is only one matrix $\mathbf{B}$ that has the above property (see Exercise 13); it is denoted by $\mathbf{A}^{-1}$ and is called the *inverse matrix* of the matrix $\mathbf{A}$.

We have the following important equivalence between the invertibility of a linear mapping and the invertibility of the matrix with which it is associated.

**THEOREM 15.7**  For a linear mapping $\mathbf{T} \colon \mathbb{R}^n \to \mathbb{R}^n$ that is associated with the $n \times n$ matrix $\mathbf{A}$, the following two assertions are equivalent:

(i) The mapping $\mathbf{T} \colon \mathbb{R}^n \to \mathbb{R}^n$ is an invertible mapping.

(ii) The matrix $\mathbf{A}$ is an invertible matrix.

**Proof**  First, suppose that the mapping $\mathbf{T} \colon \mathbb{R}^n \to \mathbb{R}^n$ is an invertible mapping. The inverse mapping $\mathbf{T}^{-1} \colon \mathbb{R}^n \to \mathbb{R}^n$ is also linear. Let the inverse mapping $\mathbf{T}^{-1} \colon \mathbb{R}^n \to \mathbb{R}^n$ be associated with the $n \times n$ matrix $\mathbf{B}$.

Now the identity mapping $\mathbf{Id} \colon \mathbb{R}^n \to \mathbb{R}^n$ is associated with the identity matrix $\mathbf{I}_n$, and the matrix associated with a composition of two mappings is the product of the matrices associated with the mappings that make up the composite. Thus, since

$$\mathbf{T} \circ \mathbf{T}^{-1} = \mathbf{Id} \colon \mathbb{R}^n \to \mathbb{R}^n \quad and \quad \mathbf{T}^{-1} \circ \mathbf{T} = \mathbf{Id} \colon \mathbb{R}^n \to \mathbb{R}^n,$$

it follows that
$$\mathbf{AB} = \mathbf{I}_n \quad and \quad \mathbf{BA} = \mathbf{I}_n,$$

so the matrix $\mathbf{A}$ is invertible.

To show that the invertibility of the matrix $\mathbf{A}$ implies the invertibility of the mapping $\mathbf{T} \colon \mathbb{R}^n \to \mathbb{R}^n$, suppose that the matrix $\mathbf{A}$ is invertible. Define $\mathbf{S} \colon \mathbb{R}^n \to \mathbb{R}^n$ to be the linear mapping associated with the matrix $\mathbf{A}^{-1}$. Now the matrix associated with the composition $\mathbf{S} \circ \mathbf{T}: \mathbb{R}^n \to \mathbb{R}^n$ is $\mathbf{A}^{-1}\mathbf{A} = \mathbf{I}_n$, which means that

$$(\mathbf{S} \circ \mathbf{T})(\mathbf{x}) = \mathbf{I}_n\mathbf{x} = \mathbf{x} \quad \text{for all } \mathbf{x} \text{ in } \mathbb{R}^n;$$

that is,
$$\mathbf{S}(\mathbf{T}(\mathbf{x})) = \mathbf{x} \quad \text{for all } \mathbf{x} \text{ in } \mathbb{R}^n. \tag{15.11}$$

Similarly, since $\mathbf{A}\mathbf{A}^{-1} = \mathbf{I}_n$,

$$\mathbf{T}(\mathbf{S}(\mathbf{x})) = \mathbf{x} \quad \text{for all } \mathbf{x} \text{ in } \mathbb{R}^n. \tag{15.12}$$

The identities (15.11) and (15.12) imply that the mapping $\mathbf{T}: \mathbb{R}^n \to \mathbb{R}^n$ is invertible (see Exercise 14) and its inverse is $\mathbf{S}: \mathbb{R}^n \to \mathbb{R}^n$. ∎

Though Theorem 15.7 asserts the equivalence between the invertibility of a linear mapping from $\mathbb{R}^n$ to $\mathbb{R}^n$ and the invertibility of its associated $n \times n$ matrix, it does not suggest how we can determine whether such a mapping is indeed invertible. For each $n \times n$ matrix $\mathbf{A}$, there is defined a number $\det \mathbf{A}$, called the *determinant* of $\mathbf{A}$. The determinant plays a central role in algebra, geometry, and analysis. One of its properties is that an $n \times n$ matrix is invertible if and only if its determinant is nonzero.

For an $n \times n$ matrix $\mathbf{A}$ and a pair of indices $i$ and $j$ such that $1 \le i \le n$ and $1 \le j \le n$, the $ij$th *minor* of $\mathbf{A}$, denoted by $\mathbf{A}^{ij}$, is the $(n-1) \times (n-1)$ matrix obtained by removing the $i$th row and the $j$th column from the matrix $\mathbf{A}$.

The determinant of a $1 \times 1$ matrix is simply the value of its single entry. Suppose that $k$ is a positive integer and the determinant has been defined for all $k \times k$ matrices. Then if $\mathbf{A}$ is a $(k+1) \times (k+1)$ matrix, the determinant of $\mathbf{A}$ is defined as follows:

$$\det \mathbf{A} \equiv \sum_{j=1}^{k+1} (-1)^{1+j} a_{1j} \det \mathbf{A}^{1j}. \tag{15.13}$$

By the Principle of Mathematical Induction, the determinant is defined for all square matrices.

**EXAMPLE 15.9**   For a $2 \times 2$ matrix $\mathbf{A} = [a_{ij}]$, the determinant of $\mathbf{A}$ is given by

$$\det \mathbf{A} = a_{11} \det \mathbf{A}^{11} - a_{12} \det \mathbf{A}^{12}$$
$$= a_{11}a_{22} - a_{12}a_{21}. \qquad \square$$

**EXAMPLE 15.10**   For a $3 \times 3$ matrix $\mathbf{A} = [a_{ij}]$, the determinant of $\mathbf{A}$ is given by

$$\det \mathbf{A} = a_{11} \det \mathbf{A}^{11} - a_{12} \det \mathbf{A}^{12} + a_{13} \det \mathbf{A}^{13}$$

$$= a_{11} \det \begin{pmatrix} a_{22} & a_{23} \\ a_{32} & a_{33} \end{pmatrix} - a_{12} \det \begin{pmatrix} a_{21} & a_{23} \\ a_{31} & a_{33} \end{pmatrix} + a_{13} \det \begin{pmatrix} a_{21} & a_{22} \\ a_{31} & a_{32} \end{pmatrix}$$

$$= a_{11}[a_{22}a_{33} - a_{23}a_{32}] - a_{12}[a_{21}a_{33} - a_{23}a_{31}] + a_{13}[a_{21}a_{32} - a_{22}a_{31}]. \qquad \square$$

It would take us too far afield to provide a full discussion of the determinant for general $n \times n$ matrices. Appendix B on linear algebra includes a discussion on the determinant of $3 \times 3$ matrices and its relation to the cross product and inner product of two vectors in $\mathbb{R}^3$. A full description of the determinant may be found in Charles Curtis's *Linear Algebra: An Introductory Approach* (Springer Verlag, 1984). In particular, the proof of the following theorem may be found in Curtis's book; the $3 \times 3$ case is proven in the appendix.

**THEOREM 15.8** **Cramer's Rule** An $n \times n$ matrix $\mathbf{A}$ is invertible if and only if $\det \mathbf{A} \neq 0$. Moreover, if $\det \mathbf{A} \neq 0$, then there is the following formula for the inverse matrix:

$$\left(\mathbf{A}^{-1}\right)_{ij} = \frac{1}{\det \mathbf{A}}\left((-1)^{i+j} \det \mathbf{A}^{ji}\right).$$

**EXAMPLE 15.11** For a $2 \times 2$ matrix $\mathbf{A} = \begin{pmatrix} a_{11} & a_{12} \\ a_{21} & a_{22} \end{pmatrix}$, Cramer's Rule implies that if

$$\det \mathbf{A} = a_{11}a_{22} - a_{12}a_{21} \neq 0,$$

then the matrix $\mathbf{A}$ has an inverse $\mathbf{A}^{-1}$, given by

$$\mathbf{A}^{-1} = \frac{1}{\det \mathbf{A}} \begin{pmatrix} a_{22} & -a_{12} \\ -a_{21} & a_{11} \end{pmatrix}.$$

In this case, direct matrix multiplication shows that this formula does in fact define the inverse matrix. $\square$

Cramer's Rule, together with the equivalence of invertibility of a mapping and the invertibility of its associated matrix, gives the following corollary.

**COROLLARY 15.9** For a linear mapping $\mathbf{T} \colon \mathbb{R}^n \to \mathbb{R}^n$ that is associated with the $n \times n$ matrix $\mathbf{A}$, the following three assertions are equivalent:

(i) $\det \mathbf{A} \neq 0$.

(ii) The matrix $\mathbf{A}$ is an invertible matrix.

(iii) The mapping $\mathbf{T} \colon \mathbb{R}^n \to \mathbb{R}^n$ is an invertible mapping.

A general continuous mapping $\mathbf{F} \colon \mathbb{R}^n \to \mathbb{R}^n$ may be one-to-one without being onto, and it may be onto without being one-to-one. However, a *linear* mapping $\mathbf{T} \colon \mathbb{R}^n \to \mathbb{R}^n$ is one-to-one if and only if it is onto. The proof of this equivalence for general $n$ can again be found in Curtis's book; the proof for $n = 3$ is in Appendix B on linear algebra. This property of linear maps between Euclidean spaces of the same dimension is necessary to prove one of the implications of the following theorem.

**THEOREM 15.10** For an $n \times n$ matrix $\mathbf{A}$, the following two assertions are equivalent:

(i) The matrix $\mathbf{A}$ is invertible.

(ii) There is a positive number $c$ such that

$$\|\mathbf{A}\mathbf{h}\| \geq c\|\mathbf{h}\| \quad \text{for all points } \mathbf{h} \text{ in } \mathbb{R}^n.$$

**Proof**  First suppose that the matrix $\mathbf{A}$ is invertible. Then observe that for each point $\mathbf{h}$ in $\mathbb{R}^n$,

$$\mathbf{h} = (\mathbf{A}^{-1}\mathbf{A})\mathbf{h} = \mathbf{A}^{-1}(\mathbf{A}\mathbf{h}),$$

and hence by the Generalized Cauchy-Schwarz Inequality,

$$\|\mathbf{h}\| = \|\mathbf{A}^{-1}(\mathbf{A}\mathbf{h})\| \leq \|\mathbf{A}^{-1}\|\|\mathbf{A}\mathbf{h}\|.$$

Thus (ii) holds where $c = 1/\|\mathbf{A}^{-1}\|$.

Conversely, suppose that (ii) holds. Then, in particular, we see that for a point $\mathbf{h}$ in $\mathbb{R}^n$, if $\mathbf{A}\mathbf{h} = 0$, then $\mathbf{h} = 0$. Let $\mathbf{T}\colon \mathbb{R}^n \to \mathbb{R}^n$ be the linear mapping associated with the matrix $\mathbf{A}$. Since for two points $\mathbf{u}$ and $\mathbf{v}$ in $\mathbb{R}^n$, $\mathbf{T}(\mathbf{u} - \mathbf{v}) = \mathbf{T}(\mathbf{u}) - \mathbf{T}(\mathbf{v})$, setting $\mathbf{h} = \mathbf{u} - \mathbf{v}$ we see that if $\mathbf{T}(\mathbf{u}) = \mathbf{T}(\mathbf{v})$, then $\mathbf{u} = \mathbf{v}$; that is, the linear mapping $\mathbf{T}\colon \mathbb{R}^n \to \mathbb{R}^n$ is one-to-one. By the remarks that preceded this theorem, the linear mapping $\mathbf{T}\colon \mathbb{R}^n \to \mathbb{R}^n$ is invertible. Thus the matrix $\mathbf{A}$ that represents it is also invertible.  ∎

To each $n \times n$ matrix $\mathbf{A}$, there corresponds another $n \times n$ matrix, called the *transpose matrix*, whose properties are closely related to those of $\mathbf{A}$. It is defined as follows.

**DEFINITION**  *For an $n \times n$ matrix $\mathbf{A} = [a_{ij}]$, the transpose matrix $\mathbf{A}^{\mathbf{T}}$ is defined to be the $n \times n$ matrix that, for indices $i$ and $j$ such that $1 \leq i \leq n$ and $1 \leq j \leq n$, has $ij$th entry equal to $a_{ji}$.*

**EXAMPLE 15.12**  For $3 \times 3$ matrices, the definition of transpose means that

$$\begin{pmatrix} a_{11} & a_{12} & a_{13} \\ a_{21} & a_{22} & a_{23} \\ a_{31} & a_{32} & a_{33} \end{pmatrix}^{\mathbf{T}} = \begin{pmatrix} a_{11} & a_{21} & a_{31} \\ a_{12} & a_{22} & a_{32} \\ a_{13} & a_{23} & a_{33} \end{pmatrix}. \qquad \square$$

For any $n \times n$ matrix $\mathbf{A}$, $\det \mathbf{A} = \det \mathbf{A}^{\mathbf{T}}$; this is proven in Curtis's book. In the cases $n = 2$ and $n = 3$, the proof is a straightforward calculation. Using Corollary 15.9 and the equality of the determinant of a matrix with the determinant of its transpose, we obtain the following important theorem.

**THEOREM 15.11**  An $n \times n$ matrix is invertible if and only if its transpose is invertible.

**EXERCISES**

1. Which of the following mappings $\mathbf{F}\colon \mathbb{R}^2 \to \mathbb{R}^2$ is linear?
   (a)  $\mathbf{F}(x, y) = (-y, e^x)$ for $(x, y)$ in $\mathbb{R}^2$.
   (b)  $\mathbf{F}(x, y) = (x - y^2, 2y)$ for $(x, y)$ in $\mathbb{R}^2$.
   (c)  $\mathbf{F}(x, y) = 17(x, y)$ for $(x, y)$ in $\mathbb{R}^2$.
2. Define

$$\mathbf{T}(x, y) = (x + y, x - y) \quad \text{for } (x, y) \text{ in } \mathbb{R}^2.$$

   (a)  Prove directly that the mapping $\mathbf{T}\colon \mathbb{R}^2 \to \mathbb{R}^2$ is one-to-one and onto.
   (b)  Use Corollary 15.9 to show that the mapping $\mathbf{T}\colon \mathbb{R}^2 \to \mathbb{R}^2$ is one-to-one and onto.
3. Define

$$\mathbf{T}(x, y, z) = (x + z, 2x - 4y + 3z, y + 6z) \quad \text{for } (x, y, z) \text{ in } \mathbb{R}^3.$$

   Prove that the mapping $\mathbf{T}\colon \mathbb{R}^3 \to \mathbb{R}^3$ is invertible.

4.  Show that there is no linear mapping $\mathbf{T}: \mathbb{R}^2 \to \mathbb{R}^2$ having the property that

$$\mathbf{T}(1,\ 1) = (4,\ 0) \quad \text{and} \quad \mathbf{T}(-2,\ -2) = (0,\ 1).$$

5.  Find a linear transformation $\mathbf{T}: \mathbb{R}^3 \to \mathbb{R}^3$ that has the property that

$$\mathbf{T}(1,\ 1,\ 1) = (0,\ 2,\ 0), \quad \mathbf{T}(1,\ 1,\ -1) = (1,\ 2,\ 0) \quad \text{and} \quad \mathbf{T}(2,\ 0,\ 0) = (1,\ 1,\ 1).$$

(*Hint:* Use linearity to determine $\mathbf{T}(\mathbf{e}_i)$ for $i = 1,\ 2,\ 3$.)

6.  For a point $(x,\ y)$ in the plane $\mathbb{R}^2$, define $\mathbf{T}(x,\ y)$ to be the point on the line $\ell = \{(x,\ y) \text{ in } \mathbb{R}^2 \mid y = 2x\}$ that is closest to $(x,\ y)$. Show that the mapping $\mathbf{T}: \mathbb{R}^2 \to \mathbb{R}^2$ is linear and find the $2 \times 2$ matrix that is associated with this mapping.

7.  For a point $(x,\ y,\ z)$ in $\mathbb{R}^3$, define $\mathbf{T}(x,\ y,\ z)$ to be the point on the plane $P = \{(x,\ y,\ z) \text{ in } \mathbb{R}^2 \mid x + y + z = 0\}$ that is closest to $(x,\ y,\ z)$. Show that the mapping $\mathbf{T}: \mathbb{R}^3 \to \mathbb{R}^3$ is linear and find the $3 \times 3$ matrix that is associated with this mapping.

8.  Define $A = \begin{pmatrix} 1 & 0 \\ 0 & 0 \end{pmatrix}$. Find all $2 \times 2$ matrices $\mathbf{B}$ that have that property that

$$\mathbf{AB} = \mathbf{BA}.$$

9.  For a number $\theta$, define the $2 \times 2$ matrix $\mathbf{A}_\theta$ by

$$\mathbf{A}_\theta = \begin{pmatrix} \cos\theta & -\sin\theta \\ \sin\theta & \cos\theta \end{pmatrix}.$$

For a point $\mathbf{x}$ in the plane $\mathbb{R}^2$ written as $\mathbf{x} = (r\cos\alpha,\ r\sin\alpha)$, use the addition properties of the cosine and the sine to verify that

$$\mathbf{A}_\theta \mathbf{x} = (r\cos(\alpha + \theta),\ r\sin(\alpha + \theta)).$$

From this formula, give a geometric interpretation of the linear mapping associated with the matrix $\mathbf{A}_\theta$. $\mathbf{A}_\theta$ is called a *rotation matrix.*

10.  Find the $2 \times 2$ matrix that is associated with the mapping in the plane that rotates points $90°$ counterclockwise about the origin.

11.  By a direct calculation of the matrix product, show that

$$\mathbf{A}_\theta \mathbf{A}_\phi = \mathbf{A}_{\theta+\phi},$$

where these matrices are rotation matrices as defined in Exercise 9. Thus the product of two rotation matrices is again a rotation matrix. Use Proposition 15.5 to explain why this product formula is expected.

12.  An $n \times n$ matrix $\mathbf{A}$ is called *upper-triangular* provided that all of the entries below the diagonal entries are 0.
   (a)  Use the inductive definition of the determinant to prove that the determinant of an upper-triangular matrix is the product of the diagonal entries.
   (b)  Prove that the product of two upper-triangular $n \times n$ matrices is again upper-triangular. What are the diagonal entries of the product?
   (c)  Use (a) and (b) to show that if $\mathbf{A}$ and $\mathbf{B}$ are upper-triangular $n \times n$ matrices, then $\det \mathbf{BA} = \det \mathbf{B} \cdot \det \mathbf{A}$.

13.  Let $\mathbf{A}$ be an $n \times n$ matrix and suppose that $\mathbf{B}$ and $\mathbf{B}'$ are two $n \times n$ matrices that have the property that

$$\mathbf{AB} = \mathbf{I}_n = \mathbf{B}'\mathbf{A}.$$

Show that $\mathbf{B} = \mathbf{B}'$ by verifying that

$$\mathbf{B} = \mathbf{I}_n \mathbf{B} = (\mathbf{B}'\mathbf{A})\mathbf{B} = \mathbf{B}'(\mathbf{AB}) = \mathbf{B}'\mathbf{I}_n = \mathbf{B}'.$$

14. Suppose that the mapping $\mathbf{T}: \mathbb{R}^n \to \mathbb{R}^n$ has the property that there is another mapping $\mathbf{S}: \mathbb{R}^n \to \mathbb{R}^n$ such that

$$\mathbf{T}(\mathbf{S}(\mathbf{x})) = \mathbf{S}(\mathbf{T}(\mathbf{x})) = \mathbf{x} \quad \text{for all } \mathbf{x} \text{ in } \mathbb{R}^n.$$

Prove that $\mathbf{T}: \mathbb{R}^n \to \mathbb{R}^n$ is invertible and that its inverse is the mapping $\mathbf{S}: \mathbb{R}^n \to \mathbb{R}^n$.

15. For an invertible $n \times n$ matrix $\mathbf{A}$ and a point $\mathbf{x}$ in $\mathbb{R}^n$, show that

$$\mathbf{A}\mathbf{x} = \mathbf{0} \quad \text{if and only if} \quad \mathbf{x} = \mathbf{0}.$$

16. Let $\mathbf{A}$ be an $n \times n$ matrix. Show that for each pair of indices $i$ and $j$ such that $1 \le i, j \le n$,

$$\langle \mathbf{A}\mathbf{e}_i, \, \mathbf{e}_j \rangle = \langle \mathbf{e}_i, \, \mathbf{A}^\mathrm{T}\mathbf{e}_j \rangle.$$

Use this and the linearity of the inner product to show that for any two points $\mathbf{u}$ and $\mathbf{v}$ in $\mathbb{R}^n$,

$$\langle \mathbf{A}\mathbf{u}, \, \mathbf{v} \rangle = \langle \mathbf{u}, \, \mathbf{A}^\mathrm{T}\mathbf{v} \rangle.$$

## 15.2 The Derivative Matrix, the Differential and First-Order Approximation

In this section, we will consider nonlinear mappings between Euclidean spaces and the manner in which they may be approximated by linear mappings. The following definition singles out the classes of mappings that we will consider.

**DEFINITION** *Let $\mathcal{O}$ be an open subset of $\mathbb{R}^n$ and consider a mapping $\mathbf{F}: \mathcal{O} \to \mathbb{R}^m$ represented in component functions as $\mathbf{F} = (\mathrm{F}_1, \ldots, \mathrm{F}_m)$.*

(i) *The mapping $\mathbf{F}: \mathcal{O} \to \mathbb{R}^m$ is said to have first-order partial derivatives at the point $\mathbf{x}$ in $\mathcal{O}$ provided that for each index $i$ such that $1 \le i \le m$, the component function $\mathrm{F}_i: \mathcal{O} \to \mathbb{R}$ has first-order partial derivatives at $\mathbf{x}$.*

(ii) *Moreover, the mapping $\mathbf{F}: \mathcal{O} \to \mathbb{R}^m$ is said to have first-order partial derivatives provided that it has first partial derivatives at every point in $\mathcal{O}$.*

(iii) *Finally, the mapping $\mathbf{F}: \mathcal{O} \to \mathbb{R}^m$ is said to be continuously differentiable provided that each component function is continuously differentiable.*

**EXAMPLE 15.13** Define the mapping $\mathbf{F}: \mathbb{R}^2 \to \mathbb{R}^3$ by

$$\mathbf{F}(x, \, y) = (x^2 + e^{xy}, \, \sin(xy), \, y + 1) \quad \text{for all } (x, \, y) \text{ in } \mathbb{R}^2.$$

Then $\mathbf{F}: \mathbb{R}^2 \to \mathbb{R}^3$ is continuously differentiable, since it is clear that each of its three component functions is continuously differentiable. $\qquad \square$

**PROPOSITION 15.12** Let $\mathcal{O}$ be an open subset of $\mathbb{R}^n$ and suppose that the mapping $\mathbf{F}: \mathcal{O} \to \mathbb{R}^m$ is continuously differentiable. Then the mapping $\mathbf{F}: \mathcal{O} \to \mathbb{R}^m$ is continuous.

**Proof**    By definition, each of the component functions of the mapping $\mathbf{F}: \mathcal{O} \to \mathbb{R}^m$ is continuously differentiable. It follows from Theorem 13.11 that each component function is continuous. Consequently, from the Componentwise Continuity Theorem (Theorem 11.4), we conclude that the mapping $\mathbf{F}: \mathcal{O} \to \mathbb{R}^m$ is itself continuous.    ∎

**DEFINITION**    *Let $\mathcal{O}$ be an open subset of $\mathbb{R}^n$ and suppose that the mapping $\mathbf{F}: \mathcal{O} \to \mathbb{R}^m$ has first-order partial derivatives at the point $\mathbf{x}$ in $\mathcal{O}$. The derivative matrix of $\mathbf{F}: \mathcal{O} \to \mathbb{R}^m$ at the point $\mathbf{x}$ is defined to be the $m \times n$ matrix $\mathbf{DF}(\mathbf{x})$, which, for each index $i$ such that $1 \le i \le m$, has $i$th row equal to $\mathbf{DF}_i(\mathbf{x})$. Thus the $ij$th entry of this derivative matrix is given by the formula*

$$(\mathbf{DF}(\mathbf{x}))_{ij} \equiv \frac{\partial F_i}{\partial x_j}(\mathbf{x}).$$

**EXAMPLE 15.14**    If the mapping $\mathbf{F}: \mathbb{R}^2 \to \mathbb{R}^2$ has first-order partial derivatives and has component representation

$$\mathbf{F}(x, y) = (u(x, y), v(x, y)) \quad \text{for } (x, y) \text{ in } \mathbb{R}^2,$$

then at the point $(x_0, y_0)$ the derivative matrix is

$$\mathbf{DF}(x_0, y_0) = \begin{bmatrix} \partial u/\partial x(x_0, y_0) & \partial u/\partial y(x_0, y_0) \\ \partial v/\partial x(x_0, y_0) & \partial v/\partial y(x_0, y_0) \end{bmatrix}. \qquad \square$$

**EXAMPLE 15.15**    Suppose that the function $f: \mathbb{R}^2 \to \mathbb{R}$ has first-order partial derivatives. Then at the point $(x_0, y_0)$ in $\mathbb{R}^2$, the derivative matrix is

$$\mathbf{D}f(x_0, y_0) = [\partial f/\partial x(x_0, y_0), \quad \partial f/\partial y(x_0, y_0)],$$

which is the $1 \times 2$ matrix corresponding to the derivative vector.    $\square$

**EXAMPLE 15.16**    Suppose that the mapping $\mathbf{F}: \mathbb{R} \to \mathbb{R}^3$ has first-order partial derivatives and has the component representation

$$\mathbf{F}(t) = (x(t), y(t), z(t)) \quad \text{for } t \text{ in } \mathbb{R}.$$

Then at the point $t_0$ in $\mathbb{R}$, the derivative matrix is the $3 \times 1$ matrix given by

$$\mathbf{DF}(t_0) = \begin{bmatrix} x'(t_0) \\ y'(t_0) \\ z'(t_0) \end{bmatrix}. \qquad \square$$

**THEOREM**
**15.13**

**The Mean Value Theorem for General Mappings**   Let $\mathcal{O}$ be an open subset of $\mathbb{R}^n$ and suppose that the mapping $\mathbf{F}: \mathcal{O} \to \mathbb{R}^m$ is continuously differentiable. Suppose that the points $\mathbf{x}$ and $\mathbf{x} + \mathbf{h}$ are in $\mathcal{O}$ and that the segment joining these points also lies in $\mathcal{O}$. Then there are numbers $\theta_1, \theta_2, \ldots, \theta_m$ in the open interval $(0, 1)$ such that

$$F_i(\mathbf{x} + \mathbf{h}) - F_i(\mathbf{x}) = \langle \mathbf{DF}_i(\mathbf{x} + \theta_i \mathbf{h}), \mathbf{h} \rangle \quad \text{for } 1 \le i \le m; \tag{15.14}$$

that is,
$$\mathbf{F}(\mathbf{x} + \mathbf{h}) - \mathbf{F}(\mathbf{x}) = \mathbf{Ah}, \tag{15.15}$$

where $\mathbf{A}$ is the $m \times n$ matrix whose $i$th row is $\mathbf{DF}_i(\mathbf{x} + \theta_i \mathbf{h})$.

**Proof**   Just apply the Mean Value Theorem for real-valued functions to each of the continuously differentiable component functions and we obtain formula (15.14). Formula (15.15) is simply a rewriting of (15.14) in matrix notation.   ∎

A natural question to ask is whether in (15.14) we can choose all the $\theta_i$'s equal. In that case, (15.15) would become

$$\mathbf{F}(\mathbf{x} + \mathbf{h}) - \mathbf{F}(\mathbf{x}) = \mathbf{DF}(\mathbf{x} + \theta \mathbf{h})\mathbf{h}, \tag{15.16}$$

and this would be a symbol-by-symbol extension of the Mean Value Theorem for real-valued functions. However, formula (15.16) is false. In general, it is not true that there is a number $\theta$ in the interval $(0, 1)$ such that (15.16) holds. The following example illustrates what can occur.

**EXAMPLE 15.17**   Define the mapping $\mathbf{F}: \mathbb{R}^2 \to \mathbb{R}^2$ by

$$\mathbf{F}(x, y) = (x^2, y^3) \quad \text{for } (x, y) \text{ in } \mathbb{R}^2.$$

Set $\psi(x, y) = x^2$ and $\phi(x, y) = y^3$ for $(x, y)$ in $\mathbb{R}^2$. Take the point $\mathbf{x}$ to be the origin $(0, 0)$ and $\mathbf{h}$ to be the point $(1, 1)$. The above extension of the Mean Value Theorem implies that there are numbers $\theta_1$ and $\theta_2$ in the interval $(0, 1)$ such that

$$\psi(1, 1) - \psi(0, 0) = \langle \mathbf{D}\psi(\theta_1, \theta_1), (1, 1) \rangle$$
$$\phi(1, 1) - \phi(0, 0) = \langle \mathbf{D}\phi(\theta_2, \theta_2), (1, 1) \rangle. \tag{15.17}$$

But $\mathbf{D}\psi(x, y) = (2x, 0)$ and $\mathbf{D}\phi(x, y) = (0, 3y^2)$ for $(x, y)$ in $\mathbb{R}^2$; hence, substituting in (15.17), we see that

$$1 = 2\theta_1 \quad \text{and} \quad 1 = 3\theta_2^2.$$

Thus $\theta_1 = 1/2$ and $\theta_2 = \sqrt{1/3}$. Therefore, we certainly cannot find $\theta_1 = \theta_2$ so that (15.17) holds.   □

Recall the First-Order Approximation Theorem for scalar-valued functions, which asserts that if $\mathcal{O}$ is an open subset of $\mathbb{R}^n$ and the function $f: \mathcal{O} \to \mathbb{R}$ is continuously differentiable, then at each point $\mathbf{x}$ in $\mathcal{O}$,

$$\lim_{\mathbf{h} \to 0} \frac{f(\mathbf{x} + \mathbf{h}) - [f(\mathbf{h}) + \langle \mathbf{D}f(\mathbf{x}), \mathbf{h} \rangle]}{\|\mathbf{h}\|} = 0. \tag{15.18}$$

There is the following extension of this result to general mappings.

**THEOREM**
**15.14**

**First-Order Approximation Theorem for Mappings**  Let $\mathcal{O}$ be an open subset of $\mathbb{R}^n$ that contains the point $\mathbf{x}$ and suppose that the mapping $\mathbf{F}: \mathcal{O} \to \mathbb{R}^m$ is continuously differentiable. Then

$$\lim_{\mathbf{h} \to 0} \frac{\|\mathbf{F}(\mathbf{x} + \mathbf{h}) - [\mathbf{F}(\mathbf{x}) + D\mathbf{F}(\mathbf{x})\mathbf{h}]\|}{\|\mathbf{h}\|} = 0. \tag{15.19}$$

**Proof**  Since $\mathcal{O}$ is open, we can choose a positive number $r$ such that the symmetric neighborhood $\mathcal{N}_r(\mathbf{x})$ is contained in $\mathcal{O}$. For a point $\mathbf{h}$ in $\mathbb{R}^n$ such that $\|\mathbf{h}\| < r$, define

$$\mathbf{R}(\mathbf{h}) = \mathbf{F}(\mathbf{x} + \mathbf{h}) - [\mathbf{F}(\mathbf{x}) + D\mathbf{F}(\mathbf{x})\mathbf{h}].$$

We must show that
$$\lim_{\mathbf{h} \to 0} \frac{\|\mathbf{R}(\mathbf{h})\|}{\|\mathbf{h}\|} = 0. \tag{15.20}$$

But if we represent the mappings $\mathbf{F}$ and $\mathbf{R}$ as $\mathbf{F} = (F_1, \ldots, F_m)$ and $\mathbf{R} = (R_1, \ldots, R_m)$, then it is clear that for each index $i$ such that $1 \leq i \leq m$,

$$R_i(\mathbf{h}) = F_i(\mathbf{x} + \mathbf{h}) - [F_i(\mathbf{x}) + \langle DF_i(\mathbf{x}), \mathbf{h} \rangle] \quad \text{for } \|\mathbf{h}\| < r.$$

Since the function $\mathbf{F}: \mathcal{O} \to \mathbb{R}^m$ is continuously differentiable, the First-Order Approximation Theorem for real-valued functions implies that

$$\lim_{\mathbf{h} \to 0} \frac{R_i(\mathbf{h})}{\|\mathbf{h}\|} = 0.$$

Since
$$\frac{\|\mathbf{R}(\mathbf{h})\|}{\|\mathbf{h}\|} = \left( \sum_{i=1}^{m} \left[ \frac{R_i(\mathbf{h})}{\|\mathbf{h}\|} \right]^2 \right)^{1/2} \quad \text{for } 0 < \|\mathbf{h}\| < r,$$

it follows that (15.20) holds. ∎

For a function $f: I \to \mathbb{R}$, where $I$ is an open interval, at the point $x$ in $I$, if there is a number $a$ such that

$$\lim_{h \to 0} \frac{f(x + h) - [f(x) + ah]}{h} = 0,$$

then since, if $h \neq 0$ and $x + h$ is in $I$,

$$\frac{f(x + h) - [f(x) + ah]}{h} = \frac{f(x + h) - f(x)}{h} - a,$$

it follows that $f: I \to \mathbb{R}$ is differentiable at $x$ and that $f'(x) = a$. This property generalizes to mappings as follows.

**THEOREM 15.15**   Let $\mathcal{O}$ be an open subset of $\mathbb{R}^n$ that contains the point $\mathbf{x}$ and consider a mapping $\mathbf{F} \colon \mathcal{O} \to \mathbb{R}^m$. Suppose that $\mathbf{A}$ is an $m \times n$ matrix that has the property that

$$\lim_{\mathbf{h} \to 0} \frac{\|\mathbf{F}(\mathbf{x} + \mathbf{h}) - [\mathbf{F}(\mathbf{x}) + \mathbf{Ah}]\|}{\|\mathbf{h}\|} = 0. \tag{15.21}$$

Then the mapping $\mathbf{F} \colon \mathcal{O} \to \mathbb{R}^m$ has first-order partial derivatives at the point $\mathbf{x}$ and

$$\mathbf{A} = \mathbf{DF}(\mathbf{x}).$$

**Proof**   Represent the mapping $\mathbf{F} \colon \mathcal{O} \to \mathbb{R}^m$ in component functions as $\mathbf{F} = (F_1, \ldots, F_m)$ and set $a_{ij} = (\mathbf{A})_{ij}$. We must show that for each pair of indices $i$ and $j$ such that $1 \le i \le m$ and $1 \le j \le n$,

$$a_{ij} = \frac{\partial F_i}{\partial x_j}(\mathbf{x}).$$

For each index $i$ such that $1 \le i \le m$, define $\mathbf{A}_i$ to be the $i$th row of the matrix $\mathbf{A}$.

Since $\mathcal{O}$ is open, we can choose a positive number $r$ such that the symmetric neighborhood $\mathcal{N}_r(\mathbf{x})$ is contained in $\mathcal{O}$. Now observe that if $1 \le i \le m$ and $\|\mathbf{h}\| < r$, then

$$F_i(\mathbf{x} + \mathbf{h}) - [F_i(\mathbf{x}) + \langle \mathbf{A}_i, \mathbf{h} \rangle] = p_i(\mathbf{F}(\mathbf{x} + \mathbf{h}) - [\mathbf{F}(\mathbf{x}) + \mathbf{Ah}])$$

so that

$$|F_i(\mathbf{x} + \mathbf{h}) - [F_i(\mathbf{x}) + \langle \mathbf{A}_i, \mathbf{h} \rangle]| \le \|\mathbf{F}(\mathbf{x} + \mathbf{h}) - [\mathbf{F}(\mathbf{x}) + \mathbf{Ah}]\|.$$

From (15.21) it follows that

$$\lim_{\mathbf{h} \to 0} \frac{F_i(\mathbf{x} + \mathbf{h}) - [F_i(\mathbf{x}) + \langle \mathbf{A}_i, \mathbf{h} \rangle]}{\|\mathbf{h}\|} = 0.$$

In particular, for an index $j$ such that $1 \le j \le n$,

$$\lim_{t \to 0} \frac{F_i(\mathbf{x} + t\mathbf{e}_j) - [F_i(\mathbf{x}) + \langle \mathbf{A}_i, t\mathbf{e}_j \rangle]}{\|t\mathbf{e}_j\|} = 0. \tag{15.22}$$

However, $\|t\mathbf{e}_j\| = |t|$, so (15.22) is equivalent to

$$\lim_{t \to 0} \frac{F_i(\mathbf{x} + t\mathbf{e}_j) - F_i(\mathbf{x})}{t} = \langle \mathbf{A}_i, \mathbf{e}_j \rangle,$$

thus proving that $\mathbf{F} \colon \mathcal{O} \to \mathbb{R}^n$ has first partial derivatives at $\mathbf{x}$ and

$$a_{ij} = \langle \mathbf{A}_i, \mathbf{e}_j \rangle = \frac{\partial F_i}{\partial x_j}(\mathbf{x}) \quad \text{for} \quad 1 \le i \le m, \ 1 \le j \le n. \qquad \blacksquare$$

The above theorem implies that for a continuously differentiable mapping, the derivative matrix is the only matrix having the first-order approximation property (15.19).

In view of the correspondence described in Section 15.1 between $m \times n$ matrices and linear mappings from $\mathbb{R}^n$ to $\mathbb{R}^m$, it is useful to introduce the following correspondent of the derivative matrix.

**DEFINITION**  *Let $\mathcal{O}$ be an open subset of $\mathbb{R}^n$ that contains the point $\mathbf{x}$ and suppose that the mapping $\mathbf{F}: \mathcal{O} \to \mathbb{R}^m$ has first-order partial derivatives at the point $\mathbf{x}$. The linear mapping*

$$d\mathbf{F}(\mathbf{x}): \mathbb{R}^n \to \mathbb{R}^m$$

*defined by* $\qquad d\mathbf{F}(\mathbf{x})(\mathbf{h}) \equiv D\mathbf{F}(\mathbf{x})\mathbf{h} \quad$ *for all $\mathbf{h}$ in $\mathbb{R}^n$*

*is called the differential of the mapping $\mathbf{F}: \mathcal{O} \to \mathbb{R}^m$ at the point $\mathbf{x}$.*

We record as a theorem the content of Corollary 15.9 in the case when the linear mapping is the differential of a nonlinear mapping at a point.

**THEOREM 15.16**  Let $\mathcal{O}$ be an open subset of $\mathbb{R}^n$ that contains the point $\mathbf{x}$ and suppose that the mapping $\mathbf{F}: \mathcal{O} \to \mathbb{R}^n$ has first-order partial derivatives at $\mathbf{x}$. Then the following three assertions are equivalent:

(i) $\det D\mathbf{F}(\mathbf{x}) \neq 0$.

(ii) The derivative matrix $D\mathbf{F}(\mathbf{x})$ is an invertible $n \times n$ matrix.

(iii) The differential $d\mathbf{F}(\mathbf{x}): \mathbb{R}^n \to \mathbb{R}^n$ is an invertible linear mapping.

The First-Order Approximation Theorem is a precise assertion of the manner in which

$$\mathbf{F}(\mathbf{x} + \mathbf{h}) \approx \mathbf{F}(\mathbf{x}) + d\mathbf{F}(\mathbf{x})(\mathbf{h})$$

when $\mathbf{h}$ is sufficiently close to $\mathbf{0}$. The general theme of the next two chapters will be expressed in descriptions of the properties that a continuously differentiable mapping inherits from the properties of its differential at a point.

**EXERCISES**

1. Define

$$\mathbf{F}(x, y) = (e^{xy} + 2x, \ y^2 + \sin(x - y)) \quad \text{for } (x, y) \text{ in } \mathbb{R}^2.$$

   Find the derivative matrix of the mapping $\mathbf{F}: \mathbb{R}^2 \to \mathbb{R}^2$ at the points $(0, 0)$ and $(\pi, 0)$.

2. Define

$$\mathbf{F}(x, y, z) = (xyz, \ x^2 + yz, \ 1 + 3x) \quad \text{for } (x, y, z) \text{ in } \mathbb{R}^3.$$

   Find the derivative matrix of the mapping $\mathbf{F}: \mathbb{R}^3 \to \mathbb{R}^3$ at the points $(1, 2, 3)$, $(0, 1, 0)$ and $(-1, 4, 0)$.

3. Suppose that the mapping $\mathbf{F}: \mathbb{R}^n \to \mathbb{R}^m$ is continuously differentiable and that the derivative matrix $D\mathbf{F}(\mathbf{x})$ at each point $\mathbf{x}$ in $\mathbb{R}^n$ has all of its entries equal to 0. Prove that the mapping $\mathbf{F}: \mathbb{R}^n \to \mathbb{R}^m$ is constant; that is, there is some point $\mathbf{c}$ in $\mathbb{R}^m$ such that

$$\mathbf{F}(\mathbf{x}) = \mathbf{c} \quad \text{for every } \mathbf{x} \text{ in } \mathbb{R}^n.$$

4. Suppose that $\mathbf{A}$ is an $m \times n$ matrix. Define the mapping $\mathbf{F}: \mathbb{R}^n \to \mathbb{R}^m$ by

$$\mathbf{F}(\mathbf{x}) = \mathbf{A}\mathbf{x} \quad \text{for every } \mathbf{x} \text{ in } \mathbb{R}^n.$$

   Prove that $D\mathbf{F}(\mathbf{x}) = \mathbf{A}$ for all $\mathbf{x}$ in $\mathbb{R}^n$.

5. Suppose that the mapping $\mathbf{F}: \mathbb{R}^n \to \mathbb{R}^m$ is continuously differentiable and that there is a fixed $m \times n$ matrix $\mathbf{A}$ so that

$$\mathbf{DF}(\mathbf{x}) = \mathbf{A} \quad \text{for every } \mathbf{x} \text{ in } \mathbb{R}^n.$$

Prove that there is some $\mathbf{c}$ in $\mathbb{R}^m$ so that

$$\mathbf{F}(\mathbf{x}) = \mathbf{Ax} + \mathbf{c} \quad \text{for every } \mathbf{x} \text{ in } \mathbb{R}^n.$$

Restate this result for the case when $n = m = 1$.

6. Define the mapping $\mathbf{F}: \mathbb{R}^2 \to \mathbb{R}^2$ by

$$\mathbf{F}(x, y) = (x^2 - y^2, 2xy) \quad \text{for } (x, y) \text{ in } \mathbb{R}^2.$$

   (a) Find the points $(x_0, y_0)$ in $\mathbb{R}^2$ at which the derivative matrix $\mathbf{DF}(x_0, y_0)$ is invertible.
   (b) Find the points $(x_0, y_0)$ in $\mathbb{R}^2$ at which the differential $\mathbf{dF}(x_0, y_0): \mathbb{R}^2 \to \mathbb{R}^2$ is an invertible linear mapping.

7. Give a proof of the First-Order Approximation Theorem that is based on the Mean Value Theorem.

8. Suppose that the mapping $\mathbf{F}: \mathbb{R}^2 \to \mathbb{R}^2$ has the property that

$$\|\mathbf{F}(x, y)\| \leq x^2 + y^2 \quad \text{for every } (x, y) \text{ in } \mathbb{R}^2.$$

   (a) Show that

$$\lim_{(x,y) \to (0,0)} \frac{\|\mathbf{F}(x, y)\|}{\|(x, y)\|} = 0.$$

   (b) Use Theorem 15.15 to verify that all the entries of the matrix $\mathbf{DF}(0)$ are 0.
   (c) Is the mapping $\mathbf{F}: \mathbb{R}^2 \to \mathbb{R}^2$ necessarily continuously differentiable?

9. Suppose that the mapping $\mathbf{F}: \mathbb{R}^n \to \mathbb{R}^n$ is continuously differentiable. Suppose also that $\mathbf{F}(\mathbf{0}) = \mathbf{0}$ and that the derivative matrix $\mathbf{DF}(\mathbf{0})$ has the property that there is some positive number $c$ such that

$$\|\mathbf{DF}(\mathbf{0})\mathbf{h}\| \geq c\|\mathbf{h}\| \quad \text{for all } \mathbf{h} \text{ in } \mathbb{R}^n.$$

Prove that there is some positive number $r$ such that

$$\|\mathbf{F}(\mathbf{h})\| \geq c/2\|\mathbf{h}\| \quad \text{if} \quad \|\mathbf{h}\| \leq r.$$

10. Suppose that the continuously differentiable mapping $\mathbf{F}: \mathbb{R}^2 \to \mathbb{R}^2$ is represented in component functions as

$$\mathbf{F}(x, y) = (\psi(x, y), \varphi(x, y)) \quad \text{for } (x, y) \text{ in } \mathbb{R}^2.$$

Define the function $g: \mathbb{R}^2 \to \mathbb{R}$ by

$$g(x, y) = \frac{1}{2}[(\psi(x, y))^2 + (\varphi(x, y))^2] \quad \text{for } (x, y) \text{ in } \mathbb{R}^2.$$

   (a) Show that

$$\mathbf{D}g(x_0, y_0) = \left[\mathbf{DF}(x_0, y_0)\right]^{\mathsf{T}} \mathbf{F}(x_0, y_0).$$

   (b) Use (a) to prove that if $(x_0, y_0)$ is a minimizer of the function $g: \mathbb{R}^2 \to \mathbb{R}$ and the matrix $\mathbf{DF}(x_0, y_0)$ is invertible, then

$$\mathbf{F}(x_0, y_0) = \mathbf{0}.$$

## 15.3 The Chain Rule

From the Chain Rule for real-valued functions of a single variable, it follows that if $\mathcal{O}$ and $\mathcal{U}$ are open sets of real numbers and the functions $f: \mathcal{O} \to \mathbb{R}$ and $g: \mathcal{U} \to \mathbb{R}$ are continuously differentiable, with $f(\mathcal{O})$ contained in $\mathcal{U}$, then the composite function

$$g \circ f: \mathcal{O} \to \mathbb{R}$$

is also continuously differentiable and, moreover, for each point $x$ in $\mathcal{O}$,

$$(g \circ f)'(x) = g'(f(x))f'(x). \tag{15.23}$$

The Chain Rule carries over to compositions of general continuously differentiable mappings, in which the derivative matrix replaces the derivative and matrix multiplication replaces scalar multiplication. The general Chain Rule follows from the following special case of the composition of a mapping with a real-valued function.

In order to clearly state the Chain Rule, it is useful to use the following notation: For an open subset $\mathcal{U}$ of $\mathbb{R}^m$ and a function $g: \mathcal{U} \to \mathbb{R}$ that has first-order partial derivatives, at each point $\mathbf{p}$ in $\mathcal{U}$ and for each index $i$ such that $1 \leq i \leq m$, we define

$$D_i g(\mathbf{p}) \equiv \lim_{t \to 0} \frac{f(\mathbf{p} + t\mathbf{e}_i) - g(\mathbf{p})}{t}.$$

This notation has the advantage that the partial derivative with respect to the $i$th component is denoted by a symbol that is independent of the notation being used for the points in the domain. Moreover, there is the formula

$$\mathbf{D}g(\mathbf{p}) = (D_1 g(\mathbf{p}), \ldots, D_m g(\mathbf{p})) \quad \text{for each point } \mathbf{p} \text{ in } \mathcal{U}.$$

**THEOREM 15.17**

**The Chain Rule** Let $\mathcal{O}$ be an open subset of $\mathbb{R}^n$ and suppose that the mapping $\mathbf{F}: \mathcal{O} \to \mathbb{R}^m$ is continuously differentiable. Suppose also that $\mathcal{U}$ is an open subset of $\mathbb{R}^m$ and that the function $g: \mathcal{U} \to \mathbb{R}$ is continuously differentiable. Finally, suppose that $\mathbf{F}(\mathcal{O})$ is contained in $\mathcal{U}$. Then the composition $g \circ \mathbf{F}: \mathcal{O} \to \mathbb{R}$ is also continuously differentiable. Moreover, for each point $\mathbf{x}$ in $\mathcal{O}$ and each index $i$ such that $1 \leq i \leq n$,

$$\frac{\partial}{\partial x_i}(g \circ \mathbf{F})(\mathbf{x}) = \sum_{j=1}^{m} D_j g(\mathbf{F}(\mathbf{x})) \frac{\partial F_j}{\partial x_i}(\mathbf{x}); \tag{15.24}$$

that is,

$$\mathbf{D}(g \circ \mathbf{F})(\mathbf{x}) = \mathbf{D}g(\mathbf{F}(\mathbf{x}))\mathbf{D}\mathbf{F}(\mathbf{x}). \tag{15.25}$$

**Proof** Let $\mathbf{x}$ be a point in $\mathcal{O}$. Since $\mathcal{O}$ is open, we can select a positive number $r$ such that the symmetric neighborhood $\mathcal{N}_r(\mathbf{x})$ is contained in $\mathcal{O}$. Moreover, since the mapping $\mathbf{F}: \mathcal{O} \to \mathbb{R}^m$ is continuous and $\mathcal{U}$ is an open subset of $\mathbb{R}^m$, we can also suppose that the segment joining the points $\mathbf{F}(\mathbf{x})$ and $\mathbf{F}(\mathbf{x} + \mathbf{h})$ lies in $\mathcal{U}$ if $\|\mathbf{h}\| < r$. For each $\mathbf{h}$ in $\mathbb{R}^n$ such that $\|\mathbf{h}\| < r$, define

$$\mathbf{R}(\mathbf{h}) = \mathbf{F}(\mathbf{x} + \mathbf{h}) - \mathbf{F}(\mathbf{x}) - \mathbf{D}\mathbf{F}(\mathbf{x})\mathbf{h}.$$

According to the First-Order Approximation Theorem for Mappings,

$$\lim_{\mathbf{h} \to 0} \frac{\|\mathbf{R}(\mathbf{h})\|}{\|\mathbf{h}\|} = 0, \tag{15.26}$$

and if $\|\mathbf{h}\| < r$, $\qquad \mathbf{F}(\mathbf{x} + \mathbf{h}) - \mathbf{F}(\mathbf{x}) = \mathbf{DF}(\mathbf{x})\mathbf{h} + \mathbf{R}(\mathbf{h}).$ (15.27)

Now for each $\mathbf{h}$ in $\mathbb{R}^n$ such that $\|\mathbf{h}\| < r$, we can apply the Mean Value Theorem to the function $g: \mathcal{U} \to \mathbb{R}$ on the segment joining the points $\mathbf{F}(\mathbf{x})$ and $\mathbf{F}(\mathbf{x} + \mathbf{h})$ in order to select a point on this segment, which we label $\mathbf{v}(\mathbf{h})$, at which

$$g(\mathbf{F}(\mathbf{x} + \mathbf{h})) - g(\mathbf{F}(\mathbf{x})) = \langle \mathbf{D}g(\mathbf{v}(\mathbf{h})), \ \mathbf{F}(\mathbf{x} + \mathbf{h}) - \mathbf{F}(\mathbf{x}) \rangle,$$

which, if we substitute (15.27), gives

$$(g \circ \mathbf{F})(\mathbf{x} + \mathbf{h}) - (g \circ \mathbf{F})(\mathbf{x}) = \langle \mathbf{D}g(\mathbf{v}(\mathbf{h})), \ \mathbf{DF}(\mathbf{x})\mathbf{h} \rangle + \langle \mathbf{D}g(\mathbf{v}(\mathbf{h})), \ \mathbf{R}(\mathbf{h}) \rangle. \quad (15.28)$$

Observe that the continuity of $\mathbf{F}: \mathcal{O} \to \mathbb{R}^m$ implies that

$$\lim_{\mathbf{h} \to 0} \mathbf{v}(\mathbf{h}) = \mathbf{F}(\mathbf{x}). \tag{15.29}$$

We will now verify (15.24). Indeed, for a number $t$ such that $0 < |t| < r$, if we define $\mathbf{h} = t\mathbf{e}_i$, then from (15.28) we obtain

$$\frac{(g \circ \mathbf{F})(\mathbf{x} + t\mathbf{e}_i) - (g \circ \mathbf{F})(\mathbf{x})}{t} = \langle \mathbf{D}g(\mathbf{v}(t\mathbf{e}_i)), \ \mathbf{DF}(\mathbf{x})\mathbf{e}_i \rangle + \left\langle \mathbf{D}g(\mathbf{v}(t\mathbf{e}_i)), \ \frac{\mathbf{R}(t\mathbf{e}_i)}{t} \right\rangle.$$

From this equality, by using (15.26) and (15.29) it follows that

$$\frac{\partial}{\partial x_i}(g \circ \mathbf{F})(\mathbf{x}) = \langle \mathbf{D}g(\mathbf{F}(\mathbf{x})), \ \mathbf{DF}(\mathbf{x})\mathbf{e}_i \rangle. \tag{15.30}$$

But $\qquad \mathbf{DF}(\mathbf{x})\mathbf{e}_i = \left( \dfrac{\partial \mathbf{F}_1}{\partial x_i}(\mathbf{x}), \dots, \dfrac{\partial \mathbf{F}_m}{\partial x_i}(\mathbf{x}) \right),$

so (15.24) is exactly (15.30). In particular, this shows that the function $g \circ \mathbf{F}: \mathcal{O} \to \mathbb{R}$ has first-order partial derivatives, and then, because of the continuity with respect to $\mathbf{x}$ of the right-hand side of formula (15.24), that $g \circ \mathbf{F}: \mathcal{O} \to \mathbb{R}$ is continuously differentiable. To conclude the proof, simply observe that (15.25) is a rewriting of (15.24) in matrix notation. ∎

We will now examine some of the forms of the Chain Rule that most commonly occur.

**EXAMPLE 15.18**   Suppose that the functions $\psi \colon \mathbb{R}^2 \to \mathbb{R}$ and $\varphi \colon \mathbb{R}^2 \to \mathbb{R}$ are continuously differentiable. Suppose also that $\mathcal{O}$ is an open subset of the plane $\mathbb{R}^2$ and that the function $f \colon \mathcal{O} \to \mathbb{R}$ is continuously differentiable. Finally, suppose that $(\psi(x, y), \varphi(x, y))$ is in $\mathcal{O}$ for all $(x, y)$ in $\mathbb{R}^2$. Then

$$\frac{\partial}{\partial x}(f(\psi(x, y), \varphi(x, y))) = D_1 f(\psi(x, y), \varphi(x, y))\frac{\partial \psi}{\partial x}(x, y)$$
$$+ D_2 f(\psi(x, y), \varphi(x, y))\frac{\partial \varphi}{\partial x}(x, y)$$

and

$$\frac{\partial}{\partial y}(f(\psi(x, y), \varphi(x, y))) = D_1 f(\psi(x, y), \varphi(x, y))\frac{\partial \psi}{\partial y}(x, y)$$
$$+ D_2 f(\psi(x, y), \varphi(x, y))\frac{\partial \varphi}{\partial y}(x, y). \qquad \square$$

**EXAMPLE 15.19**   Let the function $g \colon \mathbb{R}^3 \to \mathbb{R}$ be continuously differentiable, and define the function $\psi \colon \mathbb{R} \to \mathbb{R}$ by $\psi(x) = g(x^2, 2x, 1 - x)$ for $x$ in $\mathbb{R}$. Then for each $x$ in $\mathbb{R}$,

$$\psi'(x) = D_1 g(x^2, 2x, 1 - x)(2x)$$
$$+ D_2 g(x^2, 2x, 1 - x)(2) + D_3 g(x^2, 2x, 1 - x)(-1). \qquad \square$$

**EXAMPLE 15.20**   Suppose that each of the functions $u \colon \mathbb{R}^2 \to \mathbb{R}$, $v \colon \mathbb{R}^2 \to \mathbb{R}$, and $w \colon \mathbb{R}^2 \to \mathbb{R}$ is continuously differentiable, and that the function $g \colon \mathbb{R}^3 \to \mathbb{R}$ is also continuously differentiable. Then for each point $(x, y)$ in the plane $\mathbb{R}^2$,

$$\frac{\partial}{\partial x}(g(u(x, y), v(x, y), w(x, y))) = D_1 g(u(x, y), v(x, y), w(x, y))\frac{\partial u}{\partial x}(x, y)$$
$$+ D_2 g(u(x, y), v(x, y), w(x, y))\frac{\partial v}{\partial x}(x, y)$$
$$+ D_3 g(u(x, y), v(x, y), w(x, y))\frac{\partial w}{\partial x}(x, y).$$

$$\square$$

**Remark on Notation**   In books in which there are calculations involving partial derivatives, the reader will find a large variety of notation. For example, the latter derivative formula in Example 15.18 is often abbreviated as

$$\frac{\partial f}{\partial y} = \frac{\partial f}{\partial \psi}\frac{\partial \psi}{\partial y} + \frac{\partial f}{\partial \varphi}\frac{\partial \varphi}{\partial y}. \tag{15.31}$$

Similarly, the derivative formula in Example 15.20 is often abbreviated as

$$\frac{\partial g}{\partial x} = \frac{\partial g}{\partial u}\frac{\partial u}{\partial x} + \frac{\partial g}{\partial v}\frac{\partial v}{\partial x} + \frac{\partial g}{\partial w}\frac{\partial w}{\partial x}. \tag{15.32}$$

As another common instance of terse, but useful, notational devices we note that if the function $f: \mathbb{R}^2 \to \mathbb{R}$ is continuously differentiable and the function $g: \mathbb{R}^2 \to \mathbb{R}$ is defined by

$$g(r, \theta) = f(r \cos \theta, r \sin \theta) \quad \text{for } (r, \theta) \text{ in } \mathbb{R}^2,$$

then according to the Chain Rule, for each point $(r, \theta)$ in $\mathbb{R}^2$,

$$\frac{\partial g}{\partial r}(r, \theta) = D_1 f(r \cos \theta, r \sin \theta) \cos \theta + D_2 f(r \cos \theta, r \sin \theta) \sin \theta.$$

This last formula is frequently abbreviated as

$$\frac{\partial f}{\partial r} = \frac{\partial f}{\partial x} \cos \theta + \frac{\partial f}{\partial y} \sin \theta. \tag{15.33}$$

One must carefully interpret this formula in order to understand that it signifies the same thing as its predecessor.

Abbreviated formulas such as (15.31), (15.32), and (15.33) are very useful in compressing long equations and in shortening various calculations. But such formulas are not precise, because there is no indication of where the derivatives are to be evaluated, and there is ambiguity about what the variables are. Extra care is needed when using them.

When we analyze functions of two or three variables, especially when computing higher derivatives, it is notationally useful to denote

$$D_1 g(\mathbf{p}) \text{ by } \frac{\partial g}{\partial x}(\mathbf{p}), \quad D_2 g(\mathbf{p}) \text{ by } \frac{\partial g}{\partial y}(\mathbf{p}), \quad \text{and} \quad D_3 g(\mathbf{p}) \text{ by } \frac{\partial g}{\partial z}(\mathbf{p}),$$

even when $x$, $y$, and $z$ have not been explicitly introduced as notation for the component variables. In the following example, we will use this notational convention.

**EXAMPLE 15.21**   A function $u: \mathbb{R}^2 \to \mathbb{R}$ is said to be *harmonic* provided that it has continuous second-order partial derivatives that satisfy the identity

$$\frac{\partial^2 u}{\partial x^2}(x, y) + \frac{\partial^2 u}{\partial y^2}(x, y) = 0 \quad \text{for all } (x, y) \text{ in } \mathbb{R}^2.$$

Suppose that the function $u: \mathbb{R}^2 \to \mathbb{R}$ is harmonic. Define

$$v(x, y) = u(x^2 - y^2, 2xy) \quad \text{for all } (x, y) \text{ in } \mathbb{R}^2.$$

Then it turns out that the function $v: \mathbb{R}^2 \to \mathbb{R}$ is also harmonic. To verify this, we must show that

$$\frac{\partial^2 v}{\partial x^2}(x, y) + \frac{\partial^2 v}{\partial y^2}(x, y) = 0 \quad \text{for all } (x, y) \text{ in } \mathbb{R}^2.$$

However, for $(x, y)$ in $\mathbb{R}^2$,

$$\frac{\partial v}{\partial x}(x, y) = \frac{\partial u}{\partial x}(x^2 - y^2, 2xy)2x + \frac{\partial u}{\partial y}(x^2 - y^2, 2xy)2y,$$

so

$$\frac{\partial^2 v}{\partial x^2}(x, y) = \frac{\partial^2 u}{\partial x^2}(x^2 - y^2, 2xy)4x^2 + \frac{\partial u}{\partial x}(x^2 - y^2, 2xy)2$$

$$+ \frac{\partial^2 u}{\partial x \partial y}(x^2 - y^2, 2xy)8xy + \frac{\partial^2 u}{\partial y^2}(x^2 - y^2, 2xy)4y^2.$$

We carry out a similar computation for $\partial^2 v/\partial y^2(x, y)$, and since

$$\frac{\partial^2 u}{\partial x^2}(x^2 - y^2, 2xy) + \frac{\partial^2 u}{\partial y^2}(x^2 - y^2, 2xy) = 0 \quad \text{for all } (x, y) \text{ in } \mathbb{R}^2,$$

a calculation that is left as Exercise 4 shows that

$$\frac{\partial^2 v}{\partial x^2}(x, y) + \frac{\partial^2 v}{\partial y^2}(x, y) = 0 \quad \text{for } (x, y) \text{ in } \mathbb{R}^2. \qquad \square$$

The special case of the Chain Rule that we have just proven leads to the proof of the general case.

**THEOREM 15.18**   **The Chain Rule for General Mappings**   Let $\mathcal{O}$ be an open subset of $\mathbb{R}^n$ and suppose that the mapping $\mathbf{F}: \mathcal{O} \to \mathbb{R}^m$ is continuously differentiable. Suppose also that $\mathcal{U}$ is an open subset of $\mathbb{R}^m$ and that the mapping $\mathbf{G}: \mathcal{U} \to \mathbb{R}^k$ is continuously differentiable. Finally, suppose that $\mathbf{F}(\mathcal{O})$ is contained in $\mathcal{U}$. Then the composite mapping $\mathbf{G} \circ \mathbf{F}: \mathcal{O} \to \mathbb{R}^k$ is also continuously differentiable. Moreover, for each point $\mathbf{x}$ in $\mathcal{O}$,

$$\mathbf{D}(\mathbf{G} \circ \mathbf{F})(\mathbf{x}) = \mathbf{D}\mathbf{G}(\mathbf{F}(\mathbf{x})) \cdot \mathbf{D}\mathbf{F}(\mathbf{x}). \tag{15.34}$$

**Proof**   Represent the mapping $\mathbf{G}$ in component functions by $\mathbf{G} = (G_1, \dots, G_k): \mathcal{U} \to \mathbb{R}^k$. Then observe that the composition $\mathbf{G} \circ \mathbf{F}: \mathcal{O} \to \mathbb{R}^k$ is represented in component functions by $\mathbf{G} \circ \mathbf{F} = (G_1 \circ \mathbf{F}, G_2 \circ \mathbf{F}, \dots, G_k \circ \mathbf{F})$. For an index $j$ such that $1 \le j \le k$, the component function $G_j: \mathcal{U} \to \mathbb{R}$ is continuously differentiable, so from Theorem 15.17 it follows that

$$\mathbf{D}(G_j \circ \mathbf{F})(\mathbf{x}) = \mathbf{D}G_j(\mathbf{F}(\mathbf{x}))\mathbf{D}\mathbf{F}(\mathbf{x}) \quad \text{for all points } \mathbf{x} \text{ in } \mathcal{O}.$$

Formula (15.34) is a rewriting of this equation in matrix notation. Thus the composition $\mathbf{G} \circ \mathbf{F}: \mathcal{O} \to \mathbb{R}^k$ has first-order partial derivatives at each point, and from the continuity of the entries on the right-hand side of (15.34) we conclude that the composition is continuously differentiable. ∎

**FIGURE 15.3**

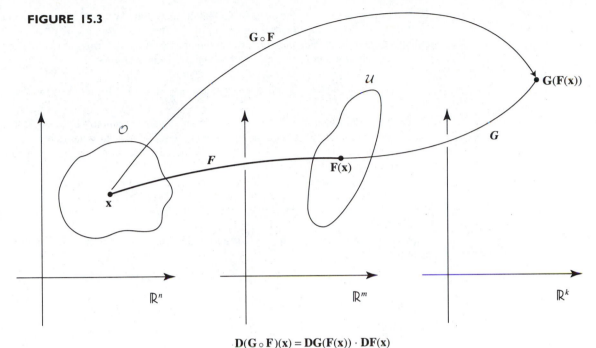

$$\mathbf{D(G \circ F)(x) = DG(F(x)) \cdot DF(x)}$$

**EXERCISES**

1. Suppose that the function $\psi: \mathbb{R}^2 \to \mathbb{R}$ is continuously differentiable. Define the function $g: \mathbb{R}^2 \to \mathbb{R}$ by

$$g(s, t) = \psi(s^2 t, s) \quad \text{for } (s, t) \text{ in } \mathbb{R}^2.$$

Find $\partial g/\partial s(s, t)$ and $\partial g/\partial t(s, t)$.

2. Suppose that the function $h: \mathbb{R}^3 \to \mathbb{R}$ is continuously differentiable. Define the function $\eta: \mathbb{R}^3 \to \mathbb{R}$ by

$$\eta(u, v, w) = (3u + 2v)h(u^2, v^2, uvw) \quad \text{for } (u, v, w) \text{ in } \mathbb{R}^3.$$

Find $D_1\eta(u, v, w)$, $D_2\eta(u, v, w)$, and $D_3\eta(u, v, w)$.

3. Suppose that the functions $g: \mathbb{R} \to \mathbb{R}$ and $h: \mathbb{R} \to \mathbb{R}$ have continuous second-order partial derivatives. Define the function $u: \mathbb{R}^2 \to \mathbb{R}$ by

$$u(s, t) = g(s - t) + h(s + t) \quad \text{for } (s, t) \text{ in } \mathbb{R}^2.$$

Prove that $\qquad \dfrac{\partial^2 u}{\partial t^2}(s, t) - \dfrac{\partial^2 u}{\partial s^2}(s, t) = 0 \quad$ for all $(s, t)$ in $\mathbb{R}^2$.

4. Carry out the calculations needed in Example (15.21) in order to verify that the function $v: \mathbb{R}^2 \to \mathbb{R}$ is harmonic.

5. Let $\mathcal{O}$ be an open subset of the plane $\mathbb{R}^2$ and let the mapping $\mathbf{F}: \mathcal{O} \to \mathbb{R}^2$ be represented by $\mathbf{F}(x, y) = (u(x, y), v(x, y))$ for $(x, y)$ in $\mathcal{O}$. Then the mapping $\mathbf{F}: \mathcal{O} \to \mathbb{R}^2$ is called a *Cauchy-Riemann mapping* provided that each of the functions $u: \mathcal{O} \to \mathbb{R}$ and $v: \mathcal{O} \to \mathbb{R}$ has continuous second-order partial derivatives and

$$\frac{\partial u}{\partial x}(x, y) = \frac{\partial v}{\partial y}(x, y) \quad \text{and} \quad \frac{\partial u}{\partial y}(x, y) = -\frac{\partial v}{\partial x}(x, y) \quad \text{for all } (x, y) \text{ in } \mathcal{O}.$$

Prove that if the function $w: \mathbb{R}^2 \to \mathbb{R}$ is harmonic and the mapping $\mathbf{F}: \mathcal{O} \to \mathbb{R}^2$ is a Cauchy-Riemann mapping, then the function $w \circ \mathbf{F}: \mathcal{O} \to \mathbb{R}$ is also harmonic.

6. Suppose that the function $w: \mathbb{R}^2 \to \mathbb{R}$ is harmonic. Use Exercise 5 to show that each of the following functions is also harmonic:
   (a) $v: \mathbb{R}^2 \to \mathbb{R}$ defined by $v(x, y) = w(e^x \cos y, e^x \sin y)$ for $(x, y)$ in $\mathbb{R}^2$
   (b) $v: \mathbb{R}^2 \to \mathbb{R}$ defined by $v(x, y) = w(x^2 - y^2, 2xy)$ for $(x, y)$ in $\mathbb{R}^2$
   (c) $v: \mathcal{O} \to \mathbb{R}$ defined by $v(x, y) = w(x/(x^2 + y^2), -y/(x^2 + y^2))$ for $(x, y)$ in $\mathcal{O}$, where $\mathcal{O} = \{(x, y)$ in $\mathbb{R}^2 \mid x^2 + y^2 > 0\}$

7. Suppose that the function $u: \mathbb{R}^2 \to \mathbb{R}$ is harmonic. Let $a$, $b$, $c$, and $d$ be real numbers such that

$$a^2 + b^2 = 1, \quad c^2 + d^2 = 1, \quad \text{and} \quad ac + bd = 0.$$

Define the function $v: \mathbb{R}^2 \to \mathbb{R}$ by

$$v(x, y) = u(ax + by, cx + dy) \quad \text{for } (x, y) \text{ in } \mathbb{R}^2.$$

Prove that the function $v: \mathbb{R}^2 \to \mathbb{R}$ is also harmonic.

8. Suppose that the functions $f: \mathbb{R} \to \mathbb{R}$ and $g: \mathbb{R} \to \mathbb{R}$ have continuous second-order partial derivatives. Also suppose that there is a number $\lambda$ such that

$$f''(x) = \lambda f(x) \quad \text{and} \quad g''(x) = \lambda g(x) \quad \text{for all } x \text{ in } \mathbb{R}.$$

Define the function $u: \mathbb{R}^2 \to \mathbb{R}$ by

$$u(x, y) = f(x)g(y) \quad \text{for } (x, y) \text{ in } \mathbb{R}^2.$$

Prove that

$$\frac{\partial^2 u}{\partial x^2}(x, y) - \frac{\partial^2 u}{\partial y^2}(x, y) = 0 \quad \text{for every } (x, y) \text{ in } \mathbb{R}^2.$$

9. Let $\mathcal{O} = \{(x, y, z)$ in $\mathbb{R}^3 \mid x^2 + y^2 + z^2 > 0\}$ and define the function $u: \mathcal{O} \to \mathbb{R}$ by

$$u(\mathbf{p}) = \frac{1}{\|\mathbf{p}\|} \quad \text{for } \mathbf{p} \text{ in } \mathcal{O}.$$

Prove that

$$\frac{\partial^2 u}{\partial x^2}(x, y, z) + \frac{\partial^2 u}{\partial y^2}(x, y, z) + \frac{\partial^2 u}{\partial z^2}(x, y, z) = 0 \quad \text{for every } (x, y, z) \text{ in } \mathcal{O}.$$

10. Suppose that the functions $f: \mathbb{R}^2 \to \mathbb{R}$, $g: \mathbb{R}^2 \to \mathbb{R}$, and $h: \mathbb{R}^2 \to \mathbb{R}$ are continuously differentiable. Express the following two limits in terms of partial derivatives of these functions:
   (a) $\displaystyle\lim_{t \to 0} \frac{f(g(1 + t, 2), h(1 + t, 2)) - f(g(1, 2), h(1, 2))}{t}$
   (b) $\displaystyle\lim_{t \to 0} \frac{f(g(1, 2) + t, h(1, 2)) - f(g(1, 2), h(1, 2))}{t}$

11. Suppose that the function $g: \mathbb{R}^n \to \mathbb{R}$ is continuously differentiable. For points $\mathbf{x}$ and $\mathbf{p}$ in $\mathbb{R}^n$, the Directional Derivative Theorem asserts that if $\psi(t) = g(\mathbf{x} + t\mathbf{p})$ for $t$ in $\mathbb{R}$, then

$$\psi'(t) = \langle \mathbf{D}g(\mathbf{x} + t\mathbf{p}), \mathbf{p} \rangle \quad \text{for every } t \text{ in } \mathbb{R}.$$

Show that this formula is a special case of the Chain Rule.

# Images and Inverses: The Inverse Function Theorem

## 16.1  Functions of a Single Variable and Maps in the Plane

Suppose that $\mathcal{O}$ is an open subset of Euclidean space $\mathbb{R}^n$ and that the nonlinear mapping $\mathbf{F}: \mathcal{O} \to \mathbb{R}^n$ is continuously differentiable. At a point $\mathbf{x}$ in $\mathcal{O}$, suppose that the differential

$$\mathbf{dF}(\mathbf{x}): \mathbb{R}^n \to \mathbb{R}^n \quad \text{is one-to-one and onto;}$$

this is equivalent to the assumption that the determinant of the derivative matrix is nonzero. Then it turns out that the nonlinear mapping inherits "local invertibility" near the point $\mathbf{x}$, in the precise sense that there is a neighborhood $U$ of the point $\mathbf{x}$ and a neighborhood $V$ of its image $\mathbf{F}(\mathbf{x})$ such that

$$\mathbf{F}: U \to V \quad \text{is one-to-one and onto}$$

and the inverse $\mathbf{F}^{-1}: V \to U$ is also continuously differentiable. This is the Inverse Function Theorem. The proof of this theorem provides the opportunity to introduce a number of ideas that are new and of independent interest. This chapter is devoted to describing these ideas, proving this theorem, and considering examples of the theorem.

In this first section we will prove the Inverse Function Theorem for functions of a single variable and state the theorem for maps in the plane; several examples will be considered. In Section 16.2, we will introduce the concept of stability of a nonlinear mapping. A linear mapping $\mathbf{T}: \mathbb{R}^n \to \mathbb{R}^n$ is stable if and only if it is invertible, and stability in a neighborhood of the point $\mathbf{x}$ is the important analytical property that a continuously differentiable nonlinear mapping inherits from the invertibility of its differential $\mathbf{dF}(\mathbf{x})\colon \mathbb{R}^n \to \mathbb{R}^n$. In Section 16.3, we will first introduce a Minimization Principle to show that a continuously differentiable stable mapping maps open sets to open sets. We will then state, prove, and discuss the General Inverse Function Theorem.

As motivation for the subject of the present chapter, we will begin by proving the following theorem about real-valued functions of a single real variable.

**THEOREM 16.1**   Let $\mathcal{O}$ be an open subset of $\mathbb{R}$ and suppose that the function $f\colon \mathcal{O} \to \mathbb{R}$ is continuously differentiable. Let $x_0$ be a point in $\mathcal{O}$ at which

$$f'(x_0) \neq 0. \tag{16.1}$$

Then there is an open interval $I$ containing the point $x_0$, and an open interval $J$ containing its image $f(x_0)$, such that the function

$$f\colon I \to J \quad \text{is one-to-one and onto.} \tag{16.2}$$

Moreover, the inverse function $f^{-1}\colon J \to I$ is also continuously differentiable, and for a point $y$ in $J$, if $x$ is the point in $I$ at which $f(x) = y$, then

$$(f^{-1})'(y) = \frac{1}{f'(x)}. \tag{16.3}$$

**FIGURE 16.1**

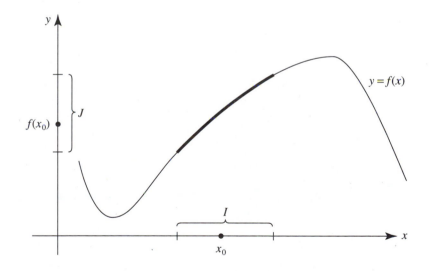

**Proof**   We will suppose that $f'(x_0) > 0$. Since $x_0$ is an interior point of $\mathcal{O}$ and the function $f'\colon \mathcal{O} \to \mathbb{R}$ is continuous, we can select a positive number $r$ such that the closed interval $[x_0 - r, x_0 + r]$ is contained in $\mathcal{O}$ and $f'(x) > 0$ for all points $x$ in the interval $[x_0 - r, x_0 + r]$. The Lagrange Mean Value Theorem implies that the function $f\colon [x_0 - r, x_0 + r] \to \mathbb{R}$ is strictly increasing. In particular, $f\colon [x_0 - r, x_0 + r] \to \mathbb{R}$ is one-to-one. Moreover, by the Intermediate Value Theorem, if the point $y$ lies strictly between $f(x_0 - r)$ and $f(x_0 + r)$, then there is some point $x$ in the open interval $(x_0 - r, x_0 + r)$ with $f(x) = y$.

Define $I = (x_0 - r, x_0 + r)$ and $J = (f(x_0 - r), f(x_0 + r))$. Then the function $f: I \to J$ is one-to-one and onto. The differentiability of the inverse and formula (16.3) have already been proven in Theorem 4.6. $\blacksquare$

The aim of this present chapter is to describe a circle of ideas that are related to the way in which we can extend Theorem 16.1 to a result about mappings $\mathbf{F}: \mathcal{O} \to \mathbb{R}^n$, where $\mathcal{O}$ is an open subset of Euclidean space $\mathbb{R}^n$. Before we describe the extension to maps in the plane, it is useful to extend the definition of symmetric neighborhood as follows:

**DEFINITION**    *An open subset of Euclidean space $\mathbb{R}^n$ that contains the point $\mathbf{x}$ is called a neighborhood of the point $\mathbf{x}$.*

For mappings of an open subset of the plane $\mathbb{R}^2$ into $\mathbb{R}^2$, we state the following theorem.

**THEOREM 16.2**    **The Inverse Function Theorem in the Plane**    Let $\mathcal{O}$ be an open subset of the plane $\mathbb{R}^2$ and suppose that the mapping $\mathbf{F}: \mathcal{O} \to \mathbb{R}^2$ is continuously differentiable. Let $(x_0, y_0)$ be a point in $\mathcal{O}$ at which the derivative matrix

$$\mathbf{DF}(x_0, y_0) \text{ is invertible.} \tag{16.4}$$

Then there is a neighborhood $U$ of the point $(x_0, y_0)$ and a neighborhood $V$ of its image $\mathbf{F}(x_0, y_0)$ such that

$$\mathbf{F}: U \to V \text{ is one-to-one and onto.} \tag{16.5}$$

Moreover, the inverse mapping $\mathbf{F}^{-1}: V \to U$ is also continuously differentiable, and for a point $(u, v)$ in $V$, if $(x, y)$ is the point in $U$ at which $\mathbf{F}(x, y) = (u, v)$, then the derivative matrix of the inverse mapping at the point $(u, v)$ is given by the formula

$$\mathbf{DF}^{-1}(u, v) = [\mathbf{DF}(x, y)]^{-1}. \tag{16.6}$$

**FIGURE 16.2**

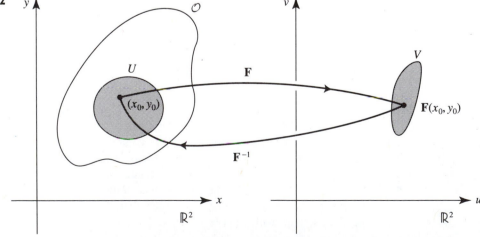

Observe that in the proof of Theorem 16.1, we used the Intermediate Value Theorem, a result that does not easily generalize to mappings whose image lies in the plane $\mathbb{R}^2$. The proof of the Inverse Function Theorem in the Plane has to be quite different. In the next two sections, we will discuss certain ideas that are of independent interest and that we will use in Section 16.3 to prove the General Inverse Function Theorem, a result that contains Theorem 16.2 as a special case. We will devote the rest of this section to a discussion of the Inverse Function Theorem in the Plane.

As we discussed in Section 15.1, an $n \times n$ matrix is *invertible* if and only if its determinant is nonzero, and when the matrix is invertible there is a formula called Cramer's Rule for the inverse matrix. For $2 \times 2$ matrices, Cramer's Rule is clear by inspection. Indeed, for a $2 \times 2$ matrix

$$\mathbf{A} = \begin{bmatrix} a_{11} & a_{12} \\ a_{21} & a_{22} \end{bmatrix},$$

if
$$\det \mathbf{A} = a_{11}a_{22} - a_{12}a_{21} \neq 0,$$

then a straightforward multiplication of matrices establishes the validity of the following formula for the inverse of $\mathbf{A}$:

$$\mathbf{A}^{-1} = \frac{1}{\det \mathbf{A}} \begin{bmatrix} a_{22} & -a_{12} \\ -a_{21} & a_{11} \end{bmatrix}.$$

In particular, for the mapping $\mathbf{F}: \mathcal{O} \to \mathbb{R}^2$ in the statement of the Inverse Function Theorem in the Plane, assumption (16.4) holds if and only if

$$\det \mathbf{DF}(x_0, \ y_0) \neq 0.$$

If the mapping $\mathbf{F}: \mathcal{O} \to \mathbb{R}^2$ is represented in component functions as

$$\mathbf{F}(x, \ y) = (\psi(x, \ y), \phi(x, \ y)) \quad \text{for } (x, \ y) \text{ in } \mathcal{O},$$

then
$$\mathbf{DF}(x, \ y) = \begin{bmatrix} \partial\psi/\partial x(x, \ y) & \partial\psi/\partial y(x, \ y) \\ \partial\phi/\partial x(x, \ y) & \partial\phi/\partial y(x, \ y) \end{bmatrix},$$

so assumption (16.4) is equivalent to the assumption that

$$\frac{\partial\psi}{\partial x}(x_0, \ y_0)\frac{\partial\phi}{\partial y}(x_0, \ y_0) - \frac{\partial\psi}{\partial y}(x_0, \ y_0)\frac{\partial\phi}{\partial x}(x_0, \ y_0) \neq 0. \tag{16.7}$$

The above explicit formula for the inverse of a $2 \times 2$ matrix permits us to use formula (16.6) to compute the partial derivatives of the component functions of the inverse mapping $\mathbf{F}^{-1}: V \to U$. Indeed, write the inverse mapping $\mathbf{F}^{-1}: V \to U$ in component functions as

$$\mathbf{F}^{-1}(u, \ v) = (g(u, \ v), \ h(u, \ v)) \quad \text{for } (u, \ v) \text{ in } V,$$

so that
$$\mathbf{DF}^{-1}(u, \ v) = \begin{bmatrix} \partial g/\partial u(u, \ v) & \partial g/\partial v(u, \ v) \\ \partial h/\partial u(u, \ v) & \partial h/\partial v(u, \ v) \end{bmatrix}.$$

For a point $(u, \ v)$ in $V$, let $(x, \ y)$ be the point in $U$ at which

$$u = \psi(x, \ y) \quad \text{and} \quad v = \phi(x, \ y).$$

For notational convenience,* set

$$J(x, y) \equiv \det \mathbf{DF}(x, y) = \frac{\partial \psi}{\partial x}(x, y)\frac{\partial \phi}{\partial y}(x, y) - \frac{\partial \psi}{\partial y}(x, y)\frac{\partial \phi}{\partial x}(x, y).$$

Then, using the above computation of the inverse of a $2 \times 2$ matrix, it follows that formula (16.6) is equivalent to

$$
\begin{aligned}
\frac{\partial g}{\partial u}(u, v) &= \frac{1}{J(x, y)} \cdot \frac{\partial \phi}{\partial y}(x, y) \\
\frac{\partial g}{\partial v}(u, v) &= -\frac{1}{J(x, y)} \cdot \frac{\partial \psi}{\partial y}(x, y) \\
\frac{\partial h}{\partial u}(u, v) &= -\frac{1}{J(x, y)} \cdot \frac{\partial \phi}{\partial x}(x, y) \\
\frac{\partial h}{\partial v}(u, v) &= \frac{1}{J(x, y)} \cdot \frac{\partial \psi}{\partial x}(x, y).
\end{aligned}
\tag{16.8}
$$

**EXAMPLE 16.1**  For a point $(x, y)$ in the plane $\mathbb{R}^2$, define

$$\mathbf{F}(x, y) = (6 + x - y + x^2 + yx^3, -3 + 2x + 3y + (x + y)^5).$$

Then the mapping $\mathbf{F}: \mathbb{R}^2 \to \mathbb{R}^2$ is continuously differentiable, since each of its component functions is continuously differentiable. At the point $(x_0, y_0) = (0, 0)$ a short computation of partial derivatives shows that

$$\mathbf{DF}(0, 0) = \begin{bmatrix} 1 & -1 \\ 2 & 3 \end{bmatrix}.$$

Observe that $\det \mathbf{DF}(0, 0) = 5 \neq 0$. We may use the Inverse Function Theorem to conclude that there are neighborhoods $U$ of the point $(0, 0)$ and $V$ of its image $\mathbf{F}(0, 0) = (6, -3)$ such that the mapping $\mathbf{F}: U \to V$ is one-to-one and onto, and that its inverse $\mathbf{F}^{-1}: V \to U$ is also continuously differentiable. Moreover, if the inverse mapping is represented in components as $\mathbf{F}^{-1}(u, v) = (g(u, v), h(u, v))$, then from (16.8) it follows that

$$\frac{\partial g}{\partial u}(6, -3) = \frac{3}{5}, \quad \frac{\partial g}{\partial v}(6, -3) = \frac{1}{5}, \quad \frac{\partial h}{\partial u}(6, -3) = -\frac{2}{5}, \quad \text{and} \quad \frac{\partial h}{\partial v}(6, -3) = \frac{1}{5}.$$

□

---

*We use the letter J here because the determinant of the derivative matrix at a point is sometimes referred to as the *Jacobian determinant*. The derivative matrix is often called the *Jacobian matrix*.

**EXAMPLE 16.2**   For a point $(x, y)$ in the plane $\mathbb{R}^2$, define

$$\mathbf{F}(x, y) = (e^{x-y} + x^2 y + x(y-1)^4, \; 1 + x^2 + x^4 + (xy)^5).$$

Again, in this example, the mapping $\mathbf{F}: \mathbb{R}^2 \to \mathbb{R}^2$ is continuously differentiable, since each of its component functions is continuously differentiable. At the point $(x_0, y_0) = (1, 1)$, a short computation of partial derivatives shows that

$$\mathbf{DF}(1, 1) = \begin{bmatrix} 3 & 0 \\ 11 & 5 \end{bmatrix}.$$

The determinant of $\mathbf{DF}(1, 1)$ is nonzero. We can use the Inverse Function Theorem to conclude that there are neighborhoods $U$ of the point $(1, 1)$ and $V$ of its image $(2, 4)$ such that the mapping $\mathbf{F}: U \to V$ is one-to-one and onto, and that the inverse mapping $\mathbf{F}^{-1}: V \to U$ is also continuously differentiable. Moreover, if the inverse is represented in components as $\mathbf{F}^{-1}(u, v) = (g(u, v), h(u, v))$, then from (16.8) it follows that

$$\frac{\partial g}{\partial u}(2, 4) = \frac{1}{3}, \quad \frac{\partial g}{\partial v}(2, 4) = 0, \quad \frac{\partial h}{\partial u}(2, 4) = \frac{-11}{15}, \quad \text{and} \quad \frac{\partial h}{\partial v}(2, 4) = \frac{1}{5}. \qquad \square$$

**EXAMPLE 16.3**   For a point $(x, y)$ in the plane $\mathbb{R}^2$, define

$$\mathbf{F}(x, y) = (x^2 - y^2, \; 2xy).$$

This mapping $\mathbf{F}: \mathbb{R}^2 \to \mathbb{R}^2$ is also continuously differentiable, since each of its component functions is obviously continuously differentiable. First consider a nonzero point $(x_0, y_0)$ in $\mathbb{R}^2$. We have

$$\mathbf{DF}(x_0, y_0) = \begin{bmatrix} 2x_0 & -2y_0 \\ 2y_0 & 2x_0 \end{bmatrix},$$

so $\det \mathbf{DF}(x_0, y_0) = 4(x_0^2 + y_0^2) \neq 0$. Once more applying the Inverse Function Theorem, it follows that there are neighborhoods $U$ of the point $(x_0, y_0)$ and $V$ of its image $(x_0^2 - y_0^2, 2x_0 y_0)$ such that the mapping $\mathbf{F}: U \to V$ is one-to-one and onto, and has an inverse $\mathbf{F}^{-1}: V \to U$ that also is continuously differentiable. If the inverse is represented in component functions as $\mathbf{F}^{-1}(u, v) = (g(u, v), h(u, v))$, then if we set $(u_0, v_0) = (x_0^2 - y_0^2, 2x_0 y_0)$, from (16.8) it follows that

$$\frac{\partial g}{\partial u}(u_0, v_0) = \frac{x_0}{2(x_0^2 + y_0^2)}, \quad \frac{\partial g}{\partial v}(u_0, v_0) = \frac{y_0}{2(x_0^2 + y_0^2)},$$

$$\frac{\partial h}{\partial u}(u_0, v_0) = \frac{-y_0}{2(x_0^2 + y_0^2)}, \quad \frac{\partial h}{\partial v}(u_0, v_0) = \frac{x_0}{2(x_0^2 + y_0^2)}.$$

Now consider the point $(x_0, y_0) = (0, 0)$. The assumptions of the Inverse Function Theorem certainly fail at this point, since

$$\mathbf{DF}(0, 0) = \begin{bmatrix} 0 & 0 \\ 0 & 0 \end{bmatrix}.$$

Moreover, the conclusion of the Inverse Function Theorem also fails at this point, because, if we observe that $\mathbf{F}(x, y) = \mathbf{F}(-x, -y)$ at all points $(x, y)$ in the plane, there is no neighborhood of the point $(0, 0)$ on which the mapping is one-to-one. $\qquad \square$

**EXAMPLE 16.4**   Define $\mathbf{F}(x, y) = (\cos(x + y^2), \sin(x + y^2))$ for $(x, y)$ in $\mathbb{R}^2$. Then the mapping $\mathbf{F} \colon \mathbb{R}^2 \to \mathbb{R}^2$ is continuously differentiable. But there is no point at which the conclusion of the Inverse Function Theorem is true. To see this, observe that if the point $(u, v)$ lies in the image of $\mathbf{F} \colon \mathbb{R}^2 \to \mathbb{R}^2$, then $u^2 + v^2 = 1$. Thus the image lies on the circle of radius 1 centered at the origin. In particular, the image does not contain an open subset of the plane, so there certainly are no open sets $U$ and $V$ in the plane such that $\mathbf{F} \colon U \to V$ is one-to-one and onto.    □

It is useful to note that the conclusion of the Inverse Function Theorem in the Plane may be refined slightly. In the conclusion of the theorem, it is asserted that the sets $U$ and $V$ are neighborhoods of the point $(x_0, y_0)$ and of its image $\mathbf{F}(x_0, y_0)$, respectively. In fact, we can choose $V$ to be a symmetric neighborhood of the point $\mathbf{F}(x_0, y_0)$. To see why this is so, first recall that a continuously differentiable mapping is continuous and that, by Theorem 11.7, the inverse image of an open set by a continuous map is again an open set. Now since $V$ is open, we can choose a positive number $r$ such that the symmetric neighborhood $\mathcal{N}_r(\mathbf{F}(x_0, y_0))$ is contained in $V$. Denote $\mathcal{N}_r(\mathbf{F}(x_0, y_0))$ by $V'$ and define $U' = \mathbf{F}^{-1}(V') \cap U$. Then $U'$ is a neighborhood of $(x_0, y_0)$. Thus, replacing $U$ by $U'$ and $V$ by $V'$, $U'$ is a neighborhood of $(x_0, y_0)$ and $V'$ is a neighborhood of $\mathbf{F}(x_0, y_0)$, and the mapping $\mathbf{F} \colon U' \to V'$ is one-to-one and onto, but now $V'$ is symmetric (see Exercise 12).

The Inverse Function Theorem has an interpretation as a statement about the solvability of systems of equations. Given two functions $\psi \colon \mathbb{R}^2 \to \mathbb{R}$ and $\phi \colon \mathbb{R}^2 \to \mathbb{R}$ and two numbers $a$ and $b$, consider the system of equations

$$\begin{aligned} \psi(x, y) &= a \\ \phi(x, y) &= b. \end{aligned} \tag{16.9}$$

We can ask whether there are any solutions of this system of equations, and, if there is a solution, whether there is only one solution. If we define the mapping $\mathbf{F} \colon \mathbb{R}^2 \to \mathbb{R}^2$ by $\mathbf{F}(x, y) = (\psi(x, y), \phi(x, y))$ for $(x, y)$ in $\mathbb{R}^2$, these two questions about the existence and the uniqueness of solutions of the system (16.9) may be rephrased as questions about the image of the mapping $\mathbf{F} \colon \mathbb{R}^2 \to \mathbb{R}^2$ and whether it has the property of being one-to-one. The following example shows how the Inverse Function Theorem provides information about systems of equations.

**EXAMPLE 16.5**   Consider the system of equations

$$\begin{aligned} e^{x-y} + x^2y + x(y - 1)^4 &= 2 \\ 1 + x^2 + x^4 + (xy)^5 &= 4. \end{aligned} \tag{16.10}$$

Observe that the point $(x, y) = (1, 1)$ is a solution of (16.10). The mapping $\mathbf{F} \colon \mathbb{R}^2 \to \mathbb{R}^2$ defined by $\mathbf{F}(x, y) = (e^{x-y} + x^2y + x(y - 1)^4, 1 + x^2 + x^4 + (xy)^5)$ for $(x, y)$ in $\mathbb{R}^2$ is precisely the mapping considered in Example 16.2. From the analysis in that example, we conclude that there is a positive number $r$ and a neighborhood $U$ of the point $(1, 1)$ such that for any numbers $a$ and $b$ with $(a - 2)^2 + (b - 4)^2 < r^2$, the system of equations

$$\begin{aligned} e^{x-y} + x^2y + x(y - 1)^4 &= a \\ 1 + x^2 + x^4 + (xy)^5 &= b, \qquad (x, y) \text{ in } U \end{aligned}$$

has exactly one solution.                                                                                  □

1.  Define the function $f: \mathbb{R} \to \mathbb{R}$ by

$$f(x) = x^3 - 3x + 1 \quad \text{for } x \text{ in } \mathbb{R}.$$

   At what points $x$ in $\mathbb{R}$ does the Inverse Function Theorem apply?

2.  Define the function $f: \mathbb{R} \to \mathbb{R}$ by

$$f(x) = x^3 + x + \cos x \quad \text{for } x \text{ in } \mathbb{R}.$$

   At what points $x$ in $\mathbb{R}$ does the Inverse Function Theorem apply? Prove that the function $f: \mathbb{R} \to \mathbb{R}$ is one-to-one and onto.

3.  Suppose that the function $f: \mathbb{R} \to \mathbb{R}$ is differentiable and is one-to-one and onto. Suppose that $f(1) = 0$, $f(0) = 1$, $f'(0) = -4$, and $f'(1) = -10$. Find $(f^{-1})'(1)$ and $(f^{-1})'(0)$.

4.  (a)  Give an example of a continuously differentiable function $f: \mathbb{R} \to \mathbb{R}$ that is one-to-one but not onto.
    (b)  Give an example of a continuously differentiable function $f: \mathbb{R} \to \mathbb{R}$ that is onto but not one-to-one.

5.  Suppose that the continuously differentiable function $f: \mathbb{R} \to \mathbb{R}$ has the property that there is some positive number $c$ such that

$$f'(x) \geq c \quad \text{for every } x \text{ in } \mathbb{R}.$$

   Show that the function $f: \mathbb{R} \to \mathbb{R}$ is both one-to-one and onto.

6.  Define $f(x) = x^3$ for $x$ in $\mathbb{R}$. Show that the function $f: \mathbb{R} \to \mathbb{R}$ is one-to-one and onto and that its inverse $f^{-1}: \mathbb{R} \to \mathbb{R}$ is continuous. At what points is the inverse differentiable?

7.  Let $\mathcal{O}$ and $V$ be open subsets of $\mathbb{R}$ and suppose that the differentiable function $f: \mathcal{O} \to V$ is one-to-one and onto. Suppose that $x$ is a point in $\mathcal{O}$ at which $f'(x) = 0$. Show that the inverse function $f^{-1}: V \to \mathbb{R}$ cannot be differentiable at the point $f(x)$. (*Hint:* Argue by contradiction and use the Chain Rule to differentiate both sides of the following identity: $f^{-1}(f(x)) = x$ for $x$ in $\mathcal{O}$.)

8.  For each of the following mappings $\mathbf{F}: \mathbb{R}^2 \to \mathbb{R}^2$, apply the Inverse Function Theorem at the point $(x_0, y_0) = (0, 0)$ and calculate the partial derivatives of the components of the inverse mapping at the point $(u_0, v_0) = \mathbf{F}(0, 0)$:
    (a)  $\mathbf{F}(x, y) = (x + x^2 + e^{x^2 y^2}, -x + y + \sin(xy))$   for $(x, y)$ in $\mathbb{R}^2$.
    (b)  $\mathbf{F}(x, y) = (e^{x+y}, e^{x-y})$   for $(x, y)$ in $\mathbb{R}^2$.

9.  Define the function $\mathbf{F}: \mathbb{R}^2 \to \mathbb{R}^2$ by

$$\mathbf{F}(x, y) = (e^x \cos y, e^x \sin y) \quad \text{for } (x, y) \text{ in } \mathbb{R}^2.$$

   (a)  Show that the Inverse Function Theorem is applicable at every point $(x_0, y_0)$ in the plane $\mathbb{R}^2$.
    (b)  Show that the function $\mathbf{F}: \mathbb{R}^2 \to \mathbb{R}^2$ is not one-to-one.
    (c)  Does (b) contradict (a)?

10.  Define the mapping $\mathbf{F}: \mathbb{R}^2 \to \mathbb{R}^2$ by

$$\mathbf{F}(r, \theta) = (r \cos \theta, r \sin \theta) \quad \text{for } (r, \theta) \text{ in } \mathbb{R}^2.$$

   (a)  At what points $(r_0, \theta_0)$ in $\mathbb{R}^2$ can we apply the Inverse Function Theorem to this mapping?

(b) Find some explicit formula for the local inverse about the point $(r, \theta) = (1, \pi/2)$. (*Hint:* The local inverse corresponds to the assignment of polar coordinates.)

11. For a pair of real numbers $a$ and $b$, consider the system of nonlinear equations

$$x + x^2 \cos y + xye^{x^3 y^2} = a$$
$$y + x^5 + y^3 - x^2 \cos(xy) = b.$$

Use the Inverse Function Theorem to show that there is some positive number $r$ such that if $a^2 + b^2 < r^2$, then this system of equations has at least one solution.

12. We have observed that the conclusion of the Inverse Function Theorem could be refined in that the neighborhood $V$ could be chosen to be symmetric. Use a similar argument to show that it is possible to choose the neighborhood $U$ to be symmetric. Show that it is generally not possible to choose both $U$ and $V$, simultaneously, to be symmetric. (*Hint:* Consider the mapping $\mathbf{F}: \mathbb{R}^2 \to \mathbb{R}^2$ defined by $\mathbf{F}(x, y) = (x, 2y)$ for $(x, y)$ in $\mathbb{R}^2$.)

13. Let the continuously differentiable mapping $\mathbf{F}: \mathbb{R}^2 \to \mathbb{R}^2$ be represented in component functions by $\mathbf{F}(x, y) = (\psi(x, y), \varphi(x, y))$ for $(x, y)$ in $\mathbb{R}^2$. Suppose that the point $(x_0, y_0)$ in $\mathbb{R}^2$ has the property that

$$\psi(x, y) \geq \psi(x_0, y_0) \quad \text{for all } (x, y) \text{ in } \mathbb{R}^2.$$

(a) Explain analytically why the hypotheses of the Inverse Function Theorem cannot hold at $(x_0, y_0)$.

(b) Explain geometrically why the conclusion of the Inverse Function Theorem cannot hold at $(x_0, y_0)$.

14. Suppose that the function $\psi: \mathbb{R}^2 \to \mathbb{R}$ is continuously differentiable and define the mapping $\mathbf{F}: \mathbb{R}^2 \to \mathbb{R}^2$ by

$$\mathbf{F}(x, y) = (\psi(x, y), -\psi(x, y)) \quad \text{for } (x, y) \text{ in } \mathbb{R}^2.$$

(a) Explain analytically why the hypotheses of the Inverse Function Theorem fail at each point $(x_0, y_0)$ in $\mathbb{R}^2$.

(b) Explain geometrically why the conclusion of the Inverse Function Theorem must fail at each point $(x_0, y_0)$ in $\mathbb{R}^2$.

## 16.2 Stability of Nonlinear Mappings

In this section we will study the concept of *stability* for nonlinear mappings. To motivate this, we will first restate the following characterization of invertibility for an $n \times n$ matrix (see Theorem 15.10).

**THEOREM 16.3** For an $n \times n$ matrix $\mathbf{A}$, the following two assertions are equivalent:

(i) The matrix $\mathbf{A}$ is invertible.

(ii) There is a positive number $c$ such that

$$\|\mathbf{Ah}\| \geq c\|\mathbf{h}\| \quad \text{for all points } \mathbf{h} \text{ in } \mathbb{R}^n.$$

As we discussed in Section 15.1, for a linear mapping $\mathbf{T}\colon \mathbb{R}^n \to \mathbb{R}^n$ that is associated with the $n \times n$ matrix $\mathbf{A}$, the assertion that the mapping $\mathbf{T}\colon \mathbb{R}^n \to \mathbb{R}^n$ is invertible is equivalent to the assertion that the matrix $\mathbf{A}$ is an invertible matrix. Thus, if we set $\mathbf{h} = \mathbf{u} - \mathbf{v}$ and observe that, because of linearity, $\mathbf{T}(\mathbf{u} - \mathbf{v}) = \mathbf{T}(\mathbf{u}) - \mathbf{T}(\mathbf{v})$, it follows from the above theorem that a linear mapping $\mathbf{T}\colon \mathbb{R}^n \to \mathbb{R}^n$ is invertible if and only if there is a positive number $c$ such that

$$\|\mathbf{T}(\mathbf{u}) - \mathbf{T}(\mathbf{v})\| \geq c\|\mathbf{u} - \mathbf{v}\| \quad \text{for all points } \mathbf{u} \text{ and } \mathbf{v} \text{ in } \mathbb{R}^n.$$

This leads us to make the following definition for general mappings.

**DEFINITION**   *Let $\mathcal{O}$ be a subset of $\mathbb{R}^n$. Then a mapping $\mathbf{F}\colon \mathcal{O} \to \mathbb{R}^n$ is called stable provided that there is some positive number $c$, called a stability constant for the mapping, such that*

$$\|\mathbf{F}(\mathbf{u}) - \mathbf{F}(\mathbf{v})\| \geq c\|\mathbf{u} - \mathbf{v}\| \quad \textit{for all points } \mathbf{u} \textit{ and } \mathbf{v} \textit{ in } \mathcal{O}. \tag{16.11}$$

The main result of this section is that if a continuously differentiable mapping has an invertible derivative matrix at a point, then there is a neighborhood of that point on which the mapping is stable. To prove this, it is useful first to establish the following result about perturbations of invertible matrices.

**LEMMA 16.4**   Let $\mathbf{A}$ be an invertible $n \times n$ matrix and $c$ be a positive number such that

$$\|\mathbf{A}\mathbf{h}\| \geq c\|\mathbf{h}\| \quad \text{for all points } \mathbf{h} \text{ in } \mathbb{R}^n.$$

If $\mathbf{B}$ is an $n \times n$ matrix such that $\|\mathbf{B} - \mathbf{A}\| \leq c/2$, then

$$\|\mathbf{B}\mathbf{h}\| \geq \frac{c}{2}\|\mathbf{h}\| \quad \text{for all points } \mathbf{h} \text{ in } \mathbb{R}^n. \tag{16.12}$$

In particular, if for each pair of indices $i$ and $j$ with $1 \leq i \leq n$ and $1 \leq j \leq n$,

$$|a_{ij} - b_{ij}| \leq \frac{c}{2n},$$

then (16.12) holds.

**Proof**   Let $\mathbf{h}$ be a point in $\mathbb{R}^n$. Since

$$\mathbf{B}\mathbf{h} = \mathbf{A}\mathbf{h} + [\mathbf{B} - \mathbf{A}]\mathbf{h},$$

from the Reverse Triangle Inequality it follows that

$$\|\mathbf{B}\mathbf{h}\| \geq \|\mathbf{A}\mathbf{h}\| - \|[\mathbf{B} - \mathbf{A}]\mathbf{h}\|.$$

By the Generalized Cauchy-Schwarz Inequality,

$$\|[\mathbf{B} - \mathbf{A}]\mathbf{h}\| \leq \|\mathbf{B} - \mathbf{A}\| \cdot \|\mathbf{h}\|,$$

so

$$\|\mathbf{B}\mathbf{h}\| \geq \{c - \|\mathbf{B} - \mathbf{A}\|\}\|\mathbf{h}\| \geq \frac{c}{2}\|\mathbf{h}\|.$$

Moreover, if the absolute value of each entry of $\mathbf{B} - \mathbf{A}$ is less than $c/2n$, then by the very definition of the norm of a matrix,

$$\|\mathbf{B} - \mathbf{A}\| = \sqrt{\sum_{1 \leq i, j \leq n} (a_{ij} - b_{ij})^2} \leq \frac{c}{2}. \qquad \blacksquare$$

**THEOREM 16.5**    **The Nonlinear Stability Theorem**    Let $\mathcal{O}$ be an open subset of $\mathbb{R}^n$ and suppose that the mapping $\mathbf{F} \colon \mathcal{O} \to \mathbb{R}^n$ is continuously differentiable. Suppose that $\mathbf{x}$ is a point in $\mathcal{O}$ at which the derivative matrix

$$\mathbf{DF}(\mathbf{x}) \text{ is invertible.}$$

Then there is a neighborhood $U$ of $\mathbf{x}$ such that the mapping $\mathbf{F} \colon U \to \mathbb{R}^n$ is stable.

**Proof**    Since the matrix $\mathbf{DF}(\mathbf{x})$ is invertible, by Theorem 16.3 we can choose a positive number $c$ such that

$$\|\mathbf{DF}(\mathbf{x})\mathbf{h}\| \geq c\|\mathbf{h}\| \quad \text{for all points } \mathbf{h} \text{ in } \mathbb{R}^n. \qquad (16.13)$$

Because the mapping $\mathbf{F} \colon \mathcal{O} \to \mathbb{R}^n$ is continuously differentiable, we can select a positive number $r$ such that the symmetric neighborhood $U = \mathcal{N}_r(\mathbf{x})$ is contained in $\mathcal{O}$ and has the property that for each point $\mathbf{z}$ in $U$ and each pair of indices $i$ and $j$ with $1 \leq i \leq n$ and $1 \leq j \leq n$,

$$\left| \frac{\partial F_i}{\partial x_j}(\mathbf{z}) - \frac{\partial F_i}{\partial x_j}(\mathbf{x}) \right| < \frac{c}{2n}. \qquad (16.14)$$

The estimates (16.13) and (16.14), together with Lemma 16.4, imply that if $\mathbf{B}$ is any $n \times n$ matrix that, for a pair of indices $i$ and $j$ with $1 \leq i \leq n$ and $1 \leq j \leq n$, has

$$b_{ij} = \frac{\partial F_i}{\partial x_j}(\mathbf{z}_{ij}) \quad \text{for some point } \mathbf{z}_{ij} \text{ in } U, \qquad (16.15)$$

then

$$\|\mathbf{Bh}\| \geq \frac{c}{2}\|\mathbf{h}\| \quad \text{for all points } \mathbf{h} \text{ in } \mathbb{R}^n.$$

To show that $\mathbf{F} \colon U \to \mathbb{R}^n$ is stable, let $\mathbf{u}$ and $\mathbf{v}$ be distinct points in $U$. According to the Mean Value Theorem for Mappings, there are $n$ points $\mathbf{z}_1, \ldots, \mathbf{z}_n$ lying on the segment between the points $\mathbf{u}$ and $\mathbf{v}$ such that if $\mathbf{B}$ is the $n \times n$ matrix whose $i$th row is $\mathbf{DF}_i(\mathbf{z}_i)$ for each index $i$ with $1 \leq i \leq n$, then

$$\mathbf{F}(\mathbf{u}) - \mathbf{F}(\mathbf{v}) = \mathbf{B}(\mathbf{u} - \mathbf{v}).$$

But $\mathbf{B}$ is a matrix of the type just described by (16.15), in which $\mathbf{z}_{ij} = \mathbf{z}_i$. Thus

$$\|\mathbf{F}(\mathbf{u}) - \mathbf{F}(\mathbf{v})\| = \|\mathbf{B}(\mathbf{u} - \mathbf{v})\| \geq \frac{c}{2}\|\mathbf{u} - \mathbf{v}\|.$$

This proves that the mapping $\mathbf{F} \colon U \to \mathbb{R}^n$ is stable. $\qquad \blacksquare$

**EXERCISES**

1.  Let $n$ be an odd positive integer and define $f(x) = x^n$ for each $x$ in $\mathbb{R}$.
    (a)  Prove that the function $f: \mathbb{R} \to \mathbb{R}$ is one-to-one and onto.
    (b)  Use the Difference of Powers Formula to show that if $n > 1$, then the function $f: \mathbb{R} \to \mathbb{R}$ is not stable.

2.  Let $I$ be an open interval and suppose that the function $f: I \to \mathbb{R}$ is differentiable. Prove that $f: I \to \mathbb{R}$ is stable, with stability constant $c$, if and only if

$$|f'(x)| \geq c \quad \text{for all } x \text{ in } I.$$

3.  Define

$$G(x, \; y) = (x^2, \; y) \quad \text{for} (x, \; y) \text{ in } \mathbb{R}^2.$$

   Show that there is no neighborhood $U$ of the point $(0, 0)$ such that the mapping $G: \mathbb{R}^2 \to \mathbb{R}^2$ is stable.

4.  Is the sum of stable mappings also a stable mapping?

5.  Suppose that the function $f: [0, \infty) \to [0, \infty)$ is continuous and stable and $f(0) = 0$. Prove that $f: [0, \infty) \to [0, \infty)$ is one-to-one and onto.

6.  Find a constant $\gamma > 0$ so that the matrix

$$\begin{bmatrix} 1 & a & 0 \\ b & 1 & c \\ c & 0 & 1 \end{bmatrix}$$

   is invertible if $|a| < \gamma$, $|b| < \gamma$ and $|c| < \gamma$.

7.  Suppose that the function $f: \mathbb{R} \to \mathbb{R}$ is stable. Show that both sets $f([0, \infty))$ and $f((-\infty, 0])$ are unbounded.

8.  (A Global Inverse Function Theorem) Suppose that the function $f: \mathbb{R} \to \mathbb{R}$ is continuous and stable. Prove that $f: \mathbb{R} \to \mathbb{R}$ is one-to-one and onto. (*Hint:* Show that both $f([0, \infty))$ and $f((-\infty, 0])$ are unbounded, with one set unbounded above and the other unbounded below.)

## 16.3   A Minimization Principle and the General Inverse Function Theorem

In order to prove the Inverse Function Theorem in the Plane and its extension to general Euclidean spaces, we must confront the question of how it is possible to show that a particular point lies in the image of a mapping. For real-valued functions of a single variable the Intermediate Value Theorem is very useful, since in order to show that a point lies in the image of a continuous function defined on an interval, it is only necessary to show that there are functional values both greater than and less than the point. The Intermediate Value Theorem does not easily generalize to mappings that have their images in $\mathbb{R}^m$ with $m > 1$. Hence we will pursue a different strategy. In order to show that a particular equation has a solution, we will introduce an auxiliary function that has the property that the minimizers of the auxiliary function are solutions of the given equation.

**PROPOSITION**
**16.6**

**A Minimization Principle**   Let $U$ be an open subset of $\mathbb{R}^n$ and suppose that the mapping $\mathbf{F}: U \to \mathbb{R}^n$ is continuously differentiable. Moreover, suppose that at each point $\mathbf{x}$ in $U$ the derivative matrix $\mathbf{DF}(\mathbf{x})$ is invertible. Let $\mathbf{y}$ be a point in $\mathbb{R}^n$. Define the function $\mathrm{E}: U \to \mathbb{R}^*$ by

$$\mathrm{E}(\mathbf{x}) = \|\mathbf{F}(\mathbf{x}) - \mathbf{y}\|^2 \text{ for } \mathbf{x} \text{ in } U.$$

Then the following two assertions are equivalent:

(i) The point $\mathbf{y}$ lies in the image of the mapping $\mathbf{F}: U \to \mathbb{R}^n$.

(ii) The function $\mathrm{E}: U \to \mathbb{R}$ attains a minimum value.

**Proof**   Since $\mathrm{E}(\mathbf{x}) \geq 0$ for each point $\mathbf{x}$ in $U$, if the point $\mathbf{y}$ lies in the image of the mapping $\mathbf{F}: U \to \mathbb{R}^n$, then the function $\mathrm{E}: U \to \mathbb{R}$ attains the value 0, and 0 is certainly the minimum value. Thus assertion (i) implies assertion (ii).

To prove the converse, suppose that the function $\mathrm{E}: U \to \mathbb{R}$ attains a minimum value. Choose a point $\mathbf{x}$ in $U$ that is a minimizer for the function $\mathrm{E}: U \to \mathbb{R}$. Then at the point $\mathbf{x}$, all of the partial derivatives of the function $\mathrm{E}: U \to \mathbb{R}$ are zero; that is,

$$0 = \frac{\partial \mathrm{E}}{\partial x_i}(\mathbf{x}) = \sum_{j=1}^{n} 2(\mathrm{F}_j(\mathbf{x}) - y_j)\frac{\partial \mathrm{F}_j}{\partial x_i}(\mathbf{x}) \quad \text{for each index } i \text{ with } 1 \leq i \leq n. \quad (16.16)$$

This, in turn, may be rewritten in matrix notation as

$$0 = \mathbf{DE}(\mathbf{x}) = 2[\mathbf{DF}(\mathbf{x})]^{\mathrm{T}}(\mathbf{F}(\mathbf{x}) - \mathbf{y}). \quad (16.17)$$

Since, by assumption, the matrix $\mathbf{DF}(\mathbf{x})$ is invertible, by Theorem 15.11, its transpose matrix $\mathbf{DF}(\mathbf{x})^{\mathrm{T}}$ also is invertible. But (16.17) means that $\mathbf{DF}(\mathbf{x})^{\mathrm{T}}(\mathbf{F}(\mathbf{x}) - \mathbf{y}) = \mathbf{0}$, and hence $\mathbf{F}(\mathbf{x}) = \mathbf{y}$.   ∎

**LEMMA 16.7**   Let $U$ be an open subset of $\mathbb{R}^n$ and suppose that the continuously differentiable mapping $\mathbf{F}: U \to \mathbb{R}^n$ is stable. Then at each point $\mathbf{x}$ in $U$, the derivative matrix $\mathbf{DF}(\mathbf{x})$ is invertible.

**Proof**   Let the positive number $c$ be a stability constant for the mapping $\mathbf{F}: U \to \mathbb{R}^n$. Let $\mathbf{x}$ be a point in $U$, and for a point $\mathbf{x} + \mathbf{h}$ in $U$, define

$$\mathbf{R}(\mathbf{h}) = \mathbf{F}(\mathbf{x} + \mathbf{h}) - \mathbf{F}(\mathbf{x}) - \mathbf{DF}(\mathbf{x})\mathbf{h}.$$

By the First-Order Approximation Theorem,

$$\lim_{\mathbf{h} \to 0} \frac{\|\mathbf{R}(\mathbf{h})\|}{\|\mathbf{h}\|} = 0.$$

Thus, we may choose a $\delta > 0$ such that

$$\|\mathbf{R}(\mathbf{h})\| \leq \frac{c}{2}\|\mathbf{h}\| \quad \text{if } 0 < \|\mathbf{h}\| < \delta.$$

---

*The symbol E is used to connote *error*, since the number $\mathrm{E}(\mathbf{x}) = \|\mathbf{F}(\mathbf{x}) - \mathbf{y}\|^2$ is a measure of how far the point $\mathbf{x}$ is from being a solution of the equation $\mathbf{F}(\mathbf{x}) = \mathbf{y}$.

Hence, by the Reverse Triangle Inequality,

$$\|\mathbf{DF(x)h}\| = \|[\mathbf{F(x+h)} - \mathbf{F(x)}] - \mathbf{R(h)}\| \geq c\|\mathbf{h}\| - \|\mathbf{R(h)}\| \geq \frac{c}{2}\|\mathbf{h}\|$$

if $0 < \|\mathbf{h}\| < \delta$. By linearity,

$$\|\mathbf{DF(x)h}\| \geq \frac{c}{2}\|\mathbf{h}\| \quad \text{for all } \mathbf{h} \text{ in } \mathbb{R}^n.$$

Theorem 16.3 implies that the derivative matrix $\mathbf{DF(x)}$ is invertible. ∎

---

**LEMMA 16.8** **The Open-Image Lemma** Let $U$ be an open subset of $\mathbb{R}^n$ and suppose that the continuously differentiable mapping $\mathbf{F}: U \to \mathbb{R}^n$ is stable. Then its image $\mathbf{F}(U)$ is also open.

**Proof** Let $\mathbf{y}_0$ be a point in $\mathbf{F}(U)$. We must show that $\mathbf{y}_0$ is an interior point of $\mathbf{F}(U)$; that is, we must find some positive number $r$ such that the symmetric neighborhood $\mathcal{N}_r(\mathbf{y}_0)$ is contained in $\mathbf{F}(U)$. To do so, let $\mathbf{x}_0$ be the point in $U$ at which $\mathbf{F}(\mathbf{x}_0) = \mathbf{y}_0$. Since $U$ is open, we can choose a positive number $r_0$ such that if $K$ is defined to be the set of points $\mathbf{x}$ in $\mathbb{R}^n$ with $d(\mathbf{x}, \mathbf{x}_0) \leq r_0$, then $K$ is contained in $U$. Define $S$ to be the set of points in $\mathbb{R}^n$ whose distance from $\mathbf{x}_0$ is equal to $r_0$, and denote $\overset{\bullet}{\mathcal{N}}_{r_0}(\mathbf{x}_0)$ by $\mathcal{N}$. Observe that $K$ is compact and $K = \mathcal{N} \cup S$.

By assumption, the mapping $\mathbf{F}: U \to \mathbb{R}^n$ is stable; we can therefore select a positive number $c$ such that

$$\|\mathbf{F(u)} - \mathbf{F(v)}\| \geq c\|\mathbf{u} - \mathbf{v}\| \quad \text{for all points } \mathbf{u} \text{ and } \mathbf{v} \text{ in } U.$$

Define $r = cr_0/2$. Then

$$\|\mathbf{F(x)} - \mathbf{y}_0\| > r \quad \text{for all } \mathbf{x} \text{ in } S. \tag{16.18}$$

Lemma 16.7 asserts that at each point in $U$, the mapping $\mathbf{F}: U \to \mathbb{R}^n$ has an invertible derivative matrix. We will use the Minimization Principle to show that the

**FIGURE 16.3**

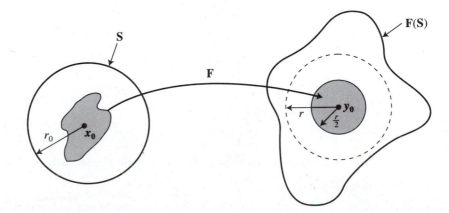

neighborhood $\mathcal{N}_r(\mathbf{y}_0)$ is contained in $\mathbf{F}(\mathcal{N})$. Indeed, let $\mathbf{y}$ belong to $\mathcal{N}_r(\mathbf{y}_0)$. Define the auxiliary function E: $U \to \mathbb{R}$ by

$$E(\mathbf{x}) = \|\mathbf{F}(\mathbf{x}) - \mathbf{y}\|^2 \quad \text{for } \mathbf{x} \text{ in } U.$$

If $\mathbf{x}$ is in $S$, then

$$\|\mathbf{F}(\mathbf{x}) - \mathbf{y}_0\| = \|\mathbf{F}(\mathbf{x}) - \mathbf{F}(\mathbf{x}_0)\| \geq cr_0,$$

and so
$$\|\mathbf{F}(\mathbf{x}) - \mathbf{y}\| \geq \|\mathbf{F}(\mathbf{x}) - \mathbf{y}_0\| - \|\mathbf{y} - \mathbf{y}_0\| > \frac{cr_0}{2} = r.$$

Then from (16.18) it follows that

$$E(\mathbf{x}_0) < r^2 < E(\mathbf{x}) \quad \text{for all } \mathbf{x} \text{ in } S. \tag{16.19}$$

However, because $K$ is compact, the restriction of the continuous function E to $K$ assumes a minimum value at some point $\mathbf{x}$ in $K$. The estimates (16.19) imply that the minimizer $\mathbf{x}$ does not belong to $S$, and hence that $\mathbf{x}$ belongs to $\mathcal{N}$. Thus $\mathbf{x}$ is a minimizer for E: $\mathcal{N} \to \mathbb{R}$, so by the Minimization Principle the point $\mathbf{y}$ belongs to $\mathbf{F}(\mathcal{N})$. ∎

**THEOREM 16.9**    **The Inverse Function Theorem**    Let $\mathcal{O}$ be an open subset of $\mathbb{R}^n$ and suppose that the mapping $\mathbf{F}: \mathcal{O} \to \mathbb{R}^n$ is continuously differentiable. Let $\mathbf{x}_0$ be a point in $\mathcal{O}$ at which the derivative matrix

$$\mathbf{DF}(\mathbf{x}_0) \text{ is invertible.} \tag{16.20}$$

**FIGURE 16.4**

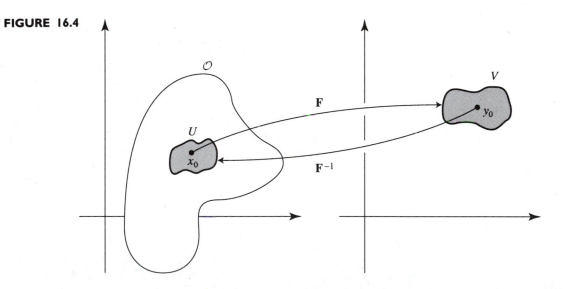

Then there is a neighborhood $U$ of the point $\mathbf{x}_0$ and a neighborhood $V$ of its image $\mathbf{F}(\mathbf{x}_0)$ such that the mapping

$$\mathbf{F}: U \to V \text{ is one-to-one and onto.} \tag{16.21}$$

Moreover, the inverse mapping $\mathbf{F}^{-1}\colon V \to U$ is also continuously differentiable, and for a point $\mathbf{y}$ in $V$, if $\mathbf{x}$ is the point in $U$ at which $\mathbf{F}(\mathbf{x}) = \mathbf{y}$, then

$$\mathbf{DF}^{-1}(\mathbf{y}) = [\mathbf{DF}(\mathbf{x})]^{-1}. \tag{16.22}$$

**Proof**    Because of the Nonlinear Stability Theorem, we may choose a neighborhood $U$ of $\mathbf{x}_0$ and a positive number $c$ such that

$$\|\mathbf{F}(\mathbf{u}) - \mathbf{F}(\mathbf{v})\| \geq c\|\mathbf{u} - \mathbf{v}\| \quad \text{for all points } \mathbf{u} \text{ and } \mathbf{v} \text{ in } U, \tag{16.23}$$

and by Lemma 16.7,

$$\mathbf{DF}(\mathbf{x}) \text{ is invertible for all points } \mathbf{x} \text{ in } U. \tag{16.24}$$

We may invoke the Open Image Lemma to conclude that the image $\mathbf{F}(U)$, which we will denote by $V$, is open. Thus $U$ is a neighborhood of $\mathbf{x}_0$ and $V$ is a neighborhood of its image $\mathbf{F}(\mathbf{x}_0) = \mathbf{y}_0$, the mapping $\mathbf{F}\colon U \to V$ is one-to-one and onto, and $\mathbf{DF}(\mathbf{x})$ is invertible for all points $\mathbf{x}$ in $U$. It remains only to prove that the inverse mapping $\mathbf{F}^{-1}\colon V \to U$ is also continuously differentiable and that formula (16.22) holds.

First, we will show that the inverse mapping is continuous; then we will establish formula (16.22). Once this has been done, the continuous differentiability of the inverse mapping will follow from the formula relating the entries of the inverse of an $n \times n$ matrix to the entries of the matrix (see Exercise 12).

Let $\mathbf{y}$ be a point in $V$ and let $\mathbf{x}$ be the point in $U$ at which $\mathbf{F}(\mathbf{x}) = \mathbf{y}$. For a point $\mathbf{y} + \mathbf{k}$ in $V$, it is convenient to define $\mathbf{h} = \mathbf{F}^{-1}(\mathbf{y} + \mathbf{k}) - \mathbf{F}^{-1}(\mathbf{y})$, so that

$$\mathbf{F}(\mathbf{x}) = \mathbf{y} \quad \text{and} \quad \mathbf{F}(\mathbf{x} + \mathbf{h}) = \mathbf{y} + \mathbf{k}. \tag{16.25}$$

Observe that from the stability estimate (16.23) we obtain the following estimate:

$$\|\mathbf{k}\| = \|\mathbf{F}(\mathbf{x} + \mathbf{h}) - \mathbf{F}(\mathbf{x})\| \geq c\|\mathbf{h}\|. \tag{16.26}$$

Since $\mathbf{h} = \mathbf{F}^{-1}(\mathbf{y} + \mathbf{k}) - \mathbf{F}^{-1}(\mathbf{y})$ this estimate may be rewritten as

$$\|\mathbf{F}^{-1}(\mathbf{y} + \mathbf{k}) - \mathbf{F}^{-1}(\mathbf{y})\| \leq \frac{1}{c}\|\mathbf{k}\|$$

from which the continuity of $\mathbf{F}^{-1}\colon V \to U$ at the point $\mathbf{y}$ follows.

We will now establish formula (16.22). To do so, define

$$\mathbf{B} = (\mathbf{DF}(\mathbf{x}))^{-1}.$$

According to Theorem 15.15, in order to prove (16.22) it is sufficient to show that

$$\lim_{\mathbf{k} \to 0} \frac{\|\mathbf{F}^{-1}(\mathbf{y} + \mathbf{k}) - [\mathbf{F}^{-1}(\mathbf{y}) + \mathbf{Bk}]\|}{\|\mathbf{k}\|} = 0. \tag{16.27}$$

Note that (16.26) provides a useful lower bound for the denominator on the left-hand side of (16.27). To find a useful upper bound for the numerator, observe that

$$\begin{aligned} \mathbf{F}^{-1}(\mathbf{y} + \mathbf{k}) - [\mathbf{F}^{-1}(\mathbf{y}) + \mathbf{Bk}] &= \mathbf{h} - \mathbf{Bk} \\ &= \mathbf{B}[\mathbf{B}^{-1}\mathbf{h} - \mathbf{k}] \\ &= \mathbf{B}[\mathbf{DF}(\mathbf{x})\mathbf{h} - \mathbf{F}(\mathbf{x} + \mathbf{h}) + \mathbf{F}(\mathbf{x})]. \end{aligned}$$

Hence, using the the Generalized Cauchy-Schwarz Inequality, it follows that

$$\|F^{-1}(y+k) - [F^{-1}(y) + Bk]\| \leq \|B\| \cdot \|F(x+h) - [F(x) + DF(x)h]\|. \quad (16.28)$$

From the estimates (16.26) and (16.28), we get the following estimate for the quotient on the left-hand side of (16.27):

$$\frac{\|F^{-1}(y+k) - [F^{-1}(y) + Bk]\|}{\|k\|} \leq \frac{\|B\|}{c} \left\{ \frac{\|F(x+h) - [F(x) + DF(x)h]\|}{\|h\|} \right\}. \quad (16.29)$$

But the estimate (16.26) implies that $h \to 0$ as $k \to 0$, so from the First-Order Approximation Theorem, the right-hand side of the estimate (16.29) converges to 0 as $k \to 0$, and hence so does the left-hand side of (16.29). This proves (16.27). ∎

Frequently, when we have a complicated problem, we first consider some simplified approximation of the problem. It is then necessary to understand the features of the approximation that are inherited by the original problem. The Inverse Function Theorem is a precise assertion of the properties of a continuously differentiable nonlinear mapping that are inherited from its linear differential at a point. Indeed, as we discussed in Section 1 of Chapter 15, for a linear mapping $T: \mathbb{R}^n \to \mathbb{R}^n$ that is associated with the $n \times n$ matrix $A$, the mapping $T: \mathbb{R}^n \to \mathbb{R}^n$ is invertible if and only if the matrix $A$ is invertible. Thus assumption (16.20) is equivalent to the assumption that the differential $dF(x_0): \mathbb{R}^n \to \mathbb{R}^n$ is an invertible linear mapping. Define the affine mapping $G: \mathbb{R}^n \to \mathbb{R}^n$ by

$$G(x) = F(x_0) + dF(x_0)(x - x_0) \quad \text{for } x \text{ in } \mathbb{R}^n.$$

The First-Order Approximation Theorem asserts that the mappings $F: \mathcal{O} \to \mathbb{R}^n$ and $G: \mathcal{O} \to \mathbb{R}^n$ are first-order approximations of one another at the point $x_0$, in the sense that

$$\lim_{x \to x_0} \frac{\|F(x) - G(x)\|}{\|x - x_0\|} = 0. \quad (16.30)$$

Since the affine mapping $G: \mathbb{R}^n \to \mathbb{R}^n$ is a constant perturbation of the differential $dF(x_0): \mathbb{R}^n \to \mathbb{R}^n$, the invertibility of one, and hence of both, of these mappings is equivalent to the invertibility of the derivative matrix $DF(x_0)$. Thus the Inverse Function Theorem asserts that if the mapping $G: \mathbb{R}^n \to \mathbb{R}^n$ is one-to-one and onto, then the mapping $F: \mathcal{O} \to \mathbb{R}^n$ inherits the same property "locally," where the conclusion of the Inverse Function Theorem makes precise what is meant by the word "locally."

---

**EXERCISES**

1. For each of the following mappings $F: \mathbb{R}^3 \to \mathbb{R}^3$, determine the points $(x_0, y_0, z_0)$ in $\mathbb{R}^3$ at which the Inverse Function Theorem applies:
   (a) $F(x, y, z) = (e^x \cos y, e^x \sin y, z^2)$ for $(x, y, z)$ in $\mathbb{R}^3$.
   (b) $F(x, y, z) = (yz, xz, xy)$ for $(x, y, z)$ in $\mathbb{R}^3$.

2. For a point $(\rho, \theta, \phi)$ in $\mathbb{R}^3$, define

$$F(\rho, \theta, \phi) = (\rho \sin \phi \cos \theta, \rho \sin \phi \sin \theta, \rho \cos \phi).$$

At what points $(\rho_0, \theta_0, \phi_0)$ in $\mathbb{R}^3$ does the Inverse Function Theorem apply to the mapping $F: \mathbb{R}^3 \to \mathbb{R}^3$?

3. Suppose that the functions $\psi: \mathbb{R}^3 \to \mathbb{R}$ and $\phi: \mathbb{R}^3 \to \mathbb{R}$ are continuously differentiable. Define

$$\mathbf{F}(x, y, z) = (\psi(x, y, z), \; \varphi(x, y, z), \; (\psi(x, y, z))^2$$
$$+(\varphi(x, y, z))^2) \quad \text{for } (x, y, z) \text{ in } \mathbb{R}^3.$$

   (a) Explain analytically why there is no point $(x_0, y_0, z_0)$ in $\mathbb{R}^3$ at which the assumptions of the Inverse Function Theorem hold for the mapping $\mathbf{F}: \mathbb{R}^3 \to \mathbb{R}^3$.
   (b) Explain geometrically why there is no point $(x_0, y_0, z_0)$ in $\mathbb{R}^3$ at which the conclusion of the Inverse Function Theorem holds for the mapping $\mathbf{F}: \mathbb{R}^3 \to \mathbb{R}^3$.

4. In the case when $n = 3$, write out explicitly, in terms of the first-order partial derivatives of the component functions of the mapping and its inverse, the meaning of the matrix formula (16.22).

5. Let $\mathbf{A}$ be an $n \times n$ matrix, and let $\mathbf{c}$ and $\mathbf{x}_0$ be points in $\mathbb{R}^n$. Define the affine mapping $\mathbf{G}: \mathbb{R}^n \to \mathbb{R}^n$ by

$$\mathbf{G}(\mathbf{x}) = \mathbf{c} + \mathbf{A}(\mathbf{x} - \mathbf{x}_0) \text{ for } \mathbf{x} \text{ in } \mathbb{R}^n.$$

   Show that the mapping $\mathbf{G}: \mathbb{R}^n \to \mathbb{R}^n$ is one-to-one and onto if and only if the matrix $\mathbf{A}$ is invertible.

6. Define $f(x) = x^2$ for $x$ in $\mathbb{R}$.
   (a) Show that if $\mathcal{O}$ is an open subset of positive numbers, then $f(\mathcal{O})$ is open.
   (b) Show that if $\mathcal{O}$ is any open subset of $\mathbb{R}$ that does not contain 0, then $f(\mathcal{O})$ is open.
   (c) Show that if $\mathcal{O}$ is an open subset of $\mathbb{R}$ that contains 0, then $f(\mathcal{O})$ is not open.

7. Give an example of a continuously differentiable mapping $\mathbf{F}: \mathbb{R}^n \to \mathbb{R}^n$ with the property that there is no open subset $U$ of $\mathbb{R}^n$ for which $\mathbf{F}(U)$ is open.

8. Let $I$ be an open interval of real numbers and suppose that the function $f: I \to \mathbb{R}$ is continuous. Let $c$ be a real number. Fix a number $x_0$ in the interval $I$ and define the auxiliary function $H: \mathbb{R} \to \mathbb{R}$ by

$$H(x) = cx - \int_{x_0}^{x} f(s) \, ds \quad \text{for } x \text{ in } I.$$

   For a point $x$ in $I$, show that $f(x) = c$ if $H'(x) = 0$. Conclude that $c$ is in the image of $f: I \to \mathbb{R}$ provided that the function $H: I \to \mathbb{R}$ has a local extreme point.

9. Give a proof of Theorem 16.1 in which, instead of the Intermediate Value Theorem, the Minimization Principle is used to show that $J$ is contained in $f(I)$.

10. Reread the proof of Darboux's Theorem and find the minimization principle used in the proof.

11. Under the assumptions of the Inverse Function Theorem, the system of equations

$$\mathbf{DF}(\mathbf{x}_0)(\mathbf{x} - \mathbf{x}_0) = \mathbf{y} - \mathbf{F}(\mathbf{x}_0)$$

   is called the *linear approximation* near the point $\mathbf{x}_0$ of the nonlinear equation

$$\mathbf{F}(\mathbf{x}) = \mathbf{y}, \; \mathbf{x} \text{ in } \mathcal{O}.$$

   For each of the following systems of equations, find the linear system of equations that approximates the system near the given point $(x_0, y_0)$:
   (a) The system

$$x + y + x^2 + xy + y^2 = u$$
$$x - y + x^4 + y^2 x = v$$

   near the point $(x_0, y_0) = (0, 0)$.
   (b) The same system as above, near the point $(x_0, y_0) = (2, 0)$.

(c)   The system

$$\cos(x - y) + e^{x^2+y} = u$$
$$\cosh(x^2 y^2) + x - y = v$$

near the point $(x_0, y_0) = (0, 0)$.

12.   Use Cramer's Rule and the continuity of the inverse mapping $\mathbf{F}^{-1}\colon V \to U$ to show that formula (16.22) implies that the inverse mapping $\mathbf{F}^{-1}\colon V \to U$ is continuously differentiable.

13.   (A Global Inverse Function Theorem) Suppose that the mapping $\mathbf{F}\colon \mathbb{R}^n \to \mathbb{R}^n$ is continuously differentiable and stable. By verifying each of the following parts, prove that the mapping $\mathbf{F}\colon \mathbb{R}^n \to \mathbb{R}^n$ is invertible.

(a)   Show that the image $\mathbf{F}(\mathbb{R}^n)$ is an open subset of $\mathbb{R}^n$. (*Hint:* Apply the Inverse Function Theorem.)

(b)   Show that the image $\mathbf{F}(\mathbb{R}^n)$ is a closed subset of $\mathbb{R}^n$. (*Hint:* Use stability and the definition of closedness.)

(c)   Show that the image $\mathbf{F}(\mathbb{R}^n) = \mathbb{R}^n$. (*Hint:* The set $\mathbf{F}(\mathbb{R}^n)$ is both open and closed in $\mathbb{R}^n$, and $\mathbb{R}^n$ is connected.)

# The Implicit Function Theorem and Its Applications

## 17.1 The Solutions of a Scalar Equation in Two Unknowns: Dini's Theorem

For $\mathcal{O}$ an open subset of the plane $\mathbb{R}^2$ and a continuously differentiable function $f: \mathcal{O} \to \mathbb{R}$, the set of solutions of the equation

$$f(x, y) = 0, \quad (x, y) \text{ in } \mathcal{O} \tag{17.1}$$

will, in general, be a very complicated subset of the plane. In particular, the set of solutions will not be the graph of a function prescribing $y$ as a function of $x$. However, if the point $(x_0, y_0)$ is a solution of this equation and $\partial f/\partial y(x_0, y_0) \neq 0$, then there is a neighborhood of the point $(x_0, y_0)$ having the property that the solutions of the above equation that are in this neighborhood make up the graph of a continuously differentiable function $g: I \to \mathbb{R}$, where $I$ is an open interval about $x_0$. Moreover, the derivative of the implicitly defined function $g: I \to \mathbb{R}$ can be computed in terms of the partial derivatives of the function $f: \mathcal{O} \to \mathbb{R}$. This assertion is called Dini's Theorem. It has an extension,

399

called the General Implicit Function Theorem, that provides a similar local description of the set of solutions of an equation of the form

$$\mathbf{F(u)} = \mathbf{0}, \quad \mathbf{u} \text{ in } \mathcal{O},$$

where $\mathcal{O}$ is an open subset of Euclidean space $\mathbb{R}^{n+k}$ and the mapping $\mathbf{F} \colon \mathcal{O} \to \mathbb{R}^k$ is continuously differentiable.

In this first section, we will provide a proof of Dini's Theorem and consider a number of examples. In Section 17.2, we will prove the General Implicit Function Theorem. The last two sections consist of applications, first to the geometry of paths and surfaces in $\mathbb{R}^3$ and then to the consideration of constrained extrema problems.

**EXAMPLE 17.1**   The set of solutions of the equation

$$\frac{x^2}{9} + \frac{y^2}{16} = 1, \quad (x, \ y) \text{ in } \mathbb{R}^2 \tag{17.2}$$

consists of the points lying on an ellipse centered at the origin. First consider the solution $(x_0, \ y_0) = (0, 4)$, which is the upper vertex of the ellipse. Then for a number $r$ between 0 and 3, define $I$ to be the open interval $(-r, \ r)$ and define the function $g \colon I \to \mathbb{R}$ by $g(x) = 4\sqrt{1 - x^2/9}$ for $x$ in $I$. Then there is a neighborhood of the solution $(0, 4)$ having the property that the set of solutions of equation (17.2) that are in this neighborhood consists of points of the form $(x, \ g(x))$ for $x$ in $I$. Now consider the second component of the solution $(0, 4)$. Observe that it is not possible to find a neighborhood $J$ of the number 4, a function $h \colon J \to \mathbb{R}$, and a neighborhood of the solution $(0, 4)$ in which the set of solutions of equation (17.2) consists of points of the form $(h(y), \ y)$ for $y$ in $J$. At each of the other vertices of the ellipse, it is possible to find a neighborhood of the vertex in which the set of solutions of equation (17.2) has a similar description. On the other hand, at a solution $(x_0, \ y_0)$ of equation (17.2) that is not a vertex of the ellipse, in a neighborhood of $(x_0, \ y_0)$ the set of solutions of equation (17.2) determines both $x$ as a function of $y$ and $y$ as a function of $x$. We leave the calculation of the specific functions as an exercise.

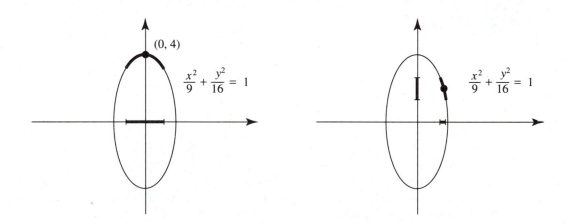

**FIGURE 17.1**

**EXAMPLE 17.2**   The set of solutions of the equation

$$y^2 - x^2 = 0, \quad (x, y) \text{ in } \mathbb{R}^2 \tag{17.3}$$

consists of the points in the plane that lie on the line $y = x$ or on the line $y = -x$. At each solution $(x_0, y_0)$ of equation (17.3) that is not equal to $(0, 0)$, there is a neighborhood of $(x_0, y_0)$ in which the set of solutions of equation (17.3) determines both $x$ as a function of $y$ and $y$ as a function of $x$. The origin $(0, 0)$ is a solution of equation (17.3), but there is no neighborhood of $(0, 0)$ in which the set of solutions coincides with the graph of a function expressing one of the components of the point $(x, y)$ as a function of the other component.                     $\square$

The equations in the above two examples are so simple that we could explicitly determine all the solutions of each of these nonlinear equations. In general, this is certainly not possible. For this reason, the following theorem is important.

---

**THEOREM 17.1**   **Dini's Theorem**   Let $\mathcal{O}$ be an open subset of the plane $\mathbb{R}^2$, and suppose that the function $f: \mathcal{O} \to \mathbb{R}$ is continuously differentiable. Let $(x_0, y_0)$ be a point in $\mathcal{O}$ at which $f(x_0, y_0) = 0$ and

$$\frac{\partial f}{\partial y}(x_0, y_0) \neq 0. \tag{17.4}$$

Then there is a positive number $r$ and a continuously differentiable function $g: I \to \mathbb{R}$, where $I$ is the open interval $(x_0 - r, x_0 + r)$, such that

(i)                     $f(x, g(x)) = 0$ for all $x$ in $I$

and

(ii)    whenever $|x - x_0| < r$,    $|y - y_0| < r$,    and   $f(x, y) = 0$,    then   $y = g(x)$.

Moreover,

(iii)         $\dfrac{\partial f}{\partial x}(x, g(x)) + \dfrac{\partial f}{\partial y}(x, g(x))g'(x) = 0$    for all $x$ in $I$.

**Proof**   We will assume that $\partial f/\partial y(x_0, y_0) > 0$. Since $\mathcal{O}$ is open and the function $\partial f/\partial y: \mathcal{O} \to \mathbb{R}$ is continuous and positive at the point $(x_0, y_0)$, we can choose positive numbers $a$ and $c$ such that the closed square $R = [x_0 - a, x_0 + a] \times [y_0 - a, y_0 + a]$ is contained in $\mathcal{O}$ and

$$\frac{\partial f}{\partial y}(x, y) \geq c \quad \text{for all points } (x, y) \text{ in } R. \tag{17.5}$$

It follows from the Mean Value Theorem for scalar functions of a single real variable that

$$f(x, y_1) < f(x, y_2) \quad \text{if} \quad |x - x_0| \leq a \quad \text{and} \quad y_0 - a \leq y_1 < y_2 \leq y_0 + a. \tag{17.6}$$

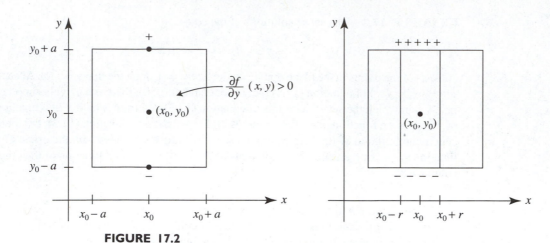

**FIGURE 17.2**

In particular, since $f(x_0, y_0) = 0$, it follows that $f(x_0, y_0 - a) < 0 < f(x_0, y_0 + a)$. Moreover, the function $f: \mathcal{O} \to \mathbb{R}$ is continuous, since it is continuously differentiable. Thus we may choose a positive number $r$ less than $a$ such that, if we let $I = (x_0 - r, x_0 + r)$,

$$f(x, y_0 - a) < 0 < f(x, y_0 + a) \quad \text{for all } x \text{ in } I.$$

Let $x$ be a point in $I$. Since $f(x, y_0 - a) < 0$ and $f(x, y_0 + a) > 0$, according to the Intermediate Value Theorem, there is some point $y$ between $y_0 - a$ and $y_0 + a$ at which $f(x, y) = 0$, and (17.6) implies that there is only one such point. Define $g(x)$ to be this point. This clearly defines a function $g: I \to \mathbb{R}$ having the properties (i) and (ii).

We will now show that $g: I \to \mathbb{R}$ is continuously differentiable and that the differentiation formula (iii) holds at the point $x_0$. Indeed, let $x_0 + h$ be a point in $I$. Then, by definition, $f(x_0 + h, g(x_0 + h)) = 0$ and $f(x_0, g(x_0)) = 0$. In particular,

$$0 = f(x_0 + h, g(x_0 + h)) - f(x_0, g(x_0)).$$

According to the Mean Value Theorem, there is some point on the segment between the points $(x_0, g(x_0))$ and $(x_0 + h, g(x_0 + h))$, which we label $\mathbf{p}(h)$, at which

$$f(x_0 + h, g(x_0 + h)) - f(x_0, g(x_0)) = \frac{\partial f}{\partial x}(\mathbf{p}(h))h + \frac{\partial f}{\partial y}(\mathbf{p}(h))[\, g(x_0 + h) - g(x_0)].$$

But the left-hand side is 0, and hence

$$g(x_0 + h) - g(x_0) = -\left[\frac{\partial f}{\partial x}(\mathbf{p}(h)) \Big/ \frac{\partial f}{\partial y}(\mathbf{p}(h))\right] h. \tag{17.7}$$

Since the function $\partial f/\partial x: \mathcal{O} \to \mathbb{R}$ is continuous, we can choose a positive number $M$ such that

$$\left|\frac{\partial f}{\partial x}(x, y)\right| \leq M \quad \text{for all points } (x, y) \text{ in } R.$$

Using this inequality, together with inequality (17.5), it follows from formula (17.7) that

$$|g(x_0 + h) - g(x_0)| \leq \frac{M}{c}|h| \quad \text{if } x_0 + h \text{ is in } I.$$

FIGURE 17.3

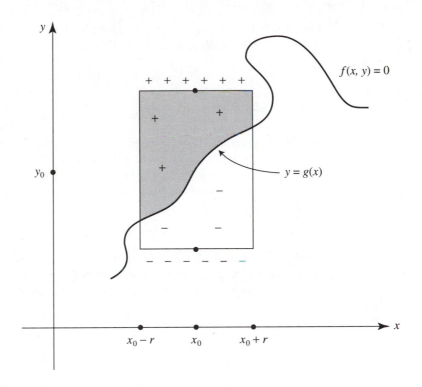

Hence the function $g: I \to \mathbb{R}$ is continuous at the point $x_0$. If we now divide (17.7) by $h$ and use the continuity of the first-order partial derivatives of $f: \mathcal{O} \to \mathbb{R}$ at the point $(x_0, y_0)$, together with the continuity of $g: I \to \mathbb{R}$ at the point $x_0$, it follows that

$$g'(x_0) = \lim_{h \to 0} \frac{g(x_0 + h) - g(x_0)}{h} = -\frac{\partial f}{\partial x}(x_0, y_0) \Big/ \frac{\partial f}{\partial y}(x_0, y_0),$$

which means that (iii) holds at $x_0$. But any other point $x$ in the interval $I$ satisfies the same assumptions as does the point $x_0$, and hence (iii) holds at all points in $I$. ∎

**EXAMPLE 17.3**   Consider the equation

$$\cos(x + y) + e^{y+x^2} + 3x - 2 - x^3 y^3 = 0, \quad (x, y) \text{ in } \mathbb{R}^2. \tag{17.8}$$

Define      $f(x, y) = \cos(x + y) + e^{y+x^2} + 3x - 2 - x^3 y^3$   for $(x, y)$ in $\mathbb{R}^2$.

Then $(x, y)$ is a solution of equation (17.8) if and only if $f(x, y) = 0$. Observe that $(0, 0)$ is a solution of (17.8) and that

$$\frac{\partial f}{\partial x}(0, 0) = 3 \quad \text{and} \quad \frac{\partial f}{\partial y}(0, 0) = 1.$$

Dini's Theorem implies that there is a positive number $r$ and a continuously differentiable function $g: I \to \mathbb{R}$, where $I$ is the open interval $(-r, r)$, such that

$$\cos(x + g(x)) + e^{g(x)+x^2} + 3x - 2 - x^3(g(x))^3 = 0 \quad \text{for all } x \text{ in } I.$$

Moreover, if $(x, y)$ is a solution of equation (17.8) with $|x| < r$ and $|y| < r$, then $y = g(x)$. Finally, $g'(0)$ is determined by the formula

$$\frac{\partial f}{\partial x}(0, 0) + \frac{\partial f}{\partial y}(0, 0)g'(0) = 0,$$

so $g'(0) = -3$. ☐

The assumption in Dini's Theorem that $\partial f / \partial y(x_0, y_0) \neq 0$ can be replaced by the assumption that $\partial f / \partial x(x_0, y_0) \neq 0$, and the conclusion remains the same except that the roles of $x$ and $y$ are interchanged. Thus if $\mathcal{O}$ is an open subset of the plane $\mathbb{R}^2$, if the function $f: \mathcal{O} \to \mathbb{R}$ is continuously differentiable, and if at the point $(x_0, y_0)$ in $\mathcal{O}$ we have $f(x_0, y_0) = 0$ and

$$\mathbf{D}f(x_0, y_0) \neq (0, 0),$$

then there is a neighborhood of the point $(x_0, y_0)$ in which the solutions of equation (17.1) make up the graph of a continuously differentiable function prescribing at least one of the components of a solution $(x, y)$ as a function of the other component.

**EXAMPLE 17.4**    Consider the equation

$$e^{x-2+(y-1)^2} - 1 = 0, \quad (x, y) \text{ in } \mathbb{R}^2. \tag{17.9}$$

Define          $f(x, y) = e^{x-2+(y-1)^2} - 1$    for $(x, y)$ in $\mathbb{R}^2$.

Then the point $(x, y)$ is a solution of equation (17.9) if and only if $f(x, y) = 0$. Observe that the point $(2, 1)$ is a solution of equation (17.9) and that

$$\frac{\partial f}{\partial x}(2, 1) = 1 \quad \text{and} \quad \frac{\partial f}{\partial y}(2, 1) = 0.$$

Dini's Theorem, with the roles of the variables $x$ and $y$ interchanged, implies that there is a positive number $r$ and a continuously differentiable function $h: J \to \mathbb{R}$, where $J$ is the open interval $(1 - r, 1 + r)$, such that

$$e^{h(y)-2+(y-1)^2} - 1 = 0 \quad \text{for all } y \text{ in } J.$$

Moreover, if $(x, y)$ is a solution of equation (17.9) with $|x - 2| < r$ and $|y - 1| < r$, then $x = h(y)$. Finally, $h'(1)$ is determined by the formula

$$\frac{\partial f}{\partial x}(2, 1)h'(1) + \frac{\partial f}{\partial y}(2, 1) = 0,$$

so $h'(1) = 0$. ☐

Of course, Dini's Theorem may be used to analyze solutions of equations of the form

$$\phi(x, y) = \eta(x, y), \quad (x, y) \text{ in } \mathcal{O}. \tag{17.10}$$

Simply define          $f(x, y) = \phi(x, y) - \eta(x, y)$    for $(x, y)$ in $\mathcal{O}$, so that the solutions of equation (17.10) are precisely the solutions of equation (17.1). In particular, for a real number $c$ we can analyze solutions of the equation

$$f(x, y) = c, \quad (x, y) \text{ in } \mathcal{O}. \tag{17.11}$$

For a given number $c$, the set of solutions of (17.11) is often called a *level curve* of the function $f: \mathcal{O} \to \mathbb{R}$. Because of Dini's Theorem, in order to justify the term *curve*, it seems reasonable to suppose that

$$\mathbf{D}f(x, y) \neq \mathbf{0} \quad \text{at each point } (x, y) \text{ in } \mathcal{O} \text{ at which } f(x, y) = c,$$

so that the set of solutions of equation (17.11) does indeed consist of the union of curves.

We conclude this section with a comment about implicit differentiation and higher-order derivatives. In the last assertion of Dini's Theorem, about the differentiability of the implicitly defined function, once we know that the function $g: I \to \mathbb{R}$ is differentiable and that

$$f(x, g(x)) = 0 \quad \text{for all } x \text{ in } I, \tag{17.12}$$

the formula (iii) for the derivative $g': I \to \mathbb{R}$ follows by differentiating each side of (17.12). This technique is known as *implicit differentiation*. Moreover, we can also find higher-order derivatives of the function $g: I \to \mathbb{R}$. For instance, if the function $f: \mathcal{O} \to \mathbb{R}$ in the statement of Dini's Theorem has continuous second-order partial derivatives, then from formula (iii) for the derivative of the implicitly defined function $g: I \to \mathbb{R}$, it follows that $g': I \to \mathbb{R}$ itself is differentiable. Thus we can differentiate each side of formula (iii) and use the Chain Rule to obtain the following formula for the second derivative of the implicitly defined function:

$$\frac{\partial^2 f}{\partial x^2}(x, g(x)) + 2\frac{\partial^2 f}{\partial x \partial y}(x, g(x))g'(x) + \frac{\partial^2 f}{\partial y^2}(x, g(x))[g'(x)]^2 + \frac{\partial f}{\partial y}(x, g(x))g''(x) = 0.$$

$$\tag{17.13}$$

---

**EXERCISES**

1.  Consider the equation

    $$\frac{x^2}{8} + \frac{y^2}{18} = 1, \quad (x, y) \text{ in } \mathbb{R}^2.$$

    (a)  Graph the set of solutions and show that Dini's Theorem applies at the solution $(2, 3)$.

    (b)  Explicitly define the function $g: I \to \mathbb{R}$ that has the property that in a neighborhood of the solution $(2, 3)$, all of the solutions are of the form $(x, g(x))$ for $x$ in $I$, and check that formula (iii) holds for the derivative $g': I \to \mathbb{R}$.

    (c)  Explicitly define the function $h: J \to \mathbb{R}$ that has the property that in a neighborhood of the solution $(2, 3)$, all of the solutions are of the form $(h(y), y)$ for $y$ in $J$.

2.  Consider the equation

    $$(x^2 + y^2 - 2)(x^2 - y^2) = 0, \quad (x, y) \text{ in } \mathbb{R}^2.$$

(a) Compute partial derivatives to show that the assumptions of Dini's Theorem do not hold at each of the following solutions: $(0, 0)$, $(1, 1)$, $(1, -1)$, $(-1, -1)$, $(-1, 1)$.

(b) By graphing the set of solutions of this equation, show that the conclusions of Dini's Theorem do not hold at each of the solutions listed in (a).

3. Consider the equation

$$x^2 + y^2 = 0, \quad (x, y) \text{ in } \mathbb{R}^2.$$

(a) Show that the assumptions of Dini's Theorem do not hold at the solution $(0, 0)$.

(b) Explain, by graphing the set of solutions of this equation, why the conclusion of Dini's Theorem does not hold at the solution $(0, 0)$.

4. Consider the equation

$$e^{2x-y} + \cos(x^2 + xy) - 2 - 2y = 0, \quad (x, y) \text{ in } \mathbb{R}^2.$$

Does the set of solutions of this equation in a neighborhood of the solution $(0, 0)$ implicitly define one of the components of the point $(x, y)$ as a function of the other component? If so, compute the derivative of this function (these functions?) at the point 0.

5. The point $(0, 0)$ is a solution of the equation

$$\ln(x^2 + y^2 + 1) = x, \quad (x, y) \text{ in } \mathbb{R}^2.$$

Find the derivative at the point 0 of the function(s) of a single variable that is (are) defined by the set of solutions of this equation in a neighborhood of $(0, 0)$.

6. Find the explicit formula for all of the solutions of the following equation, considered in Example 17.4:

$$e^{x-2+(y-1)^2} - 1 = 0, \quad (x, y) \text{ in } \mathbb{R}^2.$$

7. Let $\mathcal{O}$ be an open subset of the plane and suppose that the function $f : \mathcal{O} \to \mathbb{R}$ is continuously differentiable. At the point $(x_0, y_0)$ in $\mathcal{O}$, suppose that $f(x_0, y_0) = 0$ and that $\mathbf{D}f(x_0, y_0) \neq (0, 0)$. Show that the vector $\mathbf{D}f(x_0, y_0)$ is orthogonal to the tangent line at $(x_0, y_0)$ of the implicitly defined function.

8. Let $\mathcal{O}$ be an open subset of the plane and suppose that the function $f : \mathcal{O} \to \mathbb{R}$ is continuously differentiable. At the point $(x_0, y_0)$ in $\mathcal{O}$, suppose that $f(x_0, y_0) = 0$ and that

$$\frac{\partial f}{\partial x}(x_0, y_0) \neq 0, \quad \frac{\partial f}{\partial y}(x_0, y_0) \neq 0.$$

Show that the two functions implicitly defined by Dini's Theorem, when their domains are properly chosen, are inverses of each other.

9. Suppose that the continuously differentiable function $f : \mathbb{R}^2 \to \mathbb{R}$ may be factored as

$$f(x, y) = (x^2 + y^2)h(x, y) \quad \text{for } (x, y) \text{ in } \mathbb{R}^2,$$

where the function $h : \mathbb{R}^2 \to \mathbb{R}$ is also continuously differentiable. Show that Dini's Theorem cannot be directly applied to analyze the solutions of the equation $f(x, y) = 0$ in a neighborhood of the solution $(0, 0)$. If $h(0, 0) = 0$ and $\mathbf{D}h(0, 0) \neq (0, 0)$, use Dini's Theorem to analyze the solutions of $f(x, y) = 0$.

10. Suppose that the function $\phi : \mathbb{R} \to \mathbb{R}$ is continuously differentiable and that at the point $x_0$ in $\mathbb{R}$, $\phi'(x_0) \neq 0$. Set $y_0 = \phi(x_0)$ and define the function $f : \mathbb{R}^2 \to \mathbb{R}$ by $f(x, y) = y - \phi(x)$ for $(x, y)$ in $\mathbb{R}^2$. Apply Dini's Theorem to the function $f : \mathbb{R}^2 \to \mathbb{R}$ at the point $(x_0, y_0)$ and compare the result with the conclusion of the Inverse Function Theorem applied to the function $\phi : \mathbb{R} \to \mathbb{R}$ at the point $x_0$.

**11.** Suppose that the function $h: \mathbb{R} \to \mathbb{R}$ is differentiable and that there is a positive number $c$ such that $h'(t) \geq c$ for all points $t$ in $\mathbb{R}$. Prove that there is exactly one number $t$ at which $h(t) = 0$.

**12.** (A Global Implicit Function Theorem) Suppose that the function $f: \mathbb{R}^2 \to \mathbb{R}$ is continuously differentiable and that there is a positive number $c$ such that

$$\frac{\partial f}{\partial y}(x, \, y) \geq c \quad \text{for every } (x, \, y) \text{ in } \mathbb{R}^2.$$

Prove that there is a continuously differentiable function $g: \mathbb{R} \to \mathbb{R}$ with

$$f(x, \, g(x)) = 0 \quad \text{for every } x \text{ in } \mathbb{R}$$

and that if $f(x, \, y) = 0$, then $y = g(x)$. (*Hint:* Use Exercise 11.)

**13.** Suppose that the function $f: \mathbb{R}^3 \to \mathbb{R}$ is continuously differentiable and $(x_0, \, y_0, \, z_0)$ is a point in $\mathbb{R}^3$ at which $f(x_0, \, y_0, \, z_0) = 0$ and $\partial f / \partial z (x_0, \, y_0, \, z_0) > 0$. Follow the proof of Dini's Theorem to show that there is a positive number $r$ and a function $g: \mathcal{N}_r(x_0, \, y_0) \to \mathbb{R}$ such that

$$f(x, \, y, \, g(x, \, y)) = 0 \quad \text{for all } (x, \, y) \text{ in } \mathcal{N}_r(x_0, \, y_0).$$

**14.** In addition to the assumptions of Dini's Theorem, assume also that the function $f: \mathcal{O} \to \mathbb{R}$ has continuous second-order partial derivatives. Moreover, suppose that

$$\frac{\partial f}{\partial x}(x_0, \, y_0) = 0 \quad \text{and} \quad \frac{\partial f}{\partial y}(x_0, \, y_0) \frac{\partial^2 f}{\partial x^2}(x_0, \, y_0) > 0.$$

Prove that the graph of $g: I \to \mathbb{R}$ lies below the line $y = y_0$ if $I$ is chosen sufficiently small. (*Hint:* Use formula (17.13) to determine the sign of $g''(x_0)$.)

---

## 17.2   Underdetermined Systems of Nonlinear Equations: The General Implicit Function Theorem

Let $k$ and $n$ be positive integers, let $\mathcal{O}$ be an open subset of Euclidean space $\mathbb{R}^{n+k}$, and let the mapping $\mathbf{F}: \mathcal{O} \to \mathbb{R}^k$ be continuously differentiable. Consider the equation

$$\mathbf{F}(\mathbf{u}) = \mathbf{0}, \quad \mathbf{u} \text{ in } \mathcal{O}. \tag{17.14}$$

In the case when $n = 1$ and $k = 1$, we have already considered this equation in Section 17.1, where we proved Dini's Theorem. The object of this section is to prove the General Implicit Function Theorem, which extends Dini's Theorem to more general equations of the form (17.14). In order to emphasize the analogy between the general case and the case when $n = 1$ and $k = 1$, it is useful to introduce the following notation: For a point $\mathbf{u}$ in $\mathbb{R}^{n+k}$, we separate the first $n$ components of $\mathbf{u}$ from the last $k$ components and label them as follows:

$$\mathbf{u} = (\mathbf{x}, \, \mathbf{y}) = (x_1, \ldots, x_n, \, y_1, \ldots, y_k).$$

Then equation (17.14) can be rewritten as

$$\mathbf{F}(\mathbf{x}, \, \mathbf{y}) = \mathbf{0}, \quad (\mathbf{x}, \, \mathbf{y}) \text{ in } \mathcal{O}. \tag{17:15}$$

If the mapping $\mathbf{F}\colon \mathcal{O} \to \mathbb{R}^k$ is written in terms of its component functions, $\mathbf{F} = (F_1, \ldots, F_k)$, this equation, in turn, may be written as the following system of $k$ nonlinear scalar equations in $n + k$ scalar unknowns:

$$F_1(x_1, \ldots, x_n, y_1, \ldots, y_k) = 0$$
$$\vdots$$
$$F_i(x_1, \ldots, x_n, y_1, \ldots, y_k) = 0 \qquad (17.16)$$
$$\vdots$$
$$F_k(x_1, \ldots, x_n, y_1, \ldots, y_k) = 0$$

for $(x_1, \ldots, x_n, y_1, \ldots, y_k)$ in $\mathcal{O}$.

This system is "underdetermined" in the sense that there are fewer equations than there are variables. It is highly unlikely that we can explicitly find all the solutions of such a complicated system of nonlinear equations. Thus we seek the type of information that has already been provided by Dini's Theorem for a single scalar equation with two scalar unknowns.

Just as we have defined the partial derivatives for a function, we will now define the *partial derivative matrices* for a mapping. For a point $(\mathbf{x}, \mathbf{y})$ in $\mathcal{O}$, fix $\mathbf{x}$ and consider the mapping $\mathbf{y} \mapsto \mathbf{F}(\mathbf{x}, \mathbf{y})$, which is a mapping from an open subset of $\mathbb{R}^k$ into $\mathbb{R}^k$; we denote the $k \times k$ derivative matrix of this mapping by $\mathbf{D_y F}(\mathbf{x}, \mathbf{y})$. Similarly, if $\mathbf{y}$ is fixed, consider the mapping $\mathbf{x} \mapsto \mathbf{F}(\mathbf{x}, \mathbf{y})$, which is a mapping from an open subset of $\mathbb{R}^n$ into $\mathbb{R}^k$; we denote the $k \times n$ derivative matrix of this mapping by $\mathbf{D_x F}(\mathbf{x}, \mathbf{y})$. Displayed in terms of their entries, these matrices are

$$\mathbf{D_y F}(\mathbf{x}, \mathbf{y}) = \begin{bmatrix} \partial F_1/\partial y_1(\mathbf{x}, \mathbf{y}) \cdots \partial F_1/\partial y_j(\mathbf{x}, \mathbf{y}) \cdots \partial F_1/\partial y_k(\mathbf{x}, \mathbf{y}) \\ \cdots \quad \cdots \quad \cdots \quad \cdots \\ \partial F_i/\partial y_1(\mathbf{x}, \mathbf{y}) \cdots \partial F_i/\partial y_j(\mathbf{x}, \mathbf{y}) \cdots \partial F_i/\partial y_k(\mathbf{x}, \mathbf{y}) \\ \cdots \quad \cdots \quad \cdots \quad \cdots \quad \cdots \\ \partial F_k/\partial y_1(\mathbf{x}, \mathbf{y}) \cdots \partial F_k/\partial y_j(\mathbf{x}, \mathbf{y}) \cdots \partial F_k/\partial y_k(\mathbf{x}, \mathbf{y}) \end{bmatrix}$$

and

$$\mathbf{D_x F}(\mathbf{x}, \mathbf{y}) = \begin{bmatrix} \partial F_1/\partial x_1(\mathbf{x}, \mathbf{y}) \cdots \partial F_1/\partial x_j(\mathbf{x}, \mathbf{y}) \cdots \partial F_1/\partial x_n(\mathbf{x}, \mathbf{y}) \\ \cdots \quad \cdots \quad \cdots \quad \cdots \quad \cdots \\ \partial F_i/\partial x_1(\mathbf{x}, \mathbf{y}) \cdots \partial F_i/\partial x_j(\mathbf{x}, \mathbf{y}) \cdots \partial F_i/\partial x_n(\mathbf{x}, \mathbf{y}) \\ \cdots \quad \cdots \quad \cdots \quad \cdots \quad \cdots \\ \partial F_k/\partial x_1(\mathbf{x}, \mathbf{y}) \cdots \partial F_k/\partial x_j(\mathbf{x}, \mathbf{y}) \cdots \partial F_k/\partial x_n(\mathbf{x}, \mathbf{y}) \end{bmatrix}.$$

The following theorem, describing the solutions of Equation (17.14), is a direct generalization of Dini's Theorem.

**THEOREM 17.2**   **The General Implicit Function Theorem**   Let $n$ and $k$ be positive integers, let $\mathcal{O}$ be an open subset of $\mathbb{R}^{n+k}$, and suppose that the mapping $\mathbf{F}\colon \mathcal{O} \to \mathbb{R}^k$ is continuously differentiable. At the point $(\mathbf{x}_0, \mathbf{y}_0)$ in $\mathcal{O}$, suppose that $\mathbf{F}(\mathbf{x}_0, \mathbf{y}_0) = \mathbf{0}$ and that the $k \times k$ partial derivative matrix

$$\mathbf{D_y F}(\mathbf{x}_0, \mathbf{y}_0) \quad \text{is invertible.} \qquad (17.17)$$

Then there is a positive number $r$ and a continuously differentiable mapping $\mathbf{G}: \mathcal{N} \to \mathbb{R}^k$, where $\mathcal{N} = \mathcal{N}_r(\mathbf{x}_0)$, such that

(i) $\qquad\qquad\qquad \mathbf{F}(\mathbf{x}, \ \mathbf{G}(\mathbf{x})) = \mathbf{0} \quad$ for all points $\mathbf{x}$ in $\mathcal{N}$,

and

(ii) $\quad$ whenever $\|\mathbf{x} - \mathbf{x}_0\| < r, \ \|\mathbf{y} - \mathbf{y}_0\| < r, \ $ and $\mathbf{F}(\mathbf{x}, \ \mathbf{y}) = \mathbf{0}, \quad$ then $\mathbf{y} = \mathbf{G}(\mathbf{x})$.

Moreover,

(iii) $\qquad \mathbf{D_x F}(\mathbf{x}, \ \mathbf{G}(\mathbf{x})) + \mathbf{D_y F}(\mathbf{x}, \ \mathbf{G}(\mathbf{x})) \cdot \mathbf{DG}(\mathbf{x}) = \mathbf{0} \quad$ for all points $\mathbf{x}$ in $\mathcal{N}$.

**Proof** $\quad$ We define an auxiliary mapping

$$\mathbf{H}: \mathcal{O} \to \mathbb{R}^{n+k}$$

by $\qquad\qquad\qquad \mathbf{H}(\mathbf{x}, \ \mathbf{y}) = (\mathbf{x}, \ \mathbf{F}(\mathbf{x}, \ \mathbf{y})) \quad$ for $(\mathbf{x}, \ \mathbf{y})$ in $\mathcal{O}$.

Observe that $\qquad \mathbf{F}(\mathbf{x}, \ \mathbf{y}) = \mathbf{0} \quad$ if and only if $\quad \mathbf{H}(\mathbf{x}, \ \mathbf{y}) = (\mathbf{x}, \ \mathbf{0}). \qquad\qquad$ (17.18)

**FIGURE 17.4**

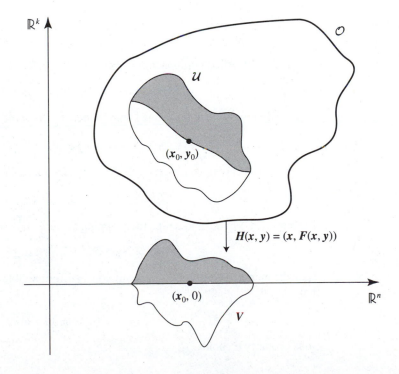

Now the mapping $\mathbf{H} \colon \mathcal{O} \to \mathbb{R}^{n+k}$ is a continuously differentiable mapping between Euclidean spaces of the same dimension, so the Inverse Function Theorem may be applied in order to analyze its image, and therefore, because of the correspondence (17.18), to analyze the points $(\mathbf{x}, \mathbf{y})$ at which $\mathbf{F}(\mathbf{x}, \mathbf{y}) = \mathbf{0}$.

The derivative matrix $\mathbf{DH}(\mathbf{x}_0, \mathbf{y}_0)$ may be partitioned as

$$\mathbf{DH}(\mathbf{x}_0, \mathbf{y}_0) = \begin{bmatrix} \mathbf{I}_n & \mathbf{0} \\ \mathbf{D}_{\mathbf{x}}\mathbf{F}(\mathbf{x}_0, \mathbf{y}_0) & \mathbf{D}_{\mathbf{y}}\mathbf{F}(\mathbf{x}_0, \mathbf{y}_0) \end{bmatrix}, \tag{17.19}$$

where $\mathbf{I}_n$ denotes the $n \times n$ identity matrix and the upper right-hand matrix is the $n \times k$ matrix all of whose entries are 0. The assumption that $\mathbf{D}_{\mathbf{y}}\mathbf{F}(\mathbf{x}_0, \mathbf{y}_0)$ is an invertible $k \times k$ matrix implies that the derivative matrix $\mathbf{DH}(\mathbf{x}_0, \mathbf{y}_0)$ is an invertible $(n+k) \times (n+k)$ matrix (see Exercise 7). Hence we may apply the Inverse Function Theorem to the mapping $\mathbf{H} \colon \mathcal{O} \to \mathbb{R}^{n+k}$ at the point $(\mathbf{x}_0, \mathbf{y}_0)$ in order to conclude that there is a neighborhood $U$ of $(\mathbf{x}_0, \mathbf{y}_0)$ in $\mathbb{R}^{n+k}$ and a neighborhood $V$ of its image $\mathbf{H}(\mathbf{x}_0, \mathbf{y}_0) = (\mathbf{x}_0, \mathbf{0})$ in $\mathbb{R}^{n+k}$ such that $\mathbf{H} \colon U \to V$ is one-to-one and onto, and the inverse mapping $\mathbf{H}^{-1} \colon V \to U$ is also continuously differentiable.

Write the inverse mapping as

$$\mathbf{H}^{-1}(\mathbf{x}, \mathbf{y}) = (\mathbf{M}(\mathbf{x}, \mathbf{y}), \mathbf{N}(\mathbf{x}, \mathbf{y})) \quad \text{for } (\mathbf{x}, \mathbf{y}) \text{ in } V.$$

The very definition of inverse mapping means that the following identities hold:

$$\begin{aligned} (\mathbf{x}, \mathbf{y}) &= \left(\mathbf{H} \circ \mathbf{H}^{-1}\right)(\mathbf{x}, \mathbf{y}) \\ &= (\mathbf{M}(\mathbf{x}, \mathbf{y}), \mathbf{F}(\mathbf{M}(\mathbf{x}, \mathbf{y}), \mathbf{N}(\mathbf{x}, \mathbf{y}))) \quad \text{for all } (\mathbf{x}, \mathbf{y}) \text{ in } V \end{aligned} \tag{17.20}$$

and

$$\begin{aligned} (\mathbf{x}, \mathbf{y}) &= \left(\mathbf{H}^{-1} \circ \mathbf{H}\right)(\mathbf{x}, \mathbf{y}) \\ &= (\mathbf{M}(\mathbf{x}, \mathbf{F}(\mathbf{x}, \mathbf{y})), \mathbf{N}(\mathbf{x}, \mathbf{F}(\mathbf{x}, \mathbf{y}))) \quad \text{for all } (\mathbf{x}, \mathbf{y}) \text{ in } U. \end{aligned} \tag{17.21}$$

But equating the first components in the identity (17.20), we see that

$$\mathbf{M}(\mathbf{x}, \mathbf{y}) = \mathbf{x} \quad \text{for all } (\mathbf{x}, \mathbf{y}) \text{ in } V,$$

which, when we equate the second components of the same identity, gives

$$\mathbf{F}(\mathbf{x}, \mathbf{N}(\mathbf{x}, \mathbf{y})) = \mathbf{y} \quad \text{for all } (\mathbf{x}, \mathbf{y}) \text{ in } V. \tag{17.22}$$

Since $U$ and $V$ are open subsets of $\mathbb{R}^{n+k}$, we can choose a positive number $r$ such that

$$\mathcal{N}_r(\mathbf{x}_0) \times \mathcal{N}_r(\mathbf{y}_0) \subseteq U \quad \text{and} \quad \mathcal{N}_r(\mathbf{x}_0) \times \{\mathbf{0}\} \subseteq V. \tag{17.23}$$

Denote $\mathcal{N}_r(\mathbf{x}_0)$ by $\mathcal{N}$ and define the mapping $\mathbf{G} \colon \mathcal{N} \to \mathbb{R}^k$ by

$$\mathbf{G}(\mathbf{x}) = \mathbf{N}(\mathbf{x}, \mathbf{0}) \quad \text{for all } \mathbf{x} \text{ in } \mathcal{N}.$$

Then the mapping $\mathbf{G} \colon \mathcal{N} \to \mathbb{R}^k$ is continuously differentiable, and from (17.22) it follows that

$$\mathbf{F}(\mathbf{x}, \mathbf{G}(\mathbf{x})) = \mathbf{0} \quad \text{for all } \mathbf{x} \text{ in } \mathcal{N}. \tag{17.24}$$

Thus property (i) holds.

To verify property (ii), note that if the point $(\mathbf{x}, \mathbf{y})$ belongs to $\mathcal{N}_r(\mathbf{x}_0) \times \mathcal{N}_r(\mathbf{y}_0)$ and $\mathbf{F}(\mathbf{x}, \mathbf{y}) = \mathbf{0}$, then, since $\mathcal{N}_r(\mathbf{x}_0) \times \mathcal{N}_r(\mathbf{y}_0) \subseteq U$, from the equality of the second components in the identity (17.21), it follows that

$$\mathbf{y} = \mathbf{N}(\mathbf{x}, \mathbf{0}) = \mathbf{G}(\mathbf{x}).$$

Thus property (ii) holds.

Finally, we will verify formula (iii) for the derivative matrix. Indeed, if we represent the mapping $\mathbf{F}$ in component functions as $\mathbf{F} = (F_1, \ldots, F_k)$, then equation (17.24) may be written in components as

$$F_i(\mathbf{x}, \mathbf{G}(\mathbf{x})) = 0 \quad \text{for all points } \mathbf{x} \text{ in } \mathcal{N} \text{ and all indices } i \text{ with } 1 \leq i \leq k.$$

Using the Chain Rule to differentiate the preceding system of equations with respect to the component $x_j$, we obtain

$$\frac{\partial F_i}{\partial x_j}(\mathbf{x}, \mathbf{G}(\mathbf{x})) + \sum_{\ell=1}^{k} \frac{\partial F_i}{\partial y_\ell}(\mathbf{x}, \mathbf{G}(\mathbf{x})) \frac{\partial G_\ell}{\partial x_j}(\mathbf{x}) = 0 \qquad (17.25)$$

for all points $\mathbf{x}$ in $\mathcal{N}$ and pairs of indices $i$ and $j$ with $1 \leq i \leq k, 1 \leq j \leq n$. However, the matrix identity (iii) is just a rewriting in matrix notation of the above system (17.25). ∎

In the statement of the Implicit Function Theorem, we singled out the first $n$ components and the last $k$ components of a point $\mathbf{u}$ in $\mathcal{O}$. But, in fact, this separation of components was rather arbitrary. A $k \times (n + k)$ matrix is said to have *maximal rank* if it has a $k \times k$ submatrix that is invertible. The Implicit Function Theorem holds for a continuously differentiable mapping $\mathbf{F} \colon \mathcal{O} \to \mathbb{R}^k$, where $\mathcal{O}$ is an open subset of $\mathbb{R}^{n+k}$, at a point $\mathbf{u}_0$ in $\mathcal{O}$ at which $\mathbf{F}(\mathbf{u}_0) = \mathbf{0}$, provided that the derivative matrix $\mathbf{DF}(\mathbf{u}_0)$ has *maximal rank*. When this is so, we select a $k \times k$ submatrix of $\mathbf{DF}(\mathbf{u}_0)$ that is invertible and has column indices $j_1, \ldots, j_k$. Then the components $u_{j_1}, \ldots, u_{j_k}$ of the solutions $\mathbf{u}$ of the equation

$$\mathbf{F}(\mathbf{u}) = \mathbf{0}, \quad \mathbf{u} \text{ in } \mathcal{O}$$

that lie in a neighborhood of the point $\mathbf{u}_0$ can be expressed as continuously differentiable functions of the remaining $n$ components.

**EXAMPLE 17.5**   Consider the system of equations

$$\begin{cases} \ln(1 + x^2 + t^2) + st + e^{s+z} - 1 = 0 \\ s^3 e^{\cos(x^2+z^2)} + s + 2z + (x + s + z)^4 = 0, \end{cases} \quad (s, x, t, z) \text{ in } \mathbb{R}^4. \qquad (17.26)$$

Observe that the point $(0, 0, 0, 0)$ is a solution of this system of equations. For a point $(s, x, t, z)$ in $\mathbb{R}^4$, define

$$\mathbf{F}(s, x, t, z) = (\ln(1 + x^2 + t^2) + st + e^{s+z} - 1, \, s^3 e^{\cos(x^2+z^2)}$$
$$+ s + 2z + (x + s + z)^4).$$

Then we can readily check that the mapping $\mathbf{F} \colon \mathbb{R}^4 \to \mathbb{R}^2$ is continuously differentiable and that its derivative matrix at the point $\mathbf{0} = (0, 0, 0, 0)$ is

$$\mathbf{DF}(\mathbf{0}) = \begin{bmatrix} 1 & 0 & 0 & 1 \\ 1 & 0 & 0 & 2 \end{bmatrix}.$$

Thus the $2 \times 2$ matrix

$$\begin{bmatrix} \partial F_1/\partial s(0,0,0,0) & \partial F_1/\partial z(0,0,0,0) \\ \partial F_2/\partial s(0,0,0,0) & \partial F_2/\partial z(0,0,0,0) \end{bmatrix} = \begin{bmatrix} 1 & 1 \\ 1 & 2 \end{bmatrix}$$

is invertible. We apply the Implicit Function Theorem to choose a positive number $r$ and continuously differentiable functions $g: \mathcal{N} \to \mathbb{R}$ and $h: \mathcal{N} \to \mathbb{R}$, where $\mathcal{N} = \mathcal{N}_r(0,0)$, such that if $x^2 + t^2 < r^2$, then $(g(x,t), x, t, h(x,t))$ is a solution of the system of equations (17.26). Moreover, if the point $(s, x, t, z)$ in $\mathbb{R}^4$ is a solution of the system (17.26), and if $s^2 + z^2 < r^2$ and $x^2 + t^2 < r^2$, then $s = g(x,t)$ and $z = h(x,t)$.  □

We will consider further examples of the Implicit Function Theorem in Section 17.3.

**EXERCISES**

For Exercises 1 through 6, use the Implicit Function Theorem to analyze the solutions of the given systems of equations near the solution **0**.

1. $\begin{cases} (x^2 + y^2 + z^2)^3 - x + z = 0 \\ \cos(x^2 + y^4) + e^z - 2 = 0 \end{cases}$   $(x, y, z)$ in $\mathbb{R}^3$.

2. $\begin{cases} a^3 + a^2 b + \sin(a+b+c) = 0 \\ \ln(1+a^2) + 2a + (bc)^4 = 0 \end{cases}$   $(a, b, c)$ in $\mathbb{R}^3$.

3. $\begin{cases} (uv)^4 + (u+s)^3 + t = 0 \\ \sin(uv) + e^{v+t^2} - 1 = 0 \end{cases}$   $(u, v, s, t)$ in $\mathbb{R}^4$.

4. $\begin{cases} x + 2y + x^2 + (yz)^2 + t^3 = 0 \\ -x + z + \sin(y^2 + z^2 + t^3) = 0 \end{cases}$   $(x, y, z, t)$ in $\mathbb{R}^4$.

5. $e^{x^2} + y^2 + z - 4xy^3 - 1 = 0$   $(x, y, z)$ in $\mathbb{R}^3$.
6. $e^{xy} + x^2 + 2y - 1 = 0$   $(x, y)$ in $\mathbb{R}^2$.
7. In the proof of the Implicit Function Theorem, it was asserted that the invertibility of the $k \times k$ matrix $\mathbf{D_y F(x_0, y_0)}$ implies the invertibility of the $(n+k) \times (n+k)$ matrix $\mathbf{DH(x_0, y_0)}$. Verify this assertion.
8. Consider the linear system of equations

$$\begin{cases} a_{11}x + a_{12}y + a_{13}z = 0 \\ a_{21}x + a_{22}y + a_{23}z = 0 \end{cases}$$   $(x, y, z)$ in $\mathbb{R}^3$.

Define $\boldsymbol{\eta}$ to be the vector $(a_{11}, a_{12}, a_{13})$ and $\boldsymbol{\beta}$ to be the vector $(a_{21}, a_{22}, a_{23})$.
(a) Show that if $\boldsymbol{\eta} \times \boldsymbol{\beta} \neq \mathbf{0}$, then the above system of equations defines two of the variables as a function of the remaining variable.
(b) Interpret (a) in the light of the geometry of lines and planes in $\mathbb{R}^3$.
9. Graph the solutions of the equation

$$y^3 - x^2 = 0, \quad (x, y) \text{ in } \mathbb{R}^2.$$

Does the Implicit Function Theorem apply at the point $(0,0)$? Does this equation define one of the components of a solution $(x, y)$ as a function of the other component?

## 17.3   Equations of Surfaces and Paths in $\mathbb{R}^3$

In this section we will obtain, in the cases of one scalar equation in three scalar unknowns and of two scalar equations in three scalar unknowns, more detailed geometric information about the graphs of the mappings that are implicitly defined by the solutions of such systems of equations.

**PROPOSITION 17.3**

Let $\mathcal{O}$ be an open subset of $\mathbb{R}^3$ and suppose that the function $f: \mathcal{O} \to \mathbb{R}$ is continuously differentiable. Assume that $(x_0,\ y_0,\ z_0)$ is a point in $\mathcal{O}$ at which

$$f(x_0,\ y_0,\ z_0) = 0 \quad \text{and} \quad \mathbf{D}f(x_0,\ y_0,\ z_0) \neq \mathbf{0}. \tag{17.27}$$

Then at the point $(x_0,\ y_0,\ z_0)$ the vector $\mathbf{D}f(x_0,\ y_0,\ z_0)$ is normal to the surface consisting of the solutions, in a neighborhood of the point $(x_0,\ y_0,\ z_0)$, of the equation

$$f(x,\ y,\ z) = 0, \quad (x,\ y,\ z) \text{ in } \mathcal{O}.$$

**FIGURE 17.5**

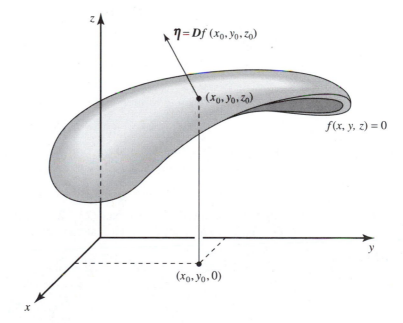

**Proof**   We can assume that it is the third component of $\mathbf{D}f(x_0,\ y_0,\ z_0)$ that is nonzero. The Implicit Function Theorem implies that there is a positive number $r$ and a continuously differentiable function $g: \mathcal{N} \to \mathbb{R}$, where $\mathcal{N} = \mathcal{N}_r(x_0,\ y_0)$, such that

$$f(x,\ y,\ g(x,\ y)) = 0 \quad \text{for all } (x,\ y) \text{ in } \mathcal{N}, \tag{17.28}$$

and moreover that in a neighborhood of the point $(x_0,\ y_0,\ z_0)$, all of the solutions belong to the graph of $g: \mathcal{N} \to \mathbb{R}$.

In Section 14.1, we showed that at the point $(x_0, y_0, z_0)$ the vector

$$\boldsymbol{\eta} = \left( -\frac{\partial g}{\partial x}(x_0, y_0), -\frac{\partial g}{\partial y}(x_0, y_0), 1 \right)$$

is normal to the surface defined by the graph of the function $g \colon \mathcal{N} \to \mathbb{R}$. On the other hand, if we differentiate (17.28), first with respect to $x$ and then with respect to $y$, we get the following two equations:

$$\frac{\partial f}{\partial x}(x_0, y_0, g(x_0, y_0)) + \frac{\partial f}{\partial z}(x_0, y_0, g(x_0, y_0))\frac{\partial g}{\partial x}(x_0, y_0) = 0$$

and

$$\frac{\partial f}{\partial y}(x_0, y_0, g(x_0, y_0)) + \frac{\partial f}{\partial z}(x_0, y_0, g(x_0, y_0))\frac{\partial g}{\partial y}(x_0, y_0) = 0,$$

to which we can adjoin the identity

$$\frac{\partial f}{\partial z}(x_0, y_0, g(x_0, y_0)) + \frac{\partial f}{\partial z}(x_0, y_0, g(x_0, y_0))(-1) = 0.$$

These last three equations can be written in vector form as

$$\mathbf{D}f(x_0, y_0, z_0) = \alpha\boldsymbol{\eta},$$

where $\alpha = \partial f/\partial z(x_0, y_0, z_0)$. Since $\alpha \neq 0$, this means that at the point $(x_0, y_0, z_0)$, the vector $\mathbf{D}f(x_0, y_0, z_0)$ is normal to the surface consisting of the solutions of the equation $f(x, y, z) = 0$ in a neighborhood of the point $(x_0, y_0, z_0)$. ∎

**EXAMPLE 17.6** The set of solutions of the equation

$$x^2 + y^2 + z^2 - 1 = 0, \quad (x, y, z) \text{ in } \mathbb{R}^3,$$

is the sphere of radius 1 centered at the origin. Define $f(x, y, z) = x^2 + y^2 + z^2 - 1$ for $(x, y, z)$ in $\mathbb{R}^3$ and observe that for a point $(x_0, y_0, z_0)$ on this sphere,

$$\mathbf{D}f(x_0, y_0, z_0) = (2x_0, 2y_0, 2z_0).$$

Proposition 17.3 implies that the vector $(2x_0, 2y_0, 2z_0)$ is normal to the sphere at the point $(x_0, y_0, z_0)$. This confirms what is already geometrically clear.

**FIGURE 17.6**

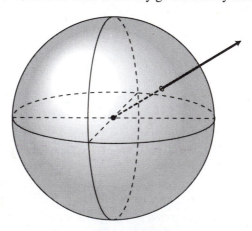

□

Let us now examine the set of solutions of a system of two equations in three unknowns. Suppose that $\mathcal{O}$ is an open subset of $\mathbb{R}^3$ and that the functions $g: \mathcal{O} \to \mathbb{R}$ and $h: \mathcal{O} \to \mathbb{R}$ are continuously differentiable. Consider the system of equations

$$\begin{cases} g(x,\ y,\ z) = 0 \\ h(x,\ y,\ z) = 0 \end{cases} \qquad (x,\ y,\ z) \text{ in } \mathcal{O}. \tag{17.29}$$

Suppose that the point $(x_0,\ y_0,\ z_0)$ is a solution of the system (17.29). Also assume that

$$\mathbf{D}g(x_0,\ y_0,\ z_0) \neq \mathbf{0} \quad \text{and} \quad \mathbf{D}h(x_0,\ y_0,\ z_0) \neq \mathbf{0}.$$

Define

$$S_1 = \{(x,\ y,\ z) \text{ in } \mathbb{R}^3 \mid g(x,\ y,\ z) = 0\}$$

and

$$S_2 = \{(x,\ y,\ z) \text{ in } \mathbb{R}^3 \mid h(x,\ y,\ z) = 0\},$$

so that the set of solutions of the system of equations (17.29) consists of the intersection of the sets $S_1$ and $S_2$. Because of Proposition 17.3, in a neighborhood of the point $(x_0,\ y_0,\ z_0)$, $S_1$ is a surface having $\mathbf{D}g(x_0,\ y_0,\ z_0)$ as a normal at $(x_0,\ y_0,\ z_0)$; $S_2$ is a surface having $\mathbf{D}h(x_0,\ y_0,\ z_0)$ as a normal at $(x_0,\ y_0,\ z_0)$. *If these two normals are not parallel, these surfaces should intersect in a path having* $\mathbf{D}g(x_0,\ y_0,\ z_0) \times \mathbf{D}h(x_0,\ y_0,\ z_0)$ *as a tangent vector at the point* $(x_0,\ y_0,\ z_0)$. These normals are nonparallel precisely when

$$\mathbf{D}h(x_0,\ y_0,\ z_0) \times \mathbf{D}g(x_0,\ y_0,\ z_0) \neq \mathbf{0}.$$

The following proposition justifies the above geometric argument:

**PROPOSITION 17.4** Let $\mathcal{O}$ be an open subset of $\mathbb{R}^3$ and suppose that the functions $g: \mathcal{O} \to \mathbb{R}$ and $h: \mathcal{O} \to \mathbb{R}$ are continuously differentiable. Let $(x_0,\ y_0,\ z_0)$ be a point in $\mathcal{O}$ at which

$$g(x_0,\ y_0,\ z_0) = h(x_0,\ y_0,\ z_0) = 0$$

and

$$\mathbf{D}g(x_0,\ y_0,\ z_0) \times \mathbf{D}h(x_0,\ y_0,\ z_0) \neq \mathbf{0}. \tag{17.30}$$

Then there is a neighborhood of the point $(x_0,\ y_0,\ z_0)$ in which the set of solutions of the system

$$\begin{cases} g(x,\ y,\ z) = 0 \\ h(x,\ y,\ z) = 0 \end{cases} \qquad (x,\ y,\ z) \text{ in } \mathcal{O} \tag{17.31}$$

consists of a *path* that at the point $(x_0,\ y_0,\ z_0)$ has the vector $\mathbf{D}g(x_0,\ y_0,\ z_0) \times \mathbf{D}h(x_0,\ y_0,\ z_0)$ as a tangent vector.

**FIGURE 17.7**

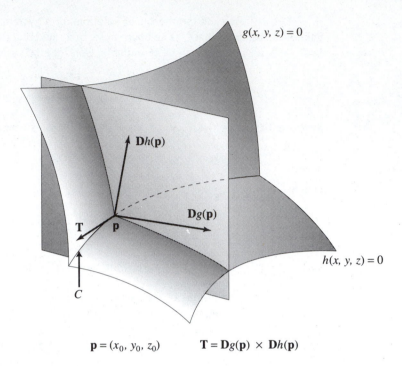

$g(x, y, z) = 0$

**D**$h$(**p**)

**D**$g$(**p**)

$h(x, y, z) = 0$

T   **p**

$C$

$$\mathbf{p} = (x_0, y_0, z_0) \qquad \mathbf{T} = \mathbf{D}g(\mathbf{p}) \times \mathbf{D}h(\mathbf{p})$$

**Proof**   Since the vector $\mathbf{D}g(x_0, y_0, z_0) \times \mathbf{D}h(x_0, y_0, z_0)$ is nonzero, at least one of its components is nonzero. We will suppose that the first component is nonzero; that is,

$$\frac{\partial g}{\partial y}(x_0, y_0, z_0)\frac{\partial h}{\partial z}(x_0, y_0, z_0) - \frac{\partial g}{\partial z}(x_0, y_0, z_0)\frac{\partial h}{\partial y}(x_0, y_0, z_0) \neq 0. \qquad (17.32)$$

Observe that (17.32) means that the $2 \times 2$ matrix

$$\begin{bmatrix} \partial g/\partial y(x_0, y_0, z_0) & \partial g/\partial z(x_0, y_0, z_0) \\ \partial h/\partial y(x_0, y_0, z_0) & \partial h/\partial z(x_0, y_0, z_0) \end{bmatrix}$$

is invertible. Thus we may apply the Implicit Function Theorem at the point $(x_0, y_0, z_0)$ to the continuously differentiable mapping $\mathbf{F}: \mathcal{O} \to \mathbb{R}^2$ defined by

$$\mathbf{F}(x, y, z) = (g(x, y, z), h(x, y, z)) \quad \text{for } (x, y, z) \text{ in } \mathcal{O}.$$

It follows that there is a symmetric neighborhood $I$ of the point $x_0$ and two continuously differentiable functions

$$\alpha: I \to \mathbb{R} \quad \text{and} \quad \beta: I \to \mathbb{R}$$

such that          $\mathbf{F}(x, \alpha(x), \beta(x)) = \mathbf{0} \quad \text{for all } x \text{ in } I,$

and in a neighborhood of the point $(x_0, y_0, z_0)$, all of the solutions of the system are of the form $(x, \alpha(x), \beta(x))$ for $x$ in $I$. Consequently, near $(x_0, y_0, z_0)$, the set of solutions of the system (17.31) coincides with the image of the parametrized path $\gamma: I \to \mathbb{R}^3$ defined by

$$\gamma(t) = (t, \alpha(t), \beta(t)) \quad \text{for } t \text{ in } I.$$

The tangent vector to this path at $t = x_0$ is

$$\mathbf{T} = \gamma'(x_0) = (1, \alpha'(x_0), \beta'(x_0)).$$

Now since $g(t, \alpha(t), \beta(t)) = 0$ and $h(t, \alpha(t), \beta(t)) = 0$ for all $t$ in $I$, we can differentiate each of these identities to obtain

$$\langle \mathbf{D}g(x_0, y_0, z_0), \mathbf{T} \rangle = 0$$

and

$$\langle \mathbf{D}h(x_0, y_0, z_0), \mathbf{T} \rangle = 0.$$

Hence $\mathbf{T}$ is orthogonal to both $\mathbf{D}g(x_0, y_0, z_0)$ and $\mathbf{D}h(x_0, y_0, z_0)$, so

$$\mathbf{T} = \alpha(\mathbf{D}g(x_0, y_0, z_0) \times \mathbf{D}h(x_0, y_0, z_0))$$

for some $\alpha \neq 0$. Thus $\mathbf{D}g(x_0, y_0, z_0) \times \mathbf{D}h(x_0, y_0, z_0)$ is tangent to the path of solutions of (17.31) at $(x_0, y_0, z_0)$. ∎

**EXAMPLE 17.7**   The set of solutions of the system of equations

$$\begin{cases} x - y = 0 \\ x^2 + y^2 + z^2 - 2 = 0 \end{cases} \quad (x, y, z) \text{ in } \mathbb{R}^3$$

consists of the intersection of a sphere and a plane. The point $(1, 1, 0)$ is a solution. By the preceding proposition and a brief computation of partial derivatives, we conclude that the vector $(0, 0, 4)$ is tangent at the point $(1, 1, 0)$ to the path of solutions of this system. □

---

**EXERCISES**

1.  Consider the system of equations

$$\begin{cases} (x - 1)^2 + (y - 1)^2 + z^2 - 2 = 0 \\ (x + 1)^2 + (y - 2)^2 + z^2 - 5 = 0 \end{cases} \quad (x, y, z) \text{ in } \mathbb{R}^3.$$

    (a)  Apply Proposition 17.4 at the point $(0, 0, 0)$ to find the tangent line to the path defined by the solutions of this system near $(0, 0, 0)$.
    (b)  Solve this system of equations near $(0, 0, 0)$ and explicitly check the tangent to the path of solutions.
2.  Explicitly find the circle that is the set of solutions of the system in Example 17.7, and directly find the tangent to this circle at the point $(1, 1, 0)$.
3.  Consider the system of equations

$$\begin{cases} (x - 1)^2 + y^2 + z^2 - 1 = 0 \\ (x - 1/4)^2 + y^2 + z^2 - 1/16 = 0 \end{cases} \quad (x, y, z) \text{ in } \mathbb{R}^3.$$

    (a)  Show that assumption (17.30) of Proposition 17.4 is not satisfied at the point $(0, 0, 0)$.
    (b)  By graphing each of the surfaces defined by the individual equations in the system, explain why there is exactly one solution of this system.

4. Construct examples demonstrating that when $1 \leq i \leq 3$ and the $i$th component of the vector $\mathbf{D}g(x_0, y_0, z_0) \times \mathbf{D}h(x_0, y_0, z_0)$ is zero, then it may be that the path of solutions of (17.31) passing through $(x_0, y_0, z_0)$ cannot be parametrized by the $i$th component.

5. Suppose that the functions $f: \mathbb{R}^3 \to \mathbb{R}$, $g: \mathbb{R}^3 \to \mathbb{R}$, and $h: \mathbb{R}^3 \to \mathbb{R}$ are continuously differentiable, and let $(x_0, y_0, z_0)$ be a point in $\mathbb{R}^3$ at which

$$f(x_0, y_0, z_0) = g(x_0, y_0, z_0) = h(x_0, y_0, z_0) = 0$$

and 
$$\langle \mathbf{D}f(x_0, y_0, z_0), \mathbf{D}g(x_0, y_0, z_0) \times \mathbf{D}h(x_0, y_0, z_0) \rangle \neq 0.$$

By considering the set of solutions of this system as consisting of the intersection of a surface with a path, explain why in a neighborhood of the point $(x_0, y_0, z_0)$, the system of equations

$$f(x, y, z) = 0$$
$$g(x, y, z) = 0 \qquad (x, y, z) \text{ in } \mathbb{R}^3$$
$$h(x, y, z) = 0$$

has exactly one solution. Also explain this by using the Inverse Function Theorem.

## 17.4 Constrained Extrema Problems and Lagrange Multipliers

A point in the domain of a real-valued function is called an *extremum*, or an *extreme point*, for the function if the function attains either a maximum or a minimum value at that point. If $\mathcal{O}$ is an open subset of $\mathbb{R}^n$ and the function $f: \mathcal{O} \to \mathbb{R}$ has first-order partial derivatives, then for the point $\mathbf{x}$ in $\mathcal{O}$ to be an extreme point for the function $f: \mathcal{O} \to \mathbb{R}$, it is necessary that

$$\mathbf{D}f(\mathbf{x}) = \mathbf{0}. \qquad (17.33)$$

The Second Derivative Test, established in Section 14.3, prescribes sufficient conditions for such a point $\mathbf{x}$ to be an extreme point for the function $f: \mathcal{O} \to \mathbb{R}$. There is no assertion regarding the *existence* of extrema for a real-valued function defined on an open subset of $\mathbb{R}^n$. However, in Section 11.2, we proved that a continuous function defined on a *compact* set attains a maximum and a minimum value.

Now suppose that $\mathcal{O}$ is an open subset of $\mathbb{R}^n$, that the function $f: \mathcal{O} \to \mathbb{R}$ has first-order partial derivatives, and that $K$ is a subset of $\mathcal{O}$. Let $\mathbf{x}$ be a point in $K$ that is an extreme point for the restricted function $f: K \to \mathbb{R}$. Such an extreme point $\mathbf{x}$ is called a *constrained extremum* for the function $f: \mathcal{O} \to \mathbb{R}$, because the values of the function $f: \mathcal{O} \to \mathbb{R}$ with which we can compare $f(\mathbf{x})$ are constrained to functional values of points in $K$; the set $K$ is called the *constraint set*. We note that if $K$ is compact, then constrained extrema do, in fact, exist.

It is not true that at a constrained extremum all of the partial derivatives are zero. The conditions that are necessarily satisfied by the partial derivatives at a constrained extremum depend on the nature of the constraint set. We shall devote this section to describing what these conditions are. Let us begin with two examples.

**EXAMPLE 17.8**   Define the function $f\colon \mathbb{R}^3 \to \mathbb{R}$ by $f(x, y, z) = z$ for all $(x, y, z)$ in $\mathbb{R}^3$. Choose as the constraint set the sphere $K = \{(x, y, z) \text{ in } \mathbb{R}^3 \mid x^2 + y^2 + z^2 = 1\}$. Then clearly the point $(0, 0, 1)$ is a maximizer for $f\colon K \to \mathbb{R}$. However, we have

$$\frac{\partial f}{\partial x}(0, 0, 1) = 0, \frac{\partial f}{\partial y}(0, 0, 1) = 0 \quad \text{and} \quad \frac{\partial f}{\partial z}(0, 0, 1) = 1, \tag{17.34}$$

so (17.33) is not satisfied at this constrained extremum, since $\partial f/\partial z(0, 0, 1) \neq 0$.   □

**EXAMPLE 17.9**   Define the function $f(x, y, z) = (x - 4y + 3z) + z^2$ for $(x, y, z)$ in $\mathbb{R}^3$. Choose as the constraint set the line $K = \{(x, y, z) \text{ in } \mathbb{R}^3 \mid x = y = z\}$. Then clearly the point $(0, 0, 0)$ is a minimizer for the function $f\colon K \to \mathbb{R}$. We have

$$\frac{\partial f}{\partial x}(0, 0, 0) = 1, \frac{\partial f}{\partial y}(0, 0, 0) = -4, \quad \text{and} \quad \frac{\partial f}{\partial z}(0, 0, 0) = 3, \tag{17.35}$$

so (17.33) is not satisfied at this constrained extremum, since in fact none of the components of $\mathbf{D}f(0, 0, 0)$ is zero.   □

In the first example, the constraint set is a surface in $\mathbb{R}^3$; in the second example, the constraint set is a path in $\mathbb{R}^3$. We will now turn to describing the conditions that must be satisfied by the partial derivatives of a function at a constrained extremum. First we consider the case when the constraint set $K$ is a surface in $\mathbb{R}^3$. Then we will consider the case when the constraint set $K$ is a path in $\mathbb{R}^3$. With these two cases understood, we will then be in a position to appreciate the significance of a general theorem on constrained extrema.

**THEOREM 17.5**   Let $\mathcal{O}$ be an open subset of $\mathbb{R}^3$, and suppose that the functions $f\colon \mathcal{O} \to \mathbb{R}$ and $g\colon \mathcal{O} \to \mathbb{R}$ are continuously differentiable. Define

$$S = \{(x, y, z) \text{ in } \mathcal{O} \mid g(x, y, z) = 0\}.$$

Suppose that the point $(x_0, y_0, z_0)$ in $S$ is an extreme point for the function

$$f\colon S \to \mathbb{R},$$

and that $$\mathbf{D}g(x_0, y_0, z_0) \neq \mathbf{0}. \tag{17.36}$$

Then there is a number $\lambda$ such that

$$\mathbf{D}f(x_0, y_0, z_0) = \lambda \mathbf{D}g(x_0, y_0, z_0). \tag{17.37}$$

**Proof**   Assumption (17.36) means that at least one of the components of the derivative vector $\mathbf{D}g(x_0, y_0, z_0)$ is nonzero. We shall suppose it is the third component; that is,

$$\frac{\partial g}{\partial z}(x_0, y_0, z_0) \neq 0.$$

According to the Implicit Function Theorem, there is a neighborhood $\mathcal{N}$ of the point $(x_0, y_0)$ in $\mathbb{R}^2$ and a continuously differentiable function $\phi: \mathcal{N} \to \mathbb{R}$ such that $\phi(x_0, y_0) = z_0$ and

$$g(x, y, \phi(x, y)) = 0 \quad \text{for all } (x, y) \text{ in } \mathcal{N}. \tag{17.38}$$

Thus the graph of the function $\phi: \mathcal{N} \to \mathbb{R}$ lies in the constraint set $S$.

Define an auxiliary function $\psi: \mathcal{N} \to \mathbb{R}$ by

$$\psi(x, y) = f(x, y, \phi(x, y)) \quad \text{for all } (x, y) \text{ in } \mathcal{N}. \tag{17.39}$$

Since the graph of the function $\phi: \mathcal{N} \to \mathbb{R}$ lies in the constraint set $S$, it follows that the point $(x_0, y_0)$ is an extreme point of the function $\psi: \mathcal{N} \to \mathbb{R}$. Thus $(x_0, y_0)$ is an *unconstrained extremum* of the function of two variables $\psi: \mathcal{N} \to \mathbb{R}$, so

$$\begin{aligned} \frac{\partial \psi}{\partial x}(x_0, y_0) &= 0 \\ \frac{\partial \psi}{\partial y}(x_0, y_0) &= 0. \end{aligned} \tag{17.40}$$

Using the Chain Rule to express the partial derivatives of the function $\psi: \mathcal{N} \to \mathbb{R}$ in terms of the partial derivatives of the function $f: \mathcal{O} \to \mathbb{R}$, we rewrite the equations (17.40) as

$$\begin{aligned} \frac{\partial f}{\partial x}(x_0, y_0, z_0) + \frac{\partial f}{\partial z}(x_0, y_0, z_0)\frac{\partial \phi}{\partial x}(x_0, y_0) &= 0 \\ \frac{\partial f}{\partial y}(x_0, y_0, z_0) + \frac{\partial f}{\partial z}(x_0, y_0, z_0)\frac{\partial \phi}{\partial y}(x_0, y_0) &= 0. \end{aligned} \tag{17.41}$$

On the other hand, by differentiating the identity (17.38), first with respect to $x$ and then with respect to $y$, we obtain

$$\begin{aligned} \frac{\partial g}{\partial x}(x_0, y_0, z_0) + \frac{\partial g}{\partial z}(x_0, y_0, z_0)\frac{\partial \phi}{\partial x}(x_0, y_0) &= 0 \\ \frac{\partial g}{\partial y}(x_0, y_0, z_0) + \frac{\partial g}{\partial z}(x_0, y_0, z_0)\frac{\partial \phi}{\partial y}(x_0, y_0) &= 0. \end{aligned} \tag{17.42}$$

Now define

$$\lambda = \frac{\partial f}{\partial z}(x_0, y_0, z_0) \Big/ \frac{\partial g}{\partial z}(x_0, y_0, z_0). \tag{17.43}$$

From the first equations in (17.41) and (17.42), respectively, we obtain

$$\frac{\partial f}{\partial x}(x_0, y_0, z_0) = \lambda \frac{\partial g}{\partial x}(x_0, y_0, z_0); \tag{17.44}$$

from the second equations in (17.41) and (17.42), respectively, we obtain

$$\frac{\partial f}{\partial y}(x_0, y_0, z_0) = \lambda \frac{\partial g}{\partial y}(x_0, y_0, z_0). \tag{17.45}$$

Finally, equations (17.43), (17.44), and (17.45) can be written in vector form as

$$\mathbf{D}f(x_0, y_0, z_0) = \lambda \mathbf{D}g(x_0, y_0, z_0). \qquad\blacksquare$$

Returning to the extremum problem in Example 17.7, if for a point $(x, y, z)$ in $\mathbb{R}^3$ we define

$$f(x, y, z) = z \quad \text{and} \quad g(x, y, z) = x^2 + y^2 + z^2 - 1,$$

at the constrained extremum $(0, 0, 1)$,

$$\mathbf{D}g(0, 0, 1) = (0, 0, 2) \quad \text{and} \quad \mathbf{D}f(0, 0, 1) = (0, 0, 1),$$

so with $\lambda = 1/2$,

$$\mathbf{D}f(0, 0, 1) = \lambda \mathbf{D}g(0, 0, 1).$$

**EXAMPLE 17.10**   Suppose we wish to find the minimum value of $x + y + 2z$ on the set

$$K = \{(x, y, z) \text{ in } \mathbb{R}^3 \mid x^2 + y^2 + z^2 \le 1\}.$$

Define $\qquad f(x, y, z) = x + y + 2z \quad$ for $(x, y, z)$ in $\mathbb{R}^3$,

and observe that since $K$ is compact, there is a minimizer $(x_0, y_0, z_0)$ for the function $f : K \to \mathbb{R}$. This minimizer cannot lie in the interior of $K$, since if it did, we would have

$$\mathbf{D}f(x_0, y_0, z_0) = (0, 0, 0).$$

But $\mathbf{D}f(x_0, y_0, z_0) = (1, 1, 2)$. Thus the point $(x_0, y_0, z_0)$ lies on the boundary of $K$; that is, $x_0^2 + y_0^2 + z_0^2 = 1$. If we define $S = \{(x, y, z) \text{ in } \mathbb{R}^3 \mid g(x, y, z) = x^2 + y^2 + z^2 - 1 = 0\}$, the point $(x_0, y_0, z_0)$ is a minimizer of the function $f : S \to \mathbb{R}$. We can apply Theorem 17.5 to assert that there is a number $\lambda$ such that

$$\mathbf{D}f(x_0, y_0, z_0) = \lambda \mathbf{D}g(x_0, y_0, z_0);$$

that is, $(1, 1, 2) = \lambda(2x_0, 2y_0, 2z_0)$. Thus $2x_0 = 2y_0 = z_0$, and since $x_0^2 + y_0^2 + z_0^2 = 1$, the minimum occurs at the point $-(1/\sqrt{6}, 1/\sqrt{6}, \sqrt{2}/\sqrt{3})$. $\qquad\square$

Theorem 17.5 provides necessary conditions for a point to be a constrained extremum of a function of three variables that is constrained to a *surface*. The next theorem gives sufficient conditions in the case when the constraint set is a *path*.

**THEOREM 17.6**   Let $\mathcal{O}$ be an open subset of $\mathbb{R}^3$ and suppose that the functions $f\colon \mathcal{O} \to \mathbb{R}$, $g\colon \mathcal{O} \to \mathbb{R}$, and $h\colon \mathcal{O} \to \mathbb{R}$ are continuously differentiable. Define

$$C = \{(x,\ y,\ z) \text{ in } \mathcal{O} \mid g(x,\ y,\ z) = h(x,\ y,\ z) = 0\}.$$

Suppose that the point $(x_0,\ y_0,\ z_0)$ in $C$ is an extreme point for the function

$$f\colon C \to \mathbb{R}$$

and that

$$\mathbf{D}g(x_0,\ y_0,\ z_0) \times \mathbf{D}h(x_0,\ y_0,\ z_0) \neq \mathbf{0}. \tag{17.46}$$

Then there are numbers $\lambda$ and $\mu$ such that

$$\mathbf{D}f(x_0,\ y_0,\ z_0) = \lambda \mathbf{D}g(x_0,\ y_0,\ z_0) + \mu \mathbf{D}h(x_0,\ y_0,\ z_0). \tag{17.47}$$

**Proof**   Since the vector $\mathbf{D}g(x_0,\ y_0,\ z_0) \times \mathbf{D}h(x_0,\ y_0,\ z_0)$ is nonzero, at least one of its components is nonzero. We will suppose it is the first component; that is,

$$\frac{\partial g}{\partial y}(x_0,\ y_0)\frac{\partial h}{\partial z}(x_0,\ y_0) - \frac{\partial g}{\partial z}(x_0,\ y_0)\frac{\partial h}{\partial y}(x_0,\ y_0) \neq 0.$$

The Implicit Function Theorem, in the form proved as Proposition 17.4, asserts that there is a neighborhood $I$ of the point $x_0$ in $\mathbb{R}$, and continuously differentiable functions $\alpha\colon I \to \mathbb{R}$ and $\beta\colon I \to \mathbb{R}$, such that $\alpha(x_0) = y_0$, $\beta(x_0) = z_0$, and

$$\begin{cases} g(x,\ \alpha(x),\ \beta(x)) = 0 \\ h(x,\ \alpha(x),\ \beta(x)) = 0 \end{cases} \qquad \text{for all } x \text{ in } I. \tag{17.48}$$

This means that if we define the parametrized path $\gamma\colon I \to \mathbb{R}$ by

$$\gamma(t) = (t,\ \alpha(t),\ \beta(t)) \quad \text{for } t \text{ in } I,$$

then the image of this parametrized path lies in the constraint set $C$. Hence if we define the auxiliary function $\psi\colon I \to \mathbb{R}$ by

$$\psi(t) = (f \circ \gamma)(t) = f(t,\ \alpha(t),\ \beta(t)) \quad \text{for } t \text{ in } I,$$

then this function $\psi\colon I \to \mathbb{R}$ attains an extreme value at the point $x_0$. But then the point $x_0$ is an *unconstrained extremum* of the function of a single variable $\psi\colon I \to \mathbb{R}$, so

$$\psi'(x_0) = 0,$$

which, because of the Chain Rule, means that

$$\psi'(x_0) = \langle \mathbf{D}f(x_0,\ y_0,\ z_0), \gamma'(x_0)\rangle = 0. \tag{17.49}$$

According to Proposition 17.4, the vector $\mathbf{D}g(x_0,\ y_0,\ z_0) \times \mathbf{D}h(x_0,\ y_0,\ z_0)$ is a nonzero multiple of the tangent vector $\gamma'(x_0)$. Hence (17.49) implies that

$$\langle \mathbf{D}f(x_0,\ y_0,\ z_0), \mathbf{D}g(x_0,\ y_0,\ z_0) \times \mathbf{D}h(x_0,\ y_0,\ z_0)\rangle = 0; \tag{17.50}$$

that is, the vector $\mathbf{D}f(x_0,\ y_0,\ z_0)$ is perpendicular to the vector cross-product $\mathbf{D}g(x_0,\ y_0,\ z_0) \times \mathbf{D}h(x_0,\ y_0,\ z_0)$. But the only vectors that are perpendicular to this nonzero cross-product are vectors that are linear combinations of the vectors $\mathbf{D}g(x_0,\ y_0,\ z_0)$ and $\mathbf{D}h(x_0,\ y_0,\ z_0)$. This means that (17.47) holds. ∎

Observe that the constrained extremum problem in Example 17.8 is of the form described by the preceding theorem. Indeed, for a point $(x, y, z)$ in $\mathbb{R}^3$, define

$$f(x, y, z) = x - 4y + 3z + z^2, \; g(x, y, z) = x - y, \; \text{and} \; h(x, y, z) = y - z.$$

Then the constraint set is given by

$$K = \{(x, y, z) \text{ in } \mathbb{R}^3 \mid x = y = z\} = \{(x, y, z) \text{ in } \mathbb{R}^3 \mid g(x, y, z) = h(x, y, z) = 0\}.$$

At the point $(0, 0, 0)$, which is a minimizer for $f \colon K \to \mathbb{R}$, we have

$$\mathbf{D}f(0, 0, 0) = (1, -4, 3), \quad \mathbf{D}g(0, 0, 0) = (1, -1, 0),$$
$$\text{and} \quad \mathbf{D}h(0, 0, 0) = (0, 1, -1),$$

so, setting $\lambda = 1$ and $\mu = -3$, we have

$$\mathbf{D}f(0, 0, 0) = \lambda \mathbf{D}g(0, 0, 0) + \mu \mathbf{D}h(0, 0, 0).$$

We arrived at our formulations of Theorems 17.5 and 17.6 aided by geometric reasoning about paths and surfaces in $\mathbb{R}^3$. The following is the general constrained extremum result, which includes these two results as particular cases.

**THEOREM 17.7**     **The General Lagrange Multiplier Theorem**   Let $\mathcal{O}$ be an open subset of $\mathbb{R}^n$, and suppose that the function $f \colon \mathcal{O} \to \mathbb{R}$ is continuously differentiable. Let $k$ be a positive integer less than $n$ and suppose that the mapping $\mathbf{G} \colon \mathcal{O} \to \mathbb{R}^k$ is also continuously differentiable. Define

$$S = \{\mathbf{x} \text{ in } \mathcal{O} \mid \mathbf{G}(\mathbf{x}) = \mathbf{0}\}.$$

Suppose that the point $\mathbf{u}$ in $S$ is an extreme point for the function

$$f \colon S \to \mathbb{R},$$

and that the $k \times n$ matrix

$$\mathbf{D}\mathbf{G}(\mathbf{u}) \text{ has maximal rank.} \tag{17.51}$$

Then there are $k$ numbers $\lambda_1, \ldots, \lambda_k$ such that

$$\mathbf{D}f(\mathbf{u}) = \sum_{i=1}^{k} \lambda_i \mathbf{D}G_i(\mathbf{u}). \tag{17.52}$$

**Proof**   Let $n = m + k$ and write points in $\mathbb{R}^n$ as $(\mathbf{x}, \mathbf{y})$, where $\mathbf{x}$ is in $\mathbb{R}^m$ and $\mathbf{y}$ is in $\mathbb{R}^k$. Since, by assumption, the $k \times n$ matrix $\mathbf{D}\mathbf{G}(\mathbf{u})$ has rank $k$, we can, by relabeling components if necessary, suppose that at the point $\mathbf{u} = (\mathbf{x}_0, \mathbf{y}_0)$ the $k \times k$ matrix

$$\mathbf{D}_\mathbf{y}\mathbf{G}(\mathbf{x}_0, \mathbf{y}_0) \text{ is invertible.} \tag{17.53}$$

According to the Implicit Function Theorem, there is a neighborhood $\mathcal{N}$ of the point $\mathbf{x}_0$ in $\mathbb{R}^m$ and a mapping $\psi \colon \mathcal{N} \to \mathbb{R}^k$ such that $\mathbf{y}_0 = \psi(\mathbf{x}_0)$ and

$$\mathbf{G}(\mathbf{x}, \psi(\mathbf{x})) = \mathbf{0} \quad \text{for } \mathbf{x} \text{ in } \mathcal{N}. \tag{17.54}$$

Thus the graph of the mapping $\psi \colon \mathcal{N} \to \mathbb{R}^k$ lies in the constraint set $S$.

Define an auxiliary function $\eta: \mathcal{N} \to \mathbb{R}$ by

$$\eta(\mathbf{x}) = f(\mathbf{x}, \boldsymbol{\psi}(\mathbf{x})) \quad \text{for } \mathbf{x} \text{ in } \mathcal{N}.$$

Then $\mathbf{x}_0$ is an *unconstrained extremum* of the function $\eta: \mathcal{N} \to \mathbb{R}$, so

$$\mathbf{D}\eta(\mathbf{x}_0) = \mathbf{0}. \tag{17.55}$$

Hence, from the definition of the function $\eta: \mathcal{N} \to \mathbb{R}$ and the Chain Rule, we have

$$\mathbf{D_x}f(\mathbf{x}_0, \mathbf{y}_0) + \mathbf{D_y}f(\mathbf{x}_0, \mathbf{y}_0)\mathbf{D}\boldsymbol{\psi}(\mathbf{x}_0) = \mathbf{0}. \tag{17.56}$$

On the other hand, differentiating the identity (17.54), it follows that

$$\mathbf{D_x}G(\mathbf{x}_0, \mathbf{y}_0) + \mathbf{D_y}G(\mathbf{x}_0, \mathbf{y}_0)\mathbf{D}\boldsymbol{\psi}(\mathbf{x}_0) = \mathbf{0}. \tag{17.57}$$

But the $k \times k$ matrix $\mathbf{D_y}G(\mathbf{x}_0, \mathbf{y}_0)$ is invertible, so (17.57) yields

$$-[\mathbf{D_y}G(\mathbf{x}_0, \mathbf{y}_0)]^{-1}\mathbf{D_x}G(\mathbf{x}_0, \mathbf{y}_0) = \mathbf{D}\boldsymbol{\psi}(\mathbf{x}_0),$$

which, when substituted in (17.56), yields

$$\mathbf{D_x}f(\mathbf{x}_0, \mathbf{y}_0) = \mathbf{D_y}f(\mathbf{x}_0, \mathbf{y}_0)[\mathbf{D_y}G(\mathbf{x}_0, \mathbf{y}_0)]^{-1}\mathbf{D_x}G(\mathbf{x}_0, \mathbf{y}_0). \tag{17.58}$$

To verify (17.52), first observe that if we define the $1 \times k$ matrix $\boldsymbol{\lambda}$ by $\boldsymbol{\lambda} = [\lambda_1, \ldots, \lambda_k]$, then (17.52) is equivalent to the matrix identity

$$\mathbf{D}f(\mathbf{x}_0, \mathbf{y}_0) = \boldsymbol{\lambda}\mathbf{D}G(\mathbf{x}_0, \mathbf{y}_0),$$

which in turn may be written in components as

$$\begin{aligned} \mathbf{D_x}f(\mathbf{x}_0, \mathbf{y}_0) &= \boldsymbol{\lambda}\mathbf{D_x}G(\mathbf{x}_0, \mathbf{y}_0) \\ \mathbf{D_y}f(\mathbf{x}_0, \mathbf{y}_0) &= \boldsymbol{\lambda}\mathbf{D_y}G(\mathbf{x}_0, \mathbf{y}_0). \end{aligned} \tag{17.59}$$

Define the $1 \times k$ matrix $\boldsymbol{\lambda}$ by

$$\boldsymbol{\lambda} = \mathbf{D_y}f(\mathbf{x}_0, \mathbf{y}_0)[\mathbf{D_y}G(\mathbf{x}_0, \mathbf{y}_0)]^{-1}. \tag{17.60}$$

This definition ensures that the second equation in the system (17.59) is satisfied. On the other hand, equation (17.58) is the assertion that this choice of $\boldsymbol{\lambda}$ also satisfies the first equation in the system (17.59). ∎

The $\lambda_i$'s in formula (17.52) are often referred to as *Lagrange multipliers*.

We have the following immediate corollary of the General Lagrange Multiplier Theorem:

**COROLLARY
17.8**

Let $\mathcal{O}$ be an open subset of $\mathbb{R}^n$ and suppose that the functions $f: \mathcal{O} \to \mathbb{R}$ and $g: \mathcal{O} \to \mathbb{R}$ are continuously differentiable. Define

$$S = \{\mathbf{x} \text{ in } \mathcal{O} \mid g(\mathbf{x}) = 0\}.$$

Suppose that the point $\mathbf{u}$ in $S$ is an extreme point for the function $f: S \to \mathbb{R}$ and that the derivative vector

$$\mathbf{D}g(\mathbf{u}) \text{ is nonzero.}$$

Then there is a number $\lambda$ such that

$$\mathbf{D}f(\mathbf{u}) = \lambda \mathbf{D}g(\mathbf{u}). \tag{17.61}$$

The above corollary has the following interesting application to the question about the existence of eigenvalues of matrices. Recall that for an $n \times n$ matrix $\mathbf{A}$ of real numbers, the real number $\lambda$ is called an *eigenvalue* of $\mathbf{A}$ provided that there is some nonzero point $\mathbf{x}$ in $\mathbb{R}^n$ such that

$$\mathbf{A}\mathbf{x} = \lambda\mathbf{x}.$$

In general, there may be no such real eigenvalues (see Exercise 10). Recall that a matrix $\mathbf{A}$ is called *symmetric* if $\mathbf{A} = \mathbf{A}^{\mathrm{T}}$. For symmetric matrices, we have the following result:

**PROPOSITION
17.9**

Every symmetric matrix has a real eigenvalue.

**Proof**   Let $\mathbf{A}$ be an $n \times n$ symmetric matrix. Define the functions $f: \mathbb{R}^n \to \mathbb{R}$ and $g: \mathbb{R}^n \to \mathbb{R}$ by

$$g(\mathbf{x}) = \langle \mathbf{x}, \mathbf{x} \rangle - 1 \quad \text{and} \quad f(\mathbf{x}) = \langle \mathbf{A}\mathbf{x}, \mathbf{x} \rangle \quad \text{for } \mathbf{x} \text{ in } \mathbb{R}^n,$$

and then define          $S = \{\mathbf{x} \text{ in } \mathbb{R}^n \mid g(\mathbf{x}) = 0\}.$

The set $S$ consists of all points in $\mathbb{R}^n$ of norm 1, so it is closed and bounded, and hence is compact. Therefore, since the function $f: \mathbb{R}^n \to \mathbb{R}$ is continuous, it follows from the Extreme Value Theorem that the function $f: S \to \mathbb{R}$ assumes a minimum value at a point $\mathbf{x}$ in $S$. From the preceding proposition, we conclude that there is a number $\lambda$ such that

$$\mathbf{D}f(\mathbf{x}) = \lambda \mathbf{D}g(\mathbf{x}). \tag{17.62}$$

However, we can explicitly compute $\mathbf{D}f(\mathbf{x})$ and $\mathbf{D}g(\mathbf{x})$. Indeed, for an index $i$ with $1 \le i \le n$, we have

$$
\begin{aligned}
\frac{\partial f}{\partial x_i}(\mathbf{x}) &= \lim_{t \to 0} \frac{f(\mathbf{x} + t\mathbf{e}_i) - f(\mathbf{x})}{t} \\
&= \lim_{t \to 0} \frac{\langle \mathbf{A}(\mathbf{x} + t\mathbf{e}_i), \ \mathbf{x} + t\mathbf{e}_i \rangle - \langle \mathbf{Ax}, \ \mathbf{x} \rangle}{t} \\
&= \lim_{t \to 0} \frac{t\langle \mathbf{Ax}, \ \mathbf{e}_i \rangle + t\langle \mathbf{Ae}_i, \ \mathbf{x} \rangle + t^2 \langle \mathbf{Ae}_i, \ \mathbf{e}_i \rangle}{t} \\
&= \langle \mathbf{Ax}, \ \mathbf{e}_i \rangle + \langle \mathbf{x}, \ \mathbf{Ae}_i \rangle \\
&= \langle \mathbf{Ax}, \ \mathbf{e}_i \rangle + \langle \mathbf{A}^{\mathrm{T}}\mathbf{x}, \ \mathbf{e}_i \rangle.
\end{aligned}
$$

But $\mathbf{A}$ is a symmetric matrix, meaning that $\mathbf{A} = \mathbf{A}^{\mathrm{T}}$, so

$$
\frac{\partial f}{\partial x_i}(\mathbf{x}) = 2\langle \mathbf{Ax}, \ \mathbf{e}_i \rangle.
$$

It follows that $$\mathbf{D}f(\mathbf{x}) = 2\mathbf{Ax}.$$

When $\mathbf{A} = \mathbf{I}_n$, the above formula reads

$$
\mathbf{D}g(\mathbf{x}) = 2\mathbf{x}.
$$

Substituting $\mathbf{D}f(\mathbf{x}) = 2\mathbf{Ax}$ and $\mathbf{D}g(\mathbf{x}) = 2\mathbf{x}$ in formula (17.62), we see that

$$
\mathbf{Ax} = \lambda\mathbf{x} \quad \text{and} \quad \|\mathbf{x}\| = 1. \qquad \blacksquare
$$

---

**EXERCISES**

1. Find the minimum of $\{x + y \mid x^2 + y^2 = 1\}$.
2. Find the maximum of $\{x^2 + y^2 \mid y^2 + x^2 + z^2 = 6\}$ by inspection and by using Lagrange multipliers.
3. Find the maximum of $\{x + y + z \mid |x| + |y| + |z| \le 1\}$ by inspection.
4. Find the maximum of $\{x^2 + y^2 + z^2 \mid 2x^2 + y^2 + 3z^2 \le 1\}$.
5. Verify the details of the applications of the Chain Rule in the proof of the General Lagrange Multiplier Theorem.
6. For numbers $a$, $b$, and $c$, find the minimum of

$$
\{ax + by + cz \mid x^2 + y^2 + z^2 \le 1\}.
$$

Give a geometric interpretation of the answer by viewing $ax + by + cz$ as the inner product of $(x, \ y, \ z)$ with $(a, \ b, \ c)$.

7. Find the point on the plane $ax + by + cz + d = 0$ that is closest to the point $(0, \ 0, \ 0)$.
8. For positive numbers $a$, $b$, and $c$, find a point on the ellipsoid

$$
S = \left\{ (x, \ y, \ z) \text{ in } \mathbb{R}^3 \ \middle|\ \frac{x^2}{a^2} + \frac{y^2}{b^2} + \frac{z^2}{c^2} = 1 \right\}
$$

that is closest to the point $(0, 0, 0)$.

9. Show, by finding an example, that Corollary 17.8 is false if we drop the assumption that $\mathbf{D}g(\mathbf{u})$ is nonzero.

**10.**   Show that the matrix

$$\begin{bmatrix} 0 & -1 \\ 1 & 0 \end{bmatrix}$$

has no real eigenvalues.

**11.**   Let $\mathbf{A}$ be an $n \times n$ symmetric matrix and define $\lambda$ to be the maximum of

$$\{\langle \mathbf{A}\mathbf{x}, \mathbf{x}\rangle \mid \langle \mathbf{x}, \mathbf{x}\rangle = 1.\}$$

Follow the proof of Proposition 17.9 to show that $\lambda$ is an eigenvalue of the matrix $\mathbf{A}$.

**12.**   Let $p$ and $q$ be numbers with $p > 1$ and $q > 1$.

(a)   Show that

$$\frac{x^p}{p} + \frac{y^q}{q} \geq \frac{1}{p} + \frac{1}{q} \qquad \text{for all } (x, y) \text{ in } \mathbb{R}^2 \text{ such that } x > 0, \ y > 0, \text{ and } xy = 1.$$

(b)   Use (a) to verify the following inequality:

If $a \geq 0$, $b \geq 0$, $p > 1$, $q > 1$,   and   $\dfrac{1}{p} + \dfrac{1}{q} = 1$,   then

$$ab \leq \frac{a^p}{p} + \frac{b^q}{q}.$$

**13.**   (a)   Show that

$$x + y + z \geq 3$$

for all $(x, y, z)$ in $\mathbb{R}^3$ such that $x > 0$, $y > 0$, $z > 0$, and $xyz = 1$.

(b)   Use (a) to verify the following Geometric Mean/Arithmetic Mean Inequality:
If $a_1$, $a_2$, and $a_3$ are positive numbers, then

$$(a_1 a_2 a_3)^{1/3} \leq \frac{a_1 + a_2 + a_3}{3}.$$

(c)   Generalize the above inequality from $n = 3$ to general positive integers $n$.

# 18

# Integration for Functions of Several Variables

## 18.1  Integration over Generalized Rectangles

Recall that if $I = [a, b]$ is a closed bounded interval of real numbers, $m$ is a positive integer, and $P = \{x_0, \ldots, x_m\}$ are $m + 1$ real numbers such that

$$a = x_0 < x_1 < \cdots < x_i < \cdots < x_m = b,$$

then $P$ is called a *partition* of $[a, b]$, and the intervals $[x_{i-1}, x_i]$, for $i$ an index between 1 and $m$, are called *intervals* in the partition P. We define the *length* of the interval $I = [a, b]$ to be $b - a$.*

Let $n$ be a positive integer and for each index $i$ between 1 and $n$, let $I_i = [a_i, b_i]$ be a closed bounded interval of real numbers. The Cartesian product of these intervals,

$$\mathbf{I} = I_1 \times \cdots \times I_i \times \cdots \times I_n = \{\mathbf{x} = (x_1, \ldots, x_n) \text{ in } \mathbb{R}^n \mid x_i \text{ in } I_i \text{ for } 1 \leq i \leq n\},$$

---

*Unless explicitly stated otherwise, it is assumed that an interval $I = [a, b]$ is nondegenerate; that is, $a < b$.

is called a *generalized rectangle*. It is convenient to refer to the interval $I_i$ as being the
$i$th *edge* of $\mathbf{I}$. We define the *volume* of $\mathbf{I}$, which is denoted by vol $\mathbf{I}$, to be the product of
the lengths of the $n$ edges; that is,

$$\text{vol } \mathbf{I} \equiv \prod_{i=1}^{n} [b_i - a_i].$$

In the case when $n = 1$, the volume is simply the length; in the case when $n = 2$, the
volume is called the *area*.

**DEFINITION**   *Given a generalized rectangle* $\mathbf{I} = I_1 \times \cdots \times I_i \times \cdots \times I_n$, *for each index i between 1
and n, let* $P_i$ *be a partition of the ith edge* $I_i$. *The collection of generalized rectangles of
the form*

$$\mathbf{J} = J_1 \times \cdots \times J_i \times \cdots \times J_n,$$

*where each* $J_i$ *is an interval in the partition* $P_i$, *is called a partition of* $\mathbf{I}$ *and is denoted
by*

$$\mathbf{P} \equiv (P_1, \ldots, P_n).$$

Consider the rectangle $[a, b] \times [c, d]$ in the plane $\mathbb{R}^2$. Let $P_1 = \{x_0, \ldots, x_m\}$ and
$P_2 = \{y_0, \ldots, y_\ell\}$ be partitions of $[a, b]$ and $[c, d]$ respectively, and define $\mathbf{P} = (P_1, P_2)$.
Then

$$\sum_{\mathbf{J} \text{ in } \mathbf{P}} \text{vol } \mathbf{J} = \sum_{j=1}^{\ell} \sum_{i=1}^{m} [x_i - x_{i-1}][y_j - y_{j-1}]$$

$$= \sum_{j=1}^{\ell} \left\{ \sum_{i=1}^{m} [x_i - x_{i-1}] \right\} [y_j - y_{j-1}]$$

$$= \sum_{j=1}^{\ell} \{[b - a]\}[y_j - y_{j-1}]$$

$$= [b - a] \sum_{j=1}^{\ell} [y_j - y_{j-1}]$$

$$= [b - a][d - c] = \text{vol } \mathbf{I}.$$

An induction argument shows that the above formula also holds in general: For each
natural number $n$, if $\mathbf{P}$ is a partition of the generalized rectangle $\mathbf{I}$ in $\mathbb{R}^n$, then

$$\text{vol } \mathbf{I} = \sum_{\mathbf{J} \text{ in } \mathbf{P}} \text{vol } \mathbf{J}.$$

Now suppose that $f : \mathbf{I} \to \mathbb{R}$ is a bounded function whose domain $\mathbf{I}$ is a generalized
rectangle, and let $\mathbf{P}$ be a partition of $\mathbf{I}$. For $\mathbf{J}$ a generalized rectangle in the partition $\mathbf{P}$,
we define

$$m(f, \mathbf{J}) \equiv \inf\{f(\mathbf{x}) \mid \mathbf{x} \text{ in } \mathbf{J}\} \quad \text{and} \quad M(f, \mathbf{J}) \equiv \sup\{f(\mathbf{x}) \mid \mathbf{x} \text{ in } \mathbf{J}\}.$$

We then define the *lower Darboux sum* for the function $f : \mathbf{I} \to \mathbb{R}$ with respect to the partition $\mathbf{P}$, which is denoted by $L(f, \mathbf{P})$, by the formula

$$L(f, \mathbf{P}) \equiv \sum_{\mathbf{J} \text{ in } \mathbf{P}} m(f, \mathbf{J}) \mathrm{vol} \, \mathbf{J},$$

and we define the *upper Darboux sum* for the function $f : \mathbf{I} \to \mathbb{R}$ with respect to the partition $\mathbf{P}$, which is denoted by $U(f, \mathbf{P})$, by the formula

$$U(f, \mathbf{P}) \equiv \sum_{\mathbf{J} \text{ in } \mathbf{P}} M(f, \mathbf{J}) \mathrm{vol} \, \mathbf{J}.$$

**DEFINITION**   *Let $\mathbf{I}$ be a generalized rectangle in $\mathbb{R}^n$ and let the function $f : \mathbf{I} \to \mathbb{R}$ be bounded. Then $f : \mathbf{I} \to \mathbb{R}$ is said to be integrable provided that there is exactly one real number $A$ with the property that*

$$L(f, \mathbf{P}) \le A \le U(f, \mathbf{P}) \quad \text{for every partition } \mathbf{P} \text{ of } \mathbf{I}.$$

*We call the number $A$ the integral of $f : \mathbf{I} \to \mathbb{R}$ and denote it by $\int_{\mathbf{I}} f$.*

These definitions of integrability and of the integral are direct extensions of the concepts defined for functions of a single variable in Section 6.2. As we will see in the remainder of this section, many of the results regarding integration of functions of a single variable extend directly to the case of functions of several variables.

**EXAMPLE 18.1**   Let $\mathbf{I}$ be a generalized rectangle in $\mathbb{R}^n$, let $c$ be a real number, and define $f : \mathbf{I} \to \mathbb{R}$ to be the constant function that assumes the value $c$ at every point. Then $f : \mathbf{I} \to \mathbb{R}$ is integrable and

$$\int_{\mathbf{I}} f = c \, \mathrm{vol} \, \mathbf{I}.$$

This follows directly from the definition. Indeed, if $\mathbf{P}$ is any partition of $\mathbf{I}$, then for each generalized rectangle $\mathbf{J}$ in $\mathbf{P}$, we have $m(f, \mathbf{J}) = c = M(f, \mathbf{J})$, so that

$$L(f, \mathbf{P}) = \sum_{\mathbf{J} \text{ in } \mathbf{P}} c \, \mathrm{vol} \, \mathbf{J} = c \, \mathrm{vol} \, \mathbf{I} = \sum_{\mathbf{J} \text{ in } \mathbf{P}} c \, \mathrm{vol} \, \mathbf{J} = U(f, \mathbf{P}). \qquad \square$$

**EXAMPLE 18.2**   For a generalized rectangle $\mathbf{I}$ in $\mathbb{R}^n$, define $f : \mathbf{I} \to \mathbb{R}$ by

$$f(\mathbf{x}) = \begin{cases} 0 & \text{if } \mathbf{x} \text{ has a rational component} \\ 1 & \text{otherwise.} \end{cases}$$

This function is not integrable. Indeed, observe that each generalized rectangle contains points with a rational component and points that have no rational component; this is a consequence of the density of the rational and the irrational numbers in $\mathbb{R}$. Thus if $\mathbf{P}$ is any partition of $\mathbf{I}$,

$$L(f, \mathbf{P}) = 0 \quad \text{and} \quad U(f, \mathbf{P}) = \mathrm{vol} \, \mathbf{I}.$$

It follows that for any number $A$ between 0 and $\mathrm{vol} \, \mathbf{I}$,

$$L(f, \mathbf{P}) \le A \le U(f, \mathbf{P}) \quad \text{for every partition } \mathbf{P} \text{ of } \mathbf{I},$$

so the function $f : \mathbf{I} \to \mathbb{R}$ is not integrable. $\qquad \square$

**LEMMA 18.1**   Let $\mathbf{I}$ be a generalized rectangle in $\mathbb{R}^n$ and let the function $f\colon \mathbf{I} \to \mathbb{R}$ be bounded. Suppose the two numbers $m$ and $M$ have the property that

$$m \le f(\mathbf{x}) \le M \quad \text{for all points } \mathbf{x} \text{ in } \mathbf{I}.$$

Then for any partition $\mathbf{P}$ of $\mathbf{I}$,

$$m\,\mathrm{vol}\,\mathbf{I} \le L(f,\mathbf{P}) \le U(f,\mathbf{P}) \le M\,\mathrm{vol}\,\mathbf{I}. \tag{18.1}$$

**Proof**   Let $\mathbf{P}$ be a partition of $\mathbf{I}$. For a generalized rectangle $\mathbf{J}$ in $\mathbf{I}$, it is clear that

$$m \le \inf\{f(\mathbf{x}) \mid \mathbf{x} \text{ in } \mathbf{J}\} = m(f,\mathbf{J}) \le M(f,\mathbf{J}) = \sup\{f(\mathbf{x}) \mid \mathbf{x} \text{ in } \mathbf{J}\} \le M,$$

so $$m\,\mathrm{vol}\,\mathbf{J} \le m(f,\mathbf{J})\mathrm{vol}\,\mathbf{J} \le M(f,\mathbf{J})\mathrm{vol}\,\mathbf{J} \le M\,\mathrm{vol}\,\mathbf{J}.$$

Summing over all the generalized rectangles $\mathbf{J}$ in the partition $\mathbf{P}$, we conclude that the inequality (18.1) holds.   ∎

Given a partition $\mathbf{P} = (P_1, \dots, P_n)$ of a generalized rectangle $\mathbf{I}$, another partition $\mathbf{P}^* = (P_1^*, \dots, P_n^*)$ of $\mathbf{I}$ is said to be a *refinement* of $\mathbf{P}$ provided that for each index $i$ between 1 and $n$, $P_i^*$ is a refinement of $P_i$. Observe that if $\mathbf{P}^*$ is a refinement of $\mathbf{P}$, then (i) each generalized rectangle $\mathbf{J}$ in $\mathbf{P}^*$ is contained in exactly one generalized rectangle in $\mathbf{P}$, and (ii) given a generalized rectangle $\mathbf{J}$ in $\mathbf{P}$, the collection of generalized rectangles in $\mathbf{P}^*$ that are contained in $\mathbf{J}$ induce a partition of $\mathbf{J}$ that we will denote by $\mathbf{P}^*(\mathbf{J})$. The following distribution formulas for the lower and upper Darboux sums follow from these two properties:

$$L(f,\mathbf{P}^*) = \sum_{\mathbf{J} \text{ in } \mathbf{P}} L(f,\mathbf{P}^*(\mathbf{J})) \quad \text{and} \quad U(f,\mathbf{P}^*) = \sum_{\mathbf{J} \text{ in } \mathbf{P}} U(f,\mathbf{P}^*(\mathbf{J})). \tag{18.2}$$

**LEMMA 18.2**   **The Refinement Lemma**   Suppose that the function $f\colon \mathbf{I} \to \mathbb{R}$ is bounded, where $\mathbf{I}$ is a generalized rectangle in $\mathbb{R}^n$. Let $\mathbf{P}$ be a partition of $\mathbf{I}$ and let $\mathbf{P}^*$ be a refinement of $\mathbf{P}$. Then

$$L(f,\mathbf{P}) \le L(f,\mathbf{P}^*) \le U(f,\mathbf{P}^*) \le U(f,\mathbf{P}). \tag{18.3}$$

**Proof**   Let $\mathbf{J}$ be a generalized rectangle in $\mathbf{P}$ and denote by $\mathbf{P}^*(\mathbf{J})$ the partition of $\mathbf{J}$ that is induced by $\mathbf{P}^*$. From Lemma 18.1, with $\mathbf{J}$ playing the role of $\mathbf{I}$, it follows that

$$m(f,\mathbf{J})\mathrm{vol}\,\mathbf{J} \le L(f,\mathbf{P}^*(\mathbf{J})) \le U(f,\mathbf{P}^*(\mathbf{J})) \le M(f,\mathbf{J})\mathrm{vol}\,\mathbf{J}.$$

If we sum these inequalities over all generalized rectangles $\mathbf{J}$ in $\mathbf{P}$ and use the distribution formulas (18.2), we arrive at the inequality (18.3).   ∎

For two partitions $P$ and $P'$ of a closed bounded interval of real numbers $I$, by taking the partition consisting of all points that are partition points in at least one of the two partitions, we obtain a partition that is a *common refinement* of the two given partitions, meaning that it is a refinement of both $P$ and $P'$. Similarly, suppose that $\mathbf{P}$ and $\mathbf{P}'$ are two partitions of a generalized rectangle $\mathbf{I}$ in $\mathbb{R}^n$ that are represented as $\mathbf{P} = (P_1, \dots, P_n)$ and $\mathbf{P}' = (P_1', \dots, P_n')$. For each index $i$ between 1 and $n$, choose $P_i''$

to be a common refinement of $P_i$ and $P_i'$ and define $\mathbf{P}'' = \left(P_1'', \ldots, P_n''\right)$. Then $\mathbf{P}''$ is a partition of $\mathbf{I}$ that is a common refinement of the partitions $\mathbf{P}$ and $\mathbf{P}'$. The existence of common refinements is what is necessary to establish the following proposition.

**PROPOSITION 18.3**    Let $\mathbf{I}$ be a generalized rectangle in $\mathbb{R}^n$ and suppose that the function $f \colon \mathbf{I} \to \mathbb{R}$ is bounded. For any two partitions $\mathbf{P}_1$ and $\mathbf{P}_2$ of $\mathbf{I}$,

$$L(f, \mathbf{P}_1) \leq U(f, \mathbf{P}_2).$$

**Proof**    Choose $\mathbf{P}$ to be a common refinement of the two partitions $\mathbf{P}_1$ and $\mathbf{P}_2$. By the Refinement Lemma,

$$L(f, \mathbf{P}_1) \leq L(f, \mathbf{P}) \leq U(f, \mathbf{P}) \leq U(f, \mathbf{P}_2). \qquad \blacksquare$$

The above proposition allows us to establish a criterion for determining when a function is integrable. Indeed, let $\mathbf{I}$ be a generalized rectangle in $\mathbb{R}^n$ and suppose that the function $f \colon \mathbf{I} \to \mathbb{R}$ is bounded. Denote by $\mathcal{L}$ the collection of all lower Darboux sums for $f \colon \mathbf{I} \to \mathbb{R}$, and denote by $\mathcal{U}$ the collection of all upper Darboux sums for $f \colon \mathbf{I} \to \mathbb{R}$. Then Proposition 18.3 amounts to the assertion that

$$\ell \leq u \quad \text{whenever } \ell \text{ is in } \mathcal{L} \text{ and } u \text{ is in } \mathcal{U}.$$

Moreover, by definition, the function $f \colon \mathbf{I} \to \mathbb{R}$ is integrable provided that there is exactly one number $A$ that has the property that

$$\ell \leq A \leq u \quad \text{whenever } \ell \text{ is in } \mathcal{L} \text{ and } u \text{ is in } \mathcal{U}.$$

It follows from a primitive theorem about sets of real numbers, the Dedekind Gap Theorem (see Section 1.1), that the function $f \colon \mathbf{I} \to \mathbb{R}$ is integrable if and only if for each $\epsilon > 0$ there are partitions $\mathbf{P}_1$ and $\mathbf{P}_2$ of $\mathbf{I}$ such that

$$U(f, \mathbf{P}_2) - L(f, \mathbf{P}_1) < \epsilon.$$

Observe, however, that if $\mathbf{P}$ is a common refinement of $\mathbf{P}_1$ and $\mathbf{P}_2$, then by the Refinement Lemma,

$$U(f, \mathbf{P}) - L(f, \mathbf{P}) < U(f, \mathbf{P}_2) - L(f, \mathbf{P}_1).$$

Thus we have the following direct extension of the Integrability Criterion for functions of a single variable.

**THEOREM 18.4**    **The Integrability Criterion**    Let $\mathbf{I}$ be a generalized rectangle in $\mathbb{R}^n$ and suppose that the function $f \colon \mathbf{I} \to \mathbb{R}$ is bounded. Then the following two assertions are equivalent:

(i) The function $f \colon \mathbf{I} \to \mathbb{R}$ is integrable.

(ii) For each positive number $\epsilon$ there is a partition $\mathbf{P}$ of $\mathbf{I}$ such that

$$U(f, \mathbf{P}) - L(f, \mathbf{P}) < \epsilon.$$

**EXAMPLE 18.3** For the generalized rectangle $\mathbf{I} = [0, 1] \times [0, 1]$ in the plane $\mathbb{R}^2$, define

$$f(x, y) = \begin{cases} 1 & \text{if } (x, y) \text{ is in } \mathbf{I} \text{ and } y > x \\ 0 & \text{if } (x, y) \text{ is in } \mathbf{I} \text{ and } y \leq x. \end{cases}$$

**FIGURE 18.1**

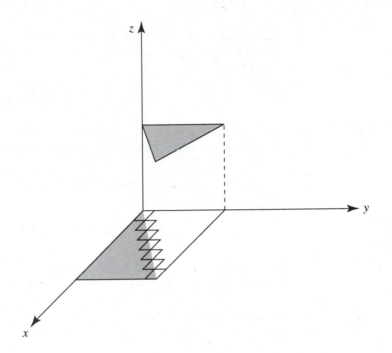

We will use the Integrability Criterion to show that the function $f: \mathbf{I} \to \mathbb{R}$ is integrable. Indeed, let $\epsilon > 0$. For a natural number $k$, let $P_k$ be the partition of the interval $[0, 1]$ into $k$ intervals of equal length $1/k$, and define the partition of $\mathbf{P}_k$ of $\mathbf{I}$ to be $(P_k, P_k)$. Observe that the only terms in the sum $U(f, \mathbf{P}_k) - L(f, \mathbf{P}_k)$ that can possibly be nonzero are those arising from the rectangles in the partition $\mathbf{P}_k$ that intersect the diagonal. Each of these contributions is $1/k^2$ and there are fewer than $3k$ of such rectangles. Hence

$$U(f, \mathbf{P}_k) - L(f, \mathbf{P}_k) < \frac{3}{k}.$$

By the Archimedean Property of $\mathbb{R}$, we can choose a natural number $k$ greater that $3/\epsilon$. Choosing $\mathbf{P} = \mathbf{P}_k$, we have

$$U(f, \mathbf{P}) - L(f, \mathbf{P}) < \epsilon. \qquad \square$$

**DEFINITION** *For a bounded subset $D$ of $\mathbb{R}^n$, the diameter of $D$, which is denoted by* diam $D$, *is defined by*

$$\text{diam } D \equiv \text{sup } \{d(\mathbf{u}, \mathbf{v}) \mid \mathbf{u} \text{ and } \mathbf{v} \text{ in } D\}.$$

Observe that for a closed bounded interval of real numbers $I = [a, b]$, the diameter of $I$ equals its length. Hence, from the definition of distance between points in $\mathbb{R}^n$, it directly follows that if the generalized rectangle $\mathbf{I}$ is the Cartesian product

$$\mathbf{I} = [a_1, b_1] \times \cdots \times [a_n, b_n],$$

then

$$\text{diam}\,\mathbf{I} = \sqrt{(b_1 - a_1)^2 + \cdots + (b_n - a_n)^2};$$

that is, the diameter of $\mathbf{I}$ is the distance between the points $(a_1, \ldots, a_n)$ and $(b_1, \ldots, b_n)$.

Recall that for a partition $P$ of an interval $[a, b]$, the *gap* of the partition, which is denoted by $\|P\|$, is defined to be the largest of the diameters of the intervals in the partition $P$. This leads us to define the gap of a partition of a generalized rectangle as follows:

**DEFINITION**    *For a partition $\mathbf{P}$ of a generalized rectangle $\mathbf{I}$ in $\mathbb{R}^n$, we define the gap of $\mathbf{P}$, denoted by $\|\mathbf{P}\|$, to be the largest of the diameters of the generalized rectangles in $\mathbf{P}$.*

It is not difficult to check that for a partition $\mathbf{P}$ of a generalized rectangle $\mathbf{I}$ in $\mathbb{R}^n$, if $\mathbf{P} = (P_1, \ldots, P_n)$, then

$$\|\mathbf{P}\| = \sqrt{\|P_1\|^2 + \cdots + \|P_n\|^2}.$$

**THEOREM 18.5**    Let $\mathbf{I}$ be a generalized rectangle in $\mathbb{R}^n$ and suppose that the function $f: \mathbf{I} \to \mathbb{R}$ is continuous. Then the function $f: \mathbf{I} \to \mathbb{R}$ is integrable.

**Proof**    We will use the Integrability Criterion. Let $\epsilon > 0$. It is necessary to find a partition $\mathbf{P}$ of $\mathbf{I}$ such that

$$U(f, \mathbf{P}) - L(f, \mathbf{P}) < \epsilon.$$

Observe that a generalized rectangle is closed and bounded, and so, by Theorem 11.11, it is compact. Consequently, by Theorem 11.15, the function $f: \mathbf{I} \to \mathbb{R}$ is uniformly continuous. Hence, since $\epsilon/\text{vol}\,\mathbf{I}$ is a positive number, we may choose a positive number $\delta$ such that

$$|f(\mathbf{u}) - f(\mathbf{v})| < \epsilon/\text{vol}\,\mathbf{I} \quad \text{if } \mathbf{u} \text{ and } \mathbf{v} \text{ are points in } \mathbf{I} \text{ such that } d(\mathbf{u}, \mathbf{v}) < \delta. \quad (18.4)$$

Choose a partition $\mathbf{P}$ of the generalized rectangle $\mathbf{I}$ such that $\|\mathbf{P}\| < \delta$. Observe that a generalized rectangle $\mathbf{J}$ in $\mathbf{P}$ is compact; thus, by the Extreme Value Theorem, the function $f: \mathbf{J} \to \mathbb{R}$ assumes a maximum and a minimum value at points in $\mathbf{J}$, and, because $\text{diam}\,\mathbf{J} < \delta$, these points are no more than the distance $\delta$ apart. In view of (18.4), this implies that

$$M(f, \mathbf{J}) - m(f, \mathbf{J}) < \epsilon/\text{vol}\,\mathbf{I}. \quad (18.5)$$

Since the sum of the volumes of the generalized rectangles in $\mathbf{I}$ equals $\text{vol}\,\mathbf{I}$, from (18.5) we conclude that

$$U(f, \mathbf{P}) - L(f, \mathbf{P}) = \sum_{\mathbf{J}\,\text{in}\,\mathbf{P}} [M(f, \mathbf{J}) - m(f, \mathbf{J})]\text{vol}\,\mathbf{J} < (\epsilon/\text{vol}\,\mathbf{I}) \sum_{\mathbf{J}\,\text{in}\,\mathbf{P}} \text{vol}\,\mathbf{J} = \epsilon. \quad \blacksquare$$

The reader can check that what we have proven so far about integration of functions of several variables that are defined on generalized rectangles in Euclidean space $\mathbb{R}^n$ follows, almost word for word, the corresponding discussion in Section 6.2 for functions of a single variable that are defined on a closed bounded interval of real numbers. Rather than continue to repeat arguments already made, and thus essentially repeat Section 6.4, we will simply state two further characterizations of the integral, one in terms of the convergence of Darboux sums and the other in terms of the convergence of Riemann sums. These are exact generalizations of corresponding results in the single-variable theory and their proofs are almost precisely the same as those for the single-variable case. It is left to the diligent reader to reread Section 6.4 and thereby verify that this is true.

**THEOREM 18.6**    **The Darboux Sum Convergence Criterion**    For a bounded function $f : \mathbf{I} \to \mathbb{R}$ defined on a generalized rectangle $\mathbf{I}$ in $\mathbb{R}^n$, and a real number $A$, the following two assertions are equivalent:

(i)  The function $f : \mathbf{I} \to \mathbb{R}$ is integrable and $\int_{\mathbf{I}} f = A$.

(ii)  If $\{\mathbf{P}_k\}$ is any sequence of partitions of $\mathbf{I}$ such that $\lim_{k \to \infty} \|\mathbf{P}_k\| = 0$, then

$$\lim_{k \to \infty} L(f, \mathbf{P}_k) = \lim_{k \to \infty} U(f, \mathbf{P}_k) = A.$$

**EXAMPLE 18.4**    For the rectangle $\mathbf{I} = [0, 1] \times [0, 1]$ in the plane $\mathbb{R}^2$, define $f : \mathbf{I} \to \mathbb{R}$ by

$$f(x, y) = x^2 y^2 \quad \text{for } (x, y) \text{ in } \mathbf{I}.$$

The function $f : \mathbf{I} \to \mathbb{R}$ is continuous, so by Theorem 18.5 it is also integrable. We will use the Darboux Sum Convergence Criterion to show that $\int_{\mathbf{I}} f = 1/9$. Indeed, for $k$ a natural number, let $P_k$ be the partition of $[0, 1]$ into $k$ intervals of equal length $1/k$ and define $\mathbf{P}_k = (P_k, P_k)$. Each rectangle $\mathbf{J}$ in $\mathbf{P}_k$ has area $1/k^2$. Moreover, we see that for two indices $i$ and $j$ between 1 and $k$, if

$$\mathbf{J} = \left[ \frac{i-1}{k}, \frac{i}{k} \right] \times \left[ \frac{j-1}{k}, \frac{j}{k} \right],$$

then
$$M(f, \mathbf{J}) = \frac{i^2 j^2}{k^4}.$$

Thus
$$U(f, \mathbf{P}_k) = \sum_{\mathbf{J} \text{ in } \mathbf{P}_k} M(f, \mathbf{J}) \text{vol } \mathbf{J}$$

$$= \sum_{1 \le i,j \le k} \frac{i^2 j^2}{k^6}$$

$$= \frac{1}{k^6} \sum_{i=1}^{k} i^2 \left[ \sum_{j=1}^{k} j^2 \right]$$

$$= \frac{1}{k^6} \sum_{i=1}^{k} i^2 \left[ \frac{k(k+1)(2k+1)}{6} \right]$$

$$= \frac{1}{k^6} \cdot \frac{k(k+1)(2k+1)}{6} \sum_{i=1}^{k} i^2$$

$$= \frac{1}{k^6} \cdot \frac{k(k+1)(2k+1)}{6} \cdot \frac{k(k+1)(2k+1)}{6}.$$

Thus, since $\lim_{k \to \infty} \|\mathbf{P}_k\| = 0$, the Darboux Sum Convergence Criterion implies that

$$\int_{\mathbf{I}} f = \lim_{k \to \infty} U(f, \mathbf{P}_k) = \lim_{k \to \infty} \left[ \frac{1}{k^6} \cdot \frac{k(k+1)(2k+1)}{6} \cdot \frac{k(k+1)(2k+1)}{6} \right] = \frac{1}{9}. \quad \square$$

**DEFINITION**  *Consider a function $f: \mathbf{I} \to \mathbb{R}$ defined on a generalized rectangle $\mathbf{I}$ in $\mathbb{R}^n$, together with a partition $\mathbf{P}$ of $\mathbf{I}$. For each generalized rectangle $\mathbf{J}$ in $\mathbf{P}$, let $c(\mathbf{J})$ be a point in $\mathbf{J}$. The sum*

$$R(f, \mathbf{P}) \equiv \sum_{\mathbf{J} \text{ in } \mathbf{P}} f(c(\mathbf{J})) vol \, \mathbf{J}$$

*is called a Riemann sum for the function $f: \mathbf{I} \to \mathbb{R}$ based on the partition $\mathbf{P}$.*

**THEOREM 18.7**  **The Riemann Sum Convergence Criterion**  For a bounded function $f: \mathbf{I} \to \mathbb{R}$ defined on a generalized rectangle $\mathbf{I}$ in $\mathbb{R}^n$, and a real number $A$, the following two assertions are equivalent:

(i) The function $f: \mathbf{I} \to \mathbb{R}$ is integrable and $\int_{\mathbf{I}} f = A$.

(ii) If $\{\mathbf{P}_k\}$ is any sequence of partitions of $\mathbf{I}$ such that $\lim_{k \to \infty} \|\mathbf{P}_k\| = 0$, and for each natural number $k$, $R(f, \mathbf{P}_k)$ is a Riemann sum for $f: \mathbf{I} \to \mathbb{R}$ based on the partition $\mathbf{P}_k$, then

$$\lim_{k \to \infty} R(f, \mathbf{P}_k) = A.$$

**THEOREM 18.8**  **Linearity of the Integral**  Suppose that the functions $f: \mathbf{I} \to \mathbb{R}$ and $g: \mathbf{I} \to \mathbb{R}$ are integrable, where $\mathbf{I}$ is a generalized rectangle in $\mathbb{R}^n$. Then for any two numbers $\alpha$ and $\beta$, the function $\alpha f + \beta g: \mathbf{I} \to \mathbb{R}$ is also integrable and

$$\int_{\mathbf{I}} [\alpha f + \beta g] = \alpha \int_{\mathbf{I}} f + \beta \int_{\mathbf{I}} g. \tag{18.6}$$

**Proof**   We will use the Riemann Sum Convergence Criterion to prove the theorem. Let $\{\mathbf{P}_k\}$ be a sequence of partitions of the generalized rectangle $\mathbf{I}$ such that

$$\lim_{k \to \infty} \|\mathbf{P}_k\| = 0.$$

For each natural number $k$, let $R(\alpha f + \beta g, \mathbf{P}_k)$ be a Riemann sum for the function $\alpha f + \beta g : \mathbf{I} \to \mathbb{R}$ based on the partition $\mathbf{P}_k$. We will show that

$$\lim_{k \to \infty} R(\alpha f + \beta g, \mathbf{P}_k) = \alpha \int_{\mathbf{I}} f + \beta \int_{\mathbf{I}} g. \tag{18.7}$$

Fix a natural number $k$. The Riemann sum $R(\alpha f + \beta g, \mathbf{P}_k)$ is defined by choosing a point in each generalized rectangle $\mathbf{J}$ in the partition $\mathbf{P}_k$; define $R(f, \mathbf{P}_k)$ and $R(g, \mathbf{P}_k)$ to be the Riemann sums for the functions $f : \mathbf{I} \to \mathbb{R}$ and $g : \mathbf{I} \to \mathbb{R}$ obtained by making the same choice of points in the generalized rectangles in $\mathbf{P}_k$. For such a choice of Riemann sums, we have

$$R(\alpha f + \beta g, \mathbf{P}_k) = \alpha R(f, \mathbf{P}_k) + \beta R(g, \mathbf{P}_k). \tag{18.8}$$

The Riemann Sum Convergence Criterion implies that

$$\lim_{k \to \infty} R(f, \mathbf{P}_k) = \int_{\mathbf{I}} f \quad \text{and} \quad \lim_{k \to \infty} R(g, \mathbf{P}_k) = \int_{\mathbf{I}} g. \tag{18.9}$$

The linearity property of convergent sequences of real numbers, together with (18.8) and (18.9), implies that

$$\lim_{k \to \infty} R(\alpha f + \beta g, \mathbf{P}_k) = \alpha \lim_{k \to \infty} R(f, \mathbf{P}_k) + \beta \lim_{k \to \infty} R(g, \mathbf{P}_k) = \alpha \int_{\mathbf{I}} f + \beta \int_{\mathbf{I}} g;$$

that is, (18.7) holds.   ∎

---

**THEOREM 18.9**   **Monotonicity of the Integral**   Suppose that the functions $f : \mathbf{I} \to \mathbb{R}$ and $g : \mathbf{I} \to \mathbb{R}$ are integrable, where $\mathbf{I}$ is a generalized rectangle in $\mathbb{R}^n$, and also suppose that

$$f(x) \leq g(x) \quad \text{for all points } \mathbf{x} \text{ in } \mathbf{I}.$$

Then

$$\int_{\mathbf{I}} f \leq \int_{\mathbf{I}} g.$$

**Proof**   Let $\{\mathbf{P}_k\}$ be any sequence of partitions of the generalized rectangle $\mathbf{I}$ such that

$$\lim_{k \to \infty} \|\mathbf{P}_k\| = 0.$$

For each natural number $k$, let $R(f, \mathbf{P}_k)$ and $R(g, \mathbf{P}_k)$ be Riemann sums for the functions $f : \mathbf{I} \to \mathbb{R}$ and $g : \mathbf{I} \to \mathbb{R}$ obtained by making the same choice of points in the generalized rectangles of $\mathbf{P}_k$. Thus

$$R(f, \mathbf{P}_k) \leq R(g, \mathbf{P}_k).$$

The order preservation property of convergent sequences of real numbers, together with the Riemann Sum Convergence Criterion, implies that

$$\int_{\mathbf{I}} f = \lim_{k \to \infty} R(f, \mathbf{P}_k) \leq \lim_{k \to \infty} R(g, \mathbf{P}_k) = \int_{\mathbf{I}} g. \qquad \blacksquare$$

**THEOREM**
**18.10**

**Additivity over Partitions**    Suppose that the function $f: \mathbf{I} \to \mathbb{R}$ is bounded, where $\mathbf{I}$ is a generalized rectangle in $\mathbb{R}^n$. Let $\mathbf{P}$ be a partition of $\mathbf{I}$. Then the function $f: \mathbf{I} \to \mathbb{R}$ is integrable if and only if for each generalized rectangle $\mathbf{J}$ in $\mathbf{P}$, $f: \mathbf{J} \to \mathbb{R}$ is integrable, in which case

$$\int_{\mathbf{I}} f = \sum_{\mathbf{J} \text{ in } \mathbf{I}} \int_{\mathbf{J}} f. \tag{18.10}$$

**Proof**    First we suppose that for each generalized rectangle $\mathbf{J}$ in $\mathbf{P}$, the function $f: \mathbf{J} \to \mathbb{R}$ is integrable. We will use the Integrability Criterion to show that $f: \mathbf{I} \to \mathbb{R}$ is integrable. Suppose that there are $m$ generalized rectangles $\mathbf{J}_1, \ldots, \mathbf{J}_m$ in $\mathbf{P}$. Let $\epsilon > 0$. For each index $i$ between 1 and $m$, we may select a partition $\mathbf{P}_i$ of $\mathbf{J}_i$ such that

$$U(f, \mathbf{P}_i) - L(f, \mathbf{P}_i) < \frac{\epsilon}{m}.$$

Choose $\mathbf{P}'$ to be a partition of $\mathbf{I}$ that contains all of the generalized rectangles that are in any one of the $\mathbf{P}_i$'s, $1 \leq i \leq m$. By the distribution formula (18.2) and the Refinement Lemma,

$$U(f, \mathbf{P}') - L(f, \mathbf{P}') = \sum_{\mathbf{J} \text{ in } \mathbf{P}} U(f, \mathbf{P}'(\mathbf{J})) - L(f, \mathbf{P}'(\mathbf{J}))$$

$$\leq \sum_{i=1}^{m} U(f, \mathbf{P}_i) - L(f, \mathbf{P}_i)$$

$$< m \, \frac{\epsilon}{m} = \epsilon.$$

Thus the function $f: \mathbf{I} \to \mathbb{R}$ is integrable.

To prove the converse, now suppose that the function $f: \mathbf{I} \to \mathbb{R}$ is integrable. For $\mathbf{J}$ a generalized rectangle in the partition $\mathbf{P}$, we will use the Integrability Criterion to show that $f: \mathbf{J} \to \mathbb{R}$ is integrable. Let $\epsilon > 0$. Since $f: \mathbf{I} \to \mathbb{R}$ is integrable, there is a partition $\mathbf{P}'$ of the generalized rectangle $\mathbf{I}$ such that $U(f, \mathbf{P}') - L(f, \mathbf{P}') < \epsilon$. Choose $\mathbf{P}^*$ to be a common refinement of $\mathbf{P}$ and $\mathbf{P}'$. The Refinement Lemma implies that

$$U(f, \mathbf{P}^*) - L(f, \mathbf{P}^*) \leq U(f, \mathbf{P}) - L(f, \mathbf{P}) < \epsilon.$$

Let $\mathbf{P}^*(\mathbf{J})$ be the partition that $\mathbf{P}^*$ induces on $\mathbf{J}$. Then

$$U(f, \mathbf{P}^*(\mathbf{J})) - L(f, \mathbf{P}^*(\mathbf{J})) \leq U(f, \mathbf{P}^*) - L(f, \mathbf{P}^*) < \epsilon.$$

Thus the function $f: \mathbf{J} \to \mathbb{R}$ is integrable.

It remains to verify formula (18.10). Let $\{\mathbf{P}_k\}$ be a sequence of partitions of $\mathbf{I}$ such that each $\mathbf{P}_k$ is a refinement of $\mathbf{P}$ and $\|\mathbf{P}_k\| < 1/k$. Then for each generalized rectangle $\mathbf{J}$ in $\mathbf{P}$, the partition $\mathbf{P}_k$ induces a partition $\mathbf{P}_k(\mathbf{J})$ of $\mathbf{J}$ with the property that $\|\mathbf{P}_k(\mathbf{J})\| < 1/k$. Moreover, the following distribution formula for the Darboux sums holds:

$$L(f, \mathbf{P}_k) = \sum_{\mathbf{J} \text{ in } \mathbf{P}_k} L(f, \mathbf{P}_k(\mathbf{J})).$$

By the Darboux Sum Convergence Criterion, for each generalized rectangle $\mathbf{J}$ in $\mathbf{P}$,

$$\lim_{k \to \infty} L(f, \mathbf{P}_k(\mathbf{J})) = \int_{\mathbf{J}} f.$$

Thus, by the sum property of convergent sequences of real numbers,

$$\lim_{k \to \infty} L(f, \mathbf{P}_k) = \lim_{k \to \infty} \sum_{\mathbf{J} \text{ in } \mathbf{P}_k} L(f, \mathbf{P}_k(\mathbf{J})) = \sum_{\mathbf{J} \text{ in } \mathbf{P}} \int_{\mathbf{J}} f.$$

Again applying the Darboux Sum Convergence Criterion, it follows that

$$\int_{\mathbf{I}} f = \lim_{k \to \infty} L(f, \mathbf{P}_k) = \sum_{\mathbf{J} \text{ in } \mathbf{P}} \int_{\mathbf{J}} f. \qquad \blacksquare$$

In preparation for our discussion, in the next section, of those aspects of integration of functions of several variables that are different from the single-variable theory, we now establish an improvement of the theorem that a continuous function defined on a generalized rectangle is integrable.

**THEOREM**
**18.11**

Let $\mathbf{I}$ be a generalized rectangle in $\mathbb{R}^n$ and suppose that the function $f : \mathbf{I} \to \mathbb{R}$ is bounded and that its restriction to the interior of $\mathbf{I}$ is continuous. Then the function $f : \mathbf{I} \to \mathbb{R}$ is integrable.

**Proof**
We will use the Integrability Criterion. First, since $f : \mathbf{I} \to \mathbb{R}$ is bounded, we can choose a number $M > 0$ such that

$$|f(\mathbf{x})| \le M \qquad \text{for all } \mathbf{x} \text{ in } \mathbf{I}. \qquad (18.11)$$

Then write $\mathbf{I}$ as the Cartesian product

$$\mathbf{I} = I_1 \times \cdots \times I_n,$$

where each $I_i = [a_i, b_i]$, and define

$$c = \max_{1 \le i \le n} [b_i - a_i]. \qquad (18.12)$$

Let $\epsilon > 0$. Define

$$\alpha = \epsilon / [8nMc^{n-1}].$$

For each index $i$ between 1 and $n$, choose an interval $I_i' = [a_i', b_i']$ that is contained in $I_i$ and such that both

$$0 < b_i - b_i' < \alpha \quad \text{and} \quad 0 < a_i' - a_i < \alpha.$$

Then define the generalized rectangle $\mathbf{I}'$ by

$$\mathbf{I}' = I_1' \times \cdots \times I_n'.$$

We have constructed $\mathbf{I}'$ to be contained in the interior of $\mathbf{I}$, so the function $f : \mathbf{I}' \to \mathbb{R}$ is continuous. Thus, by Theorem 18.5, $f : \mathbf{I}' \to \mathbb{R}$ is integrable. Since the number $\epsilon/2$ is positive, it follows from the Integrability Criterion that there is a partition $\mathbf{P}'$ of the generalized rectangle $\mathbf{I}'$ such that

$$U(f, \mathbf{P}') - L(f, \mathbf{P}') < \frac{\epsilon}{2}. \qquad (18.13)$$

Represent the partition $\mathbf{P}'$ as $\mathbf{P}' = (P_1', \ldots, P_n')$. For each index $i$, $P_i'$ is a partition of $[a_i', b_i']$ so that by adjoining the points $a_i$ and $b_i$ to $P_i'$ we obtain a partition $P_i$ of the interval $[a_i, b_i]$. Then define $\mathbf{P} = (P_1, \ldots, P_n)$. We claim that

$$U(f, \mathbf{P}) - L(f, \mathbf{P}) < \epsilon. \tag{18.14}$$

Indeed,    $\qquad U(f, \mathbf{P}) - L(f, \mathbf{P}) = U(f, \mathbf{P}') - L(f, \mathbf{P}') + E,$

where $E$ is the sum of the terms $[M(f, \mathbf{J}) - m(f, \mathbf{J})]\mathrm{vol}\,\mathbf{J}$, where $\mathbf{J}$ is a generalized rectangle in the partition $\mathbf{P}$ that is not in the partition $\mathbf{P}'$. By the estimate (18.14),

$$U(f, \mathbf{P}) - L(f, \mathbf{P}) = U(f, \mathbf{P}') - L(f, \mathbf{P}') + E < \frac{\epsilon}{2} + E. \tag{18.15}$$

Observe that for a generalized rectangle $\mathbf{J}$ that is in the partition $\mathbf{P}$ but not in the partition $\mathbf{P}'$, there is an index $i$ between 1 and $n$ such that the $i$th edge of $\mathbf{J}$ equals $[a_i, a_i']$ or equals $[b_i', b_i]$. Since, by definition, each edge of $\mathbf{I}$ has length at most $c$ and the length of the interval $[a_i, a_i']$ is less than $\alpha$, the contribution to $E$ of the sum of the terms $[M(f, \mathbf{J}) - m(f, \mathbf{J})]\mathrm{vol}\,\mathbf{J}$, where $\mathbf{J}$ is a generalized rectangle in the partition $\mathbf{P}$ that has $[a_i, a_i']$ as its $i$th edge, is less than $2M\alpha c^{n-1}$. Similarly, the contribution to $E$ of the sum of the terms $[M(f, \mathbf{J}) - m(f, \mathbf{J})]\mathrm{vol}\,\mathbf{J}$, where $\mathbf{J}$ is a generalized rectangle in the partition $\mathbf{P}$ that has $[b_i', b_i]$ as its $i$th edge, is also less than $2M\alpha c^{n-1}$. Thus the sum of all of the terms $[M(f, \mathbf{J}) - m(f, \mathbf{J})]\mathrm{vol}\,\mathbf{J}$ that are comprised by $E$ is less than $4nM\alpha c^{n-1}$. However, the number $\alpha$ was chosen so that $4nM\alpha c^{n-1} = \epsilon/2$. Hence, from (18.15), it follows that

$$U(f, \mathbf{P}) - L(f, \mathbf{P}) = U(f, \mathbf{P}') - L(f, \mathbf{P}') + E < \frac{\epsilon}{2} + E < \epsilon. \qquad \blacksquare$$

**COROLLARY 18.12**    Let $\mathbf{I}$ and $\mathbf{I}'$ be generalized rectangles in $\mathbb{R}^n$ such that $\mathbf{I}'$ is contained in $\mathbf{I}$, and suppose that the bounded function $f : \mathbf{I} \to \mathbb{R}$ has the property that

$$f(\mathbf{x}) = 0 \quad \text{for all } \mathbf{x} \text{ in } \mathbf{I} \backslash \mathbf{I}'.$$

Then $f : \mathbf{I} \to \mathbb{R}$ is integrable if and only if its restriction $f : \mathbf{I}' \to \mathbb{R}$ is integrable, in which case

$$\int_{\mathbf{I}} f = \int_{\mathbf{I}'} f.$$

**Proof**    For each index $i$ between 1 and $n$, let the $i$th edge of $\mathbf{I}$ be $[a_i, b_i]$ and the $i$th edge of $\mathbf{I}'$ be $[a_i', b_i']$, and consider the partition $P_i = \{a_i, a_i', b_i', b_i\}$ of the interval $[a_i, b_i]$. The partition $\mathbf{P} = (P_1, \ldots, P_n)$ of $\mathbf{I}$ has the property that it contains the generalized rectangle $\mathbf{I}'$. Furthermore, for each generalized rectangle $\mathbf{J}$ in $\mathbf{P}$ other than $\mathbf{I}'$, the restriction $f : \mathbf{J} \to \mathbb{R}$ is zero on the interior of $\mathbf{J}$. By Theorem 18.11, the function $f : \mathbf{J} \to \mathbb{R}$ is integrable; it is not difficult (see Exercise 7) to see that $\int_{\mathbf{J}} f = 0$. It follows from the property of additivity of integrals over partitions that the function $f : \mathbf{I} \to \mathbb{R}$ is integrable if and only if its restriction $f : \mathbf{I}' \to \mathbb{R}$ is integrable, and that

$$\int_{\mathbf{I}} f = \sum_{\mathbf{J} \text{ in } \mathbf{P}} \int_{\mathbf{J}} f = \int_{\mathbf{I}'} f. \qquad \blacksquare$$

**EXERCISES**

1. For $\mathbf{I}$ a generalized rectangle in $\mathbb{R}^n$ and a number $\delta > 0$, show that there is a partition $\mathbf{P}$ of $\mathbf{I}$ such that $\|\mathbf{P}\| < \delta$.

2. In Example 18.3, find the exact value of $U(f, \mathbf{P}_k) - L(f, \mathbf{P}_k)$.

3. Show that each generalized rectangle in $\mathbb{R}^n$ contains a point with a rational component and a point without any rational component.

4. For the generalized rectangle $\mathbf{I} = [0, 1] \times [0, 1]$ in the plane $\mathbb{R}^2$, define

$$f(x, y) = \begin{cases} 5 & \text{if } (x, y) \text{ is in } \mathbf{I} \text{ and } x > 1/2 \\ 1 & \text{if } (x, y) \text{ is in } \mathbf{I} \text{ and } x \le 1/2. \end{cases}$$

Use the Integrability Criterion to show that the function $f : \mathbf{I} \to \mathbb{R}$ is integrable.

5. For a partition $\mathbf{P} = (P_1, \ldots, P_n)$ of a generalized rectangle $\mathbf{I}$ in $\mathbb{R}^n$, verify the formula

$$\|\mathbf{P}\| = \sqrt{\|P_1\|^2 + \cdots + \|P_n\|^2}.$$

6. Let $\mathbf{I}$ be a generalized rectangle in $\mathbb{R}^n$ and suppose that the function $f : \mathbf{I} \to \mathbb{R}$ assumes the value 0 except at a single point $\mathbf{x}$ in $\mathbf{I}$. Show that $f : \mathbf{I} \to \mathbb{R}$ is integrable. Then show that $\int_{\mathbf{I}} f = 0$.

7. Let $\mathbf{I}$ be a generalized rectangle in $\mathbb{R}^n$ and suppose that the bounded function $f : \mathbf{I} \to \mathbb{R}$ has the value 0 on the interior of $\mathbf{I}$. Show that $f : \mathbf{I} \to \mathbb{R}$ is integrable and that $\int_{\mathbf{I}} f = 0$.

8. Use the Darboux Sum Convergence Criterion to evaluate the value of the integral of the function in Example 18.3.

9. For the rectangle $\mathbf{I} = [0, 1] \times [0, 1]$ in the plane $\mathbb{R}^2$, define the function $f : \mathbf{I} \to \mathbb{R}$ by

$$f(x, y) = xy \quad \text{for } (x, y) \text{ in } \mathbf{I}.$$

Use the Darboux Sum Convergence Criterion to evaluate $\int_{\mathbf{I}} f$.

10. For the rectangle $\mathbf{I} = [0, 1] \times [-1, 0]$ in the plane $\mathbb{R}^2$, define the function $f : \mathbf{I} \to \mathbb{R}$ by

$$f(x, y) = x^2 y \quad \text{for } (x, y) \text{ in } \mathbf{I}.$$

Use the Darboux Sum Convergence Criterion to evaluate $\int_{\mathbf{I}} f$.

11. For the rectangle $\mathbf{I} = [0, 2] \times [0, 1]$ in the plane $\mathbb{R}^2$, define the function $f : \mathbf{I} \to \mathbb{R}$ by

$$f(x, y) = x + 2y \quad \text{for } (x, y) \text{ in } \mathbf{I}.$$

Use the Darboux Sum Convergence Criterion to evaluate $\int_{\mathbf{I}} f$.

12. Let $\mathbf{I}$ be a generalized rectangle in $\mathbb{R}^n$ and suppose that the function $f : \mathbf{I} \to \mathbb{R}$ is integrable. Assume that $f(\mathbf{x}) \ge 0$ if $\mathbf{x}$ is a point in $\mathbf{I}$ with a rational component. Prove that $\int_{\mathbf{I}} f \ge 0$.

13. Let $\mathbf{I}$ be a generalized rectangle in $\mathbb{R}^n$ and suppose that the function $f : \mathbf{I} \to \mathbb{R}$ is continuous. Assume that $f(\mathbf{x}) \ge 0$ for all points $\mathbf{x}$ in $\mathbf{I}$. Prove that $\int_{\mathbf{I}} f = 0$ if and only if the function $f : \mathbf{I} \to \mathbb{R}$ is identically 0.

14. Let $\mathbf{I}$ be a generalized rectangle in $\mathbb{R}^n$ and suppose that the function $f : \mathbf{I} \to \mathbb{R}$ is integrable. Let the number $M$ have the property that $|f(\mathbf{x})| \le M$ for all $\mathbf{x}$ in $\mathbf{I}$. Prove that

$$\left| \int_{\mathbf{I}} f \right| \le M \cdot \text{vol} \, \mathbf{I}.$$

## 18.2 Integration over Jordan Domains

In the study of integration for functions of a single variable, we considered only those functions that have as their domain a closed bounded interval. For functions of several variables, generalized rectangles do not play a similar preeminent role: it is necessary to integrate functions of several variables that have quite general domains.

**DEFINITION** *For a bounded subset $D$ of $\mathbb{R}^n$ and a bounded function $f: D \to \mathbb{R}$, if $\mathbf{I}$ is a generalized rectangle that contains $D$, we define the zero-extension of $f: D \to \mathbb{R}$ to $\mathbf{I}$, which we denote by $\hat{f}: \mathbf{I} \to \mathbb{R}$, to be the function defined by*

$$\hat{f}(\mathbf{x}) \equiv \begin{cases} f(\mathbf{x}) & \text{if } \mathbf{x} \text{ is in } D \\ 0 & \text{if } \mathbf{x} \text{ is in } \mathbf{I} \backslash D. \end{cases}$$

**DEFINITION** *Let $D$ be a bounded subset of $\mathbb{R}^n$ and let the function $f: D \to \mathbb{R}$ be bounded. Then $f: D \to \mathbb{R}$ is said to be integrable provided that there is a generalized rectangle $\mathbf{I}$ that contains $D$ for which the zero-extension $\hat{f}: \mathbf{I} \to \mathbb{R}$ is integrable; in this case, we define*

$$\int_D f \equiv \int_{\mathbf{I}} \hat{f}. \tag{18.16}$$

It is necessary to show that the above definition is unambiguous—that is, that it is independent of the choice of generalized rectangle containing $D$. Indeed, if $\mathbf{I}_1$ and $\mathbf{I}_2$

FIGURE 18.2

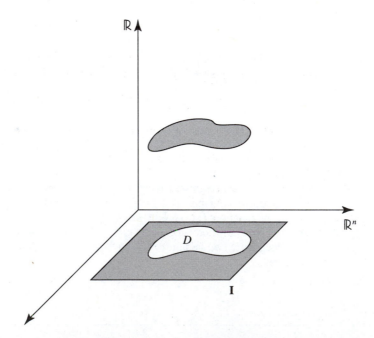

are two generalized rectangles that contain $D$, define $\mathbf{I}$ to be the intersection of $\mathbf{I}_1$ and $\mathbf{I}_2$. By Corollary 18.12, the function $\hat{f}: \mathbf{I}_1 \to \mathbb{R}$ is integrable if and only if the function $\hat{\hat{f}}: \mathbf{I} \to \mathbb{R}$ is integrable, in which case

$$\int_{\mathbf{I}} \hat{f} = \int_{\mathbf{I}_1} \hat{f},$$

and the function $\hat{f}: \mathbf{I}_2 \to \mathbb{R}$ is integrable if and only if the function $\hat{f}: \mathbf{I} \to \mathbb{R}$ is integrable, in which case

$$\int_{\mathbf{I}} \hat{f} = \int_{\mathbf{I}_2} \hat{f}.$$

Thus $\hat{f}: \mathbf{I}_1 \to \mathbb{R}$ is integrable if and only if $\hat{f}: \mathbf{I}_2 \to \mathbb{R}$ is integrable, in which case

$$\int_{\mathbf{I}_1} \hat{f} = \int_{\mathbf{I}_2} \hat{f}.$$

It follows that the above definitions of integrability and of the integral are unambiguous.

---

**THEOREM 18.13**    **Linearity of the Integral**     For $D$ a bounded subset of $\mathbb{R}^n$, suppose that the functions $f: D \to \mathbb{R}$ and $g: D \to \mathbb{R}$ are integrable. Then for any two numbers $\alpha$ and $\beta$, the function $\alpha f + \beta g: D \to \mathbb{R}$ is also integrable and

$$\int_D [\alpha f + \beta g] = \alpha \int_D f + \beta \int_D g. \tag{18.17}$$

**Proof**     Let $\mathbf{I}$ be a generalized rectangle that contains $D$. Observe that the zero-extension of $\alpha f + \beta g: D \to \mathbb{R}$ to $\mathbf{I}$ equals $\alpha \hat{f} + \beta \hat{g}: \mathbf{I} \to \mathbb{R}$. By the linearity property of integration for functions defined on rectangles (Theorem 18.8),

$$\int_D [\alpha f + \beta g] = \int_{\mathbf{I}} [\alpha \hat{f} + \beta \hat{g}] = \alpha \int_{\mathbf{I}} \hat{f} + \beta \int_{\mathbf{I}} \hat{g} = \alpha \int_D f + \beta \int_D g. \quad \blacksquare$$

---

**THEOREM 18.14**    **Monotonicity of the Integral**     For $D$ a bounded subset of $\mathbb{R}^n$, suppose that the functions $f: D \to \mathbb{R}$ and $g: D \to \mathbb{R}$ are integrable and also that

$$f(\mathbf{x}) \le g(\mathbf{x}) \quad \text{for all points } \mathbf{x} \text{ in } D.$$

Then
$$\int_D f \le \int_D g.$$

**Proof**     Choose $\mathbf{I}$ to be a generalized rectangle that contains $D$. Observe that

$$\hat{f}(\mathbf{x}) \le \hat{g}(\mathbf{x}) \quad \text{for all points } \mathbf{x} \text{ in } \mathbf{I}.$$

By the monotonicity property of integration for functions defined on generalized rectangles (Theorem 18.9),

$$\int_D f = \int_{\mathbf{I}} \hat{f} \le \int_{\mathbf{I}} \hat{g} = \int_D g. \quad \blacksquare$$

**THEOREM**
**18.15**

**Additivity over Partitions**   Let $D$ be a bounded subset of $\mathbb{R}^n$ and suppose that the function $f: D \to \mathbb{R}$ is integrable. Then, for each generalized rectangle $\mathbf{I}$ that contains $D$ and each partition $\mathbf{P}$ of $\mathbf{I}$,

$$\int_D f = \sum_{\mathbf{J} \text{ in } \mathbf{I}} \int_{\mathbf{J} \cap D} f.$$

**Proof**   By definition, the zero-extension of $f: D \to \mathbb{R}$ to $\mathbf{I}$, $\hat{f}: \mathbf{I} \to \mathbb{R}$, is integrable. According to the additivity-over-partitions property for integrable functions defined on generalized rectangles (Theorem 18.10),

$$\int_{\mathbf{I}} \hat{f} = \sum_{\mathbf{J} \text{ in } \mathbf{I}} \int_{\mathbf{J}} \hat{f}.$$

The result now follows from the observation that for each generalized rectangle $\mathbf{J}$ in $\mathbf{I}$, $\hat{f}: \mathbf{J} \to \mathbb{R}$ is the zero-extension to $\mathbf{J}$ of the function $f: \mathbf{J} \cap D \to \mathbb{R}$, and so, by definition, $\int_{\mathbf{J}} \hat{f} = \int_{\mathbf{J} \cap D} f$. ∎

For the extension of integration to functions having general domains to be significant, it is necessary to present criteria for determining when such functions are integrable. To find such criteria, it will be useful first to provide a description of the extent to which a bounded function defined on a generalized rectangle can fail to be continuous and yet still be integrable.

For a subset $S$ of $\mathbb{R}^n$, a collection $\mathcal{F}$ of subsets of $\mathbb{R}^n$ is said to *cover* $S$ provided that the union of the sets in the collection contains the set $S$—that is,

$$S \subseteq \bigcup_{F \text{ in } \mathcal{F}} F.$$

**DEFINITION**

*A bounded subset $S$ of $\mathbb{R}^n$ is said to have Jordan content $0$ provided that for each positive number $\epsilon$ there is a finite collection $\mathcal{F}$ of generalized rectangles that cover $S$ and the sum of whose volumes is less than $\epsilon$; that is, if $\mathcal{F} = \{\mathbf{I}_1, \ldots, \mathbf{I}_m\}$, then*

$$S \subseteq \bigcup_{1 \leq j \leq m} \mathbf{I}_j \quad \text{and} \quad \sum_{j=1}^{m} vol\, \mathbf{I}_j < \epsilon.$$

It is clear that if a set $D$ has Jordan content $0$, then each subset of $D$ also has Jordan content $0$. Moreover, the union of a finite number of sets, each of which has Jordan content $0$, also has Jordan content $0$. To verify this, for $k$ a positive integer, let $\{S_i\}_{1 \leq i \leq k}$ be a collection of $k$ subsets of $\mathbb{R}^n$ such that each $S_i$ has Jordan content $0$. We claim that the union $S = \bigcup_{1 \leq i \leq k} S_i$ also has Jordan content $0$. Indeed, let $\epsilon > 0$. For each index $i$ between 1 and $k$, since $\epsilon/k$ is a positive number, we can choose a finite number of generalized rectangles that cover $S_i$, the sum of whose volumes is less than $\epsilon/k$. Taking the union of these $k$ finite collections of generalized rectangles, we obtain a finite collection of generalized rectangles that covers $S$, the sum of whose volumes is less than $k[\epsilon/k] = \epsilon$.

**EXAMPLE 18.5**   Define the segment $S$ in the plane $\mathbb{R}^2$ by

$$S = \{(x, y) \mid 0 \le x \le 1, \ y = x\}.$$

**FIGURE 18.3**

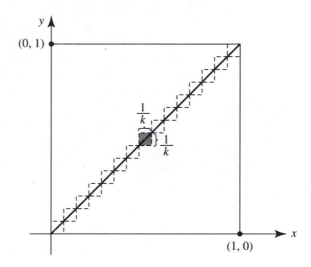

Then $S$ has Jordan content 0. To verify this, let $\epsilon > 0$. Then for each natural number $k$, let $\{\mathbf{I}_j\}_{1 \le j \le k}$ be the collection of $k$ generalized rectangles defined by

$$\mathbf{I}_j = \left[ \frac{j-1}{k}, \frac{j}{k} \right] \times \left[ \frac{j-1}{k}, \frac{j}{k} \right] \quad \text{for} \quad 1 \le j \le k.$$

Clearly, this collection of generalized rectangles covers the set $S$. Each of these rectangles has area $1/k^2$, so the sum of their areas is $1/k$. If $k$ is chosen so that $1/k < \epsilon$, then $S$ is covered by a finite collection of rectangles, the sum of whose areas is less than $\epsilon$.   □

**EXAMPLE 18.6**   Define the set $S$ in $\mathbb{R}^3$ by

$$S = \{(x, y, z) \mid 0 \le x \le 1, \ 0 \le y \le 1, z = 4\}.$$

Then $S$ also has Jordan content 0. Again, to verify this, let $\epsilon > 0$. Then the generalized rectangle

$$\mathbf{I} = [0, 1] \times [0, 1] \times [4 - \epsilon/3, 4 + \epsilon/3]$$

contains $S$ and has volume $2\epsilon/3 < \epsilon$.   □

**EXAMPLE 18.7**    For any generalized rectangle $\mathbf{I}$ in $\mathbb{R}^n$, its boundary $\partial\mathbf{I}^*$ has Jordan content 0. To verify this, we introduce an auxiliary function $f:\mathbf{I} \to \mathbb{R}$, defined by

$$f(\mathbf{x}) = \begin{cases} 1 & \text{if } \mathbf{x} \text{ is in } \partial\mathbf{I} \\ 0 & \text{if } \mathbf{x} \text{ is in int } \mathbf{I}. \end{cases}$$

Observe that the function $f:\mathbf{I} \to \mathbb{R}$ is bounded and is continuous on the interior of $\mathbf{I}$. It follows from Theorem 18.11 that $f:\mathbf{I} \to \mathbb{R}$ is integrable. Let $\epsilon > 0$. By the Integrability Criterion, there is a partition $\mathbf{P}$ of $\mathbf{I}$ such that

$$U(f, \mathbf{P}) - L(f, \mathbf{P}) < \epsilon.$$

Define $\mathcal{F}$ to be the collection of generalized rectangles in $\mathbf{P}$ that intersect the boundary of $\mathbf{I}$. From the definition of the Darboux sums and of the function $f:\mathbf{I} \to \mathbb{R}$, the above inequality means that the sum of the volumes of the generalized rectangles in $\mathcal{F}$ is less than $\epsilon$. Since the collection $\mathcal{F}$ is a cover of $\partial\mathbf{I}$, we conclude that $\partial\mathbf{I}$ has Jordan content 0. $\qquad\square$

**THEOREM 18.16**    Let $\mathbf{I}$ be a generalized rectangle in $\mathbb{R}^n$ and let the function $f:\mathbf{I} \to \mathbb{R}$ be bounded. If the set of discontinuities of $f:\mathbf{I} \to \mathbb{R}$ has Jordan content 0, then the function $f:\mathbf{I} \to \mathbb{R}$ is integrable.

**Proof**    We will use the Integrability Criterion. First, since $f:\mathbf{I} \to \mathbb{R}$ is bounded, we can choose a number $M > 0$ such that

$$|f(\mathbf{x})| \le M \qquad \text{for all } \mathbf{x} \text{ in } \mathbf{I}.$$

Let $\epsilon > 0$. Denote by $D$ the set of points in $\mathbf{I}$ at which the function $f:\mathbf{I} \to \mathbb{R}$ fails to be continuous. Since the number $\epsilon/4M$ is positive and the set $D$ has Jordan content 0, we can choose a finite collection $\mathcal{F}$ of generalized rectangles that cover $D$, the sum of whose volumes is less than $\epsilon/4M$. We can also assume that each generalized rectangle in $\mathcal{F}$ is a subset of $\mathbf{I}$.

For each index $i$ between 1 and $n$, define $P_i$ to be a partition of the $i$th edge of $\mathbf{I}$ containing all of the endpoints of the $i$th edges of the generalized rectangles in $\mathcal{F}$. Define $\mathbf{P} = (P_1, \ldots, P_n)$. The partition $\mathbf{P}$ has been constructed to have the property that each generalized rectangle in $\mathcal{F}$ is the union of rectangles in the partition $\mathbf{P}$. We divide the set of generalized rectangles in the partition $\mathbf{P}$ into those that are contained in one of the generalized rectangles in $\mathcal{F}$, which we list as $\mathbf{J}_1', \ldots, \mathbf{J}_\ell'$, and those that do not have this property, which we list as $\mathbf{J}_1, \ldots, \mathbf{J}_m$. Since the sum of the volumes of the generalized rectangles in $\mathcal{F}$ is less than $\epsilon/4M$,

$$\sum_{i=1}^{\ell} \operatorname{vol} \mathbf{J}_i' < \epsilon/4M,$$

and hence, by the choice of $M$,

$$\sum_{i=1}^{\ell} [M(f, \mathbf{J}_i') - m(f, \mathbf{J}_i')] \operatorname{vol} \mathbf{J}' < \frac{\epsilon}{2}. \tag{18.18}$$

---

*In Chapter 10 we used the notation bd $D$ to denote the boundary of a subset $D$ of $\mathbb{R}^n$. Another common notation for the boundary of $D$, particularly in the study of integration, is $\partial D$; we will use this notation in this and the next chapter.

On the other hand, for each index $i$ between 1 and $m$, because the collection of rectangles $\{\mathbf{J}'_1, \ldots, \mathbf{J}'_\ell\}$ covers $D$, the function $f : \mathbf{J}_i \to \mathbb{R}$ is continuous on the interior of $\mathbf{J}_i$. Thus, by Theorem 18.11, $f : \mathbf{J}_i \to \mathbb{R}$ is integrable, so by the Integrability Criterion we can choose a partition $\mathbf{P}_i$ of $\mathbf{J}_i$ such that

$$U(f, \mathbf{P}_i) - L(f, \mathbf{P}_i) < \frac{\epsilon}{2m}. \tag{18.19}$$

Choose $\mathbf{P}^*$ to be a refinement of the partition $\mathbf{P}$ that for each index $i$ between 1 and $m$ induces a partition of $\mathbf{J}_i$ that is a refinement of $\mathbf{P}_i$. Then, by the Refinement Lemma,

$$U(f, \mathbf{P}^*) - L(f, \mathbf{P}^*) \le \sum_{i=1}^{\ell} [M(f, \mathbf{J}'_i) - m(f, \mathbf{J}'_i)] \text{vol}\, \mathbf{J}'_i + \sum_{i=1}^{m} [U(f, \mathbf{P}_i) - L(f, \mathbf{P}_i)]$$

$$< \frac{\epsilon}{2} + m\left[\frac{\epsilon}{2m}\right] = \epsilon. \qquad \blacksquare$$

For a continuous function defined on a bounded subset $D$ of $\mathbb{R}^n$, the set of discontinuities of any zero-extension are contained in the boundary of $D$ (see Exercise 4). Thus, since a bounded function on a generalized rectangle is integrable provided that its set of discontinuities has Jordan content 0, we are led to focus on the following class of bounded subsets of $\mathbb{R}^n$.

**DEFINITION**   *A bounded subset $D$ of $\mathbb{R}^n$ is said to be a Jordan domain provided that its boundary has Jordan content 0.*

**THEOREM 18.17**   Let $D$ be a Jordan domain in $\mathbb{R}^n$ and let the function $f : D \to \mathbb{R}$ be bounded. If the set of discontinuities of $f : D \to \mathbb{R}$ has Jordan content 0, then $f : D \to \mathbb{R}$ is integrable.

**Proof**   Choose $\mathbf{I}$ to be a generalized rectangle in $\mathbb{R}^n$ that contains the domain $D$; it is necessary to show that the zero-extension $\hat{f} : \mathbf{I} \to \mathbb{R}$ is integrable. According to Theorem 18.16, to do so it suffices to show that the set of discontinuities of $\hat{f} : \mathbf{I} \to \mathbb{R}$ has Jordan content 0. But the zero-extension is clearly continuous at each point $\mathbf{x}$ in the interior of $D$ at which the function $f : D \to \mathbb{R}$ is continuous, since there is a neighborhood of $\mathbf{x}$ on which $\hat{f} : \mathbf{I} \to \mathbb{R}$ agrees with the function $f : D \to \mathbb{R}$. On the other hand, for each point $\mathbf{x}$ in $\mathbf{I}$ that is in the exterior of $D$, there is a positive number $r$ such that the function $\hat{f} : \mathbf{I} \to \mathbb{R}$ is identically 0 on $\mathcal{N}_r(\mathbf{x}) \cap \mathbf{I}$, so $\hat{f} : \mathbf{I} \to \mathbb{R}$ is certainly continuous at $\mathbf{x}$. Thus the set of discontinuities of $\hat{f} : \mathbf{I} \to \mathbb{R}$ is contained in the union of the boundary of $D$ and the set of discontinuities of $f : D \to \mathbb{R}$ that are in the interior of $D$. By assumption, each of these sets has Jordan content 0, and hence so does their union. $\qquad \blacksquare$

For the above theorem to be useful, it is necessary to provide criteria for a set to have Jordan content 0. So far, the only general result we have is that the boundary of a generalized rectangle has Jordan content 0. The following proposition and its corollary will provide many more examples.

**PROPOSITION 18.18**    Let $\mathbf{I}$ be a generalized rectangle in $\mathbb{R}^{n-1}$ and suppose that the function $g: \mathbf{I} \to \mathbb{R}$ is integrable. Then the subset of $\mathbb{R}^n$ consisting of the graph of $g: \mathbf{I} \to \mathbb{R}$,

$$\{(\mathbf{x}, g(\mathbf{x})) \text{ in } \mathbb{R}^n \mid \mathbf{x} \text{ in } \mathbf{I}\},$$

has Jordan content 0.

**Proof**    Let $\epsilon > 0$. According to the Integrability Criterion, there is a partition $\mathbf{P}$ of $\mathbf{I}$ having the property that $U(g, \mathbf{P}) - L(g, \mathbf{P}) < \epsilon$; that is,

$$\sum_{\mathbf{J} \text{ in } \mathbf{P}} [M(g, \mathbf{J}) - m(g, \mathbf{J})] \text{vol} \, \mathbf{J} < \epsilon,$$

where for each generalized rectangle $\mathbf{J}$ in $\mathbf{P}$,

$$M(g, \mathbf{J}) = \sup \{g(\mathbf{x}) \mid \mathbf{x} \text{ in } \mathbf{J}\} \quad \text{and} \quad m(g, \mathbf{J}) = \inf \{g(\mathbf{x}) \mid \mathbf{x} \text{ in } \mathbf{J}\}.$$

The graph of $g: \mathbf{I} \to \mathbb{R}$ is contained in the union of the products $\mathbf{J} \times [m(g, \mathbf{J}), M(g, \mathbf{J})]$, as $\mathbf{J}$ varies in $\mathbf{P}$. The preceding inequality means precisely that the sum of the volumes of these products is less than $\epsilon$. Thus the graph has Jordan content 0.    ∎

**COROLLARY 18.19**    Let $D$ be a Jordan domain in $\mathbb{R}^{n-1}$, let the function $f: D \to \mathbb{R}$ be bounded, and let the set of discontinuities of $f: D \to \mathbb{R}$ have Jordan content 0. Then the subset of $\mathbb{R}^n$ consisting of the graph of $f: D \to \mathbb{R}$,

$$\{(\mathbf{x}, g(\mathbf{x})) \text{ in } \mathbb{R}^n \mid \mathbf{x} \text{ in } D\},$$

has Jordan content 0.

**Proof**    Let $\mathbf{I}$ be a generalized rectangle that contains $D$. Since $f: D \to \mathbb{R}$ is integrable, the zero-extension of $f: D \to \mathbb{R}$ to $\mathbf{I}$, $\hat{f}: \mathbf{I} \to \mathbb{R}$, is also integrable. By Proposition 18.18, the graph of $\hat{f}: \mathbf{I} \to \mathbb{R}$ has Jordan content 0. Since the graph of $f: D \to \mathbb{R}$ is a subset of the graph of $\hat{f}: \mathbf{I} \to \mathbb{R}$, the graph of $f: D \to \mathbb{R}$ also has Jordan content 0.    ∎

**EXAMPLE 18.8**    The following subsets of the plane $\mathbb{R}^2$ are Jordan domains:

$$\{(x, y) \mid x^2 + y^2 < 1\}, \qquad \{(x, y) \mid x^2 + y^2 \leq 1\}.$$

Indeed, each of these sets has as its boundary the unit circle $S = \{(x, y) \mid x^2 + y^2 = 1\}$, so it is necessary to show that $S$ has Jordan content 0. But $S$ is the union of two graphs of continuous functions defined on $[-1, 1]$. The preceding corollary asserts that each of these graphs has Jordan content 0, and hence so does their union.    □

**EXAMPLE 18.9**   For $c > 0$, define the cone $C$ in $\mathbb{R}^3$ by

$$C = \left\{ \left( x, \, y, \, z \right) \mid x^2 + y^2 \leq 1, \, 0 \leq z \leq c \left( 1 - \sqrt{x^2 + y^2} \right) \right\}.$$

We claim that $C$ is a Jordan domain. Indeed, the boundary of $C$ consists of

$$\left\{ (x, \, y, \, z) \mid x^2 + y^2 \leq 1, \, z = 0 \right\} \cup \left\{ (x, \, y, \, z) \mid x^2 + y^2 \leq 1, \, z = c \left( 1 - \sqrt{x^2 + y^2} \right) \right\}.$$

Each of the two sets in this union is the graph of a continuous function of two variables defined on a domain that, by the preceding example, is a Jordan domain. The preceding corollary implies that each of these graphs has Jordan content 0, and hence so does their union.                                                                                    □

**EXAMPLE 18.10**   For $c > 0$, define the cylinder $C$ in $\mathbb{R}^3$ by

$$C = \{(x, \, y, \, z) \mid x^2 + y^2 \leq 1, \, 0 \leq z \leq c\}.$$

We claim that $C$ is a Jordan domain. Indeed, the boundary of $C$ is the union of the top, $\{(x, \, y, \, z) \mid x^2 + y^2 \leq 1, \, z = c\}$, the bottom, $\{(x, \, y, \, z) \mid x^2 + y^2 \leq 1, \, z = 0\}$, and the lateral side

$$\{(x, \, y, \, z) \mid x^2 + y^2 = 1, \, 0 \leq z \leq c\}.$$

Arguing as in Example 18.9, the top and the bottom have Jordan content 0. The lateral side may be written as

$$\left\{ (x, \, y, \, z) \mid -1 \leq x \leq 1, \, 0 \leq z \leq c, y = \pm\sqrt{1 - x^2} \right\}$$

Since this set is the union of two sets, each of which is the graph of a continuous function defined on a rectangle, the lateral side has Jordan content 0. Hence $C$ is a Jordan domain.                                                                             □

⟶ The above three examples illustrate a general way to check that a subset $D$ of $\mathbb{R}^n$ is a Jordan domain: It suffices to show that the boundary of $D$ is the union of a finite number of graphs of integrable functions of $n - 1$ variables.

**LEMMA 18.20**   Let $\mathbf{I}$ be a generalized rectangle in $\mathbb{R}^n$ and suppose that the bounded function $f: \mathbf{I} \to \mathbb{R}$ has the property that there is a subset $S$ of $\mathbf{I}$ having Jordan content 0 such that

$$f(\mathbf{x}) = 0 \quad \text{for all points } \mathbf{x} \text{ in } \mathbf{I} \backslash S.$$

Then $f: \mathbf{I} \to \mathbb{R}$ is integrable and $\int_{\mathbf{I}} f = 0$.

**Proof**   It is not difficult to see that since the set $S$ has Jordan content 0, the boundary of $S$ also has Jordan content 0 (see Exercise 14). Thus $S$ is a Jordan domain. Moreover, the function $f: \mathbf{I} \to \mathbb{R}$ is the zero-extension of the function $f: S \to \mathbb{R}$. It follows from Theorem 18.16 that the function $f: \mathbf{I} \to \mathbb{R}$ is integrable, so it is only necessary to prove that $\int_{\mathbf{I}} f = 0$. Establishing this is equivalent to showing that for each $\epsilon > 0$,

$$-\epsilon < \int_{\mathbf{I}} f < \epsilon. \tag{18.20}$$

Let $\epsilon > 0$. First, since $f: \mathbf{I} \to \mathbb{R}$ is bounded, we can choose $M > 0$ such that

$$-M \leq f(\mathbf{x}) \leq M \quad \text{for all points } \mathbf{x} \text{ in } \mathbf{I},$$

so that for each generalized rectangle $\mathbf{J}$ that is a subset of $\mathbf{I}$,

$$-M\text{vol}\,\mathbf{J} \leq m(f,\mathbf{J})\text{vol}\,\mathbf{J} \leq M(f,\mathbf{J})\text{vol}\,\mathbf{J} \leq M\text{vol}\,\mathbf{J}.$$

The construction that was made in the proof of Theorem 18.16 shows that there is a partition $\mathbf{P}$ of $\mathbf{I}$ such that the sum of the volumes of the generalized rectangles in $\mathbf{P}$ that intersect $S$ is less than $\epsilon/M$. Since for each generalized rectangle $\mathbf{J}$ in $\mathbf{P}$ that does not intersect $S$, $m(f,\mathbf{J}) = M(f,\mathbf{J}) = 0$, we see that

$$-\epsilon = -M[\epsilon/M] < L(f,\mathbf{P}) \leq \int_{\mathbf{I}} f \leq U(f,\mathbf{P}) < M[\epsilon/M] = \epsilon.$$

Thus (18.20) holds.    ∎

**THEOREM 18.21**    Let $D_1$ and $D_2$ be bounded subsets of $\mathbb{R}^n$. For $D = D_1 \cup D_2$, suppose that the function $f: D \to \mathbb{R}$ has the property that both $f: D_1 \to \mathbb{R}$ and $f: D_2 \to \mathbb{R}$ are integrable. If the intersection $D_1 \cap D_2$ has Jordan content 0, then the function $f: D \to \mathbb{R}$ is also integrable and

$$\int_D f = \int_{D_1} f + \int_{D_2} f. \tag{18.21}$$

**Proof**    Choose $\mathbf{I}$ to be a generalized rectangle that contains $D$. Define $\hat{f}_1: \mathbf{I} \to \mathbb{R}$ to be the zero-extension of $f: D_1 \to \mathbb{R}$ to $\mathbf{I}$, $\hat{f}_2: \mathbf{I} \to \mathbb{R}$ to be the zero-extension of $f: D_2 \to \mathbb{R}$ to $\mathbf{I}$, and $\hat{f}: \mathbf{I} \to \mathbb{R}$ to be the zero-extension of $f: D \to \mathbb{R}$ to $\mathbf{I}$. It is necessary to show that

$$\int_{\mathbf{I}} \hat{f} = \int_{\mathbf{I}} \hat{f}_1 + \int_{\mathbf{I}} \hat{f}_2. \tag{18.22}$$

To do so, define the auxiliary function $g: \mathbf{I} \to \mathbb{R}$ by

$$g(\mathbf{x}) = \hat{f}(\mathbf{x}) - [\hat{f}_1(\mathbf{x}) + \hat{f}_2(\mathbf{x})] \quad \text{for all points } \mathbf{x} \text{ in } \mathbf{I}.$$

Observe that the only possible points $\mathbf{x}$ in $\mathbf{I}$ at which $g(\mathbf{x}) \neq 0$ are those points in $D_1 \cap D_2$. By assumption, the set $D_1 \cap D_2$ has Jordan content 0, so it follows from Lemma 18.20 that the function $g: \mathbf{I} \to \mathbb{R}$ is integrable and $\int_{\mathbf{I}} g = 0$. By the linearity of the integral for functions defined on generalized rectangles (Theorem 18.8), the function $\hat{f}: \mathbf{I} \to \mathbb{R}$ is integrable and

$$\int_{\mathbf{I}} \hat{f} = \int_{\mathbf{I}} [\hat{f}_1 + \hat{f}_2 + g] = \int_{\mathbf{I}} \hat{f}_1 + \int_{\mathbf{I}} \hat{f}_2 + \int_{\mathbf{I}} g = \int_{\mathbf{I}} \hat{f}_1 + \int_{\mathbf{I}} \hat{f}_2;$$

that is, (18.22) holds.    ∎

We have defined the *volume* of a generalized rectangle to be the product of the lengths of its edges. It is useful to define volume for more general types of sets.

**DEFINITION**    *For a bounded subset $D$ of $\mathbb{R}^n$, suppose the function $f: D \to \mathbb{R}$ that is identically equal to 1 is integrable. Then the set $D$ is said to have volume and its volume, which is denoted by vol $D$, is defined by*

$$vol\,D = \int_D f.$$

Since a constant function is certainly continuous, it follows from Theorem 18.17 that every Jordan domain $D$ in $\mathbb{R}^n$ has volume; moreover, since by definition its boundary $\partial D$ has Jordan content 0, by Lemma 18.20, the set $\partial D$ also has volume and vol $\partial D = 0$. Thus, by the additivity of integration formula (18.21), the set $D \cup \partial D$ also has volume and

$$\text{vol}\,(D \cup \partial D) = \text{vol}\, D + \text{vol}\, \partial D = \text{vol}\, D. \tag{18.23}$$

It is useful to record for future reference an additivity of volume result.

**COROLLARY**   **Additivity of Volume**   Suppose that $D_1$ and $D_2$ are Jordan domains in $\mathbb{R}^n$ that are
**18.22**   disjoint. Then $D = (D_1 \cup D_2)$ is also a Jordan domain, as is $D \cup \partial D$, and

$$\text{vol}\,(D \cup \partial D) = \text{vol}\, D = \text{vol}\, D_1 + \text{vol}\, D_2.$$

**Proof**   It is not difficult to see that $\partial D \subseteq \partial D_1 \cup \partial D_2$. By definition, $\partial D_1$ and $\partial D_2$ have Jordan content 0, and hence so does $\partial D$. Thus $D$ is an Jordan domain. Thus, by the additivity of integration formula (18.21) and formula (18.23),

$$\text{vol}\,(D \cup \partial D) = \text{vol}\, D = \text{vol}\, D_1 + \text{vol}\, D_2. \qquad \blacksquare$$

It is not the case that all bounded subsets of $\mathbb{R}^n$ have volume. For instance, Example 18.2 shows that the set of rational numbers in the interval $[0, 1]$ does not have volume.

The following is a general mean value property for integrals, which extends a result already established for continuous functions of a single variable that are defined on closed bounded intervals.

**THEOREM**   **The Mean Value Property for Integrals**   Let $D$ be a compact, pathwise connected
**18.23**   Jordan domain in $\mathbb{R}^n$ with positive volume and suppose that the function $f : D \to \mathbb{R}$ is continuous. Then there is a point $\mathbf{x}$ in $D$ at which

$$\frac{1}{\text{vol}\, D} \int_D f = f(\mathbf{x}). \tag{18.24}$$

**Proof**   Since $D$ is compact, we may apply the Extreme Value Theorem to select points $\mathbf{x}_1$ and $\mathbf{x}_2$ in $D$ such that

$$f(\mathbf{x}_1) \le f(\mathbf{x}) \le f(\mathbf{x}_2) \quad \text{for all } \mathbf{x} \text{ in } \mathbf{I}.$$

Since $D$ is a Jordan domain, any continuous bounded function that is defined on $D$ is integrable. Thus the function $f : D \to \mathbb{R}$ is integrable, as is any constant function defined on $D$. By the monotonicity property of the integral,

$$f(\mathbf{x}_1)\text{vol}\, D = \int_D f(\mathbf{x}_1) \le \int_D f \le \int_D f(\mathbf{x}_2) = f(\mathbf{x}_2)\text{vol}\, D.$$

Dividing the entire inequality by the positive number vol $D$, we have

$$f(\mathbf{x}_1) \le \frac{1}{\text{vol}\, D} \int_D f \le f(\mathbf{x}_2).$$

Since $D$ is pathwise connected, there is a continuous function $\psi : [0, 1] \rightarrow D$ such that $\psi(0) = \mathbf{x}_1$ and $\psi(1) = \mathbf{x}_2$. Define $\phi = f \circ \psi : [0, 1] \rightarrow \mathbb{R}$. Observe that

$$\phi(0) = f(\mathbf{x}_1) \leq \frac{1}{\text{vol } D} \int_D f \leq f(\mathbf{x}_2) = \phi(1).$$

Moreover, since the composition of continuous mappings is continuous, the function $\phi : [0, 1] \rightarrow \mathbb{R}$ is continuous. Thus, by the Intermediate Value Theorem, we can select a point $c$ in the interval $[0, 1]$ at which

$$\phi(c) = \frac{1}{\text{vol } D} \int_D f.$$

For $\mathbf{x} = \psi(c)$, we see that (18.24) holds. ∎

The concept of integral that we considered in Chapters 6 and 7 and that we are considering in this chapter is often called the *Riemann integral*. There are more general concepts of integral that are also useful; the idea is to broaden the class of functions that are to be considered integrable. Any such extension also broadens the class of subsets of $\mathbb{R}^n$ that are considered to have volume. A detailed study of one such extension, called the Lebesgue integral, may be found in Chapters 10 and 15 of Apostol's book *Mathematical Analysis*, second edition, Addison-Wesley, 1974.

---

**EXERCISES**

1. For positive numbers $a$ and $b$, show that the ellipse

$$\{(x, y) \text{ in } \mathbb{R}^2 \mid x^2/a^2 + y^2/b^2 = 1\}$$

   has Jordan content 0.
2. Show that the set of real numbers $\{1/n \mid n \text{ in } \mathbb{N}\}$ has Jordan content 0.
3. Show that the ellipsoid

$$\{(x, y, z) \text{ in } \mathbb{R}^3 \mid x^2 + y^2 + z^2/4 = 1\}$$

   has Jordan content 0.
4. For a subset $S$ of $\mathbb{R}^n$, the *characteristic function* of $S$, which is denoted by $\chi : \mathbb{R}^n \rightarrow \mathbb{R}$, is defined by

$$\chi(\mathbf{x}) \equiv \begin{cases} 1 & \text{for } \mathbf{x} \text{ in } S \\ 0 & \text{for } \mathbf{x} \text{ not in } S. \end{cases}$$

   Show that the set of discontinuities of this characteristic function consists of the boundary of $S$.
5. For $\mathbf{I}$ a generalized rectangle in $\mathbb{R}^n$, define $f : \mathbf{I} \rightarrow \mathbb{R}$ to be the function with constant value 1. Find a subset $D$ of $\mathbf{I}$ such that the restriction $f : D \rightarrow \mathbb{R}$ is not integrable.
6. For a subset $S$ of $\mathbb{R}^n$ that is contained in the generalized rectangle $\mathbf{I}$, define the function $f : \mathbf{I} \rightarrow \mathbb{R}$ by

$$f(\mathbf{x}) = \begin{cases} 1 & \text{for } \mathbf{x} \text{ in } S \\ 0 & \text{for } \mathbf{x} \text{ not in } S. \end{cases}$$

   Show that $f : \mathbf{I} \rightarrow \mathbb{R}$ is integrable if $S$ has Jordan content 0.
7. Let $\mathbf{I}$ be a generalized rectangle in $\mathbb{R}^n$. By observing that there is a bounded function $f : \mathbf{I} \rightarrow \mathbb{R}$ that is not integrable, use Lemma 18.20 to conclude that $\mathbf{I}$ does not have Jordan content 0.
8. Let $A = \{x \text{ in } \mathbb{R} \mid 0 \leq x \leq 1\}$ and $B = \{(x, y) \text{ in } \mathbb{R}^2 \mid 0 \leq x \leq 1, y = 0\}$. Show that $A$ does not have Jordan content 0, whereas $B$ has Jordan content 0. Is this consistent?

9.  For two subsets $D_1$ and $D_2$ of $\mathbb{R}^n$, show that

$$\partial(D_1 \cup D_2) \subseteq \partial D_1 \cup \partial D_2.$$

Provide examples where there is equality and where there fails to be equality.

10. Prove that the union of two Jordan domains is a Jordan domain.

11. For $\mathbf{I}$ a generalized rectangle in $\mathbb{R}^n$, let $A$ be a subset of $\mathbf{I}$ of Jordan content 0, and suppose that the integrable functions $f : \mathbf{I} \to \mathbb{R}$ and $g : \mathbf{I} \to \mathbb{R}$ are such that

$$f(\mathbf{x}) = g(\mathbf{x}) \quad \text{for } \mathbf{x} \text{ in } \mathbf{I} \backslash A.$$

Show that
$$\int_{\mathbf{I}} f = \int_{\mathbf{I}} g.$$

12. Let $\mathbf{I}$ be a generalized rectangle in $\mathbb{R}^n$ and let the function $f : \mathbf{I} \to \mathbb{R}$ be integrable. Denote the interior of $\mathbf{I}$ by $D$. Show that the restriction $f : D \to \mathbb{R}$ is integrable and that

$$\int_{\mathbf{I}} f = \int_{D} f.$$

13. Let $D$ be a compact, connected Jordan domain in $\mathbb{R}^n$ with positive volume, and suppose that the function $f : D \to \mathbb{R}$ is continuous. Show that there is a point $\mathbf{x}$ in $D$ at which

$$f(\mathbf{x}) = \frac{1}{\text{vol } D} \int_{D} f.$$

14. (a)  Let $S$ and $F$ be subsets of $\mathbb{R}^n$ such that $S \subseteq F$. If $F$ is closed, show that $\partial S \subseteq F$.

(b)  Use (a) and the fact that the union of a finite number of generalized rectangles is closed to show that if $S$ has Jordan content 0, then $\partial S$ also has Jordan content 0.

---

## 18.3   Iterated Integration: Fubini's Theorem

At this point, we have no general method to actually evaluate the integral of an integrable function of several variables. This section will be devoted to reducing the evaluation of such integrals to the problem of evaluating the integrals of functions of a single variable, in which case it is often possible to use the First Fundamental Theorem of Calculus.

**THEOREM 18.24**

**Fubini's Theorem in the Plane**   Suppose that the function $f : \mathbf{I} \to \mathbb{R}$ is integrable, where $\mathbf{I} = [a, b] \times [c, d]$ is a rectangle in the plane $\mathbb{R}^2$. For each point $x$ in $[a, b]$, define the function $F_x : [c, d] \to \mathbb{R}$ by $F_x(y) = f(x, y)$ for $y$ in $[c, d]$, suppose that the function $F_x : [c, d] \to \mathbb{R}$ is integrable, and define

$$A(x) = \int_c^d f(x, y)\, dy.$$

Then the function $A : [a, b] \to \mathbb{R}$ is integrable, and

$$\int_{\mathbf{I}} f = \int_a^b A(x)\, dx = \int_a^b \left[ \int_c^d f(x, y)\, dy \right] dx. \tag{18.25}$$

**Proof**   The crucial point of the proof is to verify the following inequality that is satisfied by the Darboux sums for the two functions $f: \mathbf{I} \rightarrow \mathbb{R}$ and $A: [a, b] \rightarrow \mathbb{R}$:

$$L(f, \mathbf{P}) \leq L(A, P_1) \leq U(A, P_1) \leq U(f, \mathbf{P}) \quad \text{for every partition } \mathbf{P} = (P_1, P_2) \text{ of } \mathbf{I}.$$
(18.26)

Indeed, once this is verified, it follows immediately from the Integrability Criterion that the integrability of $f: \mathbf{I} \rightarrow \mathbb{R}$ implies the integrability of $A: [a, b] \rightarrow \mathbb{R}$. It also follows that

$$L(f, \mathbf{P}) \leq \int_a^b A(x)\, dx \leq U(f, \mathbf{P}) \quad \text{for every partition } \mathbf{P} \text{ of } \mathbf{I},$$

so since $\int_{\mathbf{I}} f$ is the only number that lies between $L(f, \mathbf{P})$ and $U(f, \mathbf{P})$ for every partition $\mathbf{P}$ of $\mathbf{I}$, formula (18.25) holds.

It remains to verify the inequality (18.26). Let $\mathbf{P} = (P_1, P_2)$ be a partition of $\mathbf{I}$, where $P_1 = \{x_0, \ldots, x_m\}$ and $P_2 = \{y_0, \ldots, y_\ell\}$ are partitions of $[a, b]$ and $[c, d]$, respectively. For a pair of indices $i$ and $j$ such that $1 \leq i \leq m$ and $1 \leq j \leq \ell$, define

$$m_{ij} = \inf \{f(x, y) \mid (x, y) \text{ in } [x_{i-1}, x_i] \times [y_{j-1}, y_j]\}$$
$$M_{ij} = \sup \{f(x, y) \mid (x, y) \text{ in } [x_{i-1}, x_i] \times [y_{j-1}, y_j]\}$$
$$m_i = \inf \{A(x) \mid x \text{ in } [x_{i-1}, x_i]\}$$

and

$$M_i = \sup \{A(x) \mid x \text{ in } [x_{i-1}, x_i]\}.$$

Then, by definition,

$$L(f, \mathbf{P}) = \sum_{i=1}^m \sum_{j=1}^\ell m_{ij}[x_i - x_{i-1}][y_j - y_{j-1}]$$

$$U(f, \mathbf{P}) = \sum_{i=1}^m \sum_{j=1}^\ell M_{ij}[x_i - x_{i-1}][y_j - y_{j-1}]$$

$$L(A, P_1) = \sum_{i=1}^m m_i[x_i - x_{i-1}]$$

and

$$U(A, P_1) = \sum_{i=1}^m M_i[x_i - x_{i-1}].$$

Fix an index $i$ between 1 and $m$ and a point $x$ in the interval $[x_{i-1}, x_i]$. Then for each index $j$ between 1 and $\ell$,

$$m_{ij} \leq f(x, y) \leq M_{ij} \quad \text{for all points } y \text{ in } [y_{j-1}, y_j].$$

The monotonicity property of the integral of functions defined on the interval $[y_{j-1}, y_j]$ implies that

$$m_{ij}[y_j - y_{j-1}] \leq \int_{y_{j-1}}^{y_j} f(x, y)\, dy \leq M_{ij}[y_j - y_{j-1}].$$

Summing this inequality for $j = 1, \ldots, \ell$ and using the additivity over intervals property of the integral, we obtain

$$\sum_{j=1}^{\ell} m_{ij}[y_j - y_{j-1}] \le \int_d^c f(x, y)\, dy \le \sum_{j=1}^{\ell} M_{ij}[y_j - y_{j-1}].$$

Since this inequality holds for each point $x$ in $[x_{i-1}, x_i]$, it follows from the definition of $m_i$ and $M_i$ that

$$\sum_{j=1}^{\ell} m_{ij}[y_j - y_{j-1}] \le m_i \le M_i \le \sum_{j=1}^{\ell} M_{ij}[y_j - y_{j-1}].$$

Multiply this inequality by $x_i - x_{i-1}$ and sum the resulting $m$ inequalities for $i = 1, \ldots, m$ to obtain

$$L(f, \mathbf{P}) \le L(A, P_1) \le U(A, P_1) \le U(f, \mathbf{P}). \qquad \blacksquare$$

**EXAMPLE 18.11** Define $f(x, y) = e^{xy}x$ for $(x, y)$ in $\mathbf{I} = [1, 2] \times [0, 1]$. Since the function $f: \mathbf{I} \to \mathbb{R}$ is continuous, it follows from Fubini's Theorem that iterated integration is permissible. By the First Fundamental Theorem of Calculus,

$$\int_{\mathbf{I}} f = \int_1^2 \left[ \int_0^1 e^{xy}x\, dy \right] dx = \int_1^2 \left[ e^x - 1 \right] dx = e^2 - e - 1. \qquad \square$$

**THEOREM 18.25** For continuous functions $h: [a, b] \to \mathbb{R}$ and $g: [a, b] \to \mathbb{R}$ that have the property that $h(x) \le g(x)$ for all points $x$ in $[a, b]$, define

$$D = \{(x, y) \mid a \le x \le b, \ h(x) \le y \le g(x)\}.$$

Suppose that the function $f: D \to \mathbb{R}$ is continuous. Then

$$\int_D f = \int_a^b \left[ \int_{h(x)}^{g(x)} f(x, y)\, dy \right] dx. \qquad (18.27)$$

**Proof** The set $D$ is a Jordan domain, since its boundary consists of the union of four graphs, each of which is the graph of a continuous function on a bounded interval. Choose an interval $[c, d]$ such that the rectangle $\mathbf{I} = [a, b] \times [c, d]$ contains $D$ and let $\hat{f}: \mathbf{I} \to \mathbb{R}$ be the zero-extension of $f: D \to \mathbb{R}$ to $\mathbf{I}$. Theorem 18.17 implies that the function $\hat{f}: \mathbf{I} \to \mathbb{R}$ is integrable. Observe that by the additivity over domains property of the integral and the integrability of bounded functions defined on an interval, which are continuous except possibly at the endpoints,

$$A(x) \equiv \int_c^d \hat{f}(x, y)\, dy = \int_{h(x)}^{g(x)} f(x, y)\, dy \quad \text{for all } x \text{ in } [a, b].$$

Formula (18.27) now follows from Fubini's Theorem in the plane. $\qquad \blacksquare$

**EXAMPLE 18.12**   For $D = \{(x, y) \mid x^2 + y^2 \le 1, y \ge 0\}$, define $f: D \to \mathbb{R}$ to be the constant function with value 1. Then

$$D = \{(x, y) \mid -1 \le x \le 1, 0 \le y \le \sqrt{1 - x^2}\},$$

so by Theorem 18.25,

$$\int_D f = \int_{-1}^1 \left[ \int_0^{\sqrt{1-x^2}} dy \right] dx = \int_{-1}^1 \left[ \sqrt{1 - x^2} \right] dx = \frac{\pi}{2}. \qquad \square$$

In the case when the function $f: \mathbf{I} \to \mathbb{R}$ fails to be continuous, some care is needed in verifying formula (18.25). As the next example shows, it is possible for the integral on the right-hand side of formula (18.25) to be properly defined and yet for the function $f: \mathbf{I} \to \mathbb{R}$ to fail to be integrable.

**EXAMPLE 18.13**   Define the function $f: \mathbf{I} \to \mathbb{R}$, where $\mathbf{I} = [0, 1] \times [0, 1]$, by

$$f(x, y) = \begin{cases} 1 & \text{if } x \text{ is rational} \\ 2y & \text{if } x \text{ is irrational.} \end{cases}$$

For each point $x$ in $[0, 1]$, $\int_0^1 f(x, y)\, dy = 1$, so that

$$\int_0^1 \left[ \int_0^1 f(x, y)\, dy \right] dx = 1.$$

But it is not difficult to see that the function $f: \mathbf{I} \to \mathbb{R}$ is not integrable (see Exercise 8). $\qquad \square$

Formula (18.25) singles out the variables $x$ and $y$ in a particular order. In fact, there is a corresponding formula if the order of integration is reversed. For an integrable function $f: [a, b] \times [c, d] \to \mathbb{R}$, for each point $y$ in $[c, d]$, define the function $F_y: [a, b] \to \mathbb{R}$ by $F_y(x) = f(x, y)$ for $x$ in $[a, b]$. Suppose that the function $F_y: [a, b] \to \mathbb{R}$ is integrable and define

$$B(y) = \int_a^b f(x, y)\, dx.$$

Then the function $B: [c, d] \to \mathbb{R}$ is integrable, and

$$\int_{\mathbf{I}} f = \int_c^d B(y)\, dy = \int_c^d \left[ \int_a^b f(x, y)\, dx \right] dy. \qquad (18.28)$$

The proof of this formula is precisely the same as the proof of Theorem 18.24, except that it requires replacing

$$\sum_{i=1}^m \sum_{j=1}^\ell m_{ij}[y_j - y_{j-1}][x_i - x_{i-1}]$$

with

$$\sum_{j=1}^\ell \sum_{i=1}^m m_{ij}[x_i - x_{i-1}][y_j - y_{j-1}].$$

**COROLLARY 18.26** Suppose that the function $f : \mathbf{I} \to \mathbb{R}$ is integrable, where $\mathbf{I} = [a, b] \times [c, d]$ is a rectangle in the plane $\mathbb{R}^2$. Then the following formula holds, provided that each side is defined:

$$\int_a^b \left[ \int_c^d f(x, y) \, dy \right] dx = \int_c^d \left[ \int_a^b f(x, y) \, dx \right] dy. \qquad (18.29)$$

In particular, this formula holds if the function $f : \mathbf{I} \to \mathbb{R}$ is continuous.

**Proof** Theorem 18.24 asserts that

$$\int_a^b \left[ \int_c^d f(x, y) \, dy \right] dx = \int_{\mathbf{I}} f,$$

provided that the integral on the left is defined. As we have discussed above, a symmetric argument shows that

$$\int_c^d \left[ \int_a^b f(x, y) \, dx \right] dy = \int_{\mathbf{I}} f,$$

provided that the integral on the left is defined. Thus each of the sides of (18.29) is equal to $\int_{\mathbf{I}} f$, provided that they are defined. ∎

The iterated integration formula (18.25) extends to the integral of functions over generalized rectangles in Euclidean space $\mathbb{R}^m$ of dimension $m \geq 2$. Moreover, once the proper notation is introduced, the statement and proof of the general result are exactly the same as for $\mathbb{R}^2$. Indeed, write $m = n + k$, where $n$ and $k$ are natural numbers. As we have done before, we write a point $\mathbf{u}$ in $\mathbb{R}^{n+k}$ as

$$\mathbf{u} = (\mathbf{x}, \mathbf{y}) \quad \text{where } \mathbf{x} \text{ is in } \mathbb{R}^n \text{ and } \mathbf{y} \text{ is in } \mathbb{R}^k.$$

Moreover, a generalized rectangle $\mathbf{I} = I_1 \times \cdots \times I_n \times I_{n+1} \times \cdots \times I_{n+k}$ in $\mathbb{R}^{n+k}$ may be represented as

$$\mathbf{I} = \mathbf{I_x} \times \mathbf{I_y},$$

where $\quad \mathbf{I_x} = I_1 \times \cdots \times I_n \quad$ and $\quad \mathbf{I_y} = I_{n+1} \times \cdots \times I_{n+k}.$

The proof that was given for Fubini's Theorem in the plane extends directly to provide a proof of the following extension.

**THEOREM 18.27** **Fubini's Theorem** Suppose that the function $f : \mathbf{I} \to \mathbb{R}$ is integrable, where $\mathbf{I} = \mathbf{I_x} \times \mathbf{I_y}$ is a rectangle in $\mathbb{R}^{n+k}$. For each point $\mathbf{x}$ in $\mathbf{I_x}$, define the function $F_{\mathbf{x}} : \mathbf{I_y} \to \mathbb{R}$ by

$$F_{\mathbf{x}}(\mathbf{y}) = f(\mathbf{x}, \mathbf{y}) \quad \text{for } \mathbf{y} \text{ in } \mathbf{I_y};$$

suppose that the function $F_{\mathbf{x}} : \mathbf{I_y} \to \mathbb{R}$ is integrable and define

$$A(\mathbf{x}) = \int_{\mathbf{I_y}} f(\mathbf{x}, \mathbf{y}) \, dy.$$

Then the function $A: \mathbf{I}_{\mathbf{x}} \to \mathbb{R}$ is integrable, and

$$\int_{\mathbf{I}} f = \int_{\mathbf{I}_{\mathbf{x}}} A(\mathbf{x}) \, d\mathbf{x} = \int_{\mathbf{I}_{\mathbf{x}}} \left[ \int_{\mathbf{I}_{\mathbf{y}}} f(\mathbf{x}, \mathbf{y}) \, d\mathbf{y} \right] d\mathbf{x}. \qquad (18.30)$$

The proof of the next theorem is exactly the same as the proof of Theorem 18.25.

**THEOREM 18.28**  For a Jordan domain $K$ in $\mathbb{R}^n$, let $h: K \to \mathbb{R}$ and $g: K \to \mathbb{R}$ be continuous bounded functions that have the property that $h(\mathbf{x}) \le g(\mathbf{x})$ for all points $\mathbf{x}$ in $K$. Define

$$D = \{(\mathbf{x}, y) \text{ in } \mathbb{R}^{n+1} \mid \mathbf{x} \text{ in } K, \; h(\mathbf{x}) \le y \le g(\mathbf{x})\}.$$

Suppose that the function $f: D \to \mathbb{R}$ is continuous and bounded. Then

$$\int_D f = \int_K \left[ \int_{h(\mathbf{x})}^{g(\mathbf{x})} f(\mathbf{x}, y) \, dy \right] d\mathbf{x}. \qquad (18.31)$$

In calculating the value of the integral of a function of a single variable, notation involving Leibnitz symbols is often useful. It is also useful for functions of several variables. For an integrable function $f: \mathbf{I} \to \mathbb{R}$ defined on a generalized rectangle $\mathbf{I} = [a_1, b_1] \times \cdots \times [a_n, b_n]$, the value of the integral is often denoted by

$$\int_{\mathbf{I}} f(\mathbf{x}) \, d\mathbf{x}$$

or by

$$\int_{a_1}^{b_1} \cdots \int_{a_n}^{b_n} f(x_1, \ldots, x_n) \, dx_n \ldots dx_1.$$

For $n = 2$ or $n = 3$, we use the notation

$$\int_{a_1}^{b_1} \int_{a_2}^{b_2} f(x, y) \, dy \, dx \quad \text{and} \quad \int_{a_1}^{b_1} \int_{a_2}^{b_2} \int_{a_3}^{b_3} f(x, y, z) \, dz \, dy \, dx.$$

In the case of $\mathbb{R}^3$ and $\mathbf{I} = [a_1, b_1] \times [a_2, b_2] \times [a_3, b_3]$, the above iterated integration formula (18.30) becomes

$$\int_{\mathbf{I}} f(x, y, z) \, dx \, dy \, dz = \int_{a_3}^{b_3} \left[ \int_{a_2}^{b_2} \int_{a_1}^{b_1} f(x, y, z) \, dx \, dy \right] dz. \qquad (18.32)$$

Moreover, if we use the two-variable iterated integration formula in the above inner integral, we obtain

$$\int_{\mathbf{I}} f(x, y, z) \, dx \, dy \, dz = \int_{a_3}^{b_3} \left[ \int_{a_2}^{b_2} \left\{ \int_{a_1}^{b_1} f(x, y, z) \, dx \right\} dy \right] dz. \qquad (18.33)$$

We emphasize that for the above formulas to be valid, it is necessary to make sure that the integrals on both sides are properly defined.

**EXERCISES**

1.  Evaluate

$$\iint_{[0,1]\times[0,1]} \sin^2 x \, \sin^2 y \, dx \, dy.$$

2.  For the following three functions, evaluate $\iint_I f$, where $I = [0, 1] \times [0, 1]$:

(a) $f(x, y) = \begin{cases} 1 - x - y & \text{if } x + y \le 1 \\ 0 & \text{otherwise.} \end{cases}$

(b) $f(x, y) = \begin{cases} x^2 + y^2 & \text{if } x^2 + y^2 \le 1 \\ 0 & \text{otherwise.} \end{cases}$

(c) $f(x, y) = \begin{cases} x + y & \text{if } x^2 \le y \le 2x^2 \\ 0 & \text{otherwise.} \end{cases}$

3.  Show that

$$\int_0^3 \left[ \int_1^{\sqrt{4-y}} (x + y) \, dx \right] dy = \int_1^2 \left[ \int_0^{4-x^2} (x + y) \, dy \right] dx = \frac{241}{60}.$$

4.  For a continuous function $f: [a, b] \times [a, b] \to \mathbb{R}$, prove Dirichlet's formula:

$$\int_a^b \left[ \int_a^x f(x, y) \, dy \right] dx = \int_a^b \left[ \int_y^b f(x, y) \, dx \right] dy.$$

5.  Suppose that the function $\phi: \mathbb{R} \to \mathbb{R}$ is continuous. Prove that for each $x \ge 0$,

$$\int_0^x \left[ \int_0^t \phi(s) \, ds \right] dt = \int_0^x (x - s)\phi(s) \, ds.$$

6.  Suppose that the function $f: [a, b] \to \mathbb{R}$ is continuous. Prove that

$$2 \int_a^b \left[ f(x) \int_x^b f(y) \, dy \right] dx = \left[ \int_a^b f(x) \, dx \right]^2.$$

7.  Follow the proof of Theorem 18.25 and thereby provide a proof of Theorem 18.28.
8.  Show that the function $f: [0, 1] \times [0, 1] \to \mathbb{R}$ defined in Example 18.13 is not integrable.
9.  For a Jordan domain $K$ in $\mathbb{R}^n$, let $h: K \to \mathbb{R}$ and $g: K \to \mathbb{R}$ be continuous bounded functions that have the property that $h(\mathbf{x}) \le g(\mathbf{x})$ for all points $\mathbf{x}$ in $K$. Define

$$D = \{(\mathbf{x}, y) \mid \mathbf{x} \text{ in } K, \, h(\mathbf{x}) \le y \le g(\mathbf{x})\}.$$

Prove that the set $D$ is also a Jordan domain.

10. Follow the proof of Theorem 18.24 and thereby provide a proof of Theorem 18.27.

## 18.4 Change of Variables

The Change of Variables Theorem for the integral of a function of a single variable was considered in Section 7.3. In this section, we will extend this theorem to functions of several variables. It will be convenient first to state the general change of variables theorem. Then we will consider some important special cases, including those of a domain in the plane $\mathbb{R}^2$ described by polar coordinates and a domain in $\mathbb{R}^3$ described by spherical coordinates. Finally, we will conclude this section with the proof of the theorem.

In Section 7.3 we proved the following result: Suppose that $\mathcal{O}$ is an open subset of $\mathbb{R}$ that contains the closed bounded interval $I = [a, b]$ and let $\psi: \mathcal{O} \to \mathbb{R}$ be a continuously differentiable function such that $\psi'(x) \neq 0$ for all $x$ in $\mathcal{O}$. Then for any continuous function $f: \psi(I) \to \mathbb{R}$,

$$\int_{\psi(a)}^{\psi(b)} f(x)\, dx = \int_a^b f(\psi(u))\psi'(u)\, du.$$

In the case when $\psi'(u) > 0$ for all $u$ in $I$, $\psi(I) = [\psi(a), \psi(b)]$; in the case when $\psi'(u) < 0$ for all $u$ in $I$, $\psi(I) = [\psi(b), \psi(a)]$. Therefore, the preceding formula may be rewritten in the following equivalent form:

$$\int_{\psi(I)} f(x)\, dx = \int_I f(\psi(u))|\psi'(u)|\, du. \tag{18.34}$$

It is in this form that the change-of-variables formula will be extended to functions of several variables.

**DEFINITION**  *Let $\mathcal{O}$ be an open subset of $\mathbb{R}^n$. A continuously differentiable mapping $\Psi: \mathcal{O} \to \mathbb{R}^n$ is called a  smooth change of variables provided that the following two properties hold:*

 (i) *The mapping $\Psi: \mathcal{O} \to \mathbb{R}^n$ is one-to-one.*

 (ii) *For each point $\mathbf{x}$ in $\mathcal{O}$, the derivative matrix $\mathbf{D}\Psi(\mathbf{x})$ is invertible.*

**THEOREM 18.29**  **The Change of Variables Theorem**  Suppose that the mapping $\Psi: \mathcal{O} \to \mathbb{R}^n$ is a smooth change of variables  on the open subset $\mathcal{O}$ of $\mathbb{R}^n$. Let $D$ be an open Jordan domain such that $K = D \cup \partial D$ is contained in $\mathcal{O}$. Then $\Psi(K)$ is a Jordan domain that has the property that for any continuous function $f: \Psi(K) \to \mathbb{R}$, the following integral transformation formula holds:

$$\int_{\Psi(K)} f(\mathbf{x})\, d\mathbf{x} = \int_K f(\Psi(\mathbf{u}))|\det \mathbf{D}\Psi(\mathbf{u})|\, d\mathbf{u}. \tag{18.35}$$

### Polar Coordinates

For each point $\mathbf{u} = (x, y) \neq (0, 0)$ in the plane $\mathbb{R}^2$, if we define $r = \sqrt{x^2 + y^2}$, then there is a unique number $\theta$ in the interval $[0, 2\pi)$ such that

$$(x, y) = (r\cos\theta, r\sin\theta).$$

**FIGURE 18.4**

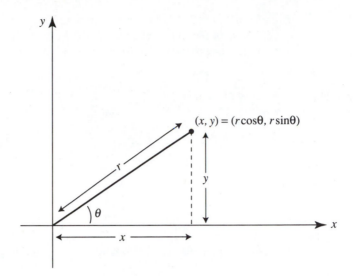

The pair of numbers $(r, \theta)$ is called a choice of *polar coordinates* for the point **u**.

Define $\mathcal{O}$ to be the subset of $\mathbb{R}^2$ consisting of points $(r, \theta)$ with $r > 0$ and $0 < \theta < 2\pi$, then define the mapping $\Psi: \mathcal{O} \to \mathbb{R}^2$ by

$$\Psi(r, \theta) = (r \cos \theta, r \sin \theta) \quad \text{for } (r, \theta) \text{ in } \mathcal{O}.$$

It is clear that the mapping $\Psi: \mathcal{O} \to \mathbb{R}^2$ is both continuously differentiable and one-to-one. Also, at each point $(r, \theta)$ in $\mathcal{O}$,

$$\det \mathbf{D}\Psi(r, \theta) = \det \begin{pmatrix} \cos \theta & -r \sin \theta \\ \sin \theta & r \cos \theta \end{pmatrix} = r \neq 0,$$

so that the derivative matrix $\mathbf{D}\Psi(r, \theta)$ is invertible. Thus the mapping $\Psi: \mathcal{O} \to \mathbb{R}^2$ is a smooth change of variables. For $0 < r_1 < r_2$ and $0 < \theta_1 < \theta_2 < 2\pi$, define $K = [r_1, r_2] \times [\theta_1, \theta_2]$. Suppose that the function $f: \Psi(K) \to \mathbb{R}$ is continuous. It follows directly from formula (18.35) and Fubini's Theorem that

$$\int_{\Psi(K)} f(x, y)\, dx\, dy = \int_K [f(r \cos \theta, r \sin \theta) r]\, dr\, d\theta$$

$$= \int_{\theta_1}^{\theta_2} \left[ \int_{r_1}^{r_2} f(r \cos \theta, r \sin \theta) r\, dr \right] d\theta. \tag{18.36}$$

We note that there is no neighborhood of the origin on which polar coordinates provide a smooth change of variables. Nevertheless, by using an approximation argument, the change-of-variables formula (18.36) for polar coordinates can be extended to Jordan domains $K$ that intersect the boundary of $\mathcal{O}$. For instance, for the above domain $K$, suppose that we allow the possibility that $r_1 = 0$, $\theta_1 = 0$, or $\theta_2 = 2\pi$. For $0 < \epsilon < \min\{r_2, [\theta_2 - \theta_1]/2\}$, define

$$K_\epsilon = \{(r, \theta) \mid \epsilon \leq r \leq r_2,\ \theta_1 + \epsilon \leq \theta \leq \theta_2 - \epsilon\}.$$

Thus formula (18.36) holds when applied to $K_\epsilon$. Now choose a number $M$ such that $|f(x, y)| \leq M$ for all $(x, y)$ in $\Psi(K)$. Then, by the addition-over-domains formula for the integrals and formula (18.36) when applied to $K_\epsilon$, we have

$$\left| \int_{\Psi(K)} f(x, y)\, dx\, dy - \int_K [f(r\cos\theta, r\sin\theta)r]\, dr d\theta \right|$$

$$= \left| \int_{\Psi(K\setminus K_\epsilon)} f(x, y)\, dx\, dy - \int_{K\setminus K_\epsilon} [f(r\cos\theta, r\sin\theta)r]\, dr\, d\theta \right|$$

$$\leq M \text{vol } \Psi(K\setminus K_\epsilon) + M r_2 \text{vol } (K\setminus K_\epsilon).$$

Since

$$\lim_{\epsilon \to 0} \text{vol } \Psi(K\setminus K_\epsilon) = \lim_{\epsilon \to 0} \text{vol } (K\setminus K_\epsilon) = 0, \tag{18.37}$$

we see that formula (18.36) also holds in this limiting case.

**EXAMPLE 18.14**   Define $D = \{(x, y) \mid x^2 + y^2 \leq 1, x \geq 0, y \geq 0\}$. We will use formula (18.36) to evaluate

$$\int_D e^{x^2+y^2}\, dx\, dy.$$

Indeed, $D = \{(r\cos\theta, r\sin\theta) \mid 0 \leq r \leq 1, 0 \leq \theta \leq \pi/2\}$, so that by formula (18.36),

$$\int_D e^{x^2+y^2}\, dx\, dy = \int_0^{\pi/2} \left[ \int_0^1 e^{r^2} r\, dr \right] d\theta = \int_0^{\pi/2} \left[ \frac{e-1}{2} \right] d\theta = \frac{\pi}{4}(e-1). \qquad \square$$

## Spherical Coordinates

We now turn to a useful change-of-variables formula in three dimensions. For each point $\mathbf{u} = (x, y, z)$ in $\mathbb{R}^3$ that does not lie on the $z$-axis, we define $\rho = \sqrt{x^2 + y^2 + z^2}$. It is not difficult to see that there are unique numbers $\theta$ in the interval $[0, 2\pi)$ and $\phi$ in the interval $(0, \pi)$ such that

$$\mathbf{u} = (x, y, z) = (\rho \sin\phi \cos\theta, \rho \sin\phi \sin\theta, \rho \cos\phi).$$

The triple of numbers $(\rho, \phi, \theta)$ is called a choice of *spherical coordinates* for the point $\mathbf{u}$. Define $\mathcal{O}$ to be the open subset of $\mathbb{R}^3$ consisting of points $(\rho, \phi, \theta)$ with $\rho > 0$, $0 < \phi < \pi$, and $0 < \theta < 2\pi$, and then define $\Psi : \mathcal{O} \to \mathbb{R}^3$ by

$$\Psi(\rho, \phi, \theta) = (\rho \sin\phi \cos\theta, \rho \sin\phi \sin\theta, \rho \cos\phi) \quad \text{for } (\rho, \phi, \theta) \text{ in } \mathcal{O}.$$

It is clear that the mapping $\Psi : \mathcal{O} \to \mathbb{R}^3$ is both continuously differentiable and one-to-one. Also, at each point $(\rho, \phi, \theta)$ in $\mathcal{O}$, the derivative matrix is given by

$$\mathbf{D}\Psi(\rho, \phi, \theta) = \begin{pmatrix} \sin\phi\cos\theta & \rho\cos\phi\cos\theta & -\rho\sin\phi\sin\theta \\ \sin\phi\sin\theta & \rho\cos\phi\sin\theta & \rho\sin\phi\cos\theta \\ \cos\phi & -\rho\sin\phi & 0 \end{pmatrix},$$

**FIGURE 18.5**

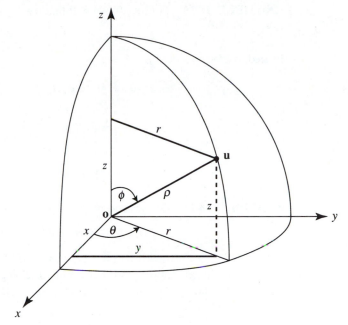

$$\mathbf{u} = (x, y, z) = (\rho \sin\phi \cos\theta, \rho\sin\phi \sin\theta, \rho\cos\phi)$$

so that $\det \mathbf{D}\Psi(\rho, \phi, \theta) = \rho^2 \sin\phi \neq 0$. Thus the derivative matrix $\mathbf{D}\Psi(\rho, \phi, \theta)$ is invertible, so $\Psi \colon \mathcal{O} \to \mathbb{R}^3$ is a smooth change of variables. For $0 < \rho_1 < \rho_2$, $0 < \phi_1 < \phi_2 < \pi$, and $0 < \theta_1 < \theta_2 < 2\pi$, define

$$K = [\rho_1, \rho_2] \times [\phi_1, \phi_2] \times [\theta_1, \theta_2].$$

Suppose that the function $f \colon \Psi(K) \to \mathbb{R}$ is continuous. Then by the integral transformation formula (18.35) and Fubini's Theorem,

$$\int_{\Psi(K)} f(x, y, z)\, dx\, dy\, dz$$

$$= \int_K [f(\rho \sin\phi \cos\theta, \rho \sin\phi \sin\theta, \rho\cos\phi)\rho^2 \sin\phi]\, d\rho\, d\phi\, d\theta$$

$$= \int_{\theta_1}^{\theta_2} \left[ \int_{\phi_1}^{\phi_2} \left\{ \int_{\rho_1}^{\rho_2} f(\rho \sin\phi \cos\theta, \rho \sin\phi \sin\theta, \rho\cos\phi)\rho^2 \sin\phi\, d\rho \right\} d\phi \right] d\theta.$$

$$(18.38)$$

Spherical coordinates do not define a smooth change of variables on any domain that intersects the boundary of $\mathcal{O}$. However, by an approximation argument similar to that used for polar coordinates, formula (18.38) may be extended to allow domains $K$ that do intersect the boundary of $\mathcal{O}$.

**EXAMPLE 18.15**   For $a > 0$, we will find the volume of the ball in $\mathbb{R}^3$ of radius $a$,

$$B_a = \{(x,\ y,\ z) \mid x^2 + y^2 + z^2 \leq a^2\}.$$

Indeed, by formula (18.38),

$$\mathrm{vol}\ B_a = \int_{B_a} 1\, dx\, dy\, dz$$

$$= \int_0^{2\pi} \left[ \int_0^\pi \left\{ \int_0^a \rho^2 \sin\phi\, d\rho \right\} d\phi \right] d\theta$$

$$= [4/3]\pi a^3. \qquad \square$$

**EXAMPLE 18.16**   For positive numbers $a$ and $b$, consider the ellipse

$$D = \{(x,\ y) \mid x^2/a^2 + y^2/b^2 \leq 1\}.$$

Define $\Psi\colon \mathbb{R}^2 \to \mathbb{R}^2$ by $\Psi(u,\ v) = (au,\ bv)$ for all $(u,\ v)$ in $\mathbb{R}^2$. Then $\Psi\colon \mathbb{R}^2 \to \mathbb{R}^2$ is an invertible linear mapping, so it is a smooth change of variables. By the change of variables formula for $\Psi$ and also for polar coordinates, we see that for any continuous function $f\colon D \to \mathbb{R}$,

$$\int_D f(x,\ y)\, dx\, dy = ab \int_{u^2+v^2 \leq 1} f(au,\ bv)\, du\, dv$$

$$= ab \int_0^{2\pi} \left[ \int_0^1 f(ar\cos\theta,\ br\sin\theta) r\, dr \right] d\theta. \qquad \square$$

We now turn to the proof of the Change of Variables Theorem. In order to prove this theorem, it is necessary to precisely compare the volume of $\Psi(\mathbf{J})$ and $d\Psi(\mathbf{x})(\mathbf{J})$,* where $\mathbf{J}$ is a generalized rectangle of small diameter in $\mathcal{O}$ that contains the point $\mathbf{x}$. The following result is the precise comparison of volume result on which the proof of the Change of Variables Theorem depends.

**THEOREM 18.30**

**The Volume Comparison Theorem**   Suppose that the mapping $\Psi\colon \mathcal{O} \to \mathbb{R}^n$ is a smooth change of variables on the open subset $\mathcal{O}$ of $\mathbb{R}^n$. Let $K$ be a compact subset of $\mathcal{O}$ and let $\epsilon$ be a positive number. Then there is a positive number $\delta$ such that if $\mathbf{J}$ is any generalized rectangle of diameter less than $\delta$ and $\mathbf{x}$ is a point in $K \cap \mathbf{J}$, then $\mathbf{J}$ is contained in $\mathcal{O}$ and

$$\mathrm{vol}\ \Psi(\mathbf{J}) = |\det D\Psi(\mathbf{x})|\mathrm{vol}\ \mathbf{J} + \mathcal{E}\mathrm{vol}\ \mathbf{J} \qquad \text{where } |\mathcal{E}| < \epsilon. \qquad (18.39)$$

It turns out that the proof of this theorem is rather technical, and it depends on properties of determinants that we will not prove in this book. Thus we prefer to describe the two fundamental ideas that underlie the theorem and provide completely elementary proofs of the Volume Comparison Theorem for polar coordinates and for spherical coordinates.

---

*Recall that $d\Psi(\mathbf{x})\colon \mathbb{R}^n \to \mathbb{R}^n$ is the linear mapping, called the *differential* of $\Psi\colon \mathcal{O} \to \mathbb{R}^n$ at the point $\mathbf{x}$, that is associated with the derivative matrix $\mathbf{D}\Psi(\mathbf{x})$ by the formula $d\Psi(\mathbf{x})(\mathbf{h}) = \mathbf{D}\Psi(\mathbf{x})\mathbf{h}$.

(i) First, for an invertible linear transformation $\mathbf{T}: \mathbb{R}^n \to \mathbb{R}^n$ that is represented by the $n \times n$ matrix $\mathbf{A}$ and a generalized rectangle $\mathbf{J}$ in $\mathbb{R}^n$, the volume of $\mathbf{T}(\mathbf{J})$ is given by

$$\text{vol } \mathbf{T}(\mathbf{J}) = |\det \mathbf{A}| \text{vol } \mathbf{J}.$$

In $\mathbb{R}^3$ this formula follows from the discussion in Appendix B, the appendix on linear algebra, of the relationship among volumes, determinants, and the cross-product. The general case follows from the product property of determinants and from a result that permits general linear mappings to be written as the composition of linear mappings of a very elementary kind. From the above volume transformation formula for linear mappings, it follows that for a point $\mathbf{x}$ in $\mathcal{O}$ and a generalized rectangle $\mathbf{J}$,

$$\text{vol } d\Psi(\mathbf{x})(\mathbf{J}) = |\det \mathbf{D}\Psi(\mathbf{x})| \text{vol } \mathbf{J}. \tag{18.40}$$

(ii) Second, recall that in Chapter 13 we proved the First-Order Approximation Theorem, from which it follows that if $\Psi: \mathcal{O} \to \mathbb{R}^n$ is a smooth change of variables, then for each point $\mathbf{x}$ in $\mathcal{O}$,

$$\lim_{\mathbf{h} \to 0} \frac{\|\Psi(\mathbf{x} + \mathbf{h}) - [\Psi(\mathbf{x}) + d\Psi(\mathbf{x})(\mathbf{h})]\|}{\|\mathbf{h}\|} = 0.$$

The First-Order Approximation Theorem is a precise assertion of the way in which in a neighborhood of a point $\mathbf{x}$ in $\mathcal{O}$, the mapping $\Psi$ is approximated by the differential of $\Psi$ at the point $\mathbf{x}$. Simply on the basis of the continuity of the mapping $\Psi$, it is not difficult to see that vol $\Psi(\mathbf{J})$ is small if diam $\mathbf{J}$ is small, so the difference

$$\text{vol } \Psi(\mathbf{J}) - |\det \mathbf{D}\Psi(\mathbf{x})| \text{vol } \mathbf{J}$$

is also small if diam $\mathbf{J}$ is small. The First-Order Approximation Theorem and formula (18.40) provide a much stronger result: Even if we divide the above difference by vol $\mathbf{J}$, the result,

$$\frac{\text{vol } \Psi(\mathbf{J})}{\text{vol } \mathbf{J}} - |\det \mathbf{D}\Psi(\mathbf{x})|, \tag{18.41}$$

remains small provided that $\mathbf{J}$ is a generalized rectangle of small diameter that contains the point $\mathbf{x}$. The Volume Comparison Theorem is a precise assertion of what is true in this respect. For the cases of polar coordinates and spherical coordinates, we will now be much more explicit.

**PROPOSITION 18.31**   **Area Transformation for Polar Coordinates**   For $0 \le r_1 < r_2$ and $0 \le \theta_1 < \theta_2 \le 2\pi$, let $\mathbf{J} = [r_1, r_2] \times [\theta_1, \theta_2]$ and define $\Psi(r, \theta) = (r\cos\theta, r\sin\theta)$ for $(r, \theta)$ in $\mathbf{J}$. Then

$$\text{area}\Psi(\mathbf{J}) = \frac{1}{2}[r_2^2 - r_1^2][\theta_2 - \theta_1] = \frac{[r_1 + r_2]}{2} \text{area } \mathbf{J}. \tag{18.42}$$

**FIGURE 18.6**

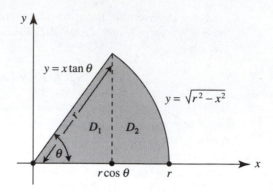

*Area D = Area D$_1$ + Area D$_2$*

**Proof** By the additivity of volume property asserted in Corollary 18.22, it suffices to prove formula (18.42) in the case when $r_1 = 0$, $\theta_1 = 0$, and $0 \le \theta_2 \le \pi/2$. Define $r = r_2$ and $\theta = \theta_2$; it is necessary to show that

$$\text{area } D = \frac{1}{2}r^2\theta.$$

Observe that $D$ is the union of the two sets $D_1 = \{(x, y) \,|\, 0 \le x \le r\cos\theta, \, 0 \le y \le \tan\theta \, x\}$ and $D_2 = \{(x, y) \,|\, r\cos\theta \le x \le r, \, 0 \le y \le \sqrt{r^2 - x^2}\}$. Thus, by the additivity property of volume and Theorem 18.25, we have

$$\text{area} D = \text{area} D_1 + \text{area} D_2$$

$$= \int_0^{r\cos\theta} \tan\theta \, x \, dx + \int_{r\cos\theta}^r \sqrt{r^2 - x^2}\,dx$$

$$= \frac{r^2}{2}\sin\theta\cos\theta + \int_{r\cos\theta}^r \sqrt{r^2 - x^2}\,dx.$$

But by the change of variables $x = r\cos t$, $0 \le t \le \theta$, it follows from the single-variable change of variables formula that

$$\int_{r\cos\theta}^r \sqrt{r^2 - x^2}\,dx = r^2\int_0^\theta \sin^2 t \, dt = r^2\int_0^\theta \left[\frac{1 - \cos 2t}{2}\right]dt = r^2\left[\frac{t}{2} - \frac{\sin 2t}{4}\right]\Bigg|_{t=0}^{t=\theta}.$$

Thus

$$\text{area } D = \frac{r^2}{2}\sin\theta\cos\theta + r^2\left[\frac{\theta}{2} - \frac{\sin 2\theta}{4}\right] = \frac{1}{2}r^2\theta. \qquad \blacksquare$$

From this area transformation formula for polar coordinates, we immediately obtain the Volume Comparison Theorem for the case of polar coordinates. Indeed, for $\mathbf{J}$ as above and any point $(r, \theta)$ in $\mathbf{J}$, it follows from formula (18.42) that

$$\text{area}\Psi(\mathbf{J}) = r \text{ area } \mathbf{J} + \mathcal{E} \text{ area } \mathbf{J}, \quad \text{where} \quad \mathcal{E} = \frac{r_1 + r_2}{2} - r.$$

Clearly, $|\mathcal{E}| < \text{diam }\mathbf{J}$, so we can let $\delta = \epsilon$ in the statement of the Volume Comparison Theorem for polar coordinates.

**PROPOSITION 18.32**

**Volume Transformation for Spherical Coordinates**   For $0 \le \rho_1 < \rho_2$, $0 \le \phi_1 < \phi_2 \le \pi$, and $0 \le \theta_1 < \theta_2 \le 2\pi$, let $\mathbf{J} = [\rho_1, \rho_2] \times [\phi_1, \phi_2] \times [\theta_1, \theta_2]$ and define

$$\Psi(\rho, \phi, \theta) = (\rho \sin\phi \cos\theta, \ \rho \sin\phi \sin\theta, \ \rho \cos\phi) \quad \text{for } (\rho, \phi, \theta) \text{ in } \mathbf{J}.$$

Then

$$\text{vol } \Psi(\mathbf{J}) = \frac{1}{3} \left[\rho_2^3 - \rho_1^3\right] \left[\cos\phi_1 - \cos\phi_2\right] \left[\theta_2 - \theta_1\right]. \tag{18.43}$$

**FIGURE 18.7**

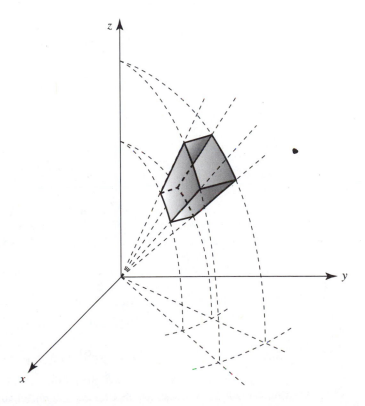

**Proof**   The additivity of volume property asserted in Corollary 18.22 implies that it is sufficient to establish formula (18.43) in the case when $\rho_1 = 0$, $0 = \phi_1 < \phi_2 \leq \pi/2$, and $\theta_1 = 0$. It is necessary to show that

$$\text{vol } \Psi(\mathbf{J}) = \frac{1}{3}\rho_2^3[1 - \cos\phi_2]\theta_2.$$

But observe that $\Psi(\mathbf{J}) = \{(x, y, z) \mid (x, y) \text{ in } D, g(x, y) \leq z \leq h(x, y)\}$, where $D = \{(x, y) = (r\cos\theta, r\sin\theta) \mid 0 \leq \theta \leq \theta_2, 0 \leq r \leq \rho_2 \sin\phi_2\}$, and for $(x, y) = (r\cos\theta, r\sin\theta)$ in $D$,

$$h(x, y) = \sqrt{\rho_2^2 - r^2} \quad \text{and} \quad g(x, y) = \frac{r}{\tan\phi_2}.$$

Using the iterated integration formula (18.31) and the change of variables formula (from Cartesian to polar coordinates) in the plane $\mathbb{R}^2$, we have

$$\text{vol } \Psi(\mathbf{J}) = \int_D [h(x, y) - g(x, y)]\, dx\, dy$$

$$= \int_0^{\theta_2} \int_0^{\rho_2 \sin\phi_2} \left[\sqrt{\rho_2^2 - r^2} - \frac{r}{\tan\phi_2}\right] r\, dr d\theta$$

$$= \int_0^{\theta_2} \left\{\int_0^{\rho_2 \sin\phi_2} \left[\sqrt{\rho_2^2 - r^2} - \frac{r}{\tan\phi_2}\right] r\, dr\right\} d\theta.$$

However, by the First Fundamental Theorem of Calculus,

$$\int_0^{\rho_2 \sin\phi_2} \left[\sqrt{\rho_2^2 - r^2} - \frac{r}{\tan\phi_2}\right] r\, dr = \left\{-\frac{1}{3}(\rho_2^2 - r^2)^{3/2} - \frac{r^3}{3}\frac{1}{\tan\phi_2}\right\}\Bigg|_{r=0}^{r=\rho_2\sin\phi_2}$$

$$= \frac{\rho_2^3}{3}\left\{-\cos^3\phi_2 - \frac{\sin^3\phi_2}{\tan\phi_2} + 1\right\}$$

$$= \frac{\rho_2^3}{3}[1 - \cos\phi_2].$$

Thus,

$$\text{vol } \Psi(\mathbf{J}) = \int_0^{\theta_2}\left\{\int_0^{\rho_2\sin\phi_2}\left[\sqrt{\rho_2^2 - r^2} - \frac{r}{\tan\phi_2}\right] r\, dr\right\} d\theta = \frac{\rho_2^3}{3}[1 - \cos\phi_2]\theta_2. \quad \blacksquare$$

Formula (18.43) implies the Volume Comparison Theorem for the case of spherical coordinates. Indeed, using the Mean Value Theorem twice, we can select points $\rho'$ in the interval $(\rho_1, \rho_2)$ and $\phi'$ in the interval $(\phi_1, \phi_2)$ such that

$$\frac{1}{3}[\rho_2^3 - \rho_1^3][\cos\phi_1 - \cos\phi_2][\theta_2 - \theta_1] = [\rho']^2[\sin\phi'][\rho_2 - \rho_1][\phi_2 - \phi_1][\theta_2 - \theta_1].$$

Thus, for a point $(\rho, \phi, \theta)$ in $\mathbf{J}$, we have

$$\text{vol } \Psi(\mathbf{J}) = [\rho^2 \sin\phi]\text{vol } \mathbf{J} + \mathcal{E} \text{ vol } \mathbf{J},$$

where

$$\mathcal{E} = \{[\rho']^2 \sin\phi' - \rho^2 \sin\phi\}.$$

Since $\det \mathbf{D}\Psi(\rho, \phi, \theta) = \rho^2 \sin\phi$, the uniform continuity on bounded sets of the function $\det \mathbf{D}\Psi: \mathbb{R}^3 \to \mathbb{R}$ shows that for each compact subset $K$ of $\mathbb{R}^3$ and $\epsilon > 0$, there is a $\delta > 0$ such that $\mathcal{E} < \epsilon$ if $\text{diam } \mathbf{J} < \delta$ and $\mathbf{J}$ contains a point of $K$.

**Proof of the Change of Variables Theorem:** Since $K$ is compact and the functions $\det \mathbf{D}\Psi : K \to \mathbb{R}$ and $f \circ \Psi : K \to \mathbb{R}$ are continuous, it follows from the Extreme Value Theorem that there is a number $M > 0$ such that

$$|f(\Psi(\mathbf{x}))| \leq M \quad \text{and} \quad |\det \mathbf{D}\Psi(\mathbf{x})| \leq M \quad \text{for all points } \mathbf{x} \text{ in } K.$$

Fix $\mathbf{I}$ to be a generalized rectangle that contains $K$. Let $\epsilon > 0$.

The Inverse Function Theorem implies that $\Psi(D)$ is an open subset of $\mathbb{R}^n$ and that the boundary of $\Psi(D)$ equals $\Psi(\partial D)$. Since, by assumption, $D$ is a Jordan domain, the set $\partial D$ has Jordan content 0. Using the Volume Comparison Theorem, it follows that $\Psi(\partial D)$ also has Jordan content 0. Thus $\Psi(K)$ is a Jordan domain, as is $\Psi(\mathbf{J})$ for any generalized rectangle $\mathbf{J}$ that is contained in $\mathcal{O}$.

A continuous function on a compact set is uniformly continuous. Hence, since $K$ is compact, the function $(f \circ \Psi) \cdot |\det \mathbf{D}\Psi| : K \to \mathbb{R}$ is uniformly continuous. Choose $\delta_1 > 0$ such that for any two points $\mathbf{u}$ and $\mathbf{v}$ in $K$,

$$\left| f(\Psi(\mathbf{u}))|\det \mathbf{D}\Psi(\mathbf{u})| - f(\Psi(\mathbf{v}))|\det \mathbf{D}\Psi(\mathbf{v})| \right| < \epsilon \quad \text{if} \quad \|\mathbf{u} - \mathbf{v}\| < \delta_1. \quad (18.44)$$

By the Volume Comparison Theorem, we can select $\delta_2 > 0$ such that if $\mathbf{J}$ is any generalized rectangle of diameter less than $\delta_2$ and $\mathbf{x}$ is a point in $K \cap \mathbf{J}$, then $\mathbf{J}$ is contained in $\mathcal{O}$ and

$$\text{vol } \Psi(\mathbf{J}) = |\det \mathbf{D}\Psi(\mathbf{x})|\text{vol } \mathbf{J} + \mathcal{E}\text{vol } \mathbf{J} \quad \text{where} \quad |\mathcal{E}| < \epsilon. \quad (18.45)$$

Define $\delta = \min\{\delta_1, \delta_2\}$. Since $\partial D$ has Jordan content 0, arguing as we have in the proof of Theorem 18.16, we can select a partition $\mathbf{P}$ of $\mathbf{I}$ such that

$$\|\mathbf{P}\| < \delta \quad \text{and} \quad \sum_{\mathbf{J} \text{ in } \mathbf{P}, \, \mathbf{J} \cap \partial D \neq \emptyset} \text{vol } \mathbf{J} < \epsilon. \quad (18.46)$$

By the additivity over domains properties of the integral, which we established in Theorems 18.10 and 18.15,

$$\int_{\Psi(K)} f(\mathbf{x})\,d\mathbf{x} - \int_K f(\Psi(\mathbf{u}))|\det \mathbf{D}\Psi(\mathbf{u})|\,d\mathbf{u}$$
$$= \sum_{\mathbf{J} \text{ in } \mathbf{P}} \int_{\Psi(K \cap \mathbf{J})} f(\mathbf{x})\,d\mathbf{x} - \sum_{\mathbf{J} \text{ in } \mathbf{P}} \int_{K \cap \mathbf{J}} f(\Psi(\mathbf{u}))|\det \mathbf{D}\Psi(\mathbf{u})|\,d\mathbf{u} \quad (18.47)$$
$$= \sum_{\mathbf{J} \text{ in } \mathbf{P}} \left[ \int_{\Psi(K \cap \mathbf{J})} f(\mathbf{x})\,d\mathbf{x} - \int_{K \cap \mathbf{J}} f(\Psi(\mathbf{u}))|\det \mathbf{D}\Psi(\mathbf{u})|\,d\mathbf{u} \right],$$

and we estimate separately the contribution to the right-hand sum from generalized rectangles that intersect $\partial D$ and from those that are contained in $D$. Since the diameter of each generalized rectangle $\mathbf{J}$ in $\mathbf{P}$ is less than $\delta$, it follows from (18.45) that for each generalized rectangle $\mathbf{J}$ in $\mathbf{I}$, vol $\Psi(K \cap \mathbf{J}) \leq (M + \epsilon)$vol $\mathbf{J}$, so

$$\left| \int_{\Psi(K \cap \mathbf{J})} f(\mathbf{x})\,d\mathbf{x} - \int_{K \cap \mathbf{J}} f(\Psi(\mathbf{u}))|\det \mathbf{D}\Psi(\mathbf{u})|\,d\mathbf{u} \right|$$

$$\le \left| \int_{\Psi(K \cap J)} f(\mathbf{x})\, d\mathbf{x} \right| + \left| \int_{K \cap J} f(\Psi(\mathbf{u})) |\det \mathbf{D}\Psi(\mathbf{u})|\, d\mathbf{u} \right|$$

$$\le M(M + \epsilon)\operatorname{vol} \mathbf{J} + M^2 \operatorname{vol} \mathbf{J}$$

$$= \{M(M + \epsilon) + M^2\}\operatorname{vol} \mathbf{J}.$$

Therefore, because the sum of the volumes of the generalized rectangles in $\mathbf{P}$ that intersect $\partial D$ has volume less than $\epsilon$, we obtain the following estimate:

$$\left| \sum_{\mathbf{J} \cap \partial D \ne \phi} \int_{\Psi(K \cap J)} f(\mathbf{x})\, d\mathbf{x} \right.$$

$$\left. - \sum_{\mathbf{J} \cap \partial D \ne \phi} \int_{K \cap J} f(\Psi(\mathbf{u})) |\det \mathbf{D}\Psi(\mathbf{u})|\, d\mathbf{u} \right| \le \{M(M + \epsilon) + M^2\}\epsilon \tag{18.48}$$

It remains to estimate the contribution to the right-hand side of (18.47) from the sum of the terms of the form

$$\int_{\Psi(J)} f(\mathbf{x})\, d\mathbf{x} - \int_{J} f(\Psi(\mathbf{u})) |\det \mathbf{D}\Psi(\mathbf{u})|\, d\mathbf{u},$$

where $\mathbf{J}$ is a generalized rectangle in $\mathbf{P}$ that is contained in $D$. Let $\mathbf{J}$ be such a generalized rectangle. Since $\mathbf{J}$ is pathwise connected, its image $\Psi(\mathbf{J})$ is also pathwise connected. It follows from the Mean Value Property of Integrals (Theorem 18.23) that there is a point $\mathbf{u}_0$ in $\mathbf{J}$ at which

$$\int_{\Psi(J)} f(\mathbf{x})\, d\mathbf{x} = f(\Psi(\mathbf{u}_0)) \operatorname{vol} \Psi(\mathbf{J}),$$

and hence, by (18.45),

$$\int_{\Psi(J)} f(\mathbf{x})\, d\mathbf{x} = f(\Psi(\mathbf{u}_0)) |\det \mathbf{D}\Psi(\mathbf{u}_0)| \operatorname{vol} \mathbf{J} + f(\Psi(\mathbf{u}_0)) \mathcal{E} \operatorname{vol} \mathbf{J} \qquad \text{where} \quad |\mathcal{E}| < \epsilon.$$

Again using the Mean Value Property of Integrals, there is a point $\mathbf{v}_0$ in $\mathbf{J}$ at which

$$\int_{J} f(\Psi(\mathbf{u})) |\det \mathbf{D}\Psi(\mathbf{u})|\, d\mathbf{u} = f(\Psi(\mathbf{v}_0)) |\det \mathbf{D}\Psi(\mathbf{v}_0)| \operatorname{vol} \mathbf{J}.$$

Since $\|\mathbf{u}_0 - \mathbf{v}_0\| \le \operatorname{diam} \mathbf{J} < \delta$, from the uniform continuity assertion (18.44) we have

$$\left| f(\Psi(\mathbf{u}_0)) |\det \mathbf{D}\Psi(\mathbf{u}_0)| - f(\Psi(\mathbf{v}_0)) |\det \mathbf{D}\Psi(\mathbf{v}_0)| \right| < \epsilon.$$

Thus,

$$\left| \int_{\Psi(J)} f(\mathbf{x})\, d\mathbf{x} - \int_{J} f(\Psi(\mathbf{u})) |\det \mathbf{D}\Psi(\mathbf{u})|\, d\mathbf{u} \right|$$

$$= \left| f(\Psi(\mathbf{u}_0)) |\det \mathbf{D}\Psi(\mathbf{u}_0)| - f(\Psi(\mathbf{v}_0)) |\det \mathbf{D}\Psi(\mathbf{v}_0)| + f(\Psi(\mathbf{u}_0)) \mathcal{E} \right| \operatorname{vol} \mathbf{J}$$

$$\le \{\epsilon + M\epsilon\} \operatorname{vol} \mathbf{J}.$$

We sum this estimate over all the generalized rectangles $\mathbf{J}$ in $\mathbf{P}$ that are contained in $D$ to obtain

$$\left| \sum_{\mathbf{J} \text{ in } \mathbf{P}, \mathbf{J} \subseteq D} \left\{ \int_{\Psi(\mathbf{J})} f(\mathbf{x})\,d\mathbf{x} - \int_{\mathbf{J}} f(\Psi(\mathbf{u}))|\det \mathbf{D}\Psi(\mathbf{u})|\,d\mathbf{u} \right\} \right| < \{\epsilon + M\epsilon\} \text{vol } \mathbf{I}.$$

Combining this with the estimate (18.48), we see that

$$\left| \int_{\Psi(K)} f(\mathbf{x})\,dx - \int_K f(\Psi(\mathbf{u}))|\det \mathbf{D}\Psi(\mathbf{u})|\,d\mathbf{u} \right|$$
$$< \{M(M+\epsilon) + M^2\}\epsilon + \{\epsilon + M\epsilon\}\text{vol } \mathbf{I}$$
$$= \epsilon\{M(M+\epsilon) + M^2 + (1+M)\text{vol } \mathbf{I}\}.$$

Since the numbers $M$ and vol $\mathbf{I}$ are fixed, and the preceding estimate holds for all positive numbers $\epsilon$, we conclude that the integral transformation formula (18.35) holds.

---

**EXERCISES**

1. Evaluate

$$\iint_{x^2+y^2 \le 1} x^2 y^2\,dx\,dy.$$

2. Evaluate

$$\iiint_V |xyz|\,dx\,dy\,dz$$

   where $\qquad V = \left\{ (x, y, z) \text{ in } \mathbb{R}^3 \,\middle|\, \dfrac{x^2}{a^2} + \dfrac{y^2}{b^2} + \dfrac{z^2}{c^2} \le 1 \right\}.$

3. Find the volume of the Jordan domain in $\mathbb{R}^3$ that is bounded by the $x$-$y$ plane and the paraboloid $z = 2 - x^2 - y^2$.

4. Show that the volume of the Jordan domain in $\mathbb{R}^3$ that is bounded by the cylinders $x^2 + y^2 = a^2$ and $x^2 + z^2 = a^2$ is equal to $16a^3/3$.

5. For $r > 0$ and $h > 0$, show that the volume of the cone $\{(x, y, z) \text{ in } \mathbb{R}^3 \mid x^2 + y^2 \le r^2, \ 0 \le z \le h/r^2(r^2 - x^2 - y^2)\}$ is equal to $[1/3]\pi r^2 h$.

6. Suppose that $\Psi: \mathcal{O} \to \mathbb{R}^n$ is a smooth change of variables on an open subset $\mathcal{O}$ of $\mathbb{R}^n$. Use the Inverse Function Theorem to show that the inverse mapping $\Psi^{-1}: \Psi(\mathcal{O}) \to \mathbb{R}^n$ is a smooth change of variables on the open subset $\Psi(\mathcal{O})$ of $\mathbb{R}^n$.

7. (Hyperbolic Coordinates) Define $\mathcal{U} = \{(x, y) \mid x > 0, \ y > 0\}$ and define $\Phi: \mathcal{U} \to \mathbb{R}^2$ by $\Phi(x, y) = (x^2 - y^2, xy)$ for $(x, y)$ in $\mathcal{U}$. Show that $\Phi: \mathcal{U} \to \mathbb{R}^2$ is a smooth change of variables. For a point $(x, y)$ in $\mathcal{U}$, the pair of numbers $(x^2 - y^2, xy) = \Phi(x, y)$ are called *hyperbolic coordinates* for $(x, y)$.

8. Define $D = \{(x, y) \mid x > 0, \ y > 0, \ 1 \le x^2 - y^2 \le 9, \ 2 \le xy \le 4\}$. For a continuous function $f: D \to \mathbb{R}$, use the hyperbolic coordinates from Exercise 7 to show that

$$\int_D [x^2 + y^2]\,dx\,dy = 8.$$

9. Verify the limits (18.37) by finding a positive number $C$ such that for each $\epsilon > 0$,

$$|\text{vol } \Psi(K \backslash K_\epsilon) + \text{vol } (K \backslash K_\epsilon)| \le C\epsilon.$$

10. Suppose that the smooth change of variables $\Psi: \mathbb{R}^2 \to \mathbb{R}^2$ has the special form $\Psi(x, y) = (x, g(x, y))$ for all $(x, y)$ in $\mathbb{R}^2$. For $K = [a, b] \times [0, 1]$, explicitly find $\Psi(K)$ and thus provide an independent proof of the change of variables formula in this particular case.

11. Show that the linear volume transformation formula (18.40) is a special case of formula (18.35).

12. Show that the volume transformation formulas (18.42) and (18.43) are special cases of formula (18.35).

13. Use the mean value property of integrals to show that the Volume Comparison Theorem may be derived as a consequence of formula (18.35).

14. Let $\mathcal{O}$ be an open subset of $\mathbb{R}^n$ that contains the compact set $K$. Show that there is a positive number $\delta$ such that if $\mathbf{J}$ is a generalized rectangle that has diameter less than $\delta$ and contains a point of $K$, then $\mathbf{J}$ is contained in $\mathcal{O}$. (*Hint:* Argue by contradiction.)

15. Suppose that $\Psi: \mathcal{O} \to \mathbb{R}^n$ is a smooth change of variables on the open subset $\mathcal{O}$ of $\mathbb{R}^n$. Show that if $A$ is a compact subset of $\mathcal{O}$ that has Jordan content 0, then its image $\Psi(A)$ also has Jordan content 0. (*Hint:* Use the Volume Comparison Theorem.)

16. Suppose that $\Psi: \mathcal{O} \to \mathbb{R}^n$ is a smooth change of variables on the open subset $\mathcal{O}$ of $\mathbb{R}^n$. Let $D$ be an open subset of $\mathbb{R}^n$ such that $D \cup \partial D$ is contained in $\mathcal{O}$. Use the Inverse Function Theorem to show that $\Psi(\partial D) = \partial \Psi(D)$—that is, the boundary of the image is the image of the boundary.

17. Suppose that $\Psi: \mathcal{O} \to \mathbb{R}^2$ is a smooth change of variables on the open subset $\mathcal{O}$ of $\mathbb{R}^n$ and that $K$ is a compact subset of $\mathcal{O}$. Prove that there is a positive number $c$ such that for any two points $\mathbf{u}$ and $\mathbf{v}$ in $K$,

$$\|\Psi(\mathbf{u}) - \Psi(\mathbf{v})\| \le c\|\mathbf{u} - \mathbf{v}\|.$$

(*Hint:* Use the Mean Value Theorem.)

18. Suppose that $\Psi: \mathcal{O} \to \mathbb{R}^n$ is a smooth change of variables on the open subset $\mathcal{O}$ of $\mathbb{R}^n$ and that $K$ is a compact subset of $\mathcal{O}$. Prove that there is a positive number $c$ such that for any two points $\mathbf{u}$ and $\mathbf{v}$ in $K$,

$$\|\Psi(\mathbf{u}) - \Psi(\mathbf{v})\| \ge c\|\mathbf{u} - \mathbf{v}\|.$$

(*Hint:* Argue by contradiction, using the Nonlinear Stability Theorem.)

# Line and Surface Integrals

## 19.1 Arclength and Line Integrals

Recall that in Section 11.3 we called a continuous mapping $\gamma: [a, b] \to \mathbb{R}^n$ a *parametrized path* whose parameter space is the interval $[a, b]$, and we called the image of this mapping a path. We emphasize that a path is a *subset* of $\mathbb{R}^n$ whereas a parametrized path is a *mapping*, and that a path will be the image of many different parametrized paths.

**DEFINITION**   *A parametrized path* $\gamma: [a, b] \to \mathbb{R}^n$ *is called smooth provided that the restriction* $\gamma: (a, b) \to \mathbb{R}^n$ *has a continuous bounded derivative such that* $\gamma'(t) \neq 0$ *for all parameter values $t$ in* $(a, b)$.

**DEFINITION**   *A parametrized path* $\gamma: [a, b] \to \mathbb{R}^n$ *is called piecewise smooth provided that there is a partition* $P = \{x_0, \ldots, x_m\}$ *of the parameter space* $[a, b]$ *such that for each index $i$ between 1 and $m$, the restriction* $\gamma: [x_{i-1}, x_i] \to \mathbb{R}^n$ *is a smooth parametrized path.*

The image of a smooth parametrized path is called a *smooth path* and the image of a piecewise smooth parametrized path is called a *piecewise smooth path.*

**473**

**EXAMPLE 19.1** Recall that for any two distinct points **p** and **q** in $\mathbb{R}^n$, the mapping $\gamma: [0, 1] \to \mathbb{R}^n$ defined by

$$\gamma(t) = t\mathbf{q} + (1 - t)\mathbf{p} \quad \text{for} \quad 0 \le t \le 1$$

is called the parametrized segment from the point **p** to the point **q**. This parametrized segment is clearly a smooth parametrized path. □

**EXAMPLE 19.2** Let $\Gamma$ be the ellipse in the plane $\mathbb{R}^2$ defined by the equation

$$\frac{x^2}{a^2} + \frac{y^2}{b^2} = 1,$$

where $a > 0$ and $b > 0$. Then $\Gamma$ is a smooth path. To verify this, we must find a smooth parametrized path that has this ellipse as its image. But define

$$\gamma(\theta) = (a \cos \theta, \, b \sin \theta) \quad \text{for} \quad 0 \le \theta \le 2\pi.$$

This is clearly a smooth parametrized path that has $\Gamma$ as its image. □

**EXAMPLE 19.3** The mapping $\gamma: [0, 4\pi] \to \mathbb{R}^3$ defined by

$$\gamma(t) = (\cos t, \, \sin t, \, t) \quad \text{for} \quad 0 \le t \le 4\pi$$

is a smooth parametrized path. The image of this smooth parametrized path is called a *helix*. □

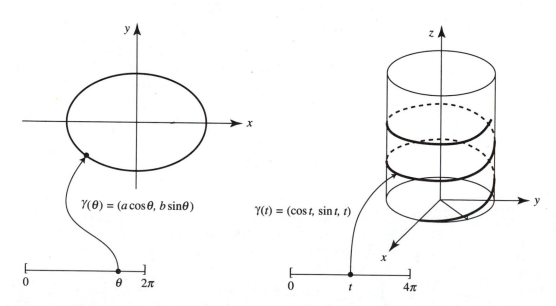

**FIGURE 19.1**

**EXAMPLE 19.4** Suppose that the continuous functions $g:[a, b] \to \mathbb{R}$ and $h:[a, b] \to \mathbb{R}$ have the property that $g(x) < h(x)$ for all $x$ in $(a, b)$ and that their restrictions $g:(a, b) \to \mathbb{R}$ and $h:(a, b) \to \mathbb{R}$ have continuous bounded derivatives. Define

$$\Omega = \{(x, y) \mid a < x < b, \ g(x) < y < h(x)\}.$$

Then the boundary of $\Omega$ is a piecewise smooth parametrized path. To see this, observe that if $g(a) < h(a)$ and $g(b) < h(b)$, then the boundary of $\Omega$ is the image of the mapping $\gamma:[0, 1] \to \mathbb{R}^2$ defined by

$$\gamma(t) = \begin{cases} (a + 4t(b-a), \ g(a + 4t(b-a))) & \text{if } 0 \le t \le 1/4 \\ (b, \ g(b) + 4(t - 1/4)(h(b) - g(b))) & \text{if } 1/4 \le t \le 1/2 \\ (b - 4(t - 1/2)(b-a), \ h(b - 4(t - 1/2)(b-a))) & \text{if } 1/2 \le t \le 3/4 \\ (a, \ h(a) + 4(t - 3/4)(g(a) - h(a))) & \text{if } 3/4 \le t \le 1. \end{cases}$$

**FIGURE 19.2**

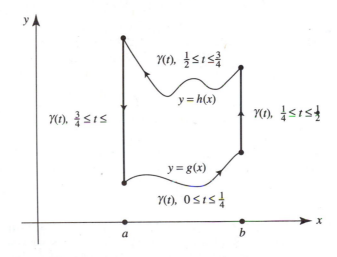

We have defined the length of the interval of real numbers $[a, b]$ to be $b - a$. Moreover, we define the length of the segment joining the points $\mathbf{p}$ and $\mathbf{q}$ in Euclidean space $\mathbb{R}^n$ to be the Euclidean distance between these points, $d(\mathbf{p}, \mathbf{q}) \equiv \|\mathbf{p} - \mathbf{q}\|$. We wish to extend this definition to the concept of *arclength* of a parametrized path. To do so, it is necessary to restrict the class of parametrized paths. The following definition of arclength is motivated by the geometric concept of polygonal approximation.

**DEFINITION** *A parametrized path $\gamma:[a, b] \to \mathbb{R}^n$ is said to have arclength (or to be rectifiable) provided that there is a number $\ell$ with the following property: For each positive number $\epsilon$ there is a positive number $\delta$ such that*

$$\left| \left[ \sum_{i=1}^{m} \|\gamma(x_i) - \gamma(x_{i-1})\| \right] - \ell \right| < \epsilon$$

*for each partition $P = \{x_0, \ldots, x_m\}$ of $[a, b]$ with $\|P\| < \delta$.*

For a parametrized path $\gamma:[a,\,b]\to\mathbb{R}^n$ and a partition $P=\{x_0,\ldots,x_m\}$ of its parameter space $[a,\,b]$, it is convenient to call the sum

$$\sum_{i=1}^{m}\|\gamma(x_i)-\gamma(x_{i-1})\|$$

the *polygonal approximation* of the arclength of $\gamma:[a,\,b]\to\mathbb{R}^n$ based on the partition $P$.

For a rectifiable parametrized path $\gamma:[a,\,b]\to\mathbb{R}^n$, there is only one number $\ell$ that has the defining polygonal approximation property (see Exercise 7) and it is called the *arclength* of the parametrized path.

We have defined the arclength of a parametrized path; it certainly seems reasonable that there should be a sense in which the arclength is a property possessed by the image of this parametrized path and does not depend on the manner of parametrization. To understand the sense in which this is so, it is necessary to consider the relationship between different parametrizations of a given path.

**DEFINITION**    *Two parametrized paths $\gamma:[a,\,b]\to\mathbb{R}^n$ and $\alpha:[c,\,d]\to\mathbb{R}^n$ are said to be equivalent provided that there is a strictly increasing parametrized path $u:[c,\,d]\to\mathbb{R}$ with image $[a,\,b]$ such that*

$$\alpha=\gamma\circ u:[c,\,d]\to\mathbb{R}.$$

**PROPOSITION 19.1**    Suppose that the parametrized path $\gamma:[a,\,b]\to\mathbb{R}^n$ is rectifiable. Then every parametrized path that is equivalent to $\gamma:[a,\,b]\to\mathbb{R}^n$ is also rectifiable and has the same arclength as $\gamma:[a,\,b]\to\mathbb{R}^n$.

**Proof**    Let $u:[c,\,d]\to\mathbb{R}$ be a continuous, strictly increasing function with image $[a,\,b]$. Then the composition $\alpha\equiv\gamma\circ u:[c,\,d]\to\mathbb{R}^n$ is equivalent to $\gamma:[a,\,b]\to\mathbb{R}^n$. We must show that $\alpha:[c,\,d]\to\mathbb{R}^n$ is rectifiable and has arclength equal to the arclength of $\gamma:[a,\,b]\to\mathbb{R}^n$, which we will denote by $\ell$.

Let $\epsilon>0$. Then there is a $\delta>0$ such that if $P=\{x_0,\ldots,x_m\}$ is a partition of the interval $[a,\,b]$ such that $\|P\|<\delta$, then

$$\left|\sum_{i=1}^{m}\|\gamma(x_i)-\gamma(x_{i-1})\|-\ell\right|<\epsilon. \tag{19.1}$$

Since the function $u:[c,\,d]\to\mathbb{R}$ is continuous, it is uniformly continuous. Thus there is a $\delta'>0$ such that $|u(t_1)-u(t_2)|<\delta$ for any two points $t_1$ and $t_2$ in $[c,\,d]$ such that $|t_1-t_2|<\delta'$. Since the function $u:[c,\,d]\to\mathbb{R}$ is strictly increasing, it is clear that a partition $P'$ of $[c,\,d]$ is mapped by $u:[c,\,d]\to\mathbb{R}$ onto a partition $P$ of $[a,\,b]$, and that the polygonal approximation for the arclength of $\alpha:[c,\,d]\to\mathbb{R}^n$ based on $P'$ is equal to the polygonal approximation for the arclength of $\gamma:[a,\,b]\to\mathbb{R}^n$ based on $P$. Thus, by the choice of $\delta$ and $\delta'$, if $\|P'\|<\delta'$, then $\|P\|<\delta$ and so the polygonal approximation of the arclength of $\alpha:[c,\,d]\to\mathbb{R}^n$ based on $P'$ differs from $\ell$ by at most $\epsilon$.   ∎

We will show that a piecewise smooth parametrized path $\gamma: [a, b] \to \mathbb{R}^n$ is rectifiable and its arclength equals $\int_a^b \|\gamma'\|$. The first step toward proving this is to establish the following approximation lemma.

**LEMMA 19.2**   A smooth parametrized path $\gamma: [a, b] \to \mathbb{R}^n$ has the following approximation property: For each subinterval $[a', b']$ of $[a, b]$, with $a < a' < b' < b$, and each positive number $\epsilon$, there is a positive number $\delta$ such that if $[c, d]$ is a subinterval of $[a', b']$ of length less than $\delta$, and $t$ is a parameter value in the open interval $(c, d)$, then

$$\left\| \frac{\gamma(d) - \gamma(c)}{d - c} - \gamma'(t) \right\| < \epsilon. \tag{19.2}$$

**Proof**   We express $\gamma: [a, b] \to \mathbb{R}^n$ in component functions as follows:

$$\gamma(t) = \big(\gamma_1(t), \ldots, \gamma_n(t)\big) \quad \text{for} \quad a \le t \le b.$$

By the very definition of a smooth parametrized path, the mapping $\gamma': [a', b'] \to \mathbb{R}^n$ is continuous. Since a continuous function defined on a closed bounded interval is uniformly continuous, for each index $i$ between 1 and $n$, the function $\gamma_i': [a', b'] \to \mathbb{R}$ is uniformly continuous.

Let $\epsilon > 0$. Define $\epsilon' \equiv \epsilon / \sqrt{n}$. For each index $i$ between 1 and $n$, we can select $\delta_i > 0$ such that if $t_1$ and $t_2$ are parameter values in the interval $[a', b']$ with $|t_2 - t_1| < \delta_i$, then $|\gamma_i'(t_2) - \gamma_i'(t_1)| < \epsilon'$. Define $\delta = \min_{1 \le i \le n} \delta_i$.

Suppose that $[c, d]$ is a subinterval of $[a', b']$ of length less than $\delta$. Then for each index $i$, we can apply the Lagrange Mean Value Theorem for scalar functions to choose a point $\eta_i$ in the open interval $(c, d)$ such that

$$\gamma_i(d) - \gamma_i(c) = \gamma_i'(\eta_i)(d - c).$$

Let $t$ be a parameter value in the open interval $(c, d)$. Then for each index $i$ with $1 \le i \le n$, $|\eta_i - t| < \delta \le \delta_i$, so by the choice of $\delta_i$,

$$|[\gamma_i(d) - \gamma_i(c)] - \gamma_i'(t)(d - c)|$$

$$= |\gamma_i'(\eta_i) - \gamma_i'(t)|(d - c) < \epsilon'(d - c) = (\epsilon / \sqrt{n})(d - c).$$

Consequently,

$$\|[\gamma(d) - \gamma(c)] - \gamma'(t)(d - c)\|$$

$$= \sqrt{\sum_{i=1}^{n} [|[\gamma_i(d) - \gamma_i(c)] - \gamma_i'(t)(d - c)|]^2} < \epsilon(d - c).$$

Dividing this inequality by $d - c$, we obtain (19.2).   ∎

**PROPOSITION 19.3**   A smooth parametrized path $\gamma: [a, b] \to \mathbb{R}^n$ is rectifiable and its arclength $\ell$ is given by

$$\ell = \int_a^b \|\gamma'\|. \tag{19.3}$$

**Proof**  Define the function $h: (a, b) \to \mathbb{R}$ by $h(t) = \|\gamma'(t)\|$ for $a < t < b$. Then, by definition, $h: (a, b) \to \mathbb{R}$ is continuous and bounded, and hence it is integrable. For each index $i$ between 1 and $n$, we can choose $M_i > 0$ such that $|\gamma_i'(t)| \le M_i$ for all $t$ in $(a, b)$. Define $M = \sqrt{M_1^2 + \cdots + M_n^2}$. Using the Mean Value Theorem for each of the component functions of the map $\gamma$ and arguing as in the proof of Lemma 19.2, it follows that for any subinterval $[c, d]$ of $[a, b]$,

$$\left| \frac{\|\gamma(d) - \gamma(c)\|}{d - c} - \|\gamma'(t)\| \right| < 2M \quad \text{for all } t \text{ in } [c, d].$$

Thus, by the monotonicity property of the integral,

$$\left| \|\gamma(d) - \gamma(c)\| - \int_c^d \|\gamma'\| \right| < 2M(d - c) \quad \text{for every subinterval } [c, d] \text{ of } [a, b].$$

$$(19.4)$$

Let $\epsilon > 0$. Choose a subinterval $[a', b']$ of $[a, b]$, with $a < a' < b' < b$, such that if we define $\eta = \max\{a' - a, b - b'\}$, then $16\eta M < \epsilon$. By Lemma 19.2, setting $\epsilon' = \epsilon/2(b' - a')$, we can choose $\delta' > 0$ such that if $[c, d]$ is a subinterval of $[a', b']$ of length less than $\delta'$, and $t$ is a parameter value in the open interval $(c, d)$, then

$$\left\| \frac{\gamma(d) - \gamma(c)}{d - c} - \gamma'(t) \right\| < \epsilon',$$

which, by the Reverse Triangle Inequality, implies that

$$\left| \frac{\|\gamma(d) - \gamma(c)\|}{d - c} - \|\gamma'(t)\| \right| < \epsilon' \quad \text{for all } t \text{ in } (c, d).$$

Hence, by the monotonicity property of the integral,

$$\left| \|\gamma(d) - \gamma(c)\| - \int_c^d \|\gamma'\| \right| < \epsilon'(d - c) \quad \text{if} \quad [c, d] \subseteq [a', b'] \quad \text{and} \quad d - c < \delta'.$$

$$(19.5)$$

Define $\delta = \min\{\delta', \eta\}$. Let $P = \{x_0, \ldots, x_m\}$ be a partition of $[a, b]$ such that $\|P\| < \delta$. By the Triangle Inequality,

$$\left| \sum_{i=1}^m \|\gamma(x_i) - \gamma(x_{i-1})\| - \int_a^b \|\gamma'\| \right| \le \sum_{i=1}^m \left| \|\gamma(x_i) - \gamma(x_{i-1})\| - \int_{x_{i-1}}^{x_i} \|\gamma'\| \right|. \quad (19.6)$$

By the estimate (19.5), for each index $i$ with $1 \le i \le m$,

$$\left| \|\gamma(x_i) - \gamma(x_{i-1})\| - \int_{x_{i-1}}^{x_i} \|\gamma'\| \right| < \epsilon'(x_i - x_{i-1}) \quad \text{if} \quad [x_{i-1}, x_i] \subseteq [a', b'].$$

Thus the contribution to the sum on the right-hand side of (19.6) from intervals $[x_{i-1}, x_i]$ contained in $[a', b']$ is less than $\epsilon'(b' - a') = \epsilon/2$. On the other hand, for each interval $[x_{i-1}, x_i]$ that is not contained in $[a', b']$, the estimate (19.4) implies that

$$\left| \|\gamma(x_i) - \gamma(x_{i-1})\| - \int_{x_{i-1}}^{x_i} \|\gamma'\| \right| < 2M(x_i - x_{i-1}),$$

and since the the sum of the lengths of such intervals is at most $4\eta$, the contribution to the sum on the right-hand side of (19.6) from intervals $[x_{i-1}, x_i]$ not contained in $[a', b']$ is at most $8\eta M < \epsilon/2$. From these two estimates it follows that

$$\left| \sum_{i=1}^{m} \|\gamma(x_i) - \gamma(x_{i-1})\| - \int_a^b \|\gamma'\| \right| < \epsilon. \qquad \blacksquare$$

**THEOREM 19.4**    A piecewise smooth parametrized path $\gamma: [a, b] \to \mathbb{R}^n$ is rectifiable and its arclength $\ell$ is given by

$$\ell = \int_a^b \|\gamma'\|. \qquad (19.7)$$

**Proof**    Let $\epsilon > 0$. It is necessary to find $\delta > 0$ such that if $P = \{x_0, \ldots, x_m\}$ is a partition of $[a, b]$ with $\|P\| < \delta$, then

$$\left| \sum_{i=1}^{m} \|\gamma(x_i) - \gamma(x_{i-1})\| - \int_a^b \|\gamma'\| \right| < \epsilon.$$

Choose a partition $\{z_0, \ldots, z_k\}$ of the interval $[a, b]$ such that the restriction of $\gamma$ to each subinterval of this partition is a smooth parametrized path. Since the mapping $\gamma: [a, b] \to \mathbb{R}^n$ is uniformly continuous, we can choose a positive number $\delta'$ such that $\|\gamma(s) - \gamma(t)\| < \epsilon/6k$ for any two parameter values $s$ and $t$ in $[a, b]$ such that $|s - t| < \delta'$.

The crucial point in the proof is the following observation: If $P$ is any partition of $[a, b]$ with $\|P\| < \delta'$, and $P'$ is the refinement of $P$ obtained by inserting all the $z_i$'s, then the difference between the polygonal approximation of the arclength of $\gamma: [a, b] \to \mathbb{R}^n$ based on $P$ and that based on $P'$ is at most $\epsilon/2$. To see this, observe that the difference between these approximations consists of fewer than $3k$ terms of the form $\|\gamma(t) - \gamma(s)\|$, where $s$ and $t$ are parameter values such that $|s - t| < \delta'$. Since each of these terms is at most $\epsilon/6k$, the polygonal approximations do indeed differ by at most $\epsilon/2$.

For each index $i$ between 1 and $k$, by Proposition 19.3, the parametrized path $\gamma: [z_{i-1}, z_i] \to \mathbb{R}^n$ is rectifiable, so we can choose a $\delta_i > 0$ such that if $P_i$ is any partition of $[z_{i-1}, z_i]$ such that $\|P_i\| < \delta_i$, then the polygonal approximation of the arclength of $\gamma: [z_{i-1}, z_i] \to \mathbb{R}^n$ differs from the integral of $\|\gamma'\|$ on $[z_{i-1}, z_i]$ by at most $\epsilon/2k$. Define $\delta = \min\{\delta', \delta_1, \ldots, \delta_k\}$. Let $P$ be any partition of $[a, b]$ with $\|P\| < \delta$. By the choice of the $\delta_i$'s, the polygonal approximation of the arclength of $\gamma: [a, b] \to \mathbb{R}^n$ based on $P'$ differs from the integral of $\|\gamma'\|$ on $[a, b]$ by at most $\epsilon/2$. By the estimate in the preceding paragraph, it follows that the polygonal approximation of the arclength of $\gamma: [a, b] \to \mathbb{R}^n$ based on $P$ differs from the integral of $\|\gamma'\|$ on $[a, b]$ by at most $\epsilon$. Thus (19.7) holds. $\blacksquare$

**EXAMPLE 19.5**    For the helix defined by

$$\gamma(t) = (\cos t, \sin t, t) \quad \text{for} \quad 0 \le t \le 4\pi,$$

observe that

$$\gamma'(t) = (-\sin t, \cos t, 1) \quad \text{for} \quad 0 \le t \le 4\pi.$$

By Theorem 19.4, this parametrized path is rectifiable and its arclength is

$$\int_0^{4\pi} \|\gamma'(t)\| \, dt = \int_0^{4\pi} \sqrt{2} \, dt = 4\sqrt{2}\pi. \qquad \square$$

Suppose that the parametrized path $\gamma:[a, b] \to \mathbb{R}^n$ is rectifiable. Then for $P$ a partition of $[a, b]$ and $P'$ a refinement of $P$, the polygonal approximation of the arclength based on $P'$ is larger than the polygonal approximation based on $P$; this is an immediate consequence of the Triangle Inequality. From this it is not difficult to show that that the arclength of a rectifiable parametrized path is the supremum of all polygonal approximations of the arclength (see Exercise 8).

In the proof that a smooth parametrized path $\gamma:[a, b] \to \mathbb{R}^n$ is rectifiable, we used the fact that for such a parametrized path the mapping $\gamma':(a, b) \to \mathbb{R}^n$ is *bounded*. In fact, as the following example shows, it is not true that a parametrized path $\gamma:[a, b] \to \mathbb{R}^n$ that has only the property that $\gamma:(a, b) \to \mathbb{R}^n$ is continuously differentiable is necessarily rectifiable.

**EXAMPLE 19.6**   First define

$$f(x) = \begin{cases} x \sin(\pi/2 + \pi/x) & \text{if } 0 < x \le 1 \\ 0 & \text{if } x = 0, \end{cases}$$

and then define $\gamma(t) = (t, \ f(t))$ for $0 \le t \le 1$. Then it is clear that $\gamma:[0, \ 1] \to \mathbb{R}^2$ is a parametrized path and that $\gamma:(0, \ 1) \to \mathbb{R}^2$ is continuously differentiable. However, the path $\gamma:[0, \ 1] \to \mathbb{R}^2$ is not rectifiable. Indeed, observe that

$$f(1/k) = \begin{cases} 1/k & \text{if } k \text{ is an even natural number} \\ -1/k & \text{if } k \text{ is an odd natural number,} \end{cases}$$

**FIGURE 19.3**

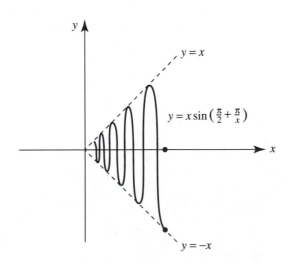

so that
$$\left\| \gamma\left(\frac{1}{k}\right) - \gamma\left(\frac{1}{(k+1)}\right) \right\| \geq \frac{1}{k} \quad \text{for each natural number } k.$$

It follows that for each natural number $n$, if we define the partition $P = \{0, 1/(n+1), 1/n, \ldots, 1/2, 1\}$, then the polygonal approximation based on $P$ is greater than $\sum_{k=1}^{n} 1/k$. Since the sequence of partial sums of the Harmonic Series $\sum_{k=1}^{\infty} 1/k$ is unbounded, we conclude that the polygonal approximations may be arbitrarily large, so the parametrized path $\gamma: [0, 1] \to \mathbb{R}^2$ is not rectifiable. $\qquad\square$

Among the equivalent parametrizations of a piecewise smooth parametrized path $\gamma: [a, b] \to \mathbb{R}^n$, there is a particular parametrization, called *parametrization by arclength*, that is often convenient. It is defined as follows: For each parameter value $t$ in $[a, b]$, define

$$u(t) \equiv \int_a^t \|\gamma'\|.$$

Then, since $\gamma'(t) \neq \mathbf{0}$, except possibly at a finite number of points in $(a, b)$, the continuous function $u: [a, b] \to \mathbb{R}$ is strictly increasing and $u([a, b]) = [0, \ell]$, where $\ell$ is the arclength of $\gamma: [a, b] \to \mathbb{R}^n$. The inverse function $u^{-1}: [0, \ell] \to [a, b]$ is also strictly increasing and continuous. The parametrized path $\alpha: [0, \ell] \to \mathbb{R}^n$ defined by the composition $\alpha = \gamma \circ u^{-1}$ is called the parametrization by arclength for the parametrized path $\gamma: [a, b] \to \mathbb{R}^n$. It has the property that for each parameter value $s$ in $[0, \ell]$, the arclength of the parametrized path $\alpha: [0, s] \to \mathbb{R}^n$ equals the parameter value $s$. The Second Fundamental Theorem of Calculus implies that except for possibly finitely many points, $u'(t) = \|\gamma'(t)\|$, so that using the Chain Rule, it follows that the parametrization by arclength $\alpha: [0, \ell] \to \mathbb{R}^n$ has the property that at each parameter value $s$ in $[0, \ell]$ at which it is differentiable, $\|\alpha'(s)\| = 1$.

**EXAMPLE 19.7**   Consider the helix defined by $\gamma(t) = (\cos t, \sin t, t)$ for $0 \leq t \leq 4\pi$. For each $t$ in $[0, 4\pi]$,

$$u(t) = \int_0^t \|\gamma'\| = \int_0^t \sqrt{2}\, dt = \sqrt{2}\,t.$$

Thus the parametrization by arclength of the helix, $\alpha: [0, 4\sqrt{2}\pi] \to \mathbb{R}^3$, is defined by

$$\alpha(s) = (\cos(s/\sqrt{2}), \sin(s/\sqrt{2}), s/\sqrt{2}) \quad \text{for} \quad 0 \leq s \leq 4\sqrt{2}\pi. \qquad\square$$

We will now consider the concept of line integral. Given a piecewise smooth parametrized path $\gamma: [a, b] \to \mathbb{R}^n$ having image $\Gamma$ and a continuous function $f: \Gamma \to \mathbb{R}$, we define the line integral of $f: \Gamma \to \mathbb{R}$ on the parametrized path $\gamma: [a, b] \to \mathbb{R}^n$ by the formula

$$\int_{\gamma:[a,b]\to\mathbb{R}^n} f \equiv \int_a^b f(\gamma(t))\|\gamma'(t)\|\, dt.$$

**THEOREM 19.5**    Suppose that $\gamma: [a, b] \to \mathbb{R}^n$ and $\alpha: [c, d] \to \mathbb{R}^n$ are two piecewise smooth parametrizations of the path $\Gamma$ that are equivalent. Then for any continuous function $f: \Gamma \to \mathbb{R}$,

$$\int_{\gamma:[a,b]\to\mathbb{R}^n} f = \int_{\alpha:[c,d]\to\mathbb{R}^n} f.$$

**Proof**    Using the addition over partitions property of the integral, it suffices to consider the case when the paths are smooth. By the definition of equivalence of parametrizations, there is a strictly increasing path $u: [c, d] \to \mathbb{R}$ such that $u([c, d]) = [a, b]$ and $\alpha = \gamma \circ u: [c, d] \to \mathbb{R}$. For each parameter value $t$ in $(c, d)$, there is an index $i$ such that $\gamma_i'(u(t)) \neq 0$, so that since $\gamma_i \circ u = \alpha_i: (c, d) \to \mathbb{R}$, from the Inverse Function Theorem and the Chain Rule it follows that $u: (c, d) \to R$ is differentiable at $t$. Thus $\alpha'(t) = \gamma'(u(t))u'(t)$ and so, in particular, $u'(t) \neq 0$, for all parameter values $t$ in $(c, d)$. Since $u: [c, d] \to \mathbb{R}$ is strictly increasing and $u'(t) \neq 0$ for all $t$ in $(c, d)$, $u'(t) > 0$ for all $t$ in $(c, d)$. By the Change of Variables Theorem and the Chain Rule,

$$\int_{\gamma:[a,b]\to\mathbb{R}^n} f = \int_a^b f(\gamma(t))\|\gamma'(t)\| \, dt$$

$$= \int_{u(c)}^{u(d)} f(\gamma(t))\|\gamma'(t)\| \, dt$$

$$= \int_c^d f(\gamma(u(t)))\|\gamma'(u(t))\|u'(t) \, dt$$

$$= \int_c^d f(\alpha(t))\|\alpha'(t)\| \, dt$$

$$= \int_{\alpha:[c,d]\to\mathbb{R}^n} f. \qquad \blacksquare$$

For a piecewise smooth parametrization $\gamma: [a, b] \to \mathbb{R}^n$ of a path $\Gamma$ and a continuous function $f: \Gamma \to \mathbb{R}$, if the arclength of this path is $\ell$ and $\alpha: [0, \ell] \to \mathbb{R}^n$ is the equivalent parametrization by arclength, then, since $\|\alpha(s)\| = 1$ except at possibly finitely many points, by Theorem 19.5,

$$\int_{\gamma:[a,b]\to\mathbb{R}^n} f = \int_0^\ell f(\alpha(s)) \, ds.$$

For this reason, it is often convenient to denote the line integral by

$$\int_\Gamma f(s) \, ds$$

and explicitly state a choice of smooth parametrization of $\Gamma$.

There are other types of line integrals associated with continuous functions defined on paths. For a parametrized path $\gamma: [a, b] \to \mathbb{R}^n$ having image $\Gamma$ and a continuous function $f: \Gamma \to \mathbb{R}$, for an index $i$ between 1 and $n$ that has the property that $\gamma_i: (a, b) \to \mathbb{R}^n$ has a continuous bounded derivative, we define

$$\int_\Gamma f(\mathbf{x}) \, dx_i = \int_a^b f(\gamma(t))\gamma_i'(t) \, dt.$$

Arguing as we did for line integrals, it follows that changing from one parametrization to another by composing with a strictly increasing smooth function does not change the value of the integral. Naturally, in $\mathbb{R}^2$ and $\mathbb{R}^3$, when points are denoted by $(x, y)$ or $(x, y, z)$ we use a more familiar notation for these integrals: In $\mathbb{R}^3$ we use

$$\int_\Gamma f(x, y, z)\,dx, \quad \int_\Gamma f(x, y, z)\,dy, \quad \text{and} \quad \int_\Gamma f(x, y, z)\,dz.$$

**EXAMPLE 19.8**   Consider the piecewise smooth parametrization $\gamma: [0, 1] \to \mathbb{R}^2$ of the path $\Gamma = \partial\Omega$ that was given in Example 19.4.

**FIGURE 19.4**

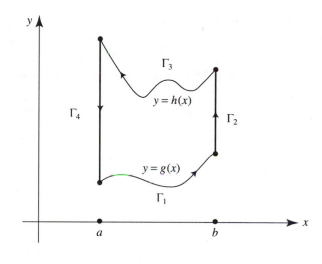

$$\int_\Gamma y\,dx = \int_{\Gamma_1} y\,dx + \int_{\Gamma_3} y\,dx$$

For this parametrization, it follows from the very definition of these line integrals and the change of variables formula for integrals of functions of a single variable that

$$\int_\Gamma y\,dx = \int_{\Gamma_1} y\,dx + \int_{\Gamma_3} y\,dx$$

$$= \int_0^{1/4} g(a + 4t(b - a))4(b - a)\,dt$$

$$+ \int_{1/2}^{3/4} h(b - 4(t - 1/2)(b - a))(-4(b - a))\,dt$$

$$= \int_a^b [g(t) - h(t)]\,dt. \qquad \square$$

1. For two points $\mathbf{p}$ and $\mathbf{q}$ in $\mathbb{R}^n$, use formula (19.3) to check that the arclength of the parametrized segment from $\mathbf{p}$ to $\mathbf{q}$ is $\|\mathbf{p} - \mathbf{q}\|$.

2. Find the arclength of the parametrized path $\gamma:[0,\, 2\pi] \to \mathbb{R}^3$ defined by

$$\gamma(t) = (3\cos t,\, 3\sin t,\, 2t), \quad 0 \le t \le 2\pi.$$

3. For the parametrized path $\gamma:[0,\, 4] \to \mathbb{R}^3$ defined by

$$\gamma(t) = (t,\, t^2,\, 1) \quad \text{for} \quad 0 \le t \le 4,$$

find a smoothly equivalent parametrization by arclength.

4. For $0 < a < b$, find an integral that equals the arclength of the ellipse $\{(x,\, y) \mid x^2/a^2 + y^2/b^2 = 1\}$. (It is necessary to use numerical approximation methods to actually evaluate this integral.)

5. Evaluate

$$\int_\gamma x \sin y\, dx + y \cos x\, dy$$

where $\gamma:[0,\, 1] \to \mathbb{R}^2$ is the parametrized segment defined by

$$\gamma(t) = (t,\, mt) \quad \text{for} \quad 0 \le t \le 1.$$

6. Evaluate

$$\int_\gamma \frac{x}{x^2 + y^2}\, dx + \frac{y}{x^2 + y^2}\, dy$$

where $\gamma:[0,\, 2\pi] \to \mathbb{R}^2$ is the parametrized path defined by

$$\gamma(t) = (e^t \cos t,\, e^t \sin t) \quad \text{for} \quad 0 \le t \le 2\pi.$$

7. Show that the definition of arclength is unambiguous in that if a parametrized path is rectifiable, then there is only one number $\ell$ that has the polynomial approximation property in the definition.

8. Show that the arclength of a rectifiable parametrized path is the supremum of all polygonal approximations of the arclength.

9. For a rectifiable parametrized path $\gamma:[a,\, b] \to \mathbb{R}^n$, define another parametrized path $\alpha:[a,\, b] \to \mathbb{R}^n$ by $\alpha(t) = \gamma(a+b-t)$ for $a \le t \le b$. Show that $\alpha:[a,\, b] \to \mathbb{R}^n$ is also rectifiable and has the same arclength as $\gamma:[a,\, b] \to \mathbb{R}^n$. Also, show that $\alpha:[a,\, b] \to \mathbb{R}^n$ is smooth if $\gamma:[a,\, b] \to \mathbb{R}^n$ is smooth.

10. Suppose that $\gamma:[a,\, b] \to \mathbb{R}^n$ and $\alpha:[c,\, d] \to \mathbb{R}^n$ are parametrized paths, each of which is one-to-one and such that $\gamma(a) = \alpha(c)$. Show that these parametrized paths are equivalent if and only if they have the same image. (*Hint:* For each parameter value $t$ in $[c,\, d]$, define $u(t) = s$ to be the unique parameter value $s$ in $[a,\, b]$ such that $\alpha(t) = \gamma(s)$.)

11. For a piecewise smooth parametrized path $\gamma:[a,\, b] \to \mathbb{R}^n$ and a continuous function $f:\Gamma \to \mathbb{R}$, where $\Gamma$ is the image of $\gamma:[a,\, b] \to \mathbb{R}^n$, prove that

$$\left| \int_\Gamma f(s)\, ds \right| \le M\ell,$$

where $M$ is such that $|f(\mathbf{p})| \le M$ for all points $\mathbf{p}$ on $\Gamma$ and $\ell$ is the arclength of $\gamma:[a,\, b] \to \mathbb{R}^n$.

## 19.2 Surface Area and Surface Integrals

In Section 19.1, we defined parametrized paths, paths, and line integrals; in this section, we will define the corresponding notions of parametrized surfaces, surfaces, and surface integrals.

We call a subset $\mathcal{R}$ of the plane $\mathbb{R}^2$ a *region* provided that it is open and is a Jordan domain—that is, it is open and bounded and its boundary has Jordan content 0. By Theorem 18.17, if $\mathcal{R}$ is a region in $\mathbb{R}^2$ and the function $g: \mathcal{R} \to \mathbb{R}$ is continuous and bounded, then its integral $\int_{\mathcal{R}} g(x, y) \, dx \, dy$ is properly defined.

**DEFINITION** *Let $\mathcal{R}$ be a region in $\mathbb{R}^2$. Then a continuously differentiable mapping $\mathbf{r}: \mathcal{R} \to \mathbb{R}^3$ is called a parametrized surface with parameter space $\mathcal{R}$ provided that the following three properties hold:*

(i) *The component functions of the mapping $\mathbf{r}: \mathcal{R} \to \mathbb{R}^3$ have bounded first-order partial derivatives.*

(ii) *The mapping $\mathbf{r}: \mathcal{R} \to \mathbb{R}^3$ is one-to-one.*

(iii) *For each point $(u, v)$ in $\mathcal{R}$,*

$$\frac{\partial \mathbf{r}}{\partial u}(u, v) \times \frac{\partial \mathbf{r}}{\partial v}(u, v) \neq \mathbf{0}.$$

**DEFINITION** *A subset $\mathcal{S}$ of $\mathbb{R}^3$ is called a surface provided that it is the image of a parametrized surface.*

For a parametrized surface $\mathbf{r}: \mathcal{R} \to \mathbb{R}^3$ and a point $(u_0, v_0)$ in the parameter space $\mathcal{R}$, since $\mathcal{R}$ is open, if the positive number $r$ is sufficiently small, then the parametrized path $\gamma: (-r, r) \to \mathbb{R}^3$ defined by

$$\gamma(t) = \mathbf{r}(u_0 + t, v_0) \quad \text{for } t \text{ in } (-r, r)$$

defines a smooth path in the surface $\mathcal{S}$ that has a tangent vector at the point $\mathbf{r}(u_0, v_0)$ given by $\partial \mathbf{r}/\partial u(u_0, v_0)$. Similarly, holding the $u$-variable constant, we get a parametrized path whose image lies in the surface $\mathcal{S}$ and that passes through the point $\mathbf{r}(u_0, v_0)$, at which point it has a tangent vector $\partial \mathbf{r}/\partial v(u_0, v_0)$. Thus, because of assumption (iii), if we define the vector $\boldsymbol{\eta}$ by

$$\boldsymbol{\eta} = \frac{\partial \mathbf{r}}{\partial u}(u_0, v_0) \times \frac{\partial \mathbf{r}}{\partial v}(u_0, v_0),$$

we conclude that the vector $\boldsymbol{\eta}$ is nonzero and, by the properties of the cross product, it is orthogonal to the tangent vectors $\partial \mathbf{r}/\partial u(u_0, v_0)$ and $\partial \mathbf{r}/\partial v(u_0, v_0)$. For this reason, we call $\boldsymbol{\eta}$, or any nonzero scalar multiple of $\boldsymbol{\eta}$, a normal to the surface $\mathcal{S}$ at the point $\mathbf{r}(u_0, v_0)$. Thus a surface has a normal at each point that, for a given parametrization, varies continuously with the parameters $(u, v)$.

**FIGURE 19.5**

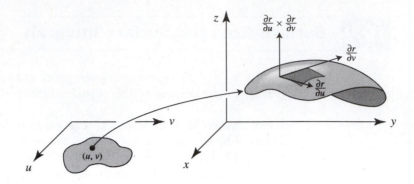

**EXAMPLE 19.9**   Fix a positive number $a$ less than 1 and consider the set

$$S = \left\{(x,\, y,\, z) \text{ in } \mathbb{R}^3 \mid x > 0,\, y > 0,\, z > 0,\, x^2 + y^2 + z^2 = 1,\, 0 < x^2 + y^2 < a^2\right\}.$$

Then the set $S$ is a surface. To verify this, we must find a parametrized surface that has image equal to $S$. Define $\mathcal{R} = \{(x,\, y) \text{ in } \mathbb{R}^2 \mid x > 0,\, y > 0,\, 0 < x^2 + y^2 < a^2\}$ and define $\mathbf{r}: \mathcal{R} \to \mathbb{R}^3$ by

$$\mathbf{r}(x,\, y) = \left(x,\, y,\, \sqrt{1 - x^2 - y^2}\right) \quad \text{for } (x,\, y) \text{ in } \mathcal{R}.$$

Then the set $\mathcal{R}$ is certainly a region, the mapping $\mathbf{r}: \mathcal{R} \to \mathbb{R}^3$ is continuously differentiable, and a computation of partial derivatives shows that the partial derivatives of the components of this mapping are bounded: For example, the estimate for the partial derivatives of the third component with respect to $x$ is

$$\left|\frac{\partial r_3}{\partial x}(x,\, y)\right| = \left|\frac{-x}{\sqrt{1 - x^2 - y^2}}\right| \leq \frac{1}{\sqrt{1 - a^2}} \quad \text{for all } (x,\, y) \text{ in } \mathcal{R}.$$

Furthermore,

$$\frac{\partial \mathbf{r}}{\partial x}(x,\, y) \times \frac{\partial \mathbf{r}}{\partial y}(x,\, y) = \left(\frac{x}{\sqrt{1 - x^2 - y^2}},\, \frac{y}{\sqrt{1 - x^2 - y^2}},\, 1\right) \neq \mathbf{0}.$$

Finally, it is clear that this mapping is one-to-one and has image equal to $S$.   □

**EXAMPLE 19.10**   In Example 19.9, if we allow $a = 1$, then the mapping $\mathbf{r}: \mathcal{R} \to \mathbb{R}^3$ defined above is not a parametrized surface, since the partial derivatives of the third component are no longer bounded. Nevertheless, the image of this mapping, $S$, is a surface. In order to verify this it is necessary to find a parametrization of $\mathcal{S}$. Define $\mathcal{R}' = \{(u, v) \text{ in } \mathbb{R}^2 \mid 0 < u < \pi/2, \, 0 < v < \pi/2\}$, then define $\mathbf{r}': \mathcal{R}' \to \mathbb{R}^3$ by

$$\mathbf{r}'(u, v) = (\sin u \cos v, \sin u \sin v, \cos u) \quad \text{for } (u, v) \text{ in } \mathcal{R}'.$$

The parameter space $\mathcal{R}'$ is a region in the plane. It is clear that this mapping is continuously differentiable and that its components have bounded partial derivatives. Moreover, geometrically interpreting the significance of the components (these are spherical coordinates), we see that the image of this map is $\mathcal{S}$ and that the map is one-to-one. Finally, a brief computation, which we leave as an exercise, shows that for each point $(u, v)$ in $\mathcal{R}$,

$$\frac{\partial \mathbf{r}}{\partial u}(u, v) \times \frac{\partial \mathbf{r}}{\partial v}(u, v) = (\sin^2 u \cos v, \sin^2 u \sin v, \sin u \cos u) \neq \mathbf{0}. \qquad \square$$

The simplest type of parametrized surface is one that has as its image the graph of a function defined on a region in a coordinate plane. Such a parametrized surface will be called a *projectionally parametrized surface*. For example, if $\mathcal{R}$ is a region in the plane $\mathbb{R}^2$ and a continuously differentiable function $g: \mathcal{R} \to \mathbb{R}$ with bounded partial derivatives is given, then each of the following three mappings describes a projectionally parametrized surface: the map $\mathbf{r}: \mathcal{R} \to \mathbb{R}^3$ defined by

$$\mathbf{r}(x, y) = (x, y, g(x, y)) \quad \text{for } (x, y) \text{ in } \mathcal{R},$$

the map

$$\mathbf{r}(x, z) = (x, g(x, z), z) \quad \text{for } (x, z) \text{ in } \mathcal{R},$$

and the map

$$\mathbf{r}(y, z) = (g(y, z), y, z) \quad \text{for } (y, z) \text{ in } \mathcal{R}.$$

Observe that in the first case a short computation shows that a normal $\boldsymbol{\eta}$ to this surface at the point $(x_0, y_0, g(x_0, y_0))$ is given by

$$\boldsymbol{\eta} = \frac{\partial \mathbf{r}}{\partial x}(x_0, y_0) \times \frac{\partial \mathbf{r}}{\partial y}(x_0, y_0) = \left( -\frac{\partial g}{\partial x}(x_0, y_0), -\frac{\partial g}{\partial y}(x_0, y_0), 1 \right).$$

This shows that the definition of normal to a surface that is a graph of a function of two variables, which we described in Section 14.1, is consistent with the general definition of normal that we are considering here.

We now wish to define surface area. To do so, we first describe some properties of the inner product and the cross product of vectors in $\mathbb{R}^3$. Recall that for two vectors $\mathbf{u}$ and $\mathbf{v}$ in $\mathbb{R}^3$, the *inner product* $\langle \mathbf{u}, \mathbf{v} \rangle$ and the *cross product* $\mathbf{u} \times \mathbf{v}$ are defined by

$$\langle \mathbf{u}, \mathbf{v} \rangle \equiv u_1 v_1 + u_2 v_2 + u_3 v_3$$

and

$$\mathbf{u} \times \mathbf{v} \equiv (u_2 v_3 - u_3 v_2, \, u_3 v_1 - u_1 v_3, \, u_1 v_2 - u_2 v_1).$$

The inner product and the cross product provide a very useful way to describe geometrical concepts in an analytical form. *

---

*In Appendix B, there is a full description of the geometric properties of the inner product and the cross product of two vectors in $\mathbb{R}^3$.

(i) Let $\mathbf{u}$ be a nonzero vector in $\mathbb{R}^3$ and let $\ell$ be the line through the origin that is parallel to $\mathbf{u}$; that is, $\ell$ consists of points of the form $t\mathbf{u}$ for $t$ in $\mathbb{R}$. Then for any point $\mathbf{q}$ in $\mathbb{R}^3$,

the point on $\ell$ that is closest to $\mathbf{q}$ is $\lambda\mathbf{u}$ where $\lambda = \dfrac{\langle \mathbf{q}, \mathbf{u} \rangle}{\langle \mathbf{u}, \mathbf{u} \rangle}$.

(The proof of this is a computation: Show that $\|\mathbf{q} - \lambda\mathbf{u}\|^2 \le \|\mathbf{q} - t\mathbf{u}\|^2$ for all $t$ in $\mathbb{R}$.)

(ii) Let the vectors $\mathbf{u}$ and $\mathbf{v}$ be linearly independent. For a point $\mathbf{p}$ in $\mathbb{R}^3$, we will define the *parallelogram* based at $\mathbf{p}$ and bounded by the vectors $\mathbf{u}$ and $\mathbf{v}$ to be the set

$$S = \{\mathbf{p} + t\mathbf{u} + s\mathbf{v} \mid 0 < t < 1, \, 0 < s < 1\}.$$

The *area* of this parallelogram is defined to be the length of the vector $\mathbf{v}$ times the distance from the point $\mathbf{p} + \mathbf{u}$ to the line through the point $\mathbf{p}$ that is parallel to the vector $\mathbf{v}$. Since the distance between points is invariant under translation, the area is independent of the choice of base point $\mathbf{p}$. Suppose $\mathbf{p} = \mathbf{0}$; then, by (i),

$$\text{area } S = \|\mathbf{v}\| \cdot \|\mathbf{u} - \lambda\mathbf{v}\|, \quad \text{where } \lambda = \frac{\langle \mathbf{u}, \mathbf{v} \rangle}{\langle \mathbf{v}, \mathbf{v} \rangle}.$$

The cross product provides a simple formula for this:

$$\text{area } S = \|\mathbf{u} \times \mathbf{v}\|. \tag{19.8}$$

The verification of this formula is a straightforward computation. Indeed,

$$
\begin{aligned}
\|\mathbf{u} \times \mathbf{v}\|^2 &= (u_2 v_3 - u_3 v_2)^2 + (u_3 v_1 - u_1 v_3)^2 + (u_1 v_2 - u_2 v_1)^2 \\
&= (u_1^2 + u_2^2 + u_3^2)(v_1^2 + v_2^2 + v_3^2) - (u_1 v_1 + u_2 v_2 + u_3 v_3)^2 \\
&= \|\mathbf{u}\|^2 \cdot \|\mathbf{v}\|^2 - \langle \mathbf{u}, \mathbf{v} \rangle^2 \\
&= \|\mathbf{v}\|^2 \left\{ \|\mathbf{u}\|^2 - \frac{\langle \mathbf{u}, \mathbf{v} \rangle^2}{\|\mathbf{v}\|^2} \right\} \\
&= \|\mathbf{v}\|^2 \langle \mathbf{u} - \lambda\mathbf{v}, \mathbf{u} \rangle \\
&= \|\mathbf{v}\|^2 \langle \mathbf{u} - \lambda\mathbf{v}, \mathbf{u} - \lambda\mathbf{v} \rangle \quad (\text{since } \langle \mathbf{u} - \lambda\mathbf{v}, \mathbf{v} \rangle = 0) \\
&= \|\mathbf{v}\|^2 \cdot \|\mathbf{u} - \lambda\mathbf{v}\|^2.
\end{aligned}
$$

**PROPOSITION 19.6**

Let $\mathbf{J}$ be an open rectangle in the plane and define the parametrized surface $\mathbf{r} : \mathbf{J} \to \mathbb{R}^3$ by

$$\mathbf{r}(x, y) = (x, y, ax + by + c) \quad \text{for } (x, y) \text{ in } \mathbf{J}.$$

Then the area of the surface $S = \mathbf{r}(\mathbf{J})$ is given by

$$\text{area } S = \sqrt{1 + a^2 + b^2} \cdot \text{area } \mathbf{J}. \tag{19.9}$$

**Proof**  Suppose that $\mathbf{J} = (x_1, x_2) \times (y_1, y_2)$. It is clear that $S$ is the parallelogram based at the point $\mathbf{p} = (x_1, y_1, ax_1 + by_1 + c)$ and bounded by the vectors $\mathbf{u} = (0, y_2 - y_1, b(y_2 - y_1))$ and $\mathbf{v} = (x_2 - x_1, 0, a(x_2 - x_1))$. By formula (19.8),

$$\text{area } S = \|\mathbf{u} \times \mathbf{v}\|$$
$$= \|(a(x_2 - x_1)(y_2 - y_1), b(x_2 - x_1)(y_2 - y_1), -(x_2 - x_1)(y_2 - y_1))\|$$
$$= \sqrt{1 + a^2 + b^2} \cdot \text{area } \mathbf{J}. \qquad \blacksquare$$

Formula (19.9) motivates the general definition of surface area. We begin with the definition of area for a projectionally parametrized surface. Let $\mathcal{R}$ be an open rectangle in the plane $\mathbb{R}^2$ and $g: \mathcal{R} \to \mathbb{R}$ be a continuously differentiable function with bounded partial derivatives. Then the mapping $\mathbf{r}: \mathcal{R} \to \mathbb{R}^3$ defined by

$$\mathbf{r}(x, y) = (x, y, g(x, y)) \quad \text{for } (x, y) \text{ in } \mathcal{R}$$

is a parametrized surface. Suppose that $\mathbf{P}$ is a partition of the parameter space $\mathcal{R}$ and that $\mathbf{J}$ is a rectangle in $\mathbf{P}$. Select a point $\mathbf{p} = (x_0, y_0)$ in the interior of $\mathbf{J}$ and let $\mathbf{T}(\mathbf{p})$ be the tangent plane to the surface $S$ at the point $\mathbf{r}(\mathbf{p})$. As we showed in Section 14.1, the tangent plane consists of points $(x, y, z)$ such that

$$z = g(\mathbf{p}) + \frac{\partial g}{\partial x}(\mathbf{p})(x - x_0) + \frac{\partial g}{\partial y}(\mathbf{p})(y - y_0).$$

According to formula (19.9), the area $\mathcal{S}(\mathbf{J})$ of the part of the tangent plane $\mathbf{T}(\mathbf{p})$ that lies above the generalized rectangle $\mathbf{J}$ is given by

$$\text{area } \mathcal{S}(\mathbf{J}) = \sqrt{1 + \left(\frac{\partial g}{\partial x}(\mathbf{p})\right)^2 + \left(\frac{\partial g}{\partial y}(\mathbf{p})\right)^2} \; \text{area } \mathbf{J}.$$

If we sum these areas over the partition $\mathbf{P}$, we obtain

$$\sum_{\mathbf{J} \text{in} \mathbf{P}} \sqrt{1 + \left(\frac{\partial g}{\partial x}(\mathbf{p})\right)^2 + \left(\frac{\partial g}{\partial x}(\mathbf{p})\right)^2} \; \text{area } \mathbf{J}. \qquad (19.10)$$

If we now take a sequence of partitions of $\mathcal{R}$, the gaps of which converge to 0, then the sums (19.10), being Riemann sums, converge to

$$\int_{\mathcal{R}} \sqrt{1 + \left(\frac{\partial g}{\partial x}(x, y)\right)^2 + \left(\frac{\partial g}{\partial y}(x, y)\right)^2} \; dx \, dy.$$

This motivates the following definition of *surface area* for a projectionally parametrized surface.

**DEFINITION**  *Suppose that $\mathcal{R}$ is a region in the plane $\mathbb{R}^2$ and that the continuously differentiable function $g: \mathcal{R} \to \mathbb{R}$ has bounded first-order partial derivatives. We define the area of the surface*

$$S = \{(x, y, g(x, y)) \text{ in } \mathbb{R}^3 \mid (x, y) \text{ in } \mathcal{R}\}$$

*by the formula*

$$area \; S \equiv \int_{\mathcal{R}} \sqrt{1 + \left( \frac{\partial g}{\partial x}(x, \; y) \right)^2 + \left( \frac{\partial g}{\partial y}(x, \; y) \right)^2} \; dx \, dy. \tag{19.11}$$

The general definition of surface area and of surface integral is as follows:

**DEFINITION**   *Suppose that $\mathcal{R}$ is a region in the plane $\mathbb{R}^2$ and let $S$ be the surface that is the image of the parametrized surface $\mathbf{r} \colon \mathcal{R} \to \mathbb{R}^3$. We define the area of $S$ by*

$$area \; S \equiv \int_{\mathcal{R}} \left\| \frac{\partial \mathbf{r}}{\partial u}(u, \; v) \times \frac{\partial \mathbf{r}}{\partial v}(u, \; v) \right\| \; du \, dv. \tag{19.12}$$

*Furthermore, for a continuous bounded function $f \colon S \to \mathbb{R}$, the surface integral of the function $f \colon S \to \mathbb{R}$ over the surface $S$, which is denoted by $\int_S f \, d\boldsymbol{\sigma}$, is defined by the formula*

$$\int_S f \, d\boldsymbol{\sigma} \equiv \int_{\mathcal{R}} f \, (\mathbf{r}(u, \; v)) \left\| \frac{\partial \mathbf{r}}{\partial u}(u, \; v) \times \frac{\partial \mathbf{r}}{\partial v}(u, \; v) \right\| du \, dv. \tag{19.13}$$

**EXAMPLE 19.11**   Consider the set

$$S = \{(x, \; y, \; z) \text{ in } \mathbb{R}^3 \mid x > 0, \; y > 0, \; z > 0, \; x^2 + y^2 + z^2 = 1.\}$$

We have shown in Example 19.10 that $S$ is a surface that is parametrized by $\mathbf{r} \colon \mathcal{R} \to \mathbb{R}^3$, where $\mathcal{R} = \{(u, \; v) \text{ in } \mathbb{R}^2 \mid 0 < u < \pi/2, \; 0 < v < \pi/2\}$, defined by

$$\mathbf{r}(u, \; v) = (\sin u \cos v, \; \sin u \sin v, \; \cos u\} \quad \text{for } (u, \; v) \text{ in } \mathcal{R}.$$

A short computation of partial derivatives and of the cross product shows that for each $(u, \; v)$ in $\mathcal{R}$,

$$\left\| \frac{\partial \mathbf{r}}{\partial u}(u, \; v) \times \frac{\partial \mathbf{r}}{\partial v}(u, \; v) \right\| = \sin u.$$

Thus, by the definition of area and Fubini's Theorem,

$$area \; S = \int_{\mathcal{R}} \left\| \frac{\partial \mathbf{r}}{\partial u}(u, \; v) \times \frac{\partial \mathbf{r}}{\partial v}(u, \; v) \right\| \; du \, dv$$

$$= \int_{\mathcal{R}} \sin u \; du \, dv$$

$$= \int_0^{\pi/2} \left[ \int_0^{\pi/2} \sin u \; du \right] dv$$

$$= \int_0^{\pi/2} 1 \; dv = \frac{\pi}{2}.$$

From this, we conclude that the surface area of the sphere about the origin of radius 1 is $4\pi$.                      $\square$

As we have already seen, a surface will always have different parametrizations. It is certainly desirable that the definition of surface integral, and in particular that of surface area, be independent of the choice of parametrization of the surface. For instance, the two definitions we have given of surface area for a projectionally parametrized surface should coincide. We will now establish the independence of parametrization of the surface integral.

**THEOREM 19.7**   Let $\mathcal{R}$ and $\mathcal{R}'$ be regions in the plane $\mathbb{R}^2$. Let $\mathbf{r}: \mathcal{R} \to \mathbb{R}^3$ and $\mathbf{r}': \mathcal{R}' \to \mathbb{R}^3$ be parametrized surfaces that have the same image $S$. Suppose that the function $f: S \to \mathbb{R}$ is continuous and bounded. Then

$$\int_{\mathcal{R}} f(\mathbf{r}(u, v)) \left\| \frac{\partial \mathbf{r}}{\partial u}(u, v) \times \frac{\partial \mathbf{r}}{\partial v}(u, v) \right\| du\, dv$$

$$= \int_{\mathcal{R}'} f(\mathbf{r}'(u', v')) \left\| \frac{\partial \mathbf{r}'}{\partial u'}(u', v') \times \frac{\partial \mathbf{r}'}{\partial u'}(u', v') \right\| du'dv'. \tag{19.14}$$

**Proof**   By the definition of parametrized surface, each of the mappings $\mathbf{r}: \mathcal{R} \to \mathbb{R}^3$ and $\mathbf{r}': \mathcal{R}' \to \mathbb{R}^3$ is one-to-one and, by assumption, each has image equal to $S$. For a point $(u, v)$ in $\mathcal{R}$, define $\mathbf{g}(u, v) = (u', v')$ to be the unique point in $\mathcal{R}'$ at which

$$\mathbf{r}'(u', v') = \mathbf{r}(u, v). \tag{19.15}$$

This defines a mapping $\mathbf{g}: \mathcal{R} \to \mathbb{R}^2$ that is one-to-one and has image equal to $\mathcal{R}'$. If we write out the parametrized surfaces and the mapping $\mathbf{g}: \mathcal{R} \to \mathbb{R}^2$ in terms of their component functions, it is clear that formula (19.15) is equivalent to the following system of identities:

$$r_1'(g_1(u, v), g_2(u, v)) = r_1(u, v)$$

$$r_2'(g_1(u, v), g_2(u, v)) = r_2(u, v) \tag{19.16}$$

$$r_3'(g_1(u, v), g_2(u, v)) = r_3(u, v).$$

We claim that the mapping $\mathbf{g}: \mathcal{R} \to \mathbb{R}^2$ is a smooth change of variables. To justify this, it is necessary to show that it is continuously differentiable and has an invertible derivative matrix at each point. Let $(u_0, v_0)$ be a point in $\mathcal{R}$. By assumption,

$$\frac{\partial \mathbf{r}'}{\partial u}(\mathbf{g}(u_0, v_0)) \times \frac{\partial \mathbf{r}'}{\partial v}(\mathbf{g}(u_0, v_0)) \neq \mathbf{0},$$

and we can suppose that it is the last component of this cross product that is nonzero. Thus we may apply the Inverse Function in the Plane to the mapping $(r_1', r_2'): \mathcal{R} \to \mathbb{R}^2$ at the point $\mathbf{g}(u_0, v_0)$ to conclude that there is a neighborhood of $\mathbf{g}(u_0, v_0)$ on which the mapping $(r_1', r_2'): \mathcal{R} \to \mathbb{R}^2$ has a continuously differentiable inverse. From the first two equations of the system (19.16), it follows that there is a neighborhood $\mathcal{N}$ of $(u_0, v_0)$ on which

$$(g_1, g_2) = (r_1', r_2')^{-1} \circ (r_1, r_2).$$

Since the composition of continuously differentiable mappings is also continuously differentiable, it follows that $\mathbf{g}: \mathcal{R} \to \mathbb{R}^2$ is continuously differentiable. It remains to verify that at each point in $\mathcal{R}$ the derivative matrix of $\mathbf{g}: \mathcal{R} \to \mathbb{R}^2$ is invertible. This will follow from the following identity: For each point $(u, v)$ in $\mathcal{R}$,

$$\left\| \frac{\partial \mathbf{r}}{\partial u}(u, v) \times \frac{\partial \mathbf{r}}{\partial v}(u, v) \right\| = \left\| \frac{\partial \mathbf{r}'}{\partial u'}(\mathbf{g}(u, v)) \times \frac{\partial \mathbf{r}'}{\partial v'}(\mathbf{g}(u, v)) \right\| \cdot \left| \det \mathbf{Dg}(u, v) \right|. \tag{19.17}$$

This identity is a consequence of the Chain Rule. Indeed, the system of identities (19.16) means that $\mathbf{r} \colon \mathcal{R} \to \mathbb{R}^3$ is the composition $\mathbf{r}' \circ \mathbf{g} \colon \mathcal{R} \to \mathbb{R}^3$. By the Chain Rule,

$$\mathbf{Dr}(u,\, v) = \mathbf{Dr}'(\mathbf{g}(u,\, v)) \cdot \mathbf{Dg}(u,\, v) \quad \text{for all } (u,\, v) \text{ in } \mathcal{R}.$$

This is an equality between two $3 \times 2$ matrices. Equating all of the $2 \times 2$ submatrices in this identity and using the product property of $2 \times 2$ determinants (see Exercise 14), it follows that for each index $i = 1,\, 2,\, 3$ and each point $(u,\, v)$ in $\mathcal{R}$,

$$\left\langle \frac{\partial \mathbf{r}}{\partial u}(u,\, v) \times \frac{\partial \mathbf{r}}{\partial v}(u,\, v),\, \mathbf{e}_i \right\rangle = \left\langle \frac{\partial \mathbf{r}'}{\partial u'}(u',\, v') \times \frac{\partial \mathbf{r}'}{\partial v'}(u',\, v'),\, \mathbf{e}_i \right\rangle \cdot \det \mathbf{Dg}(u,\, v);$$

that is,

$$\frac{\partial \mathbf{r}}{\partial u}(u,\, v) \times \frac{\partial \mathbf{r}}{\partial v}(u,\, v) = \left( \frac{\partial \mathbf{r}'}{\partial u'}(u',\, v') \times \frac{\partial \mathbf{r}'}{\partial v'}(u',\, v') \right) \cdot \det \mathbf{Dg}(u,\, v). \qquad (19.18)$$

We take the norm of each side to obtain the identity (19.17).

Finally, now that we have established that $\mathbf{g} \colon \mathcal{R} \to \mathbb{R}^2$ is a smooth change of variables, we can apply the change of variables formula (18.35) of Section 18.4 to transform the right-hand side of (19.14). Using formula (19.17) and the change of variables formula, we have

$$\int_{\mathcal{R}'} f(\mathbf{r}'(u',\, v')) \left\| \frac{\partial \mathbf{r}'}{\partial u'}(u',\, v') \times \frac{\partial \mathbf{r}'}{\partial v'}(u',\, v') \right\| du'\, dv'$$

$$= \int_{\mathcal{R}} f(\mathbf{r}'(\mathbf{g}(u,\, v))) \left\| \frac{\partial \mathbf{r}'}{\partial u'}(\mathbf{g}(u,\, v)) \times \frac{\partial \mathbf{r}'}{\partial v'}(\mathbf{g}(u,\, v)) \right\| \cdot |\det \mathbf{Dg}(u,\, v)|\, du\, dv$$

$$= \int_{\mathcal{R}} f(\mathbf{r}(u,\, v)) \left\| \frac{\partial \mathbf{r}}{\partial u}(u,\, v) \times \frac{\partial \mathbf{r}}{\partial v}(u,\, v) \right\| du\, dv. \qquad \blacksquare$$

**EXERCISES**

1. Let $a$, $b$, $c$, and $d$ be numbers with $c \neq 0$ and consider the plane

$$ax + by + cz + d = 0.$$

Parametrize the plane by $\mathbf{r}(u,\, v) = (u,\, v,\, -(au + bv + d)/c)$ and use the formula

$$\boldsymbol{\eta} = \frac{\partial \mathbf{r}}{\partial u}(u_0,\, v_0) \times \frac{\partial \mathbf{r}}{\partial v}(u_0,\, v_0)$$

to find a normal to the plane at each point $(x,\, y,\, z)$.

2. Find a unit normal vector at each point on the surface

$$S = \{(x,\, y,\, z) \mid z = x^2 + y^2,\, |x| + |y| < 4\}.$$

3. For $a > 0$ and $b > 0$, show that the cylindrical set

$$S = \{(x,\, y,\, z) \mid 0 < x < a,\, z > 0,\, y^2 + z^2 = b^2\}$$

is a surface and find its surface area.

4. For $r > 0$ and $h > 0$, show that the conical set

$$S = \{(x,\, y,\, z) \mid x > 0,\, y > 0,\, z > 0,\, z^2 = h^2(r^2 - x^2 - y^2)\}$$

is a surface and find its surface area.

5. Find the surface area of the plane
$$S = \{(x, y, z) \mid x > 0, y > 0, z > 0, 2x + y + z = 16\}.$$

6. Show that
$$\iint_S (x^2 + y^2)\, d\sigma = \frac{9\pi}{4},$$

   where     $S = \{(x, y, z) \mid x > 0, y > 0, 3 > z > 0, z^2 = 3(x^2 + y^2)\}.$

7. Compute
$$\iint_S (x^2 y^2 + y^2 z^2 + z^2 x^2)\, d\sigma,$$

   where $S$ is the portion of the cone $\{(x, y, z) \mid x^2 + y^2 = z^2, z > 0\}$ cut off by the cylinder $\{(x, y, z) \mid x^2 + y^2 - 2x = 0\}$.

8. For $0 < a < b$, a surface of the form
$$\{(u \cos v, u \sin v, g(u)) \mid a < u < b, 0 < v < 2\pi\},$$

   where the function $g\colon (a, b) \to \mathbb{R}$ has a bounded continuous derivative, is called a *surface of revolution*. Show that the area of a surface of revolution is given by the formula
$$2\pi \int_a^b u\sqrt{1 + (g'(u))^2}\, du.$$

9. Suppose that $\mathcal{R}$ is a convex region in the plane and that the function $g\colon \mathcal{R} \to \mathbb{R}$ has continuous bounded partial derivatives. Show that the surface $S = \{(x, y, g(x, y)) \mid (x, y) \text{ in } \mathcal{R}\}$ has area equal to that of $\mathcal{R}$ if and only if the function $g\colon \mathcal{R} \to \mathbb{R}$ is constant.

10. For any two vectors $\mathbf{A}$ and $\mathbf{B}$ in $\mathbb{R}^3$, show that
$$\|\mathbf{A} \times \mathbf{B}\|^2 = \|\mathbf{A}\|^2 \cdot \|\mathbf{B}\|^2 - \langle \mathbf{A}, \mathbf{B}\rangle^2.$$

   (*Hint:* Follow the verification of formula (19.8).)

11. Use formula (19.18) to show that the definition of normal to a surface is independent of the choice of parametrization.

12. For a parametrized surface $\mathbf{r}\colon \mathcal{R} \to \mathbb{R}^3$, at each point $(u, v)$ in $\mathcal{R}$ define the real-valued functions E: $\mathcal{R} \to \mathbb{R}$, F: $\mathcal{R} \to \mathbb{R}$, and G: $\mathcal{R} \to \mathbb{R}$ by
$$E(u, v) = \left\langle \frac{\partial \mathbf{r}}{\partial u}, \frac{\partial \mathbf{r}}{\partial u}\right\rangle, \quad F(u, v) = \left\langle \frac{\partial \mathbf{r}}{\partial u}, \frac{\partial \mathbf{r}}{\partial v}\right\rangle, \quad \text{and} \quad G(u, v) = \left\langle \frac{\partial \mathbf{r}}{\partial v}, \frac{\partial \mathbf{r}}{\partial v}\right\rangle.$$

   Use the identity in Exercise 10 to rewrite the formula for the surface area of the surface $S$ parametrized by $\mathbf{r}\colon \mathcal{R} \to \mathbb{R}^3$ as
$$\text{area } S = \int_{\mathcal{R}} \sqrt{EG - F^2}\, du\, dv.$$

13. Let $\mathbf{A}$, $\mathbf{B}$, $\mathbf{C}$, and $\mathbf{D}$ be vectors in $\mathbb{R}^3$. Prove Lagrange's Identity:
$$\langle \mathbf{A} \times \mathbf{B}, \mathbf{C} \times \mathbf{D}\rangle = \langle \mathbf{A}, \mathbf{C}\rangle\langle \mathbf{B}, \mathbf{D}\rangle - \langle \mathbf{A}, \mathbf{D}\rangle\langle \mathbf{B}, \mathbf{C}\rangle.$$

   (*Hint:* Prove this identity by showing that it is true if all of the four vectors are standard basis vectors and then using the linearity of the cross product and of the inner product to obtain the general case.)

14. For two $2 \times 2$ matrices $\mathbf{A}$ and $\mathbf{B}$, by explicit calculation, show that
$$\det \mathbf{AB} = \det \mathbf{A} \det \mathbf{B}.$$

15. Let **A** and **C** be $3 \times 2$ matrices and let **B** be a $2 \times 2$ matrix such that $\mathbf{AB} = \mathbf{C}$. Prove the identity

$$\|\mathbf{A}_1 \times \mathbf{A}_2\| \cdot |\det \mathbf{B}| = \|\mathbf{C}_1 \times \mathbf{C}_2\|$$

where $\mathbf{A}_i$ and $\mathbf{C}_i$ are the $i$th columns of **A** and **C**.

16. For a parametrized surface $\mathbf{r} \colon \mathcal{R} \to \mathbb{R}^3$ and a parameter value $(u_0, v_0)$ in $\mathcal{R}$, show that there is a neighborhood $\mathcal{N}$ of $(u_0, v_0)$ such that $\mathbf{r}(\mathcal{N})$ is the image of a projectionally parametrized surface. (*Hint:* Use the Inverse Function Theorem as in the proof of Theorem 19.7.)

17. Suppose that the function $h \colon \mathbb{R}^3 \to \mathbb{R}$ is continuously differentiable. Let **p** be a point in $\mathbb{R}^3$ at which the derivative vector $\mathbf{D}h(\mathbf{p})$ is nonzero and define $c = h(\mathbf{p})$. Use the Implicit Function Theorem to show that there is a neighborhood $\mathcal{N}$ of **p** such that $S = \{(x, y, z) \text{ in } \mathcal{N} \mid h(x, y, z) = c\}$ is a surface.

18. For a parametrized surface $\mathbf{r} \colon \mathcal{R} \to \mathbb{R}^3$ and a continuous bounded function $f \colon S \to \mathbb{R}$, where $S$ is the image of $\mathbf{r} \colon \mathcal{R} \to \mathbb{R}^3$, prove that

$$\left| \int_S f \right| \leq MA$$

where $M$ is such that $|f(\mathbf{p})| \leq M$ for all points **p** on $S$ and $A$ is the surface area of $S$.

## 19.3  The Integral Formulas of Green and Stokes

For a closed bounded interval $[a, b]$ in $\mathbb{R}$ and a continuous function $f \colon [a, b] \to \mathbb{R}$ that has a continuous bounded derivative on the open interval $(a, b)$, the First Fundamental Theorem of Calculus asserts that

$$\int_a^b f'(t)\, dt = f(b) - f(a). \tag{19.19}$$

There are significant generalizations of this formula for functions of several variables. In this section, we will extend this formula to the case when the open interval $(a, b)$ is replaced by certain open bounded subsets of the plane $\mathbb{R}^2$ (this is Green's Formula) and to the case when $(a, b)$ is replaced by certain surfaces in $\mathbb{R}^3$ (this is Stokes's Formula).

A parametrized path $\gamma \colon [a, b] \to \mathbb{R}^n$ is said to be *simple* provided that the mapping $\gamma \colon [a, b) \to \mathbb{R}$ is one-to-one; it is said to be *closed* provided that $\gamma(a) = \gamma(b)$. We will be concerned here with open bounded subsets $\Omega$ of the plane $\mathbb{R}^2$ that have the property that the boundary of $\Omega$, $\Gamma = \partial\Omega$, is the image of a simple, closed parametrized path.

**EXAMPLE 19.12**  Suppose that the functions $g \colon [a, b] \to \mathbb{R}$ and $h \colon [a, b] \to \mathbb{R}$ are continuous and that $g(x) < h(x)$ for all $x$ in $(a, b)$. Define

$$\Omega = \{(x, y) \mid a < x < b,\ g(x) < y < h(x)\} \quad \text{and} \quad \Gamma = \partial\Omega.$$

This set was considered in Example 19.4. In the case when $g(a) < h(a)$ and $g(b) < h(b)$, the parametrization of $\Gamma$ given in Example 19.4 is a parametrization by a simple, closed parametrized path. In the case of equality at an end, we omit the path corresponding to the end and again obtain a simple, closed parametrization of the boundary. For the obvious geometric reason, this parametrization, or a simple, closed parametrization equivalent to it, is called a *counterclockwise parametrization* of $\partial\Omega$. ☐

**EXAMPLE 19.13** Suppose that the functions $g\colon [c, d] \to \mathbb{R}$ and $h\colon [c, d] \to \mathbb{R}$ are continuous and that $g(y) < h(y)$ for all $y$ in $(c, d)$. Define

$$\Omega = \{(x, y) \mid g(y) < x < h(y), c < y < d\} \quad \text{and} \quad \Gamma = \partial\Omega.$$

Again here, $\Omega$ is a bounded open subset of $\mathbb{R}^2$ whose boundary can be parametrized by a simple, closed parametrized path. Indeed, in the case when $g(c) < h(c)$ and $g(d) < h(d)$, the following defines such a parametrization:

$$\gamma(t) = \begin{cases} (g(d + 4t(c - d)), \, d + 4t(c - d)) & \text{if } 0 \leq t \leq 1/4 \\ (g(c) + 4(t - 1/4)(h(c) - g(c)), \, c) & \text{if } 1/4 \leq t \leq 1/2 \\ (h(c - 4(t - 1/2)(c - d)), \, c - 4(t - 1/2)(c - d)) & \text{if } 1/2 \leq t \leq 3/4 \\ (h(d) + 4(t - 3/4)(g(d) - h(d)), \, d) & \text{if } 3/4 \leq t \leq 1. \end{cases}$$

**FIGURE 19.6**

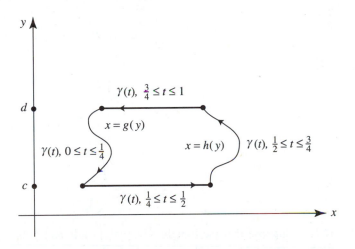

Again for the obvious geometric reason, this parametrization, or a simple, closed parametrization equivalent to it, is called a *counterclockwise parametrization* of $\partial\Omega$. ☐

**PROPOSITION 19.8**

Suppose that the functions $g\colon [a, b] \to \mathbb{R}$ and $h\colon [a, b] \to \mathbb{R}$ are continuous and that $g(x) < h(x)$ for all $x$ in $(a, b)$. Define

$$\Omega \in \{(x, y) \mid a < x < b, g(x) < y < h(x)\} \quad \text{and} \quad \Gamma \in \partial\Omega.$$

Let the function $N\colon \Omega \cup \Gamma \to \mathbb{R}$ be continuous and such that $\partial N/\partial y\colon \Omega \to \mathbb{R}$ exists and is both continuous and bounded. Then

$$\iint_\Omega \frac{\partial N}{\partial y}(x, y)\,dx\,dy \in -\int_\Gamma N(x, y)\,dx, \tag{19.20}$$

where the right-hand integral is computed with respect to a counterclockwise parametrization of the path $\Gamma$.

**Proof**    The heart of the proof consists in using Fubini's Theorem so we can apply the First Fundamental Theorem of Calculus. First, it is convenient to express $\Gamma$ as $\Gamma = \Gamma_1 \cup \Gamma_2 \cup \Gamma_3 \cup \Gamma_4$, where $\Gamma_1 = \{(x, g(x)) \mid a \leq x \leq b\}$, $\Gamma_2 = \{(b, y) \mid g(b) \leq y \leq h(b)\}$, $\Gamma_3 = \{(x, h(x)) \mid a \leq x \leq b\}$, and $\Gamma_4 = \{(a, y) \mid g(a) \leq y \leq h(a)\}$. Using Fubini's Theorem, the First Fundamental Theorem of Calculus, and the very definition of line integral, we have

$$\iint_\Omega \left[ \frac{\partial N}{\partial y}(x, y) \right] dx\, dy = \int_a^b \left\{ \int_{g(x)}^{h(x)} \frac{\partial N}{\partial y}(x, y)\, dy \right\} dx$$

$$= \int_a^b \left\{ N(x, h(x)) - N(x, g(x)) \right\} dx$$

$$= \int_a^b N(x, h(x))\, dx - \int_a^b N(x, g(x))\, dx$$

$$= -\left\{ \int_a^b N(x, g(x))\, dx - \int_a^b N(x, h(x))\, dx \right\}$$

$$= -\left\{ \int_{\Gamma_1} N\, dx + 0 + \int_{\Gamma_3} N\, dx + 0 \right\}$$

$$= -\left\{ \int_{\Gamma_1} N\, dx + \int_{\Gamma_2} N\, dx + \int_{\Gamma_3} N\, dx + \int_{\Gamma_4} N\, dx \right\}$$

$$= -\int_\Gamma N\, dx. \qquad \blacksquare$$

---

**PROPOSITION 19.9**    Suppose that the functions $g:[c, d] \to \mathbb{R}$ and $h:[c, d] \to \mathbb{R}$ are continuous and that $g(y) < h(y)$ for all $y$ in $(c, d)$. Define

$$\Omega = \{(x, y) \mid g(y) < x < h(y),\ c < y < d\} \quad \text{and} \quad \Gamma = \partial\Omega.$$

Let the function $M:\Omega \cup \Gamma \to \mathbb{R}$ be continuous and such that $\partial M/\partial x:\Omega \to \mathbb{R}$ exists and is both continuous and bounded. Then

$$\iint_\Omega \frac{\partial M}{\partial x}(x, y)\, dx\, dy = \int_\Gamma M(x, y)\, dy, \qquad (19.21)$$

where the right-hand integral is computed with respect to a counterclockwise parametrization of the path $\Gamma$.

**Proof**    As in the proof of Proposition 19.8, we will use Fubini's Theorem in order to apply the First Fundamental Theorem of Calculus. Express $\Gamma$ as $\Gamma = \Gamma_1 \cup \Gamma_2 \cup \Gamma_3 \cup \Gamma_4$, where $\Gamma_1 = \{(g(y), y) \mid c \leq y \leq d\}$, $\Gamma_2 = \{(x, c) \mid g(c) \leq x \leq h(c)\}$, $\Gamma_3 =$

$\{(h(y),\ y) \mid c \le y \le d\}$, and $\Gamma_4 = \{(x,\ d) \mid g(d) \le x \le h(d)\}$. Using Fubini's Theorem, the First Fundamental Theorem of Calculus, and the very definition of line integral, we have

$$\iint_{\Omega} \left[ \frac{\partial M}{\partial x}(x,\ y) \right] dx\, dy = \int_c^d \left\{ \int_{g(y)}^{h(y)} \frac{\partial M}{\partial x}(x,\ y)\, dx \right\} dy$$

$$= \int_c^d \left\{ M(h(y),\ y) - M(g(y),\ y) \right\} dy$$

$$= \int_c^d M(h(y),\ y)\, dy - \int_c^d M(g(y),\ y)\, dy$$

$$= \int_{\Gamma_3} M\, dy + 0 + \int_{\Gamma_1} M\, dy + 0$$

$$= \int_{\Gamma_1} M\, dy + \int_{\Gamma_2} M\, dy + \int_{\Gamma_3} M\, dy + \int_{\Gamma_4} M\, dy$$

$$= \int_{\Gamma} M\, dy. \qquad\qquad \blacksquare$$

**DEFINITION** *A region $\Omega$ in the plane $\mathbb{R}^2$ is called a Green's domain provided that its boundary $\Gamma = \partial\Omega$ is the image of a simple, closed, piecewise smooth parametrized path $\gamma: I \to \mathbb{R}^2$ for which the following property holds: Suppose that the two functions $M: \Omega \cup \Gamma \to \mathbb{R}$ and $N: \Omega \cup \Gamma \to \mathbb{R}$ are continuous and that their restrictions $M: \Omega \to \mathbb{R}$ and $N: \Omega \to \mathbb{R}$ are continuously differentiable and have bounded partial derivatives. Then*

## Green's Formula

$$\iint_{\Omega} \left[ \frac{\partial M}{\partial x}(x,\ y) - \frac{\partial N}{\partial y}(x,\ y) \right] dx\, dy = \int_{\Gamma} [M(x,\ y)\, dy + N(x,\ y)\, dx], \quad (19.22)$$

*where the right-hand integral is computed with respect to $\gamma: I \to \mathbb{R}^2$. We call such a parametrization a Green's parametrization.*

By first taking $M(x,\ y)$ identically equal to 0 and then taking $N(x,\ y)$ identically equal to 0, we see that for a domain to be a Green's domain it is only necessary to verify separately the formulas

$$\iint_{\Omega} \frac{\partial N}{\partial y}(x,\ y)\, dx\, dy = -\int_{\Gamma} N(x,\ y)\, dx$$

and

$$\iint_{\Omega} \frac{\partial M}{\partial x}(x,\ y)\, dx\, dy = \int_{\Gamma} M(x,\ y)\, dy.$$

Propositions 19.8 and 19.9 provide a means for showing that domains, with appropriate parametrizations of their boundaries, are Green's domains.

**EXAMPLE 19.14** Define $\Omega = \{(x, y) \mid x^2 + y^2 < 1\}$ so that its boundary $\Gamma$ is the circle of radius 1 centered at the origin. It is clear that $\Omega$ is a domain to which both Proposition 19.8 and Proposition 19.9 apply, so it is a Green's domain, for the counterclockwise parametrization $\gamma : [0, 2\pi] \to \mathbb{R}^2$ defined by $\gamma(\theta) = (\cos\theta, \sin\theta)$, $0 \le \theta \le 2\pi$. Thus for two continuous functions $M : \Omega \cup \Gamma \to \mathbb{R}$ and $N : \Omega \cup \Gamma \to \mathbb{R}$ whose restrictions $M : \Omega \to \mathbb{R}$ and $N : \Omega \to \mathbb{R}$ are continuously differentiable and have bounded partial derivatives, we have

$$\iint_{x^2+y^2\le 1} \left[ \frac{\partial M}{\partial x}(x, y) - \frac{\partial N}{\partial y}(x, y) \right] dx\, dy$$
$$= \int_0^{2\pi} [M(\cos\theta, \sin\theta)\cos\theta - N(\cos\theta, \sin\theta)\sin\theta]\, d\theta. \quad \square$$

**EXAMPLE 19.15** Define $\Omega = \{(x, y) = (r\cos\theta, r\sin\theta) \mid 0 < \theta < 3\pi/2, 0 < r < 1\}$. Then $\Omega$ is a Green's domain with respect to the counterclockwise parametrization, although it is not a set to which either Proposition 19.8 or Proposition 19.9 directly applies. However, setting

$$\Omega_+ = \{(x, y) \text{ in } \Omega \mid x > 0\} \quad \text{and} \quad \Omega_- = \{(x, y) \text{ in } \Omega \mid x < 0\},$$

we see that both $\Omega_+$ and $\Omega_-$ are Green's domains, since both Proposition 19.8 and Proposition 19.9 apply to each of these sets. By adding Green's Formula for each of these sets, we obtain Green's Formula for $\Omega$, because the contributions of the line integrals along the common boundary, with respect to the counterclockwise parametrization of each, cancel. $\quad \square$

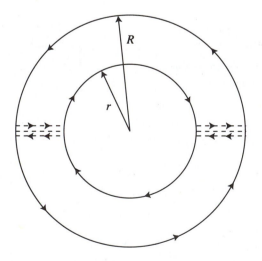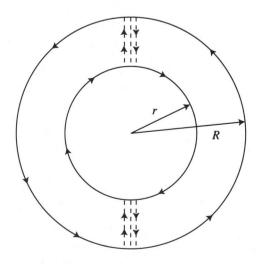

**FIGURE 19.7**

**EXAMPLE 19.16** For $0 < r < R$, consider the annulus $\Omega = \{(x, y) \mid r^2 < x^2 + y^2 < R^2\}$. The boundary of this annulus is the union of two circles. Suppose that the two functions $M: \Omega \cup \partial\Omega \to \mathbb{R}$ and $N: \Omega \cup \partial\Omega \to \mathbb{R}$ are continuous and that their restrictions $M: \Omega \to \mathbb{R}$ and $N: \Omega \to \mathbb{R}$ are continuously differentiable and have bounded partial derivatives. Then

$$\iint_{r^2 < x^2 + y^2 < R^2} \left[ \frac{\partial M}{\partial x}(x, y) - \frac{\partial N}{\partial y}(x, y) \right] dx\, dy$$

$$= \int_{x^2 + y^2 = R} [M(x, y)\, dy + N(x, y)\, dx] \qquad (19.23)$$

$$- \int_{x^2 + y^2 = r} [M(x, y)\, dy + N(x, y)\, dx],$$

where both integrals are computed with respect to a counterclockwise parametrization. To verify this, it suffices to verify the formula first when $M(x, y)$ is identically 0 and then when $N(x, y)$ is identically 0. In the case when $M(x, y)$ is identically 0, use the $x$-axis to divide the annulus into two domains $\Omega_+$ and $\Omega_-$, to each of which Proposition 19.8 applies. Apply Proposition 19.8 to each domain and sum the resulting integral equalities. Because of the choice of counterclockwise parametrization of each line integral, the contributions to the line integrals over the boundaries lying on the $x$-axis cancel out and we obtain the above formula in the case when $M(x, y)$ is identically 0. A similar computation, but now bisecting the annulus with the $y$-axis and using Proposition 19.9, establishes the above formula in the case when $N(x, y)$ is identically 0. □

For a Green's domain $\Omega$ with boundary $\Gamma$, taking $M(x, y) \equiv x$ and $N(x, y) \equiv -y$, we obtain the following formula for the area of $\Omega$:

$$\text{area } \Omega = \frac{1}{2} \int_\Gamma [x\, dy - y\, dx], \qquad (19.24)$$

where the line integral is computed with respect to the Green's parametrization of $\Gamma$.

**EXAMPLE 19.17** The elliptical region $\Omega = \{(x, y) \mid x^2/a^2 + y^2/b^2 \leq 1\}$ is a Green's domain, since it is an open bounded subset of $\mathbb{R}^2$ to which both Proposition 19.8 and Proposition 19.9 apply. The above area formula gives

$$\text{area } \Omega = \frac{1}{2} \int_0^{2\pi} [a \cos\theta\, (b \cos\theta) - b \sin\theta\, (-a \sin\theta)]\, d\theta$$

$$= \frac{ab}{2} \int_0^{2\pi} d\theta = \pi ab. \qquad □$$

The examples of Green's domains, with associated parametrizations of the boundary, that we have exhibited above have all been built out of the types of domains described in Propositions 19.8 and 19.9. There is, in fact, a very general condition for an open bounded subset of $\mathbb{R}^2$ and an associated parametrization of its boundary to be a Green's domain: Let $\Omega$ be an open bounded subset of $\mathbb{R}^2$ whose boundary is the image of a simple, closed piecewise smooth parametrized path $\gamma: [a, b] \to \mathbb{R}^2$ that has the property that

$$\int_{\gamma:[a,b]\to\mathbb{R}^2} [x\, dy - y\, dx] > 0. \tag{19.25}$$

Then $\Omega$, with this parametrization, is a Green's domain. We will not prove this general result. The intuitive indication of the reason this result is true comes from "patching" $\Omega$ together from domains to which Proposition 19.8 or Proposition 19.9 applies. It is quite a different matter to provide a rigorous proof; a completely precise proof involves considerable technical detail. We will not pursue this here. The area formula (19.24) is what motivates the assumption on the parametrization that (19.25) holds. In fact, (19.25) is taken to be the definition of what it means for a simple, closed, piecewise smooth parametrized path $\gamma: [a, b] \to \mathbb{R}^2$ to be a counterclockwise parametrization.

Suppose that $\Omega$ is a Green's domain with boundary $\Gamma$, having an associated Green's parametrization by arclength $\gamma: I \to \mathbb{R}^2$. Fix a parameter value $s$ at which $\gamma$ is differentiable. We define $\boldsymbol{\eta}(s) = (\gamma_2'(s), -\gamma_1'(s))$. Observe that

$$\|\gamma'(s)\| = 1, \quad \|\boldsymbol{\eta}(s)\| = 1, \quad \text{and} \quad \langle \gamma'(s), \boldsymbol{\eta}(s) \rangle = 0,$$

so the vector $\mathbf{N} \equiv \boldsymbol{\eta}(s)$ is a unit vector that is perpendicular to the unit tangent vector $\mathbf{T} \equiv \gamma'(s)$. The direction of this normal vector is determined by the parametrization $\gamma$, and in the specific examples we have seen, its geometric meaning is that it is the "outward-pointing" normal. It is convenient to refer to this parametrization of normals as being associated with the Green's parametrization of the boundary of $\Omega$. For a function $w: \mathcal{N} \to \mathbb{R}$ that is continuously differentiable in a neighborhood $\mathcal{N}$ of the point $\gamma(s) \equiv \mathbf{p}$, recall that in Section 13.3 we established the Directional Derivative Lemma, which provided the following formula for directional derivatives of the function $w$ at the point $\mathbf{p}$ in the direction $\mathbf{q}$:

$$\frac{\partial w}{\partial \mathbf{q}}(\mathbf{p}) \equiv \lim_{t \to 0} \frac{w(\mathbf{p} + t\mathbf{q}) - w(\mathbf{p})}{t} = \langle \nabla w(\mathbf{p}), \mathbf{q} \rangle.$$

Thus we have the following formula for the directional derivatives in the directions $\mathbf{T}$ and $\mathbf{N}$, which are called, respectively, the tangential and normal derivatives of the function $w: \mathcal{N} \to \mathbb{R}$ at the point $\mathbf{p}$:*

$$\frac{\partial w}{\partial \mathbf{T}}(\mathbf{p}) = \langle \nabla w(\mathbf{p}), \mathbf{T} \rangle \quad \text{and} \quad \frac{\partial w}{\partial \mathbf{N}}(\mathbf{p}) = \langle \nabla w(\mathbf{p}), \mathbf{N} \rangle.$$

It is customary to denote the normal derivative by $\partial w / \partial \boldsymbol{\eta}$, so that, in particular,

$$\frac{\partial x}{\partial \boldsymbol{\eta}} = \gamma_2' \quad \text{and} \quad \frac{\partial y}{\partial \boldsymbol{\eta}} = -\gamma_1'.$$

The First Fundamental Theorem of Calculus and the product formula for differentiation provided the integration by parts formula for functions of a single variable. We

---

*Until now, we have denoted the derivative vector of a differentiable function of several variables $f: D \to \mathbb{R}$ at the point $\mathbf{x}$ in $D$ by $\mathbf{D}f(\mathbf{x})$. Often, particularly when describing line and surface integrals of components of vector fields, the derivative vector is denoted by $\nabla f(\mathbf{x})$; we will use this notation for the remainder of this section.

will now use Green's Formula and the product formula for differentiation to provide the following integration by parts formula for functions of two variables.

**COROLLARY**
**19.10**

**Integration by Parts**   Let $\Omega$ be a Green's domain with boundary $\Gamma$. Suppose that the functions $a(x, y)$, $b(x, y)$, $u(x, y)$, and $v(x, y)$ have continuous bounded partial derivatives on an open set $\mathcal{O}$ containing $\Omega \cup \Gamma$. Then

$$\iint_{\Omega} \left[ \frac{\partial u}{\partial x} + \frac{\partial v}{\partial y} \right] dx\, dy = \int_{\Gamma} \left[ au\frac{\partial x}{\partial \eta} + bv\frac{\partial y}{\partial \eta} \right] ds - \iint_{\Omega} \left[ \frac{\partial a}{\partial x}u + \frac{\partial b}{\partial y}v \right] dx\, dy,$$

where the normal parametrization is that associated with the Green's parametrization of the boundary of $\Omega$.

**Proof**   Define

$$M(x, y) = a(x, y)u(x, y) \quad \text{and} \quad N(x, y) = -b(x, y)v(x, y) \quad \text{for } (x, y) \text{ in } \mathcal{O}.$$

By Green's Formula, we have

$$\iint_{\Omega} \left[ \frac{\partial (au)}{\partial x}(x, y) + \frac{\partial (bv)}{\partial y}(x, y) \right] dx\, dy$$

$$= \int_{\Gamma} [a(x, y)u(x, y)\, dy - b(x, y)v(x, y)\, dx].$$

Using the product rule for differentiation, the left-hand side of this formula becomes

$$\iint_{\Omega} \left[ \frac{\partial u}{\partial x} + \frac{\partial v}{\partial y} \right] dx\, dy + \iint_{\Omega} \left[ \frac{\partial a}{\partial x}u + \frac{\partial b}{\partial y}v \right] dx\, dy,$$

whereas for the the the right-hand side, by the definition of the line integral and the choice of normal parametrization, we have

$$\gamma_2' = \frac{\partial x}{\partial \eta} \quad \text{and} \quad -\gamma_1' = \frac{\partial y}{\partial \eta},$$

so that

$$\int_{\Gamma} [a(x, y)u(x, y)\, dy - b(x, y)v(x, y)\, dx]$$

$$= \int_0^{\ell} [a(\gamma(s))u(\gamma(s))\gamma_2'(s) - b(\gamma(s))v(\gamma(s))\gamma_1'(s)]\, ds$$

$$\equiv \int_{\Gamma} \left[ au\frac{\partial x}{\partial \eta} + bv\frac{\partial y}{\partial \eta} \right] ds.$$

From these two formulas follows the integration by parts formula.                   ∎

We will now raise Green's Formula from a surface $\mathcal{S} = \{(x, y, 0) \mid (x, y) \text{ in } \Omega\}$ that is a Green's domain in the plane $\mathbb{R}^2$ to a surface $\mathcal{S}$ in $\mathbb{R}^3$ that is parametrized by a Green's domain; the extension is called *Stokes's Formula*. In order to provide a geometric description of the integrals occurring in Stokes's Formula, it is useful to introduce the concept of a *vector-field*, which is a geometrical way of considering mappings from a subset of $\mathbb{R}^3$ into $\mathbb{R}^3$: For a subset $D$ of $\mathbb{R}^3$ and a mapping $\mathbf{F}: D \to \mathbb{R}^3$, for each point $\mathbf{p}$ in $D$ we can interpret $\mathbf{F}(\mathbf{p})$ as representing the vector associated with the segment from $\mathbf{p}$ to $\mathbf{p} + \mathbf{F}(\mathbf{p})$. Furthermore, if $\mathbf{u}: D \to \mathbb{R}^3$ is another vector-field having the property that $\|\mathbf{u}(\mathbf{p})\| = 1$ for all $\mathbf{p}$ in $D$, then for each point $\mathbf{p}$ in $D$, it follows from the geometric description of the inner product that the number $\langle \mathbf{F}(\mathbf{p}), \mathbf{u}(\mathbf{p}) \rangle$ is the component of the vector $\mathbf{F}(\mathbf{p})$ in the direction $\mathbf{u}(\mathbf{p})$. Line integrals and surface integrals of functions of the form $\mathbf{p} \mapsto \langle \mathbf{F}(\mathbf{p}), \mathbf{u}(\mathbf{p}) \rangle$ will be an essential ingredient of the extension of Green's Formula to surfaces in $\mathbb{R}^3$.

**FIGURE 19.8**

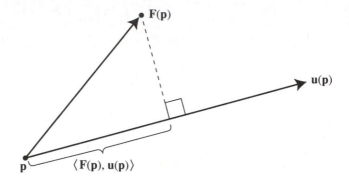

**DEFINITION**  *For a piecewise smooth parametrized path $\gamma: I \to \mathbb{R}^3$ having image $\Gamma$ and a continuous bounded mapping $\mathbf{F}: \Gamma \to \mathbb{R}^3$, we define*

$$\int_{\Gamma} \langle \mathbf{F}, \mathbf{T} \rangle \, ds \equiv \int_{I} \langle \mathbf{F}(\gamma(t)), \gamma'(t) \rangle \, dt. \tag{19.26}$$

The sign of the integral on the right-hand side of (19.26) depends on the choice of parametrization, but does not change if the given parametrization of $\Gamma$ is replaced by an equivalent parametrization. Moreover, if we choose a parametrization of $\Gamma$ by arclength, so that $I = [0, \ell]$, where $\ell$ is the arclength of $\Gamma$, then for each parameter value $s$ in $(0, \ell)$, $\mathbf{T}(s)$ is a tangent vector to $\Gamma$ at the point $\gamma(s)$ that has length 1 and

$$\int_{\Gamma} \langle \mathbf{F}, \mathbf{T} \rangle \, ds = \int_{0}^{\ell} \langle \mathbf{F}(\gamma(s)), \mathbf{T}(s) \rangle \, ds,$$

so this integral is the line integral of the tangential component of $\mathbf{F}$, where the choice of direction of the tangent is determined by the choice of parametrization of the path.

**DEFINITION**  *For a parametrized surface $\mathbf{r}: \mathcal{R} \to \mathbb{R}^3$ having image $\mathcal{S}$ and a continuous mapping $\mathbf{G}: \mathcal{S} \to \mathbb{R}^3$, we define*

$$\iint_{\mathcal{S}} \langle \mathbf{G}, \boldsymbol{\eta} \rangle \, d\sigma \equiv \iint_{\mathcal{R}} \left\langle \mathbf{G}(r(u, v)), \frac{\partial \mathbf{r}}{\partial u}(u, v) \times \frac{\partial \mathbf{r}}{\partial v}(u, v) \right\rangle \, du \, dv. \tag{19.27}$$

Observe that if for each parameter value $(u, v)$ in $\mathcal{R}$ we define

$$\boldsymbol{\eta}(u, v) = \frac{\partial \mathbf{r}/\partial u(u, v) \times \partial \mathbf{r}/\partial v(u, v)}{\|\partial \mathbf{r}/\partial u(u, v) \times \partial \mathbf{r}/\partial v(u, v)\|},$$

then $\boldsymbol{\eta}(u, v)$ is a unit normal to the surface $\mathcal{S}$ at the point $\mathbf{r}(u, v)$, and

$$\iint_{\mathcal{S}} \langle \mathbf{G}, \boldsymbol{\eta} \rangle \, d\sigma = \iint_{\mathcal{R}} \langle \mathbf{G}\left(\mathbf{r}(u, v)\right), \boldsymbol{\eta}(u, v) \rangle \left\| \frac{\partial \mathbf{r}}{\partial u}(u, v) \times \frac{\partial \mathbf{r}}{\partial v}(u, v) \right\| \, du \, dv,$$

so the integral (19.27) is the surface integral over the surface $\mathcal{S}$ of the normal component of the vector-field $\mathbf{G}$. Again, in this case, the sign of the integral on the right-hand side of (19.27) depends on the choice of parametrization.

**DEFINITION**   *Let $\mathcal{U}$ be an open subset of $\mathbb{R}^3$ and suppose that the mapping $\mathbf{F}: \mathcal{U} \to \mathbb{R}^3$ is continuously differentiable. An associated mapping, called the curl of $\mathbf{F}$ and denoted by curl $\mathbf{F}: \mathcal{U} \to \mathbb{R}^3$, is defined as follows: If we express $\mathbf{F}: \mathcal{U} \to \mathbb{R}^3$ in components as*

$$\mathbf{F}(x, y, z) = (f(x, y, z), g(x, y, z), h(x, y, z)) \quad \text{for all } (x, y, z) \text{ in } \mathcal{O},$$

*then*
$$\text{curl } \mathbf{F}(x, y, z) \equiv \left( \frac{\partial h}{\partial y} - \frac{\partial g}{\partial z}, \frac{\partial f}{\partial z} - \frac{\partial h}{\partial x}, \frac{\partial g}{\partial x} - \frac{\partial f}{\partial y} \right) \quad \text{for all } (x, y, z) \text{ in } \mathcal{O}.$$

It is not immediately apparent why the curl of a mapping should be significant. We will soon see that the curl is an essential ingredient in lifting Green's Formula out of the plane.[*]

**EXAMPLE 19.18**   Suppose that the mappings $M: \mathbb{R}^2 \to \mathbb{R}$ and $N: \mathbb{R}^2 \to \mathbb{R}$ are continuously differentiable. Define

$$\mathbf{F}(x, y, z) = (N(x, y), M(x, y), 0) \quad \text{for all } (x, y, z) \text{ in } \mathbb{R}^3.$$

Then it follows directly from the definition that

$$\text{curl } \mathbf{F}(x, y, z) = \left( 0, 0, \frac{\partial M}{\partial x}(x, y) - \frac{\partial N}{\partial y}(x, y) \right) \quad \text{for all } (x, y, z) \text{ in } \mathbb{R}^3. \quad \square$$

Directly from the definition of curl and the linearity of differentiation, it follows that the curl acts linearly; that is, for $\mathcal{U}$ an open subset of $\mathbb{R}^3$ and two continuously differentiable mappings $\mathbf{F}: \mathcal{U} \to \mathbb{R}^3$ and $\mathbf{H}: \mathcal{U} \to \mathbb{R}^3$, and any two numbers $\alpha$ and $\beta$, we have

$$\text{curl } [\alpha \mathbf{F} + \beta \mathbf{H}] = \alpha \, \text{curl } \mathbf{F} + \beta \, \text{curl } \mathbf{H}.$$

The extension of Green's Formula to surfaces in $\mathbb{R}^3$ will rely on Green's Formula itself, together with the following identity.

---

[*]In the study of physics and engineering, the curl is also an important operator that has specific physical significance in a number of different contexts. For instance, one of the most remarkable statements of physics, Maxwell's Equations, consists of assertions about the relationship between the curls of electric and magnetic vector-fields.

**LEMMA 19.11**    **Stokes's Identity**    For $\mathcal{O}$ an open subset of the plane $\mathbb{R}^2$, suppose that the components of the mapping $\mathbf{r}: \mathcal{O} \to \mathbb{R}^3$ have continuous second-order partial derivatives. Let $\mathcal{U}$ be an open subset of $\mathbb{R}^3$ containing the image of $\mathbf{r}: \mathcal{O} \to \mathbb{R}^3$ and suppose that the mapping $\mathbf{F}: \mathcal{U} \to \mathbb{R}^3$ is continuously differentiable. Then at each point $(u, v)$ in $\mathcal{O}$, we have

$$\left\langle \operatorname{curl} \mathbf{F}(\mathbf{r}(u, v)), \frac{\partial \mathbf{r}}{\partial u}(u, v) \times \frac{\partial \mathbf{r}}{\partial v}(u, v) \right\rangle$$

$$= \frac{\partial}{\partial u} \left[ \left\langle \mathbf{F}(\mathbf{r}(u, v)), \frac{\partial \mathbf{r}}{\partial v}(u, v) \right\rangle \right] - \frac{\partial}{\partial v} \left[ \left\langle \mathbf{F}(\mathbf{r}(u, v)), \frac{\partial \mathbf{r}}{\partial u}(u, v) \right\rangle \right].$$

**Proof**    First observe that both the left-hand and right-hand sides of Stokes's Identity depend linearly on the vector field $\mathbf{F}$. Thus, to verify the identity, it suffices to consider the three cases that occur when two of the component functions of the mapping $\mathbf{F}: \mathcal{U} \to \mathbb{R}^3$ are identically equal to 0. We will verify the identity in the case when

$$\mathbf{F}(x, y, z) = (g(x, y, z), 0, 0) \quad \text{for all } (x, y, z) \text{ in } \mathcal{U}. \tag{19.28}$$

The verification in the other two cases is entirely similar.

A direct computation shows that for $\mathbf{F}$ of the form (19.28),

$$\operatorname{curl} \mathbf{F}(\mathbf{r}(u, v)) = \left( 0, \frac{\partial g}{\partial z}(\mathbf{r}(u, v)), -\frac{\partial g}{\partial y}(\mathbf{r}(u, v)) \right) \quad \text{for all } (u, v) \text{ in } \mathcal{O}.$$

Therefore, writing $\partial \mathbf{r}/\partial u(u, v) \times \partial \mathbf{r}/\partial v(u, v)$ in components and taking the inner product, the left-hand side of Stokes's Identity becomes

$$\frac{\partial g}{\partial z}(\mathbf{r}(u, v)) \left[ \frac{\partial r_3}{\partial u} \frac{\partial r_1}{\partial v} - \frac{\partial r_1}{\partial u} \frac{\partial r_3}{\partial v} \right] - \frac{\partial g}{\partial y}(\mathbf{r}(u, v)) \left[ \frac{\partial r_1}{\partial u} \frac{\partial r_2}{\partial v} - \frac{\partial r_2}{\partial u} \frac{\partial r_1}{\partial v} \right]. \tag{19.29}$$

To evaluate the right-hand side of Stokes's Identity, we first observe that since the mixed second-order partial derivatives of $r_1(u, v)$ are equal, the right-hand side of Stokes's Identity equals

$$\frac{\partial}{\partial u} \left[ g(\mathbf{r}(u, v)) \right] \frac{\partial r_1}{\partial v}(u, v) - \frac{\partial}{\partial v} \left[ g(\mathbf{r}(u, v)) \right] \frac{\partial r_1}{\partial u}(u, v). \tag{19.30}$$

Thus, to prove Stokes's Identity, we have to show that (19.29) equals (19.30). To do this, observe that by the Chain Rule,

$$\frac{\partial}{\partial u} \left[ g(\mathbf{r}(u, v)) \right] \frac{\partial r_1}{\partial v} = \left[ \frac{\partial g}{\partial x}(\mathbf{r}(u, v)) \frac{\partial r_1}{\partial u} + \frac{\partial g}{\partial y}(\mathbf{r}(u, v)) \frac{\partial r_2}{\partial u} + \frac{\partial g}{\partial z}(\mathbf{r}(u, v)) \frac{\partial r_3}{\partial u} \right] \frac{\partial r_1}{\partial v}$$

and

$$\frac{\partial}{\partial v} \left[ g(\mathbf{r}(u, v)) \right] \frac{\partial r_1}{\partial u} = \left[ \frac{\partial g}{\partial x}(\mathbf{r}(u, v)) \frac{\partial r_1}{\partial v} + \frac{\partial g}{\partial y}(\mathbf{r}(u, v)) \frac{\partial r_2}{\partial v} + \frac{\partial g}{\partial z}(\mathbf{r}(u, v)) \frac{\partial r_3}{\partial v} \right] \frac{\partial r_1}{\partial u}.$$

Subtracting the second of these identities from the first, it follows that (19.29) is indeed equal to (19.30). Therefore, Stokes's Identity is verified.    ∎

**LEMMA 19.12** For $\mathcal{O}$ an open subset of the plane $\mathbb{R}^2$, suppose that the mapping $\mathbf{r}: \mathcal{O} \to \mathbb{R}^3$ is continuously differentiable. Let $\mathcal{U}$ be an open subset of $\mathbb{R}^3$ containing the image of $\mathbf{r}: \mathcal{O} \to \mathbb{R}^3$, and suppose that the mapping $\mathbf{F}: \mathcal{U} \to \mathbb{R}^3$ is also continuously differentiable. Let $\boldsymbol{\beta}: I \to \mathbb{R}^2$ be a piecewise smooth parametrized path whose image, the path $\mathcal{C}$, lies in $\mathcal{O}$, and consider the curve $\gamma$ defined by the parametrized composition $\gamma \equiv \mathbf{r} \circ \boldsymbol{\beta}: I \to \mathbb{R}^3$. Then

$$\int_\Gamma \langle \mathbf{F}, \mathbf{T} \rangle \, ds = \int_\mathcal{C} \left\langle \mathbf{F}(\mathbf{r}(u, v)), \frac{\partial \mathbf{r}}{\partial u}(u, v) \right\rangle du + \left\langle \mathbf{F}(\mathbf{r}(u, v)), \frac{\partial \mathbf{r}}{\partial v}(u, v) \right\rangle dv. \quad (19.31)$$

**FIGURE 19.9**

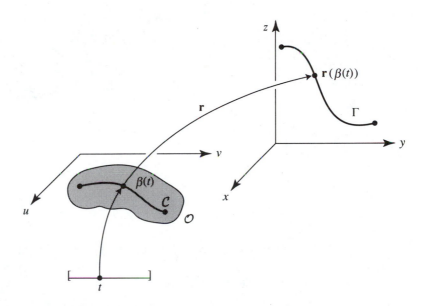

**Proof** By the Chain Rule,

$$\frac{d}{dt}(\gamma(t)) = \frac{d}{dt}(\mathbf{r}(\boldsymbol{\beta}(t))) = \frac{\partial \mathbf{r}}{\partial u}(\boldsymbol{\beta}(t))\frac{d\beta_1}{dt}(t) + \frac{\partial \mathbf{r}}{\partial v}(\boldsymbol{\beta}(t))\frac{d\beta_2}{dt}(t).$$

Hence

$$\int_\Gamma \langle \mathbf{F}, \mathbf{T} \rangle \, ds \equiv \int_I \langle \mathbf{F}(\gamma(t)), \gamma'(t) \rangle \, dt$$

$$= \int_I \left[ \left\langle (\mathbf{F} \circ \mathbf{r})(\boldsymbol{\beta}(t)), \frac{\partial \mathbf{r}}{\partial u}(\boldsymbol{\beta}(t)) \right\rangle \frac{d\beta_1}{dt}(t) + \right.$$

$$\left. \left\langle (\mathbf{F} \circ \mathbf{r})(\boldsymbol{\beta}(t)), \frac{\partial \mathbf{r}}{\partial v}(\boldsymbol{\beta}(t)) \right\rangle \frac{d\beta_2}{dt}(t) \right] dt$$

$$\equiv \int_\mathcal{C} \left\langle \mathbf{F}(\mathbf{r}(u, v)), \frac{\partial \mathbf{r}}{\partial u}(u, v) \right\rangle du + \left\langle \mathbf{F}(\mathbf{r}(u, v)), \frac{\partial \mathbf{r}}{\partial v}(u, v) \right\rangle dv.$$

**THEOREM**
**19.13**

**Stokes's Formula**   For $\mathcal{O}$ an open subset of the plane $\mathbb{R}^2$, suppose that the components of the mapping $\mathbf{r}: \mathcal{O} \to \mathbb{R}^3$ have continuous second-order partial derivatives. Let $\mathcal{U}$ be an open subset of $\mathbb{R}^3$ containing the image of $\mathbf{r}: \mathcal{O} \to \mathbb{R}^3$ and suppose that the mapping $\mathbf{F}: \mathcal{U} \to \mathbb{R}^3$ is continuously differentiable. Suppose that $\mathcal{R}$ is a Green's domain such that both $\mathcal{R}$ and its boundary are contained in $\mathcal{O}$. Let $\mathcal{S}$ be the surface defined by the parametrized surface $\mathbf{r}: \mathcal{R} \to \mathbb{R}^3$ and let $\Gamma$ be the path defined by the composition of $\mathbf{r}$ with the Green's parametrization of the boundary of $\mathcal{R}$. Then

$$\iint_{\mathcal{S}} \langle \text{curl } \mathbf{F}, \boldsymbol{\eta} \rangle \, d\sigma = \int_{\Gamma} \langle \mathbf{F}, \mathbf{T} \rangle \, ds.$$

**FIGURE 19.10**

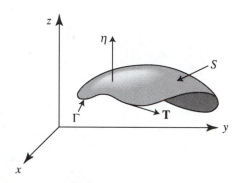

**Proof**   By the very definition of surface integral,

$$\iint_{\mathcal{S}} \langle \text{curl } \mathbf{F}, \boldsymbol{\eta} \rangle \, d\sigma = \iint_{\mathcal{R}} \left\langle \text{curl } \mathbf{F}\big(\mathbf{r}(u, v)\big), \frac{\partial \mathbf{r}}{\partial u}(u, v) \times \frac{\partial \mathbf{r}}{\partial v}(u, v) \right\rangle \, du \, dv.$$

Therefore, by Stokes's Identity,

$$\iint_{\mathcal{S}} \langle \text{curl } \mathbf{F}, \boldsymbol{\eta} \rangle \, d\sigma = \iint_{\mathcal{R}} \left\{ \frac{\partial}{\partial u}\left[ \left\langle \mathbf{F} \circ \mathbf{r}, \frac{\partial \mathbf{r}}{\partial v} \right\rangle \right] - \frac{\partial}{\partial v}\left[ \left\langle \mathbf{F} \circ \mathbf{r}, \frac{\partial \mathbf{r}}{\partial u} \right\rangle \right] \right\} \, du \, dv.$$

Since we have assumed that $\mathcal{R}$ is a Green's domain, by the preceding identity,

$$\iint_{\mathcal{S}} \langle \text{curl } \mathbf{F}, \boldsymbol{\eta} \rangle \, d\sigma = \int_{\mathcal{C}} \left\langle \mathbf{F} \circ \mathbf{r}, \frac{\partial \mathbf{r}}{\partial u} \right\rangle du + \left\langle \mathbf{F} \circ \mathbf{r}, \frac{\partial \mathbf{r}}{\partial v} \right\rangle dv,$$

where $\mathcal{C}$ is the boundary path of $\mathcal{R}$. Thus, by integral formula (19.31),

$$\iint_{\mathcal{S}} \langle \text{curl } \mathbf{F}, \boldsymbol{\eta} \rangle \, d\sigma = \int_{\Gamma} \langle \mathbf{F}, \mathbf{T} \rangle \, ds. \qquad \blacksquare$$

In the language of vector-fields, Stokes's Formula asserts that provided the parametrizations are appropriately chosen, the integral over a surface of the normal component of the curl of a vector-field equals the integral along the boundary of the surface of the tangential component of the vector-field.

**EXERCISES**

1. Use Green's Formula to evaluate

$$\int_\Gamma (2xy - x^2)\, dx + (x + y^2)\, dy$$

where $\Gamma$ is the path defined by $y = x^2$ and $x = y^2$ parametrized in the counterclockwise direction.

2. Verify Green's Formula in the case when $M = x^2 - y^2$, $N = 2xy$, and $\Omega$ is the triangle with vertices $(0, 0)$, $(2, 0)$, and $(1, 1)$.

3. Verify Green's Formula in the case when $M = x^2 - xy^3$, $N = y^2 - 2xy$, and $\Omega$ is the square with opposite vertices $(0, 0)$ and $(2, 2)$.

4. Use the area formula (19.24) to find the area of the triangle bounded by the line $x + y = 4$ and the coordinate axes.

5. Find simple, closed parametrizations of the boundary of the set $\Omega$ defined in Example 19.13 in the cases when $g(c) = h(c)$ and/or $g(d) = h(d)$.

6. Let $\mathbf{p}$ and $\mathbf{q}$ be points in $\mathbb{R}^3$. Prove that

$$\mathrm{curl}\,[(\mathbf{p} - \mathbf{q}) \times ((x,\, y,\, z) - \mathbf{q})] = 2\mathbf{p} - 2\mathbf{q}.$$

7. Verify Stokes's Formula for $\mathbf{F}(x,\, y,\, z) = (3y,\, -xz,\, yz^2)$ where $S$ is the surface of the paraboloid $2z = x^2 + y^2$ bounded by the plane $z = 2$.

8. Verify Stokes's Formula in the case when $\mathbf{F}(x,\, y,\, z) = (z,\, x,\, y)$ and $S$ is the upper hemisphere of radius 1 centered at the origin.

9. Evaluate

$$\iint_S \langle \mathbf{F},\, \boldsymbol{\eta}\rangle d\sigma,$$

where $\mathbf{F}(x,\, y,\, z) = (xz,\, yz,\, z^2)$ and $S$ is the upper hemisphere of radius 1 centered at the origin.

10. For a continuous function $f\colon [a,\, b] \to \mathbb{R}$ that has a continuous bounded derivative on the open interval $(a,\, b)$, define $\Omega = \{(x,\, y) \mid a < x < b,\, 0 < y < 1\}$ and define $M(x,\, y) = f(x)$ and $N(x,\, y) = 0$, for $(x,\, y)$ in $\Omega$. Show that Green's Formula reduces to the formula $\int_a^b f'(t)\, dt = f(b) - f(a)$.

11. Show that Green's Formula is a special case of Stokes's Formula.

12. Suppose that $\Omega$ is a Green's domain with boundary $\Gamma$. For functions $u(x,\, y)$ and $v(x,\, y)$ that are twice continuously differentiable on an open set containing $\Omega \cup \Gamma$, show that

$$\int_\Gamma \left[ u\frac{\partial v}{\partial x}\, dx + u\frac{\partial v}{\partial y}\, dy \right] \in \iint_\Omega \det \begin{vmatrix} \partial u/\partial x & \partial u/\partial y \\ \partial v/\partial x & \partial v/\partial y \end{vmatrix} dx\, dy.$$

13. For continuously differentiable vector-fields $\mathbf{F}\colon \mathbb{R}^3 \to \mathbb{R}^3$ and $\mathbf{G}\colon \mathbb{R}^3 \to \mathbb{R}^3$, show that

$$\frac{\partial}{\partial x}[\mathbf{F} \times \mathbf{G}] \in \frac{\partial \mathbf{F}}{\partial x} \times \mathbf{G} + \mathbf{F} \times \frac{\partial \mathbf{G}}{\partial x}.$$

14. Suppose that $\mathbf{F}\colon \mathbb{R}^3 \to \mathbb{R}^3$ and $f\colon \mathbb{R}^3 \to \mathbb{R}^3$ are continuously differentiable. Show that

$$\mathrm{curl}(f\mathbf{F}) \in f\,\mathrm{curl}\,\mathbf{F} + \nabla f \times \mathbf{F}.$$

15. For a function $f\colon \mathbb{R}^3 \to \mathbb{R}$ that has continuous second-order partial derivatives, show that

$$\mathrm{curl}\nabla f = \mathbf{0}.$$

16. For continuously differentiable vector-fields $\mathbf{E}\colon \mathbb{R}^3 \to \mathbb{R}^3$ and $\mathbf{H}\colon \mathbb{R}^3 \to \mathbb{R}^3$, show that

$$\nabla(\mathbf{E} \times \mathbf{H}) = \langle \mathbf{E},\, \mathrm{curl}\,\mathbf{H}\rangle - \langle \mathbf{H},\, \mathrm{curl}\,\mathbf{E}\rangle.$$

17. Under the assumptions of the integration by parts formula stated in in Corollary 19.10, suppose that $n$ and $v$ have continuous second-order partial derivatives on an open set containing $\Omega \cup \Gamma$. Use the integration by parts formula to obtain Green's First Identity:

$$\iint_{\mathcal{R}} \langle \nabla u, \nabla v \rangle \, dx \, dy = \int_C u \frac{\partial v}{\partial \eta} \, ds - \iint_{\mathcal{R}} u \Delta v \, dx \, dy.$$

By subtracting, obtain Green's Second Identity:

$$\iint_{\mathcal{R}} [u \Delta v - v \Delta u] \, dx \, dy = \int_C \left[ u \frac{\partial v}{\partial \eta} - v \frac{\partial u}{\partial \eta} \right] ds.$$

(Recall that $\Delta w$, which is called the *Laplacian* of $w$, is defined by

$$\Delta w(x, y) \equiv \frac{\partial^2 w}{\partial x^2}(x, y) + \frac{\partial^2 w}{\partial y^2}(x, y).)$$

18. Use Green's First Identity, from Exercise 17, to show that if the function $u(x, y)$ has continuous second-order partial derivatives on an open set containing $\{(x, y) \mid x^2 + y^2 \leq 1\}$, then the function that is identically 0 is the only function having the properties that

$$\begin{cases} \Delta u(x, y) = 0 & \text{for } x^2 + y^2 < 1 \\ u(x, y) = 0 & \text{for } x^2 + y^2 = 1. \end{cases}$$

# Consequences of the Field and Positivity Axioms

In the Preliminaries, we stated the Field Axioms and the Positivity Axioms for the real numbers and made various assertions regarding elementary consequences of these axioms. In this first appendix, we will verify some of these assertions.

For convenience, we will restate the Field Axioms. For each pair of real numbers $a$ and $b$, a real number called the *sum* of $a$ and $b$ is defined and is denoted by $a + b$, and a real number called the *product* of $a$ and $b$ is defined and is denoted by $ab$. These operations satisfy the following collection of axioms:

## The Field Axioms

*Commutativity of Addition*: For all real numbers $a$ and $b$,

$$a + b = b + a.$$

*Associativity of Addition*: For all real numbers $a$, $b$, and $c$,

$$(a + b) + c = a + (b + c).$$

*The Additive Identity*: There is a real number, denoted by 0, such that

$$0 + a = a \quad \text{for all real numbers } a.$$

*The Additive Inverse*: For each real number $a$, there is a real number $b$ such that

$$a + b = 0.$$

*Commutativity of Multiplication*: For all real numbers $a$ and $b$,

$$ab = ba.$$

*Associativity of Multiplication*: For all real numbers $a$, $b$, and $c$,

$$(ab)c = a(bc).$$

*The Multiplicative Identity*: There is a real number, denoted by 1, such that

$$1a = a \quad \text{for all real numbers } a.$$

*The Multiplicative Inverse*: For each real number $a \neq 0$, there is a real number $b$ such that

$$ab = 1.$$

*The Distributive Property*: For all real numbers $a$, $b$, and $c$,

$$a(b + c) = ab + ac.$$

*The Nontriviality Assumption*:

$$1 \neq 0.$$

First, observe that there is only one number that has the property asserted in the additive identity axiom. Indeed, if $0'$ also has the property that

$$0' + a = a \quad \text{for all real numbers } a,$$

then in particular we have

$$0' + 0 = 0.$$

But by the commutative property of addition and the definition of 0 as an additive identity,

$$0' + 0 = 0 + 0' = 0'.$$

Thus $0 = 0'$, so there is only one additive identity.

---

**PROPOSITION 1**    For each real number $a$,

$$a0 = 0a = 0.$$

**Proof**    Observe that

$$0 + 0 = 0.$$

Thus, by the Distributive Axiom,

$$0a + 0a = 0a.$$

If we add the additive inverse of $0a$ to each side and use the associativity of addition, we obtain $0a = 0$, and from this and the commutative property of multiplication it also follows that $a0 = 0$.    ∎

---

**PROPOSITION 2**    For any pair of real numbers $a$ and $b$, if

$$ab = 0,$$

then $a = 0$ or $b = 0$.

**Proof**    If $a = 0$, the proof is complete. So suppose that $a \neq 0$. We must show that $b = 0$. Since $a \neq 0$, by the multiplicative inverse axiom we can select a number $d$ such that $da = 1$. Since $ab = 0$, it follows from Proposition 1 that

$$d(ab) = d0 = 0.$$

On the other hand, by the associative and commutative properties of multiplication, and by the definition of 1 as the multiplicative identity, it follows that

$$d(ab) = (da)b = 1b = b.$$

Thus $b = 0$.                                                                                ■

The additive inverse axiom asserts that for each number $a$, there is a number $b$ such that $a + b = 0$. In fact, there is only one such number; it is called the *additive inverse* of $a$. To see why there is only one such number, suppose that $b'$ also has the property that $a + b' = 0$. Then

$$
\begin{aligned}
b' &= 0 + b' & \\
&= (a + b) + b' & \text{by the choice of } b \\
&= (b + a) + b' & \text{by commutativity of addition} \\
&= b + (a + b') & \text{by associativity of addition} \\
&= b + 0 & \text{by the choice of } b' \\
&= 0 + b & \text{by commutativity of addition} \\
&= b & \text{by the definition of } 0.
\end{aligned}
$$

Thus $b = b'$. Of course, we denote the additive inverse of $a$ by $-a$. The additive inverse possesses the following familiar properties:

**PROPOSITION 3**    For all real numbers $a$ and $b$,

(i)  $-(-a) = a$

(ii)  $-a = (-1)a$

(iii)  $-ab = (-a)b$

(iv)  $ab = (-a)(-b)$

(v)  $1 = (-1)(-1)$.

**Proof**    Part (i) follows from the fact that $-(-a)$ is the unique number that when added to $-a$ equals 0 and from the observation that $a$ has this property. To verify (ii), we must show that

$$a + (-1)a = 0.$$

But since 1 is the multiplicative identity,

$$a + (-1)a = 1a + (-1)a$$

$$= (1 + (-1))a \qquad \text{by the distributive property}$$

$$= 0a \qquad \text{since } -1 \text{ is the additive inverse of } 1$$

$$= 0 \qquad \text{by Proposition 1.}$$

To verify (iii), observe that

$$-ab = (-1)ab \qquad \text{by (ii)}$$

$$= ((-1)a)b \qquad \text{by associativity of multiplication}$$

$$= (-a)b \qquad \text{again by (ii)}$$

To verify (iv), observe that

$$ab = -(-ab) \qquad \text{by (i)}$$

$$= -((-a)b) \qquad \text{by (iii)}$$

$$= -(b(-a)) \qquad \text{by commutativity of multiplication}$$

$$= (-b)(-a) \qquad \text{by (iii)}$$

$$= (-a)(-b) \qquad \text{by commutativity of multiplication.}$$

Finally, observe that (v) follows from (iv) when we set $a = b = 1$.  ∎

For numbers $a$ and $b$, we define the *difference* $a - b$ by

$$a - b \equiv a + (-b).$$

Using the preceding proposition, it is not difficult to verify that for any numbers $a$, $b$, and $c$,

$$a(b - c) = ab - ac \quad \text{and} \quad -(b - c) = -b + c.$$

Let us now examine some consequences of the multiplication axioms. Just as we have shown that the additive identity is unique, a similar argument shows that the multiplicative identity is unique. Also, an argument similar to the one that showed that the additive inverse is unique shows that for a nonzero number $a$, its multiplicative inverse is unique; the multiplicative inverse of $a$ is, of course, denoted by $a^{-1}$. The multiplicative inverse possesses the following familiar properties:

**PROPOSITION 4**  For any nonzero real numbers $a$ and $b$,

(i) $(a^{-1})^{-1} = a$

(ii) $(-a)^{-1} = -a^{-1}$

(iii) $(ab)^{-1} = a^{-1}b^{-1}.$

**Proof**   To verify (i), observe that $(a^{-1})^{-1}$ is the unique number that has the property that when it is multiplied with $a^{-1}$, the product is 1, and that the number $a$ has this property. To verify (ii), we must show that

$$(-a)(-a^{-1}) = 1.$$

However, by part (iv) of Proposition 3, we have

$$(-a)(-a^{-1}) = (a)(a^{-1}) = 1.$$

Finally, to verify (iii), we must show that

$$(ab)(a^{-1}b^{-1}) = 1.$$

However, by the commutative and associative properties of multiplication,

$$(ab)(a^{-1}b^{-1}) = (aa^{-1})(bb^{-1}) = 1 \cdot 1 = 1. \qquad \blacksquare$$

For any two numbers $a$ and $b$, with $b \neq 0$, we define

$$\frac{a}{b} = ab^{-1}.$$

Directly from the definition of division and the distributive property, it follows that for any numbers $a$, $b$, and $c$, with $c \neq 0$,

$$\frac{a+b}{c} = \frac{a}{c} + \frac{b}{c}.$$

## The Positivity Axioms for the Real Numbers

In the real numbers, there is a natural notion of order—that is, of *greater than*, *less than*, and so forth. A convenient way to codify these properties is by specifying axioms that are satisfied by the set of positive numbers.

## The Positivity Axioms

There is a set of real numbers denoted by $\mathcal{P}$, called the set of *positive numbers*, that has the following two properties:

**P1**   If $a$ and $b$ are positive, then $ab$ and $a + b$ are also positive.

**P2**   For a real number $a$, exactly one of the following three alternatives is true:

$$a \text{ is positive}, \quad -a \text{ is positive}, \quad a = 0.$$

The positivity axioms lead in a natural way to an ordering of the real numbers: For real numbers $a$ and $b$, we define $a > b$ to mean that $a - b$ is positive, and $a \geq b$ to mean that $a > b$ or $a = b$. We then define $a < b$ to mean that $b > a$, and $a \leq b$ to mean that $b \geq a$.

**PROPOSITION 5**   For each real number $a \neq 0$, $a^2 > 0$. In particular, $1 > 0$.

**Proof**   Since $a \neq 0$, it follows from the second positivity axiom that either $a$ or $-a$ is positive. If $a$ is positive, then since the product of positive numbers is again positive, $a^2$ is positive. Similarly, if $-a$ is positive, so is $(-a)(-a)$. But by part (iv) of Proposition 3, $(-a)(-a) = a^2$. Thus again, in this case, $a^2$ is positive. In particular, since by the nontriviality axiom $1 \neq 0$, $1 = 1 \cdot 1$ is positive.   ∎

**PROPOSITION 6**   For each positive number $a$, its multiplicative inverse $a^{-1}$ is also positive.

**Proof**   Since $a \cdot a^{-1} = 1 \neq 0$, it follows from Proposition 2 that $a^{-1} \neq 0$. By the first positivity axiom, either $a^{-1}$ or $-a^{-1}$ is positive. But it is not possible for $-a^{-1}$ to be positive, since then $a \cdot (-a^{-1}) = -1$ would also be positive, and this contradicts Proposition 5. Thus $a^{-1}$ is positive.   ∎

**PROPOSITION 7**   If $a > b$, then

$$ac > bc \quad \text{if} \quad c > 0$$

and

$$ac < bc \quad \text{if} \quad c < 0.$$

**Proof**   The number $a - b$ is positive. If $c$ is positive, then the product $(a - b)c = ab - ac$ is also positive; that is, $ac > bc$. On the other hand, if $c < 0$, then $-c$ is positive, so $(a - b)(-c)$ also is positive. However, $(a - b)(-c) = bc - ac$, so $ac < bc$.   ∎

**EXERCISES**

1.  Prove that for any numbers $a$, $b$, and $c$,
$$a(b - c) = ab - ac \quad \text{and} \quad -(b - c) = -b + c.$$

2.  Prove that the multiplicative identity is unique.
3.  Prove that each number $a \neq 0$ has a unique multiplicative inverse.
4.  Prove that for any numbers $a$ and $b$, with $b \neq 0$,
$$-\frac{a}{b} = \frac{-a}{b} = \frac{a}{-b}.$$

# B

# Linear Algebra

In the analysis of differentiation and integration of functions of several variables, and of mappings between Euclidean spaces, we are principally concerned with the case when the functions, or mappings, are nonlinear. However, underlying the study of these nonlinear functions or mappings is an understanding of linear functions and mappings. The body of knowledge related to the study of linear functions and mappings is called *linear algebra*. The concepts of the vector sum of two vectors, the product of a number and a vector, and the inner product of two vectors were discussed in Chapter 10. The correspondence between linear mappings and matrices was established in Section 15.1, and in the same section we defined and described various properties of the determinant of a square matrix.

A full treatment of linear algebra is outside the scope of this book. In this appendix, we will first define some general concepts of linear algebra and state some general results. Then we will prove the stated results in the important special case of $\mathbb{R}^3$. Beyond the simplification that this provides, it also permits us to take a more geometric viewpoint that arises from the geometric properties of the inner product and the cross product of two vectors in $\mathbb{R}^3$.

Given $k$ vectors $\mathbf{v}_1, \ldots, \mathbf{v}_k$ in Euclidean space $\mathbb{R}^n$, a vector $\mathbf{v}$ of the form

$$\mathbf{v} = \lambda_1 \mathbf{v}_1 + \cdots + \lambda_k \mathbf{v}_k,$$

where $\lambda_1, \ldots, \lambda_k$ are numbers, is said to be a *linear combination* of $\mathbf{v}_1, \ldots, \mathbf{v}_k$.

**DEFINITION** *The $k$ vectors $\mathbf{v}_1, \ldots, \mathbf{v}_k$ in $\mathbb{R}^n$ are said to be linearly dependent provided that there are numbers $\lambda_1, \ldots, \lambda_k$, not all of which are 0, such that*

$$\lambda_1 \mathbf{v}_1 + \cdots + \lambda_k \mathbf{v}_k = \mathbf{0}.$$

*If the vectors $\mathbf{v}_1, \ldots, \mathbf{v}_k$ are not linearly dependent, they are said to be linearly independent.*

It is easy to see that the $k$ vectors $\mathbf{v}_1, \ldots, \mathbf{v}_k$ in $\mathbb{R}^n$ are linearly dependent if and only if one of these vectors is a linear combination of the remaining $k - 1$ vectors.

Moreover, if $\mathbf{v}_1, \ldots, \mathbf{v}_k$ are linearly independent and the vector $\mathbf{v}$ is a linear combination of $\mathbf{v}_1, \ldots, \mathbf{v}_k$, then there are *unique* numbers $\lambda_1, \ldots, \lambda_k$ such that

$$\mathbf{v} = \lambda_1 \mathbf{v}_1 + \cdots + \lambda_k \mathbf{v}_k.$$

**DEFINITION**   *The $k$ vectors $\mathbf{v}_1, \ldots, \mathbf{v}_k$ in $\mathbb{R}^n$ are said to span $\mathbb{R}^n$ provided that every vector $\mathbf{v}$ in $\mathbb{R}^n$ is a linear combination of $\mathbf{v}_1, \ldots, \mathbf{v}_k$.*

**DEFINITION**   *The $k$ vectors $\mathbf{v}_1, \ldots, \mathbf{v}_k$ in $\mathbb{R}^n$ are said to be a basis for $\mathbb{R}^n$ provided that for each vector $\mathbf{v}$ in $\mathbb{R}^n$, there are unique numbers $\lambda_1, \ldots, \lambda_k$ such that*

$$\mathbf{v} = \lambda_1 \mathbf{v}_1 + \cdots + \lambda_k \mathbf{v}_k.$$

The definition of linear independence may be restated by asserting that the $k$ vectors $\mathbf{v}_1, \ldots, \mathbf{v}_k$ in $\mathbb{R}^n$ are linearly independent provided that the only numbers $\lambda_1, \ldots, \lambda_k$ having the property that

$$\mathbf{0} = \lambda_1 \mathbf{v}_1 + \cdots + \lambda_k \mathbf{v}_k$$

are $\lambda_1 = 0, \ldots, \lambda_k = 0$. It follows that if the vectors $\mathbf{v}_1, \ldots, \mathbf{v}_k$ are a basis for $\mathbb{R}^n$, then they must be linearly independent. The converse is not true. There is, however, the following important theorem.

**THEOREM I**   For $n$ vectors $\mathbf{v}_1, \ldots, \mathbf{v}_n$ in $\mathbb{R}^n$, the following three assertions are equivalent:

  (i) The vectors $\mathbf{v}_1, \ldots, \mathbf{v}_n$ are a basis for $\mathbb{R}^n$.

  (ii) The vectors $\mathbf{v}_1, \ldots, \mathbf{v}_n$ span $\mathbb{R}^n$.

  (iii) The vectors $\mathbf{v}_1, \ldots, \mathbf{v}_n$ are linearly independent.

The above theorem has an immediate interpretation in terms of $n \times n$ systems of linear equations. Recall that an $n \times n$ *matrix* is a rectangular array of real numbers consisting of $n$ rows and $n$ columns. If such an $n \times n$ matrix is denoted by $\mathbf{A}$, we write

$$\mathbf{A} = [a_{ij}],$$

where for each pair of indices $i$ and $j$ such that $1 \le i \le n$ and $1 \le j \le n$, $a_{ij}$ denotes the number in the $i$th row and $j$th column of the matrix $\mathbf{A}$. For a point $\mathbf{x}$ in $\mathbb{R}^n$, by the symbol $\mathbf{Ax}$ we denote the point in $\mathbb{R}^n$ that, for each index $i$ such that $1 \le i \le n$, has $i$th component equal to the inner product of the $i$th row of $\mathbf{A}$ with $\mathbf{x}$. Thus

$$\mathbf{Ax} \equiv \mathbf{y},$$

where $$y_i \equiv \sum_{j=1}^{n} a_{ij} x_j \quad \text{for each index } i \text{ such that } 1 \le i \le n.$$

Now, for each index $i$ with $1 \le i \le n$, define the vector $\mathbf{v}_i$ by

$$\mathbf{v}_i = (a_{1i}, \ldots, a_{ni}),$$

so that $\mathbf{v}_i$ corresponds to the $i$th column of the matrix $\mathbf{A}$. Then, by the very definition of $\mathbf{Ax}$, it follows immediately that for vectors $\mathbf{x}$ and $\mathbf{y}$ in $\mathbb{R}^n$,

$$\mathbf{Ax} = \mathbf{y} \quad \text{if and only if} \quad x_1\mathbf{v}_1 + \cdots + x_n\mathbf{v}_n = \mathbf{y}.$$

This equivalence allows us to restate Theorem 1 as follows:

**THEOREM 2**    For an $n \times n$ matrix $\mathbf{A}$ and the $n \times n$ system of linear equations

$$\mathbf{Ax} = \mathbf{y}, \tag{1}$$

the following three assertions are equivalent:

(i) For each $\mathbf{y}$ in $\mathbb{R}^n$, the system of linear equations (1) has a unique solution $\mathbf{x}$.

(ii) For each $\mathbf{y}$ in $\mathbb{R}^n$, the system of linear equations (1) has a solution $\mathbf{x}$.

(iii) For $\mathbf{y} = \mathbf{0}$, the only solution of the system of linear equations (1) is $\mathbf{x} = \mathbf{0}$.

Finally, Theorem 1 (and hence also Theorem 2) has an interpretation in terms of linear mappings. The correspondence between linear mappings $\mathbf{T}: \mathbb{R}^n \to \mathbb{R}^n$ and $n \times n$ matrices is completely described in Section 15.1 and hence will not be repeated here. It should be noted that for a linear mapping $\mathbf{T}: \mathbb{R}^n \to \mathbb{R}^n$ that is associated with the $n \times n$ matrix $\mathbf{A}$ by

$$\mathbf{T}(\mathbf{x}) = \mathbf{Ax} \text{ for all } \mathbf{x} \text{ in } \mathbb{R}^n,$$

since $\mathbf{T}(\mathbf{u}) = \mathbf{T}(\mathbf{v})$ if and only if $\mathbf{T}(\mathbf{u} - \mathbf{v}) = \mathbf{0}$, we see that such a mapping is one-to-one if and only if whenever $\mathbf{Ax} = \mathbf{0}$, then $\mathbf{x} = \mathbf{0}$. This is the observation that is needed to see that Theorem 2 is equivalent to the following theorem:

**THEOREM 3**    For a linear mapping $\mathbf{T}: \mathbb{R}^n \to \mathbb{R}^n$, the following three assertions are equivalent:

(i) The mapping $\mathbf{T}: \mathbb{R}^n \to \mathbb{R}^n$ is one-to-one and has image equal to $\mathbb{R}^n$; that is, it is invertible.

(ii) The mapping $\mathbf{T}: \mathbb{R}^n \to \mathbb{R}^n$ has image equal to $\mathbb{R}^n$.

(iii) The mapping $\mathbf{T}: \mathbb{R}^n \to \mathbb{R}^n$ is one-to-one.

Although we now have the equivalence of Theorems 1, 2, and 3, we do not have a proof of any one of them. Moreover, we also lack any explicit criterion for determining when $n$ vectors in $\mathbb{R}^n$ are a basis for $\mathbb{R}^n$ or, equivalently, for determining when a linear mapping from $\mathbb{R}^n$ to $\mathbb{R}^n$ is invertible. There is a number called the *determinant*, which can be associated with any ordered $n$-tuple of vectors in $\mathbb{R}^n$; the vectors are a basis if and only if the determinant is nonzero. Equivalently, the determinant can be associated with the matrix that represents a linear mapping from $\mathbb{R}^n$ to $\mathbb{R}^n$; the determinant is nonzero if and only if the mapping is invertible. When the determinant is nonzero, its magnitude has an interpretation as a measure of volume.

As we have mentioned, we will not prove the above assertions in general Euclidean space $\mathbb{R}^n$. Rather, we will provide all details for the special but very important case of $\mathbb{R}^3$.

Not all Euclidean spaces are created equal. Of course $\mathbb{R}^1$ is special, since it is the set of real numbers, which have been described in the Preliminaries and Chapter 1 by the Field Axioms, the Positivity Axioms, and the Completeness Axiom. The plane $\mathbb{R}^2$ also is special, since it turns out that there is a concept of product called the *complex product* (or complex multiplication) that associates with a pair of points in $\mathbb{R}^2$ another point in $\mathbb{R}^2$, called the complex product of the pair. With the usual concept of sum, and with multiplication replaced by the complex product, the Field Axioms are satisfied (see Exercises 6 and 7). This has far-reaching consequences and is the basis of the subject called *complex analysis*. This topic, however, lies outside the scope of this book.* Here, we will study geometry and algebra in $\mathbb{R}^3$ by introducing a construction called the *cross product*. In contrast to the inner product, which associates with any pair of vectors in $\mathbb{R}^3$ a *number*, the cross product associates with any ordered pair of vectors in $\mathbb{R}^3$ another *vector* in $\mathbb{R}^3$. By using both the inner product and the cross product, we will obtain interesting geometric and algebraic results that have intuitive geometric interpretations.

For a point $\mathbf{p}$ and a nonzero vector $\mathbf{v}$ in $\mathbb{R}^3$, the line $\ell$ through $\mathbf{p}$ that is parallel to $\mathbf{v}$ is defined to be the set of points of the form $\mathbf{p} + t\mathbf{v}$, where $t$ is any number. For another point $\mathbf{u}$ in $\mathbb{R}^3$, it is often useful to find the point on the line $\ell$ that is closest to $\mathbf{u}$. The distance from $\mathbf{u}$ to this point is called the *distance* from the point $\mathbf{u}$ to the line $\ell$. We will now provide a formula for this point that reveals the geometric significance of the inner product.

Recall that by the linearity and symmetry of the inner product, for two vectors $\mathbf{u}$ and $\mathbf{v}$ in $\mathbb{R}^3$,

$$\langle \mathbf{u} + \mathbf{v},\, \mathbf{u} + \mathbf{v} \rangle = \langle \mathbf{u}, \mathbf{u} \rangle + \langle \mathbf{v}, \mathbf{v} \rangle + 2\langle \mathbf{u}, \mathbf{v} \rangle,$$

so that we have the following:

## The Pythagorean Identity

$$\|\mathbf{u} + \mathbf{v}\|^2 = \|\mathbf{u}\|^2 + \|\mathbf{v}\|^2 \quad \text{if and only if } \mathbf{u} \text{ and } \mathbf{v} \text{ are orthogonal.}$$

**THEOREM 4**   For a nonzero vector $\mathbf{v}$ in $\mathbb{R}^3$, let $\ell$ be the line through the origin that is parallel to $\mathbf{v}$. For a point $\mathbf{u}$ in $\mathbb{R}^3$, set $\lambda = \langle \mathbf{u}, \mathbf{v} \rangle / \langle \mathbf{v}, \mathbf{v} \rangle$. Then

(i) the vector $\mathbf{u} - \lambda\mathbf{v}$ is orthogonal to the vector $\mathbf{v}$.

(ii) $\lambda\mathbf{v}$ is the point on the line $\ell$ that is closest to $\mathbf{u}$, so the distance from $\mathbf{u}$ to $\ell$ is $\|\mathbf{u} - \lambda\mathbf{v}\|$.

**Proof**   By the linearity of the inner product and the definition of $\lambda$,

$$\langle \mathbf{u} - \lambda\mathbf{v},\, \mathbf{v} \rangle = \langle \mathbf{u}, \mathbf{v} \rangle - \lambda\langle \mathbf{v}, \mathbf{v} \rangle = 0,$$

so (i) is verified. To verify (ii), it is necessary to show that

$$\|\mathbf{u} - t\mathbf{v}\| \geq \|\mathbf{u} - \lambda\mathbf{v}\| \quad \text{for all } t \text{ in } \mathbb{R}. \tag{2}$$

---

*See, for instance, Churchill and Ward, *Complex Variables and Applications* (Fifth Edition), McGraw-Hill, 1990.

We write $\mathbf{u} - t\mathbf{v} = (\mathbf{u} - \lambda\mathbf{v}) + (\lambda - t)\mathbf{v}$. By (i), $\mathbf{u} - \lambda\mathbf{v}$ is orthogonal to $\mathbf{v}$, and hence is also orthogonal to $(\lambda - t)\mathbf{v}$. Thus, by the Pythagorean Identity,

$$\|\mathbf{u} - t\mathbf{v}\|^2 = \|(\mathbf{u} - \lambda\mathbf{v}) + (\lambda - t)\mathbf{v}\|^2 = \|\mathbf{u} - \lambda\mathbf{v}\|^2 + \|(\lambda - t)\mathbf{v}\|^2 \geq \|\mathbf{u} - \lambda\mathbf{v}\|^2,$$

so the inequality (2) holds.                                                          ■

**DEFINITION**   *For two vectors* $\mathbf{u} = (u_1, u_2, u_3)$ *and* $\mathbf{v} = (v_1, v_2, v_3)$ *in* $\mathbb{R}^3$, *the cross product of* $\mathbf{u}$ *with* $\mathbf{v}$, *which is denoted by* $\mathbf{u} \times \mathbf{v}$, *is the vector in* $\mathbb{R}^3$ *defined by the formula*

$$\mathbf{u} \times \mathbf{v} \equiv (u_2v_3 - u_3v_2, \; u_3v_1 - u_1v_3, \; u_1v_2 - u_2v_1).$$

At first glance, the geometric significance of the cross product is not at all apparent. In preparation for the discussion of its geometric significance, we will first collect some algebraic properties of the cross product.

**PROPOSITION 5**   For vectors $\mathbf{u}$, $\mathbf{v}$, and $\mathbf{w}$ in $\mathbb{R}^3$,

$$\mathbf{u} \times \mathbf{v} = -\mathbf{v} \times \mathbf{u}, \qquad \qquad (\textit{Anti-symmetry})$$

and if $\alpha$ and $\beta$ are any numbers,

$$[\alpha\mathbf{u} + \beta\mathbf{w}] \times \mathbf{v} = \alpha[\mathbf{u} \times \mathbf{v}] + \beta[\mathbf{w} \times \mathbf{v}]. \qquad (\textit{Linearity})$$

**Proof**   The proof of these identities is by inspection. By definition,

$$\mathbf{u} \times \mathbf{v} = (u_2v_3 - u_3v_2, \; u_3v_1 - u_1v_3, \; u_1v_2 - u_2v_1)$$

and

$$\mathbf{v} \times \mathbf{u} = (v_2u_3 - v_3u_2, \; v_3u_1 - v_1u_3, \; v_1u_2 - v_2u_1),$$

so $\mathbf{u} \times \mathbf{v} = -\mathbf{v} \times \mathbf{u}$. Similarly, we verify linearity.                         ■

Define $\mathbf{i} = (1, 0, 0)$, $\mathbf{j} = (0, 1, 0)$, and $\mathbf{k} = (0, 0, 1)$. For each point $\mathbf{v} = (x, y, z)$ in $\mathbb{R}^3$,

$$\mathbf{v} = x\mathbf{i} + y\mathbf{j} + z\mathbf{k},$$

so it is clear that $\mathbf{i}$, $\mathbf{j}$, and $\mathbf{k}$ form a basis for $\mathbb{R}^3$. The basis $\mathbf{i}$, $\mathbf{j}$, $\mathbf{k}$ is called the *standard basis* for $\mathbb{R}^3$. Observe that for the basis $\mathbf{i}$, $\mathbf{j}$, $\mathbf{k}$, we have

$$\mathbf{i} \times \mathbf{j} = \mathbf{k}, \quad \mathbf{k} \times \mathbf{i} = \mathbf{j}, \quad \mathbf{j} \times \mathbf{k} = \mathbf{i}.$$

The following theorem explains the significance of the length of the cross product of two vectors.

**THEOREM 6**   For vectors $\mathbf{u}$ and $\mathbf{v}$ in $\mathbb{R}^3$ such that $\mathbf{v} \neq \mathbf{0}$, set $\lambda = \langle \mathbf{u}, \mathbf{v} \rangle / \langle \mathbf{v}, \mathbf{v} \rangle$. Then

$$\|\mathbf{u} \times \mathbf{v}\| = \|\mathbf{v}\| \cdot \|\mathbf{u} - \lambda\mathbf{v}\|; \qquad (3)$$

that is, the length of $\mathbf{u} \times \mathbf{v}$ is the length of the vector $\mathbf{v}$ times the distance from the point $\mathbf{u}$ to the line through the origin that is parallel to the vector $\mathbf{v}$.

**Proof**  By part (i) of Theorem 4, the vector $\mathbf{u} - \lambda\mathbf{v}$ is orthogonal to the vector $\mathbf{v}$, and hence is also orthogonal to $\lambda\mathbf{v}$. To verify (3) we square the left-hand side and compute:

$$\|\mathbf{u} \times \mathbf{v}\|^2 = (u_2v_3 - u_3v_2)^2 + (u_3v_1 - u_1v_3)^2 + (u_1v_2 - u_2v_1)^2$$

$$= (u_1^2 + u_2^2 + u_3^2)(v_1^2 + v_2^2 + v_3^2) - (u_1v_1 + u_2v_2 + u_3v_3)^2$$

$$= \|\mathbf{u}\|^2 \cdot \|\mathbf{v}\|^2 - \langle \mathbf{u}, \mathbf{v}\rangle^2$$

$$= \|\mathbf{v}\|^2 \left\{ \|\mathbf{u}\|^2 - \frac{\langle \mathbf{u}, \mathbf{v}\rangle^2}{\|\mathbf{v}\|^2} \right\}$$

$$= \|\mathbf{v}\|^2 \langle \mathbf{u} - \lambda\mathbf{v}, \mathbf{u}\rangle \qquad \text{by the definition of } \lambda$$

$$= \|\mathbf{v}\|^2 \langle \mathbf{u} - \lambda\mathbf{v}, \mathbf{u} - \lambda\mathbf{v}\rangle \qquad \text{since } \langle \mathbf{u} - \lambda\mathbf{v}, \lambda\mathbf{v}\rangle = 0$$

$$= \|\mathbf{v}\|^2 \cdot \|\mathbf{u} - \lambda\mathbf{v}\|^2.$$

Thus formula (3) holds. The last remark in the statement of the theorem follows from part (ii) of Theorem 4. ∎

**THEOREM 7**  For two vectors $\mathbf{u}$ and $\mathbf{v}$ in $\mathbb{R}^3$, the following assertions are equivalent:

(i) The vectors $\mathbf{u}$ and $\mathbf{v}$ are linearly dependent.

(ii) $\mathbf{u} \times \mathbf{v} = \mathbf{0}$.

**Proof**  First, suppose that $\mathbf{u}$ and $\mathbf{v}$ are linearly dependent. Then one of the vectors is a scalar multiple of the other, say $\mathbf{v} = \alpha\mathbf{u}$. By the anti-symmetry of the cross product, $\mathbf{u} \times \mathbf{u} = \mathbf{0}$. Thus, using the linearity property of the cross product,

$$\mathbf{u} \times \mathbf{v} = \mathbf{u} \times \alpha\mathbf{u} = \alpha(\mathbf{u} \times \mathbf{u}) = \alpha\mathbf{0} = \mathbf{0}.$$

To prove the converse, suppose that $\mathbf{u} \times \mathbf{v} = \mathbf{0}$. If $\mathbf{v} = \mathbf{0}$, then of course $\mathbf{u}$ and $\mathbf{v}$ are linearly dependent, since $\mathbf{v} = 0\mathbf{u}$. If $\mathbf{v} \neq \mathbf{0}$, then from formula (3) we conclude that $\|\mathbf{u} - \lambda\mathbf{v}\| = 0$, and again $\mathbf{u}$ and $\mathbf{v}$ are linearly dependent, since we now have $\mathbf{u} = \lambda\mathbf{v}$. ∎

The following theorem partially reveals the significance of the direction of the cross product.

**THEOREM 8**  Let $\mathbf{u}$ and $\mathbf{v}$ be vectors in $\mathbb{R}^3$. Then

(i) The cross product $\mathbf{u} \times \mathbf{v}$ is orthogonal to both $\mathbf{u}$ and $\mathbf{v}$.

(ii) Moreover, in the case when $\mathbf{u}$ and $\mathbf{v}$ are linearly independent, if $\mathbf{w}$ is any vector in $\mathbb{R}^3$ that is orthogonal to both $\mathbf{u}$ and $\mathbf{v}$, then $\mathbf{w}$ is a scalar multiple of $\mathbf{u} \times \mathbf{v}$; that is, there is a number $\gamma$ such that $\mathbf{w} = \gamma(\mathbf{u} \times \mathbf{v})$.

**Proof**    To verify (i), it is necessary to show that

$$\langle \mathbf{u} \times \mathbf{v}, \ \mathbf{u} \rangle = \langle \mathbf{u} \times \mathbf{v}, \ \mathbf{v} \rangle = 0.$$

But by the very definition of cross product,

$$\langle \mathbf{u} \times \mathbf{v}, \ \mathbf{u} \rangle = (u_2 v_3 - u_3 v_2) u_1 + (u_3 v_1 - u_1 v_3) u_2 + (u_1 v_2 - u_2 v_1) u_3 = 0,$$

and similarly, $\langle \mathbf{u} \times \mathbf{v}, \ \mathbf{v} \rangle = 0$. We will now verify (ii). Suppose that the vectors $\mathbf{u}$ and $\mathbf{v}$ are linearly independent, and let $\mathbf{w}$ be orthogonal to both $\mathbf{u}$ and $\mathbf{v}$. We write out the orthogonality assumptions as the following system of equations:

$$\begin{aligned} u_1 w_1 + u_2 w_2 + u_3 w_3 &= 0 \\ v_1 w_1 + v_2 w_2 + v_3 w_3 &= 0. \end{aligned} \qquad (4)$$

We first eliminate $w_1$ from this system of equations by multiplying the first equation by $v_1$, multiplying the second equation by $-u_1$, and adding the resulting equations. Similarly, we eliminate $w_2$ and then $w_3$. The new system of equations is

$$\begin{aligned} w_2(u_2 v_1 - u_1 v_2) + w_3(u_3 v_1 - u_1 v_3) &= 0 \\ w_1(u_1 v_2 - u_2 v_1) + w_3(u_3 v_2 - u_2 v_3) &= 0 \\ w_1(u_1 v_3 - u_3 v_1) + w_2(u_2 v_3 - u_3 v_2) &= 0. \end{aligned} \qquad (5)$$

Now we have assumed that $\mathbf{u}$ and $\mathbf{v}$ are linearly independent. Thus, by Theorem 7, $\mathbf{u} \times \mathbf{v} \neq \mathbf{0}$, and we will suppose it is the last component of $\mathbf{u} \times \mathbf{v}$ that is nonzero; that is, $u_1 v_2 - u_2 v_1 \neq 0$. Define $\gamma = w_3 / (u_1 v_2 - u_2 v_1)$. Then, by definition, $w_3 = \gamma(u_1 v_2 - u_2 v_1)$. The first equation in (5) gives $w_2 = \gamma(u_3 v_1 - u_1 v_3)$; the second equation in (5) gives $w_1 = \gamma(u_2 v_3 - u_3 v_2)$. Consequently,

$$w_1 = \gamma(u_2 v_3 - u_3 v_2), \quad w_2 = \gamma(u_3 v_1 - u_1 v_3), \quad \text{and} \quad w_3 = \gamma(u_1 v_2 - u_2 v_1);$$

that is, $\mathbf{w} = \gamma(\mathbf{u} \times \mathbf{v})$.    ∎

---

**THEOREM 9**    Let $\mathbf{u}$ and $\mathbf{v}$ be linearly independent vectors in $\mathbb{R}^3$. Then the three vectors $\mathbf{u}$, $\mathbf{v}$, and $\mathbf{u} \times \mathbf{v}$ are a basis for $\mathbb{R}^3$.

**Proof**    Since $\mathbf{u}$ and $\mathbf{v}$ are linearly independent, $\mathbf{u} \neq \mathbf{0}$ and $\mathbf{v} \neq \mathbf{0}$. Define $\lambda = \langle \mathbf{u}, \ \mathbf{v} \rangle / \langle \mathbf{v}, \ \mathbf{v} \rangle$, and then define $\mathbf{u}' = \mathbf{u} - \lambda \mathbf{v}$. Part (i) of Theorem 4 asserts that $\mathbf{u}'$ is orthogonal to $\mathbf{v}$. Moreover, $\mathbf{u}' \neq \mathbf{0}$, since $\mathbf{u}$ and $\mathbf{v}$ are linearly independent.

Choose $\mathbf{p}$ to be a vector in $\mathbb{R}^3$. Define $\alpha' = \langle \mathbf{p}, \ \mathbf{u}' \rangle / \langle \mathbf{u}', \ \mathbf{u}' \rangle$ and $\beta' = \langle \mathbf{p}, \ \mathbf{v} \rangle / \langle \mathbf{v}, \ \mathbf{v} \rangle$. By the linearity of the inner product and the orthogonality of $\mathbf{u}'$ and $\mathbf{v}$,

$$\langle \mathbf{p} - (\alpha' \mathbf{u}' + \beta' \mathbf{v}), \ \mathbf{u}' \rangle = 0 \quad \text{and} \quad \langle \mathbf{p} - (\alpha' \mathbf{u}' + \beta' \mathbf{v}), \ \mathbf{v} \rangle = 0;$$

that is, the vector $\mathbf{p} - (\alpha' \mathbf{u}' + \beta' \mathbf{v})$ is orthogonal to both $\mathbf{u}'$ and $\mathbf{v}$. Since $\mathbf{u}'$ and $\mathbf{v}$ are nonzero and orthogonal, they are linearly independent. By part (ii) of Theorem 8, there is a number $\gamma'$ such that $\mathbf{p} - (\alpha' \mathbf{u}' + \beta' \mathbf{v}) = \gamma'(\mathbf{u}' \times \mathbf{v})$; that is,

$$\mathbf{p} = \alpha' \mathbf{u}' + \beta' \mathbf{v} + \gamma'(\mathbf{u}' \times \mathbf{v}).$$

Substituting $\mathbf{u}' = \mathbf{u} - \lambda \mathbf{v}$ in the above expression, since $\mathbf{v} \times \mathbf{v} = \mathbf{0}$, we can regroup the coefficients to find numbers $\alpha$, $\beta$, and $\gamma$ such that

$$\mathbf{p} = \alpha \mathbf{u} + \beta \mathbf{v} + \gamma(\mathbf{u} \times \mathbf{v}). \qquad (6)$$

It remains to verify that the numbers $\alpha$, $\beta$, and $\gamma$ are unique. We will suppose otherwise and derive a contradiction. Indeed, if there were two distinct triples of numbers for which (6) holds, then, by subtraction, there would be numbers $\alpha$, $\beta$, and $\gamma$, not all equal to 0, such that

$$\mathbf{0} = \alpha\mathbf{u} + \beta\mathbf{v} + \gamma(\mathbf{u} \times \mathbf{v}).$$

Taking the inner product of both sides with $\mathbf{u} \times \mathbf{v}$, we conclude that

$$0 = \gamma\|\mathbf{u} \times \mathbf{v}\|^2.$$

But $\mathbf{u}$ and $\mathbf{v}$ are linearly independent, so by Theorem 7, $\|\mathbf{u} \times \mathbf{v}\| \neq 0$. Thus $\gamma = 0$. But then

$$\mathbf{0} = \alpha\mathbf{u} + \beta\mathbf{v},$$

and either $\alpha$ or $\beta$ is nonzero. This contradicts the linear independence of $\mathbf{u}$ and $\mathbf{v}$.  ∎

With the properties of the cross product that we have so far established, we can now provide a proof of Theorem 1 (and hence also of Theorems 2 and 3) in the case when $n = 3$.

**THEOREM 10**   For the vectors $\mathbf{u}$, $\mathbf{v}$, $\mathbf{w}$ in $\mathbb{R}^3$, the following three assertions are equivalent:

(i)  The vectors $\mathbf{u}$, $\mathbf{v}$, $\mathbf{w}$ are a basis for $\mathbb{R}^3$.

(ii)  The vectors $\mathbf{u}$, $\mathbf{v}$, $\mathbf{w}$ span $\mathbb{R}^3$.

(iii)  The vectors $\mathbf{u}$, $\mathbf{v}$, $\mathbf{w}$ are linearly independent.

**Proof**   By the very definition of basis, (i) implies (ii). Now suppose (ii) holds. We will argue by contradiction to show that (iii) holds; that is, the vectors $\mathbf{u}$, $\mathbf{v}$, $\mathbf{w}$ are linearly independent. Indeed, if $\mathbf{u}$, $\mathbf{v}$, $\mathbf{w}$ are linearly dependent, then one of these vectors is a linear combination of the other two, which implies that just two of these vectors span $\mathbb{R}^3$. Suppose it is $\mathbf{u}$ and $\mathbf{v}$ that span $\mathbb{R}^3$. If $\mathbf{u} \times \mathbf{v} = \mathbf{0}$, it follows from Theorem 7 that $\mathbf{u}$ and $\mathbf{v}$ are linearly dependent, which implies that $\mathbb{R}^3$ is spanned by one of them, say $\mathbf{u}$. But this is impossible, since we can easily find a nonzero vector that is orthogonal to $\mathbf{u}$, so such a vector is certainly not a multiple of $\mathbf{u}$. It follows that $\mathbf{u} \times \mathbf{v} \neq \mathbf{0}$. This too is impossible, since $\mathbf{u} \times \mathbf{v}$ would then be a nonzero vector that is a linear combination of $\mathbf{u}$ and $\mathbf{v}$ and is orthogonal to $\mathbf{u}$ and $\mathbf{v}$. Consequently, $\mathbf{u}$, $\mathbf{v}$, $\mathbf{w}$ are linearly independent.

Now suppose that (iii) holds. Since $\mathbf{u}$, $\mathbf{v}$, $\mathbf{w}$ are linearly independent, the two vectors $\mathbf{u}$ and $\mathbf{v}$ are also linearly independent. By Theorem 9, the three vectors $\mathbf{u}$, $\mathbf{v}$, and $\mathbf{u} \times \mathbf{v}$ are a basis for $\mathbb{R}^3$. So there are real numbers $\alpha$, $\beta$, and $\gamma$ such that

$$\mathbf{w} = \alpha\mathbf{u} + \beta\mathbf{v} + \gamma(\mathbf{u} \times \mathbf{v}):$$

Since $\mathbf{u}$, $\mathbf{v}$, $\mathbf{w}$ are linearly independent, it follows that $\gamma$ is nonzero, so we have

$$\mathbf{u} \times \mathbf{v} = \frac{1}{\gamma}(\mathbf{w} - \alpha\mathbf{u} - \beta\mathbf{v}).$$

Now let $\mathbf{p}$ be a vector in $\mathbb{R}^3$. Since $\mathbf{u}$, $\mathbf{v}$, $\mathbf{u} \times \mathbf{v}$ are a basis for $\mathbb{R}^3$, there are real numbers $\alpha'$, $\beta'$, and $\gamma'$ such that

$$\mathbf{p} = \alpha'\mathbf{u} + \beta'\mathbf{v} + \gamma'(\mathbf{u} \times \mathbf{v}).$$

Then

$$\mathbf{p} = \alpha'\mathbf{u} + \beta'\mathbf{v} + \frac{\gamma'}{\gamma}(\mathbf{w} - \alpha\mathbf{u} - \beta\mathbf{v})$$

$$= \left(\alpha' - \frac{\gamma'}{\gamma}\alpha\right)\mathbf{u} + \left(\beta' - \frac{\gamma'}{\gamma}\beta\right)\mathbf{v} + \frac{\gamma'}{\gamma}\mathbf{w},$$

and hence $\mathbf{p}$ may be written as a linear combination of $\mathbf{u}$, $\mathbf{v}$, and $\mathbf{w}$. Furthermore, this representation is unique, since if

$$\mathbf{p} = \lambda_1\mathbf{u} + \lambda_2\mathbf{v} + \lambda_3\mathbf{w} = \lambda_1'\mathbf{u} + \lambda_2'\mathbf{v} + \lambda_3'\mathbf{w},$$

then we have

$$\mathbf{0} = (\lambda_1 - \lambda_1')\mathbf{u} + (\lambda_2 - \lambda_2')\mathbf{v} + (\lambda_3 - \lambda_3')\mathbf{w},$$

and since $\mathbf{u}$, $\mathbf{v}$, $\mathbf{w}$ are linearly independent, it follows that

$$\lambda_1 = \lambda_1', \quad \lambda_2 = \lambda_2', \quad \lambda_3 = \lambda_3'.$$

Hence the vectors $\mathbf{u}$, $\mathbf{v}$, $\mathbf{w}$ are a basis for $\mathbb{R}^3$.   ∎

It is useful to have a criterion for detecting when a given triple of vectors is a basis for $\mathbb{R}^3$. We will show that the three vectors $\mathbf{u}$, $\mathbf{v}$, and $\mathbf{w}$ in $\mathbb{R}^3$ are a basis for $\mathbb{R}^3$ if and only if

$$\langle \mathbf{u}, \mathbf{v} \times \mathbf{w} \rangle \neq 0.$$

The number $\langle \mathbf{u}, \mathbf{v} \times \mathbf{w} \rangle$ is called the *triple product* of the ordered triple of vectors $\mathbf{u}$, $\mathbf{v}$, $\mathbf{w}$. The dependence of the triple product on the order of the three vectors is described by the following proposition.

**PROPOSITION
11**

For vectors $\mathbf{u}$, $\mathbf{v}$, and $\mathbf{w}$ in $\mathbb{R}^3$,

$$\langle \mathbf{u}, \mathbf{v} \times \mathbf{w} \rangle = \langle \mathbf{w}, \mathbf{u} \times \mathbf{v} \rangle = \langle \mathbf{v}, \mathbf{w} \times \mathbf{u} \rangle$$

$$\langle \mathbf{u}, \mathbf{v} \times \mathbf{w} \rangle = -\langle \mathbf{v}, \mathbf{u} \times \mathbf{w} \rangle = -\langle \mathbf{w}, \mathbf{v} \times \mathbf{u} \rangle. \tag{7}$$

**Proof**    The proof is by inspection. Indeed, by definition of the inner product and the cross product,

$$\langle \mathbf{u}, \mathbf{v} \times \mathbf{w} \rangle = u_1(v_2w_3 - v_3w_2) + u_2(v_3w_1 - v_1w_3) + u_3(v_1w_2 - v_2w_1)$$

and

$$\langle \mathbf{w}, \mathbf{u} \times \mathbf{v} \rangle = w_1(u_2v_3 - u_3v_2) + w_2(u_3v_1 - u_1v_3) + w_3(u_1v_2 - u_2v_1).$$

Observe that the right-hand sides of the two above identities are equal. Similarly, it follows that $\langle \mathbf{w}, \mathbf{u} \times \mathbf{v} \rangle = \langle \mathbf{v}, \mathbf{w} \times \mathbf{u} \rangle$. Thus the first line of (7) is verified. From the anti-symmetry of the cross product and the inequalities on the first line of (7), the equalities on the second line of (7) follow; that is, the triple product of an ordered triple of vectors changes sign when two of the vectors are interchanged.   ∎

**THEOREM 12**   For vectors $\mathbf{u}$, $\mathbf{v}$, and $\mathbf{w}$ in $\mathbb{R}^3$, the following two assertions are equivalent:

(i) $\langle \mathbf{u}, \mathbf{v} \times \mathbf{w} \rangle \neq 0$.

(ii) The three vectors $\mathbf{u}$, $\mathbf{v}$, and $\mathbf{w}$ form a basis for $\mathbb{R}^3$.

Moreover, when either (and hence both) of these assertions holds, each vector $\mathbf{p}$ in $\mathbb{R}^3$ may be expressed as

$$\mathbf{p} = \alpha \mathbf{u} + \beta \mathbf{v} + \gamma \mathbf{w},$$

where the coefficients $\alpha$, $\beta$, and $\gamma$ are given by the formulas

$$\alpha = \frac{\langle \mathbf{p}, \mathbf{v} \times \mathbf{w} \rangle}{\langle \mathbf{u}, \mathbf{v} \times \mathbf{w} \rangle} \quad \beta = \frac{\langle \mathbf{u}, \mathbf{p} \times \mathbf{w} \rangle}{\langle \mathbf{u}, \mathbf{v} \times \mathbf{w} \rangle} \quad \gamma = \frac{\langle \mathbf{u}, \mathbf{v} \times \mathbf{p} \rangle}{\langle \mathbf{u}, \mathbf{v} \times \mathbf{w} \rangle}. \tag{8}$$

**Proof**   First, suppose that (i) holds. Then $\langle \mathbf{u}, \mathbf{v} \times \mathbf{w} \rangle = \langle \mathbf{w}, \mathbf{u} \times \mathbf{v} \rangle \neq 0$, so $\mathbf{u} \times \mathbf{v} \neq \mathbf{0}$. By Theorem 9, the vectors $\mathbf{u}$, $\mathbf{v}$, and $\mathbf{u} \times \mathbf{v}$ are a basis for $\mathbb{R}^3$. Thus there are numbers $\alpha$, $\beta$, and $\gamma$ such that

$$\mathbf{w} = \alpha \mathbf{u} + \beta \mathbf{v} + \gamma (\mathbf{u} \times \mathbf{v}). \tag{9}$$

Taking the inner product of each side with $\mathbf{u} \times \mathbf{v}$, we conclude that

$$\langle \mathbf{w}, \mathbf{u} \times \mathbf{v} \rangle = \gamma \|\mathbf{u} \times \mathbf{v}\|^2.$$

Thus $\gamma \neq 0$, since $\langle \mathbf{u}, \mathbf{v} \times \mathbf{w} \rangle \neq 0$. Divide equation (9) by $\gamma$. We see that $\mathbf{u} \times \mathbf{v}$ is a linear combination of $\mathbf{u}$, $\mathbf{v}$, and $\mathbf{w}$. Consequently, since every vector in $\mathbb{R}^3$ can be expressed as a linear combination of $\mathbf{u}$, $\mathbf{v}$, and $\mathbf{u} \times \mathbf{v}$, every vector in $\mathbb{R}^3$ can be expressed as a linear combination of $\mathbf{u}$, $\mathbf{v}$, and $\mathbf{w}$. It is necessary to show that each vector in $\mathbb{R}^3$ can be expressed *uniquely* as a linear combination of $\mathbf{u}$, $\mathbf{v}$, and $\mathbf{w}$. Let $\mathbf{p}$ be a point in $\mathbb{R}^3$ and suppose $\alpha$, $\beta$, and $\gamma$ are numbers such that

$$\mathbf{p} = \alpha \mathbf{u} + \beta \mathbf{v} + \gamma \mathbf{w}. \tag{10}$$

Then taking the inner product of each side of (10), first with $\mathbf{v} \times \mathbf{w}$, then with $\mathbf{w} \times \mathbf{u}$, and finally with $\mathbf{u} \times \mathbf{v}$, and using the fact that the triple product is zero if two vectors in the triple are equal, we have

$$\langle \mathbf{p}, \mathbf{v} \times \mathbf{w} \rangle = \alpha \langle \mathbf{u}, \mathbf{v} \times \mathbf{w} \rangle$$
$$\langle \mathbf{p}, \mathbf{w} \times \mathbf{u} \rangle = \beta \langle \mathbf{v}, \mathbf{w} \times \mathbf{u} \rangle$$
$$\langle \mathbf{p}, \mathbf{u} \times \mathbf{v} \rangle = \gamma \langle \mathbf{w}, \mathbf{u} \times \mathbf{v} \rangle.$$

From this, using the reordering property (7) of the triple product, we obtain (8). Thus we have proved that $\alpha$, $\beta$, and $\gamma$ are unique and, at the same time, have established formula (8) for the coefficients $\alpha$, $\beta$, and $\gamma$.

It remains to prove that (ii) implies (i). Indeed, suppose that $\mathbf{u}$, $\mathbf{v}$, and $\mathbf{w}$ are a basis for $\mathbb{R}^3$. Observe that the vectors $\mathbf{u}$ and $\mathbf{v}$ are linearly independent, since otherwise there would be two distinct ways of expressing the zero vector as a linear combination of $\mathbf{u}$, $\mathbf{v}$, and $\mathbf{w}$. Hence, by Theorem 7, $\mathbf{u} \times \mathbf{v} \neq \mathbf{0}$. Furthermore, by Theorem 9, the vectors $\mathbf{u}$, $\mathbf{v}$, and $\mathbf{u} \times \mathbf{v}$ are a basis for $\mathbb{R}^3$. In particular, there are numbers $\alpha$, $\beta$, and $\gamma$ such that

$$\mathbf{w} = \alpha \mathbf{u} + \beta \mathbf{v} + \gamma (\mathbf{u} \times \mathbf{v}). \tag{11}$$

The component $\gamma \neq 0$, since $\mathbf{u}$, $\mathbf{v}$, and $\mathbf{w}$ are a basis for $\mathbb{R}^3$. Taking the inner product of each side of (11) with $\mathbf{u} \times \mathbf{v}$, we have

$$\langle \mathbf{u}, \mathbf{v} \times \mathbf{w} \rangle = \langle \mathbf{w}, \mathbf{u} \times \mathbf{v} \rangle = \gamma \|\mathbf{u} \times \mathbf{v}\|^2 \neq 0. \qquad \blacksquare$$

Theorem 12 has an interesting interpretation in terms of $3 \times 3$ systems of linear equations. For a $3 \times 3$ matrix $\mathbf{A} = [a_{ij}]$, consider the $3 \times 3$ systems of linear equations

$$a_{11}x_1 + a_{12}x_2 + a_{13}x_3 = y_1$$
$$a_{21}x_1 + a_{22}x_2 + a_{23}x_3 = y_2 \qquad (12)$$
$$a_{31}x_1 + a_{32}x_2 + a_{33}x_3 = y_3,$$

where the triple of numbers $(y_1, y_2, y_3)$ is given and we seek a triple of numbers $(x_1, x_2, x_3)$ for which the above system of equations is satisfied. We define the three vectors $\mathbf{u}$, $\mathbf{v}$, and $\mathbf{w}$ by

$$\mathbf{u} = (a_{11}, a_{21}, a_{31}), \quad \mathbf{v} = (a_{12}, a_{22}, a_{32}), \quad \mathbf{w} = (a_{13}, a_{23}, a_{33})$$

and observe that for a given vector $\mathbf{y} = (y_1, y_2, y_3)$, the linear system of equations is equivalent to

$$x_1\mathbf{u} + x_2\mathbf{v} + x_3\mathbf{w} = \mathbf{y}. \qquad (13)$$

Now, directly from the definition of what it means for a triple of vectors to be a basis, we see that the assertion that the three vectors $\mathbf{u}$, $\mathbf{v}$, and $\mathbf{w}$ are a basis for $\mathbb{R}^3$ is equivalent to the assertion that for every triple of numbers $(y_1, y_2, y_3)$, there is a unique solution $(x_1, x_2, x_3)$ of the system of equations (12). However, Theorem 12 provides a necessary and sufficient condition for three vectors $\mathbf{u}$, $\mathbf{v}$, and $\mathbf{w}$ to be a basis for $\mathbb{R}^3$, namely that $\langle \mathbf{u}, \mathbf{v} \times \mathbf{w} \rangle \neq 0$. Since $\mathbf{u}$, $\mathbf{v}$, and $\mathbf{w}$ correspond to the first, second, and third columns, respectively, of the matrix $\mathbf{A}$, we are led to define the *determinant* of a $3 \times 3$ matrix of real numbers as follows:

**DEFINITION**    *For a $3 \times 3$ matrix*

$$\mathbf{A} = \begin{pmatrix} a_{11} & a_{12} & a_{13} \\ a_{21} & a_{22} & a_{23} \\ a_{31} & a_{32} & a_{33} \end{pmatrix},$$

*we define the determinant of $\mathbf{A}$, which is denoted by* $\det \mathbf{A}$, *by the formula*

$$\det \mathbf{A} \equiv a_{11}(a_{22}a_{33} - a_{32}a_{23}) + a_{21}(a_{32}a_{13} - a_{12}a_{33}) + a_{31}(a_{12}a_{23} - a_{22}a_{13}).$$

**THEOREM 13**    For the system of linear equations (12) that is determined by the $3 \times 3$ matrix $\mathbf{A}$, the following two assertions are equivalent:

(i) $\det \mathbf{A} \neq 0$.

(ii) For each triple of numbers $(y_1, y_2, y_3)$, there is a unique solution $(x_1, x_2, x_3)$ of the system.

Moreover, when either (and hence both) of these assertions is true, for a given triple $(y_1, y_2, y_3)$, the unique solution $(x_1, x_2, x_3)$ of the system of equations (12) is given by

$$x_1 = \frac{1}{D} \det \begin{pmatrix} y_1 & a_{12} & a_{13} \\ y_2 & a_{22} & a_{23} \\ y_3 & a_{32} & a_{33} \end{pmatrix} \quad x_2 = \frac{1}{D} \det \begin{pmatrix} a_{11} & y_1 & a_{13} \\ a_{21} & y_2 & a_{23} \\ a_{31} & y_3 & a_{33} \end{pmatrix} \quad x_3 = \frac{1}{D} \det \begin{pmatrix} a_{11} & a_{12} & y_1 \\ a_{21} & a_{22} & y_2 \\ a_{31} & a_{32} & y_3 \end{pmatrix}$$

where $D = \det \mathbf{A}$.

**Proof**    As above, define the three vectors $\mathbf{u}$, $\mathbf{v}$, and $\mathbf{w}$ by

$$\mathbf{u} = (a_{11}, a_{21}, a_{31}), \quad \mathbf{v} = (a_{12}, a_{22}, a_{32}), \quad \mathbf{w} = (a_{13}\, a_{23}\, a_{33}),$$

and for a given vector $\mathbf{y} = (y_1, y_2, y_3)$, observe that the linear system of equations is equivalent to

$$x_1\mathbf{u} + x_2\mathbf{v} + x_3\mathbf{w} = \mathbf{y}.$$

By definition, $\det \mathbf{A} = \langle \mathbf{u}, \mathbf{v} \times \mathbf{w} \rangle$. The theorem now follows from Theorem 12.    ∎

From the anti-symmetry of the cross product, it follows that the determinant of a $3 \times 3$ matrix changes sign if two columns are interchanged; from the linearity of the inner product and the cross product, it follows that if two columns are fixed, then the determinant depends linearly on the remaining column.

Now that we have understood the significance of the triple product being *nonzero*, we will turn to a description of the significance, with respect to certain volume calculations, of the *magnitude* of the triple product. For a point $\mathbf{p}$ in $\mathbb{R}^3$ and two linearly independent vectors $\mathbf{u}$ and $\mathbf{v}$ in $\mathbb{R}^3$, the *parallelogram* based at $\mathbf{p}$ and bounded by the vectors $\mathbf{u}$ and $\mathbf{v}$ is defined to be the set $\mathcal{S} = \{\mathbf{p} + \alpha\mathbf{u} + \beta\mathbf{v} \mid 0 \le \alpha \le 1, 0 \le \beta \le 1\}$; the *area* of $\mathcal{S}$ is defined to be the length of the vector $\mathbf{v}$ times the distance from the point $\mathbf{p} + \mathbf{u}$ to the line through the point $\mathbf{p}$ that is parallel to the vector $\mathbf{v}$. It follows immediately from Theorems 4 and 6 that

$$\text{area } \mathcal{S} = \|\mathbf{u} \times \mathbf{v}\|. \tag{14}$$

For a point $\mathbf{p}$ in $\mathbb{R}^3$ and a nonzero vector $\boldsymbol{\eta}$ in $\mathbb{R}^3$, the *plane* through $\mathbf{p}$ that is normal to $\boldsymbol{\eta}$ is defined to be the set $\mathcal{P}$ of points $\mathbf{p} + \mathbf{w}$ such that the vector $\mathbf{w}$ is orthogonal to $\boldsymbol{\eta}$. Given two points $\mathbf{p} + \mathbf{u}$ and $\mathbf{p} + \mathbf{v}$ in this plane such that the vectors $\mathbf{u}$ and $\mathbf{v}$ are linearly independent, it follows from part (ii) of Theorem 8 that there is a scalar $\gamma$ such that $\boldsymbol{\eta} = \gamma(\mathbf{u} \times \mathbf{v})$. Also, since the triple of vectors $\mathbf{u}$, $\mathbf{v}$, $\mathbf{u} \times \mathbf{v}$ is a basis for $\mathbb{R}^3$, it follows that $\mathcal{P} = \{\mathbf{p} + \alpha\mathbf{u} + \beta\mathbf{v} \mid \alpha \text{ in } \mathbb{R}, \beta \text{ in } \mathbb{R}\}$. The distance from a point $\mathbf{q}$ to a plane $\mathcal{P}$ is defined to be the distance from $\mathbf{q}$ to the point in $\mathcal{P}$ that is closest to $\mathbf{q}$.

**PROPOSITION 14**    For a vector $\boldsymbol{\eta}$ in $\mathbb{R}^3$ of length 1, let $\mathcal{P}$ be the plane through the origin that is normal to $\boldsymbol{\eta}$. Then for any point $\mathbf{p}$ in $\mathbb{R}^3$, the distance from $\mathbf{p}$ to the plane $\mathcal{P}$ is equal to $|\langle \mathbf{p}, \boldsymbol{\eta} \rangle|$.

**Proof**    Since, by assumption, $\langle \boldsymbol{\eta}, \boldsymbol{\eta} \rangle = 1$, by the linearity of the inner product,

$$\langle \mathbf{p} - \langle \mathbf{p}, \boldsymbol{\eta} \rangle \boldsymbol{\eta}, \boldsymbol{\eta} \rangle = \langle \mathbf{p}, \boldsymbol{\eta} \rangle - \langle \mathbf{p}, \boldsymbol{\eta} \rangle \langle \boldsymbol{\eta}, \boldsymbol{\eta} \rangle = 0;$$

that is, the vector $\mathbf{p} - \langle \mathbf{p}, \boldsymbol{\eta} \rangle \boldsymbol{\eta}$ is orthogonal to $\boldsymbol{\eta}$ and hence the point $\mathbf{p} - \langle \mathbf{p}, \boldsymbol{\eta} \rangle \boldsymbol{\eta}$ lies in $\mathcal{P}$. The distance between the point $\mathbf{p}$ and $\mathbf{p} - \langle \mathbf{p}, \boldsymbol{\eta} \rangle \boldsymbol{\eta}$ is

$$\| \langle \mathbf{p}, \boldsymbol{\eta} \rangle \boldsymbol{\eta} \| = | \langle \mathbf{p}, \boldsymbol{\eta} \rangle | \| \boldsymbol{\eta} \| = | \langle \mathbf{p}, \boldsymbol{\eta} \rangle |.$$

Thus, to prove the proposition, we must show that

$$\| \mathbf{p} - \mathbf{u} \| \geq | \langle \mathbf{p}, \boldsymbol{\eta} \rangle | \quad \text{for every point } \mathbf{u} \text{ in } \mathcal{P}.$$

However, for a point $\mathbf{u}$ in $\mathcal{P}$, write $\mathbf{p} - \mathbf{u} = (\mathbf{p} - \langle \mathbf{p}, \boldsymbol{\eta} \rangle \boldsymbol{\eta} - \mathbf{u}) + \langle \mathbf{p}, \boldsymbol{\eta} \rangle \boldsymbol{\eta}$. Since the vectors $\mathbf{p} - \langle \mathbf{p}, \boldsymbol{\eta} \rangle \boldsymbol{\eta} - \mathbf{u}$ and $\langle \mathbf{p}, \boldsymbol{\eta} \rangle \boldsymbol{\eta}$ are orthogonal, it follows from the Pythagorean Identity that

$$\| \mathbf{p} - \mathbf{u} \|^2 = \| \mathbf{p} - \langle \mathbf{p}, \boldsymbol{\eta} \rangle \boldsymbol{\eta} - \mathbf{u} \|^2 + \| \langle \mathbf{p}, \boldsymbol{\eta} \rangle \boldsymbol{\eta} \|^2 \geq \| \langle \mathbf{p}, \boldsymbol{\eta} \rangle \boldsymbol{\eta} \|^2 = | \langle \mathbf{p}, \boldsymbol{\eta} \rangle |^2,$$

so $\| \mathbf{p} - \mathbf{u} \| \geq | \langle \mathbf{p}, \boldsymbol{\eta} \rangle |$.    ∎

Now suppose that the triple of vectors $\mathbf{u}$, $\mathbf{v}$, and $\mathbf{w}$ are a basis for $\mathbb{R}^3$. The *parallelepiped* based at the point $\mathbf{p}$ and bounded by the vectors $\mathbf{u}$, $\mathbf{v}$, and $\mathbf{w}$ is defined to be the set $\mathcal{V} = \{\mathbf{p} + \alpha \mathbf{u} + \beta \mathbf{v} + \gamma \mathbf{w} \mid 0 \leq \alpha \leq 1, 0 \leq \beta \leq 1, 0 \leq \gamma \leq 1\}$; the *volume* of $\mathcal{V}$ is defined to be the area of the parallelogram based at $\mathbf{p}$ and bounded by the vectors $\mathbf{u}$ and $\mathbf{v}$ times the distance from the point $\mathbf{p} + \mathbf{w}$ to the plane containing $\mathbf{p}$, $\mathbf{p} + \mathbf{u}$, and $\mathbf{p} + \mathbf{v}$. Observe that the plane containing $\mathbf{p}$, $\mathbf{p} + \mathbf{u}$, and $\mathbf{p} + \mathbf{v}$ is the plane through the point $\mathbf{p}$ that is normal to the vector $\mathbf{u} \times \mathbf{v}$.

**THEOREM 15**    Let the vectors $\mathbf{u}$, $\mathbf{v}$, and $\mathbf{w}$ be a basis for $\mathbb{R}^3$ and let $\mathcal{V}$ be the parallelepiped based at the origin and bounded by the vectors $\mathbf{u}$, $\mathbf{v}$, and $\mathbf{w}$. Then the volume of $\mathcal{V}$ is given by the formula

$$\text{vol } \mathcal{V} = | \langle \mathbf{u}, \mathbf{v} \times \mathbf{w} \rangle | \qquad (15)$$

**Proof**    The vector $\mathbf{u} \times \mathbf{v}$ is a normal to the plane containing the origin, $\mathbf{u}$, and $\mathbf{v}$. Thus $\boldsymbol{\eta} = \mathbf{u} \times \mathbf{v} / \| \mathbf{u} \times \mathbf{v} \|$ is a normal to the plane that has length 1. Observe that

$$\langle \mathbf{w}, \boldsymbol{\eta} \rangle = \left\langle \mathbf{w}, \frac{\mathbf{u} \times \mathbf{v}}{\| \mathbf{u} \times \mathbf{v} \|} \right\rangle = \frac{\langle \mathbf{w}, \mathbf{u} \times \mathbf{v} \rangle}{\| \mathbf{u} \times \mathbf{v} \|},$$

so that by Proposition 14, the distance from the point $\mathbf{w}$ to the plane $\mathcal{P}$ equals

$$| \langle \mathbf{w}, \boldsymbol{\eta} \rangle | = \frac{| \langle \mathbf{w}, \mathbf{u} \times \mathbf{v} \rangle |}{\| \mathbf{u} \times \mathbf{v} \|}.$$

On the other hand, by Theorem 6, the area of the parallelogram based at the origin and bounded by the vectors $\mathbf{u}$ and $\mathbf{v}$ equals $\| \mathbf{u} \times \mathbf{v} \|$. Thus, by the very definition of volume,

$$\text{vol } \mathcal{V} = \frac{| \langle \mathbf{w}, \mathbf{u} \times \mathbf{v} \rangle |}{\| \mathbf{u} \times \mathbf{v} \|} \cdot \| \mathbf{u} \times \mathbf{v} \| = | \langle \mathbf{w}, \mathbf{u} \times \mathbf{v} \rangle | = | \langle \mathbf{u}, \mathbf{v} \times \mathbf{w} \rangle |.    \blacksquare$$

**COROLLARY 16**   Let $\mathcal{V}$ be the parallelepiped based at the origin that is spanned by the standard basis vectors $\mathbf{e}_1$, $\mathbf{e}_2$, $\mathbf{e}_3$. Then for an invertible linear mapping $\mathbf{T}: \mathbb{R}^3 \to \mathbb{R}^3$ that is associated with the $3 \times 3$ matrix $\mathbf{A}$, the volume of the image $\mathbf{T}(\mathcal{V})$ is given by the formula

$$\text{vol } \mathbf{T}(\mathcal{V}) = |\det \mathbf{A}| \, \text{vol } \mathcal{V}. \tag{16}$$

**Proof**   Define $\mathbf{u} = \mathbf{T}(\mathbf{e}_1)$, $\mathbf{v} = \mathbf{T}(\mathbf{e}_2)$, and $\mathbf{w} = \mathbf{T}(\mathbf{e}_3)$. Then if the point $(\alpha, \beta, \gamma)$ is in $\mathcal{V}$,

$$\mathbf{T}(\alpha, \beta, \gamma) = \alpha \mathbf{u} + \beta \mathbf{v} + \gamma \mathbf{w}.$$

Hence $\mathbf{T}(\mathcal{V})$ is the parallelepiped bounded by the vectors $\mathbf{u}$, $\mathbf{v}$, and $\mathbf{w}$. Thus

$$\text{vol } \mathbf{T}(\mathcal{V}) = |\langle \mathbf{u}, \mathbf{v} \times \mathbf{w} \rangle|. \tag{17}$$

On the other hand, by the very way in which the matrix $\mathbf{A}$ is associated with the linear mapping $\mathbf{T}: \mathbb{R}^3 \to \mathbb{R}^3$, it follows that the vectors $\mathbf{u}$, $\mathbf{v}$, and $\mathbf{w}$ correspond to the first, second, and third columns of the matrix $\mathbf{A}$. Consequently, by the definition of determinant,

$$\det \mathbf{A} = \langle \mathbf{u}, \mathbf{v} \times \mathbf{w} \rangle. \tag{18}$$

The volume formula (16) follows from the two preceding equalities, since it is clear that $\text{vol } \mathcal{V} = 1$. ∎

The triple product $\langle \mathbf{u}, \mathbf{v} \times \mathbf{w} \rangle$ of the ordered triple of vectors $\mathbf{u}$, $\mathbf{v}$, $\mathbf{w}$ is nonzero if and only if these vectors are a basis for $\mathbb{R}^3$, and when the triple product is nonzero, Theorem 15 describes the significance of the absolute value of the triple product. It is natural to inquire as to the significance of the *sign* of the triple product. The ordered triple of vectors $\mathbf{u}$, $\mathbf{v}$, $\mathbf{w}$ is defined to be *positively oriented* provided that $\langle \mathbf{u}, \mathbf{v} \times \mathbf{w} \rangle > 0$. Observe that the standard ordered basis $\mathbf{i}$, $\mathbf{j}$, $\mathbf{k}$ is positively oriented, since $\langle \mathbf{i}, \mathbf{j} \times \mathbf{k} \rangle = 1$. Moreover, the geometric significance of being positively oriented is that if the ordered triple of vectors $\mathbf{u}$, $\mathbf{v}$, $\mathbf{w}$ is positively oriented, then this basis can be continuously deformed into the standard ordered basis, in the following precise sense: There are three parametrized paths $\alpha: [0, 1] \to \mathbb{R}^3$, $\beta: [0, 1] \to \mathbb{R}^3$, and $\gamma: [0, 1] \to \mathbb{R}^3$ such that

$$\alpha(0) = \mathbf{u}, \quad \beta(0) = \mathbf{v}, \quad \gamma(0) = \mathbf{w}$$

$$\alpha(1) = \mathbf{i}, \quad \beta(1) = \mathbf{j}, \quad \gamma(1) = \mathbf{k},$$

and for each parameter value $t$ in $[0,1]$, the vectors $\alpha(t)$, $\beta(t)$, $\gamma(t)$ are a basis for $\mathbb{R}^3$.

Observe that if the vectors $\mathbf{u}$ and $\mathbf{v}$ are linearly independent, then the ordered triple of vectors $\mathbf{u}$, $\mathbf{v}$, $\mathbf{u} \times \mathbf{v}$ is positively oriented, since

$$\langle \mathbf{u}, \mathbf{v} \times (\mathbf{u} \times \mathbf{v}) \rangle = \langle \mathbf{u} \times \mathbf{v}, \mathbf{u} \times \mathbf{v} \rangle > 0.$$

The fact that the ordered basis $\mathbf{u}$, $\mathbf{v}$, $\mathbf{u} \times \mathbf{v}$ is positively oriented is what is informally described in elementary courses as the "right-hand rule."

The determinant can be defined for any $n \times n$ matrix; it has the same algebraic significance as it does in the $3 \times 3$ case. In Section 15.1, we provide a definition by induction. By inspection, we see that in the case of $3 \times 3$ matrices, this definition is consistent with the definition given in this appendix. A careful treatment of linear algebra, including a full discussion of determinants, may be found in Curtis, *Linear Algebra, an Introductory Approach*, Springer-Verlag, 1984.

---

1. Find the equation of the line $\ell$ through the origin in $\mathbb{R}^3$ that is parallel to the vector $(1, 0, 2)$. Find the distance from the point $(0, 2, 4)$ to this line.

2. Show that the set $\{\mathbf{u} = (x, y, z) \mid 2x + 3y - z = 0\}$ is the plane through the origin that is normal to $\boldsymbol{\eta} = (2, 3, -1)$. Find the distance from the point $(1, 1, 0)$ to this plane.

3. Show that the triple of vectors $\mathbf{u} = (1, 0, 1/2)$, $\mathbf{v} = (0, 2, 1)$, and $\mathbf{w} = (-4, 0, 0)$ is a basis for $\mathbb{R}^3$. Use formula (8) to write the point $(1, 0, 0)$ in this basis and also to write the point $(0, 1, 0)$ in this basis.

4. For a nonzero vector $\mathbf{u}$ in $\mathbb{R}^3$, find a nonzero vector $\mathbf{v}$ that is orthogonal to $\mathbf{u}$, and then find a nonzero vector $\mathbf{w}$ such that the triple $\mathbf{u}$, $\mathbf{v}$, $\mathbf{w}$ is a basis for $\mathbb{R}^3$.

5. Find the area of the parallelogram based at the origin and bounded by the vectors $(1, 0, 2)$ and $(0, 0, 1)$.

6. For two points $(x_1, y_1)$ and $(x_2, y_2)$ in the plane $\mathbb{R}^2$, define the *complex product*, which is denoted by $(x_1, y_1)(x_2, y_2)$, by the formula

$$(x_1, y_1)(x_2, y_2) \equiv (x_1 x_2 - y_1 y_2, \; x_1 y_2 + x_2 y_1).$$

Define the sum of two points to be the usual sum.

   (a) Show that $(1, 0)(x, y) = (x, y)$ for every point $(x, y)$ in $\mathbb{R}^2$; that is, the point $(1, 0)$ is the multiplicative identity.

   (b) Show that $(0, 0) + (x, y) = (x, y)$ for every point $(x, y)$ in $\mathbb{R}^2$; that is, the point $(0, 0)$ is the additive identity.

   (c) Finally, show that with the usual definition of sum and with the product being the complex product, the Field Axioms are satisfied.

7. Express the two points $(x_1, y_1)$ and $(x_2, y_2)$ in the plane $\mathbb{R}^2$ in polar coordinates as $(x_1, y_1) = (r_1 \cos \theta_1, r_1 \sin \theta_1)$ and $(x_2, y_2) = (r_2 \cos \theta_2, r_2 \sin \theta_2)$. Use the cosine and sine addition formulas to show that the complex product defined in the preceding exercise may be written as

$$(x_1, y_1)(x_2, y_2) = (r_1 r_2 \cos(\theta_1 + \theta_2), \; r_1 r_2 \sin(\theta_1 + \theta_2)).$$

   (a) Use this formula to provide a geometric interpretation of the complex product.

   (b) Find a geometric interpretation of the complex multiplicative inverse of a point $(x_1, y_1) \neq (0, 0)$ in $\mathbb{R}^2$.

8. Let the nonzero vectors $\mathbf{u}$ and $\mathbf{v}$ in $\mathbb{R}^3$ be orthogonal. Show that $\mathbf{u}$ and $\mathbf{v}$ are linearly independent.

9. Show that there is no vector $\mathbf{u}$ that has the property that $\mathbf{u} \times \mathbf{v} = \mathbf{v}$ for every vector $\mathbf{v}$ in $\mathbb{R}^3$.

10. Given vectors $\mathbf{u}$ and $\mathbf{v}$ in $\mathbb{R}^3$, under what conditions is there a vector $\mathbf{w}$ such that $\mathbf{u} \times \mathbf{w} = \mathbf{v}$?

11. Suppose that the triple of vectors $\mathbf{u}$, $\mathbf{v}$, $\mathbf{w}$ is a basis for $\mathbb{R}^3$. For any pair of numbers $\alpha$ and $\beta$, show that the parallelepiped based at the origin and bounded by $\mathbf{u}$, $\mathbf{v}$, and $\mathbf{w}$ has the same volume as the parallelepiped based at the origin and bounded by $\mathbf{u}$, $\mathbf{v}$, and $\alpha \mathbf{u} + \beta \mathbf{v} + \mathbf{w}$. Interpret this result geometrically.

12. Show that the system of equations

$$a_{11} x_1 + a_{12} x_2 = y_1$$

$$a_{21} x_1 + a_{22} x_2 = y_2$$

has a unique solution $(x_1, x_2)$ for each pair of numbers $(y_1, y_2)$ if and only if $a_{11}a_{22} - a_{12}a_{21} \neq 0$. *Hint*: Show that the above system of equations is equivalent to the following system:

$$a_{11}x_1 + a_{12}x_2 + 0x_3 = y_1$$
$$a_{21}x_1 + a_{22}x_2 + 0x_3 = y_2$$
$$0x_1 + 0x_2 + 1x_3 = y_3.$$

# Answers to Selected Problems

## SECTION 1.1

2. Argue by contradiction: If the set $A$ contains two points $u$ and $v$ with $u < v$, then

$$\inf A \leq u < v \leq \sup A.$$

3. Note that $x^2 - c = (x + \sqrt{c})(x - \sqrt{c})$. Thus, if $0 = x^2 - c$, then $x = \sqrt{c}$ or $x = -\sqrt{c}$.

9. Hint: Show that if $0 \leq x \leq 1$, then $x$ is in $S$, and if $x > 1$, then $x$ is not in $S$.

## SECTION 1.2

3. (a) 0 is a lower bound and the Archimedean Property asserts it is the greatest lower bound; the set has no minimum. The maximum is 1.

   (b) $\sqrt{2}$ is the least upper bound, and $-\sqrt{2}$ is the greatest lower bound. The set has no maximum and no minimum.

5. Hint: Use the density of the rationals.

7. If $a < 0 < b$, then, by the case already considered, there is a rational number and an irrational number in the interval $(0, b)$, and the interval $(0, b)$ is contained in $(a, b)$. If $a < b \leq 0$, consider the interval $(-b, -a)$ and reflect.

## SECTION 1.3

2. Hint: The Reverse Triangle Inequality implies that $|a| - |b| \leq |a - b|$.

8. Hint: Use the Difference of Powers Formula, and note that

$$a^{n-1-k}b^k \geq b^{n-1} \text{ for } 0 \leq k \leq n - 1.$$

10. (b) Hint: Use Cauchy's Inequality from Exercise 9 and the identity $abc = \sqrt{ab}\sqrt{bc}\sqrt{ca}$.

13. Use mathematical induction for both. To prove the second inequality, first show that if $a$ and $b$ are positive, them $a/b + b/a \geq 2$.

14. (a) Substitute $r = 1/(1 + x^2)$ in the first formula.

(b) Use the first formula with $n = 2$ and $r = 1 - a$.

17. Hint: Use mathematical induction, and make use of Exercise 15.

## SECTION 2.1

1. (a) Hint: Let $\epsilon > 0$. Since $\epsilon^2 > 0$, by the Archimedean Property there is a natural number $N$ such that $N > \epsilon^2$, so that $1/\sqrt{n} < \epsilon$ if $n \geq N$.

   (b) Hint: $1/(n + 5) \leq 1/n$.

6. Hint: If $a > 0$, then

$$|\sqrt{a_n} - \sqrt{a}| = |a_n - a|/(\sqrt{a_n} + \sqrt{a}\,) \leq (1/\sqrt{a})|a_n - a| \text{ for all } n.$$

   If $a = 0$, then

$$|\sqrt{a_n} - \sqrt{0}| = \sqrt{a_n} \leq a_n \text{ if } a_n \leq 1.$$

7. No: Consider the sequence $\{(-1)^n\}$.

9. Hint: Show that each $\alpha_n \geq 0$, and use the Binomial Formula as follows:

$$(1 + \alpha_n)^n = \binom{n}{0} 1^n \alpha_n^0 + \binom{n}{1} 1^{n-1} \alpha_n^1 + \binom{n}{2} 1^{n-2} \alpha_n^2 + \cdots \geq \binom{n}{0} 1^n \alpha_n^0 + \binom{n}{2} 1^{n-2} \alpha_n^2.$$

14. Hint: Use the identity $\sqrt{n+1} - \sqrt{n} = 1/(\sqrt{n+1} + \sqrt{n})$.

17. Hint: First show that, since $|a| < (1 + |a|)/2$, a natural number $N$ can be chosen such that

$$|a_n| \leq (1 + |a|)/2 < 1 \text{ for } n \geq N.$$

18. (a) For $\epsilon = 1, a = 1$, and $\{a_n\} = \{1/n\}$, this criterion holds; but $\lim_{n \to \infty} 1/n \neq 1$.

    (b) Show that this criterion holds if and only if $a_n = a$ for all natural numbers $n$.

    (c) Show that this criterion holds if and only if there is a natural number $N$ such that $a_n = a$ for $n \geq N$.

19. Hint: Use mathematical induction.

## SECTION 2.2

1. (a) monotone (b) not monotone (c) not monotone.

3. Hint: Show that for each natural number $n$,

$$1 + \frac{1}{2!} + \cdots + \frac{1}{n!} \leq 1 + \frac{1}{2} + \cdots + \frac{1}{2^n}.$$

   Now use the Geometric Sum Formula and the Monotone Convergence Theorem.

5. Hint: Use the Geometric Sum Formula to estimate the terms.

8. Hint: Consider the cases $x_1 = x_2, x_1 < x_2$, and $x_1 > x_2$ in order to verify monotonicity of $\{x_n\}$. Then show boundedness and apply the Monotone Convergence Theorem.

12. (b) The set $[0, 1)$ is not compact since $\{1 - 1/n\}$ is a sequence in $[0, 1)$ such that it, and hence all of its subsequences, converge to the number 1, which is not in $[0, 1)$.

    (c) The whole set of real numbers is not compact, since the sequence $\{n\}$ has the property that every subsequence is unbounded and hence every subsequence fails to converge.

## SECTION 3.1

2. Hint: Use Exercise 1.

4. Hint: The density of the rationals implies that for any number $x$, there is a sequence of rational number $\{x_n\}$ that converges to $x$.

5. Hint: The function is continuous at $x = 0$ since $|f(x)| = |x^2|$ for all $x$. The density of the rationals and the irrationals implies it is not continuous at any other point.

8. (a) Hint: Prove, by induction, that $f(nz) = nf(z)$, for any $z$ and any natural number $n$. In particular, $f(1) = f(n \cdot 1/n) = nf(1/n)$, so $f(1/n) = (1/n)f(1)$.

   (b) Hint: Use the density of the rationals.

## SECTION 3.2

1. (a) Show that the function $f:[0, 1] \to \mathbb{R}$ is increasing, so the maximizer is $x = 1$.

   (c) Show that the function $h:[-1, 1] \to \mathbb{R}$ is decreasing, so the maximizer is $x = -1$.

3. (e) The set of rational numbers in $[0, 1]$ is not compact, since there is a sequence in the set that converges to the irrational number $\sqrt{2}/2$ which is not in the set.

7. Hint: Define $x_0 = \inf \{x \text{ in } [0, 1] \mid f(x) = 0\}$ and prove that $x_0 > 0$.

## SECTION 3.3

1. Hint: Define $f(x) = x^9 + x^2 + 4$ for all $x$; find points at which the function $f:\mathbb{R} \to \mathbb{R}$ attains positive and negative values, and apply the Intermediate Value Theorem.

4. Hint: Define $g(x) = x - f(x)$ for $0 \le x \le 1$, and apply the Intermediate Value Theorem to the function $g:[0, 1] \to \mathbb{R}$.

7. Hint: The average of the $n$ numbers $f(x_1), \ldots, f(x_n)$ lies between the largest and the smallest of these numbers. Apply the Intermediate Value Theorem.

10. Hint: Use the density of the irrationals and the Intermediate Value Theorem.

## SECTION 3.4

1. Hint: First use the Difference of Powers formula to show that the function is strictly increasing on $(-\infty, 0]$ and on $[0, \infty)$.

6. The inverse is not continuous at $x = 1$. This does not contradict Theorem 3.10, since the domain $D$ is not an interval.

8. Hint: First show that the sequence $\{a_n\}$ is both monotone and bounded.

## SECTION 3.5

2. Hint: To verify continuity at $x = 4$, first show that for all $x > 0$,

$$|\sqrt{x} - \sqrt{4}| = |x - 4|/(\sqrt{x} + \sqrt{4}) \le (1/2)|x - 4|.$$

6. Hint: First show that for all $u$ and $v$,

$$|h(u) - h(v)| = \left|\frac{u + v}{(1 + u^2)(1 + v^2)}\right||u - v| \le 2|u - v|.$$

8. Hint: First show that for all $u$ and $v$ in the interval $[1/100, 1]$,

$$|f(u) - f(v)| = \left| \frac{u - v}{uv} \right| \le 100^2 |u - v|.$$

10. Let $\epsilon = 1$. Use the difference of cubes formula to show that no matter what $\delta > 0$ is chosen, there are points $u$ and $v$ with $|u - v| < \delta$, but $|u^3 - v^3| > 1$.

14. Uniform continuity follows from Theorem 3.14. To prove that it is not Lipschitz, show that there is no number $c$ such that

$$|\sqrt{x} - \sqrt{0}| \le c|x - 0| \text{ for all } x \text{ in } [0, 1].$$

16. Hint: Let $\epsilon = 1$. Choose $\delta > 0$ such $|f(u) - f(v)| \le 1$ for all $u$ and $v$ in $(a, b)$ such that $|u - v| < \delta$. Then observe that the function is bounded on any subinterval of length less than $\delta$. Express $(a, b)$ as the union of finitely many such subintervals.

## SECTION 3.6

2. (a) Hint: For $x \ne 1$, observe that $(x^4 - 1)/(x - 1) = x^3 + x^2 + x + 1$.

   (b) Hint: For $x > 0, x \ne 1$ observe that $(\sqrt{x} - 1)/(x - 1) = 1/(\sqrt{x} + 1)$.

5. No conclusion can be reached: for instance, set $g(x) = -f(x)$ for all $x$ in $D$, so that $\lim_{x \to x_0} [f(x) + g(x)] = 0$, but nothing is known about $\lim_{x \to x_o} f(x)$.

11. (a)

$$\lim_{x \to x_0, x < x_0} f(x) = \begin{cases} f(x_0) & \text{if } x_0 \text{ is not an integer} \\ f(x_0) - 1 & \text{if } x_0 \text{ is an integer.} \end{cases}$$

   (b)

$$\lim_{x \to x_0, x > x_0} f(x) = f(x_0) \text{ for all } x_0.$$

12. Hint: Use the Difference of Powers formula.

14. Hint at the right end-point: Choose any sequence $\{x_n\}$ in $(a, b)$ that converges to $b$. Show that $\{f(x_n)\}$ converges; define $\ell$ to be the limit of this sequence and use monotonicity to show that $\lim_{x \to b} f(x) = \ell$.

## SECTION 4.1

2. $y = 14x - 15$.

4. (a) 0.    (b) 4.

6. Hint: Show that if $x \ne 0$, then

$$\left| \frac{f(x) - f(0)}{x - 0} \right| \le |x|.$$

8. Use difference quotients directly in order to show that (a) $f'(0) = 0$, (b) $g'(0) = 0$.

13. Hint: For $h \ne 0$, write

$$\frac{f(x_0 + h) - f(x_0 - h)}{h} = \frac{f(x_0 + h) - f(x_0)}{h} + \frac{f(x_0 - h) - f(x_0)}{-h}.$$

14. Hint: Write $xf(x_0) - x_0 f(x) = x[f(x_0) - f(x)] + f(x)[x - x_0]$.

15. If $g(0) \neq 0$, then the function $|g|: \mathbb{R} \to \mathbb{R}$ is differentiable at $x = 0$. If $g(0) = 0$, then the function $|g|: \mathbb{R} \to \mathbb{R}$ is differentiable at $x = 0$ if and only if $g'(0) = 0$.

## SECTION 4.2

1. Use the composition property of limits by noting that for $x_0 > 0$ and $x > 0, x \neq x_0$,

$$\frac{g(x) - g(x_0)}{x - x_0} = c \left[ \frac{f(cx) - f(cx_0)}{cx - cx_0} \right].$$

4. No: Consider $f(x) = x$ for all $x$.

8. Hint: By the Bolzano-Weierstrass Theorem, a subsequence of $\{x_n\}$ converges to a number $x_0$. First, show that $f(x_0) = 0$, and then show that $f'(x_0) = 0$.

## SECTION 4.3

4. Hint: Define $f(x) = x^5 + 5x + 1$ for $-1 < x < 0$. Show that the derivative of $f: (-1, 0) \to \mathbb{R}$ is positive and that this function attains positive and negative values.

5. Hint: To show there are at most two solutions, define $f(x) = x^4 + 2x^2 - 6x + 2$ for all $x$. Since $f''(x) > 0$ for all $x$, Rolle's Theorem implies that the equation cannot have three solutions.

8. No: Define $f(x) = g(x) = 0$ for all $x > 0$, $f(x) = 0$ for $x < 0$, and $g(x) = 1$ for $x < 0$.

12. Define $g(x) = -\sqrt{1 - x^2} + 26$ for $-1 < x < 1$. The function $g: (-1, 1) \to \mathbb{R}$ is a solution, and the Identity Criterion implies it is the only solution.

13. Define $h(x) = f(x)/g(x)$ for all $x$. Show that $h: \mathbb{R} \to \mathbb{R}$ is constant, since its derivative is identically 0.

## SECTION 4.4

2. Hint: Use Theorem 4.21.

4. (b) Show that for $0 < c < 1$, if $f(1) - f(0) = f'(c)(1 - 0)$, then $c = 1/2$. On the other hand, if $g(1) - g(0) = g'(c)(1 - 0)$, then $c = \sqrt{1/3}$.

6. Hint: Use Theorem 4.21.

9. Hint: Prove Leibnitz' Formula by induction, using the product rule for differentiation together with the following identity for binomial coefficients:

$$\binom{n+1}{k} = \binom{n}{k-1} + \binom{n}{k}.$$

## SECTION 4.5

1. By the Identity Criterion, $F(x) = x^2 + x^4/4 + 7/4$ for all $x$. Thus $F(-1) = 3$.

4. The Identity Criterion implies that there are constants $c_1$ and $c_2$ such that $F(x) = x^2/2 + c_1$ for $x > 0$, and $F(x) = x^3/3 + c_2$ for $x < 0$. Since $F: \mathbb{R} \to \mathbb{R}$ is continuous at $x = 0$, $F(0) = c_1 = c_2$, and since $F(2) = 4$, $c_1 = c_2 = 4 - 8/3$.

5. Hint: The density of the irrationals and Darboux's Theorem imply that the derivative of $F: \mathbb{R} \to \mathbb{R}$ is constant. Now apply the Identity Criterion.

## SECTION 5.1

2. Hint: Use the very definition of the derivative at $x = 0$ of the function $f \colon \mathbb{R} \to \mathbb{R}$ defined by $f(x) = a^x$ for all $x$: $\lim_{n \to \infty} n[f(1/n) - f(0)] = f'(0)$.

3. Hint: Apply the Lagrange Mean Value Theorem to the natural logarithm on the interval $[a, b]$.

7. Hint: Define $f(x) = xe^x - 2$ for $0 < x < 1$. First show that the function $f \colon (0, 1) \to \mathbb{R}$ is strictly increasing. Then show that the function attains both positive $(e > 2)$ and negative values.

9. Hint: First show that $h(0) = 1$, and then, for any $x$ and any $t \neq 0$, observe that

$$\frac{h(x + t) - h(x)}{t} = h(x)\left[\frac{h(t) - h(0)}{t}\right].$$

## SECTION 5.2

4. Hint: Define $f(x) = e^{2x} + \cos x + x$ for all $x$. Show that the derivative of the function $f \colon \mathbb{R} \to \mathbb{R}$ is always positive and that $f \colon \mathbb{R} \to \mathbb{R}$ attains both positive and negative values.

8. Hint: Define $g(x) = k^2[f(x)]^2 + [f'(x)]^2$ for all $x$. Show that the derivative of $g \colon \mathbb{R} \to \mathbb{R}$ is identically 0 and $g(0) = 0$. Then note that $0 \leq k^2[f(x)]^2 \leq g(x)$ for all $x$.

10. Hint: First consider the case $a \geq 0$ and $b \geq 0$, and $0 \leq \theta \leq \pi/2$. Do the other cases based on the oddness of the sine and the evenness of the cosine.

12. Hint: Choose $\theta_0$ such that $\cos \theta_0 = c_1$ and $\sin \theta_0 = c_2$.

14. No; otherwise, since $f'(0) > 0$, there would be a symmetric neighborhood of 0 on which the derivative is positive, and so on which the function is strictly increasing.

## SECTION 5.3

1. Hint: Use the Identity Criterion.

6. Hint: Use the Mean Value Theorem.

## SECTION 6.2

2. Hint: Since $f \colon [0, 1] \to \mathbb{R}$ is increasing, for each $i$, $m_i = (i - 1)^2/n^2$ and $M_i = i^2/n^2$, so that

$$L(f, P_n) - U(f, P_n) = (1/n^3)\left[\sum_{i=1}^{n} i^2 - \sum_{i=1}^{n}(i - i)^2\right].$$

4. For the partition $P = \{a, b\}$, $m_1 \geq 0$. Thus, $L(f, P) = m_1(b-a) \geq 0$. Hence, $0 \leq L(f, P) \leq \int_a^b f$.

7. Hint: For $P = \{x_1, \ldots, x_n\}$ a partition of $[a, b]$, and $1 \leq i \leq n$, by the density of the rationals, $m_i \leq 0 \leq M_i$, so $L(f, P) \leq 0 \leq U(f, P)$. Now use the very definition of the integral.

13. In general, no. However, the sum of monotone increasing (resp. decreasing) functions is monotone increasing (decreasing). If one of the functions only attains values of one sign, the product of monotone functions is monotone.

16. Hint: Argue by contradiction. If there is a point $x_0$ in $[a, b]$ at which $f(x_0) > 0$, then since the function is continuous, there is an interval $[c, d]$ containing $x_0$ and contained in $[a, b]$, with $c < d$, such that $f(x) > f(x_0)/2$ for all $x$ in $[c, d]$. From the definition of the integral it follows that $\int_c^d f \geq (f(x_0)/2)(d - c) > 0$.

## SECTION 6.3

1. (b) For each $n$, $L(f, P_n) = 1 + 4(n-1)n/2n^2$ and $U(f, P_n) = 1 + 4n(n+1)/2n^2$ :

$$\lim_{n \to \infty} \left[1 + 4(n-1)n/2n^2\right] = \lim_{n \to \infty} \left[1 + 4n(n+1)/2n^2\right] = 3.$$

3. (c) Use the identity $x\sqrt{10-x} = 10\sqrt{10-x} - (10-x)^{3/2}$ and the First Fundamental Theorem.

## SECTION 6.4

3. This is the limit of Riemann sums for the function defined by $f(x) = 1/(1+x)$ for $0 \le x \le 1$, based on regular partitions of $[0, 1]$. By the Riemann Sum Convergence Criterion and the First Fundamental Theorem, the limit is $\ln 2$.

5. The limit equals $\lim_{n \to \infty} 1/\sqrt{n \cdot n} + \lim_{n \to \infty} R_n$, where the second limit is the limit of Riemann sums for $f(x) = 1/\sqrt{1+x}$ for $0 \le x \le 1$, based on regular partitions of $[0.1]$. By the Riemann Sum Convergence Criterion and the First Fundamental Theorem, the limit is $2(\sqrt{2} - 1)$.

8. Hint: For an interval $[x_{i-1}, x_i]$ in the partition $P$ and a point $c_i$ in $[x_{i-1}, x_i]$, show that for all $x$ in $[x_{i-1}, x_i]$,

$$f(c_i) - c\|P\| \le f(x) \le f(c_i) + c\|P\|.$$

## SECTION 6.5

4. Hint: For all real numbers $\lambda$, by the monotonicity and linearity of the integral,

$$0 \le \int_a^b [\lambda g - f]^2 = \lambda^2 \int_a^b g^2 - 2\lambda \int_a^b gf + \int_a^b f^2 \equiv a\lambda^2 + b\lambda + c.$$

For this quadratic polynomial in the variable $\lambda$ to be nonnegative, its discriminant must be nonpositive.

5. (a) By the Cauchy-Schwarz Inequality for Integrals,

$$\int_0^1 \sqrt{1 + x^4} \, dx = \int_0^1 \sqrt{1 + x^4} \cdot 1 \, dx \le \sqrt{\int_0^1 [1 + x^4] \, dx} \cdot \sqrt{\int_0^1 1 \, dx} = \sqrt{\frac{6}{5}}.$$

7. Define $h(x) = x - \sin x$ and $g(x) = \sin x - (2/\pi)x$ for all $x$. Since $h(0) = 0$ and $h'(x) > 0$ for $0 < x < \pi/2$, $h(x) > 0$ for $0 < x < \pi/2$. This proves the right-hand inequality. Since $g(0) = g(\pi/2) = 0$ and $g''(x) = -\sin x \ne 0$, for $0 < x < \pi/2$, it follows from the Mean Value Theorem that $g(x) \ne 0$ for $0 < x < \pi/2$. But $g'(0) > 0$, so $g(x) > 0$ for $x > 0$ sufficiently small. Thus, by the Intermediate Value Theorem, $g(x) > 0$ for all for $0 < x < \pi/2$. This proves the left-hand inequality. The integral inequality follows from the monotonicity of the integral.

## SECTION 7.1

1. (a)

$$\frac{d}{dx}\left(\int_0^x x^2 t^2 \, dt\right) = \frac{d}{dx}\left(x^2 \int_0^x t^2 \, dt\right) = (5/3)x^4.$$

(c)

$$\frac{d}{dx}\left(\int_{-x}^x e^{t^2} \, dt\right) = \frac{d}{dx}\left(\int_0^x e^{t^2} \, dt - \int_0^{-x} e^{t^2} \, dt\right) = 2e^{x^2}.$$

8. Consider the function $f: [0, 1] \to \mathbb{R}$ that has the value $-1$ on $[0, 1/2]$ and the value $1$ on the interval $(1/2, 1]$. Then $\int_0^1 f = 0$, but there is no point $x$ in $[0, 1]$ at which $f(0) = 0$.

9. Show that $\int_0^1 p(x)\,dx = 0$ and then apply the Mean Value Theorem for Integrals.

## SECTION 7.2

1. (a) By Theorem 7.10 and the First Fundamental Theorem,

$$F(x) = e^{-x} + e^{-x} \int_0^x e^t t \, dt = 2e^{-x} + x - 1.$$

5. Show that $f'(x) - f(x) = 0$ for all $x$, and that $f(0) = 0$; then apply Proposition 7.9.

7. Hint: Show that $h'(x) = 1/g(x)$ for all $x$ and that $f'(x) = 1/h'\big(f(x)\big)$ for all $x$ in $J$.

## SECTION 7.3

1. (c) $4e^9 - (3/2)e^4$.

2. (b) $-1 - 2\ln(4/3)$.

4. Observe that the formula holds at $x = a$. Then differentiate each side of the formula and apply the Identity Criterion.

9. (a) Hint: (A diagram will help) Choose $c$ such that $f(c) = b$. Consider the cases $a = c, a < c$, and $a > c$. For instance, if $a < c$, then $f(x) \leq b$ for $a \leq x \leq c$, so using Exercise 8,

$$ab = cf(c) - (c - a)b \leq F(a) + G(b) + \int_a^c f - (a - c)b \leq F(a) + G(b).$$

## SECTION 7.4

1. (c) $\int_0^1 x^4\,dx = 1/5$; the Trapezoid Approximation based on the partition $\{0, 1\}$ is $1/2$; the error estimate from Theorem 7.14 is 1. For Simpson's Rule based on the partition $\{0, 1\}$ the value is $5/24$; the error estimate from Theorem 7.17 is $24/2880$.

6. The Simpson Rule approximation of $\ln 2 = \int_1^2 1/t\,dt$ based on the partition $\{1, 3/2, 2\}$ is

$$1/12\big[f(1) + 4f(5/4) + 2f(3/2) + 4f(7/4) + f(2)\big] = 1747/2520,$$

where $f(t) = 1/t$. Since $\|P\| = 1/2$ and $|f^{(4)}(t)| \leq 24$ for $1 \leq t \leq 2$, the error estimate from Theorem 7.19 is $24/(2^4 \cdot 2880)$.

## SECTION 8.1

1. (a) Highest order of contact is 0. (c) Highest order of contact is 2.

2. (a) $p_3(x) = x - x^3/3$.

4. Hint: Show that $f(0) > 0, f'(0) > 0$, and $f''(0) < 0$, and then use continuity of the function and its first two derivatives at $x = 0$.

## SECTION 8.2

1. Hint: Find the first Taylor polynomial for $f(x) = \sqrt{1 + x}$ at $x_0 = 0$ and estimate the remainder.

5. Hint: Find the first Taylor polynomial for $f(x) = \sin x$ at $x_0 = a$, and use this to estimate $\sin(a + b)$.

7. Hint: A polynomial of degree $n$ is identically equal to its $n$th Taylor polynomial at $x = x_0$.

## SECTION 8.3

1. In (a) and (b), the estimate (8.12) holds with $M = 1$.

4. Hint: Argue by induction that the function $F: \mathbb{R} \to \mathbb{R}$ has derivatives of all orders and that for each natural number $n$,
$$F^{(n+2)}(x) - F^{(n+1)}(x) - F^{(n)}(x) = 0.$$

Use this formula to show that on each interval $[-r, r]$ the estimate (8.12) holds where $M = 2(1 + c)$ and $c$ is chosen so that $|F'(x)| \le c$ and $|F''(x)| \le c$, for $|x| \le r$.

## SECTION 8.4

5. Hint: By the Lagrange Remainder Theorem, for $x > -1, x \ne 0$, there is a point $c$, strictly between 0 and $x$, such that
$$\ln(1 + x) - p_n(x) = \frac{(-1)^{n+1}}{(n+1)(1+c)^{n+1}} x^{n+1} = \frac{(-1)^{n+1}}{(n+1)} \left(\frac{x}{1+c}\right)^{n+1}.$$

Show that $|x|/(1 + c) \le 1$ if $-1/2 \le x \le 1$. Thus, the Lagrange Remainder Theorem implies the validity of the expansion for $-1/2 \le x \le 1$.

## SECTION 8.5

4. Hint: Let $x_m$ and $x_M$ be minimizers and maximizers of $g: [a, b] \to \mathbb{R}$. Then, by the monotonicity of the integral,
$$g(x_m) \int_a^b h(x)\, dx \le \int_a^b g(x) h(x)\, dx \le g(x_M) \int_a^b h(x)\, dx.$$

If $\int_a^b h(x)\, dx > 0$, divide the above inequality by the integral and apply the Intermediate Value Theorem.

7. Hint: Show that for $0 < c < x < 1$, since $(1 + c)^{\beta - k} \le (1 + c)^\beta$,
$$\left| \frac{f^{(k)}(c)}{k!} x^k \right| = \left| (1 + c)^{\beta - k} \binom{\beta}{k} x^k \right| \le (1 + c)^\beta \binom{\beta}{k} |x|^k,$$

and use Lemma 8.18.

## SECTION 8.6

2. Argue by contradiction: if there is such an estimate, then from Theorem 8.11 it follows that the Taylor series expansion converges to the function at every point.

4. Hint: Argue by induction. By assumption, $g(0) = 0$. Let $n$ be a natural number, and suppose that for $0 \le k \le n - 1$, $g^{(k)}(0) = 0$. Then by the Lagrange Remainder Theorem, for each $x$ there is a point $c$ between 0 and $x$ such that $g(x) = (g^{(n)}(c)/n!)x^n$. The estimate $|g(x)| \le c_{n+1}|x|^{n+1}$ for $|x| < \delta_n$, and the continuity of the $n$th derivative, imply that $g^{(n)}(0) = 0$.

## SECTION 8.7

4. Hint: Argue by contradiction. Suppose, for instance, that there is a polynomial $p: \mathbb{R} \to \mathbb{R}$ such that $|e^x - p(x)| \le 1$ for all $x$. Let the degree of this polynomial be $k$. Show that this is inconsistent with the fact that $e^x \ge x^{k+1}/(k+1)!$ for all $x > 0$.

7. By the Weierstrass Approximation Theorem, for each natural number $k$, we may choose a polynomial $q_k(x)$ such that $|q_k(x) - f(x)| \leq 1/k$ for all $x$ in $[-1, 1]$. For $x$ in $[-1, 1]$, define $p_1(x) = q_1(x)$ and $p_n(x) = q_n(x) - q_{n-1}(x)$ if $n > 1$. Show that $f(x) = \sum_{n=1}^{\infty} p_n(x)$ for all $x$.

## SECTION 9.1

1. (d) Converges; either find the exact formula for the partial sums or compare with $\sum_{k=1}^{\infty} 1/k^2$. (e) Converges; the Integral Test. (g) Diverges; the terms do not converge to 0.

6. The sequence of partial sums of the series $\sum_{k=1}^{\infty}(a_{2k} + a_{2k-1})$ is a *subsequence* of the sequence of partial sums of the series $\sum_{k=1}^{\infty} a_k$.

## SECTION 9.2

1. The sequence converges pointwise to the function $f: \mathbb{R} \to \mathbb{R}$ with $f(x) = 1$ for $|x| < 1$, $f(x) = -1$ for $|x| > 1$ and $f(\pm 1) = 0$.

3. The sequence converges pointwise to the function that is identically 0.

7. Hint: Choose $n$ such that $|f(x) - f_n(x)| \leq 1$ for all $x$ in $\mathbb{R}$. Thus, $|f(x)| \leq |f_n(x)| + 1$ for all $x$ in $\mathbb{R}$. The boundedness of $f_n: \mathbb{R} \to \mathbb{R}$ implies the boundedness of $f: \mathbb{R} \to \mathbb{R}$.

10. Hint: For all natural numbers $n$ and $k$,

$$|f_{n+k}(x) - f_n(x)| \leq \sum_{j=n+1}^{n+k} \frac{a_j}{j!} r^j \text{ for } |x| \leq r.$$

This, together with the Weierstrass Uniform Convergence Criterion and the convergence of $\sum_{j=1}^{\infty}(a_j/j!)r^j$, implies uniform convergence.

## SECTION 9.3

2. Hint: Since $e^t \geq t^2/2$ for all $t > 0$, $|f_n(x)| \leq (2/n|x|^3)$ for all $x$ in $(0, 1]$ and all natural numbers $n$. This implies pointwise convergence to the function $f: [0, 1] \to \mathbb{R}$ that is identically 0. For each $n$, on the other hand, $\int_0^1 f_n = (1/2)[1 - e^{-n}]$. Thus, $\lim_{n \to \infty} \int_0^1 f_n = 1/2 \neq \int_0^1 f$. There is no contradiction, because the convergence is not uniform.

4. For a natural number $n$ and any $x$ in $(-1, 1)$, define $f_n(x) = (1/n)\sin(n^2 x)$. Then $|f_n(x)| \leq 1/n$ for all $n$, and all $x$ in $(-1, 1)$, so the sequence $f_n: (-1, 1) \to \mathbb{R}$ converges pointwise to the function that is identically 0. Observe that $f'(0) = n$ for each $n$.

## SECTION 9.4

3. Apply Theorem 9.24 to justify term-by-term differentiation of the series in Exercise 2.

5. Substitute $r = (1 - x)/x$ in the Geometric Series.

9. Show convergence of each of these series for all $x$, and apply Theorem 9.24 to justify term-by-term differentiation.

11. Use Exercise 10.

14. (a) If $|x| < 1/\alpha$, set $r = [1 + |\alpha x|]/2$. Show that since $|\alpha x| < r$, there is a natural number $N$ such that $|a_n x^n| \leq r^n$ for $n \geq N$. Since the Geometric Series $\sum_{n=1}^{\infty} r^n$ converges, show that the sequence of partial sums of $\sum_{n=1}^{\infty} a_n x^n$ is a Cauchy sequence.

## SECTION 9.5

2. First show that the interval $[x_0/r - 1/2, x_0/r + 1/2]$ contains an integer $m$. If $m$ is in $[x_0/r - 1/2, x_0/r]$, then $[x_0, x_0 + r]$ is contained in $[mr, (m+1)r]$, while if $m$ is in $[x_0/r, x_0/r + 1/2]$, then $[x_0 - r, x_0]$ is contained in $[(m-1)r, mr]$,

6. Define $F(x) = \int_0^x f$ for each $x$, where $f \colon \mathbb{R} \to \mathbb{R}$ is the function defined in Theorem 9.27. The Second Fundamental Theorem implies that $F'(x) = f(x)$ for all $x$.

## SECTION 10.1

1. $||\mathbf{u}|| = \sqrt{14}$, $||\mathbf{v}|| = \sqrt{24}$, and $\langle \mathbf{u}, \mathbf{v} \rangle = 0$. The equality follows from the Pythagorean Identity.

2. The Cauchy-Schwarz Inequality implies that

$$x + 2y + 3z \leq \sqrt{x^2 + y^2 + z^2}\sqrt{14} \text{ for all } (x, y, z) \text{ in } \mathbb{R}^3.$$

Thus, the maximum value of $\sqrt{14}$ is attained at the point $(1, 2, 3)$.

7. Hint: Express $a_1 + \cdots + a_n$ as an inner product of two vectors in $\mathbb{R}^n$, and apply the Cauchy-Schwarz Inequality.

## SECTION 10.2

3. Hint: Use the Componentwise Convergence Criterion and apply the properties of real convergent sequences to the sequence $\{||\mathbf{u}_k||\}$.

5. Hint: By the Reverse Triangle Inequality, $||\mathbf{u} - \mathbf{u}_k|| \geq ||\mathbf{u}|| - ||\mathbf{u}_k||$ for all $k$.

## SECTION 10.3

1. (a), (c) open. (d), (e) closed. (b) neither open nor closed.

2. (a) open. (b), (d) closed. (c) neither open nor closed.

9. (b) In general, bd A is not contained in bd B. One counterexample:

$$A = \{\mathbf{x} \text{ in } \mathbb{R}^n \mid ||\mathbf{x}|| \leq 1\}, B = \{\mathbf{x} \text{ in } \mathbb{R}^n \mid ||\mathbf{x}|| \leq 2\}.$$

## SECTION 11.1

4. Hint: For $\mathbf{u}$ in $\mathbb{R}^n$, use the density of the rationals in $\mathbb{R}$ to find a sequence $\{\mathbf{u}_k\}$ in $\mathbb{R}^n$ converging to $\mathbf{u}$ such that the first component of $\mathbf{u}_k$ is rational for all natural numbers $k$. Then use the continuity of the function $f \colon \mathbb{R}^n \to \mathbb{R}$ at the point $\mathbf{u}$.

5. (c) Define $f(x, y) = x^2 - y$ for $(x, y)$ in $\mathbb{R}^2$. The function $f \colon \mathbb{R}^2 \to \mathbb{R}$ is continuous, so, by Corollary 11.8, the set $\{(x, y) \mid f(x, y) < 0\}$ is open.

8. Hint: $\{\mathbf{u} \text{ in } \mathbb{R}^n \mid u_n > 0\} = \{\mathbf{u} \text{ in } \mathbb{R}^n \mid p_n(\mathbf{u}) > 0\}$.

10. $\{\mathbf{v} \text{ in } \mathbb{R}^n \mid f(\mathbf{v}) = c\} = \{\mathbf{v} \text{ in } \mathbb{R}^n \mid f(\mathbf{v}) \geq c\} \cap \{\mathbf{v} \text{ in } \mathbb{R}^n \mid f(\mathbf{v}) \leq c\}$. Apply Corollary 11.8 and Theorem 10.15.

11. Hint: Use the $\epsilon - \delta$ criterion for continuity at the point $\mathbf{u}$ with $\epsilon = f(\mathbf{u})/2$. Or use Corollary 11.8.

## SECTION 11.2

1. None of the three subsets of $\mathbb{R}$ is compact. (a), (b) not closed. (c) not bounded since $\{x \text{ in } \mathbb{R} \,|\, e^x - x^2 \le 2\}$ contains the negative integers.

4. Not necessarily: Consider $A = \{\mathbf{x} \text{ in } \mathbb{R}^n \,|\, 0 < \|\mathbf{x}\| < 1\}$, and the function $f: A \to \mathbb{R}$ defined by $f(\mathbf{x}) = 1/\|\mathbf{x}\|$ for $\mathbf{x}$ in $A$.

6. Hint: Apply Theorem 11.11.

8. Define the function $f: A \to \mathbb{R}$ by $f(\mathbf{u}) = d(\mathbf{u}, \mathbf{v})$ for $\mathbf{u}$ in $A$. Show that $f: A \to \mathbb{R}$ is continuous and apply the Extreme Value Theorem.

## SECTION 11.3

5. Define $\mathcal{U} = \{x \text{ in } \mathbb{R} \,|\, x < \sqrt{2}\}$ and $\mathcal{V} = \{x \text{ in } \mathbb{R} \,|\, x > \sqrt{2}\}$. Show that $\mathcal{U}$ and $\mathcal{V}$ separate $\mathbb{Q}$.

9. (a) Hint: Define $h: A \to \mathbb{R}$ by $h(x, y, z) = y$ for $(x, y, z)$ in $A$. By Theorem 11.21, the image of $h: A \to \mathbb{R}$ is an interval.

   (b) Hint: Define $g: A \to \mathbb{R}$ by $g(x, y, z) = \sqrt{x^2 + y^2 + z^2}$ for $(x, y, z)$ in $A$. By Theorem 11.21, the image of $g: A \to \mathbb{R}$ is an interval.

14. Hint: Assume the contrary. Then $\mathcal{U}$ and $\mathcal{V}$ will separate $B$, and this contradicts the connectedness of $B$.

18. The compact, connected subsets of $\mathbb{R}$ are precisely the closed, bounded intervals $[a, b]$, where $a \le b$ (and, of course, the empty-set).

## SECTION 12.1

2. (a) $d(x, \cos x) = 1$. (b) $d(4x^3, 6x^2 - 3x) = 1$.

4. Hint: Use the Mean Value Theorem to show that for a natural number $k$ and $x$ in $[0, 1]$,
$$|\cos(x/k) - 1| = |\cos(x/k) - \cos(0)| \le 1/k.$$

8. (c) Show that convergence of a sequence with respect to the metric $d^*$ is equivalent to componentwise convergence.

11. (c) By part (b), uniform convergence implies convergence with respect to the metric $d^*$. The converse is false: consider $f_k(x) = x^k$ for $0 \le x \le 1$, and $k$ a natural number.

## SECTION 12.2

2. (a) $0 \le \alpha \le 4$. (b) $0 \le \alpha < 1$.

9. The metric space $C([0, 1], \mathbb{R})$ is complete. Thus, this sequence is not Cauchy in $C([0, 1], \mathbb{R})$, since it does not converge uniformly.

17. Hint: Use the Triangle Inequality to show $T(K) \subseteq K$. If $d(T(p_0), p_0) = 0$, then $T(p_0) = p_0$, so the proof is complete. If $d(T(p_0), p_0) > 0$, show that $c < 1$ and apply the Contraction Mapping Principle.

## SECTION 12.3

1. (a) $f(x) = x \sin x + \cos x$ for $x$ in $\mathbb{R}$.

   (b) $f(x) = x + x^4/4 + 11/4$ for $x$ in $\mathbb{R}$.

   (c) $f(x) = \int_0^x e^{t^2}\, dt$ for $x$ in $\mathbb{R}$.

2. (a) $f(x) = 2e^x - 1$ for $x$ in $\mathbb{R}$.

   (b) $f(x) = x + 1$ for $x$ in $\mathbb{R}$.

   (c) $f(x) = e^{2x-1} - e^x$ for $x$ in $\mathbb{R}$.

3. (c) The maximal interval is $I = (-1, \infty)$, and $f(x) = -1/(1+x)$ for $x$ in $I$.

6. The maximal interval is $I = (-\pi\epsilon/2, \pi\epsilon/2)$, and $f(x) = \tan(x/\epsilon)$ for $x$ in $I$.

## SECTION 12.4

6. Hint: Use the density of the rationals in $\mathbb{R}$.

7. In general, this is false. One counterexample: Define $f : \mathbb{R} \to \mathbb{R}$ by $f(x) = x^2$ for $x$ in $\mathbb{R}$. Then $(-1, 1)$ is open in $\mathbb{R}$, but $f((-1, 1)) = [0, 1)$ is not open in $\mathbb{R}$.

8. Hint: First show that $X \backslash f^{-1}(C) = f^{-1}(Y \backslash C)$. Then use the Complementing Characterization and Theorem 12.20.

## SECTION 12.5

4. Hint: $D$ and $X \backslash D$ separate $X$.

9. (b) Hint: Use Theorem 12.23 and part (a).

11. Hint: Fix $p$ in $X$. Show that $\{p\}$ and $X \backslash \{p\}$ separate $X$.

## SECTION 13.1

1. Show that $|x^3 y/(x^2 + y^2)| \leq |xy|$ for $(x, y) \neq (0, 0)$. So the limit is 0.

2. (a) 1. (b) 0.

3. (a) Does not exist. (b) Does not exist. (c) 1.

5. Consider a set containing a single point.

9. Define $f(\mathbf{x}) = \mathbf{x}$ for all $\mathbf{x}$ in $\mathbb{R}^n$.

## SECTION 13.2

1. (c) $\partial f/\partial x = -y \cos xy \sin xy / \sqrt{1 + \cos^2 xy}$; $\partial f/\partial z = 0$.

7. Hint: For a sufficiently small open interval $J$ about 0, define $\psi(t) = g(f(\mathbf{x} + t\mathbf{e}_i))$ if $t$ is in $J$. Apply the Chain Rule for functions of a single variable to $\psi : J \to \mathbb{R}$ at $t = 0$.

8. (a) $\partial h/\partial x(x, y) = y^2 g'(xy^2 + 1)$, $\partial h/\partial y(x, y) = 2xy g'(xy^2 + 1)$.

   (b) $\partial h/\partial u(u, v) = 4g'(4u + 7v)$, $\partial h/\partial v(u, v) = 7g'(4u + 7v)$.

   (c) $\partial h/\partial t(t, s) = g'(t - s)$, $\partial h/\partial s = -g'(t - s)$.

11. (a), (b) harmonic. (c) not harmonic.

14. Hint: Choose $\epsilon > 0$ such that $\mathcal{N}_\epsilon(x_0, y_0)$ is contained in $\mathcal{U}$. Let $r = \epsilon/3$.

## SECTION 13.3

1. (a) $\mathbf{D}f(\mathbf{x}) = 2\mathbf{x} e^{||\mathbf{x}||^2}$ for $\mathbf{x}$ in $\mathbb{R}^n$.

   (c) $\mathbf{D}f(\mathbf{x}) = -2\mathbf{x}/||\mathbf{x}||^4$ for $\mathbf{x}$ in $\mathbb{R}^n$.

3. $g'(f(\mathbf{x}))\mathbf{D}f(\mathbf{x})$.

6. $\theta = (\sqrt{13} - 2)/3$.

10. (a) implies (b), (a) implies (c); (b) implies (c), (b) does not imply (a); (c) does not imply (b), (c) does not imply (a).

11. (a) Hint: $K$ is compact. (b) Hint: Assume the contrary, that the minimizer $\mathbf{x}$ in (a) has norm 1. Use the fact that $\langle \mathbf{D}f(\mathbf{x}), \mathbf{x} \rangle > 0$ to show that $\mathbf{x}$ cannot be a minimizer.

## SECTION 14.1

1. $\psi(x, y) = e(1 + 2x + 4y)$ for $(x, y)$ in $\mathbb{R}^2$.

2. The tangent plane is parallel to the $x$-$y$ plane only at the point $(x, y) = (-4/7, -1/7)$.

6. $g(x, y, z) = z$ for $(x, y, z)$ in $\mathbb{R}^3$.

13. Hint: Consider the function evaluated on the sequence $\{(1/n, 0)\}$.

## SECTION 14.2

1. $\phi''(0) = 44$.

2. $\phi''(0) = -9/4$.

3. (a) $\phi(t) = \sqrt{1 + 3t^2 - 3t^3}$.

5. For each unit vector $\mathbf{h}$, there is an $r = r(\mathbf{h}) > 0$ such that $f(t\mathbf{h}) > 0$ for $0 < |t| < r$. But $r$ depends on $\mathbf{h}$.

8. (a) $A = \begin{pmatrix} 1 & 0 \\ 0 & -1 \end{pmatrix}$. (b) $A = \begin{pmatrix} 1 & 4 \\ 4 & 1 \end{pmatrix}$.

9. Hint: For $\mathbf{u}$ in $\mathbb{R}^n$, let A be the $n \times n$ matrix whose first row is $\mathbf{u}$ and whose other entries are zero. Then if $\mathbf{v}$ is in $\mathbb{R}^n$, then A$\mathbf{v}$ is a point in $\mathbb{R}^n$ whose only nonzero component is equal to $\langle \mathbf{u}, \mathbf{v} \rangle$.

11. $Q: \mathbb{R} \to \mathbb{R}$ is not a quadratic function.

## SECTION 14.3

1. (a) $(0, 2)$ is strict local minimizer.

   (b) $(0, 2, 0)$ is strict local minimizer. Hint: Consider part (a).

   (c) $(0, 0)$ is strict local minimizer.

3. The largest value of $f: K \to \mathbb{R}$ is $f(1, 0) = 1/e$. The smallest value is $f(-1, 0) = -e$.

7. Hint: Show that for $t > 0$ sufficiently small, $f(t\mathbf{u}) > f(\mathbf{0}) > f(t\mathbf{v})$.

11. (a) Does not exist. (b) 0. (c) Does not exist. (d) 0.

## SECTION 15.1

5. $\mathbf{T}(x, y, z) = (x/2 - z/2, x/2 + 3y/2, x/2 - y/2)$ for $(x, y, z)$ in $\mathbb{R}^3$.

6. $\begin{pmatrix} 1/5 & 2/5 \\ 2/5 & 4/5 \end{pmatrix}$.

8. All $2 \times 2$ diagonal matrices, $\begin{pmatrix} a & 0 \\ 0 & b \end{pmatrix}$.

10. $\begin{pmatrix} 0 & -1 \\ 1 & 0 \end{pmatrix}$.

## SECTION 15.2

1. $\mathbf{DF}(0,0) = \begin{pmatrix} 2 & 0 \\ 1 & -1 \end{pmatrix}$, $\mathbf{DF}(\pi,0) = \begin{pmatrix} 2 & \pi \\ -1 & 1 \end{pmatrix}$.

2. $\mathbf{DF}(1,2,3) = \begin{pmatrix} 6 & 3 & 2 \\ 2 & 3 & 2 \\ 3 & 0 & 0 \end{pmatrix}$.

6. Since $\det \mathbf{DF}(x,y) = 4(x^2 + y^2)$, the matrix $\mathbf{DF}(x,y)$ is invertible if $(x,y) \neq (0,0)$. Same answer for (a) and (b). (Why?)

## SECTION 15.3

1. $\partial g/\partial s(s,t) = D_1\psi(s^2t,s)2st + D_2\psi(s^2t,s)$, $\partial g/\partial t(s,t) = D_1\psi(s^2t,s)s^2$.

2. $D_1\eta(u,v,w) = 3h(u^2,v^2,uvw) + (3u+2v)\left[D_1h(u^2,v^2,uvw)2u + D_3h(u^2,v^2,uvw)vw\right]$.

   $D_2\eta(u,v,w) = 2h(u^2,v^2,uvw) + (3u+2v)\left[D_2h(u^2,v^2,uvw)2v + D_3h(u^2,v^2,uvw)uw\right]$.

   $D_3\eta(u,v,w) = (3u+2v)D_3(u^2,v^2,uvw)uv$.

10. (a) $D_1f\left(g(1,2),h(1,2)\right)\partial g/\partial x(1,2) + D_2f\left(g(1,2),h(1,2)\right)\partial h/\partial x(1,2)$.

    (b) $D_1f(g(1,2),h(1,2))$.

## SECTION 16.1

1. $x \neq 1, -1$.

2. All $x$ in $\mathbb{R}$. To show that $f\colon \mathbb{R} \to \mathbb{R}$ is one-to-one, show it is strictly increasing; to show $f(\mathbb{R}) = \mathbb{R}$, use the Intermediate Value Theorem.

5. Hint: To verify that $f(\mathbb{R}) = \mathbb{R}$, show that $f(x) \geq f(0) + cx$, if $x > 0$, and that $f(x) \leq f(0) + cx$, if $x < 0$.

8. (a) $\partial g/\partial u(1,0) = 1$, $\partial g/\partial v(1,0) = 0$, $\partial h/\partial u(1,0) = 1$, $\partial h/\partial v(1,0) = 1$.

   (b) $\partial g/\partial u(1,1) = \frac{1}{2}$, $\partial g/\partial v(1,1) = \frac{1}{2}$, $\partial h/\partial u(1,1) = \frac{1}{2}$, $\partial h/\partial v(1,1) = -\frac{1}{2}$.

13. (a) Hint: Since $(x_0, y_0)$ is a minimizer for $\psi\colon \mathbb{R}^2 \to \mathbb{R}$, $\partial \psi/\partial x(x_0, y_0) = 0$ and $\partial \psi/\partial y(x_0, y_0) = 0$. Thus, $\mathbf{DF}(x_0, y_0)$ is not invertible.

    (b) Hint: Since $\psi(x,y) \geq \psi(x_0, y_0)$ for all $(x,y)$ in $\mathbb{R}^2$, the image of $\mathbf{F}\colon \mathbb{R}^2 \to \mathbb{R}^2$ does not contain a neighborhood of the point $\mathbf{F}(x_0, y_0)$.

## SECTION 16.2

2. Hint: To verify stability, use the Mean Value Theorem. For the converse, express the derivative as the limit of difference quotients.

3. Hint: Show that there are not $c > 0$ and $r > 0$ such that

$$\|\mathbf{G}(x,0) - \mathbf{G}(0,0)\| \geq c|x| \text{ for all } x \text{ in } (-r,r).$$

6. $\gamma = \frac{1}{6}$ is one correct answer, which follows from Lemma 16.4 with

$$A = \begin{pmatrix} 1 & 0 & 0 \\ 0 & 1 & 0 \\ 0 & 0 & 1 \end{pmatrix} \text{ and } B = \begin{pmatrix} 1 & a & 0 \\ b & 1 & c \\ c & 0 & 1 \end{pmatrix}.$$

## SECTION 16.3

1. (a) The Inverse Function Theorem applies at all points $(x, y, z)$ such that $z \neq 0$, since

$$\det \mathbf{DF}(x, y, z) = \det \begin{pmatrix} e^x \cos y & -e^x \sin y & 0 \\ e^x \sin y & e^x \cos y & 0 \\ 0 & 0 & 2z \end{pmatrix} = 2ze^x.$$

2. The Inverse Function Theorem applies for $(\rho_0, \theta_0, \phi_0)$ in $\mathbb{R}^3$ with $\rho_0 \neq 0$ and $\phi_0 \neq n\pi$ for any integer $n$.

3. (a) Show that the derivate matrix is always noninvertible, since the last row is a linear combination of the first two rows. (b) The image of the mapping is contained in the paraboloid $\{(x, y, z) \mid z = x^2 + y^2\}$, so it does not contain an open subset of $\mathbb{R}^3$.

6. (c) The image contains 0, but fails to contain any negative number.

7. Consider any constant mapping.

11. $x + y = u$

    $x - y = v.$

## SECTION 17.1

4. Define $f: \mathbb{R}^2 \to \mathbb{R}$ by $f(x, y) = e^{2x-y} + \cos(x^2 + xy) - 2 - 2y$ for $(x, y)$ in $\mathbb{R}^2$. Since $\partial f/\partial y(0, 0) \neq 0$, by Dini's Theorem there exists $r_1 > 0$ and a continuously differentiable function $\phi: (-r_1, r_1) \to \mathbb{R}$ such that $f(x, \phi(x)) = 0$ for $|x| < r_1$. Also, since $\partial f/\partial x(0, 0) \neq 0$, there exists $r_2 > 0$ and a continuously differentiable function $\psi: (-r_2, r_2) \to \mathbb{R}$ such that $f(\psi(y), y) = 0$ for $|y| < r_2$. Moreover, $\phi'(0) = \frac{2}{3}$ and $\psi'(0) = \frac{3}{2}$.

10. By Dini's Theorem, there is an $r > 0$ and a function $\psi: (y_0 - r, y_0 + r) \to \mathbb{R}$ that is continuously differentiable and such that $0 = f(\psi(y), y) = y - \phi(\psi(y))$, so $\phi(\psi(y)) = y$, for $|y - y_0| < r$. Note that $\phi$ and $\psi$ are inverses of each other when restricted to appropriate domains. Dini's Theorem also asserts that

$$\psi'(y) = \frac{1}{\phi'(\psi(y))} \text{ whenever } |y - y_0| < r,$$

which is the formula for the derivative of the inverse asserted in the Inverse Function Theorem.

9. Hint: Observe that $f(x, y) = 0$ if and only if $h(x, y) = 0$, but that $\mathbf{D}f(0, 0) \neq \mathbf{D}h(0, 0)$.

14. Hint: Since $f(x, g(x)) = 0$ for all $x$ in $I$, by differentiating we obtain

$$\partial f/\partial x(x, g(x)) + \partial f/\partial y(x, g(x))g'(x) = 0 \text{ for all } x \text{ in } I.$$

Hence $g'(x_0) = 0$, and, differentiating once more, we obtain

$$\partial^2 f/\partial x^2(x_0, y_0) + \partial f/\partial y(x_0, y_0)g''(x_0) = 0,$$

so that $g''(x_0) < 0$.

## SECTION 17.2

4. Hint: Define the continuously differentiable mapping $\mathbf{F}: \mathbb{R}^4 \to \mathbb{R}^2$ by

$$\mathbf{F}(x, y, z, t) = (x + 2y + x^2 + (yz)^2 + t^3, -x + z + \sin(y^2 + z^2 + t^3)) \text{ for } (x, y, z, t) \text{ in } \mathbb{R}^4,$$

and use the fact that three $2 \times 2$ submatrices of $\mathbf{DF}(0, 0, 0, 0)$ are nonsingular.

8. (a) Define $\mathbf{F}\colon \mathbb{R}^3 \to \mathbb{R}^2$ by

$$\mathbf{F}(x,y,z) = (a_{11}x + a_{12}y + a_{13}z, a_{21}x + a_{22}y + a_{23}z) \text{ for } (x,y,z) \text{ in } \mathbb{R}^3.$$

Note that $\mathbf{F}(0,0,0) = \mathbf{0}$ and that $\mathbf{DF}(0,0,0)$ is the $2 \times 3$ matrix with first row $\boldsymbol{\eta}$ and second row $\boldsymbol{\beta}$. One component of $\boldsymbol{\eta} \times \boldsymbol{\beta}$ is nonzero; without loss of generality, assume the first component is nonzero. Then it follows that the $2 \times 2$ submatrix

$$\begin{pmatrix} \partial F_1/\partial y(0,0,0) & \partial F_1/\partial z(0,0,0) \\ \partial F_2/\partial y(0,0,0) & \partial F_2/\partial z(0,0,0) \end{pmatrix}$$

of $\mathbf{DF}(0,0,0)$ is invertible. Apply the Implicit Function Theorem to define the second and third components of a solution $(x,y,z)$ as functions of the first component. (The cases when the second and/or third component of $\boldsymbol{\eta} \times \boldsymbol{\beta}$ are nonzero are handled similarly.)

(b) Note that the solutions of the given system of equations are precisely those points $(x,y,z)$ in $\mathbb{R}^3$ which are orthogonal to both $\boldsymbol{\eta}$ and $\boldsymbol{\beta}$. Since $\boldsymbol{\eta} \times \boldsymbol{\beta} \neq \mathbf{0}$, Theorem 8 of the appendix on linear algebra asserts that the solutions are precisely the multiples of $\boldsymbol{\eta} \times \boldsymbol{\beta}$. The functions obtained in part (a) give only a small portion of these solutions.

## SECTION 17.3

4. One example: Define $g, h\colon \mathbb{R}^3 \to \mathbb{R}$ by $g(x,y,z) = x$, $h(x,y,z) = y$ for $(x,y,z)$ in $\mathbb{R}^3$. Then the first component of $\mathbf{D}g(0,0,0) \times \mathbf{D}h(0,0,0)$ is zero. Also, since the set of solutions of (17.31) consists of the $z$-axis, the solutions cannot be parametrized by the first component.

## SECTION 17.4

1. The minimum is $\sqrt{2}$.

6. The minimum is $-\sqrt{a^2 + b^2 + c^2}$.

13. (a) Use Lagrange multipliers to find the minimum of $\{x + y + z \mid xyz = 1\}$.

(b) Set

$$x = \frac{a_1}{(a_1 a_2 a_3)^{1/3}}, \quad y = \frac{a_2}{(a_1 a_2 a_3)^{1/3}}, \quad z = \frac{a_3}{(a_1 a_2 a_3)^{1/3}},$$

and apply part (a).

## SECTION 18.1

6. Hint: Let $\epsilon > 0$. Choose a partition $\mathbf{P}$ such that $\|\mathbf{P}\|^n < \epsilon/(2|f(\mathbf{x})|)$ and $\mathbf{x}$ is contained in exactly one generalized rectangle in $\mathbf{P}$. Estimate $L(f, \mathbf{P})$ and $U(f, \mathbf{P})$.

8. $\frac{1}{2}$.

10. $-1/6$.

11. 4.

14. Use Lemma 18.1 with $m = -M$, and the definition of the integral.

## SECTION 18.2

1. The ellipse is the union of the graphs of continuous functions defined on intervals.

5. One example: Let $D$ be the set of points in $\mathbf{I}$ with a rational first component.

8. There is no inconsistency, since $A$ is a subset of $\mathbb{R}$, whereas $B$ is a subset of $\mathbb{R}^2$.

**13.** Hint: Follow the proof of Theorem 18.23, but note that since $D$ connected, it has the Intermediate Value Property.

## SECTION 18.3

**1.** $(\frac{1}{2} - \frac{1}{4}\sin 2)^2$.

**2.** (a) 1/6. (b) $\pi/8$. (c) $\int_0^{1/\sqrt{2}}\left[\int_{x^2}^{2x^2}(x+y)\,dy\right]dx + \int_{1/\sqrt{2}}^{1}\left[\int_{x^2}^{1}(x+y)\,dy\right]dx$.

**5.** Hint: Differentiate each side with respect to $x$.

## SECTION 18.4

**1.** $\pi/24$.

**3.** $2\pi$.

**7.** First show that $\psi: D \to \mathbb{R}$ is one-to-one. Then show that $\det \mathbf{D}\psi(x,y) = 2(x^2+y^2) \neq 0$ for all $(x,y)$ in $D$, and apply the Inverse Function Theorem.

**17.** Hint: As $\mathbf{x}$ varies in $K$, the norm $\|\mathbf{DF}(\mathbf{x})\|$ assumes a maximum value.

## SECTION 19.1

**2.** $2\pi\sqrt{13}$.

**6.** $2\pi$.

**8.** Hint: Use the Triangle Inequality to show that polygonal approximations increase when the partition is refined.

## SECTION 19.2

**1.** $\boldsymbol{\eta} = (a/c, b/c, 1)$.

**8.** For $\mathbf{r}(u,v) = (u\cos v, u\sin v, g(u))$, show that $\|\partial\mathbf{r}/\partial u \times \partial\mathbf{r}/\partial v\| = \|(-ug'(u)\cos v, ug'(u)\sin v, u)\|$.

**9.** Hint: Use formula 19.11 to show that $\partial g/\partial x(x,y) = \partial g/\partial y(x,y) = 0$ for $(x,y)$ in $\mathcal{R}$. What does this say about $g: \mathcal{R} \to \mathbb{R}$?

## SECTION 19.3

**1.** 1/30.

**7.** $-20\pi$.

**8.** $\pi$.

**18.** Hint: Use Exercise 17 to show that if the function $u: \{(x,y) \mid x^2 + y^2 \leq 1\} \to \mathbb{R}$ has the stated properties, then

$$\int_{x^2+y^2\leq 1}\langle \nabla u, \nabla u\rangle\,dxdy = 0.$$

# Index